Guiding, Diffraction, and Confinement of Optical Radiation

Guiding, Diffraction, and Confinement of Optical Radiation

Salvatore Solimeno
UNIVERSITY OF NAPLES
NAPLES, ITALY

Bruno Crosignani
UNIVERSITY OF ROME, "LA SAPIENZA"
ROME, ITALY

Paolo DiPorto
UNIVERSITY OF L'AQUILA
L'AQUILA, ITALY

1986

ACADEMIC PRESS, INC.
Harcourt Brace Jovanovich, Publishers
Orlando San Diego New York Austin
London Montreal Sydney Tokyo Toronto

ACADEMIC PRESS, INC.
Orlando, Florida 32887

United Kingdom Edition published by
ACADEMIC PRESS INC. (LONDON) LTD.
24–28 Oval Road, London NW1 7DX

Library of Congress Cataloging in Publication Data

Solimeno, S. (Salvatore)
Guiding, diffraction, and confinement of optical radiation.

Includes bibliographies and index.
1. Beam optics. 2. Diffraction. 3. Radiation.
I. Crosignani, Bruno. II. Di Porto, Paolo.
III. Title.
QC389.S65 1986 535.5 84-18534
ISBN 0–12–654340–2 (hardcover) (alk. paper)
ISBN 0–12–654341–0 (paperback) (alk. paper)

PRINTED IN THE UNITED STATES OF AMERICA

86 87 88 89 9 8 7 6 5 4 3 2 1

To our beloved parents

Contents

Chapter III Plane-Stratified Media

Chapter IV Fundamentals of Diffraction Theory

Chapter V Asymptotic Evaluation of Diffraction Integrals

Chapter VI Aperture Diffraction and Scattering from Metallic and Dielectric Obstacles

Chapter VII Optical Resonators and Fabry–Perot Interferometers

Chapter VIII Propagation in Optical Fibers

Preface

Following the advent of the laser in the early sixties, a host of devices have been developed in the course of the past twenty years that are capable of handling and manipulating electromagnetic radiation at optical frequencies. Although a number of excellent books have been written that deal, at various levels of description, with specific classes of these optical devices (for example, optical resonators and optical fibers), we felt that there was space for a text in which the reader could become acquainted with the general principles on which they are based and, in the spirit of this unifying purpose, find the illustration of a large variety of them, usually spread among several monographs. In this approach, the reader is introduced to a number of analytic techniques (many of them being in general mastered by people working in classical electromagnetism more than by those active in the field of optics) that provide him with the basis for a full comprehension of the single topic.

The present volume is the result of the scientific experience of the authors, who have worked in fields strictly related to many of the covered subjects, and of their teaching activity at the Universities of L'Aquila (Di Porto, solid state physics), Rome (Crosignani, optoelectronics), and Naples (Solimeno, electromagnetic fields and optics). This has led to a text that can be used as both a textbook and an advanced research book, the emphasis being shifted to the former by the presence of a large number of problems.

Chapter I concerns the general features of electromagnetic propagation and introduces the basic concepts pertaining to the description of the electromagnetic field and its interaction with matter.

Chapter II is mainly devoted to asymptotic methods of solution of the wave equation, with particular relevance being given to the asymptotic representation of the field in the form of the Luneburg–Kline series (of which geometrical optics constitutes the lowest-order approximation). In particular, a number of optical systems characterized by different refractive index distributions are investigated by relying on the eikonal equation.

Chapter III deals with stratified media (e.g., multilayered thin films, metallic and dielectric reflectors, and interference filters), a subject extremely significant in the frame of integrated optics, and it includes propagation through periodic structures.

In Chapter IV, the problem of propagation is faced by means of diffraction theory, which basically consists of evaluating the field in the region of interest through the contributions arising from the field itself on a reference surface. This approach, which can be considered as a direct consequence of Huygens's principle, poses as its main problem the evaluation of diffraction integrals, a task accomplished by the systematic use of the techniques developed in Chapter V.

Chapter VI is essentially devoted to scattering from obstacles and includes the description of metallic and dielectric gratings. The formalisms of diffraction matrix and *S* scattering matrix, seldom to be found in other texts on optics, are also dealt with in some detail.

The enhancement of the field in certain finite regions of space, taking place under particular combinations of wavelength and medium inhomogeneities and giving rise to an effect that can be termed radiation confinement, is investigated in the last two chapters. In particular, Chapter VII describes passive and active resonators employed in connection with laser sources for producing a confinement near the axis of an optical cavity and Fabry–Perot interferometers and mainly relies on the use of diffraction theory. In Chapter VIII, the analytic approach to the study of transverse confinement near the axis of a dielectric waveguide hinges on the introduction of modal solutions of the wave equation. This allows one to treat in a unified way up-to-date topics associated with such nonlinear optical effects as self-phase modulation and soliton propagation.

The book is primarily directed toward graduate students in optics and electromagnetism and researchers interested in the problems of propagation and confinement of optical radiation. It requires, as a prerequisite, a basic knowledge of electromagnetic theory at the level of a good undergraduate course and, as concerns the mathematical apparatus, a solid background in special functions, Fourier series, and differential equations. We believe that, owing to the broadness of the topics covered and to the rigor of the analytic description, it should also serve as a reference text for physicists and engineers active in the field of quantum electronics.

We are grateful to Mrs. C. Cutillo for her competent typing, to M. Sansone for preparing the original illustrations, and to A. Torre for a careful reading of the manuscript. Two of us (Crosignani and DiPorto) wish to thank the Fondazione Ugo Bordoni (Istituto Superiore Poste e Telecomunicazioni), Rome, for its kind hospitality during part of the writing of the book. Bruno Crosignani owes a great debt to his wife, Maria, and his daughters, Ginevra and Viera, for their encouragement in preparing this book. The thanks of Salvatore Solimeno go to the Consiglio Nazionale della Ricerche and the Instituto Nazionale di Fisica Nucleare for their support and to R. Bruzzese for useful suggestions.

Chapter I

General Features of Electromagnetic Propagation

1 Maxwell's Equations

The behavior of the electromagnetic field in a region of space in which the physical properties of the medium are continuous is characterized by four vector quantities, that is, $\mathbf{E}(\mathbf{r}, t)$, $\mathbf{B}(\mathbf{r}, t)$, $\mathbf{D}(\mathbf{r}, t)$, $\mathbf{H}(\mathbf{r}, t)$, which satisfy Maxwell's equations:

$$\nabla \times \mathbf{E}(\mathbf{r}, t) = (-\partial/\partial t)\mathbf{B}(\mathbf{r}, t), \tag{I.1.1}$$

$$\nabla \times \mathbf{H}(\mathbf{r}, t) = \mathbf{J}(\mathbf{r}, t) + (\partial/\partial t)\mathbf{D}(\mathbf{r}, t), \tag{I.1.2}$$

$$\nabla \cdot \mathbf{B}(\mathbf{r}, t) = 0, \tag{I.1.3}$$

$$\nabla \cdot \mathbf{D}(\mathbf{r}, t) = \rho(\mathbf{r}, t). \tag{I.1.4}$$

Throughout this book we will use mksa units, so that the *electric field* $\mathbf{E}$ is measured in volts per meter, the *magnetic induction* $\mathbf{B}$ in webers per square meter, the *electric induction* $\mathbf{D}$ in coulombs per square meter, the *magnetic field* $\mathbf{H}$ in amperes per meter, the *volume density of charge* ρ in coulombs per cubic meter, and the *current density* $\mathbf{J}$ in amperes per square meter.

The current density $\mathbf{J}$ appearing in Eq. (I.1.2) may be due to the presence of a conducting material (viz., a metal or a semiconductor) or may represent a source (such as the one associated with a magnetic or an electric dipole or with a moving electron). In some cases, $\mathbf{J}$ is not known *a priori*; for example, the current circulating on the surface of a metallic body enveloped by an electromagnetic wave depends in a complex way on the incident and scattered field that it generates. Since the solution of these problems goes beyond the scope of this book, which is meant to further the analysis of optical problems, $\mathbf{J}$ will usually represent a prescribed source. The charge density is produced only by a nonvanishing divergence of $\mathbf{J}$, according to the relation

$$\partial\rho/\partial t = -\nabla \cdot \mathbf{J}, \tag{I.1.5}$$

which follows from Eqs. (I.1.2) and (I.1.4).

In general, at optical frequencies one deals with media whose *magnetic permeability* μ coincides approximately with that of the vacuum, μ_0 ($=4\pi \times 10^{-7}$ H/m), so that one can put

$$\mathbf{B}(\mathbf{r}, t) = \mu_0 \mathbf{H}(\mathbf{r}, t) \tag{I.1.6}$$

(cases in which this relation is not satisfied are actually encountered in the frame of moving media, for which the reader is referred to Section I.7).

The relation connecting $\mathbf{E}(\mathbf{r}, t)$ and $\mathbf{D}(\mathbf{r}, t)$ is in general more complicated (see the next section), but if we restrict ourselves to the case of a *monochromatic field* oscillating at the angular frequency ω, i.e., $\mathbf{E}(\mathbf{r}, t) = \mathbf{e}(\mathbf{r}) \cos[\omega t + \Phi(\mathbf{r})]$ and $\mathbf{D}(\mathbf{r}, t) = \mathbf{d}(\mathbf{r}) \cos[\omega t + \Psi(\mathbf{r})]$, then between the *complex representation* (see Section I.8) $\hat{\mathbf{E}}(\mathbf{r}, t) = \mathbf{e}(\mathbf{r}) \exp[i\Phi(\mathbf{r}) + i\omega t] \equiv \mathbf{E}(\mathbf{r}) \exp(i\omega t)$ and $\hat{\mathbf{D}}(\mathbf{r}, t) = \mathbf{d}(\mathbf{r}) \exp[i\Psi(\mathbf{r}) + i\omega t] \equiv \mathbf{D}(\mathbf{r}) \exp(i\omega t)$ a relation similar to Eq. (I.1.6) holds; that is,

$$\hat{\mathbf{D}}(\mathbf{r}, t) = \varepsilon_0(1 + \chi_\omega)\hat{\mathbf{E}}(\mathbf{r}, t), \tag{I.1.7}$$

where ε_0 ($\sim 8.85 \times 10^{-12}$ F/m) is the vacuum *dielectric permittivity* and χ_ω the generally complex *dielectric susceptibility* of the medium. Equation (I.1.7) describes *homogeneous isotropic* media but can be generalized to include homogeneous *anisotropic* media (see Section I.4) by letting χ_ω depend on the direction of propagation of the electromagnetic field (assumed to be a plane wave).

Limiting ourselves to the isotropic case, it is possible to derive, by employing Eq. (I.1.7) together with Maxwell's equations and assuming $\mathbf{J} = \rho = 0$, a single equation for $\mathbf{E}(\mathbf{r})$:

$$\nabla^2 \mathbf{E}(\mathbf{r}) + k_0^2(1 + \chi_\omega)\mathbf{E}(\mathbf{r}) = 0, \tag{I.1.8}$$

where $k_0 = \omega/c$ and $c = 1/(\varepsilon_0\mu_0)^{1/2}$ is the *velocity of light* in vacuum, or equivalently:

$$\nabla^2 \mathbf{E}(\mathbf{r}) + k_0^2 \tilde{n}^2(\omega)\mathbf{E}(\mathbf{r}) = 0, \tag{I.1.9}$$

where $\tilde{n}(\omega) = (1 + \chi_\omega)^{1/2}$ is the (generally complex) *refractive index* of the medium. This is accomplished by applying the operator $\nabla \times$ on both sides of Eqs. (I.1.1) and (I.1.2), written in the complex representation, and taking advantage of the vector identity (A.13) (see the Appendix) and of Eq. (I.1.4).

For an isotropic *nonhomogeneous* medium, the electric susceptibility becomes a function of position and Eq. (I.1.7) takes the form

$$\hat{\mathbf{D}}(\mathbf{r}, t) = \varepsilon_0[1 + \chi_\omega(\mathbf{r})]\hat{\mathbf{E}}(\mathbf{r}, t). \tag{I.1.10}$$

In this case the equation obeyed by $\mathbf{E}(\mathbf{r})$ reads

$$\nabla^2 \mathbf{E}(\mathbf{r}) - \nabla\nabla \cdot \mathbf{E}(\mathbf{r}) + k_0^2 \tilde{n}^2(\mathbf{r}, \omega)\mathbf{E}(\mathbf{r}) = 0, \tag{I.1.11}$$

where the refractive index, $\tilde{n}(\mathbf{r}, \omega) = \sqrt{1 + \chi_\omega(\mathbf{r})}$, is now a function of $\mathbf{r}$. By taking advantage of Eqs. (I.1.4) and (A.7) we can write

$$\nabla \cdot \mathbf{D} = \nabla \cdot (\varepsilon_0 \tilde{n}^2 \mathbf{E}) = \varepsilon_0 \mathbf{E} \cdot \nabla \tilde{n}^2 + \varepsilon_0 \tilde{n}^2 \nabla \cdot \mathbf{E} = 0,$$

from which it follows that $\nabla \cdot \mathbf{E}(\mathbf{r}) = -2\mathbf{E}(\mathbf{r}) \cdot \nabla \tilde{n}/\tilde{n}$, a relation that in turn allows us to neglect the second term of Eq. (I.1.11) whenever $\tilde{n}(\mathbf{r}, \omega)$ is a slowly varying function of $\mathbf{r}$ and/or $\mathbf{E}$ is perpendicular to $\nabla \tilde{n}$. In this case, the generic cartesian component $u(\mathbf{r})$ of $\mathbf{E}(\mathbf{r})$ satisfies the *scalar wave equation*

$$\nabla^2 u(\mathbf{r}) + k_0^2 \tilde{n}^2(\mathbf{r}, \omega) u(\mathbf{r}) = 0. \tag{I.1.12}$$

A great deal of the mathematical apparatus presented in this book is devoted to the solution of this equation. In this connection, recall that in regions of space where the properties of the medium change abruptly (see, for example, Chapters III and VIII), Maxwell's equations and the associated wave equations must be supplemented by the appropriate relations describing the transition of **E**, **H**, **B**, and **D** across a discontinuity surface. These relations state that, for dielectric media, in the absence of external charges and currents, the normal components of the magnetic and electric induction are continuous across the discontinuity surface, as are the tangential components of the electric and magnetic field [1].

1.1 *Vector and Scalar Potentials*

Maxwell's equations contain six scalar functions $[E_i, B_i; (i = x, y, z)]$ that can be reduced to four by expressing **E** and **B** in terms of the *vector potential* $\mathbf{A}(\mathbf{r}, t)$ and the *scalar potential* $\Phi(\mathbf{r}, t)$ as

$$\mathbf{H} = \nabla \times \mathbf{A}, \qquad \mathbf{E} = -\nabla \Phi - \partial \mathbf{A}/\partial t, \tag{I.1.13}$$

these forms being, respectively, suggested by Eq. (I.1.3) and the vector identity (A.15) and by Eq. (I.1.1) and the vector identity (A.14).

For a stationary, homogeneous, anisotropic time–space dispersive medium, for which (see Section I.5)

$$\mathbf{D}(\mathbf{r}, t) = \iiint_{-\infty}^{+\infty} d\mathbf{r}' \int_{-\infty}^{t} dt' \, \chi(\mathbf{r} - \mathbf{r}', t - t') \cdot \mathbf{E}(\mathbf{r}', t') \equiv \tilde{\varepsilon} \cdot \mathbf{E}, \tag{I.1.14}$$

where the tensor $\tilde{\varepsilon}$ is a linear operator and $d\mathbf{r}' \equiv dx'\, dy'\, dz'$, we can use Eq. (I.1.6), the vector identity (A.13), and the time invariance of $\tilde{\varepsilon}$ [that is, $(\partial/\partial t)\tilde{\varepsilon} = \tilde{\varepsilon}(\partial/\partial t)$] to show that **A** and Φ satisfy the set of equations

$$\nabla \cdot (\tilde{\varepsilon} \cdot \nabla \Phi) + \frac{\partial}{\partial t} \nabla \cdot (\tilde{\varepsilon} \cdot \mathbf{A}) = -\rho, \tag{I.1.15a}$$

$$\nabla^2\mathbf{A} - \mu_0\left(\frac{\partial}{\partial t}\tilde{\boldsymbol{\varepsilon}}\right)\cdot\frac{\partial}{\partial t}\mathbf{A} = -\mu_0\mathbf{J} + \boldsymbol{\nabla}\boldsymbol{\nabla}\cdot\mathbf{A} + \mu_0\left(\frac{\partial}{\partial t}\tilde{\boldsymbol{\varepsilon}}\right)\cdot\boldsymbol{\nabla}\Phi, \tag{I.1.15b}$$

which are equivalent to Maxwell's equations.

It is noteworthy that $\mathbf{A}$ and Φ are defined apart from an arbitrary function $\Psi(\mathbf{r}, t)$, since Eqs. (I.1.13) furnish the same field vectors $\mathbf{E}$ and $\mathbf{B}$ if $\mathbf{A}$ and Φ are replaced by $\mathbf{A}' = \mathbf{A} + \boldsymbol{\nabla}\Psi$ and $\Phi' = \Phi - \partial\Psi/\partial t$. This property can be exploited to choose $\mathbf{A}'$ in such a way as to satisfy the *Coulomb condition*

$$\boldsymbol{\nabla}\cdot(\tilde{\boldsymbol{\varepsilon}}\cdot\mathbf{A}') = 0, \tag{I.1.16}$$

which is equivalent to letting Ψ satisfy

$$\boldsymbol{\nabla}\cdot(\tilde{\boldsymbol{\varepsilon}}\cdot\boldsymbol{\nabla}\Psi) = -\boldsymbol{\nabla}\cdot(\tilde{\boldsymbol{\varepsilon}}\cdot\mathbf{A}) \equiv -g(\mathbf{r}, t) \tag{I.1.17}$$

with $g(\mathbf{r}, t)$ a generic function. This equation can be easily integrated by taking its space–time Fourier transform, which yields

$$(\mathbf{q}\cdot\tilde{\boldsymbol{\varepsilon}}_{\mathbf{q},\omega}\cdot\mathbf{q})\Psi_{\mathbf{q},\omega} = g_{\mathbf{q},\omega}, \tag{I.1.18}$$

where we have defined

$$g_{\mathbf{q},\omega} \equiv \iiint_{-\infty}^{+\infty} d\mathbf{r}\int_{-\infty}^{+\infty} dt\, e^{i\mathbf{q}\cdot\mathbf{r} - i\omega t} g(\mathbf{r}, t) \tag{I.1.19}$$

by solving Eq. (I.1.18) to obtain $\Psi_{\mathbf{q},\omega}$ and then by taking the inverse transform of the resulting expression.

In the *Coulomb gauge* described by Eq. (I.1.16), the set of Eqs. (I.1.15) simplifies to

$$\boldsymbol{\nabla}\cdot(\boldsymbol{\varepsilon}\cdot\boldsymbol{\nabla}\Phi) = -\rho, \tag{I.1.20a}$$

$$\nabla^2\mathbf{A} - \boldsymbol{\nabla}(\boldsymbol{\nabla}\cdot\mathbf{A}) - \mu_0\left(\frac{\partial}{\partial t}\tilde{\boldsymbol{\varepsilon}}\right)\cdot\frac{\partial}{\partial t}\mathbf{A} = -\mu_0\mathbf{J} + \mu_0\left(\frac{\partial}{\partial t}\tilde{\boldsymbol{\varepsilon}}\right)\cdot\boldsymbol{\nabla}\Phi \equiv -\mu_0\mathbf{J}^{(t)}, \tag{I.1.20b}$$

the first of which shows how the scalar potential $\Phi(\mathbf{r}, t)$ reacts to the variation of the charge density $\rho(\mathbf{r}, t)$. More precisely, by taking its time Fourier transform, it is immediately seen that $\Phi_\omega(\mathbf{r})$ is linearly related to $\rho_\omega(\mathbf{r})$ through the equation (see Problems 2 and 3)

$$\varepsilon_0\Phi_\omega(\mathbf{r}) = \iiint_{-\infty}^{+\infty} d\mathbf{r}'\, G_{\mathrm{C}}(\mathbf{r} - \mathbf{r}', \omega)\rho_\omega(\mathbf{r}'), \tag{I.1.21}$$

where the *Green's function* G_{C} is given by

$$G_{\mathrm{C}}(\mathbf{r}, \omega) = \frac{\varepsilon_0}{(2\pi)^3}\iiint_{-\infty}^{+\infty} d\mathbf{q}\frac{e^{-i\mathbf{q}\cdot\mathbf{r}}}{\mathbf{q}\cdot\tilde{\boldsymbol{\varepsilon}}_{\mathbf{q},\omega}\cdot\mathbf{q}}. \tag{I.1.22}$$

In an analogous way, by taking the Fourier transform of Eqs. (I.1.20), it is possible to obtain

$$(\omega^2\mu_0\tilde{\boldsymbol{\varepsilon}}_{\mathbf{q},\omega} + \mathbf{qq} - q^2\mathbf{1})\cdot\mathbf{A}_{\mathbf{q},\omega} = -\mu_0\left(1 - \frac{\tilde{\boldsymbol{\varepsilon}}_{\mathbf{q},\omega}\cdot\mathbf{qq}}{\mathbf{q}\cdot\tilde{\boldsymbol{\varepsilon}}_{\mathbf{q},\omega}\cdot\mathbf{q}}\right)\cdot\mathbf{J}_{\mathbf{q},\omega}$$

$$\equiv -\mu_0\mathbf{J}^{(t)}_{\mathbf{q},\omega}, \tag{I.1.23}$$

where advantage has been taken of the *charge conservation* condition [Eq. (I.1.5)] in its Fourier-transformed form

$$\mathbf{q}\cdot\mathbf{J}_{\mathbf{q},\omega} = \omega\rho_{\mathbf{q},\omega}. \tag{I.1.24}$$

Note that, since $\mathbf{J}^{(t)}_{\mathbf{q},\omega}\cdot\mathbf{q} = 0$, $\mathbf{J}^{(t)}_{\mathbf{q},\omega}$ can be identified with the transverse component of $\mathbf{J}_{\mathbf{q},\omega}$.

When the medium is *isotropic* and *homogeneous*, we can require that the potentials **A** and Φ obey the so-called *Lorentz gauge*, that is

$$\nabla\cdot\mathbf{A} + \mu_0\tilde{\varepsilon}\frac{\partial\Phi}{\partial t} = 0. \tag{I.1.25}$$

In this case, **A** and Φ satisfy the *nonhomogeneous Helmholtz equations*

$$\nabla^2\Phi - \mu_0\tilde{\varepsilon}(\partial^2/\partial t^2)\Phi = -\rho/\tilde{\varepsilon}, \tag{I.1.26a}$$

$$\nabla^2\mathbf{A} - \mu_0\tilde{\varepsilon}(\partial^2/\partial t^2)\mathbf{A} = -\mu_0\mathbf{J}, \tag{I.1.26b}$$

which contain, respectively, ρ and **J**.

Lastly, for a homogeneous anisotropic medium, we can express $\mathbf{E}_\omega(\mathbf{r})$ by means of the convolution integral

$$\mathbf{E}_\omega(\mathbf{r}) = -i\omega\mu_0\iiint_{-\infty}^{+\infty} d\mathbf{r}'\,\boldsymbol{\Gamma}(\mathbf{r}-\mathbf{r}')\cdot\mathbf{J}_\omega(\mathbf{r}'), \tag{I.1.27}$$

where $\boldsymbol{\Gamma}(\mathbf{r})$ is the *dyadic Green's function* (see Problems 4, 5, and 7).

2 Propagation in Time-Dispersive Media

In the preceding section, the time dependence of the electromagnetic field was accounted for mainly by the presence of a factor $\exp(i\omega t)$ in its complex representation. This does not constitute a limitation if the medium is static and (approximately) nondispersive over the range of frequencies spanned by the field, since in this case the general solution can be expressed as a superposition of monochromatic solutions. While the superposition principle still applies for a dispersive linear medium, this is no longer true when nonlinear phenomena become relevant. It is worthwhile to note that, due to the (usually satisfied) assumption $\delta\omega/\omega_0 \ll 1$ between the field bandwidth $\delta\omega$ and its mean frequency ω_0, it is often possible to factorize the field as the product of $\exp(i\omega_0 t)$ times a slowly varying amplitude. This is sometimes feasible also in the space domain whenever the field can be approximately represented by a plane wave (see, for example, Section II.1).

The interaction between radiation field and material medium is usually described by means of the *induced polarization vector* **P**, defined as the electric polarization per unit volume induced by the electric field **E** [1]. Its influence is accounted for through the relation

$$\mathbf{D} = \varepsilon_0 \mathbf{E} + \mathbf{P}, \tag{I.2.1}$$

where **P** is, in general, a complicated function of **E**. For most of the subjects covered in this book the relation between **P** and **E** is of the simple form (for notational convenience we omit the space dependence of the quantities under consideration)

$$\mathbf{P}(t) = \varepsilon_0 \int_{-\infty}^{+\infty} \chi(t')\mathbf{E}(t - t')\,dt', \tag{I.2.2}$$

where the function $\chi(t')$, which depends only on the characteristics of the medium, must vanish for $t' < 0$ because of the *causality principle*. By inserting Eq. (I.2.2) into Eq. (I.2.1) and taking the time Fourier transform of both sides, we immediately obtain the *constitutive relation*

$$\mathbf{D}_\omega = \varepsilon_0(1 + \chi_\omega)\mathbf{E}_\omega \equiv \tilde{\varepsilon}(\omega)\mathbf{E}_\omega. \tag{I.2.3}$$

The symbol f_ω indicates the time Fourier transform of $f(t)$ and is defined as

$$f_\omega = \int_{-\infty}^{+\infty} e^{-i\omega t} f(t)\,dt. \tag{I.2.4}$$

Equation (I.2.3) allows us to express the *dielectric permeability* $\tilde{\varepsilon}(\omega)$ in terms of the *electric susceptibility* $\chi_\omega \equiv \chi'_\omega - i\chi''_\omega$ and the *refractive index* $\tilde{n}(\omega)$ through the relation

$$\tilde{\varepsilon}(\omega)/\varepsilon_0 = 1 + \chi_\omega = \tilde{n}^2(\omega), \tag{I.2.5}$$

where the dependence of $\tilde{\varepsilon}$ on ω is usually called its *dispersion law*.

The simple linear relation (I.2.2) represents the first-order approximation of a general expression that, for a given material, reduces to the linear form at relatively low material densities and field intensities [2]. Whenever the medium presents time fluctuations that are slow compared with the characteristic time scale of $\chi(t)$, Eq. (I.2.3) can be generalized by allowing χ_ω to exhibit a parametric dependence on time [3], which amounts to substituting χ_ω with the slowly varying time-dependent susceptibility $\chi_\omega(t)$. (This circumstance is, for example, responsible for the broadening of monochromatic radiation propagating in a fluctuating medium.) A situation in which the above consideration applies concerns the optical kerr effect (Section VIII.19).

The dependence of the electric susceptibility, and consequently of the refractive index $\tilde{n}$, on ω is usually referred to as the *material dispersion* (see Fig. I.1). It is responsible, as shown in great detail in Chapter VIII in connection with propagation in optical fibers, for signal distortion in *transparent media* [where $\tilde{n}(\omega)$ can be considered real; see Section I.3]. The underlying mechanism involves the different *group velocities*

$$v(\omega) = \left\{ \frac{d}{d\omega} \left[\frac{\omega}{c} n(\omega) \right] \right\}^{-1} \tag{I.2.6}$$

where $n(\omega)$ is the real refractive index of transparent media, at which different frequency components of the field travel. It should be noted that while in optics it is customary to identify frequency regions of *normal* and *anomalous* dispersion in which the refractive index increases and decreases, respectively, with frequency ($dn/d\omega > 0$ and $dn/d\omega < 0$), in the current literature of optical fibers the two terms often refer, respectively, to regions in which $(d/d\omega)[1/v(\omega)] > 0$ and $(d/d\omega)[1/v(\omega)] < 0$, that is, to regions in which $v(\omega)$ is a decreasing and increasing function, respectively, of ω (see Section VIII.19).

As shown in Fig. I.1, a region of anomalous dispersion is confined between a maximum and a minimum of $\chi'(\omega)$; in a small interval around the center ω_0 of this region (which corresponds to a *resonance frequency* of the elementary systems of which the medium is made up), the medium exhibits strong absorption, that is, the electronic susceptibility acquires a large imaginary component χ''_ω. The investigation of this region requires a microscopic level of description, which is given in Section I.2.2 in connection with propagation in a two-level resonant system.

Whenever the strength of the electric field is such that the nonlinear contributions cannot be completely neglected, the simple expression furnished by Eq. (I.2.2) is modified, and in most situations the concept of refractive index becomes meaningless. As a consequence, the functional dependence of **P** on **E**

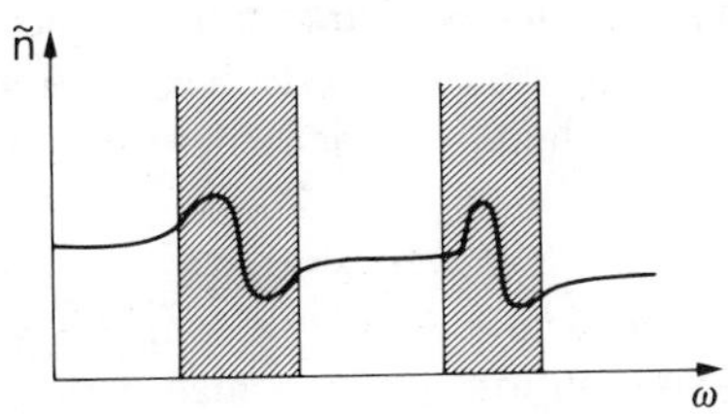

Fig. I.1. Qualitative behavior of refractive index versus frequency. The dashed areas represent regions of resonance.

must be determined for any specific nonlinear process and introduced into Maxwell's equations through Eq. (I.2.1). Examples of this situation will be considered in the following section, but we wish to remark here that, in relation to the *optical Kerr effect* (see Section VIII.19), the concept of refractive index retains its validity notwithstanding the intrinsic nonlinear nature of the effect.

2.1 *Nonlinear Wave-Propagation*

Let us consider Maxwell's equations (I.1.1) and (I.1.2) supplemented by Eq. (I.2.1) and by the differential form of *Ohm's law*, that is,

$$\mathbf{J} = \sigma \mathbf{E}, \tag{I.2.7}$$

where σ is the *electric conductivity* of the medium. Equation (I.2.7) holds true for a conductor in the low-frequency limit $\omega \to 0$. In the present form, which is strictly valid for a nearly monochromatic field centered around a frequency ω_0, $\sigma \equiv \sigma(\omega_0)$ has no direct physical meaning and is introduced only to take into account the losses associated with the imaginary part of the dielectric susceptibility. By a procedure analogous to that leading to Eq. (I.1.8), it is not difficult to derive the equation

$$\nabla^2 \mathbf{E} - \mu_0 \sigma \frac{\partial \mathbf{E}}{\partial t} - \frac{1}{c^2}\frac{\partial^2 \mathbf{E}}{\partial t^2} = \mu_0 \frac{\partial^2 \mathbf{P}}{\partial t^2}, \tag{I.2.8}$$

where we have made use of the approximate relation $\nabla \cdot \mathbf{E} = 0$ (rigorously valid only when the medium can be considered perfectly homogeneous) to get rid of the term $\nabla\nabla \cdot \mathbf{E}$ [see Eq. (I.1.11)].

If we now write the induced polarization vector $\mathbf{P}$ as the sum of a linear part $\mathbf{P}^{(L)}$ [obeying Eq. (I.2.2)] and a nonlinear part $\mathbf{P}^{(NL)}$, $\mathbf{P} = \mathbf{P}^{(L)} + \mathbf{P}^{(NL)}$, Eq. (I.2.8) becomes

$$\nabla^2 \mathbf{E} - \mu_0 \sigma \frac{\partial \mathbf{E}}{\partial t} - \left(\frac{1}{c^2}\frac{\partial^2 \mathbf{E}}{\partial t^2} + \mu_0 \frac{\partial^2 \mathbf{P}^{(L)}}{\partial t^2}\right) = \mu_0 \frac{\partial^2 \mathbf{P}^{(NL)}}{\partial t^2}. \tag{I.2.9}$$

At this stage, before proceeding and facing the problem of solving Eq. (I.2.9), it is necessary to derive the expression for $\mathbf{P}^{(NL)}$ for the particular nonlinear process being considered. This task can be accomplished at various levels of description, ranging from a completely quantum mechanical microscopic approach to a purely phenomenological one. In the next section we will give an example of the first procedure; here we briefly describe the second one.

The ith component of the polarization vector $\mathbf{P}(t)$ can be written as the sum of first-, second-, and third-order polarizabilities and so on in the form [4] (we use here, for simplicity, the convention of summation over repeated indices)

$$P_i(t) = \varepsilon_0 \int_{-\infty}^{+\infty} \chi_{ij}(t - t')E_j(t')\,dt' + \varepsilon_0 \iint_{-\infty}^{+\infty} \chi_{ijk}(t - t', t - t'')E_j(t')E_k(t'')\,dt'\,dt''$$

$$+ \varepsilon_0 \iiint_{-\infty}^{+\infty} \chi_{ijkl}(t - t', t - t'', t - t''')E_j(t')E_k(t'')E_l(t''')\,dt'\,dt''\,dt''' + \cdots,$$

$$i, j, k, l = x, y, z, \qquad \text{(I.2.10)}$$

where the tensor character of χ_{ij} allows us to generalize Eq. (I.2.2) to include anisotropic media.

The general expression furnished by Eq. (I.2.10) may assume a simpler form in particular situations, in connection with either the symmetry properties or the time response of the medium. As an example, the second term on the right side of Eq. (I.2.10) is zero in media possessing inversion symmetry; besides, under suitable conditions for the field frequency and its bandwidth, the response of the medium can be considered instantaneous, which amounts to assuming a δ-function type behavior for the χ's (see, for example, Section VIII.19).

Most often, nonlinear optical propagation is investigated by assuming that the field is a superposition of several monochromatic waves. This point of view has internal consistency, since the structure of Eq. (I.2.10) then implies the generation of waves in the form of a discrete superposition of monochromatic fields. Accordingly, the electric field is often written as

$$\mathbf{E}(t) = \sum_{l=1}^{n} (\mathbf{E}_{\omega_l} e^{i\omega_l t} + \mathbf{E}^*_{\omega_l} e^{-i\omega_l t}). \qquad \text{(I.2.11)}$$

Example: Second-Harmonic Generation. Let us consider, as an example, the *second-harmonic generation* (SHG) process for which second-order polarizability is responsible. In order to describe the process, the electric field is assumed to consist of the sum of a term vibrating at angular frequency ω_1 and a term vibrating at angular frequency $\omega_2 = 2\omega_1$ [see Eq. (I.2.11)].

After introducing this expression in the second term on the right side of Eq. (I.2.10), it is necessary to first extract the two contributions vibrating at ω_1 and $2\omega_1$, so that

$$P_i^{(\mathrm{NL})}(t) = p_{\omega_1 i} e^{i\omega_1 t} + q_{2\omega_1 i} e^{i2\omega_1 t} + \mathrm{cc}, \qquad \text{(I.2.12)}$$

where cc stands for "complex conjugate," and

$$p_{\omega_1 i} = \varepsilon_0 \chi_{-\omega_1, 2\omega_1 ijk} E^*_{\omega_1 j} E_{2\omega_1 k} + \varepsilon_0 \chi_{2\omega_1, -\omega_1 ijk} E_{2\omega_1 j} E^*_{\omega_1 k}, \qquad \text{(I.2.13)}$$

$$q_{2\omega_1 i} = \varepsilon_0 \chi_{\omega_1, \omega_1 ijk} E_{\omega_1 j} E_{\omega_1 k} \qquad \text{(I.2.14)}$$

with the symbol $\chi_{\omega, \omega' ijk}$ indicating the double time Fourier transform of

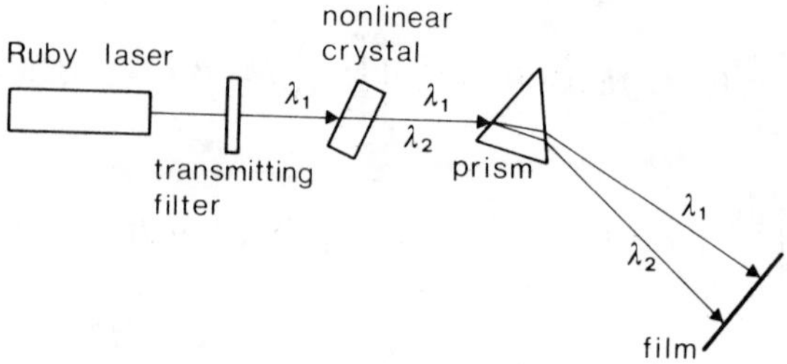

Fig. I.2. Experimental arrangement for the demonstration of second-harmonic light generation.

$\chi_{ijk}(t', t'')$. Inserting Eq. (I.2.12) into Eq. (I.2.9), we obtain

$$\nabla^2 E_{\omega_1 i} - i\mu_0 \sigma \omega_1 E_{\omega_1 i} + \frac{\omega_1^2}{c^2} \frac{\varepsilon_{\omega_1 ij}}{\varepsilon_0} E_{\omega_1 j} = -\mu_0 \omega_1^2 p_{\omega_1 i}, \tag{I.2.15a}$$

$$\nabla^2 E_{2\omega_1 i} - 2i\mu_0 \sigma \omega_1 E_{2\omega_1 i} + \frac{(2\omega_1)^2}{c^2} \frac{\varepsilon_{2\omega_1 ij}}{\varepsilon_0} E_{2\omega_1 j} = -4\mu_0 \omega_1^2 q_{2\omega_1 i}. \tag{I.2.15b}$$

having introduced the *dielectric permeability tensor*

$$\varepsilon_{\omega ij} = \varepsilon_0(\delta_{ij} + \chi_{\omega ij}). \tag{I.2.16}$$

The coupled set of Eqs. (I.2.15) provides the analytical description of SHG and its solution, for which the reader is referred to more specialized texts [5], allowing one to draw general conclusions about the efficiency and the characteristics of the process. In particular, its efficiency depends critically on the difference between the phase velocities of the pump (field at ω_1) and SH (field at $2\omega_1$) waves (see Fig. I.2). This coincidence (*phase-matching* condition) can be achieved in uniaxial crystals by exploiting the different velocities of the ordinary and extraordinary waves (see Section I.4).

2.2 *Pulse Propagation in a Two-Level Resonant System*

In this section we give an example of explicit evaluation of the polarization vector **P** by the microscopic approach. The intereaction of an electromagnetic field with a two-level system provides a general model for a number of situations involving the *coherent interaction* of radiation with matter.

The problem of expressing **P** as a function of **E**, a quasi-monochromatic plane wave whose midfrequency coincides with the central emission frequency $\omega_0 = (E_a - E_b)/\hbar$ of the system at rest (see Fig. I.3), is associated with solving the Schrödinger equation in the presence of an external perturbation. We follow a semiclassical approach in which the electric field is a classical quantity (which is consistent with neglecting the contribution to the electromagnetic field arising from the spontaneous emission decay), while the atomic system is

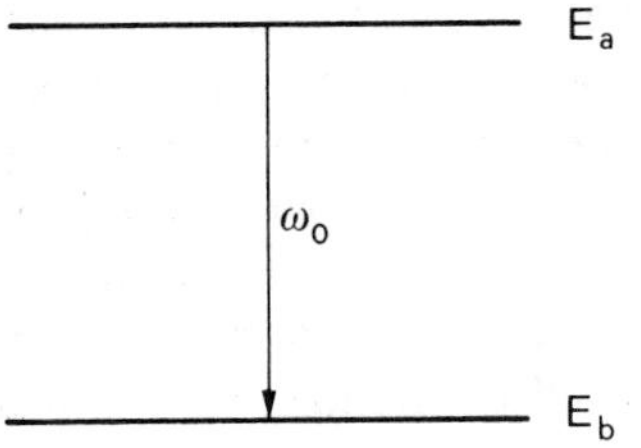

Fig. I.3. Two-level energy system.

quantized, its state $|\Psi\rangle$ being represented as a superposition of the energy eigenstates $|u_a\rangle$ and $|u_b\rangle$, corresponding to the eigenvalues E_a and E_b, in the form

$$|\Psi\rangle = a(t)|u_a\rangle + b(t)|u_b\rangle. \tag{I.2.17}$$

The state $|\Psi\rangle$ must obey the time-dependent Schrödinger equation:

$$H|\Psi\rangle = (H_0 + V)|\Psi\rangle = i\hbar\,\partial|\Psi\rangle/\partial t, \tag{I.2.18}$$

where H_0 is the Hamiltonian of the unperturbed system and V the external perturbation, which describes the interaction between the radiation field and the atom given, in the *dipole approximation* corresponding to a wavelength much larger than the dimension of the atomic system, by the expression $V = -e\mathbf{r}\cdot\mathbf{E}$. Once Eq. (I.2.18) is solved, that is, the time behavior of $a(t)$ and $b(t)$ is determined in terms of $\mathbf{E}$, the dipole moment $\mathbf{p}$ of the elementary system is given by

$$\mathbf{p} = \mathbf{p}_0(ab^* + a^*b), \tag{I.2.19}$$

where

$$p_0 = -e\langle u_a|\mathbf{r}|u_b\rangle \tag{I.2.20}$$

is the magnitude of the electric-dipole matrix element, and the polarization vector $\mathbf{P}$ is obtained by summing the contributions of all the systems present in the unit volume.

To proceed, let us assume that, at time $t = 0$, each atom is in either state a or state b and let us label it according to its velocity v_z along the propagation direction z of the electromagnetic field or, equivalently, by its central emission frequency $\omega = \omega_0 + k_0 v_z$ as seen in the laboratory frame. If we now indicate by $g(\omega)$ the probability distribution of ω, arising from atomic motion in a gas or from local crystal inhomogeneities in a solid, and by N_{a0} and N_{b0} the number of atoms per unit volume in the upper and lower states at time $t = 0$, we have

$$P(z,t) = \frac{p_0}{\sqrt{3}}\int_{-\infty}^{+\infty} d\omega\, g(\omega) \sum_{\alpha=a,b} N_{\alpha 0}[a(z,t,\omega,\alpha)b^*(z,t,\omega,\alpha) + \mathrm{cc}], \tag{I.2.21}$$

where $a(z,t,\omega,\alpha)$ and $b(z,t,\omega,\alpha)$ refer to an atom having central emission frequency ω, occupying position z at time t, and being in state $|u_\alpha\rangle$ at $t = 0$, while the factor $1/\sqrt{3}$ is due to the averaging over all possible orientations of the atomic systems, which have been assumed to be isotropically polarizable.

Actually, the existence of a statistical uncertainty concerning the phases of $a(t)$ and $b(t)$, due to the random collisions taking place among the atoms, requires a suitable time-averaging operation to be performed on **P**, as expressed by Eq. (I.2.21), in order to evaluate the polarizability **P** to be used in connection with the macroscopic Maxwell's equations [in particular, with Eq. (I.2.8)]. Thus, we need to know $\langle ab^* \rangle$ and $\langle a^*b \rangle$ (where $\langle \cdots \rangle$ indicates ensemble average), which are the off-diagonal elements of the *density matrix* [6]

$$\boldsymbol{\rho} = \begin{bmatrix} \langle |a|^2 \rangle & \langle ab^* \rangle \\ \langle a^*b \rangle & \langle |b|^2 \rangle \end{bmatrix} \equiv \begin{bmatrix} \rho_{aa} & \rho_{ab} \\ \rho_{ba} & \rho_{bb} \end{bmatrix}, \tag{I.2.22}$$

whose time evolution can be easily obtained starting from the Schrödinger equation and introducing some phenomenological damping terms. In this way, we obtain the *Bloch equations*

$$\begin{aligned} \dot{\rho} &= -i\omega\rho_{ab} - \gamma_{ab}\rho_{ab} + \frac{i}{\hbar}V(z,t)(\rho_{aa} - \rho_{bb}), \\ \dot{\rho}_{aa} &= -\gamma_a(\rho_{aa} - \rho^{\text{eq}}{}_{aa}) + \frac{i}{\hbar}V(z,t)(\rho_{ab} - \rho_{ba}), \\ \dot{\rho}_{bb} &= -\gamma_b(\rho_{bb} - \rho^{\text{eq}}{}_{bb}) - \frac{i}{\hbar}V(z,t)(\rho_{ab} - \rho_{ba}), \end{aligned} \tag{I.2.23}$$

where the dot indicates differentiation with respect to time and $\gamma_{ab} = \frac{1}{2}(\gamma_a + \gamma_b) + 1/T_c$. The inverses of the natural lifetimes of states a and b and the characteristic atomic collision time, given, respectively, by γ_a, γ_b, and T_c, have been inserted in Eq. (I.2.23) in a phenomenological way. Equations (I.2.23) can be recast such as to resemble those for a magnetic dipole undergoing precession in a magnetic field, as originally shown by Feynman, Vernon, and Hellwarth. (See also Shen [5] and Sargent *et al.*[6].)

Since we are looking for a solution to Maxwell's equations corresponding to a linearly polarized (see Section I.3), narrowband plane wave centered around a given frequency ω_0, it is natural to set

$$E(z,t) = |\mathscr{E}(z,t)| \cos[\omega_0 t - kz + \Phi(z,t)] \tag{I.2.24}$$

and

$$\begin{aligned} P(z,t) &= S(z,t)\sin[\omega_0 t - kz + \Phi(z,t)] + C(z,t)\cos[\omega_0 t - kz + \Phi(z,t)], \\ &= \left(\frac{P_0}{\sqrt{3}}\right) \sum_{\alpha=a,b} N_{\alpha 0} \int_{-\infty}^{+\infty} g(\omega)[\rho_{ab}(z,t,\omega,\alpha) + \text{cc}]\, d\omega \end{aligned} \tag{I.2.25}$$

where $\mathscr{E} = |\mathscr{E}|e^{i\Phi}$ is the field envelope and the real quantities S and C are assumed to vary on a temporal and a spatial scale, respectively, much larger than $1/k_0$ and $1/\omega_0$. It is noteworthy that in the electric dipole approximation according to Eq. (I.2.24) the perturbing energy V coincides with $-\frac{1}{2}p_0\mathscr{E}e^{i(\omega-\Omega)t}/\sqrt{3}$. By inserting Eqs. (I.2.24) and (I.2.25) into Eq. (I.2.8), we obtain, under the condition just mentioned,

$$\left(\frac{\partial}{\partial z} + \frac{n_1}{c}\frac{\partial}{\partial t} + \alpha\right)|\mathscr{E}| = \frac{\omega_0\mu_0 c}{2n_1}S \tag{I.2.26}$$

and

$$|\mathscr{E}|\left(\frac{\partial}{\partial z} + \frac{n_1}{c}\frac{\partial}{\partial t}\right)\Phi = -\frac{\omega_0\mu_0 c}{2n_1}C \tag{I.2.27}$$

where $\alpha = \mu_0(\sigma/2n_1)$ is the loss per unit length and $n_1 = \sqrt{\varepsilon_1/\varepsilon_0}$ the refractive index of the inert background supporting the active atoms.

We must now add equations describing the evolution of $S(z,t)$ and $C(z,t)$, which can be worked out by starting from Eqs (I.2.23). After some lengthy calculations (for details the reader is referred to Hopf and Scully [7]), it is possible to show that, for a $g(\omega)$ symmetrical around $\omega = \omega_0$, $C = \Phi = 0$. If we assume that the contributions to the electric field due to spontaneous emission are negligible, propagation can then be described in terms of the two quantities $\mathscr{E}(z,t)$ and the *complex susceptibility integral*

$$\chi(z,T,t) = \frac{1}{2\pi g(\omega_0)}\sum_{\alpha=a,b}\frac{N_{\alpha 0}}{N_{a0}-N_{b0}}\int_{-\infty}^{+\infty} g(\omega)\cos[(\omega-\omega_0)T]$$
$$\times\,[\rho_{aa}(z,t,\omega,\alpha) - \rho_{bb}(z,t,\omega,\alpha)]\,d\omega \tag{I.2.28}$$

where the density matrix elements ρ_{aa} and ρ_{bb} refer to an atom possessing a central emission frequency ω_0 occupying the position z at time t and in the state u_α at $t = 0$. The quantity $\exp(-\gamma_{ab}t')\chi(z,t',t-t')$ plays, for the slowly varying amplitudes $S(z,t)$ and $\mathscr{E}(z,t)$, the same role as χ(t) in Eq. (I.2.2), as it enters the relation

$$S(z,t) = -\frac{2c\varepsilon_1}{\omega_0}d\int_0^{+\infty} e^{-t'\gamma_{ab}}\chi(z,t',t-t')\mathscr{E}(z,t-t')\,dt' \tag{I.2.29}$$

with $d = \frac{1}{3}p_0^2\omega_0(N_{a0} - N_{b0})\pi g(\omega_0)/(cn_1\varepsilon_0\hbar)$. More precisely, the propagation of a signal in the two-level system is described by the two coupled integro-differential equations

$$\frac{\partial\mathscr{E}}{\partial z} + \frac{n_1}{c}\frac{\partial\mathscr{E}}{\partial t} + \alpha\mathscr{E} = d\int_{-\infty}^{t} e^{-(t-t')\gamma_{ab}}\chi(z,t-t',t')\mathscr{E}(z,t')\,dt'$$

and
$$\frac{\partial}{\partial t}\chi(z,T,t)=\frac{\gamma_a^2}{\gamma_{ab}}\chi(z,T,-\infty)-\gamma_{ab}\chi(z,T,t)$$
$$-\frac{p_0^2}{6\hbar^2}\mathscr{E}(z,t)\int_{-\infty}^{t}e^{-(t-t')\gamma_{ab}}[\chi(z,T+t-t',t')$$
$$+\chi(z,T-t+t',t')]\mathscr{E}(z,t')\,dt' \tag{I.2.31}$$

with the boundary condition $\mathscr{E}(z=0,t)=\mathscr{E}_i$ for the electric field and the initial condition

$$\chi(z,T,t=-\infty)=\frac{1}{2\pi g(\omega_0)}\int_{-\infty}^{+\infty}g(\omega)\cos[(\omega-\omega_0)T]\,d\omega. \tag{I.2.32}$$

Let us now consider some limiting situations in which the set of equations assumes a much simpler form and, in some cases, becomes analytically tractable.

2.2.a *Rate Equation Regime for Long Pulses*

If T_c becomes negligible with respect to all other significant times ($1/\gamma_a$, $1/\gamma_b$, the inverse $\tilde{T}_2$ of the bandwidth of $g(\omega)$, and the pulse duration T_p), Eqs. (I.2.30,31) reduce to

$$\frac{\partial\mathscr{E}}{\partial z}+\frac{n_1}{c}\frac{\partial\mathscr{E}}{\partial t}+\alpha\mathscr{E}=-dT_2\chi(z,0,t)\mathscr{E}(z,t) \tag{I.2.33}$$

and

$$\frac{\partial\chi}{\partial t}(z,0,t)=-\frac{p_0^2}{3\hbar^2}T_2\chi(z,0,t)\mathscr{E}^2(z,t), \tag{I.2.34}$$

while Eq. (I.2.28) yields

$$\chi(z,0,t)=\frac{1}{2\pi g(\omega_0)}\frac{N_a(z,t)-N_b(z,t)}{N_{a0}-N_{b0}}, \tag{I.2.35}$$

where $N_a(z,t)$ and $N_b(z,t)$ are the atom densities in the upper and lower states [in particular, $N_a(z,-\infty)=N_{a0}$, $N_b(z,-\infty)=N_{b0}$]. Thus, Eqs. (I.2.33,34) represent two coupled equations connecting the instantaneous intensity of the field [which is proportional to $\mathscr{E}^2(z,t)$; see Section I.8] and the density of the *population inversion* $\Delta N=N_a-N_b$. These are, for example, the equations that are usually employed to study the behavior of a laser oscillator in the stationary regime [8] (see Section VII.19), the *rate equations* being in general able to describe situations in which the temporal variations of the field amplitude take place on a time scale long compared with T_2. The opposite situation, which is referred to as *coherent propagation*, is described in the next subsection.

The set of Eqs. (I.2.33,34) has, in the limit $\alpha=0$ corresponding to the absence of losses, an exact analytical solution [9] that is useful for investigating pulse propagation in a chain of laser oscillators.

2.2b *Self-Induced Transparency*

In the absence of dissipation, for a sharp line at exact resonance ($\omega_0 = \Omega$), it can be shown that

$$\frac{p_0}{3^{1/2}\hbar}\left(\frac{d}{dz} + \alpha\right)\int_{-\infty}^{+\infty} \mathscr{E}(z,t)\,dt = \frac{d}{2}\sin\left\{\frac{p_0}{3^{1/2}\hbar}\int_{-\infty}^{+\infty} \mathscr{E}(z,t)\,dt\right\}. \quad \text{(I.2.36)}$$

Therefore, if we indicate with $\Theta(z)$ the quantity in square brackets proportional to the field envelope, we have

$$d\Theta/dz + \alpha\Theta = \tfrac{1}{2}d\sin\Theta. \quad \text{(I.2.37)}$$

Equation (I.2.37) has, in the limit $\alpha = 0$, the general solution

$$\Theta(z) = 2\arctan(\tan(\tfrac{1}{2})\Theta_0)e^{zd/2}), \quad \text{(I.2.38)}$$

which leads to the so-called *area theorem* [10], according to which the values $\Theta(z) = (2m+1)\pi$ and $\Theta(z) = 2m\pi$ (with m an integer) are, respectively, stable solutions in the case of an amplifier ($d > 0$, $N_{a0} > N_{b0}$) and an attenuator ($d < 0$) (see Fig. I.4). In particular, for a given initial value Θ_0, $\Theta(z)$ evolves toward the nearest odd or even multiple of π, respectively, for the amplifier and the attenuator.

For $\alpha = 0$, Eq. (I.2.30) implies that the *hyperbolic-secant pulse*[10]

$$\mathscr{E}(z,t) = \frac{3^{1/2}\hbar}{p_0 T_p}\operatorname{sech}\left\{\frac{t - z/v}{T_p}\right\}, \quad \text{(I.2.39)}$$

(with $T_p \ll \tilde{T}_2$) propagates without changing its shape with group velocity

$$\frac{1}{v} = \frac{n_1}{c} + \frac{T_p^2 d}{2\pi g(\omega_0)}. \quad \text{(I.2.40)}$$

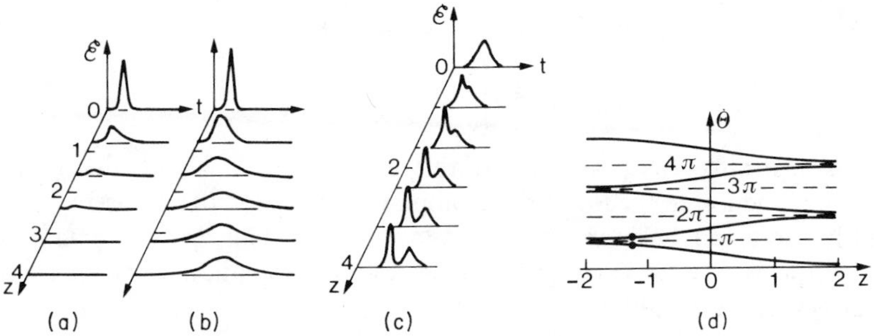

Fig. I.4. Computer plots of the evolution of pulses with initial areas (a) $\Theta_0 = 0.9\pi$, (b) $\Theta_0 = 1.1\pi$, and (c) $\Theta_0 = 4\pi$ for a positive attenuation parameter α. The distance z is measured in absorption length units of π/α. Note that for $\Theta_0 = 4\pi$ the pulse splits into two separate 2π pulses. Shown in (d) is the evolution of the pulse area as described by Eq. (I.2.37). (From McCall and Hahn [10].)

The circumstance according to which a short pulse may propagate with anomalously low energy losses when interacting with a two-level system of absorbers, provided its power is adjusted to a optical level, is known as *self-induced transparency* [10].

2.2.c General Case

In the general situation represented by Eqs. (I.2.30) and (I.2.31), it is usually necessary to resort to numerical methods in order to obtain a solution. To this end, it is convenient to put the equations in a suitable differential form. More precisely, by employing Eqs. (I.2.28) and (I.2.29) it is possible to show that for $\tilde{T}_2 > T_2$ (see Section I.2.2a) they are equivalent to the set of Eqs. (I.2.26) and

$$\frac{\partial}{\partial t}S = -\frac{S}{T_2} - \frac{p_0^2}{3\hbar}\mathscr{E}(N_a - N_b), \tag{I.2.41a}$$

$$\hbar\frac{\partial}{\partial t}(N_a - N_b) = -\mathscr{E}S, \tag{I.2.41b}$$

supplemented by the conditions $N_a(z, t = 0) - N_b(z, t = 0) = N_{a0} - N_{b0}$, $\mathscr{E}(z = 0, t) \equiv \mathscr{E}_i(t)$, and $S(z, t = 0) = 0$.

This is the system of equations that describes very short pulse ($T_p \lesssim 1$ ns) propagation in high-power laser amplifiers used in inertial confinement experiments.

2.3 *Kramers–Kronig Relations and Monochromatic Waves.*

The causality principle, by requiring that the polarization **P** of the medium at a given instant be influenced by the values of the electric field at previous instants only [see Eq. (I.2.2)], has relevance for the possible analytical behavior of χ_ω. In fact, the vanishing of $\chi(t)$ for negative values of the argument allows one to write

$$\chi_\omega = \int_0^\infty e^{-i\omega t}\chi(t)\,dt, \tag{I.2.42}$$

which, regarding ω as a complex variable ($\omega = \omega' + i\omega''$), implies that χ_ω is an analytic function in the lower half complex plane ($\omega'' < 0$) for passive media and in the upper half for active ones (e.g., lasers). This, in turn, entails the derivation (see, e.g., Yariv [8]) of the following relations, known as the *Kramers–Kronig relations*:

$$\chi'_\omega = \frac{1}{\pi}P\int_{-\infty}^{+\infty}\frac{\chi''_{\omega'}}{\omega' - \omega}\,d\omega', \tag{I.2.43a}$$

$$\chi''_\omega = -\frac{1}{\pi} P \int_{-\infty}^{+\infty} \frac{\chi'_{\omega'}}{\omega' - \omega} d\omega'. \qquad \text{(I.2.43b)}$$

(where P denotes Cauchy's principal value of the integral) between the real and imaginary parts of $\chi_\omega = \chi'_\omega - i\chi''_\omega$.

When a dispersive medium is in thermal equilibrium it is possible to demonstrate (see Landau and Lifshitz [1], Sect. 61) that in this case χ''_ω is always positive (negative) for $\omega > 0$, ($\omega < 0$) while at $\omega = 0$ it assumes the value zero as a consequence of the relation

$$\chi_\omega = \chi^*_{-\omega}, \qquad \text{(I.2.44)}$$

which immediately follows from Eq. (I.2.42). This, of course, does not forbid χ''_ω to assume very small values in a certain interval of frequencies, a dielectric being *transparent* in the frequency range over which $|\chi''_\omega| \ll |\chi'_\omega|$.

Using the fact that, according to Eq. (I.2.43b), χ''_ω is an odd function of ω, it is possible to rewrite Eq. (I.2.43a) in the form

$$\chi'_\omega = \frac{2}{\pi} P \int_0^\infty \frac{\omega'\chi''_{\omega'}}{\omega'^2 - \omega^2} d\omega' = \lim_{\varepsilon \to 0} \frac{2}{\pi} \int_{0 - i\varepsilon}^{\infty - i\varepsilon} \frac{\omega'\chi''_{\omega'}}{\omega'^2 - \omega^2} d\omega' - i2\chi''_\omega \qquad \text{(I.2.45)}$$

with $\varepsilon > 0$. This expression can be easily justified by replacing the integration domain $(0, \omega - \varepsilon) \cup (\omega + \varepsilon, \infty)$, on which the Cauchy principal value is calculated, with the half-line $(0 - i\varepsilon, \infty - i\varepsilon)$ deprived of the half-circle of radius ε and center in ω. The quantity ε must be so small as to exclude the presence of poles of χ''_ω between the old and the new integration paths.

If we differentiate Eq. (I.2.45) with respect to ω we obtain

$$\frac{d\chi'_\omega}{d\omega} = \lim_{\varepsilon \to 0} \frac{4\omega}{\pi} \int_{0 - i\varepsilon}^{\infty - i\varepsilon} \frac{\omega'\chi''_{\omega'}}{(\omega'^2 - \omega^2)^2} d\omega' - i2\frac{d\chi''_\omega}{d\omega}. \qquad \text{(I.2.46)}$$

When $d\chi''_\omega/d\omega$ is negligible, the right side of Eq. (I.2.46) is a positive quantity for $\chi'' > 0$ (passive-medium) and χ'_ω is a monotonically increasing function of ω together with the real part n_ω of the refractive index, $n_\omega = (1 + \chi'_\omega)^{1/2}$, and the *dielectric constant* $\varepsilon_\omega = \varepsilon_0(1 + \chi'_\omega)$. Recalling the definition given in Section I.2, this implies that the regions of constant losses of the spectrum are regions of normal dispersion.

The above considerations apply equally well to a dielectric and to a metal at high frequencies. In the limit of very high frequencies, there is not even a quantitative difference between the two cases, as confirmed by the fact that in most cases the dielectric constant (see, e.g., Landau and Lifshitz [1], Sect. 60) takes the form

$$\varepsilon_\omega = \varepsilon_0\left(1 - \frac{Ne^2}{m\varepsilon_0\omega^2}\right) \equiv \varepsilon_0\left(1 - \frac{\omega_p^2}{\omega^2}\right), \qquad \text{(I.2.47)}$$

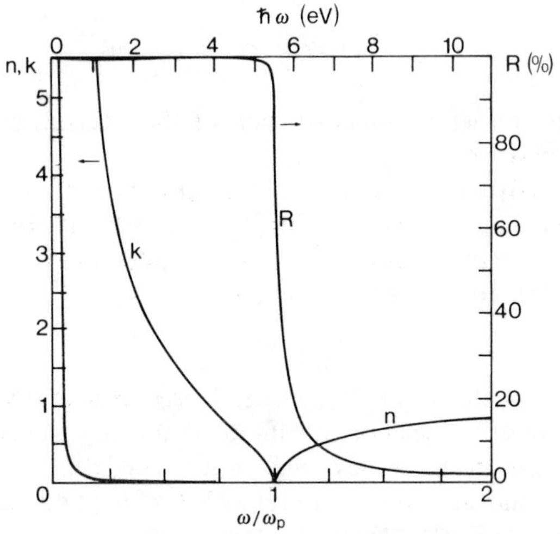

Fig. I.5. Spectral dependence of the real (n) and imaginary (k) parts of the complex dielectric constant $\tilde{n} = n - i\kappa$ of a metal whose dielectric constant is described by a law similar to Eq. (I.2.47) with a plasma frequency $\hbar\omega_p = \hbar(Ne^2/m\varepsilon_0)^{1/2} = 4.7$ eV. In addition, the term ω^2 is replaced by $\omega(\omega - i\gamma)$ to take into account damping effects represented by the coefficient γ, equal in the present case to $4 \times 10^{-3}\,\omega_p$. The curve $R(\omega)$ represents the reflectivity given by the Fresnel formulas discussed in Chapter III. (After F. Wooten [10a].)

where ω_p is the *plasma frequency*, e and m the electron charge and mass, and N the total number of electrons per unit volume present in the medium (see Fig. 1.5).

2.3.a *The Dielectric Susceptibility in a Two-Level Resonant System*

The results of Section I.2 allow us to explicitly evaluate, by referring to a simple but realistic model, the real and imaginary parts of the dielectric susceptibility χ_ω in the vicinity of a resonance frequency ω_0. Note that, by recalling Eqs. (I.2.24,25) and the definition of analytic signal (see Section I.8), we can write

$$\hat{E}(z,t) \cong e^{i\omega_0 t - ikz}\mathscr{E}(z,t), \qquad \hat{P}(z,t) \cong -ie^{i\omega_0 t - ikz}S(z,t), \tag{I.2.48}$$

so that from Eq. (I.2.29) it follows that

$$\hat{P}(z,t) = -i\int_0^{+\infty} dt'\,\xi(z,t',t-t')\hat{E}(z,t-t')\,dt', \tag{I.2.49}$$

where we have set

$$\xi(z,t',t-t') = -(2c\varepsilon d/\omega_0)e^{-\gamma_{ab}t'+i\omega_0 t'}\chi(z,t',t-t'). \tag{I.2.50}$$

Let us now assume that the field intensity is small enough not to modify the population distribution of the two-level atomic system, which is assumed to be in thermodynamic equilibrium. We then have (see Eq. I.2.28) $\chi(z,t',t-t') \cong \chi(z,t',-\infty)$, so that by taking the time Fourier transform of both sides of Eq. (I.2.49) we obtain

$$\hat{P}_\omega(z) = -i\xi_\omega \hat{E}_\omega(z), \tag{I.2.51}$$

where ξ_ω is the Fourier transform of $\xi(z,t',0)$. On the other hand, from the definition of $\chi(t)$ [see Eq. (I.2.2)] it follows that, for $\omega > 0$,

$$\hat{P}_\omega(z) = \varepsilon_0 \chi_\omega \hat{E}_\omega(z), \tag{I.2.52}$$

so that by comparing Eqs. (I.2.51,52) we have

$$\chi_\omega = -(i/\varepsilon_0)\xi_\omega, \tag{I.2.53}$$

that is,

$$\chi_\omega = \frac{2icn_1^2 d}{\omega_0}\int_0^{+\infty} e^{-\gamma_{ab}t'-i(\omega-\omega_0)t'}\chi(z,t',-\infty)\,dt'. \tag{I.2.54}$$

If we take into account Eq. (I.2.30c), we finally obtain

$$\chi_\omega = f\frac{e^2}{\varepsilon_0 m 2\omega_0}\Delta N_0 i\int_{-\infty}^{+\infty}\frac{g(\omega')}{\gamma_{ab}+i(\omega-\omega')}d\omega',$$

where

$$f = \frac{2m\omega_0 p_0^2}{3\hbar e^2}, \qquad \Delta N_0 = N_{a0} - N_{b0}, \tag{I.2.56}$$

m being the electron mass, ΔN_0 the population inversion per unit volume, and f the so-called *oscillator strength.* For $g = \delta(\omega-\omega_0) + \delta(\omega+\omega_0)$,

$$\chi'_\omega = f\frac{\Delta N_0 e^2}{2\varepsilon_0 m\omega_0}\frac{\omega-\omega_0}{\gamma_{ab}^2+(\omega-\omega_0)^2}, \qquad \chi''_\omega = -f\frac{\Delta N_0 e^2}{2\varepsilon_0 m\omega_0}\frac{\gamma_{ab}}{\gamma_{ab}^2+(\omega-\omega_0)^2}.$$

These are precisely the expressions employed in drawing the plots of

$$\frac{\chi'_\omega}{\chi''_{\omega_0}} = -\zeta\mathscr{L}, \qquad \frac{\chi''_\omega}{\chi''_{\omega_0}} = \mathscr{L}, \tag{I.2.58}$$

shown in Fig. I.6, where $\zeta = (\omega-\omega_0)/\gamma_{ab}$ is the so-called *detuning parameter* and $\mathscr{L} = (1+\zeta^2)^{-1}$ the *lorentzian function.* For additional details the reader is referred to Section VII.19.

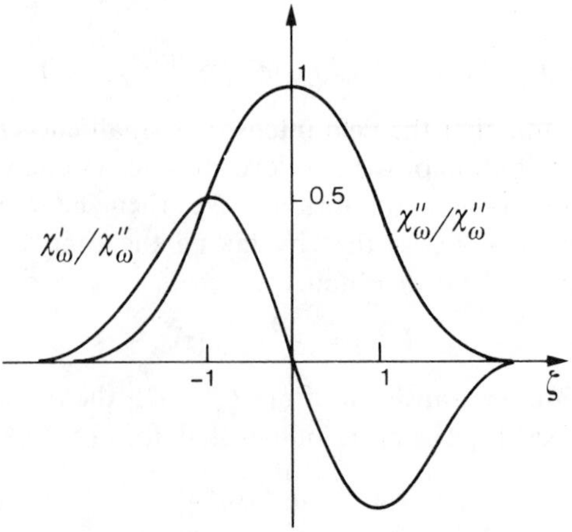

Fig. I.6. Real χ'_ω and imaginary χ''_ω parts of the susceptibility versus the detuning parameter ζ for a Lorentzian line.

2.3.b *Propagation of a Monochromatic Wave in a Homogeneous Medium*

In the high-frequency range we are considering, the Maxwell equations sufficient to describe propagation of a single-frequency wave in a dispersive medium in the absence of external charges and currents read

$$\nabla \times \mathbf{E}(\mathbf{r}) = -i\omega\mu_0\mathbf{H}(\mathbf{r}), \tag{I.2.59a}$$

$$\nabla \times \mathbf{H}(\mathbf{r}) = i\omega\varepsilon_0\tilde{n}^2(\omega)\mathbf{E}(\mathbf{r}). \tag{I.2.59b}$$

If we assume for both $\mathbf{E}(\mathbf{r})$ and $\mathbf{H}(\mathbf{r})$ a space dependence of the type $\exp(-i\mathbf{k}\cdot\mathbf{r})$, where $\mathbf{k} = \mathbf{k}' - i\mathbf{k}''$ is permitted to be a complex quantity, we can immediately derive from Eqs. (I.2.59) the relations

$$\mathbf{k} \times \mathbf{E} = \omega\mu_0\mathbf{H}, \qquad \mathbf{k} \times \mathbf{H} = -\omega\varepsilon_0\tilde{n}^2(\omega)\mathbf{E}. \tag{I.2.60}$$

From these equations it follows, by taking the scalar product of both sides with $\mathbf{k}$, that

$$\mathbf{k}\cdot\mathbf{E} = \mathbf{k}\cdot\mathbf{H} = 0, \tag{I.2.61}$$

while by eliminating $\mathbf{E}$ (or $\mathbf{H}$) between them and employing the vector identity (A.2), we obtain

$$k^2 = k'^2 - k''^2 - 2i\mathbf{k}'\cdot\mathbf{k}'' = k_0^2\tilde{n}_\omega^2.$$

If $\mathbf{k}'$ and $\mathbf{k}''$ are parallel (say to the z axis) the spatial dependence of $\mathbf{E}$ and $\mathbf{H}$ is of the form of a *plane evanescent wave* (the surfaces of constant field are orthogonal to the propagation direction along which the wave attenuates);

that is,

$$\mathbf{E}, \mathbf{H} \propto e^{-ik'z - k''z}, \tag{I.2.62}$$

where

$$k' - ik'' = k_0\tilde{n} \equiv k_0(n - i\kappa), \tag{I.2.63}$$

the quantity κ being called the *absorption* (or *extinction*) *coefficient*.

3 State of Polarization of the Electromagnetic Field

We wish to recall in this section some basic concepts concering the polarization of electromagnetic radiation. Historically, the first intuition concerning a property of light beams associated with their state of polarization is due to Newton, who inferred it in connection with the interpretation of an experiment performed by Huygens. More precisely, Huygens observed that two rays, *ordinary* and *extraordinary*, which are obtained when a single ray penetrates a birefringent crystal (see Section I.4), behave differently from a ray that has not undergone double refraction when they are allowed to propagate in a second birefringent crystal.

Let us first consider propagation in a homogeneous medium, where the electromagnetic vector lies in a plane orthogonal to the wave propagation direction z. The tip of the electric vector $\mathbf{E} = (E_x, E_y, 0)$ associated with a wave whose analytic expression reads, in complex notation,

$$\hat{\mathbf{E}}(x, y, z, t) = a_1(x, y)e^{i\omega t - ikz}\hat{x} + a_2(x, y)e^{i\omega t - ikz}\hat{y} \tag{I.3.1}$$

describes, at any fixed position, a closed curve that can be easily proved to be an ellipse by eliminating the parameter t between E_x and E_y. By doing that, we obtain

$$\frac{E_x^2}{|a_1|^2} + \frac{E_y^2}{|a_2|^2} - 2\frac{E_x E_y}{|a_1 a_2|}\cos\delta = \sin^2\delta, \tag{I.3.2}$$

where we have set $a_1 = |a_1|\exp(i\delta_1)$, $a_2 = |a_2|\exp(i\delta_2)$, and $\delta = \delta_2 - \delta_1$. This equation degenerates into a circumference for $|a_1| = |a_2|$ and $\delta = \pm\pi/2$ (*circular polarization*) and a straight line for $\delta = 0, \pm\pi$ (*linear polarization*).

These considerations can be extended [11] to a generic field $\mathbf{E} = (E_x, E_y, E_z)$ having a complex representation of the kind

$$\hat{\mathbf{E}}(\mathbf{r}, t) = [\mathbf{p}(\mathbf{r}) + i\mathbf{q}(\mathbf{r})]e^{i\omega t}, \tag{I.3.3}$$

where $\mathbf{p}(\mathbf{r})$ and $\mathbf{q}(\mathbf{r})$ are two real vectors. At any fixed position $\mathbf{r}$, the tip of the vector $\mathbf{E}$ describes a curve that, due to its periodicity, is closed and lies in the plane specified by $\mathbf{p}(\mathbf{r})$ and $\mathbf{q}(\mathbf{r})$. Since in this plane it is always possible to choose two real mutually orthogonal vectors **a** and **b** such that

$$\mathbf{p} + i\mathbf{q} = (\mathbf{a} + i\mathbf{b})e^{-i\eta}, \tag{I.3.4}$$

η being a real quantity given by

$$\tan 2\eta = \frac{2\mathbf{p}\cdot\mathbf{q}}{q^2 - p^2}, \tag{I.3.5}$$

we can rewrite Eq. (I.3.3) in the form

$$\hat{\mathbf{E}}(\mathbf{r}, t) = \mathbf{a}\exp(i\omega t - i\eta) + i\mathbf{b}\exp(i\omega t - i\eta),$$

which, after identifying the directions of **a** and **b** with those of $\hat{x}$ and $\hat{y}$, is immediately seen to be similar to Eq. (I.3.1). Accordingly, in the plane we are considering the tip of **E** describes an ellipse whose equation is obtained from Eq. (I.3.2) by putting $\delta = \pi/2$:

$$\frac{E_x^2}{a^2} + \frac{E_y^2}{b^2} = 1. \tag{I.3.6}$$

3.1 *Stokes Parameters and Jones Matrix*

As we have seen above, the electric field of a monochromatic wave changes in time at a given point **r** by remaining parallel to a plane. If we identify it with the xy plane, the general expression for the field has the form

$$\mathbf{E}(t) = a_x\cos(\omega t + \Phi_x)\hat{x} + a_y\cos(\omega t + \Phi_y)\hat{y}, \tag{I.3.7}$$

or in complex notation

$$\hat{\mathbf{E}}(t) = \tilde{a}_x e^{i\omega t}\hat{x} + \tilde{a}_y e^{i\omega t}\hat{y}, \tag{I.3.8}$$

where $\tilde{a}_x = a_x\exp(i\Phi_x)$, $\tilde{a}_y = a_y\exp(i\Phi_y)$. The tip of the vector **E** describes an ellipse whose major axis forms an angle ψ with the x axis given by

$$\psi = \frac{1}{2}\arctan\left(\frac{2a_x a_y}{a_x^2 - a_y^2}\cos\delta\right) \tag{I.3.9}$$

with $\delta = \Phi_x - \Phi_y$. In addition, the ratio b/a between the minor semiaxis and the major semiaxis of the ellipse can be expressed as a function of an auxiliary angle χ, such that $b/a = \tan\chi$, where

$$\chi = \frac{1}{2}\arcsin\left(\frac{2a_x a_y}{a_x^2 + a_y^2}\sin\delta\right). \tag{I.3.10}$$

Thus, the field in **r** is specified by the plane containing its polarization ellipse, by the angles χ and ψ defined above, and by its intensity. Alternatively, the field can be characterized by the so-called *Stokes parameters* s_0, s_1, s_2, and s_3, given by

$$\begin{aligned} s_0 &= a_x^2 + a_y^2, \\ s_1 &= a_x^2 - a_y^2 = s_0 \cos 2\chi \cos 2\psi, \\ s_2 &= 2a_x a_y \cos\delta = s_0 \cos 2\chi \sin 2\psi = 2\,\mathrm{Re}(\tilde{a}_x \tilde{a}_y^*), \\ s_3 &= 2a_x a_y \sin\delta = 2\,\mathrm{Im}(\tilde{a}_x \tilde{a}_y^*) = s_0 \sin 2\chi. \end{aligned} \tag{I.3.11}$$

Obviously, s_0 is proportional to the field intensity, while s_1, s_2, and s_3 can be interpreted [11] as the cartesian coordinates of a point on a sphere of radius s_0 (see Fig. I.7), known as the *Poincaré sphere*, whose longitude and latitude are 2ψ and 2χ, respectively. In particular, the north and south poles correspond respectively to left- and right-handed circular polarization, while linearly polarized fields correspond to points on the equator.

When dealing with quasi-monochromatic waves, the amplitudes a_x and a_y and the phase are time-dependent random quantities and the Stokes parameters are replaced by the ensemble averages (see Section I.8)

$$\begin{aligned} s_0 &= \langle a_x^2(t)\rangle + \langle a_y^2(t)\rangle, \\ s_1 &= \langle a_x^2(t)\rangle - \langle a_y^2(t)\rangle, \\ s_2 &= 2\langle a_x(t)a_y(t)\cos\delta(t)\rangle, \\ s_3 &= 2\langle a_x(t)a_y(t)\sin\delta(t)\rangle. \end{aligned} \tag{I.3.12}$$

In particular, if the field consists of a superposition of several statistically independent fields, the Stokes parameters are the sums of the Stokes parameters of the individual fields. In view of this, van de Hulst formulated the following *principle of optical equivalence* [12]: "It is impossible by means of any instrument to distinguish between various incoherent sums of simple waves that may together form a beam with the same Stokes parameters."

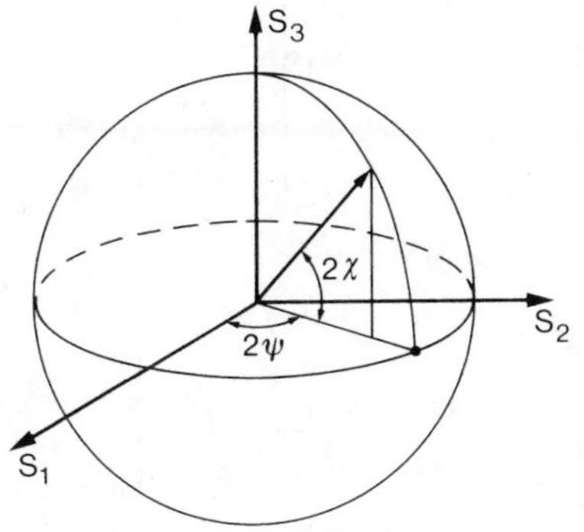

Fig. I.7. The Poincaré sphere.

A wave for which $s_0 \neq 0$ and $s_1 = s_2 = s_3 = 0$ is said to be *unpolarized*. A given quasi-monochromatic wave is decomposable uniquely into an unpolarized part and a polarized part; in fact, in view of the additivity of the Stokes parameters, we can write

$$s_0 = s_0^{(1)} + s_0^{(2)}, \qquad s_i = s_i^{(2)} \qquad (i = 1, 2, 3), \tag{I.3.13}$$

indicating with the superscript (1) the contribution of the polarized wave and with (2) that of the polarized one. Since [see Eq. (I.3.11)] $(s_0^{(2)})^2 = (s_1^{(2)})^2 + (s_2^{(2)})^2 + (s_3^{(2)})^2$, if we define *degree of polarization* as the ratio m between the intensity $s_0^{(2)}$ of the polarized component and the intensity s_0 of the total wave, we have

$$m = [(s_1^{(2)})^2 + (s_2^{(2)})^2 + (s_3^{(2)})^2]^{1/2}/s_0. \tag{I.3.14}$$

It is sometimes important to know the change in the Stokes parameters of a beam traveling in a given medium or scattered by some obstacle. In these cases, the output field $E'_{x'}$, $E'_{y'}$, referred to a particular choice of cartesian coordinates x', y', is related to the input field E_x, E_y by a linear transformation characterized by the so-called *Jones matrix* $\mathbf{A}$, that is [13],

$$\begin{bmatrix} E'_{x'} \\ E'_{y'} \end{bmatrix} = \begin{bmatrix} A_2 & A_3 \\ A_4 & A_1 \end{bmatrix} \cdot \begin{bmatrix} E_x \\ E_y \end{bmatrix}. \tag{I.3.15}$$

If we define the real quantities

$$M_k = A_k A_k^*, \qquad S_{kj} = S_{jk} = \mathrm{Re}(A_j A_k^*)$$
$$(j, k = 1, 2, 3, 4), \tag{I.3.16}$$
$$D_{jk} = -D_{kj} = -\mathrm{Im}(A_j A_k^*)$$

then it can be shown by simple algebra that

$$\begin{bmatrix} \frac{1}{2}s'_0 + \frac{1}{2}s'_1 \\ \frac{1}{2}s'_0 - \frac{1}{2}s'_1 \\ s'_2 \\ s'_3 \end{bmatrix} = \mathbf{F} \cdot \begin{bmatrix} \frac{1}{2}s_0 + \frac{1}{2}s_1 \\ \frac{1}{2}s_0 - \frac{1}{2}s_1 \\ s_2 \\ s_3 \end{bmatrix}, \tag{I.3.17}$$

where

$$\mathbf{F} = \begin{bmatrix} M_2 & M_3 & S_{23} & -D_{23} \\ M_4 & M_1 & S_{41} & -D_{41} \\ 2S_{24} & 2S_{31} & S_{21} + S_{34} & -D_{21} + D_{34} \\ 2D_{24} & 2D_{31} & D_{21} + D_{34} & S_{21} - S_{34} \end{bmatrix}. \tag{I.3.18}$$

Further details about the $\mathbf{F}$ matrix can be found in Section VI.13.

3.2 *Polarization of Evanescent Waves*

Let us consider an evanescent wave of the form encountered in Section I.2. It is possible to show that, as a consequence of Eq. (I.2.61) and of the complex nature of **k**, the field is elliptically polarized. To this aim, consider a transverse-magnetic (TM) wave (see Section III.7) with electric field components

$$E_x(\mathbf{r}) = E_0 \cos\theta e^{-i\mathbf{k}\cdot\mathbf{r}}, \tag{I.3.19a}$$

$$E_z(\mathbf{r}) = E_0 \sin\theta e^{-i\mathbf{k}\cdot\mathbf{r}}, \tag{I.3.19b}$$

where $\theta = \theta' - i\theta''$ is the complex angle between **k** and $\hat{z}$. The above relations imply the validity of $\mathbf{E}\cdot\mathbf{k} = 0$ as well as $H_x = H_z = 0$ (TM wave).

If we assume, without loss of generality, that $E_0 = 1$ and neglect the nonessential common phase factor $\exp(-i\mathbf{k}\cdot\mathbf{r})$, the time dependence of the electric field reads

$$\hat{E}_x(t) = \mathrm{Re}(\cos\theta e^{i\omega t}) = \cos\theta' \cosh\theta'' \cos\omega t - \sin\theta' \sinh\theta'' \sin\omega t \tag{I.3.20a}$$

$$\hat{E}_z(t) = \mathrm{Re}(\sin\theta e^{i\omega t}) = \sin\theta' \cosh\theta'' \cos\omega t + \cos\theta' \sinh\theta'' \sin\omega t, \tag{I.3.20b}$$

and by eliminating t between the above equations, we easily obtain

$$\hat{E}_x^2(\sin^2\theta' + \sinh^2\theta'') + \hat{E}_z^2(\cos^2\theta' + \sinh^2\theta'') - \hat{E}_x\hat{E}_z \sin 2\theta' = \tfrac{1}{4}\sinh^2 2\theta''. \tag{I.3.21}$$

Therefore, the vector $\hat{E}_x(t)\hat{x} + \hat{E}_y(t)\hat{y}$ describes an ellipse whose major axis is inclined with respect to the x direction by an angle θ', the lengths of the two axes being proportional to $\cosh\theta''$ and $|\sinh\theta''|$. Thus, a TM wave is associated with an elliptically polarized electric field parallel to the xz plane (see Fig. I.8), while the magnetic field lies along the y direction. For $\theta'' = 0$

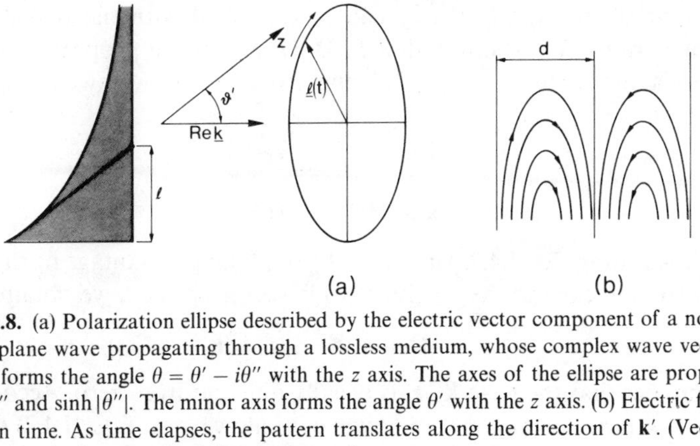

Fig. I.8. (a) Polarization ellipse described by the electric vector component of a nonhomogeneous plane wave propagating through a lossless medium, whose complex wave vector $\mathbf{k} = \mathbf{k}' - i\mathbf{k}''$ forms the angle $\theta = \theta' - i\theta''$ with the z axis. The axes of the ellipse are proportional to $\cosh\theta''$ and $\sinh|\theta''|$. The minor axis forms the angle θ' with the z axis. (b) Electric field lines at a given time. As time elapses, the pattern translates along the direction of $\mathbf{k}'$. (Vectors are underlined in the figure and boldface in the text.)

(propagating waves), the ellipse degenerates into a straight line. For transverse-electric (TE) waves, the vector **H** turns out to be elliptically polarized while **E** is orthogonal to the plane of incidence.

4 Propagation in Anisotropic Media

Many macroscopic media are anisotropic, a circumstance that introduces preferential directions in the geometry of the propagation problem and requires a generalization of the concepts introduced up to now [13]. In particular, it is obvious that, from a macroscopic point of view, the medium will no longer be represented by scalar dielectric and magnetic permeabilities ε and μ, the obvious generalization to the constitutive relations connecting $\mathbf{E}_\omega$ and $\mathbf{H}_\omega$ to $\mathbf{D}_\omega$ and $\mathbf{B}_\omega$ being, as already envisaged in Eq. (I.2.3), of the kind

$$D_{\omega i} = \tilde{\varepsilon}_{\omega ij} E_{\omega j}, \tag{I.4.1a}$$

$$B_{\omega i} = \tilde{\mu}_{\omega ij} H_{\omega j}, \tag{I.4.1b}$$

with $\tilde{\varepsilon}_{\omega ij}$ and $\tilde{\mu}_{\omega ij}$ the elements of a second-rank tensor.

Equations (I.4.1) apply to optical propagation in a crystal, μ usually being a scalar quantity nearly equal to μ_0 at the frequencies of interest. The medium is thus described by a dielectric permeability tensor $\tilde{\boldsymbol{\varepsilon}}_\omega$ whose elements are given by Eq. (I.2.16), which expresses the fact that the medium responds in different ways to different directions of application of the electric field (**D** is not necessarily parallel to **E**). This, in turn, implies that the characteristics of propagation depend on the directions of propagation and polarization of the wave (*birefringence*).

4.1 *Birefringence*

Let us consider the propagation of plane monochromatic waves in a transparent crystal (for which $\tilde{\boldsymbol{\varepsilon}}_\omega$ can be identified with the real dielectric constant tensor $\boldsymbol{\varepsilon}_\omega$). Assuming that **E**, **D**, **H**, and **B** are proportional to the common factor $\exp(i\omega t - i\mathbf{k}\cdot\mathbf{r})$, $\mathbf{k}$ being a real vector, Maxwell's equations (I.2.60) take the form

$$\mathbf{k} \times \mathbf{E} = \omega\mu_0 \mathbf{H}, \tag{I.4.2a}$$

$$\mathbf{k} \times \mathbf{H} = -\omega \mathbf{D}. \tag{I.4.2b}$$

By substituting Eq. (I.4.2a) into (I.4.2b) and taking advantage of the vector identity (A.2) to expand the resulting triple vector product, we obtain

$$\mu_0 \omega^2 \mathbf{D} = k^2 \mathbf{E} - (\mathbf{k}\cdot\mathbf{E})\mathbf{k}, \tag{I.4.3}$$

which implies that the term $\mathbf{k}\cdot\mathbf{E}$ is usually different from zero. Accordingly, the vector **k** is not orthogonal to **E**, while it follows from Eqs. (I.4.2) that it is

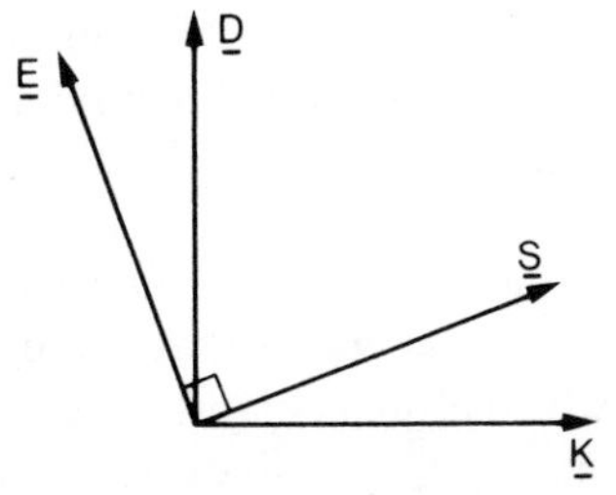

Fig. I.9. Vector geometry for propagation of normal modes in anisotropic media.

orthogonal to both **D** and **H**. The situation is depicted in Fig. I.9, which in particular shows that **k** and the Poynting vector **S** (see Section I.6) are in general not parallel and that, consequently, the direction of the power flow is not necessarily the same as the direction of the wave front normal.

In order to proceed, let us assume that the direction of propagation $\hat{s} = \mathbf{k}/k$ of the beam in the crystal is assigned; it is then possible to show that the set of Eqs. (I.4.2) is consistent with two possible choices for the magnitude of **k** and, accordingly, for the refractive index $n = k/k_0$. To this end, it is convenient to introduce a system of cartesian axes coinciding with the *principal axes* $(\bar{x}\bar{y}, \bar{z})$ of the dielectric constant tensor $\boldsymbol{\varepsilon}_\omega$, in which Eq. (I.4.1a) takes the diagonal form

$$D_{\omega i} = \bar{\varepsilon}_{\omega i} E_{\omega i} \tag{I.4.4}$$

the $\bar{\varepsilon}_{\omega i}$ being called the *principal dielectric constants*. In this reference system, after introducing Eq. (I.4.4) into Eq. (I.4.3), we easily obtain

$$E_{\omega i} = \frac{n^2_{\omega,s}\hat{s} \cdot \mathbf{E}_\omega}{n^2_{\omega,s} - \bar{\varepsilon}_{\omega i}/\varepsilon_0} s_i, \tag{I.4.5}$$

and multiplying both sides of equation by s_1 and summing over i, we obtain

$$\hat{s} \cdot \mathbf{E}_\omega = \hat{s} \cdot \mathbf{E}_\omega \sum_{i=1}^{3} \frac{n^2_{\omega,s} s_i^2}{n^2_{\omega,s} - \mu_0 \bar{\varepsilon}_{\omega,i}}, \tag{I.4.6}$$

from which the *Fresnel equation* follows:

$$\frac{s_1^2}{n^2_{\omega,\hat{s}} - \bar{\varepsilon}_{\omega 1}/\varepsilon_0} + \frac{s_2^2}{n^2_{\omega,\hat{s}} - \varepsilon_{\omega 2}/\varepsilon_0} + \frac{s_3^2}{n^2_{\omega,\hat{s}} - \varepsilon_{\omega 3}/\varepsilon_0} = \frac{1}{n^2_{\omega,\hat{s}}}. \tag{I.4.7}$$

This equation is biquadratic in the unknown $n_{\omega,\hat{s}}$ and consequently admits two pairs of solutions $\pm n_1$ and $\pm n_2$. Neglecting the trivial degeneracy associated with the $\pm$ sign (corresponding to waves propagating in opposite directions), this implies that, in correspondence to a fixed direction $\hat{s}$, two distinct plane waves propagate with unequal phase velocities c/n_1 and c/n_2. It is possible to show that each of the waves is linearly polarized and that the polarization directions (that is, the directions of the vector **E**) are mutually orthogonal, so that, for any given $\hat{s}$, two plane waves with orthogonal linear polarization can propagate inside the anisotropic medium, each "seeing" a different refractive index n_1 or n_2.

If we plot along each direction $\hat{t}$ of the Poynting vector of the plane wave of wave vector $k\hat{s}$, starting from the origin, a segment of length n^{-1}, we obtain the so-called *ray surface* or *Fresnel wave surface*, which gives a complete picture of the distribution of the ray velocities $v_{1,2} = c/n_{1,2}$ for all possible directions. In general, we obtain a two-sheeted surface, one sheet of which corresponds to c/n_1, the other one to c/n_2. These two sheets intersect in two (uniaxial) or four (biaxial) points, which define the optic axes of the crystal.

When we plot the quantities $n_{1,2}^{-1}$ as radius vectors in the direction $\hat{s}$ from the origin of a coordinate system, we obtain the two-sheeted *normal surface*. In particular, for a uniaxial crystal the normal surface reduces to a sphere, corresponding to the O-wave, and a fourth-order revolution surface, corresponding to the E-wave. These two surfaces touch each other at the points corresponding to the optic axis.

It is worth noting at this point that it is not easy to reproduce a situation in which the two waves share the same propagation vector **k** since each of them will be refracted, if entering from outside the crystal, at a different angle (see Born and Wolf [11], Chapter XIV).

In any uniaxial crystal, one of the principal dielectric axes coincides with the axis of symmetry (*optic axis*, usually identified with the z-axis) while the other two are perpendicular to it, but otherwise arbitrary (it is usual to set $\bar{\varepsilon}_1 = \varepsilon_x = \hat{\varepsilon}_2 = \varepsilon_y$).

In general we can solve the propagation problem by resorting to a geometric representation known as the *index ellipsoid* (see Born and Wolf [11], Chapter XIV) defined as

$$(x^2/n_x^2) + (y^2/n_y^2) + (z^2/n_z^2) = 1 \tag{I.4.8}$$

where $n_{x,y,z}$ are the principal refractive indexes. The index ellipsoid, also known as the *optical indicatrix*, can be used to find the two refractive indexes n_1, and n_2 associated with the two independent linearly polarized plane waves which can propagate along an arbitrary direction $\hat{s}$ in a crystal. This is done by finding the intersection ellipse between the plane normal to $\hat{s}$ through the origin and the index ellipsoid. The two semiaxes of the intersection ellipse are equal to the refractive indexes n_1 and n_2 of the two normal modes. These axes are in turn parallel, respectively, to the directions of the vectors $\mathbf{D}_{1,2}$ of the two modes. The electric fields $\mathbf{E}_{1,2}$ are parallel to the normals to the index ellipsoid at its intersection with the ellipse axes.

It turns out that the value assumed by one of the two solutions of the Fresnel equation depends on the angle between $\hat{s}$ and z. More precisely, while

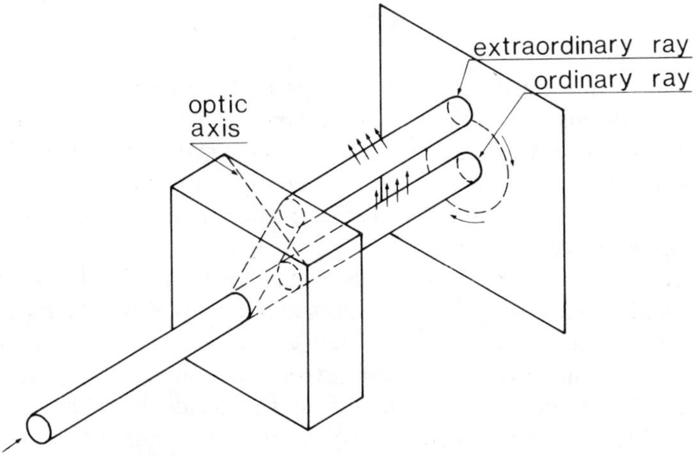

Fig. I.10. Double refraction of ordinary and extraordinary rays.

one of the two waves, termed *ordinary*, exhibits an effective refractive index independent from θ and given by $n_1 \equiv n_o \equiv \sqrt{\bar{\varepsilon}_x/\varepsilon_0}$, the other one (*extraordinary*) has a refractive index $n_2 \equiv n_2(\theta)$ which depends on θ and whose value ranges between n_o and $\sqrt{\bar{\varepsilon}_z/\varepsilon_0} \equiv n_e$ (see Problem 13).

The difference between the values assumed by the effective refractive indices of the ordinary and extraordinary rays has as an obvious consequence the fact that the two rays undergo a different refraction when entering the crystal (see Fig. I.10), which justifies the term "birefringence" for the phenomenon described in this section.

It should be noted that the Poynting vector $\frac{1}{2}\mathbf{E} \times \mathbf{H}$ is generally tilted with respect to the wave vector of the normal mode. When we consider the propagation of a ray bundle, for example, a Gaussian laser beam, the direction of flight of the pencil *does not* coincide with the propagation vector of the central component of the plane waves composing the beam. S. M. Rytov has shown that the ray pencil propagates along the direction of the *Poynting vector* relative to the central component of the plane-wave packet. This result can be easily proved by representing the field with a diffraction integral (see Chapter IV), which is in turn calculated by applying the *stationary-phase method* discussed in Chapter V.

4.1.a *The Electrooptic Effect*

The application of an electric field $\mathbf{E}^{(0)}$ along some direction in a crystal changes the values of the elements ε_{ij} of the dielectric constant tensor, the new components ε_{ij} being expressible in terms of the old ones by a relation of the kind

$$\varepsilon'_{ij} = \varepsilon_{ij} + t_{ijk}E_k^{(0)} + T_{ijkl}E_k^{(0)}E_l^{(0)}, \qquad \text{(I.4.8')}$$

t_{ijk} and T_{ijkl} being third- and fourth-order rank tensors that characterize, respectively, the *linear electrooptic effect* (known as the *Pockels effect*) and the *quadratic electro-optic effect* (known as the *Kerr effect*). While the Pockels effect exists only in crystals that lack a center of symmetry (i.e., that do not possess an inversion symmetry), the Kerr effect is also observed in centrosymmetric crystals.

According to Eq. (I.4.8), the application of an electric field $\mathbf{E}^{(0)}$ changes both the principal axes and the principal dielectric constants of the crystal in a way that depends on the direction and strength of the field. Hence, by varying the second of this quantities, one can expect to control the propagation characteristics of an electromagnetic field $\mathbf{E}^{(1)}$ incident on the crystal. A *Pockels cell*, for example, consists of a crystal plate, with conducting electrodes deposited on the working faces of the plate, placed between two mutually orthogonal polarizers (see Fig. I.11). By applying a voltage V to the crystal, the phase retardance Γ between two rays propagating through the cell can be altered by a desired amount.

Let us consider, in particular, a uniaxial crystal of potassium dihydrogen phosphate (KH_2PO_4; usually known as KDP) cut in the form of a plate with two faces perpendicular to the quadruple symmetry axis (optic axis z) and with two electrodes made of transparent metal oxide coatings. With no voltage applied, one of the principal axes is the optic axis z and the other two can be chosen arbitrarily in a plane orthogonal to it. Application of the voltage V removes this degeneracy and, while one of the principal axes still coincides with z, the other two are given by the two double symmetry axes of the crystal rotated by an angle of $\pi/4$ [13]. The relative principal electric constants $\bar{\varepsilon}_x$ and $\bar{\varepsilon}_y$ can be expressed in terms of the ordinary refractive index n_o, in the absence of voltage, and the applied field $E^{(0)}$ in the form

$$\varepsilon_0/\bar{\varepsilon}_x = (1/n_o^2) + rE^{(0)}, \qquad \text{(I.4.9a)}$$

$$\varepsilon_0/\bar{\varepsilon}_y = (1/n_o^2) - rE^{(0)}, \qquad \text{(I.4.9b)}$$

where r is a constant typical of the crystal. Equations (I.4.9) imply that the two effective refractive indices are approximately given by (usually, $rE^{(0)} \ll 1$)

$$n_1 = n_o - \tfrac{1}{2}rn_o^3E^{(0)}, \qquad n_2 = n_o + \tfrac{1}{2}rn_o^3E^{(0)}, \qquad \text{(I.4.10)}$$

so that the two components of a linearly polarized field $\mathbf{E}^{(1)}$ propagating along the z axis will acquire, over the crystal length L, a *phase retardation* Γ given by

$$\Gamma = k_0(n_2 - n_1)L = n_o^3k_0rLE^{(0)} = k_0n_o^3rV, \qquad \text{(I.4.11)}$$

where $V = LE^{(0)}$.

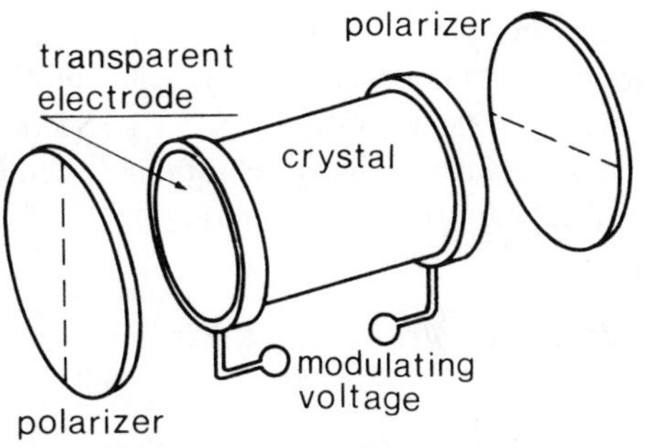

Fig. I.11. Pockels cell.

It is often convenient to introduce the quantity

$$V_\pi = \lambda/2n_o^3 r, \tag{I.4.12}$$

which represents the voltage that induces a phase delay π. For $\lambda = 0.54\ \mu$m, KDP is characterized by $V_\pi = 7.9$ kv, while in KD*P, where the hydrogen has been replaced by deuterium, $V_\pi \simeq 3.4$ kV. If the input polarization analyzer is oriented in such a way that $\hat{E}_x^{(1)} = \hat{E}_y^{(1)}$ and the output polarization analyzer is rotated by $\pi/2$ with respect to it, then the output intensity is given, apart from a proportionally factor, by

$$I_{\text{out}} = \tfrac{1}{2}|\hat{E}_x^{(1)}(e^{i\Gamma} - 1)|^2, \tag{I.4.13}$$

and the transmittance T of the Pockels cell depends on the applied voltage V through the relation

$$T = I_{\text{out}}/I_{\text{in}} = \sin^2(\Gamma/2) = \sin^2(\pi V/2V_\pi). \tag{I.4.14}$$

5 Propagation in Spatially Dispersive Media

The relation between the electric induction **D** and the electric field **E** given in Section I.2 [see Eqs. (I.2.1) and (I.2.2)] is nonlocal in the time argument; that is, the value of **D** at time t depends on the value of **E** at time t and at other earlier times $t' < t$. This nonlocalization in time is unavoidably associated with a nonlocalization in space [completely neglected in Eq. (I.2.2), where the value of **P** at a given position **r** depends only on the value of **E** at the same point]. In fact, if the polarization vector **P** at a given instant t and position **r** is determined by the value of **E** at the same point at time $t - \Delta t$, during the time interval Δt the electric disturbance will travel a distance $v\,\Delta t$ (v being its typical velocity); as a consequence, the polarization **P** and **r** will also be influenced by the value of the electric field at time t and at adjacent positions **r**$'$ such that $|\mathbf{r}' - \mathbf{r}| = v\,\Delta t$. Thus, whenever $v\,\Delta t$ (which represents a characteristic length l of the medium, e.g., molecular dimension, lattice constant, or Debye radius) becomes comparable with the field wavelength λ, *spatial dispersion* must be taken into account. For example, this might occur near a resonance frequency, where the refractive index n may become very large and the wavelength $\lambda = \lambda_0/n$ in the medium may scale accordingly (see, e.g., Section I.2.2).

These considerations lead us to replace the usual relation between **P** and **E** in the time domain [see Eq. (I.2.2)] with the more general one

$$\mathbf{P}(\mathbf{r}, t) = \varepsilon_0 \iiint_{-\infty}^{+\infty} d\mathbf{r}' \int_{-\infty}^{t} dt' \, \chi(\mathbf{r} - \mathbf{r}', t - t')\mathbf{E}(\mathbf{r}', t'). \tag{I.5.1}$$

By using the time–space Fourier transform [see Eq. (I.1.19)] the constitutive relation (I.2.3) becomes

$$\mathbf{D}_{\mathbf{q},\omega} = \varepsilon_0(1 + \chi_{\mathbf{q},\omega})\mathbf{E}_{\mathbf{q},\omega} = \tilde{\varepsilon}_{\mathbf{q},\omega}\mathbf{E}_{\mathbf{q},\omega}, \tag{I.5.2}$$

where **q** and ω must be considered independent variables. As a consequence of spatial dispersion, the *dielectric permittivity* $\tilde{\varepsilon}_{\mathbf{q},\omega}$ in general acquires a tensorial character so that Eq. (I.5.2) reads

$$\mathbf{D}_{\mathbf{q},\omega} = \tilde{\boldsymbol{\varepsilon}}_{\mathbf{q},\omega} \cdot \mathbf{E}_{\mathbf{q},\omega'}, \tag{I.5.3}$$

where $\tilde{\boldsymbol{\varepsilon}}_{\mathbf{q},\omega}$ is a tensor of rank two, also for a medium that would normally be considered isotropic (in the zeroth-order approximation in l/λ). In addition, if the medium is spatially dispersive but macroscopically inhomogeneous, $\tilde{\boldsymbol{\varepsilon}}_{\mathbf{q},\omega}$ exhibits, in complete analogy with the case of temporally dispersive media (see Section I.2), a parametric dependence on the position **r**.

The main differences between propagation in ordinary and spatially dispersive media can be appreciated by comparing monochromatic plane-wave solutions of Maxwell's equations (I.1.1,2), written for $\mathbf{J} = 0$, and with the help of Eq. (I.1.6) it is possible to derive

$$\nabla \times (\nabla \times \mathbf{E}) + \mu_0 \frac{\partial^2}{\partial t^2}\mathbf{D} = 0. \tag{I.5.4}$$

If we now look for solutions of this equation in the form of *normal modes*, that is,

$$\mathbf{E}'(\mathbf{r}, t) = \mathbf{E}_{\mathbf{q},\omega}e^{i\omega t - i\mathbf{q}\cdot\mathbf{r}}, \qquad \mathbf{D}'(\mathbf{r}, t) = \mathbf{D}_{\mathbf{q},\omega}e^{i\omega t - i\mathbf{q}\cdot\mathbf{r}}, \tag{I.5.5}$$

we have, after taking advantage of Eqs. (I.5.3) and (I.1.4) (written for $\rho = 0$,

$$\mathbf{q} \times (\mathbf{q} \times \mathbf{E}_{\mathbf{q},\omega}) + \mu_0\omega^2\tilde{\boldsymbol{\varepsilon}}_{\mathbf{q},\omega} \cdot \mathbf{E}_{\mathbf{q},\omega} = 0 \tag{I.5.6}$$

and

$$\mathbf{q} \cdot \tilde{\boldsymbol{\varepsilon}}_{\mathbf{q},\omega} \cdot \mathbf{E}_{\mathbf{q},\omega} = 0. \tag{I.5.7}$$

It is then possible to distinguish between two possibilities, according to whether $\tilde{\boldsymbol{\varepsilon}}_{\mathbf{q},\omega} \neq 0$ or $\tilde{\boldsymbol{\varepsilon}}_{\mathbf{q},\omega} = 0$. In the first case we must have $\mathbf{q} \cdot \mathbf{D}_{\mathbf{q},\omega} = 0$ (*transverse waves*), and Eq. (I.5.6) implies (after using the vector identity A.2)

$$D(\mathbf{q}, \omega) \equiv \det(\mu_0\omega^2\tilde{\boldsymbol{\varepsilon}}_{\mathbf{q},\omega} + \mathbf{qq} - q^2\mathbf{1}) = 0, \tag{I.5.8}$$

representing the relation (*dispersion equation*) that enables one to express $\mathbf{q}$ in terms of ω (or vice versa) by means of the dispersion function $D(\mathbf{q}, \omega)$. In particular, for a spatially nondispersive medium $\tilde{\boldsymbol{\varepsilon}}_{\mathbf{q},\omega}$ is independent of $\mathbf{q}$ and Eq. (I.5.8) is quadratic in q^2. In the following we will indicate with $\mathbf{k} = \mathbf{q}(\omega)$ the value of $\mathbf{q}$ that satisfies the above dispersion equation for an assigned ω. In particular, if we put $\mathbf{q} = nk_0\hat{s}$, we can show that for $\tilde{\boldsymbol{\varepsilon}}$ independent of $\mathbf{q}$, Eq. (I.5.8) is equivalent to the Fresnel equation (I.4.6).

In the second case ($\tilde{\boldsymbol{\varepsilon}}_{\mathbf{q},\omega} = 0$), it follows from Eq. (I.5.6) that

$$\mathbf{q} \times \mathbf{E}_{\mathbf{q},\omega} = 0 \tag{I.5.9}$$

so that the waves are *longitudinal.*

5.1 *Natural Optical Activity*

An effect produced by spatial dispersion is *natural optical activity*, namely the rotation of the plane of polarization of linearly polarized light propagating through certain media. In order to understand this process we note that, since spatial dispersion is expected to play a role only when the wavelength λ of the field is so small as to become comparable with the characteristic length l of the medium (the nondispersive theory being the zero-order approximation in l/λ), we can expand the induction vector $\mathbf{D}$ in the form

$$D_i = \tilde{\varepsilon}_{ij}E_j + \gamma_{ijk}\frac{\partial}{\partial x_j}E_k, \tag{I.5.10}$$

where for simplicity, we omit the subscript ω. In this equation $\tilde{\varepsilon}_{ij}$ is the dielectric constant tensor in the limit $q \to 0$ and γ_{ijk} an antisymmetric tensor. For a plane wave, with wave vector $\mathbf{q} = k_0\hat{s}$, we have $\nabla\mathbf{E}_\omega = -ik_0\hat{s}\mathbf{E}_\omega$, so that the dielectric constant tensor can be rewritten as

$$\tilde{\varepsilon}_{\mathbf{q},\omega ij} = \tilde{\varepsilon}_{0,\omega ij} - ik_0\gamma_{ijk}s_k. \tag{I.5.11}$$

It is convenient at this point to introduce the *gyration vector* g by means of the relation

$$k_0\gamma_{ijk}s_k = e_{ijk}g_k, \tag{I.5.12}$$

where e_{ijk} is the *antisymmetric unit tensor of rank three* ($e_{ijk} \neq 0$ for $i \neq j \neq k$, $e_{ijk} = -e_{jik} = -e_{ikj} = -e_{kji}, e_{123} = 1$) (Levi–Civita tensor), and write accordingly

$$\tilde{\varepsilon}_{\mathbf{q},\omega ij} = \tilde{\varepsilon}_{0,\omega ij} - ie_{ijk}g_k, \tag{I.5.13}$$

which in turn, when inserted in Eq. (I.5.3), implies

$$\mathbf{D}_{\mathbf{q},\omega} = \tilde{\boldsymbol{\varepsilon}}_\omega \cdot \mathbf{E}_{\mathbf{q},\omega} + i\mathbf{g} \times \mathbf{E}_{\mathbf{q},\omega}, \tag{I.5.14}$$

where $\tilde{\boldsymbol{\varepsilon}}_\omega = \tilde{\boldsymbol{\varepsilon}}_{0,\omega}$.

In particular, in an isotropic material $\boldsymbol{\varepsilon}_\omega$ is diagonal and $\mathbf{g} = g\hat{s}$; that is, the gyration vector is parallel to the direction of propagation of the wave. This fact has as an immediate consequence that since $\mathbf{D}_{\mathbf{q},\omega} \cdot \hat{s} = 0$ (*transverse waves*), $\mathbf{E}_{\mathbf{q},\omega} \cdot \hat{s} = 0$, as immediately seen from Eq. (I.5.14). By inserting this equation into Eq. (I.5.4) it is possible to obtain, after choosing the propagation direction as the z axis the solution [1]

$$\begin{aligned} \hat{E}_x(z,t) &= E_0 e^{ikz - i\omega t} \cos \chi z, \\ \hat{E}_y(z,t) &= E_0 e^{ikz - i\omega t} \sin \chi z, \end{aligned} \tag{I.5.15}$$

where $k = \frac{1}{2}k_0(n_+ + n_-)$, $\chi = \frac{1}{2}k_0(n_+ - n_-)$, and n_+ and n_- are obtainable from

$$n_\pm^2 = n_0^2 \pm g n_0, \tag{I.5.16}$$

where n_0 is the refractive index of the isotropic medium in the absence of spatial dispersion. Materials for which $g \neq 0$ are said to possess *natural optical activity*.

According to Eqs. (I.5.15), when the wave has traveled a distance L inside the medium, the ratio $\hat{E}_y/\hat{E}_x = \tan(\chi L)$ is still real and the field is linearly polarized but the direction of polarization is changed. This is precisely the phenomenon discovered by Arago in 1811 in quartz crystals and by Biot in 1815 in liquids (such as oil of turpentine and aqueous solutions of tartaric acid). In 1810 Fresnel explained optical activity as the result of *circular birefringence*, i.e., of different propagation velocities in a medium of right and left circularly polarized light, and in 1848 Pasteur attributed it to *enantiomorphism* (the existence of mirror-image forms of the same molecule), thus originating the branch of chemistry now called *stereochemistry*.

6 Energy Relations

The *instantaneous Poynting vector* is defined as

$$\mathbf{S}(\mathbf{r},t) = \mathbf{E}(\mathbf{r},t) \times \mathbf{H}(\mathbf{r},t). \tag{I.6.1}$$

If we now apply the operator $\nabla\cdot$ on both sides of this equation, use the vector identity (A.9), and take advantage of Maxwell's equations (I.1.1) and (I.1.2), we immediately obtain

$$\nabla \cdot \mathbf{S} = \mathbf{H} \cdot \nabla \times \mathbf{E} - \mathbf{E} \cdot \nabla \times \mathbf{H} = -\mathbf{H} \cdot \frac{\partial}{\partial t}\mathbf{B} - \mathbf{E} \cdot \frac{\partial}{\partial t}\mathbf{D} - \mathbf{J} \cdot \mathbf{E}. \tag{I.6.2}$$

If we integrate both sides of this equation over an arbitrary volume V and apply the Gauss theorem to express its left-hand side as an integral over the surface ∂V enclosing it, we get, indicating by $\hat{n}$ the unit outward normal,

$$-\oiint_{\partial V} \mathbf{S} \cdot \hat{n}\, dS = \iiint_V \left(\mathbf{J} \cdot \mathbf{E} + \mathbf{H} \cdot \frac{\partial}{\partial t} \mathbf{B} + \mathbf{E} \cdot \frac{\partial}{\partial t} \mathbf{D} \right) dV \qquad \text{(I.6.3)}$$

or, recalling Eqs. (I.1.6) and (I.2.1),

$$-\oiint_{\partial V} \mathbf{S} \cdot \hat{n}\, dS = \iiint_V \left(\frac{\partial}{\partial t} w_0 + \mathbf{J} \cdot \mathbf{E} + \mathbf{E} \cdot \frac{\partial}{\partial t} \mathbf{P} \right) dV, \qquad \text{(I.6.4)}$$

where

$$w_0 = \tfrac{1}{2}\varepsilon_0 \mathbf{E} \cdot \mathbf{E} + \tfrac{1}{2}\mu_0 \mathbf{H} \cdot \mathbf{H}. \qquad \text{(I.6.5)}$$

Equation (I.6.5) may be interpreted as the *energy law* of the electromagnetic field by assuming that the flux of the Poynting vector (changed in sign) represents the total power flowing through into the volume V, w_0 the electromagnetic energy density stored *in vacuo*, $-\mathbf{J} \cdot \mathbf{E}$ the power density transferred from $(\mathbf{J} \cdot \mathbf{E} < 0)$ [or to $(\mathbf{J} \cdot \mathbf{E} > 0)$] the sources to (or from) the electromagnetic field, and $\mathbf{E} \cdot (\partial \mathbf{P}/\partial t)$ the power density expended by the field on the dielectric dipoles.

6.1 *Poynting Relations for a Quasi-Monochromatic Ray Bundle in a Spatially Dispersive Medium*

When the medium is temporally and spatially dispersive, the term $\mathbf{E} \cdot (\partial \mathbf{D}/\partial t)$ cannot be interpreted as the time derivative of an energy density, which by its nature must depend on the local instantaneous value of the field. However, if we consider a linear medium and quasi-monochromatic light possessing a spatial distribution similar to that of a plane wave, we can rearrange the first two terms of the right-hand side of Eq. (I.6.2) as the sum of two contributions representing, respectively, the power transferred reversibly and irreversibly to (or from) the medium. In order to see this, let us consider a narrow beam directed along a generally complex wave vector $\mathbf{k}$, which can be represented in the form

$$\mathbf{E}(\mathbf{r}, t) = \mathrm{Re}[\mathscr{E}(\mathbf{r}, t) e^{i\omega t - i\mathbf{k} \cdot \mathbf{r}}] = \mathrm{Re}\, \hat{\mathbf{E}}(\mathbf{r}, t), \qquad \text{(I.6.6)}$$

where ω and $\mathbf{k}$ satisfy the dispersion equation (I.5.8) and

$$\mathscr{E}(\mathbf{r}, t) = \frac{1}{(2\pi)^4} \iiint_{-\infty}^{+\infty} d\mathbf{q}' \int_{-\infty}^{+\infty} d\omega'\, \mathbf{E}_{\mathbf{k}+\mathbf{q}', \omega+\omega'} e^{i\omega' t - \mathbf{q}' \cdot \mathbf{r}}, \qquad \text{(I.6.7)}$$

with $\mathbf{E}_{\mathbf{k}+\mathbf{q}', \omega+\omega'}$ sensibly different from zero for ω' and the *real* vector $\mathbf{q}'$ sufficiently small (*ray bundle*). Under these assumptions, the electric induction

D can be expressed in the form of Eq. (I.6.6), that is,

$$\mathbf{D}(\mathbf{r}, t) = \mathrm{Re}[\mathscr{D}(\mathbf{r}, t)e^{i\omega t - i\mathbf{k}\cdot\mathbf{r}}] = \mathrm{Re}\,\hat{\mathbf{D}}(\mathbf{r}, t), \tag{I.6.8}$$

with

$$\begin{aligned}\mathscr{D}(\mathbf{r}, t) &= \frac{1}{(2\pi)^4}\iiint_{-\infty}^{\infty} d\mathbf{q}' \int_{-\infty}^{+\infty} d\omega' \tilde{\boldsymbol{\varepsilon}}_{\mathbf{k}+\mathbf{q}',\omega+\omega'} \cdot \mathbf{E}_{\mathbf{k}+\mathbf{q}',\omega+\omega'} e^{i\omega' t - i\mathbf{q}'\cdot\mathbf{r}} \\ &\cong \frac{1}{(2\pi)^4}\iiint_{-\infty}^{+\infty} d\mathbf{q}' \int_{-\infty}^{+\infty} d\omega' \left[\tilde{\boldsymbol{\varepsilon}}_{\mathbf{k},\omega} + \mathbf{q}' \cdot (\nabla_{\mathbf{k}}\tilde{\boldsymbol{\varepsilon}}_{\mathbf{k},\omega}) + \omega'\left(\frac{\partial}{\partial\omega}\tilde{\boldsymbol{\varepsilon}}_{\mathbf{k},\omega}\right)\right] \\ &\quad \cdot \mathbf{E}_{\mathbf{k}+\mathbf{q}',\omega+\omega'} e^{i\omega' t - i\mathbf{q}'\cdot\mathbf{r}}\end{aligned} \tag{I.6.9}$$

that is equivalent to

$$\mathscr{D} = \tilde{\boldsymbol{\varepsilon}}_{\mathbf{k},\omega}\mathscr{E} + i(\nabla_{\mathbf{k}}\tilde{\boldsymbol{\varepsilon}}_{\mathbf{k},\omega}) \cdot \nabla\mathscr{E} - i\left(\frac{\partial}{\partial\omega}\tilde{\boldsymbol{\varepsilon}}_{\mathbf{k},\omega}\right)\frac{\partial}{\partial t}\mathscr{E}. \tag{I.6.10}$$

Now suppose we perform a time average (indicated by $\langle\cdots\rangle_t$) over a period long compared with $2\pi/\omega$ so that we can write approximately (see Section I.8)

$$\left\langle \mathbf{E}\cdot\frac{\partial}{\partial t}\mathbf{D}\right\rangle_t = \frac{1}{2}\mathrm{Re}\left(\hat{\mathbf{E}}\cdot\frac{\partial}{\partial t}\hat{\mathbf{D}}^*\right). \tag{I.6.11}$$

On the other hand, according to Eqs. (I.6.8) and (I.6.10) we have

$$\frac{\partial\hat{\mathbf{D}}}{\partial t} \cong e^{i\omega t - i\mathbf{k}\cdot\mathbf{r}}\left[i\omega\tilde{\boldsymbol{\varepsilon}}_{\mathbf{k},\omega}\cdot\mathscr{E} - \omega(\nabla_{\mathbf{k}}\tilde{\boldsymbol{\varepsilon}}_{\mathbf{k},\omega}) : \nabla\mathscr{E} + \frac{\partial}{\partial\omega}(\omega\tilde{\boldsymbol{\varepsilon}}_{\mathbf{k},\omega})\cdot\frac{\partial\mathscr{E}}{\partial t}\right] \tag{I.6.12}$$

so that, after writing $\boldsymbol{\varepsilon}_{\mathbf{k},\omega} = \boldsymbol{\varepsilon}'_{\mathbf{k},\omega} - i\boldsymbol{\varepsilon}''_{\mathbf{k},\omega}$, where the dyadics $\boldsymbol{\varepsilon}'_{\mathbf{k},\omega}$ and $\boldsymbol{\varepsilon}''_{\mathbf{k},\omega}$ are both Hermitian, we obtain

$$\begin{aligned}\mathrm{Re}\left(\hat{\mathbf{E}}^*\cdot\frac{\partial\hat{D}}{\partial t}\right) &= \omega\boldsymbol{\varepsilon}''_{\mathbf{k},\omega} : \mathscr{E}\mathscr{E}^* - \left(\frac{\omega}{2}\right)(\nabla_{\mathbf{k}}\boldsymbol{\varepsilon}'_{\mathbf{k},\omega}) \vdots \nabla(\mathscr{E}\mathscr{E}^*) \\ &\quad + \frac{1}{2}\frac{\partial(\omega\boldsymbol{\varepsilon}'_{\mathbf{k},\omega})}{\partial\omega} : \frac{\partial(\mathscr{E}\mathscr{E}^*)}{\partial t} \\ &\quad + \frac{i\omega}{2}\Bigg\{\nabla_{\mathbf{k}}\boldsymbol{\varepsilon}''_{\mathbf{k},\omega}) \vdots (\nabla\mathscr{E})\mathscr{E}^* - \mathscr{E}\,\nabla\mathscr{E}^*)) \\ &\quad - \frac{\partial}{\partial\omega}(\omega\boldsymbol{\varepsilon}''_{\mathbf{k},\omega}) : \left(\left(\frac{\partial}{\partial t}\mathscr{E}\right)\mathscr{E}^* - \mathscr{E}\frac{\partial\mathscr{E}^*}{\partial t}\right)\Bigg\},\end{aligned} \tag{I.6.13}$$

and the energy law takes the form

$$\frac{\partial}{\partial t}\langle w\rangle_t + \frac{1}{2}\omega\boldsymbol{\varepsilon}''_{\mathbf{k},\omega}:\mathscr{E}\mathscr{E}^* + \left(\frac{i\omega}{2}\right)\left\{\nabla_{\mathbf{k}}\boldsymbol{\varepsilon}''_{\mathbf{k}}:((\nabla\mathscr{E})\mathscr{E}^* - \mathscr{E}\,\nabla\mathscr{E}^*)\right.$$
$$\left. - \frac{\partial(\omega\boldsymbol{\varepsilon}''_{\mathbf{k},\omega})}{\partial\omega}:\left(\left(\frac{\partial}{\partial t}\mathscr{E}\right)\mathscr{E}^* - \mathscr{E}\frac{\partial\mathscr{E}^*}{\partial t}\right)\right\}$$
$$= -\nabla\cdot(\mathbf{S}^{(0)} + \mathbf{S}^{(1)}) - \langle\mathbf{J}\cdot\mathbf{E}\rangle_t, \tag{I.6.14}$$

where

$$\langle w\rangle_t = \frac{1}{4}\frac{\partial}{\partial\omega}(\omega\boldsymbol{\varepsilon}'_{\mathbf{k},\omega}):\mathscr{E}\mathscr{E}^* + \frac{1}{4}\mu_0|\mathscr{H}|^2, \tag{I.6.15}$$

$$S^{(0)} = \frac{1}{2}\mathrm{Re}(\mathscr{E}\times\mathscr{H}^*), \tag{I.6.16}$$

$$S^{(1)} = -\frac{\omega}{4}\nabla_{\mathbf{k}}\boldsymbol{\varepsilon}'_{\mathbf{k},\omega}:\mathscr{E}\mathscr{E}^*. \tag{I.6.17}$$

For our wave packet, the Poynting vectors $S^{(0)} + S^{(1)}$ must be equal to the product of the energy density $\langle w\rangle_t$ times the group velocity, i.e.,

$$S^{(0)} + S^{(1)} = \langle w\rangle_t\mathbf{v}_g. \tag{I.6.18}$$

In particular, for an anisotropic non-space-dispersive medium the group velocity is given by

$$\mathbf{v}_g = \frac{2\,\mathrm{Re}(\mathscr{E}\times\mathscr{H}^*)}{[\partial(\omega\boldsymbol{\varepsilon}'_{\mathbf{k},\omega})/\partial\omega]:\mathscr{E}\mathscr{E}^* + \mu_0|\mathscr{H}|^2}. \tag{I.6.19}$$

7 Propagation in Moving Media

When an homogeneous, isotropic, nondispersive medium moves with uniform velocity $\mathbf{v} = c\boldsymbol{\beta}$ with respect to a reference frame F, the vectors **D** and **B** are related to **E** and **H** by the *Minkowski constitutive relations* [15]

$$\mathbf{D} = \varepsilon\mathsf{A}\cdot\mathbf{E} + \mathbf{b}\times\mathbf{H}, \qquad \mathbf{B} = \mu\mathsf{A}\cdot\mathbf{H} - \mathbf{b}\times\mathbf{E}, \tag{I.7.1}$$

where A is the dyadic

$$\mathsf{A} = a(\mathsf{I} - \mathsf{V}) + \mathsf{V}, \tag{I.7.2}$$

I the unit dyadic, $\mathsf{V} = \hat{v}\hat{v}$, and

$$a = \frac{1-\beta^2}{1-n\beta^2}, \qquad \mathbf{b} = b\hat{v} = \frac{\beta}{c}\frac{n^2-1}{1-n^2\beta^2}\hat{v}. \tag{I.7.3}$$

Instead of dealing with an electromagnetic field characterized by the complicated constitutive relations expressed by Eq. (I.7.1), we can introduce a fictitious field $\mathbf{E}'$, $\mathbf{H}'$ produced by sources ρ', $\mathbf{J}'$ in a space spanned by the vector $\mathbf{r}'$, related to the corresponding physical quantities by the string of relations [16]

$$\mathbf{r} = \mathsf{A}^{1/2}\cdot\mathbf{r}', \qquad \rho(\mathbf{r}) = \frac{T}{a}\rho'(\mathbf{r}'), \qquad \mathbf{J}(\mathbf{r}) = \frac{T}{a}\mathsf{A}^{1/2}\cdot\mathbf{J}'(\mathbf{r}'),$$

$$\mathbf{E}(\mathbf{r}) = T\mathsf{A}^{-1/2}\cdot\mathbf{E}'(\mathbf{r}'), \qquad \mathbf{H}(\mathbf{r}) = T\mathsf{A}^{-1/2}\cdot\mathbf{H}'(\mathbf{r}'), \tag{I.7.4}$$

where T is the linear operator $\exp(b\hat{v}\cdot\mathbf{r}'\partial/\partial t)$ mapping any function $f(\mathbf{r}',t)$ into $f(\mathbf{r}',t+b\hat{v}\cdot\mathbf{r}')$. For $a>0$, the space spanned by $\mathbf{r}'$ corresponds to the physical space in which the coordinates perpendicular to $\hat{v}$ are scaled by $1/\sqrt{a}$. For $a<0$, $\mathbf{r}'$ spans a space with complex coordinates, and consequently some precautions are required when dealing with the corresponding fields $\mathbf{E}'$ and $\mathbf{H}'$. It can be shown [16] that they satisfy the following equations:

$$\begin{aligned}
\nabla\cdot\mathbf{E}' &= \frac{1}{\varepsilon a}(\rho' + \mathbf{b}\cdot\mathbf{J}'),\\
\nabla\cdot\mathbf{H}' &= 0,\\
\nabla\times\mathbf{H}' &= \mathbf{J}' + \varepsilon a\frac{\partial}{\partial t}\mathbf{E}',\\
\nabla\times\mathbf{E}' &= -\mu a\frac{\partial}{\partial t}\mathbf{H}',
\end{aligned} \tag{I.7.5}$$

so that they can be interpreted as solutions of Maxwell's equations for a medium at rest characterized by a dielectric and magnetic permeability, respectively, given by εa and μa, while the charge density is given by $\rho' + \mathbf{b}\cdot\mathbf{J}'$.

In the absence of sources, Eqs. (I.7.5) admit as solutions the set of plane waves

$$\mathbf{E}'(\mathbf{r}',t) = \mathbf{E}'_0 e^{i\omega t - i\mathbf{k}'\cdot\mathbf{r}'}, \tag{I.7.6}$$

with $\mathbf{E}'_0\cdot\mathbf{k}' = 0$ and $k' = \omega a\sqrt{\varepsilon\mu}$, so that the true field $\mathbf{E}$ is

$$\mathbf{E}(\mathbf{r}) = \mathsf{A}^{-1/2}\cdot\mathbf{E}'_0 e^{-i\mathbf{k}\cdot\mathbf{r}}, \tag{I.7.7}$$

where

$$\mathbf{k} = \mathsf{A}^{-1/2}\cdot\mathbf{k}' - \omega b\hat{v}. \tag{I.7.8}$$

In particular, if θ is the angle between **k** and $\hat{v}$, the amplitude of **k** is given by [17]

$$k = k_0 \frac{[1 + \gamma^2(n^2 - 1)(1 - \beta^2 \cos^2\theta)]^{1/2} - \beta\gamma^2(n^2 - 1)\cos\theta}{1 - \gamma^2(n^2 - 1)\beta^2\cos^2\theta}$$

$$\equiv \frac{\omega}{v_{\text{ph}}}, \qquad \text{(I.7.9)}$$

where $\gamma^2 = (1 - \beta^2)^{-1}$, which for $\beta \ll 1$ reduces to

$$k = k_0[n - (n^2 - 1)\beta\cos\theta]. \qquad \text{(I.7.10)}$$

Rigorously speaking, this result applies only to the case in which the medium velocity is uniform all over the space. However, this condition can be relaxed by requiring **v** to be uniform only over a volume of a few wavelengths [18], and consequently Eq. (I.7.10) can also be used when the medium is rotating. In the most general case, the phase velocity v_{ph} differs from that relative to the medium at rest by a quantity proportional to the component of **v** along the direction of propagation, viz.

$$v_{\text{ph}} = \frac{c}{n} + \left(1 - \frac{1}{n^2}\right)\mathbf{v}\cdot\hat{k}. \qquad \text{(I.7.11)}$$

This is the well-known Fresnel formula, experimentally verified by Fizeau, and $1 - 1/n^2$ is called the *Fresnel drag coefficient.*

8 Coherence Properties of the Electromagnetic Field

The developments of the preceding sections were mainly based on a deterministic description of the electromagnetic field, which was assumed to be a prescribed quantity, even when possessing a finite bandwidth (which can be ascribed to amplitude or phase modulation). Actually, there is a certain amount of statistical uncertainty associated with every electromagnetic field (including those generated by the best amplitude-stabilized single-mode lasers), which must be accounted for. In the spirit of the statistical approach, this is done by introducing appropriate quantities that are suitable ensemble (or time) averages of the (in practice) inaccessible ones. The introduction of these averages and their connection with experimentally observable quantities constitute the subject of the *coherence theory* of electromagnetic radiation. In the rest of this book we will mainly refer to deterministic monochromatic fields (except for problems connected with incoherent imagery; see Section IV.15). It is, however, important for the reader to become acquainted with some basic elements of coherence theory in order to understand the procedure that allows us, once the single deterministic realization of the field is known, to evaluate its significant statistical averages.

Let us start by introducing the concept of the *analytic signal* $\hat{f}(t)$ of a given time-dependent quantity $f(t)$ [11], a concept that turns out to be particularly useful for the development of the theory. If a *real signal* $f(t)$ admits the Fourier expansion

$$f(t) = \frac{1}{2\pi} \int_{-\infty}^{+\infty} f_\omega e^{i\omega t}\, d\omega, \tag{I.8.1}$$

then the analytic signal $\hat{f}(t)$ is defined by the relation

$$\hat{f}(t) = \frac{1}{\pi} \int_{0}^{\infty} f_\omega e^{i\omega t}\, d\omega, \tag{I.8.2}$$

from which it immediately follows that $f(t) = \mathrm{Re}\, \hat{f}(t)$. If the signal is monochromatic, that is, $f(t) = f_0 \cos(\omega t + \Phi)$, then

$$\hat{f}(t) = f_0 e^{i\Phi + i\omega t} \qquad (\omega > 0), \tag{I.8.3}$$

an expression that is also referred to as *complex representation.*

From a conceptual point of view, the usefulness of the analytic signal stems from the way in which, at optical frequencies, a fast detector performs measurements of the *instantaneous optical intensity*. More precisely, we refer to a realistic detector having a response time T_d small compared with the inverse of the field bandwidth $\delta\omega$ but large with respect to the inverse of its mean frequency ω_0. It gives a response proportional to [19]

$$I(t) = \frac{1}{T_d} \int_{t - T_d/2}^{t + T_d/2} E^2(t')\, dt', \tag{I.8.4}$$

the assumed condition

$$1/\omega_0 \ll T_d \ll 1/\delta\omega \tag{I.8.5}$$

justifying the term instantaneous intensity applied to the above integral. If we now write

$$E(t) = \frac{1}{\pi} \mathrm{Re}\left(e^{i\omega_0 t} \int_0^\infty E_\omega e^{i(\omega - \omega_0)t}\, d\omega \right) = \mathrm{Re}[e^{i\omega_0 t} \mathscr{E}(t)], \tag{I.8.6}$$

where $\mathscr{E}(t) = e^{-i\omega_0 t} \hat{E}(t)$ can be assumed to vary on a time scale that is long compared with T_d we have as a good approximation that the instantaneous optical intensity is the square modulus of the analytic signal, that is,

$$I(t) = |\hat{E}(t)|^2 = |\mathscr{E}|^2. \tag{I.8.7}$$

In the same way, we can show that the flux of the instantaneous Poynting vector (see Section I.6), averaged over a few periods of the electromagnetic field, is most conveniently expressed as the flux of the real part of the *complex Poynting vector*

$$\tilde{\mathbf{S}} = \tfrac{1}{2} \hat{\mathbf{E}} \times \hat{\mathbf{H}}^*. \tag{I.8.8}$$

For a single plane wave, the modulus of the Poynting vector can be expressed as

$$|\tilde{\mathbf{S}}| = \tfrac{1}{2}|\tilde{\varepsilon}/\mu_0|^{1/2}|\hat{\mathbf{E}}|^2 = \tfrac{1}{2}\tilde{n}\zeta_0 I \tag{I.8.9}$$

which is proportional to the instantaneous optical intensity I (for a comparison of $|\tilde{\mathbf{S}}|$ and $\hat{\mathbf{E}} \cdot \hat{\mathbf{E}}^*$, see, e.g., Marathay [19]) through the free-space impedance ζ_0 and the refractive index $\tilde{n}$. In general, the instantaneous optical intensity exhibits a fluctuating behavior, and thus $I(t)$ constitutes the single realization of a statistical ensemble. Accordingly, a complete description requires, in principle, the introduction of a hierarchy of ensemble averages of the kind

$$\langle I(t)\rangle, \langle I^2(t)\rangle, \ldots, \langle I^n(t)\rangle \tag{I.8.10}$$

and

$$\langle I(t)I(t')\rangle, \ldots, \langle I(t_1)\cdots I(t_n)\rangle, \tag{I.8.11}$$

the ensemble averaging operation being equivalent, in stationary situations, to a time average over an interval long compared with $1/\delta\omega$. Note that the necessity for introducing higher-order averages was scarcely felt before the advent of the lasers (early 1960s), because there were no optical sources having a *coherence time* $T_c = 2\pi/\delta\omega$ longer than the response time of the available detectors. In fact, if T_d is larger than T_c, $I(t)$, as given by Eq. (I.8.4), becomes for all practical purposes a deterministic variable and all averages factorize into products of the lowest-order ones.

If we now wish to take into account more general situations in which two or more beams are made to interfere (as, for example, in the Young or Michelson interference experiments; see Fig. I.12) it is natural to introduce the correlation function [20] (also known as the *mutual coherence function*) (cf. Section IV.15)

$$G^{(1,1)}(\mathbf{r}\, t; \mathbf{r}', t') = \langle \hat{E}^*(\mathbf{r}, t)\hat{E}(\mathbf{r}', t')\rangle, \tag{I.8.12}$$

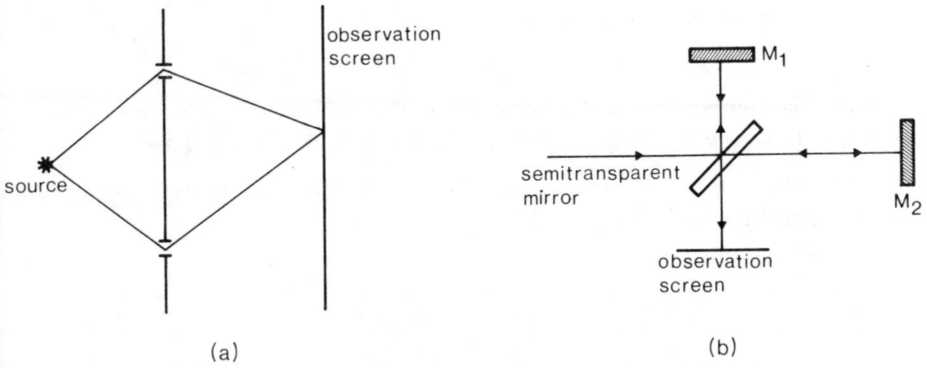

Fig. I.12. Schematic of (a) Young's and (b) Michelson's interferometers.

which represents the contribution of the interference term to the signal falling on the detector, together with higher-order averages of the type

$$\begin{aligned} &G^{(n,m)}(\mathbf{r}_1, t_1, \ldots, \mathbf{r}_n, t_n, \ldots, \mathbf{r}_{n+m}, t_{n+m}) \\ &\quad = \langle \hat{E}^*(\mathbf{r}_1, t_1) \cdots \hat{E}^*(\mathbf{r}_n, t_n) \hat{E}(\mathbf{r}_{n+1}, t_{n+1}) \cdots \hat{E}(\mathbf{r}_{n+m}, t_{n+m}) \rangle. \end{aligned} \tag{I.8.13}$$

From an analytical point of view, the statistical description introduced above completely characterizes a stochastic electromagnetic field. In practice, averages of the electromagnetic field of order $n + m$ higher than the second are connected either with higher-order interference experiments (such as the stellar interferometer experiments of Hanbury–Brown and Twiss) or with photon-counting experiments [21].

Problems

Section 1

1. Show that, for a homogeneous stationary medium, the Fourier transforms $\mathbf{A}_{\mathbf{q},\omega}, \Phi_{\mathbf{q},\omega}$ of the potentials satisfy the relation

$$(q^2\mathbf{1} - \mathbf{qq} - \omega^2\mu_0\boldsymbol{\varepsilon}_{q,\omega}) \cdot \mathbf{A}_{\mathbf{q},\omega} + \omega\mu_0\boldsymbol{\varepsilon}_{q,\omega} \cdot \mathbf{q}\Phi_{\mathbf{q},\omega} = \mu_0\mathbf{J}_{\mathbf{q},\omega}.$$

2. Show that the space–time Fourier transform of the scalar potential Φ in the Coulomb gauge [see Eq. (I.1.20a)] is related to $\rho_{\mathbf{q},\omega}$ by

$$\Phi_{\mathbf{q},\omega} = \rho_{\mathbf{q},\omega}/\mathbf{q} \cdot \boldsymbol{\varepsilon}_{q,\omega} \cdot \mathbf{q}.$$

3. Show that $G_{\mathrm{C}}(\mathbf{r}, \omega)$ [see Eq. (I.1.22)] for an anisotropic medium is given by

$$G_{\mathrm{C}}(\mathbf{r}, \omega) = \varepsilon_0/(\bar{\varepsilon}_1 x^2 + \bar{\varepsilon}_2 y^2 + \bar{\varepsilon}_3 z^2)^{1/2},$$

$\bar{\varepsilon}_{1,2,3}$ being the components of the dielectric tensor relative to the principal axes (see Section I.4).

4. Show that the space–time Fourier transform of the dyadic Green's function for an anisotropic medium is

$$\boldsymbol{\Gamma}_{\mathbf{q},\omega} = (\omega^2\mu_0\boldsymbol{\varepsilon}_{\mathbf{q},\omega} + \mathbf{qq} - q^2)^{-1} = \mathbf{A}'/D(\mathbf{q}, \omega),$$

where the components of the tensor $\mathbf{A}'$ are the algebraic complements of the tensor $\mathbf{A} = \omega^2\mu_0\boldsymbol{\varepsilon}_{\mathbf{q},\omega} + \mathbf{qq} - q^2$, and $D(\mathbf{q}, \omega) = \det \mathbf{A}$ [see Eq. (I.5.9)].

5. Show that the dyadic Green's function $\boldsymbol{\Gamma}$ for an isotropic and homogeneous medium is given by

$$\boldsymbol{\Gamma}(\mathbf{r}) = \frac{1}{(2\pi)^3} \iiint_{-\infty}^{+\infty} \left(\mathbf{1} - \frac{\mathbf{qq}}{k^2}\right) \frac{e^{-i\mathbf{q}\cdot\mathbf{r}}}{k^2 - q^2} d^3q$$

$$= \left(\mathbf{1} + \frac{\nabla\nabla}{k^2}\right)\frac{1}{(2\pi)^3}\iiint \frac{e^{-i\mathbf{q}\cdot\mathbf{r}}}{k^2 - q^2}\, d^3q = \left(\mathbf{1} + \frac{\nabla\nabla}{k^2}\right) G(\mathbf{r}),$$

where $G(\mathbf{r})$ is the scalar Green's function.

6. Using the Fourier integral representation of the scalar Green's function of the above problem, show that the Green's function is

$$G(\mathbf{r}) = \frac{1}{(2\pi)^3}\iiint_{-\infty}^{+\infty} \frac{e^{-i\mathbf{q}\cdot\mathbf{r}}}{k^2 - q^2}\, d^3q = \frac{e^{-ikr}}{4\pi r}.$$

7. Using the asymptotic expression for $\mathbf{\Gamma}$,

$$\mathbf{\Gamma}(\mathbf{r}) \sim (\mathbf{1} - \hat{r}\hat{r})G(\mathbf{r}),$$

show that the far field radiated by a source distribution is

$$\mathbf{E}(\mathbf{r}) \sim -i\omega\mu_0 G(\mathbf{r}) \iiint_{-\infty}^{+\infty} (\mathbf{1} - \hat{n}\hat{n}) \cdot \mathbf{J}(\mathbf{r}') e^{ik\hat{n}\cdot\mathbf{r}'}\, dV',$$

$\hat{n}$ being the direction parallel to $\mathbf{r} - \mathbf{r}'$.

8. Let us consider the time Fourier transform $\mathbf{E}_1, \mathbf{H}_1$ and $\mathbf{E}_2, \mathbf{H}_2$ of the fields radiated by the two monochromatic sources

$$\mathbf{J}_1(\mathbf{r}, t) = \mathbf{J}_1(\mathbf{r})e^{i\omega t}, \qquad \mathbf{J}_2(\mathbf{r}, t) = \mathbf{J}_2(\mathbf{r})e^{i\omega t}.$$

Then, making use of the vector identity (A.9) and of Maxwell's equations, prove the following relation:

$$\iiint_V (\mathbf{E}_2 \cdot \mathbf{J}_1 - \mathbf{E}_1 \cdot \mathbf{J}_2)\, dV = \oiint_S (\mathbf{E}_1 \times \mathbf{H}_2 - \mathbf{E}_2 \times \mathbf{H}_1) \cdot \hat{n}\, dS$$

$$= +\, i\omega \iiint_V E_{1i} E_{2j} (\varepsilon_{ij} - \varepsilon_{ji})\, dV$$

for an anisotropic medium. In particular, show that when $\varepsilon_{ij} = \varepsilon_{ji}$ and the volume is limited by a metallic surface, we have

$$\iiint_V \mathbf{E}_2 \cdot \mathbf{J}_1\, dV = \iiint_V \mathbf{E}_1 \cdot \mathbf{J}_2\, dV,$$

which defines the *reciprocity theorem.*

Section 2

9. Show that $n(\omega)$ and $\kappa(\omega)$ are connected by the following Kramers–Kronig relations:

$$n(\omega) - 1 = \frac{2}{\pi} P \int_0^\infty \frac{\omega' \kappa(\omega')}{\omega'^2 - \omega^2} d\omega', \qquad \kappa(\omega) = \frac{2\omega}{\pi} P \int_0^\infty \frac{n(\omega') - 1}{\omega'^2 - \omega^2} d\omega',$$

after folding the integrals into the positive axis, using the property $\tilde{n}^*(\omega) = \tilde{n}(-\omega)$. Furthermore, write the dielectric constant in the form

$$\varepsilon_{ij}(\omega) = \varepsilon_0 [\delta_{ij} - i\sigma_{ij}(\omega)/(\omega\varepsilon_0)],$$

and prove the following Kramers–Kronig relations:

$$\frac{1}{\omega^2} \operatorname{Re} \sigma_{ij} = -\frac{2}{\pi} P \int_0^{+\infty} \frac{\operatorname{Im} \sigma_{ij}}{\omega'^2 - \omega^2} \frac{1}{\omega'} d\omega',$$

$$\frac{1}{\omega} \operatorname{Im} \sigma_{ij} = \frac{2}{\pi} P \int_0^{+\infty} \frac{\operatorname{Re} \sigma_{ij}}{\omega'^2 - \omega^2} d\omega'.$$

10. Calculate the susceptibility χ_ω of an ensemble of two-level oscillators characterized by the susceptibility χ of Eq. (I.2.58), with the transition frequencies $\omega_0 + \delta$ distributed with a Gaussian law.

$$\chi(\omega) \propto -\frac{-if}{\pi^{1/2}\sigma} \int_{-\infty}^{+\infty} \frac{e^{-\delta^2/2\sigma^2}}{\gamma_{ab} + i(\omega - \omega_0 - \delta)} d\delta$$

11. Show that for $\sigma \gg \gamma_{ab}$ the imaginary component χ'' of the above system reduces to

$$\chi'' \propto (f/\sigma)\pi^{1/2} e^{-(\omega - \omega_0)^2/2\sigma^2}.$$

Then use the Kramers–Kronig relations of Problem 8 to derive χ',

$$\chi' = -\frac{2}{\pi} P \int_0^\infty \frac{\omega' \chi''(\omega')}{\omega'^2 - \omega^2} d\omega',$$

and show that χ' can be expressed by means of Dawson's integral [see Eq. (VII.19.9)].

Section 4

12. The normal modes of an anisotropic medium are characterized by a refractive index $\tilde{n}$ defined as

$$\mathbf{q} = (\omega/c)\tilde{n}(\omega, \mathbf{q})\hat{s},$$

where ω and $\mathbf{q}$ satisfy the dispersion relation given by Eq. (I.5.8). Show that $\tilde{n}$ obeys the Fresnel equation

$$\varepsilon_0^2 \tilde{n}^4 \varepsilon_{ij} s_i s_j - \tilde{n}^2 \varepsilon_0 (\varepsilon_{ij} s_i s_j \varepsilon_{ll} - \varepsilon_{il} \varepsilon_{lj} s_i s_j) + \det \boldsymbol{\varepsilon}_{\mathbf{q},\omega} = 0.$$

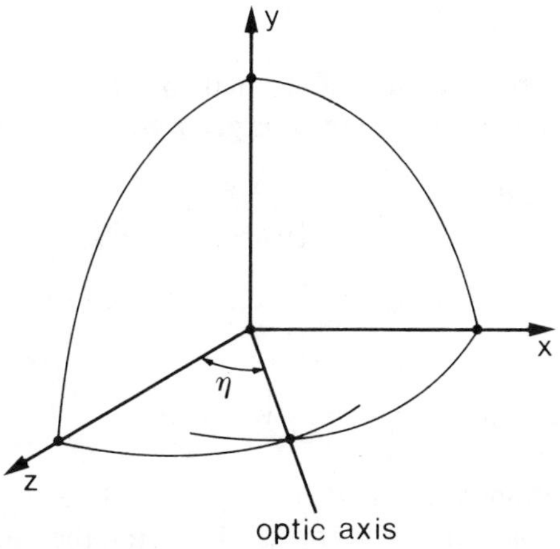

Fig. I.13. Wave vector surface for a biaxial crystal.

13. Show that for a uniaxial crystal the refractive index of the extraordinary ray is given by

$$\tilde{n}_{\hat{e}}(\theta) = (\cos^2\theta/n_0^2 + \sin^2\theta/n_e^2)^{-1/2}$$

(see Problem 15 for the definition of uniaxial crystal).

14. If we plot $\tilde{n}$ as a radius vector in the direction $\hat{s}$ from the origin, we obtain a two-sheeted surface, called a *wave vector surface,* one sheet of which corresponds to $\tilde{n}_1$, the other to $\tilde{n}_2$. Using the second equation of Problem 12, deduce that the intersection of the wave vector surface with a plane perpendicular to a principal axis (say $\hat{z}$) is represented by

$$\tilde{n}_1 = (\bar{\varepsilon}_z/\varepsilon_0)^{1/2}, \qquad \tilde{n}_2 = [\bar{\varepsilon}_x\bar{\varepsilon}_y/\varepsilon_0(\bar{\varepsilon}_x s_x^2 + \bar{\varepsilon}_y s_y^2)]^{1/2}$$

15. The directions in crystals for which $\tilde{n}_1 = \tilde{n}_2$ are known as *optic axes.* Using the second equation of Problem 12, show that a crystal has at most two such directions (biaxial) (if the two directions coincide the crystal is referred to as *uniaxial*). In particular, show that for $\bar{\varepsilon}_z > \bar{\varepsilon}_y > \bar{\varepsilon}_x$ the two optic axes lie in the x–z plane (Fig. I.13) and form with the z axis an angle η given by

$$\tan\eta = [\bar{\varepsilon}_z(\bar{\varepsilon}_y - \bar{\varepsilon}_x)/\bar{\varepsilon}_x(\bar{\varepsilon}_z - \bar{\varepsilon}_y)]^{1/2}.$$

16. Show that, for a uniaxial crystal, the wave vector surface separates into a sphere (ordinary ray) and an ellipsoid (extraordinary ray). For $n_e > n_0$ the sphere lies inside the ellipsoid and the crystal is referred to as *positive uniaxial.* For $n_0 > n_e$ it is termed *negative uniaxial.*

Section 5

17. Discuss the propagation of normal modes through a magnetically biased plasma, whose dielectric tensor is given by

$$\tilde{\varepsilon}_{xx} = \tilde{\varepsilon}_{yy} = \varepsilon_0 \left\{ 1 - \frac{\omega_p^2(\omega - i\nu)}{\omega[(\omega - i\nu)^2 - \omega_g^2]} \right\},$$

$$\tilde{\varepsilon}_{xy} = -\tilde{\varepsilon}_{yx} = i\varepsilon_0 \frac{\omega_p^2 \omega_g}{\omega(\omega + \omega_g - i\nu)(\omega - \omega_g - i\nu)},$$

$$\tilde{\varepsilon}_{zz} = \varepsilon_0 \left(1 - \frac{\omega_p^2}{\omega^2 - i\omega\nu} \right),$$

the remaining component being identically zero. Here, $\omega_g = (e/m)B_0\hat{z}$ represents the *gyrofrequency* of the electrons, $B_0\hat{z}$ being the applied field, and ω_p and ν are, respectively, the plasma frequency and the collision frequency.

18. Discuss the propagation through a polar crystal characterized by a scalar dielectric constant given by

$$\varepsilon(\omega) = \varepsilon_b(\omega)\left(1 + \frac{\Omega^2}{\omega_T^2 - \omega^2 + i\omega\gamma} \right) \equiv \varepsilon'(\omega) - i\varepsilon''(\omega),$$

where $\varepsilon_b(\omega)$ is the contribution from the electronic polarizability of the atoms, ω_T is the *transverse resonance frequency* of the oscillator assembly, and $\Omega^2 = Ne^2/[m_{\text{eff}}\varepsilon_b(\omega)]$, N being the number density of the ions contributing to the polarizability and m_{eff} a characteristic mass, while γ takes into account damping effects. The frequency at which $\varepsilon'(\omega_L) = 0$ is called the *longitudinal resonance frequency*; ω_L and ω_T are related to $\varepsilon_S = \varepsilon(0)$ and $\varepsilon_\infty = \varepsilon(\infty)$ by the *Liddane–Sachs–Teller relation* $\omega_L^2/\omega_T^2 = \varepsilon_S/\varepsilon_\infty$.

19. Consider a gyrotropic medium characterized by the dielectric tensor

$$\tilde{\boldsymbol{\varepsilon}} = \begin{bmatrix} \varepsilon_1 & ig & 0 \\ -ig & \varepsilon_1 & 0 \\ 0 & 0 & \varepsilon_3 \end{bmatrix}.$$

Find the normal modes and show that those propagating along the z axis are circularly polarized.

20. A wave packet travels with a group velocity $\mathbf{v}_g = \partial\omega/\partial k_x, \partial\omega/\partial k_y, \partial\omega/\partial k_z$. Show that $\mathbf{v}_g$ can be obtained in terms of the dispersion function $D(\mathbf{q}, \omega)$ through the relation

$$\mathbf{v}_g = -\left(\frac{\partial D}{\partial \omega}\right)^{-1} \nabla_q D.$$

In particular, show that for a nondispersive uniaxial crystal the group velocity of the extraordinary ray is

$$\mathbf{v}_g = (\bar{\varepsilon}_1 \bar{\varepsilon}_3 \mu_0)^{-1/2} (\boldsymbol{\varepsilon} : \mathbf{qq})^{-1/2} \boldsymbol{\varepsilon} \cdot \mathbf{q},$$

and forms with $\mathbf{q}$ an angle δ given by

$$\delta = \arctan \frac{(n_e^2 - n_0^2) \sin\theta \cos\theta}{n_e^2 \cos^2\theta + n_0^2 \sin^2\theta},$$

where θ is the angle formed by $\mathbf{q}$ with the optic axis.

21. Show that the dispersion function $D(\mathbf{q}, \omega)$ for an anisotropic crystal can be written as

$$D(\mathbf{q}, \omega) = \omega^2 \mu_0 \varepsilon_{ij} s_i s_j (q^2 - k_0^2 \tilde{n}_1^2)(q^2 - k_0^2 \tilde{n}_2^2),$$

$\tilde{n}_{1,2}$ being the refractive indices relative to the normal waves propagating along the direction $\hat{s}$.

References

1. Landau, L. D., and Lifshitz, E. M., "Electrodynamics of Continuous Media." Pergamon, Oxford, 1960.
2. Van Kranendonk, J., and Sipe, J. E., *Prog. Opt.* **15**, 245–350 (1977).
3. Mandel, L., and Wolf, E., *Opt. Commun.* **8**, 95 (1973).
4. Owyoung, A., Ph.D. Thesis, California Inst. Technol., Pasadena, 1971 (Clearing House Fed. Sci. Tech. Inf. Rep. AFOSR-TR-71-3132).
5. Shen, Y. R., "Principles of Nonlinear Optics." Wiley, New York, 1984.
6. Sargent, M., Scully, M. O., and Lamb, W. E., "Laser Physics." Addison-Wesley, Reading, Massachusetts, 1974.
7. Hopf, F. A., and Scully, M. O., *Phys. Rev.* **179**, 399 (1969).
8. Yariv, A., "Quantum Electronics," 2nd ed. Wiley, New York, 1975.
9. Frantz, L. M., and Nodvik, J. S., *J. Appl. Phys.* **34**, 2346 (1963).
10. McCall, S. L., and Hahn, E. L., *Phys. Rev.* **183**, 457 (1969).

10a. Wooten, F., "Optical Properties of Solids." Academic Press, New York, 1972.

11. Born, M., and Wolf, E., "Principles of Optics." Pergamon, Oxford, 1970.
12. van de Hulst, H. C., "Light Scattering by Small Particles," Wiley, New York, 1957.
13. Yariv, A., and Yeh, P., "Optical Waves in Crystals," Wiley, New York, 1983.
14. Agranovich, V. M., and Ginzburg, V. L., "Crystal Optics with Spatial Dispersion," Springer-Verlag, New York, 1984.
15. Sommerfeld, A., "Electrodynamics." Academic Press, New York, 1952.
16. Solimeno, S., *Alta Freq*, **43**, 1005 (1974); *J. Math. Phys.* **16**, 218 (1975).
17. Papas, C. H., "Theory of Electromagnetic Wave Propagation." McGraw-Hill, New York, 1965.
18. Censor, D., *IEEE Trans. Microwave Theory Tech.* **MTT-16**, 565 (1968).
19. Marathay, A. S., "Elements of Optical Coherence Theory." Wiley, New York, 1982.
20. Mandel, L., and Wolf, E., *Rev. Mod. Phys.* **37**, 231 (1965).
21. Peřina, J., "Quantum Statistics of Linear and Nonlinear Optical Phenomena." Reidel Publ., Dordrecht, Netherlands, 1984.

Bibliography

Akhmanov, S. A., and Khokhlov, R. V., "Problems of Nonlinear Optics." Gordon and Breach, New York, 1972.

Allen, L., and Eberly, J. H., "Optical Resonances and Two-Level atoms." Wiley, New York 1975.

Bloembergen, N., "Nonlinear Optics." Benjamin-Cummings, Menlo Park, California, 1965.

Brewer, R. G., *Phys. Today* **30**, 50 (1977).

Chow, W. W., Gea-Banacloche, J., Pedrotti, L. M., Sanders, V. E., Schleich, W., and Scully, M. O., *Rev. Mod. Phys.* **57**, 61 (1985).

Clarke, D., and Grainger, J. F., "Polarized Light and Optical Measurement." Pergamon, Oxford, 1971.

Feynman, R. P., Vernon F. L., and Hellwarth R. W., *J. Appl. Phys.* **28**, 49 (1957).

Haken H., "Light." Vols. 1, and 2. North-Holland Publ., Amsterdam, 1981, and 1985.

Hange, P. S., Muller, R. H., and Smith, C. G., *Surf. Sci.* **96,** 81 (1980).

Hecht, E., and Zajac, A., "Optics." Addison-Wesley, Reading, Massachusetts, 1969.

Loudon, R., "The Quantum Theory of Light." Oxford Univ. Press (Clarendon), London and New York, 1978.

Mo, T. C., *J. Math Phys.* **11**, 2589 (1970).

Nussenzveig, H. M., "Causality and Dispersion Relations." Academic Press, New York, 1972.

Pekar, S. I., "Crystal Optics and Additional Light Waves." Benjamin-Cummings, Menlo Park, California, 1983.

Rabin, H., and Tang, C. L., eds., "Quantum Electronics," Vol. 1, Parts A and B. Academic Press, New York, 1975.

Ramachandran, G. N., and Ramasehan, S., *in* "Handbuch der Physik" (S. Flugge, ed.), Vol. 25, Part 1, pp. 1–217. Springer-Verlag, Berlin, 1961.

Toraldo di Francia, G., "Electromagnetic Waves." Wiley (Interscience), New York, 1955.

Van Bladel, J., *Proc. IEEE* **64**, 301 (1976).

Whitham, G. B., "Linear and Nonlinear Waves." Wiley, New York, 1973.

Chapter II

Ray Optics

1 Approximate Representation of the Electromagnetic Field

Maxwell's equations in a medium characterized by a uniform refractive index $n(\omega)$ result in monochromatic plane wave solutions whose complex representation is $\mathbf{E}(\mathbf{r}, t) = E_0 \exp(-ik_0 n\hat{s} \cdot \mathbf{r} + i\omega t)$. When n depends on $\mathbf{r}$, no plane-wave solution exists (we omit the dependence of n on ω, unless strictly necessary). As a first approximation, we can discuss the possibility of describing the field in terms of "local" plane waves of the form

$$\mathbf{E}(\mathbf{r}, t) = \mathbf{E}_0(\mathbf{r}) \exp[-ik_0 S(\mathbf{r}) + i\omega t] \tag{II.1.1}$$

where $\mathbf{E}_0(\mathbf{r})$ is a slowly varying function and $S(\mathbf{r})$ a generic function of $\mathbf{r}$, which respectively reduce to a constant and to $n\hat{s} \cdot \mathbf{r}$ for a homogeneous medium. If we introduce Eqs. (II.1.1) into Maxwell's equations, written for $\mathbf{J} = \rho = 0$ and $\mathbf{B} = \mu_0 \mathbf{H}$, we immediately obtain, after taking advantage of Eqs. (A.7) and (A.8),

$$\nabla S \times \mathbf{E}_0 - \zeta_0 \mathbf{H}_0 + (i/k_0) \nabla \times \mathbf{E}_0 = 0, \tag{II.1.2a}$$

$$\nabla S \times \mathbf{H}_0 + (n^2/\zeta_0) \mathbf{E}_0 + (i/k_0) \nabla \times \mathbf{H}_0 = 0, \tag{II.1.2b}$$

$$\nabla S \cdot \mathbf{H}_0 + (i/k_0) \nabla \cdot \mathbf{H}_0 = 0, \tag{II.1.2c}$$

$$\nabla S \cdot \mathbf{E}_0 + [i/(k_0 n^2)] \nabla \cdot (n^2 \mathbf{E}_0) = 0, \tag{II.1.2d}$$

where $\zeta_0 = (\mu_0/\varepsilon_0)^{1/2}$ ($\cong 377$ ohms) is the *vacuum impedance*. In the following we will indicate with $\zeta \equiv \zeta_0/n$ the impedance of a medium with refractive index n.

If we now let $k_0 \to \infty$, we can neglect the terms in $1/k_0$, having chosen both $\mathbf{E}_0$ and S independent of k_0. After doing this and multiplying vectorially

Eq. (II.1.2a) by ∇S, we obtain, with the help of Eqs. (II.1.2b), (II.1.2d), and (A.2),

$$\begin{aligned}\nabla S \times (\nabla S \times \mathbf{E}_0) - n^2\mathbf{E}_0 &= \nabla S(\nabla S \cdot \mathbf{E}_0) - \mathbf{E}_0(\nabla S)^2 + n^2\mathbf{E}_0 \\ &= [n^2 - (\nabla S)^2]\mathbf{E}_0 = 0.\end{aligned} \tag{II.1.3}$$

Accordingly, the function $S(\mathbf{r})$, known as the *eikonal*, must be chosen to satisfy the *eikonal equation*

$$(\nabla S)^2 = n^2(\mathbf{r}). \tag{II.1.4}$$

For finite values of ω, we cannot neglect *a priori* the terms in $1/k_0$ appearing in Eq. (II.1.2), since we are not able to appreciate the error deriving from this approximation, once we accept as an ansatz the representation of the field given by Eq. (II.1.1). An answer to this problem can be found by resorting to a more accurate representation of the field in terms of *asymptotic series*, the subject of the following section.

2 Asymptotic Solution of the Scalar Wave Equation

An approach to the search for approximate solutions of the wave equation is based on the introduction, for the generic component of the electric field, of the complex representation u that locally obeys the plane-wave type relation

$$u(\mathbf{r} + d\mathbf{r}) \sim A(\mathbf{r})e^{-i\mathbf{k}\cdot d\mathbf{r}} \tag{II.2.1}$$

Equation (II.2.1) implies that, for a generic displacement $d\mathbf{r}$, the phase of u changes by an amount $-\mathbf{k}(\mathbf{r}) \cdot d\mathbf{r}$, where $\mathbf{k}(\mathbf{r})$ is the $\mathbf{r}$-dependent local wave vector. Thus, it defines a generalized plane wave whose direction and velocity vary during propagation, and whose complex amplitude

$$A(\mathbf{r}) = |u(\mathbf{r})|e^{i\phi(\mathbf{r})} \tag{II.2.2}$$

has a small relative variation over a distance of the order of the wavelength. The symbol $\sim$ stands for *asymptotically equal* and implies

$$\lim_{k_0 \to \infty} [u(\mathbf{r} + d\mathbf{r}) - A(\mathbf{r})e^{-i\mathbf{k}\cdot d\mathbf{r}}] = 0, \tag{II.2.3}$$

so that u can be substituted by its asymptotic expression only for $k_0 \to \infty$.

By superimposing solutions of the kind described by Eq. (II.2.1), we obtain the most general asymptotic representation of the field in the form

$$u(\mathbf{r} + d\mathbf{r}) \sim \sum_n A_n(\mathbf{r})e^{-i\mathbf{k}_n\cdot d\mathbf{r}}, \tag{II.2.4}$$

describing a situation in which many waves are passing through a given point $\mathbf{r}$, such as happens when the various waves reflected, refracted, or diffracted by discontinuities are superimposed on a principal wave.

The validity of Eq. (II.2.1) is limited to a small region around **r**. The extension of this region depends on **r** itself, and it shrinks to zero for **r** varying in some zones (*caustics*). There the concept of local plane wave expressed by Eq. (II.2.1) is obviously meaningless. The opposite case arises for an ideal plane wave, which is represented in all space by Eq. (II.2.1) with $\sim$ replaced by the equality sign.

In order to place on more rigorous grounds the intuitive approach of Eq. (II.2.1), let us assume that

$$u(\mathbf{r}) \sim e^{-ik_0 S(\mathbf{r})} \sum_{m=0}^{\infty} \frac{A_m(\mathbf{r})}{(-ik_0)^m} \tag{II.2.5}$$

known as the *asymptotic series of Luneburg and Kline* [1, 2], where the symbol $\sim$ now means that, for every integer N,

$$u(\mathbf{r}) = e^{-ik_0 S(\mathbf{r})} \sum_{m=0}^{N} \frac{A_m(\mathbf{r})}{(-ik_0)^m} + \mathrm{o}(k_0^{-N}), \tag{II.2.6}$$

Landau's symbol $\mathrm{o}(k_0^{-N})$ indicating a function that vanishes more rapidly than k_0^{-N}, while

$$\mathbf{k}(\mathbf{r}) = k_0 \nabla S. \tag{II.2.7}$$

The asymptotic series of Eq. (II.2.5) is also termed the *ray optical* (*RO*) *representation*, since the eikonal $S(\mathbf{r})$ (introduced by Burns in 1895) leads to the intuitive concept of a ray, as will become apparent in Section II.4. A general feature of Eq. (II.2.5) is that it furnishes a physical picture of electromagnetic propagation more complete than the one associated with the traditional geometrical optics. An example of this will be explicitly considered in Section II.7 in connection with the introduction of complex values of $S(\mathbf{r})$.

In order to determine $S(\mathbf{r})$ and the A_m's, let us introduce Eq. (II.2.6) into the wave equation [Eq. (I.1.12)], thus obtaining [3]

$$\sum_{m=0}^{N} \frac{Q_m(\mathbf{r})}{(-ik_0)^{m-2}} = \mathrm{o}(k_0^{-N}), \tag{II.2.8a}$$

where

$$Q_0(\mathbf{r}) = (\nabla S)^2 - n^2(\mathbf{r}), \tag{II.2.8b}$$

$$Q_1(\mathbf{r}) = (\nabla^2 S + 2\nabla S \cdot \nabla) A_0, \tag{II.2.8c}$$

$$Q_m(\mathbf{r}) = (\nabla^2 S + 2\nabla S \cdot \nabla) A_{m-1} + \nabla^2 A_{m-2} \qquad (m = 2, 3, 4, \ldots). \tag{II.2.8d}$$

Equation (II.2.8a) is satisfied for every N only if

$$Q_m(\mathbf{r}) = 0 \qquad (m = 0, 1, 2, \ldots), \tag{II.2.9}$$

which yields the desired set of equations for $S(\mathbf{r})$ and the A_m's.

Example: Expansion of the Hankel Function $H_0^{(2)}$. The meaning of asymptotic series can be found by considering, as a particular example, the field radiated *in vacuo* by a *current line source*. In this case, the exact solution is known:

$$u \propto H_0^{(2)}(k_0 \rho), \tag{II.2.10}$$

where ρ is the distance from the source and $H_0^{(2)}$ is the Hankel function of the second kind and zeroth order, which tends asymptotically to [4]

$$H_0^{(2)}(k_0 \rho) \sim [2/(\pi k_0 \rho)]^{1/2} e^{-i(k_0 \rho - \pi/4)}. \tag{II.2.11}$$

In order to improve the accuracy of representation of the field, we can represent $H_0^{(2)}$ as an asymptotic series having as leading term the right-hand side of Eq. (II.2.11). The higher-order terms are generated by using Eqs. (II.2.9). This procedure yields

$$H_0^{(2)}(k_0 \rho) \sim \left(\frac{2}{\pi k_0 \rho}\right)^{1/2} e^{-i(k_0 \rho - \pi/4)} \sum_{m=0}^{\infty} \frac{(0, m)}{(-2ik_0 \rho)^m}, \tag{II.2.12}$$

where

$$(0, m) = [(2m)!/(2^m m!)]^2 / 2^{2m} m!; \tag{II.2.13}$$

which coincides with the Hankel asymptotic expansion of $H_0^{(2)}$, as it must.

If Eq. (II.2.5) is convergent for large k_0, then it reduces to the Taylor's series expansion in the wave number giving the exact solution of Eq. (I.1.12), and there is no novelty in using an asymptotic series. In most cases, asymptotic expansions are divergent series exhibiting the following features:

(i) the error made by halting the series with the nth term is less than or equal to the mth term with $m = n + 1$;
(ii) as the order increases, the terms first decrease and then increase;
(iii) for a given k_0, there is a term such that the best approximation is achieved by truncating the series at that term.

Accordingly, while taking an infinite number of terms would cause u to diverge, a limited number will give a good approximation. Since the error is of the order of the first term neglected, the most accurate sum is obtained by cutting off at a value N such that the successive term is a minimum. A surprising feature of these expansions is their almost unexpected success even for values of k_0 that are not very large.

Before proceeding further, it is worth noting that the above form [Eq. (II.2.5)] of the asymptotic representation of the field is not unique. Some authors have found it advantageous to assume that

$$u = e^{-ik_0 \psi(\mathbf{r})}, \tag{II.2.14}$$

with

$$\psi(\mathbf{r}) \sim S(\mathbf{r}) + \sum_{n=0}^{\infty} (-ik_0)^{-n-1} B_n(\mathbf{r}). \tag{II.2.15}$$

We have now developed the apparatus of ray optics to be used in the following sections in order to describe light propagation in physical systems.

3 The Eikonal Equation

It is easy to see that S and the A_m's can be obtained by recurrence through the set of Eqs. (II.2.9). In particular, the relation $Q_0(\mathbf{r}) = 0$, that is, in cartesian coordinates,

$$(\partial S/\partial x)^2 + (\partial S/\partial y)^2 + (\partial S/\partial z)^2 = n^2(\mathbf{r}), \tag{II.3.1}$$

is known as the *eikonal equation* or the *Hamilton–Jacobi equation.* The surfaces on which the eikonal S is constant are called *wave fronts.* If the eikonal S is known on a generic surface Σ, its value on a nearby surface Σ' is obtained from the relation (see Fig. II.1).

$$\partial S/\partial \tau = [n^2(\mathbf{r}) - (\partial S/\partial \sigma)^2]^{1/2}, \tag{II.3.2}$$

where $\partial/\partial\tau$ and $\partial/\partial\sigma$ indicate, respectively, the derivatives normal and tangent to Σ (of course, $\partial S/\partial\sigma = 0$ if Σ is a wave front). Thus, whenever S (and, as a consequence, $\partial S/\partial\sigma$) is given on a surface Σ, it is possible to evaluate its variation (apart from the sign) when moving to Σ', so that a sufficient

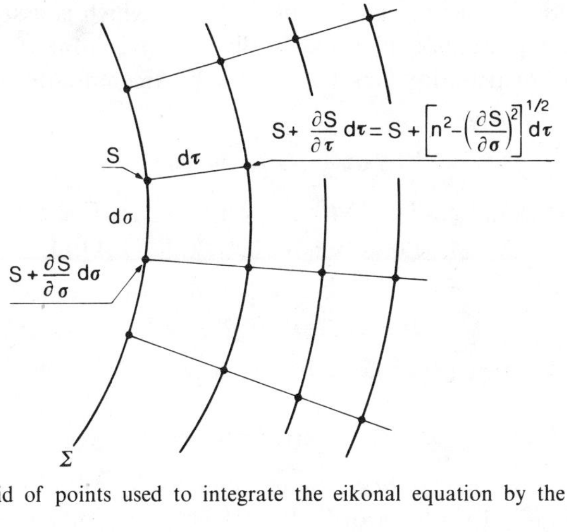

Fig. II.1. Grid of points used to integrate the eikonal equation by the finite-difference method.

condition for solving Eq. (II.3.1) is the knowledge of S on a surface and the direction in which it increases.

Let us now consider a nondispersive medium (refractive index independent of ω) and introduce the function $\Phi(\mathbf{r}, t)$ defined as

$$\Phi(\mathbf{r}, t) = S(\mathbf{r}) - ct, \tag{II.3.3}$$

which, with the help of Eq. (II.3.1), is immediately seen to obey the relation

$$(\nabla\Phi)^2 - (n^2(\mathbf{r})/c^2)(\partial\Phi/\partial t)^2 = 0. \tag{II.3.4}$$

If we remember that the *characteristic equation* of a second-order partial differential equation [5] is obtained by substituting the second derivatives with the product of the corresponding first derivatives, we observe that Eq. (II.3.4) is the characteristic equation of the time-dependent wave equation

$$\nabla^2 u(\mathbf{r}, t) - (n^2(\mathbf{r})/c^2)(\partial^2 u(\mathbf{r}, t)/\partial t^2) = 0. \tag{II.3.5}$$

Since Φ is constant on the surfaces of discontinuity of $u(\mathbf{r}, t)$, we can analyze the evolution of the fronts of discontinuity (for example, those delimiting a space region where the field abruptly vanishes) in terms of $\Phi(\mathbf{r}, t)$. If $\Phi = 0$ on a given front of discontinuity, the position of this front at time t is [see Eq. (II.3.3)] $S(\mathbf{r}) = ct$.

The above discussion furnishes a physical interpretation of the concept of the eikonal and shows its relevance. It is now useful to determine the condition under which an assigned family of surfaces

$$F(\mathbf{r}, \tilde{c}) = 0, \tag{II.3.6}$$

where $\tilde{c}$ is a continuous parameter any value of which is associated with a given surface, represents a possible family of wave fronts for an assigned refractive index distribution $n(\mathbf{r})$. To this end, by differentiating Eq. (II.3.6) we obtain

$$\nabla F \cdot d\mathbf{r} + (\partial F/\partial\tilde{c})d\tilde{c} = 0, \tag{II.3.7}$$

where ∇ indicates the gradient with respect to x, y, z. For a displacement $d\mathbf{r}$ normal to the surface passing through $\mathbf{r}$ (i.e., $d\mathbf{r}$ parallel to ∇F), we have, with the help of Eq. (II.3.2),

$$d\mathbf{r} = (\nabla F/|\nabla F|)[dS/n(\mathbf{r})]. \tag{II.3.8}$$

By inserting Eq. (II.3.8) into Eq. (II.3.7), we can write

$$dS/d\tilde{c} = -n(\mathbf{r})(\partial F/\partial\tilde{c})/|\nabla F|. \tag{II.3.9}$$

Thus, since S depends only $\tilde{c}$, the right-hand side of Eq. (II.3.9) must be a function of $\tilde{c}$ only. In other words, Eq. (II.3.6) represents a family of wave fronts if the right-hand side of Eq. (II.3.9) remains constant on each surface of the family.

4 The Ray Equation

Let us consider a single-valued eikonal $S(\mathbf{r})$ and define the unit vector (see Fig. II.2)

$$\hat{s}(\mathbf{r}) = \nabla S/|\nabla S| = \nabla S/n(\mathbf{r}), \tag{II.4.1}$$

where $\hat{s}(\mathbf{r})$ is perpendicular to the wave fronts and points in the propagation direction, and let us introduce the trajectories (*rays*) $\mathbf{r}(s)$ tangent to $\hat{s}(\mathbf{r})$ at each $\mathbf{r}$. Whenever S is not single-valued, the region will be spanned by a multiplicity of ray families. If s is made to coincide with the curvilinear abscissa along the ray, we can write

$$n(\mathbf{r})\, d\mathbf{r}/ds = n(\mathbf{r})\hat{s}(\mathbf{r}) = \nabla S = \mathbf{k}/k_0 \tag{II.4.2}$$

[see Eq. (II.2.7)] and, by differentiating both sides with respect to s,

$$\frac{d}{ds}\left[n(\mathbf{r})\frac{d\mathbf{r}}{ds}\right] = \frac{d}{ds}\nabla S = \left(\frac{d\mathbf{r}}{ds}\cdot\nabla\right)\nabla S = \frac{1}{n(\mathbf{r})}(\nabla S\cdot\nabla)\nabla S. \tag{II.4.3}$$

On the other hand, Eq. (II.3.1) gives

$$2n(\mathbf{r})\,\nabla n = \nabla(\nabla S\cdot\nabla S) = 2(\nabla S\cdot\nabla)\nabla S, \tag{II.4.4}$$

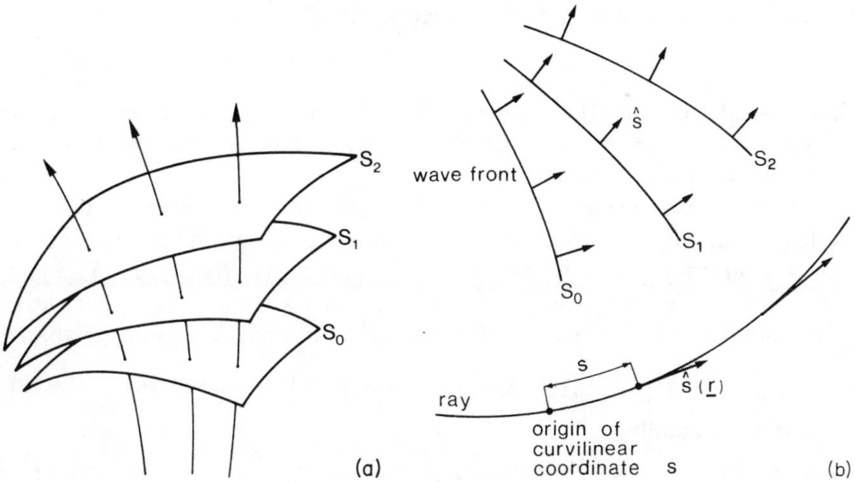

Fig. II.2. (a) Wavefront family and (b) the curvilinear coordinate s and direction $\hat{s}$ along a generic trajectory (ray) perpendicular to the wave front family.

the last equality resulting from the vector identities (A.11) and (A.14). By inserting Eq. (II.4.4) into Eq. (II.4.3), we finally obtain

$$\frac{d}{ds}\left[n(\mathbf{r})\frac{d\mathbf{r}}{ds}\right] = \nabla n, \tag{II.4.5}$$

or equivalently,

$$[\hat{s}(\mathbf{r}) \cdot \nabla][n(\mathbf{r})\hat{s}(\mathbf{r})] = \nabla n, \tag{II.4.6}$$

which is known as the *vector ray equation* (for uniaxial crystals, a particular case of an anisotropic medium, see Stavroudis [6] and Section II.14.1). It is interesting to observe that if we replace s with the parameter

$$\tau = \int^{s} \frac{ds'}{n(s')}, \tag{II.4.7}$$

the ray equation can be rewritten as

$$d^2\mathbf{r}/d\tau^2 = \nabla(\tfrac{1}{2}n^2), \tag{II.4.8}$$

according to which the rays are formally equivalent to the trajectories of a unit-mass particle moving through a potential $V = -n^2/2$.

4.1 *Malus–Dupin Theorem*

Equation (II.4.5) represents a *necessary* condition for a ray bundle to be orthogonal to a family of wave fronts, but it does not always imply the existence of an eikonal. The eikonal exists only if

$$\nabla \times [n(\mathbf{r})\hat{s}(\mathbf{r})] = 0, \tag{II.4.9}$$

as obtained from Eqs. (II.4.2) and (A.14). In this case, the ray bundle is referred to as a *normal congruence* or *orthotomic system*, and its behavior can be described in terms of the eikonal equation.

We now wish to show that Eq. (II.4.9) is verified everywhere, provided it holds true at one point. We observe that imposing the vanishing of the curl of $n\hat{s}$ at a given point is equivalent to assuming that the *dyadic* $\nabla(n\hat{s})$ is symmetrical. We will prove that all-order derivatives of $\nabla(n\hat{s})$ are symmetrical, if this holds true for $\nabla(n\hat{s})$. In fact, from Eq. (II.4.6) we have the dyadic relations

$$\nabla\nabla(\tfrac{1}{2}n^2) = \nabla[(n\hat{s} \cdot \nabla)n\hat{s}] = [\nabla(n\hat{s})]^2 + (n\hat{s} \cdot \nabla)[\nabla(n\hat{s})], \tag{II.4.10}$$

and, more generally,

$$(n\hat{s} \cdot \nabla)^m[\nabla(n\hat{s})] = (n\hat{s} \cdot \nabla)^{m-1}[\nabla\nabla(\tfrac{1}{2}n^2)] - (n\hat{s} \cdot \nabla)^{m-1}[\nabla(n\hat{s})]^2. \tag{II.4.11}$$

Since the first dyadic on the right-hand side of Eq. (II.4.11) is clearly symmetrical and the second one as well, if all the derivatives of $\nabla(n\hat{s})$ of order lower than m are symmetrical, then the left-hand side is also symmetrical. In conclusion, we have shown by induction that all the derivatives of $\nabla(n\hat{s})$ are symmetrical at the point considered, which implies that $\nabla(n\hat{s})$ is symmetrical along the whole ray (provided that n is an analytic function of r) and, equivalently, that $\nabla \times (n\hat{s}) = 0$ everywhere.

It can be shown that the above result, i.e., the validity of Eq. (II.4.9) along the whole ray if it is valid at one point, holds true even when the refractive index is discontinuous on a refracting or reflecting surface. This result is known as the *Malus–Dupin theorem* (*see Chapter I, Born and Wolf* [11]). An intuitive proof can be obtained if we imagine the rays as limiting trajectories associated with a continuous refractive index distribution that changes gradually up to a sudden variation of n on the discontinuity surface. Since $\nabla \times (n\hat{s}) = 0$ for all the rays obtained with a regular index distribution, it must vanish when the discontinuity is reached.

4.2 *Curvature and Torsion of Rays*

The geometry of a ray is conveniently described in terms of the three orthogonal unit vectors $\hat{s}$, $\hat{n}$, and $\hat{b} = \hat{s} \times \hat{n}$, which are respectively directed as the *tangent*, *normal*, and *binormal* to $\mathbf{r}(s)$ (see Fig. II.3). The *Frénet equations*

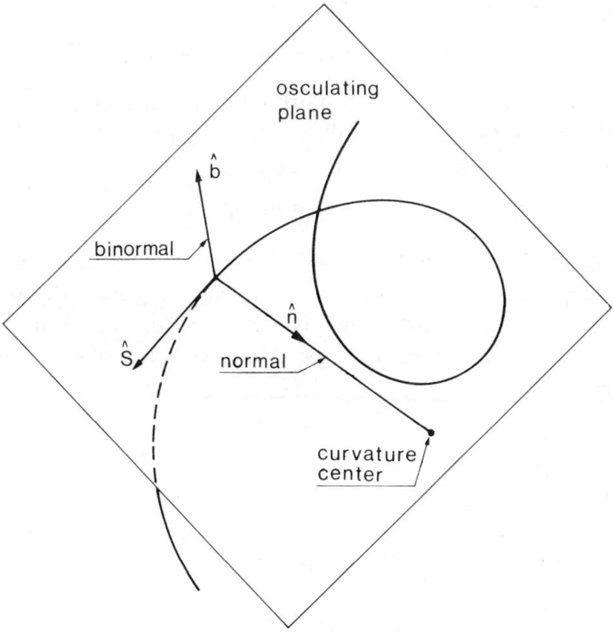

Fig. II.3. Tangent, normal, and binormal unit vectors and osculating plane associated with a space curve.

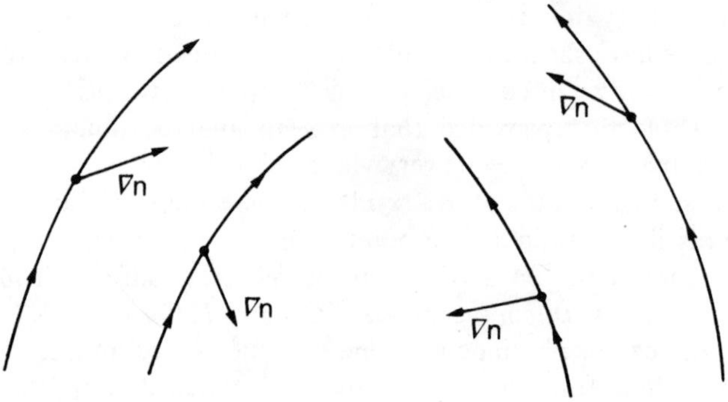

Fig. II.4. Ray bending due to the gradient of the refractive index.

[7], valid for a generic curve, are

$$(d/ds)\hat{s} = \hat{n}/\rho \equiv \mathbf{K}, \tag{II.4.12a}$$

$$(d/ds)\hat{n} = -\hat{s}/\rho + \hat{b}/\tilde{\tau}, \tag{II.4.12b}$$

$$(d/ds)\hat{b} = -\hat{n}/\tilde{\tau}, \tag{II.4.12c}$$

where ρ and $\tilde{\tau}$ indicate, respectively, the *curvature radius* and the *torsion*, which measures the deviation of the curve from planarity. From Eqs. (II.4.5) and (II.4.12a) it follows that

$$\nabla n = (d/ds)(n\hat{s}) = (dn/ds)\hat{s} + n\mathbf{K}, \tag{II.4.13}$$

so that the rays lie locally in the osculating plane containing $\hat{s}$ and ∇n. From Eqs. (II.4.12a) and (II.4.13) it also follows that the *curvature* $K = 1/\rho$ is related to the refractive index through the relation

$$K = \hat{n} \cdot \nabla(\ln n), \tag{II.4.14}$$

according to which a ray tends to bend in such a way that ∇n points into the side where the center of curvature lies (see Fig. II.4). As an example, a ray impinging on a refraction surface will be deflected "up" or "down" according to whether $\mu = n/n'$ is smaller or larger than one, where n' is the index of the refracting medium.

4.3 *Construction of a Ray*

A simple computational scheme for evaluating the trajectory of a ray [8] can be immediately obtained by means of the relation $\hat{s} = d\mathbf{r}/ds$ and Eq. (II.4.12a). In fact,

$$\mathbf{r}(s_1) = \mathbf{r}(s_0) + \hat{s}(s_0)\,\Delta s + (1/2!)\mathbf{K}(s_0)(\Delta s)^2 + \cdots, \tag{II.4.15}$$

where $\Delta s = s_1 - s_0$ is a finite increment. By differentiation,

$$\hat{s}(s_1) = \hat{s}(s_0) + \mathbf{K}(s_0)\,\Delta s + \cdots. \qquad \text{(II.4.16)}$$

The computational scheme proceeds as follows. For a given initial position $\mathbf{r}_0$ and normal $\hat{n}(s_0)$ of a ray, Eq. (II.4.14) allows us to obtain the initial curvature vector $\mathbf{K}(s_0)$. Thus, if we know $\hat{s}(s_0)$, we are able to evaluate $\mathbf{r}(s_1)$ and $\hat{s}(s_1)$ by means of Eqs. (II.4.15) and (II.4.16) to the desired degree of approximation. Then the new vector $\mathbf{K}(s_1)$ is obtained by using Eq. (II.4.14) with $\hat{n}(s_1)$ determined by means of the osculating plane containing $\hat{s}(s_0)$ and $\hat{s}(s_1)$, and the extrapolation process continues.

5 Field-Transport Equation for A_0.

We now wish to analyze in some detail the evolution of the field amplitudes A_0. By taking the product of Eq. (II.2.9) (for $m = 1$) with A_0, we obtain the *transport equation*

$$A_0^2 \nabla^2 S + 2A_0 \nabla S \cdot \nabla A_0 = 0, \qquad \text{(II.5.1)}$$

which, with the help of the vector relation (A.7), yields

$$\nabla \cdot (A_0^2 n \hat{s}) = 0, \qquad \text{(II.5.2)}$$

where use has been made of Eq. (II.4.2).

If the quantity $A_0^2 n \hat{s}$ is considered analogous to the Poynting vector for the scalar field $A_0(\mathbf{r}) \exp[-ik_0 S(\mathbf{r})]$, Eq. (II.5.2) can be interpreted as the conservation of the power flux. Once it is written in integral form by applying Gauss's theorem to a small volume made up by the rays contiguous to the trajectory $\mathbf{r}(s)$ (see Fig. II.5), it determines the law of variation of $A_0(s)$ along a

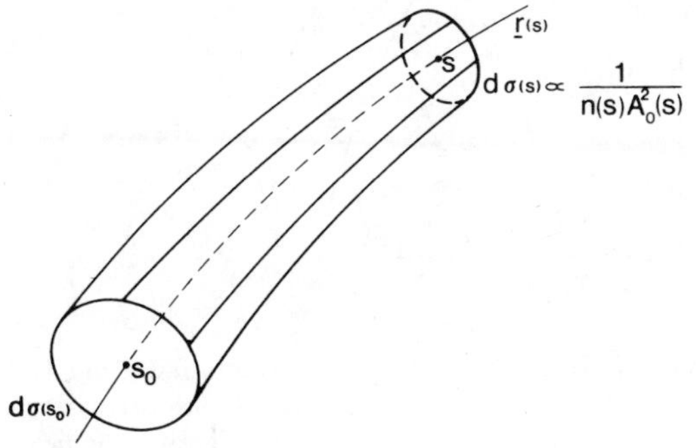

Fig. II.5. Geometry of a ray bundle.

generic ray according to the relation

$$A_0(s) = A_0(s_0)\left[\frac{n(s_0)\,d\sigma(s_0)}{n(s)\,d\sigma(s)}\right]^{1/2} \tag{II.5.3}$$

where $d\sigma(s_0)$ and $d\sigma(s)$ represent the sections of the ray tube relative to s_0 and s.

Observing now that Eq. (II.5.1), which can be written, with the help of Eq. (II.4.1), in the form

$$\nabla^2 S + 2n(d/ds)(\ln A_0) = 0, \tag{II.5.4}$$

gives the integral

$$A_0(s) = A_0(s_0)\exp\left\{-\frac{1}{2}\int_{s_0}^{s}\frac{\nabla^2 S}{n(s')}\,ds'\right\}, \tag{II.5.5}$$

we can rewrite Eq. (II.5.3) as

$$\frac{n(s_0)\,d\sigma(s_0)}{n(s)\,d\sigma(s)} = \left(\frac{A_0(s)}{A_0(s_0)}\right)^2 = \exp\left[-\int_{s_0}^{s}\frac{\nabla^2 S}{n(s')}\,ds'\right]. \tag{II.5.6}$$

As a consequence, when $d\sigma \to 0$, $\nabla^2 S \to -\infty$. This fact is connected with the concept of caustic (see Section II.10.1b) as the locus of points where the Laplacian of the eikonal becomes a divergent quantity (see Fig. II.6).

According to the previous results, the field amplitude along a ray depends on the way in which $d\sigma$ or, equivalently, $\nabla^2 S$ varies. Thus, the knowledge of the trajectory of a single ray is not sufficient to deduce the amplitude A_0, but it is necessary to determine all the rays in the proximity of the reference trajectory.

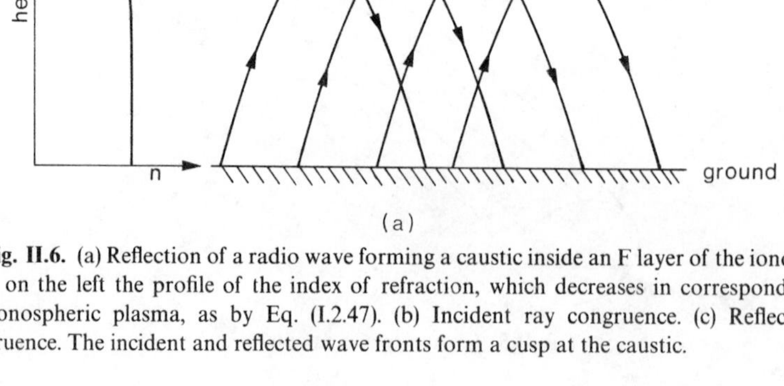

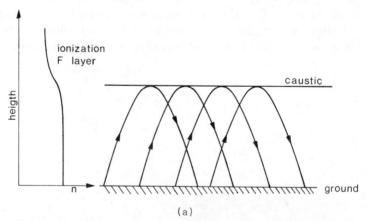

Fig. II.6. (a) Reflection of a radio wave forming a caustic inside an F layer of the ionosphere. Note on the left the profile of the index of refraction, which decreases in correspondence to the ionospheric plasma, as by Eq. (I.2.47). (b) Incident ray congruence. (c) Reflected ray congruence. The incident and reflected wave fronts form a cusp at the caustic.

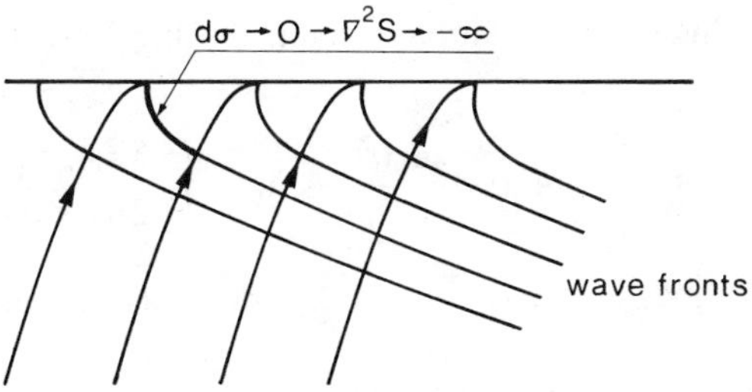

(b)

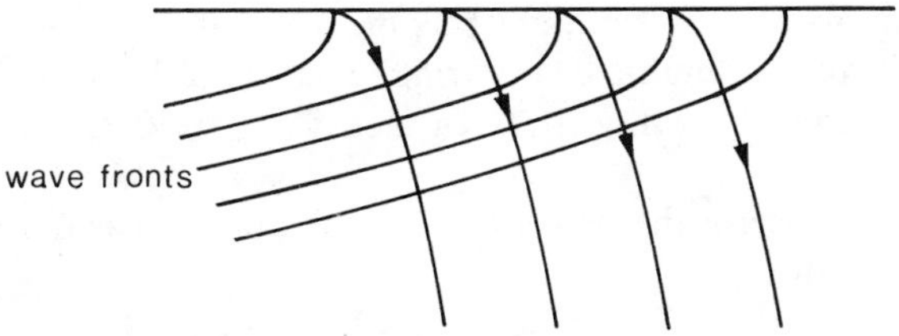

(c)

Fig. II.6. (*continued*)

In a homogeneous medium (n independent of $\mathbf{r}$), the single trajectory is a straight line, a circumstance that notably simplifies the law of variation of $d\sigma$.

6 Field-Transport Equations for the Higher-Order Terms A_m

We now wish to look at the behavior of the A_ms ($m > 0$). To this end, it is convenient to introduce the quantities A'_m defined as

$$A_m(\mathbf{r}) = \exp\left[-\frac{1}{2}\int_{so}^{s} \frac{\nabla^2 S}{n(s')}\,ds'\right] A'_m(\mathbf{r}), \tag{II.6.1}$$

so that Eq. (II.2.9), written in the form

$$\left(\frac{\nabla^2 S}{2n} + \frac{d}{ds}\right) A_m = \frac{-1}{2n}\nabla^2 A_{m-1} \qquad (m = 1, 2, 3, \ldots) \tag{II.6.2}$$

(where use has been made of the relation $\nabla S \cdot \nabla = n\, d/ds$), is equivalent to the *transport equation*

$$\frac{dA'_m}{ds} = -\exp\left[\frac{1}{2}\int_{so}^{s} \frac{\nabla^2 S}{n(s')}\,ds'\right] \frac{\nabla^2 A_{m-1}}{2n(s)}. \tag{II.6.3}$$

Integrating this equation along a ray and using Eqs. (II.5.5) and (II.6.1), we obtain

$$A_m(s) = A_m(s_0)\frac{A_0(s)}{A_0(s_0)} - A_0(s)\int_{s_0}^{s}\frac{\nabla^2 A_{m-1}(s')}{2A_0(s')n(s')}\,ds', \tag{II.6.4}$$

which determines the variation of A_m along a generic ray in terms of A_0 and A_{m-1}.

If $s_0 = 0$ and $A_m(0) = 0$, we have

$$A_m(s) = -\frac{1}{2}A_0(s)\int_0^s \frac{\nabla^2 A_{m-1}(s')}{A_0(s')n(s')}\,ds', \tag{II.6.5}$$

so that the higher-order contributions become relevant for increasing s, as if they were produced by "multiple scattering" related to the variation of the A_m on the wave fronts (see Fig. II.7). In fact, $\nabla^2 A_m \cong \nabla_t^2 A_m$, where $\nabla^2 = \nabla_t^2 + \partial^2/\partial s^2$.

In order to summarize the preceding considerations, it is convenient to introduce the operator

$$\hat{L} = -A_0(s)\int_0^s \frac{\nabla^2}{2A_0(s')n(s')}\,ds', \tag{II.6.6}$$

so that Eq. (II.6.5) yields

$$A_m(s) = \hat{L}A_{m-1}(s) = \hat{L}^m A_0(s). \tag{II.6.7}$$

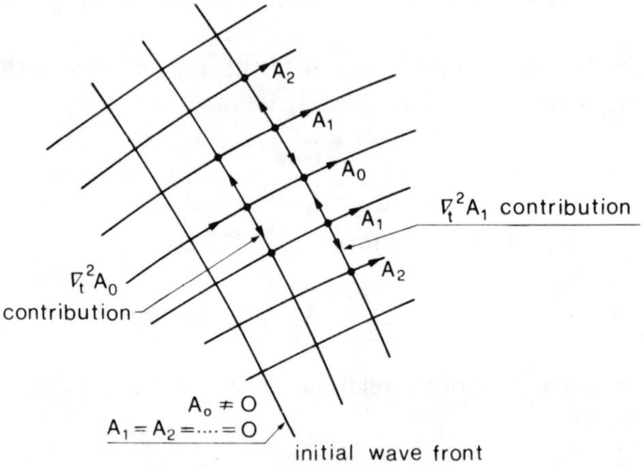

Fig. II.7. Generation of higher-order terms.

Finally, with the help of Eqs. (II.6.7) and (II.2.5) {written as $u(s) \sim A(s)\exp[-ik_0 S(s)]$}, we have

$$A(s) = i\hat{L}A(s)/k_0 + A_0(s), \tag{II.6.8}$$

which yields, with the help of Eq. (II.6.6),

$$A(s) = \frac{-iA_0(s)}{k_0}\int_0^s \frac{\nabla^2 A(s')}{2A_0(s')n(s')}\,ds' + A_0(s) \tag{II.6.9}$$

or, after differentiation,

$$\frac{d}{ds}\frac{A(s)}{A_0(s)} = \frac{-i}{k_0}\frac{\nabla^2 A(s)}{2A_0(s)n(s)}. \tag{II.6.10}$$

6.1 *Fock–Leontovich Parabolic Wave Equation*

Whenever a field propagating *in vacuo* ($n = 1$) can be approximately described by means of an ideal congruence of parallel rays, Eq. (II.5.3) ensures the independence of A_0 from s, so that Eq. (II.6.10) reads

$$(\partial/\partial z)A(\mathbf{r}) = (-i/2k_0)\nabla^2 A(\mathbf{r}), \tag{II.6.11}$$

where the common direction of the rays has been made to coincide with the z axis. Equation (II.6.11) is approximated by the *Fock–Leontovich parabolic wave equation*

$$(\partial/\partial z)A(\mathbf{r}) = (-i/2k_0)\nabla_t^2 A(\mathbf{r}), \tag{II.6.12}$$

where we neglect the longitudinal contribution to ∇^2. Thus, electromagnetic propagation is reduced to an irreversible diffusive process similar to those associated with heat diffusion and wave function evolution in quantum mechanics.

Finally, we observe that the asymptotic development of the solution of Eq. (II.6.12) coincides with that of Eq. (II.6.11) up to terms of order k_0^{-2}. The proof is left as a problem.

7 Evanescent Waves and Complex Eikonals

In the preceding sections we have tacitly assumed that S is a real function, with which we have associated (real) rays. However, in order to deal with fields whose amplitude undergoes a nonnegligible variation over a distance approximately equal to λ, we can still rely on the ray-optical formalism if we

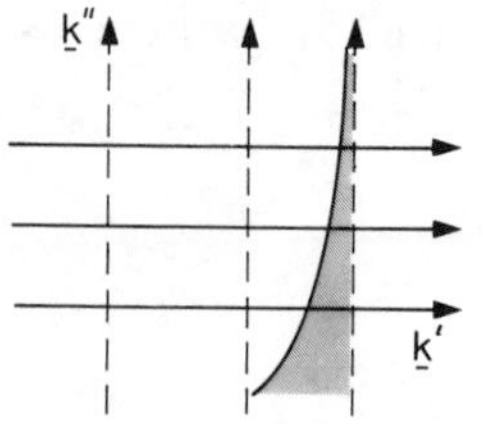

Fig. II.8. Plane evanescent wave. The shaded area represents an exponentially decreasing amplitude distribution. (Vectors are underlined in the figure and boldface in the text.)

introduce a *complex eikonal.* To this end, let us consider the simple case of a free-space plane *evanescent* wave (Fig. II.8) in the form

$$u \propto e^{-i\mathbf{k}'\cdot\mathbf{r}-\mathbf{k}''\cdot\mathbf{r}} = e^{-ik_0S(\mathbf{r})}, \tag{II.7.1}$$

where $k'^2 - k''^2 = k_0^2$ and $\mathbf{k}' \cdot \mathbf{k}'' = 0$.

This example has lead Felsen to look in general for complex solutions [3]

$$S(\mathbf{r}) = R(\mathbf{r}) - iI(\mathbf{r}) \tag{II.7.2}$$

of the eikonal equation [Eq. (II.3.1)], so that $R(\mathbf{r})$ and $I(\mathbf{r})$ obey the equation

$$(\nabla R)^2 - (\nabla I)^2 \equiv \beta^2 - \alpha^2 = n^2(\mathbf{r}), \tag{II.7.3}$$

with $\nabla R \cdot \nabla I = 0$ (*equiphase and equiamplitude surfaces* mutually orthogonal). Fields associated with complex eikonals are usually referred to as *homogeneous waves.* Similarly, if we set

$$A_0(\mathbf{r}) = e^{w(\mathbf{r})-iv(\mathbf{r})}, \tag{II.7.4}$$

the transport equation [Eq. (II.5.1)] gives

$$\tfrac{1}{2}\nabla^2 R + \nabla R \cdot \nabla w - \nabla I \cdot \nabla v = 0, \qquad \tfrac{1}{2}\nabla^2 I + \nabla R \cdot \nabla v + \nabla I \cdot \nabla w = 0. \tag{II.7.5}$$

Let us now consider two ray congruences that are perpendicular to the families of surfaces $R(\mathbf{r}) = \text{const}$ and $I(\mathbf{r}) = \text{const}$. The trajectories of the first congruence normal to the *phase fronts* (*equiphase surfaces*) are called *phase paths* and are characterized by a constant value of I. The trajectories of the second congruence (*attenuation paths* or *equiphase contours*), normal to the equiamplitude surfaces $I(\mathbf{r}) = \text{const}$, lie on phase fronts. A procedure analogous to that leading to Eq. (II.4.5) yields the trajectory equations

$$(d/ds)(\beta\hat{s}) = \nabla\beta, \qquad (d/dt)(\alpha\hat{t}) = \nabla\alpha, \tag{II.7.6}$$

where ds and dt are respectively, the length elements of the phase and attenuation paths (see Fig. II.9) and the unit vectors $\hat{s}$ and $\hat{t}$ determine the corresponding directions. Equations (II.7.6) show that α and β play the formal role of refractive indices.

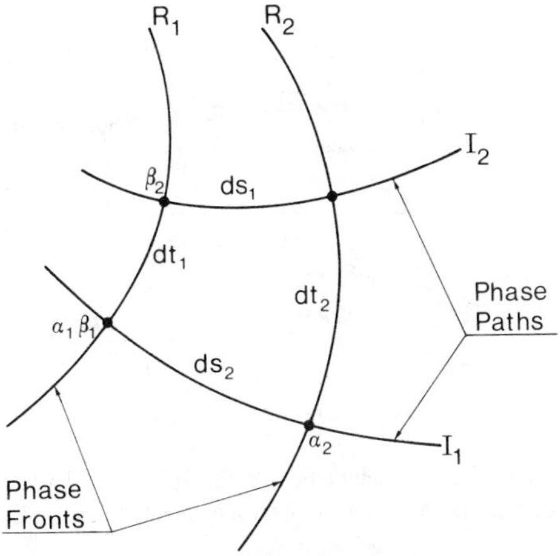

Fig. II.9. Geometry of the mutually orthogonal phase fronts ($R = \text{const}$) and phase paths ($I = \text{const}$).

By specifying that $\nabla R \cdot \nabla = \beta\, d/ds$ and $\nabla I \cdot \nabla = \alpha\, d/dt$, the transport equations [Eqs. (II.7.5)] become

$$\begin{aligned} &\tfrac{1}{2}\nabla \cdot (\beta \hat{s}) + \beta(dw/ds) - \alpha(dv/dt) = 0, \\ &\tfrac{1}{2}\nabla \cdot (\alpha \hat{t}) + \beta(dv/ds) + \alpha(dw/dt) = 0. \end{aligned} \tag{II.7.7}$$

The *curvatures* K_s and K_t are found by extending the procedure leading to Eq. (II.4.14):

$$K_s = \hat{t} \cdot \nabla(\ln \beta) = (1/\beta)(d\beta/dt), \qquad K_t = \hat{s} \cdot \nabla(\ln \alpha) = (1/\alpha)(d\alpha/ds). \tag{II.7.8}$$

The above considerations show that the Luneburg–Kline asymptotic series can be generalized to investigate fields that do not possess a real S, provided a suitable complex eikonal function is introduced.

7.1 *Two-Dimensional Evanescent Fields*

Simple examples of evanescent waves are obtained by considering two-dimensional fields. If we assign a family of phase paths, then the orthogonal attenuation paths are uniquely determined. Using suitable orthogonal curvilinear coordinates μ, ν to label the two families, we have $R = R(\mu)$, $I = I(\nu)$, so that Eq. (II.7.3) transforms into

$$(dR/d\mu)^2(1/h_\mu^2) - (dI/d\nu)^2(1/h_\nu^2) = n^2(\mu, \nu), \tag{II.7.9}$$

h_μ and h_ν being the *scale factors* of the chosen coordinates.

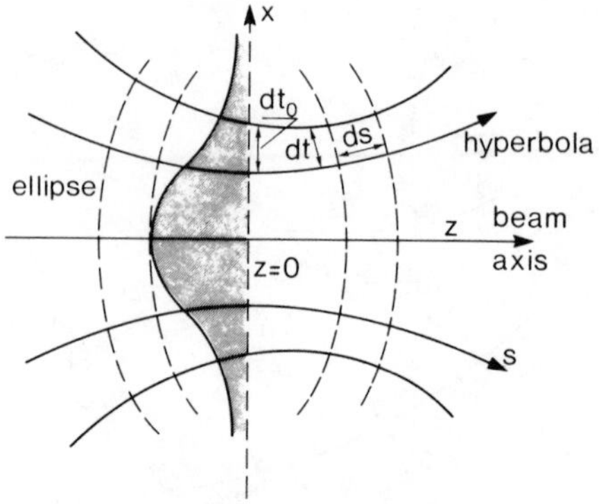

Fig. II.10. Geometry of an evanescent field whose phase paths form a family of confocal hyperbolas. The field amplitude varies along the x axis according to a Gaussian law.

A case of interest is that in which the phase paths are confocal hyperbolas and the attenuation paths ellipses (see Fig. II.10). Thus, it is convenient to use elliptic coordinates defined by

$$\begin{aligned} x &= b \sin \nu \cosh \mu \\ z &= b \cos \nu \sinh \mu \end{aligned} \qquad (0 \leqq \nu \leqq 2\pi, \quad (0 \leqq \mu \leqq \infty), \qquad \text{(II.7.10)}$$

with $h_\mu = h_\nu = b(\cosh^2 \mu - \sin^2 \nu)^{1/2}$, and Eq. (II.7.9) for a field *in vacuo* reads

$$(dR/d\mu)^2 - (dI/d\nu)^2 = b^2(\cosh^2 \mu - \sin^2 \nu). \qquad \text{(II.7.11)}$$

Let us consider the half-plane $z > 0$ ($\cos \nu > 0$) and the integral of Eq. (II.7.11) $I(\nu) = b(1 - \cos \nu)$, $R(\mu) = b \sinh \mu = z/\cos \nu$. If we concentrate on the region $z \gg |x|$ ($\cos \nu \cong 1$), we have

$$I(\nu) = \frac{b}{2} \frac{x^2}{(z^2 + b^2)}, \qquad \mathrm{R}(\mu) = z\left(1 + \frac{x^2}{2(z^2 + b^2)}\right), \qquad \text{(II.7.12)}$$

so that [see Eq. (II.7.2)]

$$e^{-ik_0 S} = e^{-x^2/w^2 - ik_0[z + x^2/(2\rho)]}, \qquad \text{(II.7.13)}$$

where

$$w^2(z) = 2(z^2 + b^2)/bk_0, \qquad \rho(z) = z + b^2/z. \qquad \text{(II.7.14)}$$

These kinds of fields present a Gaussian distribution with width $w(z)$, while the wave fronts have a curvature radius $\rho(z)$. They are called *Gaussian beams*, as they describe fields having the form of very narrow pencils (see Section VII.7).

8 Ray Optics of Maxwell Vector Fields

8.1 *Asymptotic Expansion of the Electric Field*

The electric field **E** in a dielectric inhomogeneous medium satisfies the *vector wave equation* [Eq. (I.1.11)] written in the form

$$\nabla^2\mathbf{E} + k_0^2 n^2\mathbf{E} + 2\nabla[\mathbf{E}\cdot\nabla(\ln n)] = 0. \tag{II.8.1}$$

In complete analogy with the scalar case represented by Eq. (II.2.5), we look for a ray optical representation of the electric field by replacing $A_m(\mathbf{r})$ with $\mathbf{E}_m(\mathbf{r})$.

By inserting the right-hand side of Eq. (II.2.5) into Eq. (II.8.1) and equating to zero the coefficients of each power of k_0, we obtain the recursive system

$$\begin{gathered}(\nabla S)^2 - n^2(\mathbf{r}) = 0,\\ (\nabla^2 S + 2\nabla S\cdot\nabla)\mathbf{E}_0 + 2[\mathbf{E}_0\cdot\nabla(\ln n)]\,\nabla S = 0,\\ [\nabla^2 S + 2\nabla S\cdot\nabla + 2\nabla S\nabla(\ln n)\,\cdot]\mathbf{E}_m + \nabla^2\mathbf{E}_{m-1} + 2\nabla[\mathbf{E}_{m-1}\cdot\nabla(\ln n)] = 0,\end{gathered} \tag{II.8.2}$$

the last equation referring to $m > 0$. If we compare Eqs. (II.8.2) with Eqs. (II.2.9), we observe that the vector theory reduces to the scalar one only if the $\mathbf{E}_m$ are perpendicular to the gradient of the refractive index. In general, E_x, E_y, and E_z mix because of the terms containing $\nabla(\ln n)$, so that an initially linearly polarized field does not maintain its polarization during propagation.

In order to study the lowest-order term of the asymptotic series, it is convenient to replace $\mathbf{E}_0(\mathbf{r})$ with $\mathbf{E}'(\mathbf{r})$ defined as [cf. Eq. (II.6.1)]

$$\mathbf{E}_0(\mathbf{r}) = \mathbf{E}'(\mathbf{r})\exp\left[-\frac{1}{2}\int_0^s \frac{\nabla^2 S}{n(s')}\,ds'\right], \tag{II.8.3}$$

so that, with the help of Eq. (II.4.1), Eq. (II.8.2b) reduces to

$$n(d/ds)\mathbf{E}' + \hat{s}(\mathbf{E}'\cdot\nabla n) = 0. \tag{II.8.4}$$

If we multiply (scalarly) the above relation by $\hat{s}$ and use Eq. (II.4.6), we obtain

$$n\hat{s}\cdot(d/ds)\mathbf{E}' + \mathbf{E}'\cdot\nabla n = n\hat{s}\cdot(d/ds)\mathbf{E}' + \mathbf{E}'\cdot(d/ds)(n\hat{s}) = 0, \tag{II.8.5}$$

which implies the constancy of the scalar product $n\hat{s}\cdot\mathbf{E}'$ along a ray.

In particular, if $\mathbf{E}'$ is perpendicular to $\hat{s}$ at one point, it remains perpendicular along the whole ray path. Furthermore, scalar multiplication of Eq. (II.8.4) by $\mathbf{E}'$ yields

$$n\mathbf{E}'\cdot(d/ds)\mathbf{E}' = \tfrac{1}{2}n(d/ds)(\mathbf{E}'\cdot\mathbf{E}') = 0, \tag{II.8.6}$$

if $\mathbf{E}' \cdot \hat{s} = 0$. Thus $\mathbf{E}'$ is a constant amplitude vector orthogonal to $\hat{s}$, if it is such at one point.

8.2 *Asymptotic Expansion of the Magnetic Field*

In order to complete the description of the vector field, we must now look for an asymptotic expression for the magnetic field $\mathbf{H}$. We observe that Eq. (II.4.1), with the help of the vector relation (A.8), allows us to obtain the asymptotic representation of $\nabla \times \mathbf{E}$,

$$\nabla \times \mathbf{E} \sim e^{-ik_0 S} \sum_{m=-1}^{\infty} \frac{(n\hat{s} \times \mathbf{E}_{m+1} + \nabla \times \mathbf{E}_m)}{(-ik_0)^m}, \qquad \mathbf{E}_{-1} \equiv 0, \tag{II.8.7}$$

so that for $\mu = \mu_0$ we obtain from Eq. (I.1.1)

$$\mathbf{H} \sim \frac{e^{-ik_0 S(\mathbf{r})}}{\zeta_0} \sum_{m=0}^{\infty} \frac{n\hat{s} \times \mathbf{E}_m + \nabla \times \mathbf{E}_{m-1}}{(-ik_0)^m} \equiv e^{-ik_0 S(\mathbf{r})} \sum_{m=0}^{\infty} \frac{\mathbf{H}_m(\mathbf{r})}{(-ik_0)^m}. \tag{II.8.8}$$

In particular, $\zeta_0 \mathbf{H}_0 = \hat{s} \times \mathbf{E}_0$ ($\zeta \equiv \zeta_0/n$). Accordingly, if $\mathbf{E}_0$ is orthogonal to $\hat{s}$, then $\hat{s}$, $\mathbf{E}_0$, and $\mathbf{H}_0$ are mutually orthogonal, and the field is a transverse electromagnetic (TEM) wave at the zeroth order in k_0^{-1}.

8.3 *Asymptotic Expansion of the Poynting Vector*

By means of the above expansions the complex Poynting vector [see Eq. (I.6.8)] $\mathbf{S} = \frac{1}{2}\mathbf{E} \times \mathbf{H}^*$ possesses the asymptotic representation

$$\mathbf{S} \sim \sum_{m=0}^{\infty} \mathbf{S}_m k_0^{-m}, \tag{II.8.9}$$

with

$$\mathbf{S}_m = \frac{1}{2} \sum_{m'=0}^{m} \mathbf{E}_{m'} \times \mathbf{H}^*_{m-m'} (-i)^{-m'} (i)^{m'-m}$$

$$= \frac{i^{-m}}{2\zeta_0} \sum_{m'=0}^{m} (-1)^{m'} \mathbf{E}_{m'} \times (n\hat{s} \times \mathbf{E}^*_{m-m'} + \nabla \times \mathbf{E}^*_{m-m'-1}). \tag{II.8.10}$$

In particular, because of the vector relation (A.2), we can write, whenever $\mathbf{E}_0 \cdot \hat{s} = 0$,

$$\mathbf{S}_0 = \frac{1}{2\zeta} \mathbf{E}_0 \times (\hat{s} \times \mathbf{E}_0^*) = \hat{s} \frac{1}{2\zeta} |\mathbf{E}'|^2 \exp\left[-\int_0^s \frac{\nabla^2 S}{n(s')} ds' \right], \tag{II.8.11a}$$

$$\mathbf{S}_1 = \frac{-i}{2\zeta_0} [-\mathbf{E}_1 \times (n\hat{s} \times \mathbf{E}_0^* + \mathbf{E}_0 \times (n\hat{s} \times \mathbf{E}_1^* + \nabla \times \mathbf{E}_0^*)]$$

$$= \frac{-i}{2\zeta_0} [(-\mathbf{E}_1 \cdot \mathbf{E}_0^* + \mathbf{E}_1^* \cdot \mathbf{E}_0) n\hat{s} + \mathbf{E}_0^* (n\hat{s} \cdot \mathbf{E}_1) + \mathbf{E}_0 \times (\nabla \times \mathbf{E}_0^*)]. \tag{II.8.11b}$$

Equation (II.8.11a) ensures that $\mathbf{S}_0$ is a real vector. On the other hand, the orthogonality of $\mathbf{E}_0$ and $\hat{s}$ implies that $\frac{1}{2}\varepsilon\mathbf{E}_0 \cdot \mathbf{E}_0^* = \frac{1}{2}\mu_0\mathbf{H}_0 \cdot \mathbf{H}_0^*$. Consequently, the equality at the zeroth order between electric and magnetic energy densities and the reality of $\mathbf{S}_0$ are related phenomena.

The preceding considerations imply a local plane wave at the zeroth order in k_0^{-1}, while this is not the case for the first-order field. In fact, if we neglect the terms in $\mathbf{E}_1$, Eq. (II.8.11b) becomes $i2\zeta_0\mathbf{S}_1 = \mathbf{E}_0 \times (\nabla \times \mathbf{E}_0^*)$, so with the help of Eqs. (II.8.3), (A.8), and (A.2) and the constancy of $|\mathbf{E}'|$ we get

$$\mathbf{S}_1 = \frac{i}{2\zeta_0}|\mathbf{E}_0|^2 \frac{\nabla^2 S}{2n}\hat{s} = \frac{i}{2n^2}(\nabla^2 S)\mathbf{S}_0, \tag{II.8.12}$$

where we have considered, as a first approximation, only the variation of $\mathbf{E}_0$ along a ray. The purely imaginary value of $\mathbf{S}_1$ signals an unbalance between electric and magnetic energy (see Chapter I, Papas [17]).

8.4 *The Luneburg–Kline Series at the Interface between Two Media*

A peculiar property of the plane wave $\mathbf{E}_0$ arises in connection with the presence of a discontinuity. More precisely, let us consider a wave traveling in a stratified medium in a direction $\hat{z}$, parallel to ∇n, so that $\mathbf{E}_0$ and $\mathbf{H}_0$ are both perpendicular to ∇n. For a refractive index discontinuous on a given plane $z =$ const, $\mathbf{E}_0$ and $\mathbf{H}_0$ cannot both be continuous functions of z. This result is in contrast with the fact that, according to Maxwell's equations, the components of $\mathbf{E}$ and $\mathbf{H}$ parallel to the discontinuity surface of the refractive index must be continuous. In order to restore this situation, we must consider a second wave arising from the discontinuity surface and traveling in the direction $-\hat{z}$. The amplitudes of the reflected and transmitted waves will be derived in Chapter III, to which the reader is referred for an exhaustive analysis of propagation through plane-stratified media.

9 Differential Properties of Wave Fronts

We wish to investigate some relevant differential properties of the wave fronts in a *homogeneous* medium. If we expand S in proximity to the point $\mathbf{r}_0$, we obtain

$$S(\mathbf{r}_0 + \Delta\mathbf{r}) = S(\mathbf{r}_0) + \Delta\mathbf{r} \cdot \nabla S + \tfrac{1}{2}\Delta\mathbf{r}\,\Delta\mathbf{r} : \mathbf{Q} + \cdots, \tag{II.9.1}$$

where $\mathbf{Q} \equiv \nabla\nabla S$ is a tensor representing the curvature of the wave front, as will become clear in the following, and the "product" symbol ":" indicates the operation yielding the scalar quantity $\Sigma_{i,j}\Delta r_i \Delta r_j(\partial/\partial x_i)(\partial/\partial x_j)S$. Since

$\nabla S \cdot \nabla S = n^2 = $ const in a homogeneous medium, the vector identities (A.11) and (A.4) yield

$$0 = \nabla(\nabla S \cdot \nabla S) = 2(\nabla S \cdot \nabla)\nabla S + 2\nabla S \times (\nabla \times \nabla S) = 2\hat{s} \cdot \mathbf{Q} \qquad \text{(II.9.2)}$$

(where we have assumed without loss of generality that $n = 1$).

The vanishing of the product $\mathbf{V} \cdot \mathbf{A}$ between a nonvanishing vector $\mathbf{V}$ and a nonvanishing tensor $\mathbf{A}$, yielding by definition the vector $B_i = \Sigma_j V_j A_{ji}$, $(i, j = 1, 2, 3)$, implies the following properties, as easily seen by inspection: at least one of the three components of the diagonal representation of $\mathbf{A}$ vanishes, and the components of $\mathbf{V}$ along the *principal directions* of $\mathbf{A}$, corresponding to the eigenvalues different from zero, vanish as well. Thus, in view of Eq. (II.9.2), the tensor $\mathbf{Q}$ can be represented as

$$\mathbf{Q} = q_1 \hat{t}_1 \hat{t}_1 + q_2 \hat{t}_2 \hat{t}_2, \qquad \text{(II.9.3)}$$

with $\hat{t}_1$ and $\hat{t}_2$ (*principal directions*) orthogonal to ∇S. We shall now see that $\mathbf{Q}$ describes the *curvature* of the wave front at $\mathbf{r}_0$[9]. (In particular, when $\mathbf{Q} = 0$, the wave is locally a plane wave, while $q_1 = q_2 = 0$ for a spherical wave and $q_1 = 0$, $q_2 \neq 0$ for a cylindrical one.) To this end, we insert Eq. (II.9.3) into Eq. (II.9.1) and obtain, with the help of Eq. (II.4.2),

$$S(\mathbf{r}_0 + \Delta\mathbf{r}) = S_0 + \Delta\mathbf{r} \cdot \hat{s}_0 + \tfrac{1}{2}(\Delta\mathbf{r} \cdot \hat{t}_1)^2 q_1 + \tfrac{1}{2}(\Delta\mathbf{r} \cdot \hat{t}_2)^2 q_2 + \cdots, \qquad \text{(II.9.4)}$$

where the quantities on the right side are evaluated at $\mathbf{r}_0$. By differentiation

$$\hat{s}(\mathbf{r}_0 + \Delta\mathbf{r}) = \hat{s}_0 + \Delta\mathbf{r} \cdot \hat{t}_1 \hat{t}_1 q_1 + \Delta\mathbf{r} \cdot \hat{t}_2 \hat{t}_2 q_2 + \cdots. \qquad \text{(II.9.5)}$$

If we apply Eq. (II.9.4) for $\Delta\mathbf{r}$ lying on the cross sections of the wave front $S = S(\mathbf{r}_0) \equiv$ const with a plane passing through $\mathbf{r}_0$ and parallel to $\hat{s}_0$ and $\hat{t}_1$ and with a plane parallel to $\hat{s}_0$ and $\hat{t}_2$, we obtain, respectively (see Fig. II.11)

$$\begin{aligned} \hat{s}_0 \cdot \Delta\mathbf{r} + \tfrac{1}{2} q_1 (\Delta\mathbf{r} \cdot \hat{t}_1)^2 + \cdots &= 0, \\ \hat{s}_0 \cdot \Delta\mathbf{r} + \tfrac{1}{2} q_2 (\Delta\mathbf{r} \cdot \hat{t}_2)^2 + \cdots &= 0. \end{aligned} \qquad \text{(II.9.6)}$$

Thus, these cross sections are locally parabolas with curvature, respectively, equal to q_1 and q_2. In addition, a curvature is considered positive if the corresponding curvature center lies on the opposite side with respect to the ray direction $\hat{s}_0$. Accordingly, the *principal curvature radii*

$$\rho_1 \equiv q_1^{-1}, \qquad \rho_2 \equiv q_2^{-1} \qquad \text{(II.9.7)}$$

of a wave front that appears concave to an observer approaching it along $\hat{s}_0$ are both positive. The opposite occurs for a convex wave front. In conclusion, according to our convention, ρ_1 and ρ_2 are positive for a diverging wave and negative for a converging one.

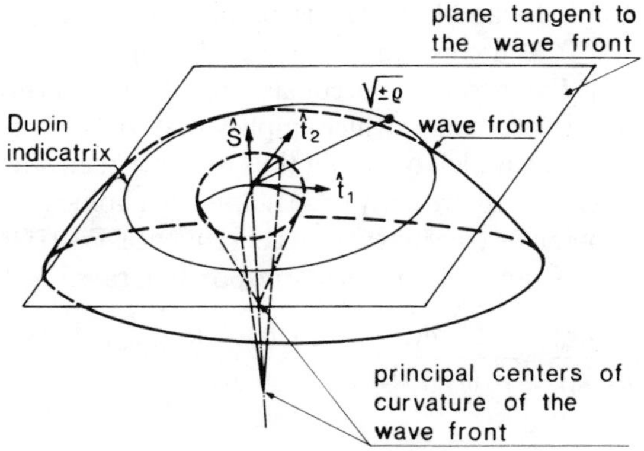

Fig. II.11. Differential properties of a wave front. The *Dupin indicatrix* is the locus of the points lying on the tangent plane at a distance proportional to $|\rho|^{1/2}$ from the point of tangency, ρ being the curvature radius of the cross section of the surface with a plane normal to it and passing through the indicatrix point.

We are now in the position of giving an explicit form for **Q**. To this end, note that for an infinitely small increment $d\mathbf{r}$, Eqs. (II.9.4) and (II.9.5) reduce to

$$S(\mathbf{r}_0 + d\mathbf{r}) = S_0 + d\mathbf{r} \cdot \hat{s}_0, \tag{II.9.8}$$

$$\hat{s}(\mathbf{r}_0 + d\mathbf{r}) = \hat{s}_0 + d\mathbf{r} \cdot (\hat{t}_1\hat{t}_1 q_1 + \hat{t}_2\hat{t}_2 q_2). \tag{II.9.9}$$

In particular, the last equation implies that $\hat{s}_0$, $d\mathbf{r}$, and $\hat{s}(\mathbf{r}_0 + d\mathbf{r})$ are coplanar when $\mathbf{r}$ is parallel to $\hat{t}_1$ or $\hat{t}_2$. We now wish to prove by simple geometric considerations (see Fig. II.12) that the unit vectors $\hat{t}_1$ and $\hat{t}_2$ are constant along a ray. Let us consider two points $\mathbf{r}_0$ and $\mathbf{r}'_0 = \mathbf{r}_0 + \hat{s}_0\, ds$, a vector $d\mathbf{r}_0$ parallel to $\hat{t}_1$, and a ray passing through $\mathbf{r} \equiv \mathbf{r}_0 + d\mathbf{r}_0$. Since $d\mathbf{r}_0$ is parallel to $\hat{t}_1$, $\hat{s} \equiv \hat{s}(\mathbf{r})$ is coplanar with $\hat{s}_0$ and $d\mathbf{r}_0$. Consider now the point $\mathbf{r}' = \mathbf{r} + \hat{s}\, ds$ and the vector $d\mathbf{r}'_0 = d\mathbf{r}_0 + \hat{s}\, ds - \hat{s}_0\, ds$. Since $\mathbf{r}'_0$ and $\mathbf{r}'$ are equidistant (ds) from the wave front passing through $\mathbf{r}_0$ and $\mathbf{r}$, they lie on

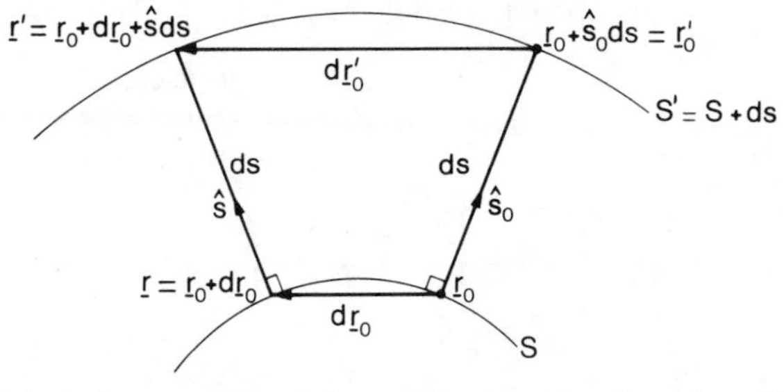

Fig. II.12. Principal cross section of the wave front passing through $\mathbf{r}_0$. The vector $d\mathbf{r}'_0$ is parallel to $d\mathbf{r}_0$ and coplanar with $\hat{s}$ and $\hat{s}_0$, so that it is parallel to the principal direction of the wave front passing through $\mathbf{r}'_0$.

a same wave front. In addition, $d\mathbf{r}'_0$ is parallel to $d\mathbf{r}_0$ since $d\mathbf{r}'_0 \times d\mathbf{r}_0 = ds(\hat{s} - \hat{s}_0) \times d\mathbf{r}_0 = 0$, as a consequence of Eq. (II.9.9) and of the parallelism between $d\mathbf{r}_0$ and $\hat{t}_1$. Furthermore, the coplanarity of $\hat{s}$, $\hat{s}_0$, and $d\mathbf{r}_0$ ensures that $d\mathbf{r}'_0$, $\hat{s}$, and $\hat{s}_0$ are also coplanar, which implies that $d\mathbf{r}'_0$ is directed as $\hat{t}_1(\mathbf{r}'_0)$. Therefore, since $d\mathbf{r}'_0$ is parallel to $d\mathbf{r}_0$ and thus to $\hat{t}_1(\mathbf{r}_0)$, then $\hat{t}_1(\mathbf{r}'_0) = \hat{t}_1(\mathbf{r}_0)$. In conclusion, $\hat{t}_1$ is *constant* along a ray propagating in a homogeneous medium, and the same is obviously true for $\hat{t}_2$. As a consequence, Eq. (II.9.9) yields, due to the constancy of $\hat{s}_0$ and $\hat{s}$ along the corresponding rays,

$$\frac{\hat{t}_1 \cdot d\mathbf{r}_0}{\rho_1(s_0)} = \frac{\hat{t}_1 \cdot d\mathbf{r}'_0}{\rho_1(s_0 + ds)} = \frac{\hat{t}_1 \cdot d\mathbf{r}_0 + (\hat{s} - \hat{s}_0) \cdot \hat{t}_1\, ds}{\rho_1(s_0 + ds)}, \tag{II.9.10}$$

which entails

$$d\rho = \frac{(\hat{s} - \hat{s}_0) \cdot \hat{t}_1 \rho_1\, ds}{\hat{t}_1 \cdot d\mathbf{r}_0} = ds. \tag{II.9.11}$$

The above argument applies as well to $\hat{t}_2$, so that $\mathbf{Q}(\mathbf{r})$ has the simple matrix form (in the reference frame formed by $\hat{t}_1$, $\hat{t}_2$, and $\hat{s}$)

$$\mathbf{Q}(\mathbf{r}) = \begin{bmatrix} \dfrac{1}{\rho_1(0) + s} & 0 \\ 0 & \dfrac{1}{\rho_2(0) + s} \end{bmatrix}. \tag{II.9.12}$$

Here, $\rho_1(0)$ and $\rho_2(0)$ are the principal curvature radii of the wave front passing through the origin O (Fig. II.13), where $s = 0$. The corresponding centers of curvature (*principal centers of curvature*) are at $s = -\rho_1(0)$ and $s = -\rho_2(0)$.

A remarkable expression describing the evolution of A_0 along a ray can now be given. In fact, since

$$\nabla^2 S = \operatorname{Tr} \nabla \nabla S = \operatorname{Tr} \mathbf{Q} = \frac{1}{\rho_1(s)} + \frac{1}{\rho_2(s)} = \frac{1}{\rho_1(0) + s} + \frac{1}{\rho_2(0) + s}, \tag{II.9.13}$$

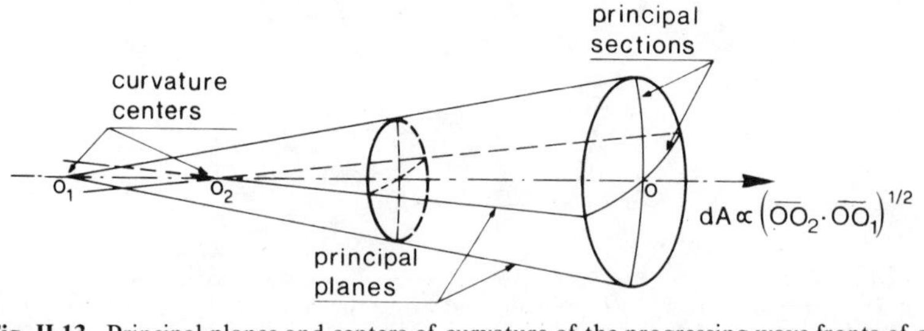

Fig. II.13. Principal planes and centers of curvature of the progressing wave fronts of a ray bundle propagating through a homogeneous medium.

Eq. (II.5.5) becomes

$$A_0(s) = A_0(0)\left[\frac{\rho_1(0)}{\rho_1(0)+s}\right]^{1/2}\left[\frac{\rho_2(0)}{\rho_2(0)+s}\right]^{1/2}, \qquad \text{(II.9.14)}$$

where the square roots must be taken as real positive or imaginary positive. For a spherical wave ($\rho_1 = \rho_2$) the phase along a ray undergoes a jump of π in passing through the focus. This effect was observed a century ago by Gouy and has been referred to since then as *phase anomaly*. As will be shown rigorously in Chapter IV for a spherical wave of finite aperture, the phase undergoes a rapid but continuous change of π. Along the axis, the phase anomaly fluctuates periodically between $-\frac{1}{2}\pi$ and $\frac{3}{2}\pi$ (see Fig. IV.23).

An intuitive argument justifying the phase jump can be given in the following way. The phase along a ray is approximately of the form

$$\phi(s) = -\int^{s} \kappa(s')\,ds' \qquad \text{(II.9.15)}$$

where $\kappa(s')$ represents the "effective" component of the wave vector along the propagation direction. When the wave possesses an amplitude distribution almost uniform on a plane perpendicular to $\hat{s}$, one has $\kappa \cong k_0$. Near the centers of curvature the wave is strongly inhomogeneous and $\kappa = (k_0^2 - k_t^2)^{1/2} < k_0$, where k_t represents an effective transverse wave number. Accordingly, the phase $\phi(s)$ undergoes a positive increment whenever the ray crosses a center of curvature. More precisely, $\phi(s)$ is larger than $\phi(0)$ by an amount $n\pi/2$, where n is the number of centers of curvature between 0 and s. This holds true only if no source is located at the centers of curvature; otherwise, one would obviously have to consider two distinct rays along any straight line crossing the centers.

10 Caustics and Wave Fronts

The field amplitude predicted by ray optics becomes infinite on certain surfaces called *caustics*, where $\nabla^2 S = -\infty$ [Eq. (II.5.6)]. Furthermore, at special points (*cusps*) of a caustic, it diverges in a manner different from the way in which it becomes infinite in the rest of the caustic. In a homogeneous medium, the caustic is the locus of the principal centers of curvature (*foci*) of the wave front, as inferred from Eq. (II.9.14). Since, in most cases, there exist two principal centers of curvature in correspondence to each point of a given wave front, the caustic can be considered as a two-sheeted surface. In a nonhomogeneous medium, the concept of center of curvature is meaningless and the caustic turns out to coincide with the envelope of the ray congruence. In fact, each ray is tangent to this surface at some point, which implies the shrinkage of an elementary ray tube containing the ray up to a segment at the point itself and, as a consequence, the singularity of the field [Eq. (II.5.3)].

More rigorously, let us consider a two-dimensional ray congruence and a triangular contour composed by a ray element δ, a caustic segment, and a wave front segment Δ. The flux of $\hat{s}$ through this contour is clearly equal to $-\Delta$. The area limited by the contour goes to zero approximately as $\delta\Delta$. As a consequence, the ratio between flux and area diverges as $-1/\delta$. This proves, with the help of Gauss's theorem, that $\nabla \cdot \hat{s}$ tends to $-\infty$ on the caustic. Since $\nabla^2 S = \nabla n \cdot \hat{s} + n\nabla \cdot \hat{s}$ [Eq. (II.4.1)], $\nabla^2 S$ also tends to $-\infty$, which entails the singularity of A_0 [Eq. (II.5.6)].

The envelope surface coincides with the locus of the principal centers of curvature for a homogeneous medium and reduces to a single point for a spherical wave (Fig. II.14). As an example, when the wave fronts are cylindrical surfaces, one sheet of the caustic is a cylinder, while the other sheet is at infinity. Another example is furnished by wave fronts in a homogeneous medium exhibiting rotational symmetry; the principal centers of curvature relative to the meridional sections are generally separated from those that lie on the rotation axis (*focal line*), so that the caustic is composed of a rotation surface (possessing a cusp on the rotation axis) and a segment of the axis itself (Fig. II.15). In the two cases considered, the caustics are completely determined by curves that are, respectively, their sections with a plane orthogonal to the cylindrical wave fronts and with a plane containing the rotation axis.

Let us consider a two-dimensional problem. Figure II.16 shows a point P in proximity to a caustic with two rays passing through P (we refer hereafter to

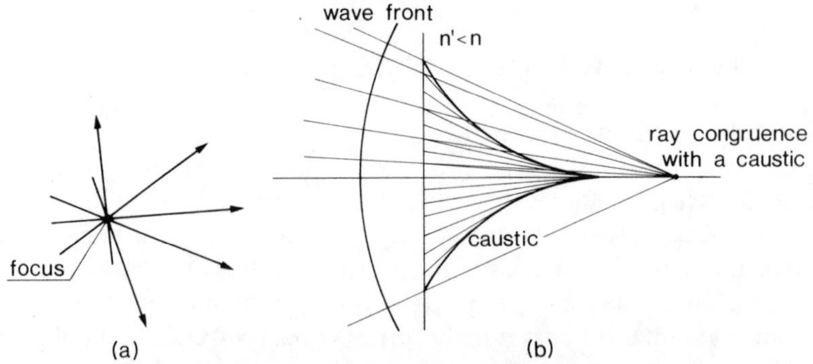

Fig. II.14. Examples of caustics. (a) Caustic reducing to a point for a spherical wave. Notice the difference from the case in which a point source *emits* rays in all directions. While in the former case the field remains finite in the focus, the same does not hold true in the latter one. (b) Caustic formed when a spherical wave enters a less dense dielectric through a plane interface.

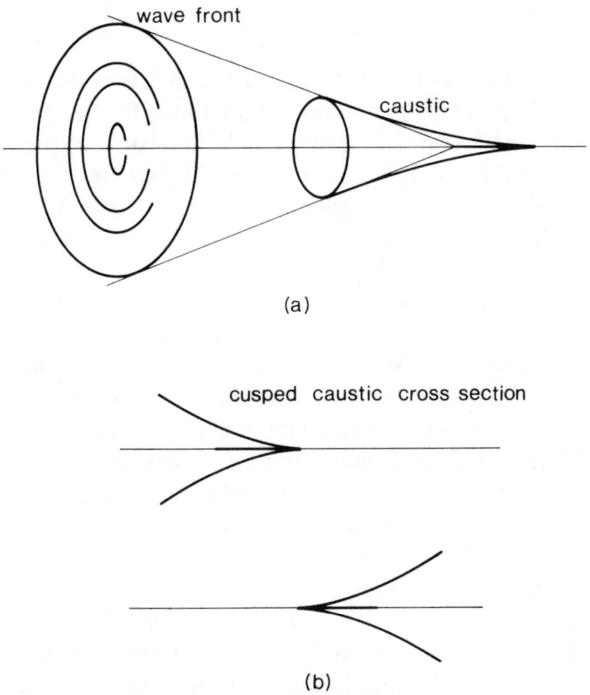

Fig. II.15. (a) The two-sheeted caustic surface associated with a wave front exhibiting rotational symmetry. The cross sections of these caustics with a plane passing through the rotational symmetry axis are shown in (b). The cusped sections can be oppositely oriented according to the wave front behavior.

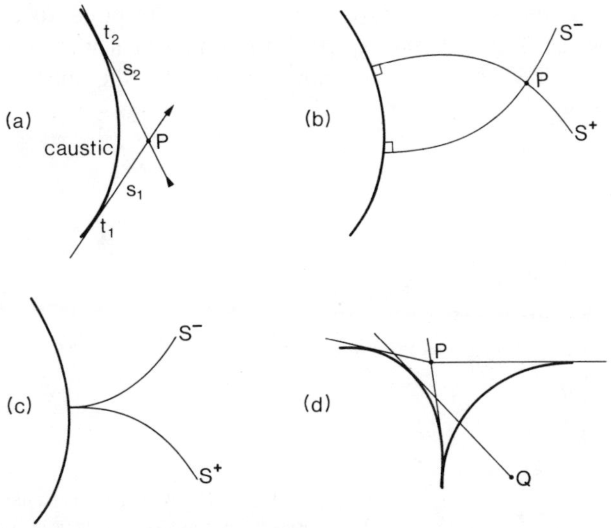

Fig. II.16. (a) Ray crossing in proximity to a regular caustic point. (b) Wave fronts relative to case (a). (c) Cusp formed by the wave fronts on the caustic. (d) Ray crossing in proximity to a cusp. In this case more than two rays can pass through a point, thus giving rise to a complex interference figure. Also indicated is the point Q lying on the dark side of one branch of the caustic and on the lit side of the other branch.

homogeneous media). Each point close to the caustic lies on a ray that has left the caustic (corresponding to an arc length t_1 along the caustic and to a distance s_1 between the point of tangency and P) and on a ray approaching the caustic (corresponding to t_2 and s_2). Thus, the field in P is given by the superposition of two ray fields with respective eikonals S^+ and S^-. This implies an oscillation of the field amplitude in proximity to the caustic due to the interference of the ray fields present there.

The wave fronts $S^{\pm} = \text{const}$ can be traced by the end of a string that is unwound from the caustic. In fact, the center of rotation of the straight section of the string coincides with the point of tangency, so that the end of the string moves perpendicularly to the ray itself, that is, tangentially to a wave front. In other words, if we consider a reference point Q on the caustic, we can imagine that the field is generated by a source in Q. The rays initially follow a curved trajectory along the caustic and then escape tangentially from it. The optical path calculated along this trajectory is constant when P moves on a wave front. This interpretation suggests a simple extension to the caustics of ray congruences in three dimensions. In this case it can be shown [10] that the wave fronts can be constructed by unwinding a bundle of strings that are stretched along the geodesic lines of the caustic surface. This procedure is the direct generalization of the one already described for plane geometry, and allows us to obtain the wave fronts from the caustic, that is, to solve the problem inverse to the one of building up the caustic from the wave fronts.

The wave fronts passing through two positions lying along the same ray, on opposite sides with respect to the point of contact of the caustic and the ray itself, possess opposite curvatures, since they correspond to ingoing and outgoing waves (Fig. II.17). In general, the phase of a wave undergoes an increment of $\pi/2$ when the corresponding ray touches the caustic (see the end of Section II.9).

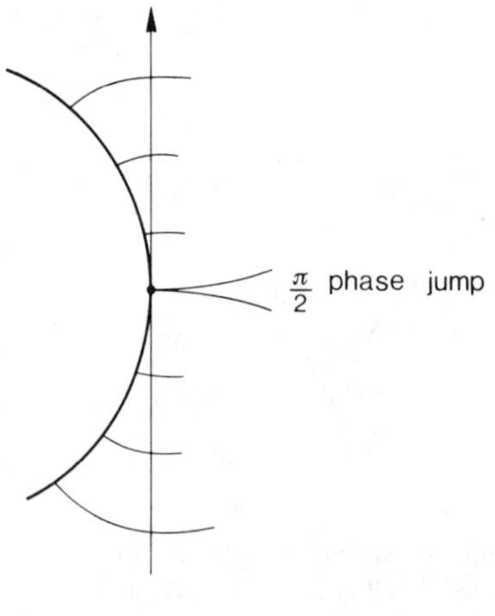

Fig. II.17. Wave front behavior in proximity to the point of tangency of a ray with a caustic.

10.1 *Analytic Properties of Ray Congruences, Wave Fronts, and Caustics*

Let us now consider from the analytical point of view the connection between ray congruences, wave fronts, and caustics. A congruence of straight rays, such as those propagating in a homogeneous medium, can be defined by the following pair of parametric equations:

$$x = \alpha z + f(\alpha, \beta), \qquad y = \beta z + g(\alpha, \beta). \tag{II.10.1}$$

The corresponding tangent unit vectors $\hat{s}$ are given by

$$\begin{aligned}\hat{s} &= \frac{\alpha}{(1 + \alpha^2 + \beta^2)^{1/2}}\hat{x} + \frac{\beta}{(1 + \alpha^2 + \beta^2)^{1/2}}\hat{y} + \frac{1}{(1 + \alpha^2 + \beta^2)^{1/2}}\hat{z} \\ &\equiv p\hat{x} + q\hat{y} + (1 - p^2 - q^2)^{1/2}\hat{z}.\end{aligned} \tag{II.10.2}$$

According to Eq. (II.4.9), a congruence is normal, and thus defines a wave front family, if

$$\nabla \times \hat{s} = 0. \tag{II.10.3}$$

Now, if we make the change of variables $(x, y, z) \to (\alpha, \beta, z)$, we have

$$\frac{\partial}{\partial x} = \left[\left(z + \frac{\partial g}{\partial \beta}\right)\frac{\partial}{\partial \alpha} - \frac{\partial g}{\partial \alpha}\frac{\partial}{\partial \beta}\right] \Bigg/ \left[z^2 + z\left(\frac{\partial f}{\partial \alpha} + \frac{\partial g}{\partial \beta}\right) + \frac{\partial f}{\partial \alpha}\frac{\partial g}{\partial \beta} - \frac{\partial f}{\partial \beta}\frac{\partial g}{\partial \alpha}\right],$$

$$\frac{\partial}{\partial y} = \left[\left(z + \frac{\partial f}{\partial \alpha}\right)\frac{\partial}{\partial \beta} - \frac{\partial f}{\partial \beta}\frac{\partial}{\partial \alpha}\right] \Bigg/ \left[z^2 + z\left(\frac{\partial f}{\partial \alpha} + \frac{\partial g}{\partial \beta}\right) + \frac{\partial f}{\partial \alpha}\frac{\partial g}{\partial \beta} - \frac{\partial f}{\partial \beta}\frac{\partial g}{\partial \alpha}\right],$$

$$\left.\frac{\partial}{\partial z}\right|_{x,y} = \frac{\left[\beta\frac{\partial f}{\partial \beta} - \alpha\left(z + \frac{\partial g}{\partial \beta}\right)\right]\frac{\partial}{\partial \alpha} + \left[\alpha\frac{\partial g}{\partial \alpha} - \beta\left(z + \frac{\partial f}{\partial \alpha}\right)\right]\frac{\partial}{\partial \beta} + \left.\frac{\partial}{\partial z}\right|_{\alpha,\beta}}{z^2 + z\left[\frac{\partial f}{\partial \alpha} + \frac{\partial g}{\partial \beta}\right] + \frac{\partial f}{\partial \alpha}\frac{\partial g}{\partial \beta} - \frac{\partial f}{\partial \beta}\frac{\partial g}{\partial \alpha}}. \tag{II.10.4}$$

Consequently, using the condition that $\nabla \times \hat{s}$ vanishes, we get

$$(1 + \alpha^2)\frac{\partial g}{\partial \alpha} + \alpha\beta\frac{\partial g}{\partial \beta} = \alpha\beta\frac{\partial f}{\partial \alpha} + (1 + \beta^2)\frac{\partial f}{\partial \beta}, \tag{II.10.5}$$

or equivalently, $\partial g/\partial p = \partial f/\partial q$.

10.1.a *Wave Fronts*

In order to obtain the wave fronts associated with the normal ray congruence discussed above, let us indicate with $z = z(\alpha, \beta, \tilde{c})$ the z coordinate of a wave front defined by the parameter $\tilde{c}$. Then, the other two coordinates of the wave front points will be given by

$$x = \alpha z(\alpha, \beta, \tilde{c}) + f(\alpha, \beta), \qquad y = \beta z(\alpha, \beta, \tilde{c}) + g(\alpha, \beta). \tag{II.10.6}$$

We now impose the orthogonality between the displacement of a wave front point ($\tilde{c}$ = const) and the ray direction $\hat{s}$, which yields, with the help of Eq. (II.10.2),

$$\begin{aligned} &\left(z + \alpha\frac{\partial z}{\partial \alpha} + \frac{\partial f}{\partial \alpha}\right)\alpha + \left(\beta\frac{\partial z}{\partial \alpha} + \frac{\partial g}{\partial \alpha}\right)\beta + \frac{\partial z}{\partial \alpha} = 0. \\ &\left(\alpha\frac{\partial z}{\partial \beta} + \frac{\partial f}{\partial \beta}\right)\alpha + \left(z + \beta\frac{\partial z}{\partial \beta} + \frac{\partial g}{\partial \beta}\right)\beta + \frac{\partial z}{\partial \beta} = 0; \end{aligned} \tag{II.10.7}$$

that is,

$$\begin{aligned} \alpha z + (\alpha^2 + \beta^2 + 1)(\partial z/\partial \alpha) &= -\alpha(\partial f/\partial \alpha) - \beta(\partial g/\partial \alpha), \\ \beta z + (\alpha^2 + \beta^2 + 1)(\partial z/\partial \beta) &= -\alpha(\partial f/\partial \beta) - \beta(\partial g/\partial \beta). \end{aligned} \tag{II.10.8}$$

If we set

$$\zeta = z(1 + \alpha^2 + \beta^2)^{1/2}, \tag{II.10.9}$$

we have

$$\begin{aligned} \frac{\partial \zeta}{\partial \alpha} &= -(1 + \alpha^2 + \beta^2)^{-1/2}\left(\alpha\frac{\partial f}{\partial \alpha} + \beta\frac{\partial g}{\partial \alpha}\right), \\ \frac{\partial \zeta}{\partial \beta} &= -(1 + \alpha^2 + \beta^2)^{-1/2}\left(\alpha\frac{\partial f}{\partial \beta} + \beta\frac{\partial g}{\partial \beta}\right), \end{aligned} \tag{II.10.10}$$

which can be integrated, since the relation $\partial g/\partial p = \partial f/\partial q$ implies

$$\frac{\partial^2 \zeta}{\partial \alpha\, \partial \beta} = \frac{\partial^2 \zeta}{\partial \beta\, \partial \alpha}. \tag{II.10.11}$$

In this way, we are able to determine the function

$$z = z(\alpha, \beta, \tilde{c}), \tag{II.10.12}$$

which, with the help of Eqs. (II.10.6), allows us to eliminate the parameters α and β to obtain the equation of the wave fronts $F(x, y, z, \tilde{c}) = 0$. Different values of the parameter $\tilde{c}$ correspond to different solutions of Eqs. (II.10.10), so that the most general solution is associated with the most general wave front.

10.1.b *Caustics*

Let us now look for the caustic associated with the congruence we are considering. It is convenient to consider the subfamily of straight lines defined as

$$x - \alpha z - f[\alpha, \beta(\alpha)] = 0, \qquad y - \beta(\alpha)z - g[\alpha, \beta(\alpha)] = 0, \tag{II.10.13}$$

where $\beta(\alpha)$ is a function that implies the existence of an envelope curve Γ of the subfamily. The point (x_c, y_c, z_c) of contact between the ray associated with a fixed value of α and Γ is determined by imposing

$$z_c + \frac{\partial f}{\partial \alpha} + \frac{\partial f}{\partial \beta}\frac{d\beta}{d\alpha} = 0, \qquad z_c \frac{d\beta}{d\alpha} + \frac{\partial g}{\partial \alpha} + \frac{\partial g}{\partial \beta}\frac{d\beta}{d\alpha} = 0. \tag{II.10.14}$$

This condition corresponds to the intuitive fact that any point of the envelope is an accumulation point for the rays, which implies the vanishing of $dx_c/d\alpha$ and $dy_c/d\alpha$ for fixed z_c. As a consequence of Eqs. (II.10.14), we obtain

$$\frac{d\beta}{d\alpha} = \frac{(\partial g/\partial \beta) - (\partial f/\partial \alpha) \pm \{[(\partial f/\partial \alpha) - (\partial g/\partial \beta)]^2 + 4(\partial f/\partial \beta)(\partial g/\partial \alpha)\}^{1/2}}{2\,\partial f/\partial \beta}. \tag{II.10.15}$$

Solving this differential equation yields the function $\beta(\alpha)$, for which the above subfamily of straight rays defines an envelope Γ. In general, $\beta(\alpha)$ depends on a parameter γ, so that Γ depends on γ and can be indicated with the symbol Γ_γ. Letting γ change continuously, Γ_γ will describe the caustic. Since Eq. (II.10.15) has two roots, there are in general two disconnected surfaces (two-sheeted caustic). It can be shown [10] that the Γ_γ are the geodesic lines of the caustic.

Example: Determination of the Caustic of a Ray Congruence. Let us consider the case in which

$$f = p, \qquad g = q \tag{II.10.16}$$

[a common factor of the dimension of a length is tacitly assumed in the right-hand sides of Eqs. (II.10.16)], where p and q are implicitly defined in Eq. (II.10.2). We refer to a ray congruence possessing rotational symmetry, since any ray describes a rotation surface around the z axis when α and β change in all possible ways keeping $\alpha^2 + \beta^2$ constant.

Equations (II.10.15,16) give

$$d\beta/d\alpha = -\alpha/\beta, \qquad d\beta/d\alpha = \beta/\alpha, \tag{II.10.17}$$

so that, correspondingly,

$$\beta = (\gamma^2 - \alpha^2)^{1/2}, \qquad \beta = \gamma\alpha. \tag{II.10.18}$$

If we now use Eqs. (II.10.14) and (II.10.13), we obtain

$$\begin{aligned} z_c &= -(1 + \alpha^2 + \beta^2)^{-1/2}; \qquad x_c = y_c = 0, \\ z_c &= -(1 + \alpha^2 + \beta^2)^{-3/2}; \qquad \frac{x_c}{\alpha} = \frac{y_c}{\beta} = \frac{\alpha^2 + \beta^2}{(\alpha^2 + \beta^2 + 1)^{3/2}}, \end{aligned} \tag{II.10.19}$$

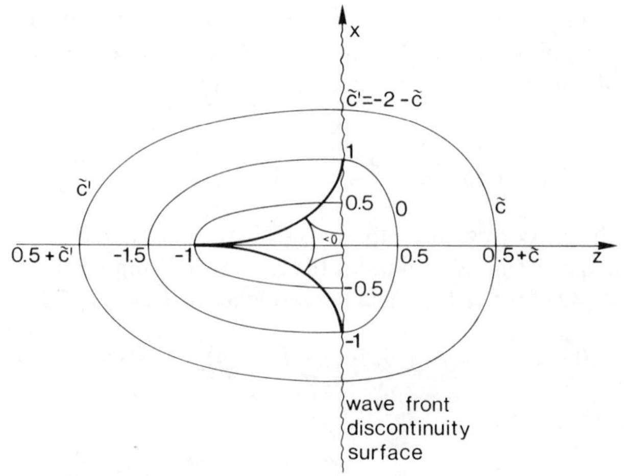

Fig. II.18. Caustic and wave fronts relative to the ray congruence of Eq. (II.10.16). The eikonal is discontinuous on the x axis (wavy line) and the wave fronts intersect perpendicularly the z and x axes and the caustic represented by the thick line connecting the points $(0, \pm 1)$ with $(-1, 0)$.

which represents a caustic (segment $-1, 0$ on the symmetry axis plus rotation surface) of the kind expected in a case of rotational symmetry (Fig. II.18). The rotation surface can be viewed as a horn with the cusp in $(0, 0, -1)$ and the rim tangent to the plane $z = 0$ along the circumference $x^2 + y^2 = 1$. We note that in the present situation the caustic is contained in a finite volume, although the rays span the whole space.

In order to determine the wave fronts, we observe that Eq. (II.10.10) yields, with the help of Eqs. (II.10.16),

$$\partial\zeta/\partial\alpha = -\alpha(1 + \alpha^2 + \beta^2)^{-2}, \qquad \text{(II.10.20)}$$

whose integral is [see Eq. (II.10.9)]

$$z(\alpha, \beta, \tilde{c}) = \tfrac{1}{2}(1 + \alpha^2 + \beta^2)^{-3/2} + \tilde{c}(1 + \alpha^2 + \beta^2)^{-1/2}, \qquad \text{(II.10.21)}$$

where the parameter $\tilde{c}$ identifying the wave front is given by

$$\tilde{c} = z(0, 0, \tilde{c}) - \frac{1}{2}. \qquad \text{(II.10.22)}$$

Finally, using Eqs. (II.10.6), (II.10.16), and (II.10.21), we obtain

$$(x^2 + y^2)^{1/2} = \left| \frac{1}{2} \frac{(\alpha^2 + \beta^2)^{1/2}}{(1 + \alpha^2 + \beta^2)^{3/2}} + (\tilde{c} + 1) \frac{(\alpha^2 + \beta^2)^{1/2}}{(1 + \alpha^2 + \beta^2)^{1/2}} \right|. \qquad \text{(II.10.23)}$$

Equations (II.10.21) and (II.10.23) are the parametric equations (the parameter being $\alpha^2 + \beta^2$) of the wave fronts relative to the ray congruence associated with Eqs. (II.10.16). It is worth noting that, due to Eq. (II.10.22) and the orthogonality between the wave fronts and the z axis, $\tilde{c}$ represents the eikonal to within an additive constant.

11 Reflection and Refraction of a Wave Front at the Curved Interface of Two Media

One of the most relevant problems in optics consists in determining the variations undergone by a wave front during passage through a sequence of lenses. This requires the solution of the canonical problem of the refraction of a ray congruence at the curved interface between two uniform media having refractive indices n and n', respectively. The questions to be answered concern the deviation of the direction of the incident ray and the change of the local wave front, where by "local" we mean the small portion of wave front that the ray is passing through. Because of its small extension, this portion can be analytically represented by a quadric surface characterized by its principal curvature radii. Thus, we can reformulate the problem by saying that we wish to establish a relation between the principal curvature radii of the wave front immediately before and after the passage through the surface of discontinuity. Kneisly [11] has obtained simple relations that are illustrated in the next section together with a short introduction to the differential properties of a surface.

11.1 *Local Matching of the Incident and Refracted Fields along a Curved Interface*

We consider a field hitting a curved surface of discontinuity $f(\mathbf{r}) = 0$ of the refractive index. In order to guarantee the continuity of the electric and magnetic components tangent to the surface implied by Maxwell's equations, the reflected and refracted fields (indicated, respectively, by $''$ and $'$) must be introduced (see Fig. II.19). Then considering the ray-field approximation $\mathbf{E}_0(\mathbf{r})\exp[-ik_0 S(\mathbf{r})]$, we have on $f(\mathbf{r}) = 0$ the matching conditions

$$(\mathbf{1} - \hat{n}\hat{n}) \cdot \mathbf{E}_0(\mathbf{r}) + (\mathbf{1} - \hat{n}\hat{n}) \cdot \mathbf{E}_0''(\mathbf{r}) e^{-ik_0[S''(\mathbf{r}) - S(\mathbf{r})]}$$
$$= (\mathbf{1} - \hat{n}\hat{n}) \cdot \mathbf{E}_0'(\mathbf{r}) e^{-ik_0[S'(\mathbf{r}) - S(\mathbf{r})]}, \tag{II.11.1}$$

and

$$(\mathbf{1} - \hat{n}\hat{n}) \cdot [\mathbf{k} \times \mathbf{E}_0(\mathbf{r})] + (\mathbf{1} - \hat{n}\hat{n}) \cdot [\mathbf{k}'' \times \mathbf{E}_0''(\mathbf{r})] e^{-ik_0[S''(\mathbf{r}) - S(\mathbf{r})]}$$
$$= (\mathbf{1} - \hat{n}\hat{n}) \cdot [\mathbf{k}' \times \mathbf{E}_0'(\mathbf{r})] e^{-ik_0[S'(\mathbf{r}) - S(\mathbf{r})]}, \tag{II.11.2}$$

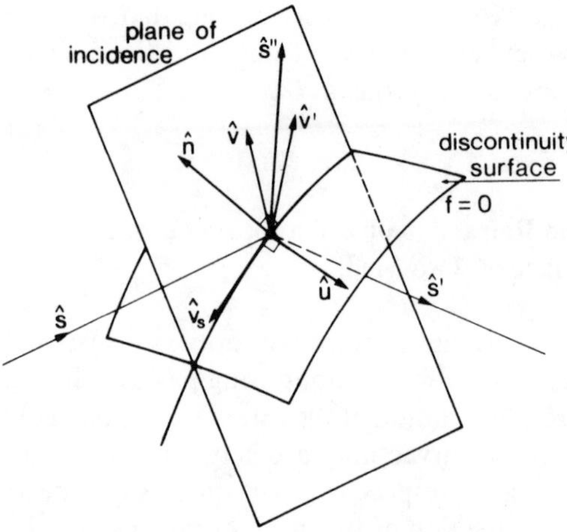

Fig. II.19. Reflection and refraction on a surface of discontinuity of the refractive index. The unit vectors $\hat{v}$, $\hat{v}'$, and $\hat{v}_s$ lie on the plane of incidence and are normal to $\hat{u}$.

yielding, respectively, the continuity of the electric and magnetic tangential components (see Section II.8.2). In fact, the symbol $\mathbf{1}$ denotes the unit tensor and $\hat{n}$ is a unit vector orthogonal to the surface, so that $(\mathbf{1} - \hat{n}\hat{n}) \cdot \mathbf{V}$ represents the projection of the vector $\mathbf{V}$ on the surface.

While $\mathbf{E}_0$, $\mathbf{E}'_0$, and $\mathbf{E}''_0$ are slowly varying functions, this is not the case for the phase factors $\exp(-ik_0 S)$, so that the above vector equations can be satisfied only if the increments $\Delta(S'' - S)$ and $\Delta(S' - S)$ vanish identically on the discontinuity surface. By using Eqs. (II.4.1) and (II.9.1), these conditions can be written as (hereafter n is the refractive index relative to the incident and reflected rays, and n' the one relative to the refracted ray)

$$\Delta(S'' - S) = n\,\Delta\mathbf{r} \cdot (\hat{s}'' - \hat{s}) + \tfrac{1}{2}\Delta\mathbf{r}\,\Delta\mathbf{r} : \nabla\nabla(S'' - S) + \cdots = 0, \qquad \text{(II.11.3a)}$$

$$\Delta(S' - S) = \Delta\mathbf{r} \cdot (n'\hat{s}' - n\hat{s}) + \tfrac{1}{2}\Delta\mathbf{r}\,\Delta\mathbf{r} : \nabla\nabla(S' - S) + \cdots = 0, \qquad \text{(II.11.3b)}$$

with the additional constraint on $\mathbf{r}$ of belonging to the surface $f = 0$, i.e.,

$$\Delta\mathbf{r} \cdot \hat{n} + \tfrac{1}{2}\Delta\mathbf{r}\,\Delta\mathbf{r} : \mathbf{s} + \cdots = 0, \qquad \text{(II.11.4)}$$

where

$$\mathbf{s} = \nabla\nabla f / |\nabla f|. \qquad \text{(II.11.5)}$$

If we now add Eq. (II.11.4) (multiplied by a generic factor $-n'\gamma$) to Eq. (II.11.3b), we obtain

$$\Delta\mathbf{r} \cdot (n'\hat{s}' - n\hat{s} - n'\gamma\hat{n}) + \tfrac{1}{2}\Delta\mathbf{r}\,\Delta\mathbf{r} : [\nabla\nabla(S' - S) - n'\gamma\mathbf{s}] + \cdots = 0, \qquad \text{(II.11.6)}$$

which reduces, for infinitely small increments of $\mathbf{r}$, to

$$d\mathbf{r} \cdot (n'\hat{s}' - n\hat{s} - n'\gamma\hat{n}) = 0. \tag{II.11.7}$$

In order to satisfy Eq. (II.11.7) for any $d\mathbf{r}$ tangent to the surface, that is, orthogonal to $\hat{n}$, the vector $n'\hat{s}' - n\hat{s}$ must be parallel to $\hat{n}$. This condition yields *Snell's law*

$$n' \sin \theta' = n \sin \theta, \tag{II.11.8}$$

relating the *angles of incidence and refraction* θ and θ', defined, respectively, as the angle between $\hat{n}$ and $-\hat{s}$ and the angle between $\hat{n}$ and $-\hat{s}'$, the direction of $\hat{n}$ being such that $\theta < \pi/2$ and $\theta' < \pi/2$. Furthermore, the parallelism between $\hat{n}$ and $n'\hat{s}' - n\hat{s}$ implies the existence of a quantity $\bar{\gamma}$ for which

$$\hat{s}' = \mu\hat{s} + \bar{\gamma}\hat{n}, \tag{II.11.9}$$

with

$$\mu = n/n', \tag{II.11.10}$$

so that the *plane of incidence* $\hat{n}\hat{s}$ coincides with the *plane of refraction* $\hat{n}\hat{s}'$. This statement, together with Eq. (II.11.8), constitutes the well-known *law of refraction* (Snell's law), which was arrived at in 1621 by the Dutch mathematician Willebrord Snell of the University of Leiden and, independently, the French philosopher and mathematician René Descartes (see, e.g., Herzberger [12]).

Scalar multiplication of Eq. (II.11.9) by $\hat{n}$ yields

$$\bar{\gamma} = \mu \cos \theta - \cos \theta'. \tag{II.11.11}$$

For $\gamma = \bar{\gamma}$ the first term of Eq. (II.11.6) vanishes, and we have

$$d\mathbf{r}\, d\mathbf{r} : [\nabla\nabla(S' - S) - n'\bar{\gamma}\mathbf{s}] = 0, \tag{II.11.12}$$

which implies that the element of discontinuity surface $(x, y, z - x + dx, y + dy, z + dz)$ lies on a cone with the vertex on $\mathbf{r}$, whenever the tensor in square brackets does not vanish. Since $\mathbf{r}$ is supposed to be a regular point of the surface, we must conclude that

$$\nabla\nabla S' = \nabla\nabla S + n'\bar{\gamma}\mathbf{s}. \tag{II.11.13}$$

In particular, we have

$$\begin{aligned} \bar{\kappa}' &= \mathrm{Tr}(\nabla\nabla S')/n' = \mathrm{Tr}(\nabla\nabla S)/n' + \mathrm{Tr}(\bar{\gamma}\mathbf{s}) \equiv \mu\bar{\kappa} + \mathrm{Tr}(\bar{\gamma}\mathbf{s}), \\ \kappa' &= \det(\nabla\nabla S')/n'^2 = \det(\nabla\nabla S + n'\bar{\gamma}\mathbf{s})/n'^2, \end{aligned} \tag{II.11.14}$$

where $\bar{\kappa}'$ (and $\bar{\kappa}$) and κ' are, respectively, the *average* and *Gaussian curvature* of the corresponding wave front, Tr stands for trace, and det denotes the determinants of the two-dimensional matrices formed by the first two rows of the three-dimensional matrices of Eq. (II.11.13), where we have chosen the third coordinate axis orthogonal to the principal axes $\hat{t}'_1$, $\hat{t}'_2$ [for the factors in the definitions of $\bar{\kappa}'$, $\bar{\kappa}$, and κ' see the discussion after Eqs. (II.11.23)].

11.2 *Principal Directions and Curvature Radii of the Refracted Wave Front*

In order to find the curvature radii of the refracted wave front, we introduce a unit vector $\hat{u}$ perpendicular to the plane of incidence $\hat{n}\hat{s}$ (see Figs. II.19,20). Since Eq. (II.9.3) ensures that the principal directions of $\nabla\nabla S$ are orthogonal to ∇S, and thus to $\hat{s}$, we can write

$$\nabla\nabla S = \left(\frac{\hat{u}\hat{u}}{\rho_u} + \frac{\hat{v}\hat{v}}{\rho_v} + \frac{\hat{u}\hat{v} + \hat{v}\hat{u}}{\sigma}\right)n, \tag{II.11.15}$$

where

$$\hat{v} = \hat{u} \times \hat{s} \tag{II.11.16}$$

is a unit vector parallel to the plane of incidence. The parameter σ goes to infinity when $\hat{u}$ and $\hat{v}$ happen to be the principal directions of the incident ray.

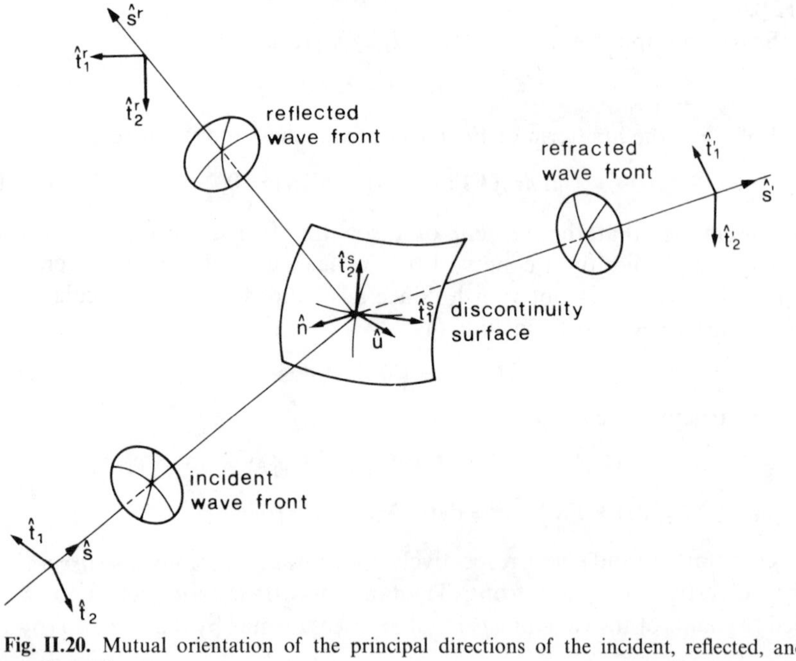

Fig. II.20. Mutual orientation of the principal directions of the incident, reflected, and refracted wave fronts.

We now observe that the function f whose vanishing determines the discontinuity surface can always be chosen in such a way that $\nabla f \cdot \nabla f = \text{const.}$ This implies the "orthogonality" between ∇f and $\nabla\nabla f$ [see Eq. (II.9.2)], which allows us to write, with the help of Eq. (II.11.5),

$$\mathbf{s} = \frac{\hat{u}\hat{u}}{\rho_u^s} + \frac{\hat{v}_s\hat{v}_s}{\rho_v^s} + \frac{1}{\sigma_s}(\hat{u}\hat{v}_s + \hat{v}_s\hat{u}), \tag{II.11.17}$$

with $\hat{v}_s = \hat{u} \times \hat{n}$, in full analogy with Eq. (II.11.15). Finally, we have

$$\nabla\nabla S' = \left(\frac{\hat{u}\hat{u}}{\rho_u'} + \frac{\hat{v}'\hat{v}'}{\rho_v'} + \frac{\hat{u}\hat{v}' + \hat{v}'\hat{u}}{\sigma'}\right)n', \tag{II.11.18}$$

where

$$\hat{v}' = \hat{u} \times \hat{s}' \tag{II.11.19}$$

Thus, Eq. (II.11.13) can be put in the form

$$\frac{\hat{u}\hat{u}}{\rho_u'} + \frac{\hat{v}'\hat{v}'}{\rho_v'} + \frac{\hat{u}\hat{v}' + \hat{v}'\hat{u}}{\sigma'} = \hat{u}\hat{u}\left(\frac{\mu}{\rho_u} + \frac{\bar{\gamma}}{\rho_u^s}\right) + \frac{\mu\hat{v}\hat{v}}{\rho_v} + \bar{\gamma}\frac{\hat{v}_s\hat{v}_s}{\rho_v^s}$$
$$+ \mu\left(\frac{\hat{u}\hat{v} + \hat{v}\hat{u}}{\sigma}\right) + \bar{\gamma}\left(\frac{\hat{u}\hat{v}_s + \hat{v}_s\hat{u}}{\sigma_s}\right). \tag{II.11.20}$$

On the other hand, vector-multiplying Eq. (II.11.9) by $\hat{u}$ yields

$$\hat{v}' = \mu\hat{v} + \bar{\gamma}\hat{v}_s, \tag{II.11.21}$$

with which it can be easily proved [11,13] that Eq. (II.11.13) is satisfied only if

$$1/\rho_u' = \mu/\rho_u + \bar{\gamma}/\rho_u^s, \tag{II.11.22a}$$

$$(\cos\theta')/\sigma' = (\mu\cos\theta)/\sigma + \bar{\gamma}/\sigma_s, \tag{II.11.22b}$$

$$(\cos^2\theta')/\rho_v' = (\mu\cos^2\theta)/\rho_v + \bar{\gamma}/\rho_v^s, \tag{II.11.22c}$$

which allow us to determine the quantities ρ_u', ρ_v', and σ'. In turn, the principal directions $\hat{t}_1'$ and $\hat{t}_2'$ of the refracted ray are obtained by diagonalizing $\nabla\nabla S'$, that is, by expressing it in the form $(\hat{t}_1'\hat{t}_1'/\rho_1' + \hat{t}_2'\hat{t}_2'/\rho_2')n'$ starting from Eq. (II.11.18), so that [13]

$$1/\rho_u' = (\cos^2\phi')/\rho_1' + (\sin^2\phi')/\rho_2', \tag{II.11.23a}$$

$$1/\rho_v' = (\sin^2\phi')/\rho_1' + (\cos^2\phi')/\rho_2', \tag{II.11.23b}$$

$$2/\sigma' = (1/\rho_1' - 1/\rho_2')\sin(2\phi'), \tag{II.11.23c}$$

where ϕ' represents the angle through which $\hat{u}$ and $\hat{v}'$ must be rotated in the plane $\hat{u}\hat{v}'$ in order to coincide, respectively, with $\hat{t}_1'$ and $\hat{t}_2'$. The principal curvature radii of the refracted ray are ρ_1' and ρ_2'. We observe that the concept of curvature radius, as introduced in Section II.9, is connected with the relation $\nabla S \cdot \nabla S = 1$, while in the present section $\nabla S \cdot \nabla S = n^2$ and $\nabla S' \cdot \nabla S' = n'^2$. This discrepancy is accounted for through the factors n and n' in Eqs. (II.11.15) and (II.11.18).

To summarize, the principal directions and the curvature radii of a wave front immediately after passage through a refractive surface can be obtained in three steps:

1. transformation of ρ_1, ρ_2 into ρ_u, ρ_v, and σ by means of Eq. (II.11.15),
2. application of the Kneisly formulas [Eqs. (II.11.22)],
3. transformation of ρ'_u, ρ'_v, and σ' into ρ'_1, ρ'_2 by means of Eqs. (II.11.23).

The above considerations can be repeated for the *reflected* wave front. In particular, since $\mu = n/n'' = 1$, $\theta'' = \pi - \theta$, and $\bar{\gamma} = 2\cos\theta$. Furthermore, the plane of incidence coincides with the *plane of reflection* $\hat{n}\hat{s}''$. The first and third statements constitute the *law of reflection.*

11.3 *Spherical Refracting Surfaces*

Let us consider the particular example of a spherical discontinuity surface (e.g., the front surface of a spherical lens) with ray $R^s = \rho_u^s = \rho_v^s$. In this case, any couple of axes that are mutually orthogonal (and orthogonal to $\hat{n}$) diagonalizes Eq. (II.11.17), so that $1/\sigma_s = 0$ and

$$\begin{aligned} 1/\rho'_u &= \mu/\rho_u + \bar{\gamma}/R^s, \\ 1/\sigma' &= (\mu\cos\theta)/(\sigma\cos\theta'), \\ 1/\rho'_v &= (\mu\cos^2\theta)/(\rho_v\cos^2\theta') + \bar{\gamma}/(R^s\cos^2\theta'). \end{aligned} \tag{II.11.24}$$

If the incident wave front is also spherical ($1/\sigma = 0$), Eq. (II.11.24b) implies that $1/\sigma' = 0$, so that $\phi' = 0$ [Eq. (II.11.23c)] and, as a consequence, $\hat{u}$ and $\hat{v}'$ are the principal directions of the refracted ray (Fig. II.21). Thus,

$$\rho'_1 = \rho'_u, \qquad \rho'_2 = \rho'_v, \tag{II.11.25}$$

so that

$$1/\rho'_1 = \mu/R + \bar{\gamma}/R^s, \tag{II.11.26a}$$

$$1/\rho'_2 = (\mu\cos^2\theta)/(R\cos^2\theta') + \bar{\gamma}/(R^s\cos^2\theta'), \tag{II.11.26b}$$

where R is the curvature radius of the incident wave front. Thus, the curvature radii ρ'_1 and ρ'_2 of the refracted wave front are in general different, unless $\theta = \theta' = 0$ (*normal incidence*). This implies a certain astigmatism in the bundle of refracted rays, whose degree increases with the departure from normal incidence.

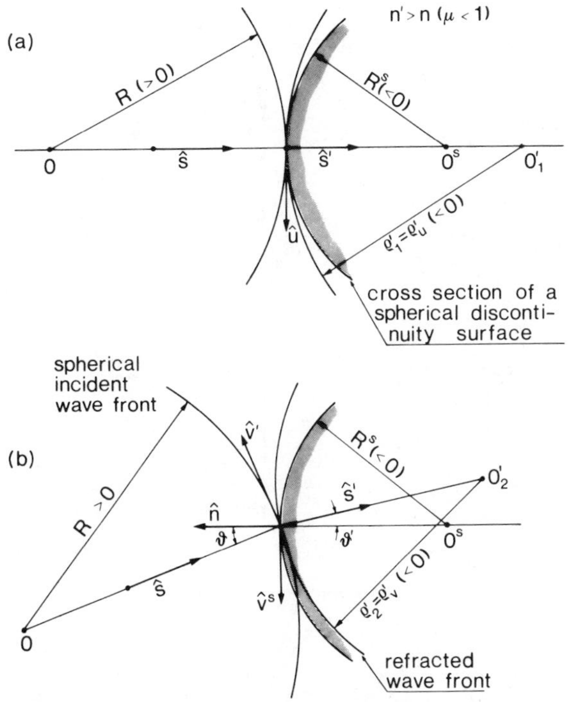

Fig. II.21. Change of the principal curvature radii of an initially spherical wave front passing through a spherical discontinuity surface. (a) Section normal to the plane of incidence. (b) Section parallel to the plane of incidence.

11.4 *Paraxial Limit*

For normal incidence, Eq. (II.11.11) yields $\bar{\gamma} = \mu - 1$, so that, if we set $\rho'_1 = \rho'_2 = -d'$, Eq. (II.11.26a) reduces to the well-known relation

$$n'/d' + n/d = (n' - n)/R^s = n/f = n'/f' \tag{II.11.27}$$

(see Fig. II.22), where $d \equiv R$ and f and f' represent the *focal distances.* Equation (II.11.27) allows us to connect the distance d' between the image point O' and the vertex V of a spherical discontinuity surface with the distance d between the object O and V. It is worth noting that in the examples of Fig. II.22 R^s has to be chosen as a positive quantity. In fact, Eq. (II.11.4) implies that ∇f is directed like the normal $\hat{n}$, which in turn is oriented in the opposite direction with respect to the center of curvature of the surface. This entails a positive value of R^s, in the same way in which an outgoing spherical wave front is shown to have a positive curvature radius (see Section II.9).

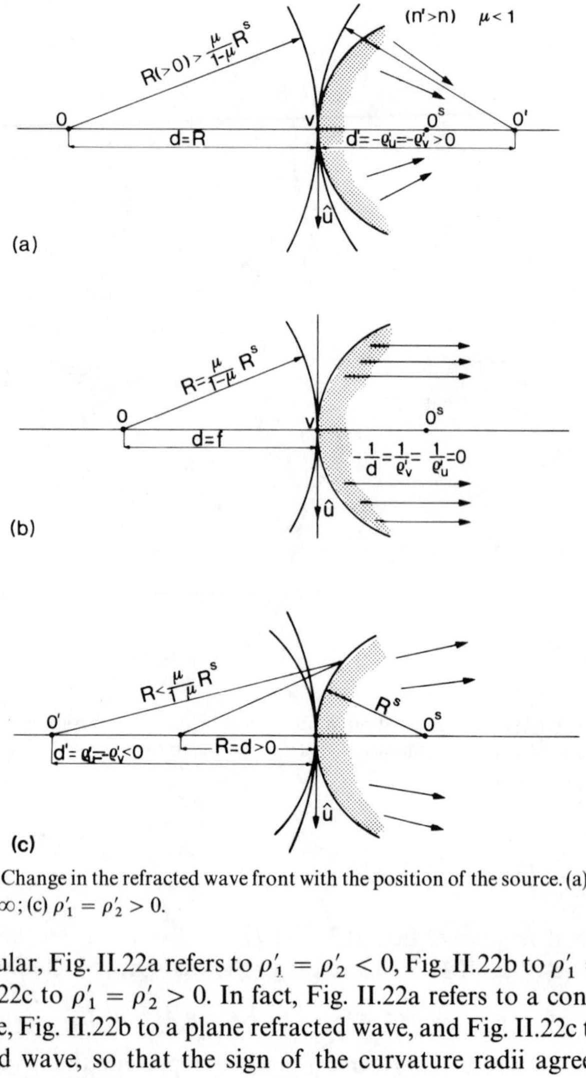

Fig. II.22. Change in the refracted wave front with the position of the source. (a) $\rho_1' = \rho_2' < 0$; (b) $\rho_1' = \rho_2' = \infty$; (c) $\rho_1' = \rho_2' > 0$.

In particular, Fig. II.22a refers to $\rho_1' = \rho_2' < 0$, Fig. II.22b to $\rho_1' = \rho_2' = \infty$, and Fig. II.22c to $\rho_1' = \rho_2' > 0$. In fact, Fig. II.22a refers to a converging refracted wave, Fig. II.22b to a plane refracted wave, and Fig. II.22c to a diverging refracted wave, so that the sign of the curvature radii agrees with the results of Section II.9.

11.5 *Spherical Reflecting Surfaces*

Another example is furnished by a spherical wave reflected by a spherical surface. In this case, Eq. (II.11.11) with the substitution $\theta' \to \theta''$ yields

$$\bar{\gamma} = 2\cos\theta, \tag{II.11.28}$$

because of the relations $\mu = n/n'' = 1$ and $\theta'' = \pi - \theta$, so that by analogy with Eqs. (II.11.26) we have

$$1/\rho_2'' = 1/\rho_1'' + (2\tan\theta\sin\theta)/R^s, \tag{II.11.29}$$

which shows a degree of astigmatism in the bundle of reflected rays, vanishing for normal incidence.

Finally, we observe that ray optics is not able to determine the transmission (or reflection) coefficient, and thus the ratio between transmitted (or reflected) and incident electromagnetic fields. This can be accomplished by means of the Fresnel formulas (see Section III.8), which are rigorously valid only for a plane discontinuity surface.

12 Solution of the Eikonal Equation by the Method of Separation of Variables

The most powerful method for solving the Helmholtz wave equation consists of finding a suitable system of orthogonal coordinates x_1, x_2, x_3 so that the field can be expressed as a product of functions of the single variables x_i, that is, $u = u_1(x_1)u_2(x_2)u_3(x_3)$. This method (*separation of variables*) has a counterpart for the eikonal equation consisting of expressing S as a *sum* of functions of the single variables x_i. The existence of these additional coordinates depends on the form taken by the function $n^2(\mathbf{r})$. In the following we will examine some cases of practical interest.

12.1 *Cartesian Coordinates*

In many cases of practical interest, the square of the refractive index is separable for some choice of coordinate system [14]. As a preliminary example, let us consider the situation in which n^2 is separable in *cartesian coordinates*, that is,

$$n^2(\mathbf{r}) = f(x) + g(y) + h(z). \tag{II.12.1}$$

If we look for eikonals of the form

$$S(\mathbf{r}) = X(x) + Y(y) + Z(z), \tag{II.12.2}$$

Eq. (II.3.1) splits into

$$\begin{aligned} (dX/dx)^2 &= a + f(x), \\ (dY/dy)^2 &= b + g(y), \\ (dZ/dz)^2 &= c + h(z), \end{aligned} \tag{II.12.3}$$

a, b, and c being three constants subject to the condition

$$a + b + c = 0. \tag{II.12.4}$$

Thus, we can write

$$S(\mathbf{r}) = \int^x [f(x') + a]^{1/2}\,dx' + \int^y [g(y') + b]^{1/2}\,dy' + \int^z [h(z') + c]^{1/2}\,dz', \tag{II.12.5}$$

which yields

$$\nabla^2 S = \frac{1}{2}[f(x) + a]^{-1/2}\frac{df}{dx} + \frac{1}{2}[g(y) + b]^{-1/2}\frac{dg}{dy} + \frac{1}{2}[h(z) + c]^{-1/2}\frac{dh}{dz} \tag{II.12.6}$$

and

$$\begin{aligned} \nabla S &= [f(x) + a]^{1/2}\hat{x} + [g(y) + b]^{1/2}\hat{y} + [h(z) + c]^{1/2}\hat{z} \\ &= (f + g + h)^{1/2}\hat{s}, \end{aligned} \tag{II.12.7}$$

the latter equality being a consequence of Eqs. (II.4.1) and (II.12.1). Since $\hat{s}\cdot\hat{x}\,ds = dx$ and

$$n\hat{s}\cdot\hat{x} = \partial S/\partial x = [f(x) + a]^{1/2}, \tag{II.12.8}$$

we have

$$ds/n(s) = dx/[f(x)+a]^{1/2} = dy/[g(y)+b]^{1/2} = dz/[h(z)+c]^{1/2}, \tag{II.12.9}$$

from which we obtain, with the help of Eq. (II.12.6),

$$\int^s \frac{\nabla^2 S}{n(s')}\,ds' = \frac{1}{2}\ln[(f + a)(g + b)(h + c)] + \text{const}, \tag{II.12.10}$$

$$\exp\left[-\frac{1}{2}\int^s \frac{\nabla^2 S}{n(s')}\,ds'\right] \propto (f + a)^{-1/4}(g + b)^{-1/4}(h + c)^{-1/4}. \tag{II.12.11}$$

Thus, Eq. (II.5.5) allows us to write the ray field as

$$\begin{aligned} u(\mathbf{r}) &= \exp[-ik_0 S(\mathbf{r})]A_0(\mathbf{r}) \\ &= F(\mathbf{r})[f(x) + a]^{-1/4}[g(y) + b]^{-1/4}[h(z) + c]^{-1/4} \\ &\quad \times \exp\left\{-ik_0\left[\int^x [f(x') + a]^{1/2}\,dx' + \int^y [g(y') + b]^{1/2}\,dy'\right.\right. \\ &\quad \left.\left. + \int^z [h(z') + c]^{1/2}\,dz'\right]\right\}, \end{aligned} \tag{II.12.12}$$

$F(\mathbf{r})$ being a function that remains constant along each ray. In this way, the field is determined by the geometry of the ray congruence and by its distribution on a wave front or, more generally, on a surface crossing all rays.

We observe that Eq. (II.12.12) yields diverging values of the field when one of the factors $f + a$, $g + b$, or $h + c$ vanishes, so that the relations

$$f(x_c) = -a, \qquad g(y_c) = -b, \qquad h(z_c) = -c = a + b \tag{II.12.13}$$

determine the possible plane faces of the caustic $x = x_c$, $y = y_c$, $z = z_c$.

12.1.a *Axially Symmetric Refractive Index*

A relevant case is that of an *axially symmetric* profile having the form

$$n^2(\mathbf{r}) = -\alpha(x^2 + y^2) + n_0^2, \tag{II.12.14}$$

from which

$$f(x) = -\alpha x^2, \qquad g(y) = -\alpha y^2, \qquad h(z) = n_0^2. \tag{II.12.15}$$

The corresponding eikonal is [Eq. (II.12.5)]

$$\begin{aligned} S(\mathbf{r}) = {} & \tfrac{1}{2}x(-\alpha x^2 + a)^{1/2} - (ia/2\alpha^{1/2})\ln[ix(\alpha/a)^{1/2} + (1 - x^2\alpha/a)^{1/2}] \\ & + \tfrac{1}{2}y(-\alpha y^2 + b)^{1/2} - (ib/2\alpha^{1/2})\ln[iy(\alpha/b)^{1/2} + (1 - y^2\alpha/b)^{1/2}] \\ & + (n_0^2 + c)^{1/2}z + \text{const.} \end{aligned} \tag{II.12.16}$$

In particular, for $x^2 \gg |a|/\alpha$ and $y^2 \gg |b|/\alpha$, that is, far away from the optical z axis, Eq. (II.12.12) gives

$$\begin{aligned} u(\mathbf{r}) = {} & F(\mathbf{r})(-\alpha x^2 + a)^{-1/4}(-\alpha y^2 + b)^{-1/4}(n_0^2 + c)^{-1/4} \\ & \times \exp[-k_0\alpha^{1/2}(x^2 + y^2)/2 - ik_0(n_0^2 + c)^{1/2}z] \\ & \times x^{k_0a/(2\alpha^{1/2})}y^{k_0b/(2\alpha^{1/2})}, \end{aligned} \tag{II.12.17}$$

which shows that, for $\alpha > 0$, the field decays radially according to a Gaussian law. Equation (II.12.17) corresponds to a guided mode propagating along the z axis with propagation constant

$$\beta = k_0(n_0^2 + c)^{1/2}. \tag{II.12.18}$$

12.1.b *Lenslike Media*

When the refractive index is a decreasing function of the distance from the optical axis, we call the medium *lenslike*. In order to remove the nonphysical singularities on the caustic, it is necessary to impose the conditions (see Section III.3.4, where we refer to propagation along the x-direction)

$$\begin{aligned} k_0[X(|x_c|) - X(-|x_c|)] &= (p + \tfrac{1}{2})\pi, \\ k_0[Y(|y_c|) - Y(-|y_c|)] &= (q + \tfrac{1}{2})\pi, \end{aligned} \tag{II.12.19}$$

so that we can write, with the help of Eq. (II.12.16) and the relations $x_c^2 = a/\alpha$ and $y_c^2 = b/\alpha$,

$$k_0 a/2\alpha^{1/2} - 1/2 = p, \qquad k_0 b/2\alpha^{1/2} - 1/2 = q, \tag{II.12.20}$$

which entails a discrete set of values for a and b and, in turn, for β. More precisely, Eqs. (II.12.18) and (II.12.4) give

$$\begin{aligned}\beta = \beta_{pq} &= k_0[n_0^2 - 2\alpha^{1/2}(p + \tfrac{1}{2})/k_0 - 2\alpha^{1/2}(q + \tfrac{1}{2})/k_0]^{1/2} \\ &\cong k_0 n_0 - \alpha^{1/2}(p + q + 1)/n_0.\end{aligned} \tag{II.12.21}$$

If we compare this with the result of the wave analysis of parabolic profile multimode optical fibers we find perfect agreement, which clearly shows the potential of the ray-optical analysis (see Eq. (VIII.7.19)).

12.1.c *Physical Properties of Complex Eikonals*

A few comments are now necessary to clarify the physical meaning of complex eikonal. Simple inspection of Eq. (II.12.16) shows a nonvanishing imaginary part of $S(\mathbf{r})$ whenever at least one of the quantities $a - \alpha x^2$ or $b - \alpha y^2$ is negative. This suggests to us that all space is divided into two domains: the *lit side* of the caustic, for which

$$a - \alpha x^2 > 0, \qquad b - \alpha y^2 > 0, \tag{II.12.22}$$

and the *dark side*, for which the two conditions are not both satisfied.

The lit side is characterized by a real eikonal, the dark side by a complex eikonal giving rise to an exponential decay of the electric field for increasing distances from the caustic, as shown by Eq. (II.12.17) (see also the example of Section II.7). Correspondingly, a ray coming from the lit zone lands tangentially on the caustic, and gives rise to a back-reflected real ray and to a *complex ray* penetrating the dark zone.

In the dark region, the unit vector $\hat{s}$ takes on complex values. More precisely, the component of $\hat{s}(\mathbf{r})$ normal to the caustic is purely imaginary for $\mathbf{r}$ varying on the caustic itself, so that there is no energy flow in the corresponding direction, and the trasmitted field does not drain any power from the incident one. However, the nonvanishing field corresponds to *reactive energy* accumulated in the dark region, which is associated with the jump of $\pi/2$ undergone by the phase of the field after crossing the caustic as discussed in Section II.10 for homogeneous media (see Fig. II.17).

In the lit region, the unit vector $\hat{s}$ is a real quantity. As we will see in connection with diffraction theory, the field near the caustic is well described by the Airy function, which tends, while penetrating in the lit domain, to assume the form of two traveling waves as represented by the ray optical field.

The above results hold true for a generic form of the refractive index distribution $n(\mathbf{r})$. In any event, ray optics alone cannot provide the reflection and transmission coefficients for a wave touching a caustic. This problem can be solved by introducing suitable transmission functions to match the field on both sides (Section III.3).

12.2 *Cylindrical Coordinates*

Let us now consider the case of n^2 separable in cylindrical coordinates, viz.

$$n^2(\mathbf{r}) = f(\rho) + g(\phi)/\rho^2 + h(z). \tag{II.12.23}$$

Eikonals of the form

$$S(\mathbf{r}) = R(\rho) + \Phi(\phi) + Z(z) \tag{II.12.24}$$

obey Eq. (II.3.1) if

$$(dR/d\rho)^2 + 1/\rho^2(d\Phi/d\phi)^2 + (dZ/dz)^2 = f(\rho) + g(\phi)/\rho^2 + h(z), \tag{II.12.25}$$

which yields

$$\begin{aligned} R(\rho) &= \int^{\rho}\left[f(\rho') - \frac{b}{\rho'^2} - c\right]^{1/2} d\rho', \\ \Phi(\phi) &= \int^{\phi}[g(\phi') + b]^{1/2}\, d\phi', \\ Z(z) &= \int^{z}[h(z') + c]^{1/2}\, dz', \end{aligned} \tag{II.12.26}$$

where b and c are constants and Φ fulfills the periodicity conditions

$$\Phi(\phi + 2\pi) = \Phi(\phi) + 2m\pi/k_0 \qquad (m \text{ integer}) \tag{II.12.27}$$

giving rise to a single-valued function $\exp[-ik_0 S(\mathbf{r})]$. Thus, if $g = 0$, we have

$$\Phi(\phi) = m\phi/k_0 + \text{const}, \tag{II.12.28}$$

$$b = m^2/k_0^2. \tag{II.12.29}$$

By proceeding in the same way as for the cartesian case, it can be shown that

$$\begin{aligned} u(\mathbf{r}) = {} & \frac{F(\mathbf{r})}{\rho^{1/2}}\left[f(\rho) - c - \frac{b}{\rho^2}\right]^{-1/4}[g(\phi) + b]^{-1/4}[h(z) + c]^{-1/4} \\ & \times \exp\left\{-ik_0\left[\int^{\rho}\left(f(\rho') - c - \frac{b}{\rho'^2}\right)^{1/2} d\rho' \right.\right. \\ & \left.\left. + \int^{\phi}(g(\phi') + b)^{1/2}\, d\phi' + \int^{z}(h(z') + c)^{1/2}\, dz'\right]\right\}, \end{aligned} \tag{II.12.30}$$

with $F(\mathbf{r})$ constant along each ray.

12.2.a *Axially Symmetric Profile*

For $g = 0$ and $h = $ const, that is, for an axially symmetric refractive index distribution, Eq. (II.12.30) is of the form

$$u(\mathbf{r}) = \frac{F(\mathbf{r})}{\rho^{1/2}}\left[n^2(\rho) - \frac{\beta^2}{k_0^2} - \frac{m^2}{(k_0^2\rho^2)}\right]^{-1/4}$$
$$\times \exp\left\{-ik_0\int^{\rho}\left[n^2(\rho') - \frac{\beta^2}{k_0^2} - \frac{m^2}{(k_0^2\rho'^2)}\right]^{1/2} d\rho' - i\beta z - im\phi\right\}, \tag{II.12.31}$$

where use has been made of Eqs. (II.12.23) and (II.12.29) and of the definition

$$\beta = k_0(h + c)^{1/2}. \tag{II.12.32}$$

The caustic is a two-sheeted circular cylinder with radii $\rho_{\max}$ and $\rho_{\min}$ corresponding to two consecutive zeros of the factor $[n^2(\rho) - \beta^2/k_0^2 - m^2/(k_0^2\rho^2)]$ in Eq. (II.12.31), and, in complete analogy with the cartesian case, we impose the condition

$$k_0[R(\rho_{\max}) - R(\rho_{\min})] = (v + \tfrac{1}{2})\pi \qquad (v \text{ integer}), \tag{II.12.33}$$

which allows us to determine a discrete set of values β_{vm} for the propagation constant. In particular, Eq. (II.12.31) represents the propagation modes of an optical fiber having a refractive index distribution $n(\rho)$. The parabolic profile of Eq. (II.12.14) is obtained for

$$n^2(\rho) = -\alpha\rho^2 + n_0^2. \tag{II.12.34}$$

12.3 *Spherical Coordinates*

For n^2 separable in *spherical coordinates* and symmetrical with respect to the z axis, that is,

$$n^2(\mathbf{r}) = h(r) + g(\theta)/r^2, \tag{II.12.35}$$

we set $S(\mathbf{r}) = R(r) + \Theta(\theta) + \Phi(\phi)$, so that

$$R(r) = \int^r \left[h(r') + \frac{c}{r'^2}\right]^{1/2} dr',$$
$$\Theta(\theta) = \int^{\theta}\left[g(\theta') - c - \frac{m^2}{(k_0^2 \sin^2\theta')}\right]^{1/2} d\theta', \tag{II.12.36}$$
$$\Phi(\phi) = \frac{m\phi}{k_0}$$

as easily seen by expressing Eq. (II.3.1) in spherical coordinates. In the usual way, it is possible to obtain

$$u(\mathbf{r}) = \frac{F(\mathbf{r})}{r}\left[h(r) + \frac{c}{r^2}\right]^{-1/4} [(g(\theta) - c)k_0^2 \sin^2\theta - m^2]^{-1/4}$$
$$\times \exp\left\{-ik_0\left[\int^r \left(h(r') + \frac{c}{r'^2}\right)^{1/2} dr' + \int^\theta \left(g(\theta') - c - \frac{m^2}{(k_0^2 \sin^2\theta')}\right)^{1/2} d\theta'\right] - im\phi\right\}, \quad \text{(II.12.37)}$$

$F(\mathbf{r})$ being constant along each ray.

Example: Scattering from a Finite Body. If we choose $c = m = 0$, Eq. (II.12.37) reduces, for a medium homogeneous outside a sphere of radius R, to the spherical wave (spherical wave fronts)

$$u(\mathbf{r}) = f(\theta, \phi)e^{-ik_0 r}/r, \quad \text{(II.12.38)}$$

where, without loss of generality, we have chosen $h = 1$. Starting from this equation, we can now construct the relative Luneburg–Kline series. To this end, note that when we express ∇^2 in spherical coordinates and observe that $A_0(s) = A_0(r) \propto 1/r$, the operator $\hat{L}$ of Eq. (II.6.6) is

$$\begin{aligned}\hat{D}^2 &= \frac{-1}{2r}\int_\infty^r \frac{(\partial/\partial r')r'^2(\partial/\partial r') + \hat{D}^2}{r'}dr' \\ &= \frac{-1}{2}\left(\frac{1}{r} + \frac{\partial}{\partial r} + \frac{\hat{D}^2}{r}\int_\infty^r \frac{dr'}{r'}\right),\end{aligned} \quad \text{(II.12.39)}$$

where

$$\hat{D}^2(\theta, \phi) = \frac{1}{\sin\theta}\frac{\partial}{\partial\theta}\left(\sin\theta\frac{\partial}{\partial\theta}\right) + \frac{1}{\sin^2\theta}\frac{\partial^2}{\partial\phi^2}. \quad \text{(II.12.40)}$$

The lower limit in the integral of Eq. (II.12.39) is ∞, since we suppose $A_m(\infty) = 0$ for $m \neq 0$ [see the procedure leading to Eq. (II.6.6)], which implies that Eq. (II.12.38) is the *exact expression* of the field for $r \to \infty$. Now, if we set

$$(i/k_0)^n \hat{L}^n f(\theta, \phi)/r = f_n(\theta, \phi)/r^{n+1}, \quad \text{(II.12.41)}$$

the relation

$$A_0(r) = f(\theta, \phi)/r \quad \text{(II.12.42)}$$

implies, with the help of Eq. (II.6.7), that Eq. (II.2.5) can be cast in the form

$$u(\mathbf{r}) \sim \frac{e^{-ik_0 r}}{r}\sum_{m=0}^{\infty}\frac{f_m(\theta, \phi)}{r^m}. \quad \text{(II.12.43)}$$

The f_n are seen to satisfy the recursive system

$$-2ik_0(n+1)f_{n+1} = [n(n+1) + \hat{D}^2]f_n, \qquad \text{(II.12.44)}$$

with $f_0 \equiv f$, which can be conveniently solved by expanding the f_n into series of spherical harmonics $Y_l^m(\theta, \phi)$, which are eigenfunctions of the operator $\hat{D}^2$. In this way, the series of Eq. (II.12.43) turns out to be a combination of terms given by products of Y_l^m and spherical Hankel functions $h_n(k_0 r)$ (see Section VI.12).

From an intuitive point of view, the above procedure implies that the field can be calculated at any point (at least in the Luneburg–Kline sense), once its *far-field pattern* $f(\theta, \phi)$ is known. This property is strictly connected with the possibility of obtaining an integral representation of the field by using the values taken on a surface (see Section IV.2.2).

If we consider the flux of the vector $|u|^2\hat{r}$ through a sphere, we have

$$\int_{4\pi} |u|^2 r^2 \, d\Omega \underset{r\to\infty}{=} \int_{4\pi} |f(\theta, \phi)|^2 \, d\Omega. \qquad \text{(II.12.45)}$$

Thus, $|f|^2 \, d\Omega$ is proportional to the power radiated in the solid angle $d\Omega$, while the integral is proportional to the total power radiated to infinity.

We can also consider the case of a medium inhomogeneous for $r < R$ illuminated by a plane wave of *unit amplitude*. In this situation, the quantity $\exp(-ik_0 r)f(\theta, \phi)/r$ represents the field scattered at infinity. The *scattering amplitude* [3] is represented by f (see Section VI.11), while

$$|f|^2 \equiv \sigma(\theta, \phi) \qquad \text{(II.12.46)}$$

is the *differential cross section*, which depends on the structure of the scattering medium. As we will see in next section, σ can be calculated by tracing the field u along each ray of the collimated congruence representing the incident plane wave. The calculation is notably simplified whenever the inhomogeneous medium has a center of symmetry.

13 Ray Paths Obtained by the Method of Separation of Variables

The method of separation of variables proves in many cases to be quite useful for finding a parametric representation of the ray trajectories. Let μ, ν, σ be orthogonal curvilinear coordinates with *scale factors* h_μ, h_ν, h_σ, so that

$$(dx)^2 + (dy)^2 + (dz)^2 = h_\mu^2(d\mu)^2 + h_\nu^2(d\nu)^2 + h_\sigma^2(d\sigma)^2. \qquad \text{(II.13.1)}$$

If we indicate with μ, ν, σ the coordinates of the points of a given ray, projecting Eq. (II.4.1) on the coordinate axis μ yields

$$1 + \frac{h_\nu^2}{h_\mu^2}\left(\frac{d\nu}{d\mu}\right)^2 + \frac{h_\sigma^2}{h_\mu^2}\left(\frac{d\sigma}{d\mu}\right)^2 = \frac{n^2 h_\mu^2}{(\partial S/\partial \mu)^2}, \tag{II.13.2}$$

where ν and σ are considered functions of μ along the ray. Analogous equations hold true if we interchange μ with ν and with σ.

13.1 *Axially Symmetric Media*

If

$$n^2(\mathbf{r}) = f(\rho) + h(z), \tag{II.13.3}$$

Eq. (II.13.2) can be written [see the expression for S implicitly given in Eq. (II.12.31)]

$$1 + \rho^2\left(\frac{d\phi}{d\rho}\right)^2 + \left(\frac{dz}{d\rho}\right)^2 = \frac{n^2 k_0^2 \rho^2}{n^2 k_0^2 \rho^2 - \beta^2(z)\rho^2 - m^2}, \tag{II.13.4}$$

where we have chosen cylindrical coordinates $(\rho, \phi, z) = (\mu, \nu, \sigma)$. If we allow h to depend on z, then β is also a function of z, which can be determined by applying Eq. (II.12.33) at each section $z = \text{const}$. By interchanging ρ and z, we obtain

$$1 + \rho^2(d\phi/dz)^2 + (d\rho/dz)^2 = n^2 k_0^2/\beta^2(z). \tag{II.13.5}$$

The system of Eqs. (II.13.4) and (II.13.5) yields

$$\rho\, d\phi/d\rho = m/[n^2 k_0^2 \rho^2 - \beta^2(z)\rho^2 - m^2]^{1/2}, \tag{II.13.6}$$

$$dz/d\rho = \beta\rho/[n^2 k_0^2 \rho^2 - \beta^2(z)\rho^2 - m^2]^{1/2}, \tag{II.13.7}$$

so that

$$d\phi/dz = m/\beta(z)\rho^2. \tag{II.13.8}$$

We observe that the above equations can be obtained directly from the ray equation. In this way, the quantity m is not necessarily an integer. Here, we are considering a congruence of rays whose wave fronts extend over the whole space. This particular situation refers to an eikonal of the kind given in Eq. (II.12.24), leading to an integer value of m. On the other hand, if we try to span the whole space with a set of rays having a noninteger m, we are unable to find a single-valued eikonal. Accordingly, we can accept a noninteger m only when dealing with ray pencils having small cross sections.

Usually, m is referred to as *skewness invariant*, since it can be connected with the transverse direction cosines p' and q' through the relation

$$m = k_0 \frac{\partial S}{\partial \phi} = k_0 \left(x \frac{\partial S}{\partial y} - y \frac{\partial S}{\partial x} \right) = k_0 n(xq' - yp'). \tag{II.13.9}$$

If n does not depend on z, β is a constant of motion and

$$\beta = k_0\, \partial S/\partial z = k_0 n(\rho) \cos \gamma, \tag{II.13.10}$$

where $\cos \gamma$ is the direction cosine relative to the z axis.

Meridional rays ($m = 0$) lie in planes containing the z axis [see Eq. (II.13.8)], while Eq. (II.13.7) yields

$$d^2\rho/dz^2 = \tfrac{1}{2}(k_0/\beta)^2\, d(n^2)/d\rho, \tag{II.13.11}$$

whose general solution determines all possible meridional trajectories.

Example 1: Meridional Rays in Graded-Index Fibers. As an example, we consider a refractive index distribution

$$n^2(\rho) = n_0^2[1 - \delta(\rho/\rho_0)^2 + \alpha_2 \delta^2 (\rho/\rho_0)^4 + \alpha_3 \delta^3 (\rho/\rho_0)^6 + \cdots], \tag{II.13.12}$$

where $\alpha_i (i = 1, 2, 3, \ldots)$, δ, and ρ_0 are arbitrary constants. Then if we use normalized coordinates

$$R = \rho/\rho_0, \tag{II.13.13}$$

$$Z = z\delta^{1/2} n_0 k_0/(\beta\rho_0), \tag{II.13.14}$$

Eq. (II.13.11) becomes

$$\ddot{R} + R = R \sum_{j=2}^{\infty} j\alpha_j \delta^{j-1} R^{2(j-1)}, \tag{II.13.15}$$

where the double dot indicates the second derivative with respect to Z. This is a nonlinear differential equation, which for $\alpha_j = 0$ reduces to the equation of the harmonic oscillator, so that the rays describe sinusoidal paths with spatial period

$$\Lambda = 2\pi\rho_0\beta/(\delta^{1/2} n_0 k_0), \tag{II.13.16}$$

which depends through β on both the initial ($Z = 0$) radial position ρ_{in} and the angle γ_{in} with respect to the z axis. In general, the solution of Eq. (II.13.15) is a periodic function of Z, i.e., $R(Z + 2\pi/\Omega) = R(Z)$, where Ω is an angular frequency that depends on the nonlinearity parameter δ, the set of coefficients α_j, and the initial conditions R_0 and $\dot{R}_0$. Consequently, $R(Z)$ can be expressed in the form of a Fourier series (*Linstedt method* [15], see also *Krylov, Bogoliubov, Mitropolski method* [16])

$$R(Z) = a\cos(\Omega Z + \psi_{in}) + \sum_{\substack{k=1 \\ j \geq k}}^{\infty} \frac{a^{2j+1}\delta^j}{2^{4j}} A_{j,\,2k+1} \times \cos[(2k+1)(\Omega Z + \psi_{in})]. \tag{II.13.17}$$

The constants a, Ω, and ψ_{in} can be conveniently expressed as power series in δ,

$$a = \sum_{j=0}^{\infty} a_j \delta^j, \tag{II.13.18}$$

$$\Omega = 1 - \sum_{j=1}^{\infty} \frac{\delta^j a^{2j}}{2^{4j-2}} B_j, \tag{II.13.19}$$

$$\psi_{in} = \sum_{j=0}^{\infty} \psi_j \delta^j. \tag{II.13.20}$$

If we insert the above expressions in Eq. (II.13.17), Eq. (II.13.15) yields

$$A_{1,3} = -\alpha_2, \qquad B_1 = 3\alpha_2, \tag{II.13.21}$$

$$a_j = \frac{a_0^{2j+1}}{2^{4j}} \sum_{k=0}^{2j} C_{j,2k} \cos(2k\psi_0), \tag{II.13.22}$$

$$\psi_j = \frac{a_0^{2j}}{2^{4j}} \sum_{k=1}^{2j} S_{j,2k} \sin(2k\psi_0), \tag{II.13.23}$$

where a_0 and ψ_0 are determined through the initial conditions

$$R_{in} = a_0 \cos\psi_0, \qquad \dot{R}_{in} = -a_0 \sin\psi_0. \tag{II.13.24}$$

On the other hand, the $C_{j,2k}$, $S_{j,2k}$ depend on the α_i. In particular,

$$\begin{aligned} &C_{1,0} = 6\alpha_2, \qquad C_{1,2} = -4\alpha_2, \qquad C_{1,4} = -\alpha_2, \\ &S_{1,2} = 8\alpha_2, \qquad S_{1,4} = \alpha_2. \end{aligned} \tag{II.13.25}$$

If we truncate the series expansion of Eq. (II.13.17) to $j = 1$, the procedure described gives

$$\begin{aligned} R(Z) &\cong a\cos\psi(Z) + \delta A_{1,3} \frac{a^3}{16} \cos[3\psi(Z)] \\ &= a\cos\psi(Z) - \delta\alpha_2 \frac{a^3}{16} \cos[3\psi(Z)], \end{aligned} \tag{II.13.26}$$

with

$$a = a_0 + \frac{\delta a_0^3}{16} \alpha_2 [6 - 4\cos(2\psi_0) - \cos(4\psi_0)], \tag{II.13.27}$$

$$\psi(Z) = \left(1 - \frac{3\delta a^2 \alpha_2}{4}\right) Z + \psi_0 + \delta a_0^2 \frac{\alpha_2 [8\sin(2\psi_0) + \sin(4\psi_0)]}{16}. \tag{II.13.28}$$

Comparing Eq. (II.13.28) with Eq. (II.13.19), we obtain, with the help of Eq. (II.13.14) in which β is expressed through Eq. (II.13.10) specialized to the initial conditions,

$$\Omega Z \cong \left(1 - \frac{3\delta a_0^2 \alpha_2}{4}\right)(1 - \delta R_{\rm in}^2)^{-1/2}\left(1 + \delta \dot{R}_{\rm in}^2 \frac{n_0^2 k_0^2}{\beta^2}\right)^{1/2} \frac{\delta^{1/2} z}{\rho_0}, \tag{II.13.29}$$

where use has been made of Eqs. (II.13.17) and (II.13.10), both specialized to the initial values. In turn, Eqs. (II.13.24) yield

$$\Omega Z = \left[1 + \frac{\delta}{2}\left(1 - \frac{3\alpha_2}{2}\right)(R_{\rm in}^2 + \dot{R}_{\rm in}^2)\right]\frac{\delta^{1/2} z}{\rho_0} + \text{higher-order terms in } \delta, \tag{II.13.30}$$

which shows that the phase ΩZ is independent of the initial conditions for $\alpha_2 = \frac{2}{3}$.

Example 2: Selfoc Fibers. For a *sech profile* the refractive index is given by

$$n^2(\rho)/n_0^2 = \operatorname{sech}^2(\delta^{1/2}\rho/\rho_0) = 1 - \delta(\rho/\rho_0)^2 + (2\delta^2/3)(\rho/\rho_0)^4 + \cdots. \tag{II.13.31}$$

Equation (II.13.11) can be solved exactly, giving

$$\begin{aligned} \sinh(\delta^{1/2}\rho/\rho_0) &= \sinh(\delta^{1/2}\rho_{\rm in}/\rho_0)\cos(\delta^{1/2}z/\rho_0) \\ &\quad + \tan\gamma_{\rm in}\cosh(\delta^{1/2}\rho_{\rm in}/\rho_0)\sin(\delta^{1/2}z/\rho_0), \end{aligned} \tag{II.13.32}$$

which yields approximately, for small argument $\delta^{1/2}\rho/\rho_0$,

$$\rho = \rho_{\rm in}\cos(\delta^{1/2}z/\rho_0) + \tan\gamma_{\rm in}\sin(\delta^{1/2}z/\rho_0)\rho_0/\delta^{1/2}, \tag{II.13.33}$$

so that the period is independent of the initial position and slope of the ray.

13.2 *Spherically Symmetric Media*

If the refractive index has a spherically symmetric distribution $n(r)$, it is convenient to adopt spherical coordinates $(r, \theta, \phi) = (\mu, \nu, \sigma)$. In this case, simple symmetry considerations ensure that any ray describes a plane trajectory, which lies on the plane determined by the initial direction of the ray $\hat{s}_{\rm in}$ and the initial position vector $\mathbf{r}_{\rm in}$.

Equation (II.13.2) and the analogous equation with r and θ interchanged are, with the help of Eqs. (II.12.36a) and (II.12.36b),

$$1 + r^2\left(\frac{d\theta}{dr}\right)^2 + r^2\sin^2\theta\left(\frac{d\phi}{dr}\right)^2 = \frac{n^2}{n^2 + c/r^2}, \tag{II.13.34}$$

$$1 + \frac{1}{r^2}\left(\frac{dr}{d\theta}\right)^2 + \sin^2\theta\left(\frac{d\phi}{d\theta}\right)^2 = \frac{-n^2 r^2}{c + m^2/(k_0^2\sin^2\theta)}, \tag{II.13.35}$$

which yield

$$\left(\frac{d\theta}{dr}\right)^2 = \frac{-ck_0^2 \sin^2\theta - m^2}{(c + n^2r^2)k_0^2r^2 \sin^2\theta}, \tag{II.13.36}$$

$$\left(\frac{d\phi}{dr}\right)^2 = \frac{m^2}{(c + n^2r^2)k_0^2r^2 \sin^4\theta}. \tag{II.13.37}$$

As a consequence, if we indicate with ψ the angle between the tangent to the ray and the radius vector from the origin (see Fig. II.23),

$$\cos^2\psi = \left(\frac{dr}{ds}\right)^2 = \frac{n^2r^2 + c}{n^2r^2}. \tag{II.13.38}$$

We now observe that, whenever we are concerned with a single ray, we can always choose our system of coordinates in such a way that the plane of the trajectory contains the z axis, which corresponds to set $m = 0$ [see Eq. (II.13.37)]. With this choice, Eqs. (II.13.36) and (II.13.38) yield

$$n^2r^4\left(\frac{d\theta}{ds}\right)^2 = n^2r^4\left(\frac{d\theta}{dr}\right)^2 \cos^2\psi = -c \qquad (c \le 0). \tag{II.13.39}$$

In particular, a straight ray passing through the origin corresponds to $c = 0$. Since $\sin\psi\, ds = r\, d\theta$, Eq. (II.13.39) is equivalent to

$$nr\sin\psi = (-c)^{1/2} = \text{const along any trajectory.} \tag{II.13.40}$$

The above relation is a generalization of Snell's law applied to spherically symmetric media known as *Bouguer's theorem*, and it can be interpreted as the law of conservation of angular momentum of the photons moving through the medium.

It is worth noting that a ray congruence refers to an eikonal of the form $S(\mathbf{r}) = R(r) + \Theta(\theta) + \Phi(\phi)$ only if all the rays share common values of c

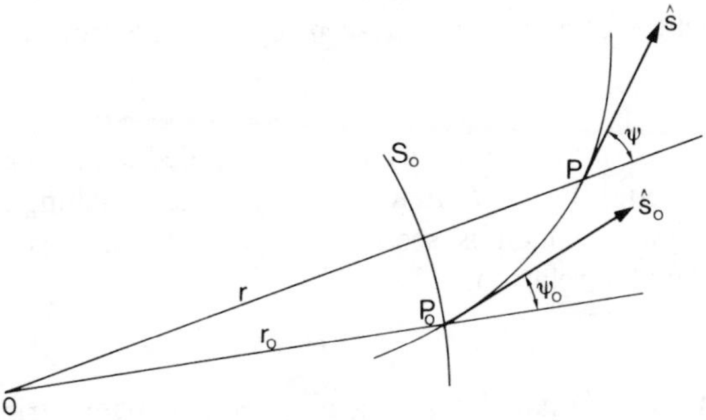

Fig. II.23. Ray path in a spherical medium of varying refractive index.

and m. If this condition is not verified, the eikonal is not expressible in separable form.

For $m = 0$, Eq. (II.13.36) reduces to

$$(d\theta/dr)^2 = -c/(c + n^2 r^2) r^2, \tag{II.13.41}$$

which is singular when

$$c + n^2 r^2 = 0. \tag{II.13.42}$$

Equation (II.13.42) defines the caustic relative to the ray field composed of the rays with $m = 0$ and a common value of c, which thus turns out to be composed of a number of spherical surfaces with center at the origin. The lit zone, that is, the portion of space where real trajectories take place, is immediately obtained from Eq. (II.13.41) as

$$c + n^2 r^2 \geqq 0. \tag{II.13.43}$$

Circular trajectories centered at the origin can take place only if, for some $r = \bar{r}$, the ray curvature radius coincides with $\bar{r}$ itself, that is [see Eq. (II.4.14)],

$$\frac{1}{\bar{r}} = -\left.\frac{d \ln n}{dr}\right|_{r=\bar{r}} \tag{II.13.44}$$

Furthermore, since in this case $d\theta/dr = \infty$, we have from Eq. (II.13.41)

$$c = -n^2 \bar{r}^2. \tag{II.13.45}$$

Other relevant trajectories are associated with radial index distributions $n(r)$ such that the function $n(r)r$ has a relative maximum. In this case, for suitable values of c, Eq. (II.13.42) has two roots r_1 and r_2 and a ray describes a path between the two circles of radius r_1 and r_2 coplanar with the ray (see Fig. II.24). In particular, congruences composed of closed trajectories correspond to *oscillation modes* of the medium.

Let us now consider a ray bundle coming from $z = +\infty$, $\theta = 0$, that is, initially parallel to the z axis. For a given ray, we have from Eq. (II.13.40)

$$c = -n_\infty^2 l^2, \tag{II.13.46}$$

l being the initial distance from the z axis (also called the *impact parameter* in scattering theory and the *height* in lens theory). After reaching the distance r^* of closest approach, which is seen, with the help of Eqs. (II.13.40) and (II.13.46), to obey the relation

$$n^2 l^2 = n^2(r^*) r^{*2}, \tag{II.13.47}$$

the ray returns to infinity, describing a trajectory symmetrical to the one described during the approach and asymptotically forming an angle θ_s with

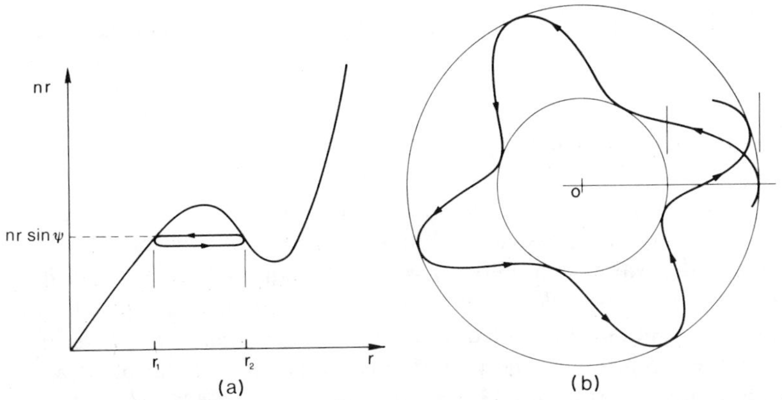

Fig. II.24. Ray trajectories in a medium with radial symmetry. (a) Plot of nr versus r. (b) The ray bounces back and forth between the two spherical caustics.

the negative z axis. We can calculate θ_s by integrating Eq. (II.13.41):

$$\theta_s = \pi - 2\int_{r^*}^{\infty} \frac{n_\infty l}{(n^2r^2 - n_\infty^2 l^2)^{1/2} r}\, dr. \tag{II.13.48}$$

In general, the *scattering angle* $\theta_s(l)$ is a complicated function of l. For $n(r)$ continuous together with its derivatives, θ_s tends to vanish as $l \to 0$ and $l \to \infty$. In other situations, for instance, if n is supposed to be singular at the origin, $\theta_s \to \pi$ as l vanishes. This means that rays with a small impact parameter tend to be back-reflected.

When $n(r)$ is continuous for $r = R$ and the medium is uniform (n independent of r) for $r > R$, a straight ray with impact parameter $l = R$ can be trapped by the gradient index region. More precisely, if

$$1/R > -d \ln n/dr, \tag{II.13.49}$$

the straight trajectory goes on unperturbed, while if

$$1/R < -d \ln n/dr, \tag{II.13.50}$$

the ray curvature radius is smaller than R, so that the ray enters the region $r < R$, reaches the distance of closest approach, and returns to infinity.

For an incident plane wave, a ray bundle with impact parameter l and annular section $2\pi l\, dl$ leaves the region of inhomogeneity $r < R$ within a solid angle

$$d\Omega = 2\pi \sin\theta_s |d\theta_s/dl|\, dl. \tag{II.13.51}$$

Accordingly (see Section II.12.3), it is easy to show that

$$\sigma(\theta_s) = (l/\sin\theta_s)|dl/d\theta_s|, \tag{II.13.52}$$

which is valid whenever there is a one-to-one correspondence between θ_s and l.

13.3 *Gradient Index Elements*

Another application of the formalism introduced in this section is connected with the possibility of manufacturing lens elements whose refractive index varies continuously within the material. Such *gradient index elements* [17, 18] are effective in reducing aberrations and achieving special results. An example of a graded-index lens is the *crystalline lens* present in the human eye. According to the Woinow measurement, its index of refraction in an adult man varies from 1.4387 at the inner core to 1.4005 at the less dense cortex. Interest in lenses with spherical index gradients stemmed originally from microwave applications. In particular, when

$$n = a/(b + r^2), \qquad a > b > 0, \tag{II.13.53}$$

the medium, known as *Maxwell's fish-eye*, exhibits the important property of focusing every point of the whole space. In fact, it can be shown that all the rays emerging from a given source intersect at a second focus aligned with the origin ($r = 0$) and the source [13]. The rays form a system of circles with centers on the plane perpendicular to the line connecting the source and the focus and equidistant from them (see Fig. II.25). The distances of the source and the focus from the origin satisfy the relation

$$r_s r_f = b. \tag{II.13.54}$$

The source and the focus, whose roles can obviously be interchanged, are said to be *conjugate points.*

Another interesting optical system is the *Luneburg lens*, characterized by the refractive index profile

$$n^2 = 2 - r^2/R^2, \tag{II.13.55}$$

inside a sphere of radius R. When a source is placed on the surface of the sphere, the rays describe ellipses [18] determined by

$$x^2 + y^2(1 + 2\cot^2\alpha) - 2xy\cot\alpha = R^2, \tag{II.13.56}$$

where α represents the angle between the ray emerging from the source and the diameter passing through the source itself (see Fig. II.26). It can be shown that the tangents to the ellipses are all parallel to the x axis on the surface of the sphere. As a consequence, if the refractive index is uniformly equal to one for $r > R$, the ray system transmitted outside is parallel to the x axis (see Fig. II.27), that is, to the diameter passing through the source.

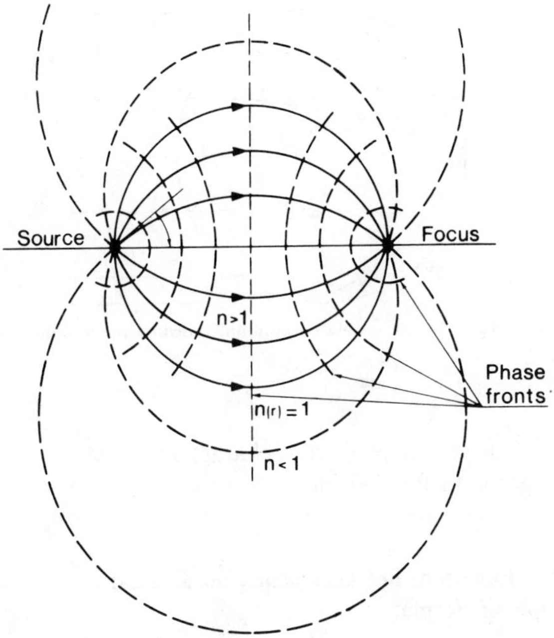

Fig. II.25. Ray trajectories in Maxwell's fish-eye lens.

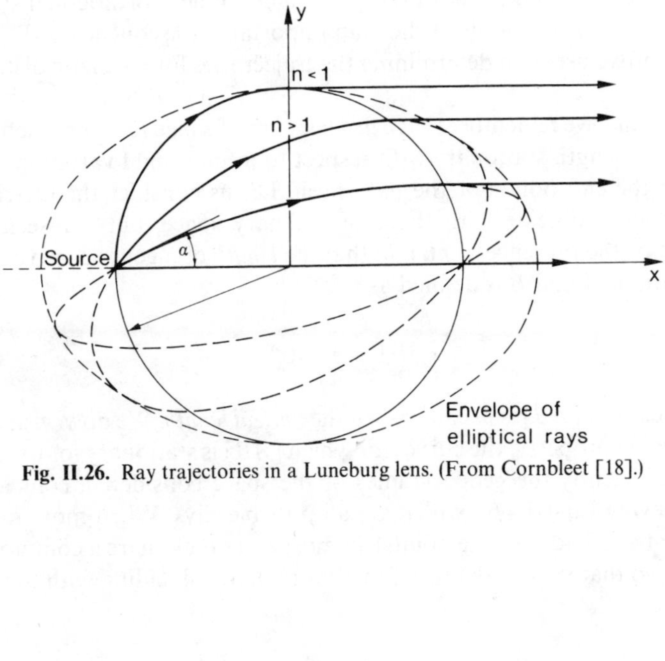

Fig. II.26. Ray trajectories in a Luneburg lens. (From Cornbleet [18].)

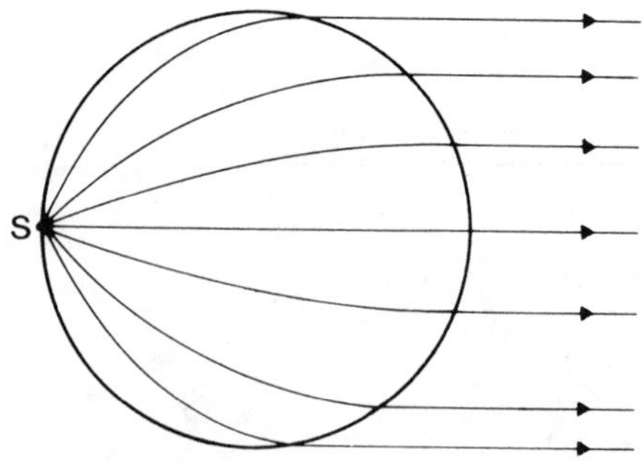

Fig. II.27. Luneburg lens used to obtain a collimated beam from a point source placed on its surface.

The proofs of the above properties characterizing Maxwell's fish-eye and the Luneburg lens are left as problems.

14 Scalar Ray Equations in Curvilinear Coordinates: the Principle of Fermat

The method of separation of variables can be conveniently applied to determine the ray trajectories whenever the refractive index distribution $n(\mathbf{r})$ possesses some symmetry with respect to a curvilinear orthogonal system of coordinates. We now wish to show an important consequence of Eq. (II.4.6) that can prove useful in determining the trajectories for a generic distribution $n(\mathbf{r})$.

To this end, we remember that a *geodesic line* of a *metric space* is defined as a path with a length stationary with respect to infinitesimal variations that do not alter the end points of the path itself. Let us consider the metric space whose points coincide with those of ordinary space, but whose metric is modified by the presence of $n(\mathbf{r})$, so that the *length* of a generic curve between two positions A and B is defined as

$$[AB] = \int_A^B n(r)\,ds. \tag{II.14.1}$$

The quantity $[AB]$ is usually called the *optical length*. We now wish to show the following property: the curves for which $[AB]$ is stationary for fixed A and B, or equivalently the geodesic lines of the space considered, coincide with those obeying Eq. (II.4.6), which are all possible rays. When more rays pass through A and B (conjugate points), in most cases they share a common value of $[AB]$, so that we can define a function $V(A, B)$ coinciding with the optical length of the rays. This function is called the *point characteristic* of the medium and will be discussed further in Section II.15.

In the language of the *calculus of variations* the above property means that the optical path is an *extremal* compared with all other paths that do not follow the laws of optics. In order to show this property, we can calculate the variation δ of $[AB]$ by using for mathematical convenience a cartesian coordinate, say z, as the integration variable:

$$\delta \int_A^B n(x, y, z)(1 + \dot{x}^2 + \dot{y}^2)^{1/2}\, dz \equiv \delta \int_A^B F(x, \dot{x}, y, \dot{y}, z)\, dz, \qquad \text{(II.14.2)}$$

where the dot denotes the derivative with respect to z. A curve connecting the two fixed points A and B corresponds to a particular couple of functions $x(z)$ and $y(z)$ selected in the set of continuous functions with assigned values at the extremes of the interval (z_A, z_B). A variation of the path corresponds to a change of these functions, $x(z) \to x(z) + \delta x(z)$ and $y(z) \to y(z) + \delta y(z)$, so that

$$\delta[AB] = \int_A^B \left(\frac{\partial F}{\partial x} \delta x + \frac{\partial F}{\partial \dot{x}} \delta \dot{x} + \frac{\partial F}{\partial y} \delta y + \frac{\partial F}{\partial \dot{y}} \delta \dot{y} \right) dz. \qquad \text{(II.14.3)}$$

Integrating by parts the second and fourth terms, and taking into account the possible discontinuity surfaces of $n(\mathbf{r})$, we obtain

$$\delta[AB] = \frac{\partial F}{\partial \dot{x}_j} \delta x_j \bigg|_A^B - \sum_S \left(\Delta \frac{\partial F}{\partial \dot{x}_j} \right) \delta x_j - \sum_S \int_{A_{S-1}}^{A_S} \left(\frac{d}{dz} \frac{\partial F}{\partial \dot{x}_j} - \frac{\partial F}{\partial x_j} \right) \delta x_j\, dz. \qquad \text{(II.14.4)}$$

Here the second term is a sum over the surfaces S of discontinuity of $n(\mathbf{r})$, Δ being related to sudden changes of n in the sense of increasing s, and the third term is a sum of contributions related to the regions of continuity. Two equal indices represent a sum over x and y. Since A and B are fixed, $\delta x(A) = \delta x(B) = \delta y(A) = \delta y(B) = 0$, so that the first term on the right side of Eq. (II.14.4) vanishes. On the other hand, we are looking for geodesic lines, characterized by the condition

$$\delta[AB] = \delta \int_A^B n(\mathbf{r})\, ds = 0 \qquad (A, B \text{ fixed}) \qquad \text{(II.14.5)}$$

for all possible values of δx and δy. Equation (II.14.4) ensures that this condition is satisfied only if

$$\frac{d}{dz} \frac{\partial F}{\partial \dot{x}} - \frac{\partial F}{\partial x} = 0, \qquad \frac{d}{dz} \frac{\partial F}{\partial \dot{y}} - \frac{\partial F}{\partial y} = 0 \qquad \text{(II.14.6)}$$

(*Euler's equations*), supplemented with the *saltus conditions*

$$\Delta(\partial F/\partial \dot{x}) = \Delta(\partial F/\partial \dot{y}) = 0 \tag{II.14.7}$$

on each discontinuity surface. By using the relations $ds = dz(1 + \dot{x}^2 + \dot{y}^2)^{1/2}$, $s_x = \dot{x}/(1 + \dot{x}^2 + \dot{y}^2)^{1/2}$, $s_y = \dot{y}/(1 + \dot{x}^2 + \dot{y}^2)^{1/2}$, Eqs. (II.14.6) are immediately seen to coincide with the scalar components of the ray equation [Eq. (II.4.5)] along the x and y axes (the same obviously holding true for the z component). The proof of the coincidence between rays and geodesic lines is complete if we observe that Eqs. (II.14.7) are equivalent to Snell's law, as we easily see if, for instance, we choose the z axis orthogonal to the discontinuity surface and the plane x–z as the incidence plane. A more rigorous derivation of Snell's law by means of the calculus of variations can be found in Zatzkis [19].

Usually, Eq. (II.14.5) is referred to as the *principle of Fermat*. More precisely, this principle asserts that the optical length of a ray between two points A and B is smaller than the optical length of any other curve joining A and B and lying in the neighborhood of the ray. Strictly speaking, Fermat's principle is valid, that is, yields curves satisfying Eq. (II.4.6) and the laws of refraction (and reflection), whenever A and B are not too far away. An example is provided by the concave spherical mirror [1] (see Fig. II.28). Let us consider the ray AQB, where the points A and B are symmetrically aligned with the center O of the spherical mirror γ. The ellipse ε with foci in A and B lies to the right of γ, so that $[AQB] = [AQ'B] > [AQ''B]$, and $[AQB]$ is a relative *maximum* with respect to displacements of Q on the mirror surface. The opposite situation arises if we consider two points A' and B' close enough to Q. Because of its limited validity, Fermat's principle has been reformulated (by Carathéodory) as follows: for each point of a light ray, there is a finite neighborhood for which Fermat's principle holds true. (For the extension of Fermat's principle to diffracted rays see Section VI.7.)

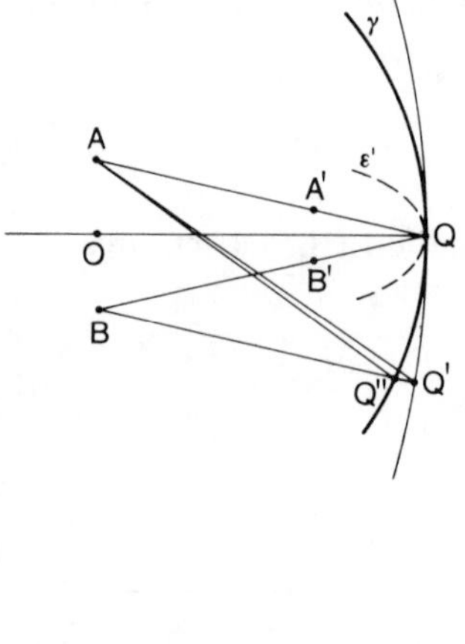

Fig. II.28 Illustration of Carathéodory's formulation of Fermat's principle (see Luneburg [1].

14.1 *Anisotropic Media*

The treatment described above can be generalized to the case in which n depends on both $\mathbf{r}$ and $\hat{s}$, that is, the dielectric constant is a tensor [20, 21] (see Section I.4). If we define

$$F(x, \dot{x}, y, \dot{y}, z) = n(x, y, z, \dot{x}, \dot{y})(1 + \dot{x}^2 + \dot{y}^2)^{1/2}, \tag{II.14.8}$$

we can look for a suitable generalization of Eq. (II.4.6). In fact, Eq. (II.14.6) yields

$$\frac{d}{dz}\frac{\partial F}{\partial \dot{x}} - \frac{\partial F}{\partial x} = (1 + \dot{x}^2 + \dot{y}^2)^{1/2} \frac{d}{ds}\left[\frac{\partial n}{\partial \dot{x}}(1 + \dot{x}^2 + \dot{y}^2)^{1/2} + n\frac{\dot{x}}{(1 + \dot{x}^2 + \dot{y}^2)^{1/2}}\right] - \frac{\partial n}{\partial x}(1 + \dot{x}^2 + \dot{y}^2)^{1/2} = 0, \tag{II.14.9}$$

that is,

$$\frac{d}{ds}\left[(1 + \dot{x}^2 + \dot{y}^2)^{1/2}\frac{\partial n}{\partial \dot{x}} + ns_x\right] = \frac{\partial n}{\partial x}. \tag{II.14.10}$$

Since

$$\frac{\partial}{\partial \dot{x}} = \frac{\partial s_x}{\partial \dot{x}}\frac{\partial}{\partial s_x} + \frac{\partial s_y}{\partial \dot{x}}\frac{\partial}{\partial s_y} = \frac{1 + \dot{y}^2}{(1 + \dot{x}^2 + \dot{y}^2)^{3/2}}\frac{\partial}{\partial s_x} - \frac{\dot{x}\dot{y}}{(1 + \dot{x}^2 + \dot{y}^2)^{3/2}}\frac{\partial}{\partial s_y}, \tag{II.14.11}$$

Eq. (II.14.10) is

$$\frac{d}{ds}\left[\frac{\partial n}{\partial s_x} + ns_x - s_x\left(s_x\frac{\partial n}{\partial s_x} + s_y\frac{\partial n}{\partial s_y}\right)\right] = \frac{\partial n}{\partial x}. \tag{II.14.12}$$

This equation has been derived by considering n as a function of $\mathbf{r}$, s_x, and s_y, which is justified by the relation $s_z = (1 - s_x^2 - s_y^2)^{1/2}$. We can also express n as an explicit function of s_x, s_y, and s_z, provided the following relations are taken into account:

$$\left.\frac{\partial}{\partial s_x}\right|_{s_y = \text{const}} = \left.\frac{\partial}{\partial s_x}\right|_{s_y = \text{const},\, s_z = \text{const}} - \left.\frac{s_x}{s_z}\frac{\partial}{\partial s_z}\right|_{s_x = \text{const},\, s_y = \text{const}},$$

$$\left.\frac{\partial}{\partial s_y}\right|_{s_x = \text{const}} = \left.\frac{\partial}{\partial s_y}\right|_{s_x = \text{const},\, s_z = \text{const}} - \left.\frac{s_y}{s_z}\frac{\partial}{\partial s_z}\right|_{s_x = \text{const},\, s_y = \text{const}}.$$

In this way, Eq. (II.14.12) and the analogous equations for the y and z axes relative to the surface having normal $\hat{n}_0$ can be rewritten in the form

$$\frac{d}{ds}\left[ns_i + \frac{\partial n}{\partial s_i} - s_i\left(s_j\frac{\partial n}{\partial s_j}\right)\right] = \frac{\partial n}{\partial x_i} \qquad (i = 1, 2, 3). \tag{II.14.14a}$$

$$\Delta[\{(n - \hat{s}\cdot\nabla_s n)\hat{s} + \nabla_s n\}\cdot(\mathbf{1} - \hat{n}_0\hat{n}_0)] = 0, \tag{II.14.14b}$$

Equations (II.14.14) yield the extension of Eq. (II.4.6) to *anisotropic media*, where the refractive index depends on the direction of propagation.

We note that the above relation furnishes the trajectory described by a narrow ray bundle. In an anisotropic medium, a narrow ray bundle travels parallel to the Poynting vector (see Section I.6), while it is tilted with respect to the wave front normal. The latter property marks the difference from the isotropic situation, in which the rays coincide with the trajectories normal to the wave fronts.

As a further clarification, we observe that, according to Eqs. (II.14.14), a ray describes a straight line in a homogeneous anisotropic medium. In contrast, the wave fronts produced by a point source inside the same medium are spheroids, and the curves normal to the wave front family are generally curvilinear.

Example 1: *An Application to Electron Optics.* An interesting analogy is furnished by an electron moving under the action of a combination of *static* electric and magnetic fields. The trajectory of the electron can be determined by introducing an equivalent refractive index n_{eq} given by

$$n_{\mathrm{eq}} = \frac{mv}{(1-\beta^2)^{1/2}} - \mu_0 e\mathbf{A}\cdot\hat{s} \qquad \text{(apart from an arbitrary constant factor)}, \tag{II.14.15}$$

where $-e$ is the electron charge, m the electron rest mass, $\mathbf{A}$ the vector potential, and $\beta = v/c$. The electron velocity v is related to the scalar potential Φ through the relativistic energy conservation law

$$\frac{mc^2}{(1-\beta^2)^{1/2}} - e\Phi = \text{const.} \tag{II.14.16}$$

Equation (II.14.15) is the *fundamental equation of electron optics*.

If we introduce Eq. (II.14.15) into Eqs. (II.14.14), we obtain

$$\frac{d}{ds}\left[-\mu_0 e A_i + \frac{mvs_i}{(1-\beta^2)^{1/2}}\right] = \frac{\partial n_{\mathrm{eq}}}{\partial x_i}, \tag{II.14.17}$$

Example 2: *Refraction of Extraordinary (E) Ray in a Uniaxial Crystal.* Some caution must be adopted in choosing the correct equation expressing the dependence of n on $\hat{s}$. Since $\hat{s}$ represents the tangent to the ray trajectory, the ray-surface equation (see Section I.4.1) must be used. In particular, for a uniaxial crystal

$$n^2(\hat{s}) = (\hat{s}\cdot\hat{c})^2(n_0^2 - n_e^2) + n_e^2,$$

where $\hat{c}$ and $\hat{s}$ represent the optic axis (forming an angle α with the normal $\hat{n}_0$ to the surface) and the ray (parallel to the Poynting vector) directions respectively. Accordingly,

$$\nabla_{\hat{s}} n = n^{-1}(n_0^2 - n_e^2)(\hat{s}\cdot\hat{c})\hat{c}, \tag{II.14.18}$$

and, for a ray incident normally, Eq. (II.14.14b) yields

$$\tan\theta' = -\frac{\sin\alpha\cos\alpha(n_0^2 - n_e^2)}{n_e^2\cos^2\alpha + n_0^2\sin\alpha}. \qquad \text{(II.14.19)}$$

14.2 *Geodesic Lenses*

An interesting application of Fermat's principle is the *geodesic lens*, which consists of thin dielectric layers of constant n deposited on a substrate having a shallow depression (see Fig. II.29). This kind of two-dimensional structure possesses guiding properties, so that a ray initially tangent to the depression surface changes its direction, following the surface tangentially [22]. For suitable types of surface, the system can act like a lens on a light beam traveling into this optical waveguide.

In general, a geodesic lens consists of a surface of revolution (see Fig. II.30). In cylindrical coordinates, the lens profile is described by the function $z(\rho)$ and the optical path is given by

$$L = \int\left[1 + \left(\frac{dz}{d\rho}\right)^2 + \rho^2\left(\frac{d\phi}{d\rho}\right)^2\right]^{1/2} d\rho, \qquad \text{(II.14.20)}$$

where we have set $n = 1$, which is justified by the supposed constancy of the refractive index.

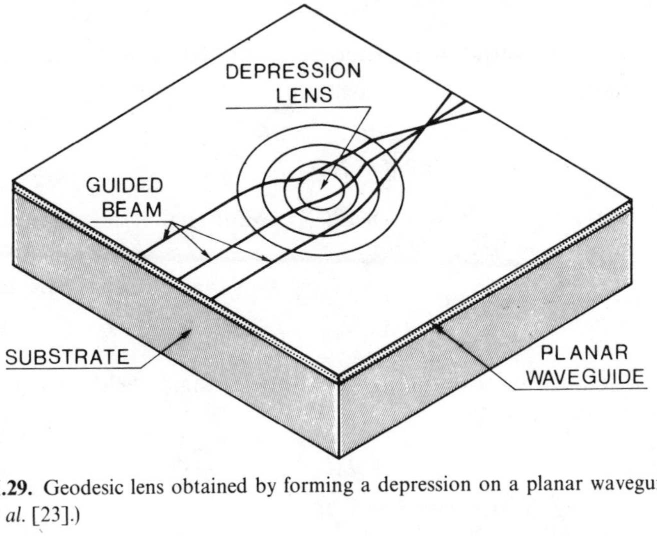

Fig. II.29. Geodesic lens obtained by forming a depression on a planar waveguide. (From Righini *et al.* [23].)

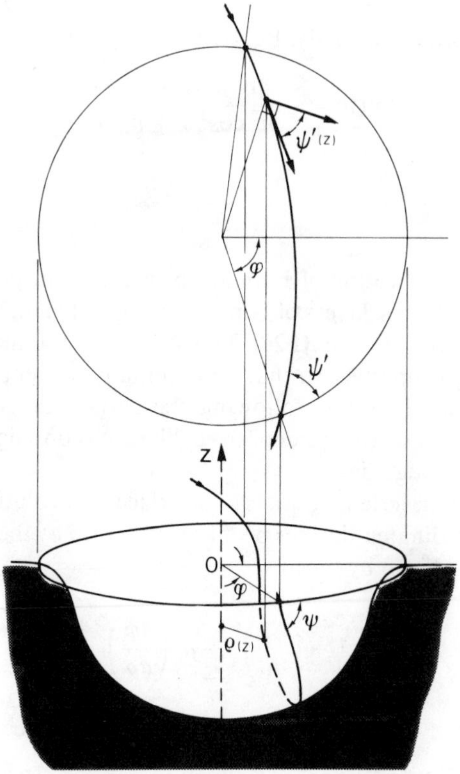

Fig. II.30. Ray trajectories in a geodesic lens. Here ψ' is the "projection" of ψ on the plane z = const. (From Righini *et al.* [23].)

If we apply the variational criterion of Eq. (II.14.5), the relative Euler equation reads

$$\frac{d}{d\rho}\frac{\partial F}{\partial(d\phi/d\rho)} - \frac{\partial F}{\partial \phi} = 0, \tag{II.14.21}$$

where F represents the integrand of Eq. (II.14.20), so that, taking into account the relation $\partial F/\partial\phi = 0$, we obtain

$$\frac{\rho^2}{[1 + (dz/d\rho)^2 + \rho^2(d\phi/d\rho)^2]^{1/2}}\frac{d\phi}{d\rho} = C, \tag{II.14.22}$$

where C is a constant along the path. Solving for $\dot{\phi}$ yields the differential equation of a geodesic

$$\dot{\phi} = C(1 + \dot{z}^2)^{1/2}/\rho(\rho^2 - C^2)^{1/2}, \tag{II.14.23}$$

the dot indicating the derivative with respect to ρ.

If we denote by $\pi/2 - \psi$ the angle between the ray path and the meridional line ($\phi = \text{const}$) on the surface and by dL the arc length, Eqs. (II.14.20) and (II.14.22) lead to

$$\rho \cos \psi = \rho^2 |d\phi/dL| = |C|. \qquad \text{(II.14.24)}$$

This equation, known as *Clairaut's theorem*, can be considered a generalization of Snell's law [see also Bouguer's theorem, Eq. (II.13.40)] applicable to homogeneous media.

14.3 *Absolute Differential Calculus Formalism*

The equivalence between rays and geodesic lines allows us to exploit the standard formalism of *absolute differential calculus*, usually adopted in general relativity, in order to obtain differential equations furnishing the optical trajectories in any coordinate system.

As already observed, we consider the three-dimensional space defined by the metric

$$d\psi^2 \equiv n^2 ds^2 = g_{ij} dx^i dx^j \qquad \text{(II.14.25)}$$

[the symbol $d\psi$ used hereafter has nothing to do with the angle ψ of Eq. (II.14.24)], where the sum over repeated indices is adopted (*Einstein's convention*). The g_{ij} depend on the particular coordinate system. For example, for orthogonal coordinates they are

$$g_{ij} = \begin{bmatrix} n^2 h_1^2 & 0 & 0 \\ 0 & n^2 h_2^2 & 0 \\ 0 & 0 & n^2 h_3^2 \end{bmatrix}, \qquad \text{(II.14.26)}$$

the h_i being suitable scale factors.

The geodesic lines relative to the metric of Eq. (II.14.25) obey the equations [24]

$$\frac{d^2 x_i}{d\psi^2} + \Gamma^i_{jk} \frac{dx^j}{d\psi} \frac{dx^k}{d\psi} = 0, \qquad \text{(II.14.27)}$$

where the Γ^i_{jk} (*Christoffel symbols of the second kind*) are defined as

$$\Gamma^i_{jk} = \tfrac{1}{2} g^{im} (g_{mj,k} + g_{mk,j} - g_{jk,m}), \qquad \text{(II.14.28)}$$

the suffix ",i" indicating partial derivative with respect to x^i. The tensor g^{im} is the *dual tensor* of g_{im} and its components satisfy the relation

$$g^{im} g_{mj} = \delta^i_j, \qquad \text{(II.14.29)}$$

where

$$\delta^i_j = 1 \quad (i = j), \qquad \delta^i_j = 0 \quad (i \neq j) \qquad \text{(II.14.30)}$$

is the *Kronecker tensor.* For orthogonal coordinates corresponding to Eq. (II.14.26), we obtain immediately

$$g^{ij} = \begin{bmatrix} 1/(n^2h_1^2) & 0 & 0 \\ 0 & 1/(n^2h_2^2) & 0 \\ 0 & 0 & 1/(n^2h_2^3) \end{bmatrix}. \tag{II.14.31}$$

In spite of their formal elegance, the *scalar ray equations in curvilinear coordinates* [Eqs. (II.14.27)] are not simple to use, because of the dependence of the coefficients Γ^i_{jk} on the distribution of the refractive index. To put them in a simpler form it is convenient to introduce the quantities $\hat{g}_{ij}$, defined as

$$\hat{g}_{ij} = g_{ij}/n^2, \tag{II.14.32}$$

which depend only on the system of coordinates. If we introduce Eqs. (II.14.32) into Eqs. (II.14.28), we have

$$\Gamma^i_{jk} = \hat{\Gamma}^i_{jk} + (1/n)\hat{g}^{im}(n_{,k}\hat{g}_{mj} + n_{,j}\hat{g}_{mk} - n_{,m}\hat{g}_{jk}), \tag{II.14.33}$$

so that we can write, with the help of the duality relations defining the g^{ij},

$$\Gamma^i_{jk} = \hat{\Gamma}^i_{jk} + (1/n)(\delta^i_j n_{,k} + \delta^i_k n_{,j} - n_{,m}\hat{g}^{im}\hat{g}_{jk}). \tag{II.14.34}$$

The symbol $\hat{\Gamma}^i_{jk}$ indicates the quantity given in terms of the $\hat{g}$'s analogous to Γ^i_{jk} in terms of the g's. We now observe that

$$\begin{aligned} \frac{d^2x^i}{d\psi^2} &= \frac{1}{n}\frac{d}{ds}\left(\frac{1}{n}\frac{dx^i}{ds}\right) = \frac{1}{n^2}\frac{d^2x^i}{ds^2} - \frac{1}{n^3}\frac{dn}{ds}\frac{dx^i}{ds} \\ &= \frac{1}{n^2}\frac{d^2x^i}{ds^2} - \frac{1}{n^3}\frac{\partial n}{\partial x^j}\frac{dx^j}{ds}\frac{dx^i}{ds}, \end{aligned} \tag{II.14.35}$$

and that, from Eqs. (II.14.25) and (II.14.32),

$$\hat{g}_{ij}\,dx^i\,dx^j/ds^2 = 1. \tag{II.14.36}$$

These relations allow us to put Eqs. (II.14.27) in the form

$$\frac{d^2x^i}{ds^2} + \hat{\Gamma}^i_{jk}\frac{dx^j}{ds}\frac{dx^k}{ds} + \frac{d\ln n}{ds}\frac{dx^i}{ds} - \frac{1}{n}\hat{g}^{ij}\frac{\partial n}{\partial x^j} = 0. \tag{II.14.37}$$

The second term is directly connected with the particular system of coordinates, while the third and fourth terms are associated with the inhomogeneity medium.

Example: Cylindrical coordinates. As an example, let us write Eqs. (II.14.37) for cylindrical coordinates $x^1 = \rho$, $x^2 = \phi$, $x^3 = z$. It can be easily proved, and is left as a problem, that

$$\hat{\Gamma}^1_{22} = -\rho, \qquad \hat{\Gamma}^2_{12} = \hat{\Gamma}^2_{21} = 1/\rho, \tag{II.14.38}$$

all other $\hat{\Gamma}$ vanishing, so that Eqs. (II.14.37) yield

$$\frac{d^2\rho}{ds^2} - \rho\left(\frac{d\phi}{ds}\right)^2 + \frac{d\ln n}{ds}\frac{d\rho}{ds} = \frac{1}{n}\frac{\partial n}{\partial \rho},$$

$$\frac{d^2\phi}{ds^2} + \frac{2}{\rho}\frac{d\rho}{ds}\frac{d\phi}{ds} + \frac{d\ln n}{ds}\frac{d\phi}{ds} = \frac{1}{n\rho^2}\frac{\partial n}{\partial \phi}, \qquad \text{(II.14.39)}$$

$$\frac{d^2 z}{ds^2} + \frac{d\ln n}{ds}\frac{dz}{ds} = \frac{1}{n}\frac{\partial n}{\partial z}.$$

15 Elements of Hamiltonian Optics

An optical instrument K can be envisaged as a medium whose refractive index is a prescribed function of the coordinates. In general, the design is such that K will produce, as nearly as possible, an image of a given characteristic of the object considered. Conventionally, the region in which the object is situated is referred to as *object space*, while the term *image space* is reserved for the region in which one observes the rays from the object after their passage through K. Additional simplifications are introduced when the medium is homogeneous in both the image and object spaces. In this case, the initial and final parts of a ray are straight lines.

In the most common situation, we wish K to produce a sharp image of a plane object geometrically similar to the object itself. Less frequently, we might want K to produce a square image of a rectangular object, as in lenses needed in making wide-screen motion pictures, where the extra-large horizontal field of view must be condensed into the regular film format. In other cases, we might wish a plane object to have a sharp image lying on some surface other than a plane.

Perfect imagery is achieved when all the rays from any point O of the object pass through the corresponding image point O'. In real systems, rays from O will, in general, fail to pass through the appropriate point O' of the image space. For systems imaging a plane object on Π_o into a plane image on Π_i, if a particular ray from O intersects Π_i in O'_1, the vector $O'O'_1$ connecting the desired image O' to O'_1 is a measure of the *ray aberration*.

When the object and image spaces are homogeneous, a homocentric pencil centered around the ray R and having focus in O is transformed by K into an *astigmatic* pencil characterized by two focal lines separated by the *astigmatic focal distance*. Simple geometric considerations show that, for focal segments of equal length a, the rays of the bundle pass through a circular path with diameter $a/2$. In this case, the best point image O' lies halfway between the focal lines, at the center of the *disk of least confusion*.

In general, the astigmatic focal length cannot be made to vanish by orienting the central ray of the bundle, except for particular surfaces of the object field. In the latter situation, a sharp image of these points is obtained by limiting the aperture of the ray pencil. For special systems, there are surfaces that are perfectly imaged by wide-angle beams. For example, for a homogeneous sphere of radius r and refractive index n immersed in a medium of index n', the spheres of radius rn/n' and rn'/n are perfectly imaged into one another the proof is left as a problem. In 1840–1850 G. Amici used spherical lenses based on this property in his microscope objectives (see Problem 13).

To be more specific, let us consider an *axisymmetric system* K consisting of surfaces of revolution with a common *optic axis*. The object point O and the optic axis define the *meridional plane*. A ray tangent to a meridional plane must lie totally in it. A nonmeridional ray is called *skew* and it never crosses the optic axis. The focal lines (in the image space) of a meridional ray are, respectively, normal and parallel to the meridional plane, which can be proved by using Eqs. (II.11.22). As a consequence, they are called *sagittal* and *tangential* focal lines. The same property does not hold for skew rays. In particular, when O lies on the optic axis, every ray from it is meridional, and the caustic corresponding to a wide-angle pencil is composed of a sagittal focal surface of revolution around the optic axis and of a tangential focal surface coincident with a segment of the optic axis (see the example in Section II.10.1.b). These surfaces reduce to a point for a moderate aperture, when O coincides with the *aplanatic point* of the lens. In the language of aberration theory, the finiteness of the caustic of an axial point source is mainly due to the presence of spherical aberration, which is minimized for a particular position of the object.

15.1 *Point, Angle, and Mixed Characteristics*

In order to analyze the quality of an image produced by a system made of a sequence of curved refraction surfaces, it would be necessary to trace a sufficiently large number of rays by integrating the ray equation written in the most convenient coordinate system. In addition, it would be necessary to repeatedly apply Eqs. (II.11.22) to obtain the centers of curvature of the pencils relative to the single homogeneous regions. This program can be performed very quickly by using special computer codes. However, for an initial assessment of the parameters of a lens, an approximate analytic evaluation of the aberrations is necessary. To this end, an elegant theory of aberrations originally due to William Hamilton is of great help. The power of his method lies in its ability to yield excellent results by taking into account only the symmetries of the system.

The main ingredient of the Hamiltonian method, known as *Hamiltonian optics*, is the optical distance $[P_0, P_1]$ between two generic points P_0 and P_1 of

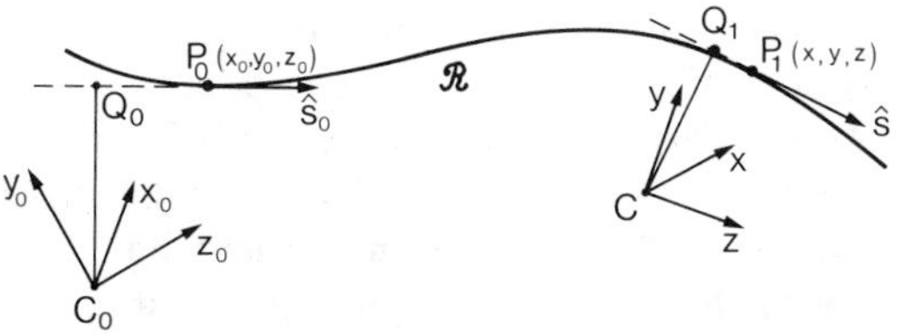

Fig. II.31. Optical distance between two points P_0 and P_1, respectively in the object and image space.

the system K (see Fig. II.31). This distance is called the *point characteristic* and will be indicated by the symbol $V(P_0, P_1)$. It coincides with the eikonal in P_1 of the ray field produced by a source in P_0, or vice versa. If we bear in mind the eikonal equation [Eq. (II.3.1)], we immediately see that the gradient of V with respect to $P_0(P_1)$ is a vector directed as the ray R through $P_0(P_1)$ and having an amplitude coincident with the refractive index $n(P_0)$ $[n(P_1)]$. The orientation of the gradient coincides with that of R for points lying in the image space, the opposite holding in the object space. In the following, we will assume that P_0 is in the object space and P_1 in the image space, so that

$$\nabla_0 V(P_0, P_1) = -n_0\hat{s}(P_0), \qquad \nabla_1 V(P_0, P_1) = n_1\hat{s}(P_1). \qquad \text{(II.15.1)}$$

It is now convenient to introduce the *optical direction cosines* $p = ns_x$, $q = ns_y$, and $r = ns_z$ so that the above equations can be recast in scalar form as

$$\begin{aligned} p_0 &= -\partial V/\partial x_0, & q_0 &= -\partial V/\partial y_0, & r_0 &= -\partial V/\partial z_0, \\ p_1 &= \partial V/\partial x_1, & q_1 &= \partial V/\partial y_1, & r_1 &= \partial V/\partial z_1, \end{aligned} \qquad \text{(II.15.2)}$$

where x_0, y_0, z_0 are the cartesian coordinates of P_0 in a system C_0, while x_1, y_1, z_1 are those of P_1 referred in general to a different system. Indicating with a dot the derivative with respect to z_0, we obtain from the ray equation [Eq. (II.4.6)]

$$\begin{aligned} \dot{p}_0 &= \partial r_0/\partial x_0, & \dot{q}_0 &= \partial r_0/\partial y_0, \\ \dot{x}_0 &= -\partial r_0/\partial p_0, & \dot{y}_0 &= -\partial r_0/\partial q_0. \end{aligned} \qquad \text{(II.15.3)}$$

Then p_0, q_0 and x_0, y_0 are conjugate variables with respect to the *Hamiltonian function* $H(z_0, x_0, y_0, p_0, q_0) = r_0 = [n^2(x_0, y_0, z_0) - p_0^2 - q_0^2]^{1/2}$, which is in general an explicit function of z_0. Accordingly, the dynamics of a ray is identical to that of a time-dependent system with two degrees of freedom. Consequently, we can introduce a *phase space* formed by the points of coordinates p_0, q_0, x_0, and y_0. For systems governed by an Hamiltonian, the *Liouville theorem* [25, 26] states that *a volume element of the phase space remains constant during its evolution.* In our case, the quantity $dx_0\,dy_0\,dp_0\,dq_0$ is constant along a ray. As a consequence of this property, the determinant of the ray matrix that will be introduced in this section must be unity. Thus, the region occupied by the ray bundle in the phase space changes in shape as it travels through the optical instrument, but its volume does not change. It is noteworthy that in the paraxial limit this property gives the*Lagrange invariant* [Eq. (II.15.27)].

Analysis of propagation in phase space is a common practice in the design of particle beam transport systems, where each element (e.g., a quadrupole) is characterized by an acceptance area. In this case, only the beams whose phase space volume is contained in the acceptance areas of the single components can be transmitted [27].

The six coordinates x_0, y_0, z_0, x_1, y_1, and z_1 are sufficient to determine R, as are the direction cosines p_0, q_0, p_1, and q_1 together with z_0 and z_1, unless we consider *afocal systems* such as those used as beam expanders. In general, we can establish a one-to-one correspondence between the two sets of variables $(x_0, y_0, z_0, x_1, y_1, z_1) \leftrightarrow (z_0, z_1, p_0, q_0, p_1, q_1)$. In order to use the second set, it is necessary to introduce a different characteristic function T called the *angle characteristic*, which is related to V through the equation

$$T(P_0, P_1) = V(P_0, P_1) + x_0 P_0 + y_0 q_0 + z_0 r_0 - x_1 p_1 - y_1 q_1 - z_1 r_1. \tag{II.15.4}$$

For small displacements of P_0 and P_1, and using Eqs. (II.15.2), the differential of T is

$$dT = x_0\,dp_0 + y_0\,dq_0 + z_0\,dr_0 - x_1\,dp_1 - y_1\,dq_1 - z_1\,dr_1. \tag{II.15.5}$$

Then, taking into account the relations between r_0, r_1 and p_0, q_0, p_1, q_1, we have

$$\begin{aligned} x_0 &= \partial T/\partial p_0 + z_0(p_0/r_0), & y_0 &= \partial T/\partial q_0 + z_0(q_0/r_0), \\ x_1 &= -\partial T/\partial p_1 - z_1(p_1/r_1), & y_1 &= -\partial T/\partial q_1 - z_1(q_1/r_1). \end{aligned} \tag{II.15.6}$$

Simple geometric considerations show that T represents the optical distance between the base points Q_0 and Q_1 of the perpendiculars dropped onto the ray from the origins of the coordinate systems relative to the object and image spaces. In particular, for axisymmetric systems and for reference systems C_0 and C_1 having z axes coincident with the optical axis, T depends only on the following combination of direction cosines:

$$u = p_0^2 + q_0^2, \qquad v = p_1^2 + q_1^2, \qquad w = 2(p_0 p_1 + q_0 q_1). \tag{II.15.7}$$

In some problems, such as those encountered in the diffraction analysis of optical instruments discussed in Section IV.13, it is convenient to use the *mixed characteristic* W, defined as

$$W(x_0, y_0, z_0; z_1, p_1, q_1) = V(P_0, P_1) - x_1 p_1 - y_1 q_1 - z_1 r_1, \tag{II.15.8a}$$

for which

$$p_0 = -\partial W/\partial x_0, \qquad q_0 = -\partial W/\partial y_0,$$
$$x_1 = -\partial W/\partial p_1 - z_1(p_1/r_1), \qquad y_1 = -\partial W/\partial q_1 - z_1(q_1/r_1). \tag{II.15.8b}$$

15.2 *Angle Characteristic of a Surface of Revolution*

Composite lenses are made of a sequence of surfaces of revolution separating homogeneous media. In general, if we indicate with $z = f(x, y)$ the equation of a single surface, then we have as an immediate consequence of Fermat's principle (Section II.14)

$$\begin{aligned} V(x_0, y_0, z_0; x_1, y_1, z_1) = g(P_0, P_1; x, y, z) \\ \equiv n_0[(x - x_0)^2 + (y - y_0)^2 + (z - z_0)^2]^{1/2} \\ + n_1[(x_1 - x)^2 + (y_1 - y)^2 + (z_1 - z)^2]^{1/2}, \end{aligned} \tag{II.15.9}$$

subject to the condition

$$\frac{\partial g}{\partial x} + \frac{\partial g}{\partial z}\frac{\partial f}{\partial x} = 0, \qquad \frac{\partial g}{\partial y} + \frac{\partial g}{\partial z}\frac{\partial f}{\partial y} = 0. \tag{II.15.10}$$

As a consequence, if we use as a reference for object and image planes the plane passing through the vertex of the surface, $z_0 = z_1 = z_v = f(0, 0)$, then we have for the relative angle characteristic T_0, by virtue of its geometric interpretation (Fig. II.32),

$$T_0 = (p_0 - p_1)x + (q_0 - q_1)y + (r_0 - r_1)f(x, y), \tag{II.15.11}$$

where $x, y, f(x, y)$ is the point P of intersection of the ray R with the refraction

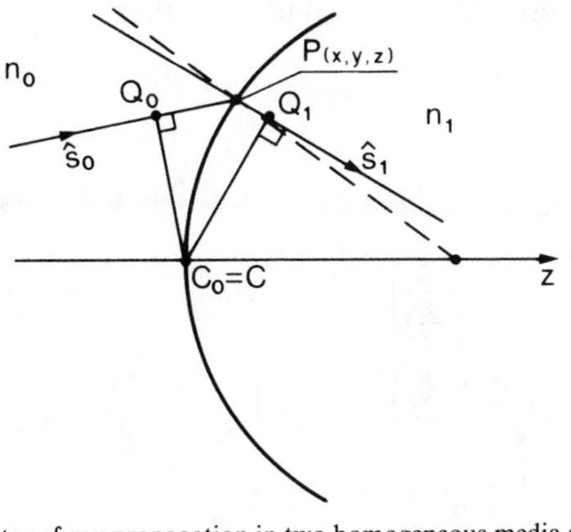

Fig. II.32. Geometry of ray propagation in two homogeneous media separated by a surface of revolution around the z-axis.

surface. On the other hand, Snell's law implies

$$p_0 - p_1 = -(r_0 - r_1)\,\partial f/\partial x, \qquad q_0 - q_1 = -(r_0 - r_1)\,\partial f/\partial y, \tag{II.15.12}$$

so that

$$T_0 = (r_1 - r_0)[x(\partial f/\partial x) + y(\partial f/\partial y) - f]. \tag{II.15.13}$$

Next, drawing on Luneburg [1], we introduce the variables $\alpha = \partial f/\partial x$ and $\beta = \partial f/\partial y$ and the function $\Omega(\alpha, \beta)$, defined as

$$\Omega(\alpha, \beta) = x\frac{\partial f}{\partial x} + y\frac{\partial f}{\partial y} - f = x(\alpha, \beta) + y(\alpha, \beta) - f(\alpha, \beta). \tag{II.15.14}$$

Then, taking into account Eqs. (II.15.12, 13, 14), we finally obtain

$$T_0 = (r_1 - r_0)\Omega\left(-\frac{p_0 - p_1}{r_0 - r_1}, -\frac{q_0 - q_1}{r_0 - r_1}\right). \tag{II.15.15}$$

For a surface of revolution $z = f(\rho)$, we can easily verify that, if we set $\Omega = \rho\,\partial f/\partial\rho - f_1$, the above relation becomes

$$T_0 = (r_1 - r_0)\Omega\left[\left(\frac{u + v - w}{r_1 - r_0}\right)^{1/2}\right], \tag{II.15.16}$$

where u, v, and w are defined by Eqs. (II.15.7). For a spherical surface of radius R (>0 if convex, <0 otherwise), Eq. (II.15.16) takes the simple form

$$\begin{aligned} T_0 &= R\,\mathrm{sgn}(n_1 - n_0)(n_0^2 + n_1^2 - w - 2r_1 r_0)^{1/2} + R(r_0 - r_1) \\ &= R\,\mathrm{sgn}(n_1 - n_0)|n_0\hat{s}_0 - n_1\hat{s}_1| + R(r_0 - r_1). \end{aligned} \tag{II.15.17}$$

We can now immediately find the angle characteristic for $z_0 \neq z_1 \neq z_v$ by noting that in general T changes linearly with z_0 and z_1, the proportionality coefficients being, respectively, $-r_0$ and r_1, so that

$$\begin{aligned} T(z_0, z_1; p_0, q_0, p_1, q_1) &= R\,\mathrm{sgn}(n_1 - n_0)(n_0^2 + n_1^2 - w - 2r_1 r_0)^{1/2} \\ &\quad + (z_1 - z_v - R)r_1 - (z_0 - z_v - R)r_0. \end{aligned} \tag{II.15.18}$$

In particular, by assuming $z_v = 0$ and defining $\Delta = n_1 - n_0$ we get, for $u, v, w \ll 1$

$$\begin{aligned} T &\sim n_1 z_1 - n_0 z_0 + \frac{1}{2\Delta}\left[-Rw + Ru + \frac{\Delta}{n_0}z_0 u + Rv - \frac{\Delta}{n_1}z_1 v\right] \\ &\quad + \frac{R}{8}\left[\frac{1}{\Delta}\left(\frac{u}{n_0^2} + \frac{v}{n_1^2}\right)^2 + \frac{n_1^2 n_0^2}{\Delta^3}\left(\frac{w}{n_0 n_1} - \frac{u}{n_0^2} - \frac{v}{n_1^2}\right)^2 \right. \\ &\quad \left. - \frac{v^2}{n_1^3} + \frac{u^2}{n_0^3}\right] + \frac{z_1}{8n_1^3}v^2 - \frac{z_0}{8n_0^3}u^2 \end{aligned} \tag{II.15.19}$$

15.3 *Ray Matrix*

Let us introduce the transverse vectors $\boldsymbol{\rho}_0 = \hat{x}x_0 + \hat{y}y_0$ and $\boldsymbol{\rho}_1 = \hat{x}x_1 + \hat{y}y_1$ for an axisymmetric system. In this case, we can recast Eqs. (II.15.6) in vector form

$$\begin{aligned} \boldsymbol{\rho}_0 &= 2\mathbf{p}_0\, \partial T/\partial u + 2\mathbf{p}_1\, \partial T/\partial w, \\ \boldsymbol{\rho}_1 &= -2\mathbf{p}_0\, \partial T/\partial w - 2\mathbf{p}_1\, \partial T/\partial v, \end{aligned} \tag{II.15.20}$$

where $\mathbf{p}_0 = \hat{x}p_0 + \hat{y}q_0$, $\mathbf{p}_1 = \hat{x}p_1 + \hat{y}q_1$. In particular, from the above relations it follows that the vector $\mathbf{p} \times \boldsymbol{\rho}$ remains constant along a ray propagating through an axisymmetric system. The invariance of its modulus m, termed *skewness* (see also *skewness invariant*, Section II.13.1), can be used to simplify the ray tracing in centered systems, as originally done by T. Smith in 1821. For example, the skewness invariance implies the *optical sine theorem* [28]. In general, it is convenient to write Eqs. (II.15.20) in matrix form, viz.

$$\begin{bmatrix} \boldsymbol{\rho}_1 \\ \mathbf{p}_1 \end{bmatrix} = \begin{bmatrix} A & B \\ C & D \end{bmatrix} \cdot \begin{bmatrix} \boldsymbol{\rho}_0 \\ \mathbf{p}_0 \end{bmatrix} \equiv \mathbf{S} \cdot \begin{bmatrix} \boldsymbol{\rho}_0 \\ \mathbf{p}_0 \end{bmatrix}, \tag{II.15.21}$$

where

$$A = -\frac{\partial T/\partial v}{\partial T/\partial w}, \qquad D = -\frac{\partial T/\partial u}{\partial T/\partial w}, \qquad C = \frac{1}{2(\partial T/\partial w)}, \qquad B = \frac{AD - 1}{C}. \tag{II.15.22}$$

In view of the above expressions, the determinant of the *ray matrix* $\mathbf{S}$ is equal to unity. In particular, for the spherical surface described by the characteristic of Eq. (II.15.16) we can write

$$\begin{aligned} A &= r_0/r_1 - (z_1 - z_v - R)K_1/r_1, \\ D &= r_1/r_0 + (z_v + R - z_0)K_1/r_0, \\ B &= (z_1 - z_v - R)/r_0 - (z_v + R - z_0)/r_1 + K_1(z_1 - z_v - R)(z_v + R - z_0)/(r_1 r_0), \\ C &= -|n_0\hat{s}_0 - n_1\hat{s}_1|/[R\,\mathrm{sgn}(n_1 - n_0)] \equiv -K_1, \end{aligned} \tag{II.15.23}$$

where K_1 is called the *skew power* of the surface. It is, in general, a function of u, v, and w.

For $\boldsymbol{\rho}_0$ and $\mathbf{p}_0$ sufficiently small, the coefficients A, B, C, and D become independent of $\mathbf{p}_0$ and $\mathbf{p}_1$. This is the domain of *Gaussian optics* (so called in honor of K. F. Gauss, who undertook the analysis of a refractive sphere by using power series expansions in his celebrated memoir *Dioptrische Untersuchungen*, published in Göttingen in 1841), in which the optical instruments are described by a system matrix $\mathbf{S}_{1,0}$ formed by the *Gaussian constants* A, B, C, and D. For example, for a spherical refracting surface, Eqs. (II.15.23) yield, for vanishing $\mathbf{p}_0$ and $\mathbf{p}_1$,

$$\mathbf{S}_{1,0} = \begin{bmatrix} 1 - \dfrac{K(z_1 - z_v)}{n_1} & -\dfrac{z_v - z_0}{n_0} + \dfrac{z_1 - z_v}{n_1} + \dfrac{K(z_1 - z_v)(z_v - z_0)}{n_1 n_0} \\ -K & 1 + \dfrac{K(z_v - z_0)}{n_0} \end{bmatrix} \tag{II.15.24}$$

where $K = (n_1 - n_0)/R$ is the limit value of K_1 for small angles and is usually termed *refracting power*.

If we displace the planes $z = z_0$ and $z = z_1$, the new matrix can be obtained by exploiting the simple linear dependence of the angular characteristic T on z_0 and z_1, which gives

$$\begin{aligned} \mathbf{S}(z'_0, z'_1) &= \begin{bmatrix} 1 & (z'_1 - z_1)/n_1 \\ 0 & 1 \end{bmatrix} \cdot \mathbf{S}(z_0, z_1) \cdot \begin{bmatrix} 1 & (z_0 - z'_0)/n_0 \\ 0 & 1 \end{bmatrix} \\ &\equiv \mathbf{T}(z'_1, z_1) \cdot \mathbf{S}(z_0, z_1) \cdot \mathbf{T}(z_0, z'_0), \end{aligned} \tag{II.15.25}$$

where the translation matrices $\mathbf{T}$ account for the translation of the reference planes.

In general, when the coefficient B vanishes, all the rays from P_0 pass through P_1, independent of their initial direction. This means that P_1 is the *paraxial image* of P_0. If this property holds true in the paraxial limit for the points of the planes $z = z_0$ and $z = z_1$, these planes are called *conjugate planes*. Thus, a condition for two planes to be conjugate is the vanishing of B. For two generic values $z = z_0$, $z = z_1$, the plane conjugate of $z = z_0$ can be found by translating the reference plane in the image space, that is, by changing z'_1 in Eq. (II.15.25) until the coefficient B does not vanish.

Analogously, when the system is illuminated by a collimated beam, the vector $\boldsymbol{\rho}_1$ will be independent of $\boldsymbol{\rho}_0$ if $z = z_1$ coincides with the *focal plane* of the instrument. Its position can be found by repeating the procedure of displacing the plane $z = z'_1$ until A does not vanish. Every system admits a couple of *principal* (or *unit*) *planes* such that each point $\boldsymbol{\rho}_0$ is imaged in a point $\boldsymbol{\rho}_1 = \boldsymbol{\rho}_0$. These planes can be found by imposing the conditions $A = 1$ and $B = 0$. The axial point such that an initial ray from that point makes an angle with the optic axis equal to that formed by the final ray is called a *nodal point* and can be determined by imposing the conditions $B = 0$, $D = n_1/n_0$. When $n_1 = n_0$, the nodal points (in the object and image space) lie on the unit planes.

In general, an optical system is characterized by its focal and principal planes. The image of an object O is obtained by means of the geometric

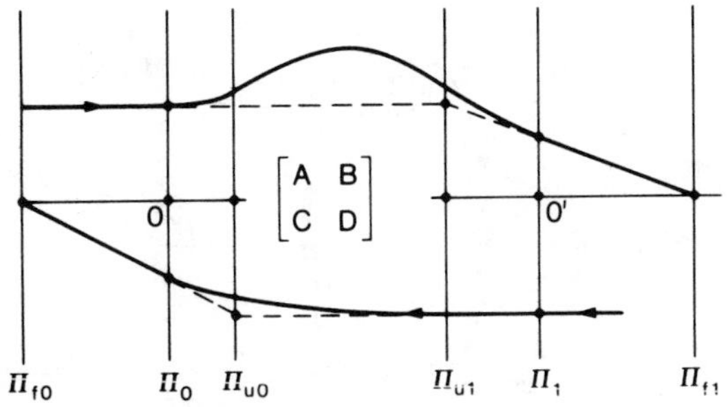

Fig. II.33. Location of the unit planes Π_{u0} and Π_{u1} and of the focal planes Π_{f0} and Π_{f1} of a lens characterized by a ray matrix referred to the generic planes Π_0 and Π_1.

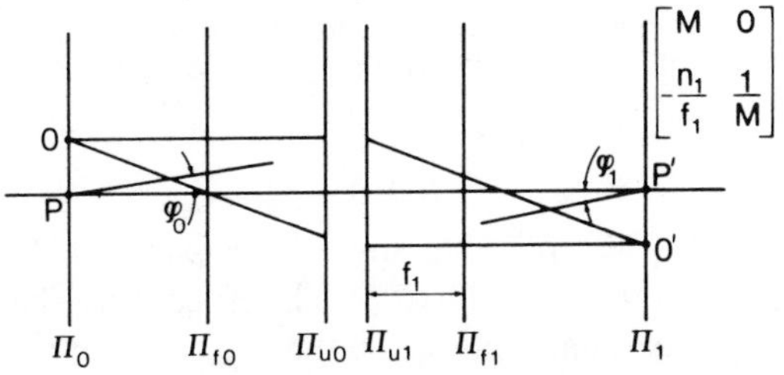

Fig. II.34. Ray matrix relative to two conjugate planes Π_0 and Π_1; M represents the linear magnification of the object on Π_0, while f_1 is the distance of the focal plane Π_{f1} from the unit plane Π_{u1}; f_1 is positive if Π_{f1} is to the right of Π_{u1}.

construction illustrated in Figs. II.33 and II.34. The corresponding matrix is

$$\mathbf{S} = \begin{bmatrix} M & 0 \\ -n_1/f_1 & 1/M \end{bmatrix}, \tag{II.15.26}$$

where M represents the *transverse magnification* of the object and f is the distance of the image focal plane from the relative unitary plane; f is positive if Π_{f1} lies to the right of Π_{u1}. Now, let us indicate with ϕ_0 and ϕ_1 the initial and final angles formed with the optic axis by a paraxial ray passing through P and P'. Then, in view of the above expression for $\mathbf{S}$, we have $\phi_1 = \phi_0 n_1/(M n_0)$. On the other hand, $\rho_1 = M\rho_0$. Combining these two relations yields the so-called *Smith–Helmholtz* or *Lagrange* invariant (see Chapter I, Born and Wolf [11]):

$$n_0 \rho_0 \phi_0 = n_1 \rho_1 \phi_1. \tag{II.15.27}$$

Most often, the optical system will contain a stop that limits the bundles of rays capable of passing through K. In practice, this limitation is sometimes imposed by other obstacles such as lens rims, in which case one speaks of

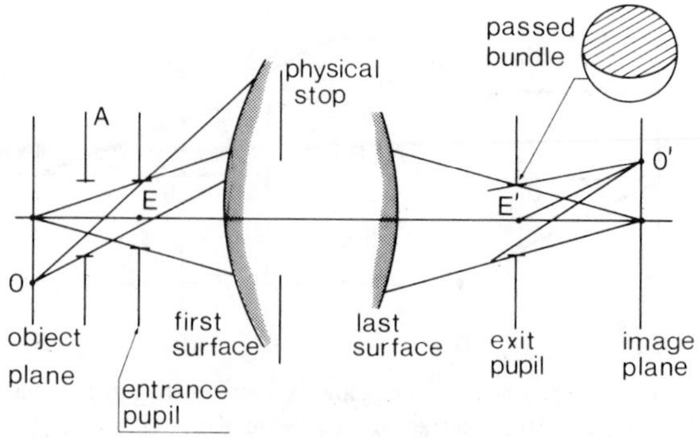

Fig. II.35. Entrance and exit pupils of an optical system. The ray passing through E' and O' is called the principal ray. Another aperture of the system, A, does not limit the bundles of rays from the axial points. When the object O is too far from the axis, the exit pupil appears partially illuminated, as shown in the insert on the right top. This partial illumination is known as *vignetting.*

vignetting (Fig. II.35). The paraxial images by parts of K preceding and following it form the *entrance* and *exit* pupil, respectively. The rays touching the pupil edge are called marginal, while the ray through the object and the center of the exit pupil is called the *principal ray.* This ray plays an important role in the analysis of aberrations.

A quantity frequently used to specify the amount of light power crossing the system is the *numerical aperture* NA. This is defined in the paraxial limit as the product of the refractive index n_0 times the half-angle of the cone of rays collected by the input pupil for an axial object point in O. An analogous quantity can be defined for the image space. By combining the wave vector with the numerical aperture and the coordinates x_0, y_0, z_0 of an object P_0, we can define P_0 through the following nondimensional quantities, called *optical coordinates*:

$$v_x = k_0 x_0 \mathrm{NA}_0, \qquad v_y = k_0 y_0 \mathrm{NA}_0, \qquad \bar{u} = k_0 z_0 \mathrm{NA}_0^2. \quad \text{(II.15.28)}$$

Analogous definitions hold for points in the image space, with NA_0 replaced by $\mathrm{NA}_1 = (n_0/n_1)\mathrm{NA}_0/|M|$, M being the transverse magnification. If the planes $z_0 = 0$ and $z_1 = 0$ are conjugate, it is easy to show that the moduli of the optical coordinates of two conjugate points, respectively close to O and O', are equal.

An optical instrument is characterized by the dimension of the object that it is able to reproduce more or less faithfully. If we indicate by A the area of the object domain and by Ω the instrument angle, we call the quantity

$$U = A\Omega \qquad \text{(II.15.29)}$$

the *étendue* or *light-gathering power*. This is a parameter of great importance in spectroscopic instruments (see Section VII.21.2). In addition, U/λ^2 measures the number of points of the object that can be resolved by the image furnished by an instrument of a given étendue (see Section IV.15.5).

Having shown how far we can go by using Gaussian optics as a reference scheme for analyzing optical systems, we conclude by reminding the reader that the system ray matrix can be constructed by decomposing a lens with a sequence of planes tangent to the vertices of the refracting surfaces. Multiplying in sequence the matrices related to the refracting surfaces and the translation matrices connecting two adjacent planes, we end up with the final matrix of the system. This method of analysis is not new, having been exploited for years in order to design and analyze systems used in accelerator technology for the transport of charged particles. An interesting description of this application is given in Steffen [27].

15.4 *Wave Front Aberrations*

In the analysis of aberrations, initiated in 1856 by L. Seidel, it is convenient to use particular coordinates for the object and image planes and the entrance and exit pupils. They are chosen in such a way that, in the paraxial limit, the coordinates of the intersection points of a ray with the above planes are all coincident. In this way, the variations of these coordinates (*Seidel's coordinates*) for a finite ray are a measure of the deviation from the ideal paraxial trajectory.

We have already defined the aberration of an optical system K as its failure to let a ray from an object point O pass through an image point O'. In order to describe the diffraction effects of the deviation of K from ideal behavior, it is convenient to describe the aberrations as a departure of the wave fronts from some ideal surfaces. In particular, when we consider the imaging of a point O by an axisymmetric system K, we locate its Gaussian image O' at the intersection of the principal ray from O with the conjugate plane of the object plane. Then we consider a wave front W passing through the center E' of the exit pupil and a Gaussian reference sphere W_s of radius R, also passing through E' and having its center in O' (see Fig. II.36). A ray R from O will intersect W and W_s at two points, say A and B, respectively, whose signed distance $[AB] \equiv W_0$ is taken as a measure of the deformation of W and is called the *aberration function* or *retardation of the wavefront* (see Buchdahl [29], p. 106).

We are now faced with the problem of evaluating W_0 by using the characteristic functions of K. In order to simplify the calculations, it is useful to compare the single quantities occurring in the expression for W_0 with the

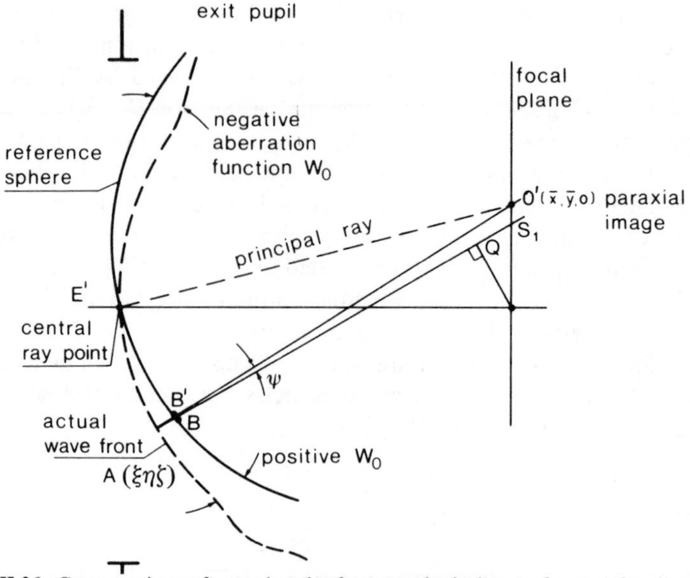

Fig. II.36. Cross sections of actual and reference spherical wave fronts (of radius R). The aberration function W_0 measures the distance $[AB]$ between these two surfaces. The exit pupil determines the domain of definition of the function W_0, which, in particular, vanishes for the principal ray.

parameters p and q of the ray R. In particular, once expanded in a power series of p and q, the mixed characteristic of K takes the form $W = W^{(0)} + W^{(2)} + W^{(4)}$, where $W^{(2n)}$ is a polynomial of order $2n$ in p and q. Because of axial symmetry, no odd terms are present. While $W^{(2)}$ corresponds to paraxial optics, the aberrations are due to terms of order no less than 4. This circumstance is expressed by saying that the part of W that contributes to the aberrations is $O(4)$. Analogously, we say that $[S_1 O']$ is $O(3)$ and so on (see Fig. II.36).

After these preliminaries, we observe that

$$[AQ] - R + p\bar{x} + q\bar{y} = [AS_1] - R - p\,\delta x - q\,\delta y$$
$$= [AS_1] - [BS_1] + O(6) = W_0 + O(6). \qquad \text{(II.15.30)}$$

In fact, $p\,\delta x + q\,\delta y = O(4)$, while $R = [BS_1] - p\,\delta x - q\,\delta y + O(\delta x^2 + \delta y^2)$. Therefore, we can make W_0 coincident with the left side of the above equation for calculating aberrations of the third and fifth order, viz.

$$W_0(X, Y) = W(z_0, z_1; x_0, y_0, p, q) + p\bar{x} + q\bar{y} - R - V(O, E')$$
$$= T(z_0, z_1; p_0, q_0, p, q) - p_0 x_0 - q_0 y_0$$
$$+ p\bar{x} + q\bar{y} - R - V(O, E'), \qquad \text{(II.15.31)}$$

where X and Y are the coordinates of the exit pupil coinciding with $X = Rp/r$ and $Y = Rq/r$.

Now, we can expand T in a power series of $s_0 \equiv u$, $s_1 \equiv v$, and $s_2 \equiv w$. We have already remarked that W_0 must be at least of the fourth order, so that $W_0 = W_0^{(4)} + W_0^{(6)} + \cdots$, where, according to Eq. (II.15.31), $W_0^{(4)} = T^{(4)}$, viz.

$$W_0^{(4)} = \frac{1}{2} \sum_{i,j=0}^{2} T_{ij} s_i s_j$$

By introducing the quantities $\bar{u} = x_0^2 + y_0^2$, $v = p_1^2 + q_1^2$ and $\bar{w} = 2(x_0 p_1 + y_0 q_1)$, Eq. (II.15.32) can be recast as

$$W_0^{(4)} = -\tfrac{1}{4}Bv^2 - \tfrac{1}{4}C\bar{w}^2 - \tfrac{1}{2}D\bar{u}v + \tfrac{1}{2}E\bar{u}\bar{w} + \tfrac{1}{2}Fv\bar{w}, \qquad \text{(II.15.33)}$$

where B, C, D, E, F are known as *primary* or *Seidel's aberration coefficients* [30, 31]. They are associated with defects of the image known as *spherical aberration* (B), *coma* (F), *astigmatism* (C), *curvature of the field* (D) and *distortion* (E).

Problems

Section 2

1. Show that the terms of the Hankel asymptotic series of $H^{(2)}(k_0\rho)$ [Eq. (II.2.12)] satisfy the recurrence relations of Eq. (II.2.9). *Hint*: $\nabla^2 = d^2/d\rho^2 + \rho^{-1} d/d\rho$.

Section 4

2. Show that in the *paraxial limit* (ray bundles slightly deviating from a reference direction) the ray equation reduces to

$$\frac{d^2x}{dz^2} + \frac{\partial \ln n}{\partial z}\frac{dx}{dz} = \frac{\partial}{\partial x}\ln n,$$

$$\frac{d^2y}{dz^2} + \frac{\partial \ln n}{\partial z}\frac{dy}{dz} = \frac{\partial}{\partial y}\ln n.$$

3. Show that in the paraxial limit a ray describes a sinusoidal path in a medium (*lenslike medium*) whose refractive index is given by

$$n^2 = n_0^2 - a(x^2 + y^2).$$

Section 6

4. Show that the parabolic wave equation [Eq. (II.6.12)] can be solved asymptotically by setting

$$A(\mathbf{r}) \cong \sum_{m=0}^{\infty} \frac{A_m(r)}{(-ik_0)^m},$$

where the A_m satisfy the recurrence relations

$$(\partial/\partial z)A_m = -\tfrac{1}{2}\nabla_t^2 A_{m-1}.$$

5. Prove that the asymptotic solutions of Eqs. (II.6.11) and (II.6.12) coincide up to the second order in k_0^{-1}. *Hint*: See Problem 4.

Section 8

6. Consider a current line source $\mathbf{J}(\rho, t) = e^{i\omega t}[\delta(\rho)/2\pi\rho]\hat{z}$ flowing along the z axis. The relative vector potential $\mathbf{A}$ is proportional to $\hat{z}H_0^{(2)}(k_0\rho)$. Calculate $\mathbf{E}$, $\mathbf{H}$, and $\mathbf{S}$ and expand them in asymptotic series by using Eq. (II.2.12).

Section 9

7. Consider a refractive index distribution such that a straight ray can propagate along the z axis. Very near it, $n(\mathbf{r})$ can be approximated by

$$n^2(\mathbf{r}) = n_0^2(z) + f(z)x^2 + g(z)y^2.$$

Show that the eikonal can be expanded in a power series

$$S(\mathbf{r}) = S_0(z) + \tfrac{1}{2}a(z)x^2 + \tfrac{1}{2}b(z)y^2 + c(z)xy + \text{higher powers of } x \text{ and } y,$$

where

$$\frac{dS_0}{dz} = n_0(z), \qquad \frac{dc}{dz} = -(a+b)\frac{c}{n_0},$$

$$\frac{da}{dz} = -\frac{a^2}{n_0} + \frac{f - c^2}{n_0} \quad \text{(Riccati's equation)}, \qquad \frac{db}{dz} = -\frac{b^2}{n_0} + \frac{g - c^2}{n_0}.$$

8. Derive the law of variation of the principal curvature radii *in vacuo* [Eq. (II.9.12)], relying on the transport equations of Problem 7.

Section 10

9. Consider a point source in a medium with refractive index n at a distance t from a plane interface with a medium of index n'. Show that the caustic of the refracted rays is described by the equation

$$\mu^{2/3}z^{2/3} + (\mu^2 - 1)^{1/3}\rho^{2/3} = t^{2/3} \qquad (\mu = n/n'),$$

where z and $\rho = (x^2 + y^2)^{1/2}$ are relative to a coordinate system with origin at the interface and z axis passing through the source perpendicularly to the interface. Show that the caustic is the evolute of an ellipse when $\mu \geqq 1$. Analyze the shape of the wave fronts. *Hint*: See Section X.4 of Ref. [13].

Section 11

10. Calculate the field reflected by a metallic sphere illuminated by a plane wave. Assume a reflection coefficient for the scalar field equal to -1, which means a vanishing total field on the surface.

11. Consider a point source in a dielectric sphere of radius R and refractive index n_1 contained in a medium of refractive index n_0 ($<n_1$). Prove that, when the distance of the source S from the center O of the sphere is equal to Rn_0/n_1, the rays that form an angle $\leq \pi/2$ with the direction SO give rise to refracted rays having a virtual focus at S' at a distance Rn_1/n_0 from O. The perfect conjugate points S and S' are termed *aplanatic points* of the sphere. *Hint*: Use Eqs. (II.11.22) in conjunction with Snell's law.

12. Consider a point source and a ray bundle perpendicular to a window or to a cylindrical lens of given refractive indices and dimensions (see Fig. II.37). Calculate the principal curvature radii along the ray bundle while it crosses the lens and after it does so.

13. Consider *Amici's meniscus* (see Fig. II.38), obtained from a dielectric sphere of radius R and refractive index n_1 immersed in a medium of refractive index n_0 ($<n_1$) by digging a sphere having its center at the aplanatic point S (see Problem 11). Calculate the relation between the *numerical aperture* NA of the incident congruence and that of the refracted one emerging from the meniscus, when the source is in S. The NA is defined as $n \sin\theta_{max}$, where θ_{max} is the angle between the optical axis of the lens and the most inclined ray of the congruence.

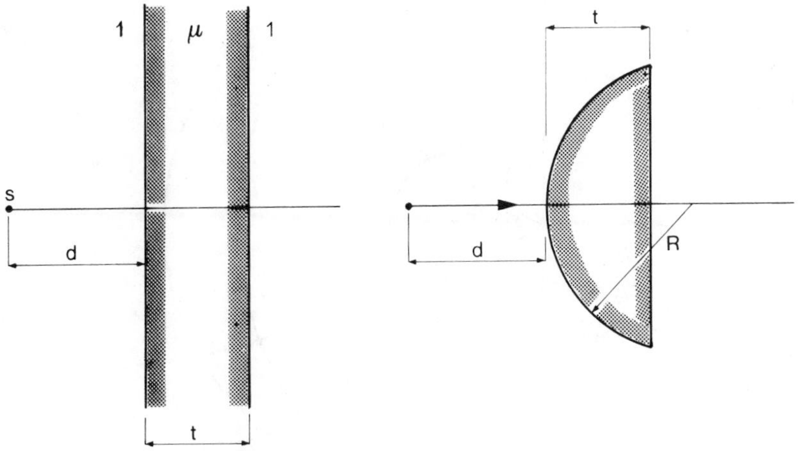

Fig. II.37. Geometry of Problem 12.

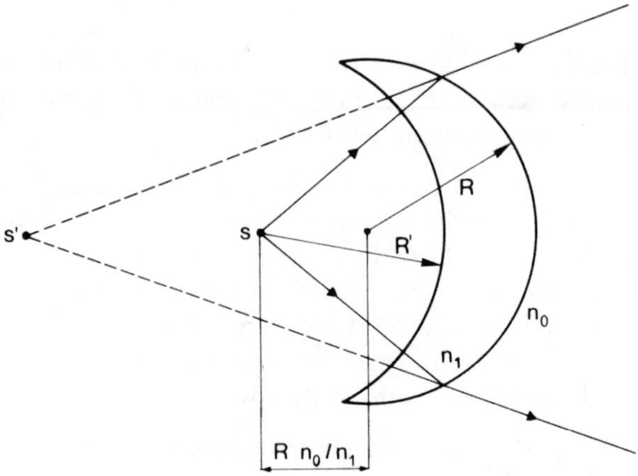

Fig. II.38. Amici's meniscus.

14. Consider the *oil-immersion microscope objective* of Fig. II.39. The object is immersed in a fluid whose refractive index n_{oil} matches that of the spherical first element. The source is an aplanatic point of this sphere whose image is located at the center of curvature of R_2 and coincides with an aplanatic point of R_3. Calculate the numerical aperture of the congruence emerging from Amici's meniscus (see Problem 13) as a function of the input $\text{NA} = n_{\text{oil}} \sin\theta$.

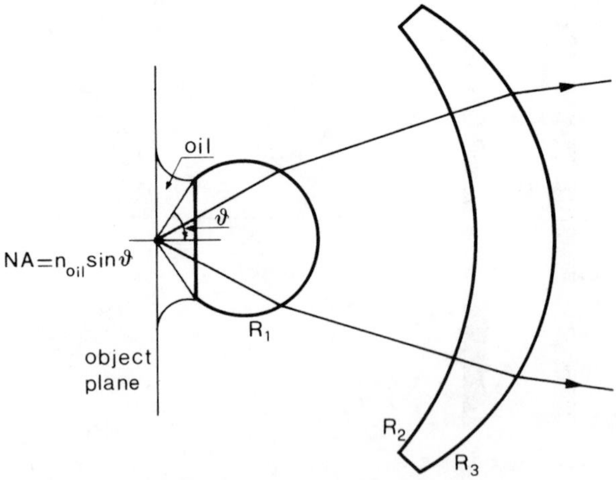

Fig. II.39. Oil-immersion microscope objective.

Section 12

15. Write the ray equation in elliptical cylindrical coordinates (y, μ, ν) [see Eqs. (II.7.10)] by assuming a refractive index constant on each ellipse of the family and independent of y. Find the constant of motion corresponding to the skewness in rotationally symmetrical fibers for an assigned ray. Analyze the modes propagating along the y axis through this elliptically graded optical fiber by studying the relative caustics and the dependence of the propagation constant β on the modal indices.

Section 13

16. Find the circular trajectories with center at the origin for the Maxwell fish-eye [Eq. (II.13.53)] and Luneburg [Eq. (II.13.55)] lenses by using Eq. (II.13.45). Show that the radii are given by $b^{1/2}$ and R, respectively.

17. Consider a ray traveling along the x axis in a Luneburg lens. Using the transport equations of Problem 7, show that the curvature radius ρ of a rotationally symmetric wave front along the x axis obeys the equation

$$\frac{1}{\rho(x)} = \frac{1}{R}\frac{1}{[2-(x/R)^2]^{1/2}}\tan\left(-\sin^{-1}\frac{x}{2^{1/2}R} + \phi\right)$$

where ϕ is an angle related to the initial value of ρ and x is measured from the center of the lens.

18. Prove that the input and output curvature radii of a rotationally symmetric wave front along a diameter of a Luneburg lens fulfill the relation

$$\rho_{\text{in}}\rho_{\text{out}} = -R^2.$$

Section 14

19. Solve Eq. (II.14.19) for a uniform magnetic field (**A** is not uniform!). Show that the trajectory is a helix with axis parallel to **H**. Find the frequency of rotation around **H** (*cyclotron frequency*).

20. Consider a periodically bent *nematic liquid crystal*, in which the components of the dielectric tensor depend only on the z coordinate,

$$\boldsymbol{\varepsilon}(z) = \varepsilon\begin{bmatrix} 1+\delta\cos(\beta z) & 0 & \delta\sin(\beta z) \\ 0 & 1+\delta & 0 \\ \delta\sin(\beta z) & 0 & 1-\delta\cos(\beta z) \end{bmatrix}$$

where $|\delta| \ll 1$ and $\beta = 2\pi/L$, L being the spatial period of the distortion. Show that $\boldsymbol{\varepsilon}$ represents a uniaxial crystal and calculate the direction of the optic axis as a function of z. Then, assuming $L \gg \lambda$ use Eq. (II.14.14a) for calculating the trajectory of an E-ray. *Hint*: See Example 2 of Section II.14.1 (see also Ong and Meyer, bibliography).

21. Consider a ray entering a homogeneous uniaxial crystal and propagating as an E-ray. Assuming the optic axis coplanar with the plane of incidence and using Eq. (II.14.19), obtain the angle of refraction as a function of the angle of incidence. *Hint*: Show that Eq. (II.14.19) generalizes into

$$\tan\theta' = -C/B + (\sin\theta/B)\{(AB - C^2)/(B - \sin^2\theta)\}^{1/2},$$

where

$$A = n_e^2 \sin^2\alpha + n_o^2\cos^2\alpha,\ B = n_o^2\sin^2\alpha + n_e^2\cos^2\alpha,\ C = \sin\alpha\cos\alpha(n_o^2 - n_e^2).$$

22. Use the Christoffel symbols for spherical coordinates to derive the trajectories in a dielectric whose refractive index depends on r. Solve for the Maxwell fish-eye.

23. Analyze the space trajectories relative to a plane wave deflected by the static gravitational field of a star, using spherical coordinates r, θ, ϕ. *Hint*: The *space–time* trajectory of a photon is a *null geodesic* of the metric

$$ds^2 = e^{\lambda(r)}\,dr^2 + r^2\,d\theta^2 + r^2\sin^2\theta\,d\phi^2 - e^{\nu(r)}c^2\,dt^2,$$

where λ and ν are suitable functions describing the effect of the gravitational field [32] ($\lambda = \nu = 0$ in *flat space-time*). As a consequence, the space trajectories can be obtained by introducing an equivalent medium with refractive index

$$n_{\text{eq}}^2(r) = e^{-\nu(r)},$$

and a metric tensor having the components

$$g_{rr} = e^{\lambda(r) - \nu(r)}, \qquad g_{\theta\theta} = r^2 e^{-\nu(r)}, \qquad g_{\phi\phi} = r^2\sin^2\theta\, e^{-\nu(r)}.$$

24. Relying on the results of Problem 27, calculate the trajectory of a ray deviated from a star of mass M representing a *black hole.* Set

$$g_{rr} = (1 - r_g/r)^{-2}, \qquad g_{\theta\theta} = g_{\phi\phi}\sin^{-2}\theta = r^2(1 - r_g/r)^{-1},$$

where $r_g = 2GM/c^2$ and G is the gravitational constant. Show that when the impact parameter $l > 3^{3/2}r_g/2 = l_{\text{crit}}$ the rays are deflected by M, while for $l < l_{\text{crit}}$ they are captured by the gravitational field [32] (see Fig. II.40).

Section 15

25. Calculate the angle characteristic of a spherical lens at the second order in u, v, w. *Hint*: choose the reference planes Π_0 ($z = 0$) and Π_1 ($z = d$) tangent to the spheres, introduce an auxiliary plane Π' coincident, for example, with Π_1, and express T_{lens} in the form

$$T_{\text{lens}} = T(0, d; p_0, q_0, p', q') + T_0(d, d; p', q', p_1, q_1).$$

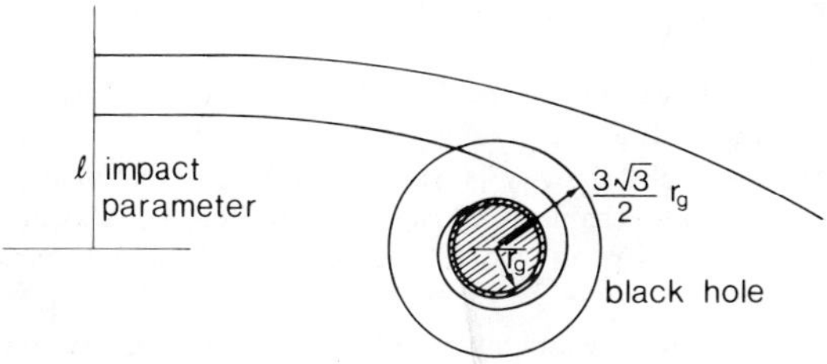

Fig. II.40. Ray trajectories in the presence of a black hole.

Since the intersection with Π' of the ray coming from the left coincides with the intersection with the same plane of the ray coming from the right, then T_{lens} is stationary with respect to p' and q'. These conditions allow us to eliminate p', q' and thus to obtain T_{lens} as an explicit function of p_0, q_0, p_1, q_1 by means of Eq. (II.15.17).

References

1. Luneburg, L. K., "Mathematical Theory of Optics." Univ. of California Press, Berkeley, 1964.
2. Kline, M., and Kay, I. W., "Electromagnetic Theory and Geometrical Optics." Wiley (Interscience), New York, 1965.
3. Felsen, L. B., and Marcuvitz, M., "Radiation and Scattering of Waves." Prentice-Hall, Englewood Cliffs, New Jersey, 1973.
4. Abramowitz, M., and Stegun, I. A., "Handbook of Mathematical Functions." Dover, New York, 1965.
5. Courant, R., and Hilbert, D., "Methods of Mathematical Physics," Vol. 2. Wiley (Interscience), New York, 1962.
6. Stavroudis, O. N., *J. Opt. Soc. Am.* **52**, 187 (1962).
7. Laugwitz, D., "Differential and Riemannian Geometry." Academic Press, New York, 1965.
8. Montagnino, L., *J. Opt. Soc. Am.* **58**, 1667 (1968).
9. Deschamps, G. A., *Proc. IEEE* **60**, 1022 (1972).
10. Ludwig. D., *Commun. Pure Appl. Math.* **19**, 215 (1966).
11. Kneisly, J. A., II, *J. Opt. Soc. Am.* **54**, 229 (1964).
12. Herzberger, M., *Appl. Opt.* **5**, 1383 (1966).
13. Stavroudis, O. N., "The Optics of Rays, Wavefronts and Caustics." Academic Press, New York, 1972.
14. Buchdahl, H. A., *J. Opt. Soc. Am.* **63**, 46 (1973).
15. Streifer, W., and Paxton, K. B., *Appl. Opt.* **10**, 769 (1971); **10**, 1164 (1971).
16. Bogoliubov, N. N., Mitropolsky, Yu. A., and Samoilenko, A. M., "Methods of Accelerated Convergence in Nonlinear Mechanics." Springer-Verlag, Berlin and New York, 1976.
17. Marchand, E. W., "Gradient Index Optics." Academic Press, New York, 1978.
18. Cornbleet, S., "Microwave Optics." Academic Press, New York, 1976.
19. Zatzkis, H., *J. Opt. Soc. Am.* **55**, 59 (1965).
20. Budden, K. G., "Radio Waves in the Ionosphere." Cambridge Univ. Press, London and New York, 1961.

21. Brandstatter, J. J., "An Introduction to Waves, Rays and Radiation in Plasma Media." McGraw-Hill, New York, 1963.
22. Righini. G. C., Russo, V., Sottini, S., and Toraldo di Francia, G., *Appl. Opt.* **12**, 1477 (1973).
23. Sottini, S., Russo, V., and Righini, G. C., *J. Opt. Soc. Am.* **69**, 1248 (1979).
24. Möller, C., "The Theory of Relativity." Oxford Univ. Press, London and New York, 1972.
25. Goldstein, H., "Classical Mechanics." Addison-Wesley, Reading, Massachusetts, 1980.
26. Marcuse, D., "Light Transmission Optics." Van Nostrand-Reinhold, Princeton, New Jersey, 1972.
27. Steffen, K. G., "High Energy Beam Optics." Wiley (Interscience), New York, 1965.
28. Welford, W. T., "Aberrations of the Symmetrical Optical System." Academic Press, New York, 1974.
29. Buchdahl, H. A., "An Introduction to Hamiltonian Optics." Cambridge Univ. Press, London and New York, 1970.
30. Kingslake, R., "Lens Design Fundamentals." Academic Press, New York, 1978.
31. Cagnet, M., Françon, M., and Thrierr, J. C., "Atlas of Optical Phenomena." Springer-Verlag, Berlin and New York, 1962.
32. Zeldovich, Y. B., and Novikov, L. D., "Relativistic Astrophysics," Vol. 1. Chicago Univ. Press, Chicago, Illinois, 1971.

Bibliography

Brouwer, W., "Matrix Methods in Optical Instruments Design," Benjamin, New York, 1964.
Chrétien, H., "Calcul des Combinaisons Optiques," Masson, Paris, 1980.
Cornbleet, S., "Microwave and Optical Ray Geometry." Wiley, New York, 1984.
Flugger, S., ed., "Handbuch der Physik," Vol. 24. Springer-Verlag, Berlin, 1967.
Focke, J., *Prog. Opt.* **4**, 1–36 (1965).
Gerrard, A., and Burch, J. M., "Introduction to Matrix Methods in Optics," Wiley, New York.
Herzberger, M., "Modern Geometrical Optics." Wiley (Interscience), New York, 1958.
Kravtsov, Yu. A., and Orlov, Yu. I., *Sov. Phys. Usp.* (*Engl. Transl.*) **26**, 1083 (1983).
Mu, G., and Zhan, Y., "Optics," The People's Education Publishing House, Peking, 1978. (Chinese.)
Ong, H. L., and Meyer, R. B., *J. Opt. Soc. Am.* **A2**, 198 (1985).
Shaomin, W., *Opt. Quantum Electron.* **17**, 1 (1985).
Slyusarev, G. G., "Aberration and Optical Design Theory." Adam Hilger, Bristol, 1984.
Smith, W. J., "Modern Optical Engineering." McGraw-Hill, New York, 1966.
Welford, W. T., and Winston, R., "The Optics of Nonimaging Concentrators," Academic Press, New York, 1980.

Chapter III

Plane-Stratified Media

1 Introduction

Refractive index discontinuities are the most recurrent example of inhomogeneities encountered in optics. They produce reflected waves that interfere with the incident one and produce complicated interference patterns. In most cases it is necessary to account for an infinite number of multiple reflections undergone by an incident beam to calculate the amplitude of the waves reflected or transmitted by an optical system. In the language of ray optics this amounts to representing the total field as the superposition of an infinite sequence of ray fields. This circumstance marks the main difference from the cases considered before. In particular, it means that new methods must be used that encompass the difficulty of dealing with an infinite sequence of ray fields and answer the main questions about the amplitudes of the reflected and transmitted beams. In order to give an idea of the practical relevance of these methods, we list below a few examples of existing applications, in which a modulated dielectric constant produces reflected or transmitted waves having amplitudes sensitive to the operation frequency.

Coating glass surfaces with thin dielectric films [1] in order to improve the performance of optical instruments by increasing their transparency and reducing ghost images is a common practice. The advantages of this antireflection treatment become more pronounced as the number of surfaces increases. Semitransparent thin films are also used to increase the light reflected by a glass surface [2]. These dielectric mirrors play an important role in laser cavities and interferometers. They also make it possible to design filters that limit the transmission to a narrow frequency spectrum.

The development in recent years of synchrotron radiation sources has required the fabrication of mirrors for the far-ultraviolet region of the spectrum [3]. It is known that below 300 Å no material has a practically useful reflectivity unless it is used at grazing incidence. In this case only stacks of thin films can be used as reflectors. Dielectric mirrors for wavelengths down to a few angstroms have been designed, and have shown a reflectivity greater than 50% [4].

Photographic material may present a periodic structure imposed on it by exposure to two (or more) interfering monochromatic beams and subsequent processing, so that the film ends up with a periodicity that depends on the wavelength and the angle formed by the waves [5]. These systems give resonant-type reflection or transmission and can replace traditional gratings by virtue of their higher efficiency and resolution [6] (see discussion on holographic gratings, Section VI.10).

The above technique is used for recording the three-dimensional shape of objects in *thick holograms* by exposing a thick emulsion of a high-resolution photographic plate to the superposition of a *reference* spherical beam and the coherent field scattered by the object to be reconstructed.

The simplest example of an inhomogeneous medium is that in which the refractive index n changes only along one direction [7, 8]. In this case the medium is stratified, its dielectric properties being constant on every plane perpendicular to the stratification axis. More complex situations occur where the dielectric constant is constant on families of coaxial cylinders, which happens in graded-index fibers, discussed in Chapter VIII. However, plane-stratified media are by far the most common structures. This is why the present chapter is devoted only to them. In particular, the following sections are dedicated to media exhibiting *slow*, *stepwise*, and *sinusoidal* modulations of the index.

Although the focus of this chapter is on propagation in the stratification z direction, some attention will be paid to lateral waves that propagate *parallel* to a thin film (x or y direction, see Fig. III.1). These *surface* and *leaky* waves show that the structure can be designed to allow the propagation of waves

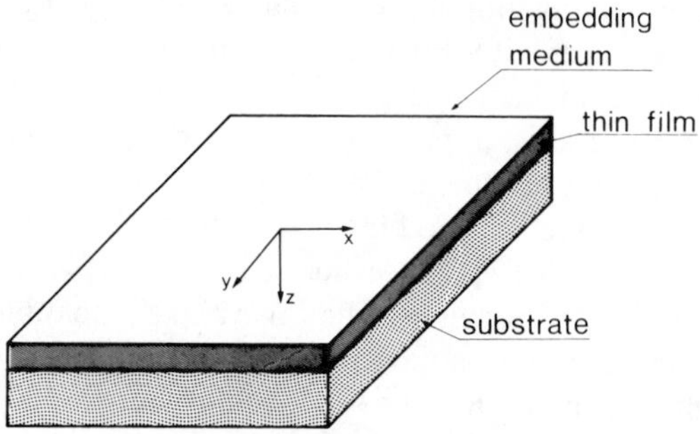

Fig. III.1. Planar film-guide structure.

confined inside the stratified region, which can have a thickness of a few wavelengths. This property is currently used in making optical planar waveguides and other devices used in *integrated optics* [9].

Propagation of surface waves can also be used as a tool for investigating the physical properties of the thin films. This is particularly true for metals, where the parameters characterizing the *surface plasmons* can be determined by exciting and analyzing the surface waves supported by thin films [10].

This chapter is essentially divided into two parts, the shorter of which deals with media exhibiting a scale of variation of n large compared with the wavelength, while the longer one discusses the opposite situation. The first part completes the analysis of graded-index media begun in Chapter II, by studying the representation of the field in proximity to the critical regions (*caustics* or *turning points*). Practical examples are provided at optical frequencies by graded-index multimode optical fibers. The second part is essentially dedicated to the analysis of piecewise-constant refractive index profiles.

Very little is said about the synthesis of optical filters, a problem requiring a mathematical apparatus that is irreducible to the size of the present chapter. The interested reader should consult the specialized books, some of which are listed in the references.

2 Ray Optics for Stratified Media

We wish to consider a refractive index distribution independent of x and y, corresponding to an *unbounded stratified medium*. The refractive index $n(z)$ tends to the constant values n_0 and n_1, respectively, as z approaches $-\infty$ and $+\infty$ (see Fig. III.2a). We assume a field originating at $z = -\infty$ initially represented by a plane wave u_i forming an angle θ with the z axis:

$$u_i = e^{-ik_0 n_0 (z\cos\theta + x\sin\theta)}, \tag{III.2.1}$$

having adopted a unitary initial amplitude without loss of generality with respect to the considerations that follow.

The standard way to study the associated propagation problem consists of looking for solutions of the wave equation in the frame of some approximation scheme, exact analytical solutions being obtainable only for some particular classes of $n(z)$. In the Wentzel–Kramers–Brillouin (*WKB*) method [11], for example, one looks for an approximate solution that is asymptotic in the parameter $\varepsilon = (dn/dz)/(k_0/n)$, whose smallness expresses the condition of slow variation of the refractive index on a scale comparable to the wavelength. In this section we adopt, as an application of the formalism developed in Chapter II, the ray optics approach, which leads, to the lowest significant order in $1/k_0$, to the same results as the WKB method.

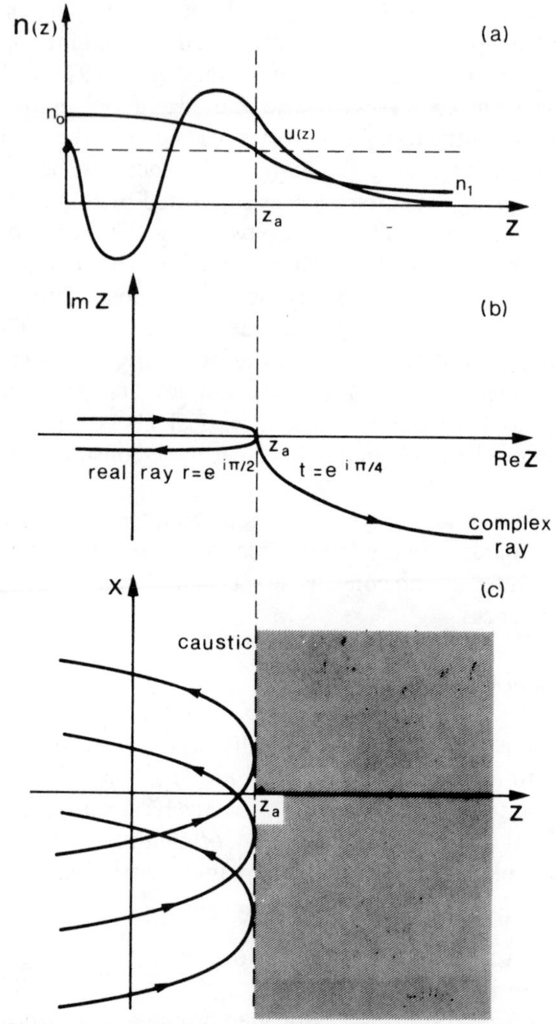

Fig. III.2. Reflection and transmission of a plane wave at the caustic of a plane-stratified medium. (a) Refractive index profile and field distribution in proximity of the caustic at $z = z_a$; (b) complex ray trajectories; (c) trajectories of the rays in the x,z plane.

The ray equation in the form of Eq. (II.4.10) can be solved by using cartesian coordinates, which yields

$$d^2z/d\tau^2 = (d/dz)(\tfrac{1}{2}n^2), \qquad d^2x/d\tau^2 = d^2y/d\tau^2 = 0. \tag{III.2.2}$$

Thus, $n\,dx/ds$ and $n\,dy/ds$ are constants of motion, and we can conveniently choose the direction of the x axis in such a way that all the rays of an initially plane wave lie on planes orthogonal to the y axis, as we implicitly did in writing Eq. (III.2.1). Therefore, $dy = 0$ and

$$d\tau \equiv ds/n = (dx^2 + dz^2)^{1/2}/n. \tag{III.2.3}$$

On the other hand, Eq. (III.2.2) gives $d\tau = dx/a$ (a being a constant), so that it reduces to

$$dx/dz = a/(n^2 - a^2)^{1/2}. \tag{III.2.4}$$

For $z \to -\infty$, Eqs. (III.2.4) and (III.2.1) imply that $\tan\theta = a(n_0^2 - a^2)^{-1/2}$, which yields

$$a = n_0 \sin\theta. \tag{III.2.5}$$

Equation (III.2.4) determines any ray in the form

$$x = x_0 + \int_{z_0}^{z} \frac{a}{(n^2 - a^2)^{1/2}}\, dz', \tag{III.2.6}$$

where x_0 and z_0 are the coordinates of a generic point of the trajectory.

The eikonal reads

$$S = S_0 + a(x - x_0) + \int_{z_0}^{z} (n^2 - a^2)^{1/2}\, dz', \tag{III.2.7}$$

S_0 being a constant. In fact, this relation agrees with Eq. (II.12.5) and describes a wave whose propagation direction forms an angle θ with the z axis at $z = -\infty$. According to Eq. (III.2.7),

$$\nabla^2 S = \frac{1}{(n^2 - a^2)^{1/2}} \frac{d}{dz}\left(\frac{1}{2} n^2\right). \tag{III.2.8}$$

Furthermore, since $ds/n\,dz = (n^2 - a^2)^{1/2}$ [see Eqs. (III.2.3,4) and (II.5.6)], we can write

$$A_0(\mathbf{r}) = \exp\left[-\frac{1}{4}\int_{z_0}^{z} \frac{1}{(n^2 - a^2)}\left(\frac{d}{dz'} n^2\right) dz'\right] = \frac{(n_0 \cos\theta)^{1/2}}{(n^2 - a^2)^{1/4}}, \tag{III.2.9}$$

having put $z_0 = -\infty$.

Finally, the ray field is easily obtained with the help of Eq. (III.2.7) in the form

$$u(\mathbf{r}) = A_0(\mathbf{r}) e^{-ik_0 S(\mathbf{r})} = \frac{(n_0 \cos\theta)^{1/2}}{(n^2 - a^2)^{1/4}}\, \bar{u}_i(\mathbf{r}) e^{-i\delta_i}, \tag{III.2.10}$$

where $\bar{u}_i(\mathbf{r})$ represents the plane wave of Eq. (III.2.1), supposed to travel unperturbed through a uniform medium with refractive index n_0, and δ_i is the phase delay of u with respect to $\bar{u}_i$ associated with the refractive index variation

$$\delta_i(z) = k_0 \int_{-\infty}^{z} [(n^2 - a^2)^{1/2} - n_0 \cos\theta]\, dz'. \tag{III.2.11}$$

The above field is singular when $n^2 - a^2 = 0$. The corresponding value z_a, for which $n^2(z_a) = a^2 = n_0^2 \sin^2\theta$, is a *turning point* for the wave equation. The plane $z = z_a$, if any, is the caustic associated with the congruence of all the rays sharing a common value of $\sin\theta$.

In order to study the field near z_a, let us go back to the wave equation. Our initial conditions allow us to look for a solution of Eq. (I.1.12) of the kind

$$u = e^{-ik_0 n_0 x \sin\theta} f(z), \tag{III.2.12}$$

which, when inserted into Eq. (I.1.12) yields

$$\partial^2 u/\partial z^2 + k_0^2(n^2 - a^2)u = 0. \tag{III.2.13}$$

For our purpose, it is convenient to put the solution of Eq. (III.2.13) in the form

$$u = u_+ \exp\left[-ik_0 \int^{z} (n^2 - a^2)^{1/2}\, dz' - ik_0 n_0 x \sin\theta\right]$$

$$+ u_- \exp\left[ik_0 \int^{z} (n^2 - a^2)^{1/2}\, dz' - ik_0 n_0 x \sin\theta\right], \tag{III.2.14}$$

where u_+ and u_- are two functions of z that obey the relation

$$\pm \frac{i}{k_0}\frac{d^2 u_\pm}{dz^2} + 2(n^2 - a^2)^{1/2}\frac{du_\pm}{dz} + u_\pm \frac{d}{dz}(n^2 - a^2)^{1/2} = 0, \tag{III.2.15}$$

as easily seen by inserting Eq. (III.2.14) into Eq. (III.2.13). As $k_0 \to \infty$, we obtain

$$2(n^2 - a^2)^{1/2}\frac{du_\pm}{dz} + u_\pm \frac{d}{dz}(n^2 - a^2)^{1/2} = 0, \tag{III.2.16}$$

so that u_+ and u_- are two slowly varying functions of z [all fast dependence having been included in the exponential of Eq. (III.2.14)] of the type $u_\pm \propto (n^2 - a^2)^{-1/4}$.

Thus we recognize that the limit $k_0 \to \infty$ in Eq. (III.2.14), represents a ray field, provided that the term in u_- is identified with a wave traveling in the negative z direction (WKB solution).

3 Matched Asymptotic Expansion: Langer's Method

In order to eliminate the nonphysical behavior of the field at $z = z_a$, we can resort to the method of *matched asymptotic expansions* [11] for constructing a global approximation to the solution of a differential equation having turning points. The method consists of joining together various WKB (ray optics) approximations that hold true in their respective regions of validity. In Chapter V we will discuss an analogous approach for the evaluation of the diffraction integrals in the *transition regions*.

To begin with, let us replace $k_0^2(n^2 - a^2)$ in Eq. (III.2.13) with

$$-\gamma(z - z_a) \equiv k_0^2(z - z_a)\left.\frac{dn^2}{dz}\right|_{z=z_a}, \tag{III.3.1}$$

so that Eq. (III.2.13) reduces, in proximity to z_a, to the *Airy equation* (see Chapter II, ref. [4])

$$\partial^2 u/\partial z^2 - \gamma(z - z_a)u = 0. \tag{III.3.2}$$

This is a particular example of a general class of equations with turning points of order m

$$\partial^2 u/\partial z^2 - \gamma(z - z_a)^m u = 0, \tag{III.3.3}$$

whose solution is of the type

$$u(z) \propto \xi^{1/2}\gamma(z - z_a)^{-1/4}J_{\pm n}(\xi), \qquad \xi = \int_{z_a}^{z} \gamma^{1/2}(z' - z_a)^{m/2}\,dz', \tag{III.3.4}$$

where J_n is a Bessel function of fractional order $n = 1/(m + 2)$. For $m = 1$, Eq. (III.3.4) reduces to Eq. (III.3.2) and its solution, if we remember Eq. (III.2.14), is

$$u = c_1 e^{-ik_0 n_0 x \sin\theta}\,\mathrm{Ai}[k_0\eta^{1/3}(z - z_a)] + c_2 e^{-ik_0 n_0 x \sin\theta}\,\mathrm{Bi}[k_0\eta^{1/3}(z - z_a)], \tag{III.3.5}$$

where c_1 and c_2 are constant quantities, $\eta = \gamma/k_0^3$ is a nondimensional parameter, and Ai and Bi are the *Airy functions*.

We observe that our assumption of a wave coming from $z = -\infty$ implies that the half-space $z < z_a$ is the lit zone corresponding to a real eikonal. Thus, for $z < z_a$, $n^2(z) > a^2 = n^2(z_a)$, so that γ and η are positive quantities.

While both *transition functions* Ai and Bi oscillate for negative arguments, for positive arguments Ai decreases and Bi increases monotonically. Thus, for physical reasons, we set $c_2 = 0$ and u takes the form depicted in the top diagram of Fig. III.2.

Let us now compare the two expressions for u given by Eqs. (III.2.14) and (III.3.5). The first one represents the ray field and provides a satisfactory approximation far enough from the caustic $z = z_a$; the second one is a solution of the wave equation whose validity is limited to values of $|z - z_a|$ small enough to justify Eq. (III.3.1). If the regions of validity overlap, which happens for realistic distributions $n(z)$, we can match the two solutions and obtain relevant information about the behavior of our field. More precisely, if we use the asymptotic expression for Ai for large arguments (see Chapter II, Abramowitz and Stegun [4])

$$\mathrm{Ai}(\zeta) = \frac{|\zeta|^{-1/4}}{2\sqrt{\pi}} \begin{cases} e^{-(2/3)\zeta^{3/2}} & (\zeta > 0), \quad \text{(III.3.6a)} \\ 2\sin[(2/3)|\zeta|^{3/2} + \pi/4] & (\zeta < 0), \quad \text{(III.3.6b)} \end{cases}$$

Eq. (III.3.5) yields

$$u \sim \frac{c_1}{2\sqrt{\pi}} \frac{e^{-ik_0 n_0 x \sin\theta}}{[\eta^{1/3} k_0 |z - z_a|]^{1/4}} \begin{cases} e^{-(2/3)(\eta^{1/2} k_0^{3/2})(z - z_a)^{3/2}} & (z > z_a) \quad \text{(III.3.7a)} \\ e^{-i\pi/4}\{e^{i(2/3)\eta^{1/2}[k_0(z_a - z)]^{3/2}} \\ \quad + e^{-i(2/3)\eta^{1/2}[k_0(z_a - z)]^{3/2}} e^{i\pi/2}\} & (z < z_a), \end{cases}$$

(III.3.7b)

which represent the field in the region where the two approximations provide the same result. Accordingly, when a wave approaches a turning point, a reflected wave represented by the second term on the right side of Eq. (III.3.7b) is always present. This field has the same amplitude as the incident one and is anticipated in time by $\pi/2\omega$ (see Section II.10 and Fig. II.17).

3.1 *Transmitted and Reflected Fields*

In agreement with the above discussion, the reflected and transmitted ray fields are obtained by using the following rules: when a real ray approaches the caustic $z = z_a$, it becomes tangent to the caustic and splits into a real reflected ray and a complex one that penetrates the dark region $z > z_a$ and corresponds to an evanescent wave (see Fig. III.2b); the reflected amplitude is initially multiplied by $e^{i\pi/2}$ and the transmitted one by $e^{i\pi/4}$. We observe that the incident and reflected rays are symmetrical with respect to the z axis, so that, in analogy with Eq. (III.2.14), the reflected ray field is given by

$$u_r(\mathbf{r}) = \frac{(n_0 \cos\theta)^{1/2}}{(n^2 - a^2)^{1/4}} \bar{u}_r(\mathbf{r}) e^{-i\delta_r} r, \qquad \text{(III.3.8)}$$

where $\bar{u}_r$ represents a uniform wave that is "specularly symmetrical" to $\bar{u}_i$ with respect to the caustic plane and has the same value (amplitude and phase) as $\bar{u}_i$, $r = e^{i\pi/2}$ plays the role of a reflection coefficient (see Fig. III.2b), and δ_r is the phase delay accumulated by the field with respect to $\bar{u}_i$ and $\bar{u}_r$ in traveling from $-\infty$ to the caustic and back to z. For δ_r we obtain, in analogy with

Eq. (III.2.11),

$$\begin{aligned}\delta_r(z) &= k_0 \int_{-\infty}^{z} [(n^2 - a^2)^{1/2} - n_0 \cos\theta]\, dz' \\ &\quad + 2k_0 \int_{z}^{z_a} [(n^2 - a^2)^{1/2} - n_0 \cos\theta]\, dz' \\ &= \delta_i(z) + 2k_0 \int_{z}^{z_a} [(n^2 - a^2)^{1/2} - n_0 \cos\theta]\, dz'. \end{aligned} \tag{III.3.9}$$

In an analogous way, the evanescent transmitted ray field u_t for $z > z_a$ obeying Eq. (III.3.7a) in the overlapping region is

$$u_t(\mathbf{r}) = \frac{(n_0 \cos\theta)^{1/2}}{(a^2 - n^2)^{1/4}} \bar{u}_i(\mathbf{r}) e^{-i\delta_t} t, \tag{III.3.10}$$

where $t = e^{i\pi/4}$ and $\delta_t = \delta_t' - i\delta_t''$ is given by an expression similar to that for δ_r. The quantity δ_t'' corresponds to the exponentially decreasing behavior of the field in the dark region (see also Jacobson [12]).

3.2 *Transition from the Forbidden Zone to the Permitted One*

Let us consider the opposite case of transition from the forbidden zone to the permitted one. More precisely, we assume that the half-space $z > z_a$ is the lit region and that an evanescent wave originates in the dark region $z < z_a$. Since only one progressive wave exists for $z > z_a$, the solution of the Airy equation must be expressed by a suitable combination of Ai and Bi, which reads for large arguments (see Chapter II, Abramowitz and Stegun [4])

$$\mathrm{Bi}(\zeta) \sim (1/\pi^{1/2})|\zeta|^{-1/4} \cos[(2/3)|\zeta|^{3/2} + \pi/4] \qquad (\zeta < 0), \tag{III.3.11}$$

and the correct combination furnishing the transition function is

$$u \propto -i\,\mathrm{Ai}[k_0 \eta^{1/3}(z - z_a)] + \mathrm{Bi}[k_0 \eta^{1/3}(z - z_a)], \tag{III.3.12}$$

which is immediately obtained with the help of Eqs. (III.3.5,6a). (In the present situation the parameter η is negative, and thus $\zeta < 0$ corresponds to the lit zone $z > z_a$). From Eqs. (III.3.12) and (III.3.6a) and (III.3.11) we derive an initial phase factor $e^{-i\pi/4}$ in the progressive transmitted wave. On the other hand, for large arguments

$$\mathrm{Bi}(\zeta) \sim (1/\pi^{1/2})|\zeta|^{-1/4} e^{(2/3)\zeta^{3/2}} \qquad (\zeta > 0). \tag{III.3.13}$$

Since the wave originating in the dark region is represented in the transition zone by Bi, Eq. (III.3.12) implies that a second complex ray originates by reflection, its initial amplitude being $\frac{1}{2}e^{-i\pi/2}$ times that of the evanescent field Bi (see Fig. III.3, in which the lit region is $z < z_a$).

3.3 *Dark Barrier*

We can now consider the case in which there are two turning points z_a and $z'_a > z_a$, the *dark barrier* being the interval $z_a < z < z'_a$, where $n(z)$ has a dip. The above results allow us to look for a ray field in the form (see Fig. III.4)

$$u(\mathbf{r}) = \frac{(n_0 \cos\theta)^{1/2}}{(n^2 - a^2)^{1/4}}$$

$$\times \begin{cases} \bar{u}_r(\mathbf{r})e^{-i\delta_r}r_b + \bar{u}_i(\mathbf{r})e^{-i\delta_i} & (z < z_a) \quad \text{(III.3.14a)} \\ \bar{u}_i(\mathbf{r})e^{-i\delta'_t}t_b & (z > z'_a) \quad \text{(III.3.14b)} \\ \bar{u}_i(\mathbf{r})e^{-i\delta'_t}t_b[e^{i\pi/4+\alpha-\delta''_t} + \frac{1}{2}e^{-i\pi/4-\alpha+\delta''_t}] & (z_a < z < z'_a) \quad \text{(III.3.14c)} \end{cases}$$

with

$$\alpha = \delta''_t(z'_a) = k_0 \int_{z_a}^{z'_a} (a^2 - n^2)^{1/2}\, dz'.$$

The assumption of the absence of a wave coming from $z = +\infty$, which justifies the form of Eq. (III.3.14b), implies a ray field in the dark region in the form of Eq. (III.3.14c), because this behavior is consistent with the previous results concerning the transition between the forbidden and the permitted zone. In order to determine the ray field for $-\infty < z < z_a$, we observe that the term $[e^{i\pi/4+\alpha-\delta''_t} + \frac{1}{2}e^{-i\pi/4-\alpha+\delta''_t}]$ appearing in Eq. (III.3.14c) must be substituted, near the first turning point, by a term containing the transition function $e^{i\pi/4+\alpha}\mathrm{Ai} + (1/4)e^{-i\pi/4-\alpha}\mathrm{Bi}$, as shown by the expressions for Ai and Bi for large arguments [Eqs. (III.3.6a) and (III.3.13)]. Following the method already used in this section, we move continuously to the left of z_a. Next, we express Ai and Bi again as in Eqs. (III.3.6b,11), and we impose the coincidence of this asymptotic approximation with Eq. (III.3.14a). In doing so, we find that r_b and t_b (*reflection and transmission coefficients*) are given by

$$r_b = e^{i\pi/2}(4 - e^{-2\alpha})/(4 + e^{-2\alpha}), \qquad t_b = 4e^{-\alpha}/(4 + e^{-2\alpha}) \qquad \text{(III.3.15)}$$

and obey the relation $|r_b|^2 + |t_b|^2 = 1$, which expresses the conservation of energy flux along the z propagation direction.

For completeness, we note that the same result can be obtained by considering the reflected and transmitted beams as the sums of infinite

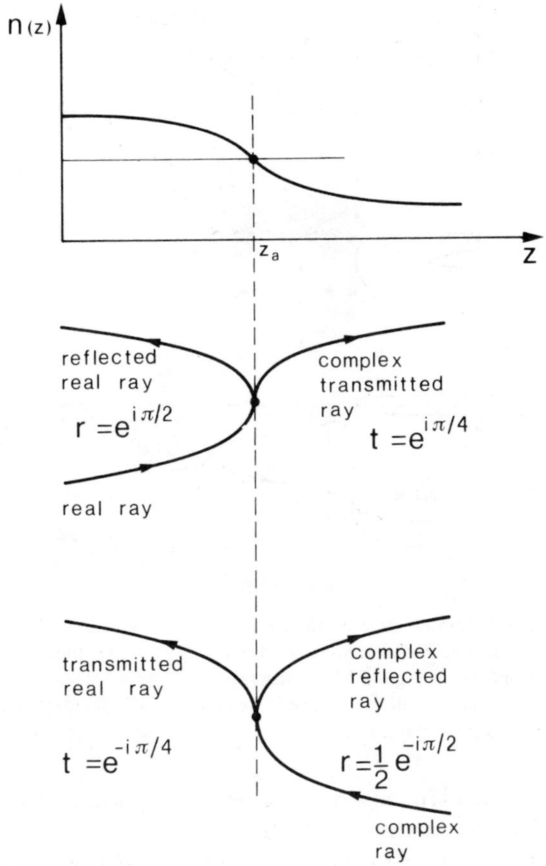

Fig. III.4. Intensity distribution (lower curve) in proximity to a refractive-index well (upper curve) hit by a plane wave from the left.

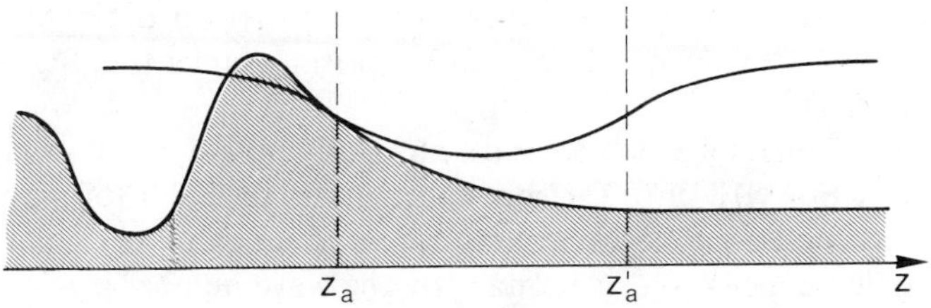

Fig. III.4. Intensity distribution (lower curve) in proximity to a refractive-index well (upper curve) hit by a plane wave from the left.

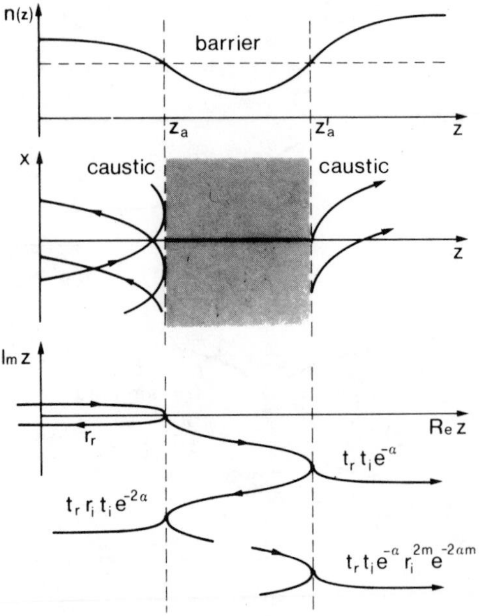

Fig. III.5. Schematic representation of the changes produced in the field distribution by a refractive-index well (barrier). The bottom diagram represents the multiple reflections undergone by a ray entering from the lit region. The transmitted and reflected field amplitudes can be obtained by summing the contributions of the infinite sequence of fields labeled with the index *m*. The subscripts r and i refer to z_a and z'_a.

contributions due to a kind of bouncing between z_a and z'_a of the evanescent wave "trapped" in the barrier (Fig. III.5).

Of course, if $z'_a \to \infty$, we have $\alpha \to \infty$, i.e., $t_b = 0$ and $r_b = \exp(i\pi/2)$, in agreement with Eq. (III.3.8). In general, t_b is a nonvanishing quantity; this is due to a partial leakage of energy through the dark region, which parallels the well-known *tunneling effect* of quantum mechanics. The amount of this effect depends on the quantity α, which measures the strength of the gap between the two sheets of the caustic $z = z_a$ and $z = z'_a$. The leakage gives rise to the so-called *frustrated total reflection*, exploited in integrated optics for exciting waves in a thin-film waveguide by using a prism coupler (see Fig. III.6 and Section III.20.1).

We observe that, for $\alpha = 0$, we must obviously obtain $r_b = 0$ and $t_b = 1$, in contrast with Eqs. (III.3.15). This is connected with the failure of the matching approach whenever the dark barrier is not large enough.

The explicit dependence of r_b and t_b on the wave number is related to the role of Ai and Bi in connection with the peculiar situation near the turning points.

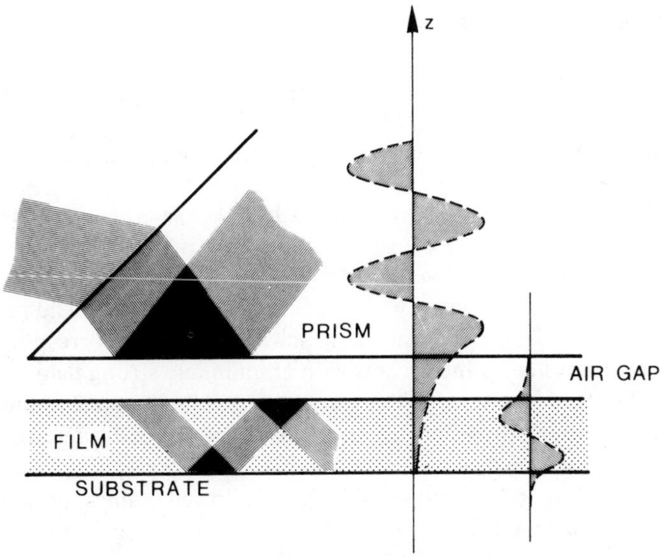

Fig. III.6. Prism coupler used to excite a guided wave on a film substrate by exploiting the frustrated total reflection effect.

Finally, we wish to consider the case of an *evanescent* wave $u^{(e)}$ originating at $z = -\infty$. This assumption is accounted for by the substitution $\theta \to \pi/2 + i\theta''$, from which the reader can derive

$$u^{(e)}(\mathbf{r}) = \frac{(n_0 \sinh \theta'')^{1/2}}{(n_0^2 \cosh^2 \theta'' - n^2)^{1/4}} \bar{u}_i^{(e)}(\mathbf{r}) e^{-i\delta^{(e)}}, \tag{III.3.16}$$

where $\bar{u}_i^{(e)}(\mathbf{r}) = \exp(-k_0 n_0 z \sinh \theta'' - ik_0 n_0 x \cosh \theta'')$, the phase delay $\delta^{(e)}$ being

$$\delta^{(e)}(z) = -ik_0 \int_{-\infty}^{z} [(a^{(e)2} - n^2)^{1/2} - n_0 \sinh \theta''] \, dz', \tag{III.3.17}$$

with $a^{(e)} = n_0 \cosh \theta''$.

In conclusion, when a complex ray approaches a turning point z_a, it splits into a complex reflected ray and a real transmitted one.

3.4 *Duct*

Whenever there are two turning points z_a and z_a' limiting a duct—that is, an interval where $n(z)$ has a bump—a situation comparable to that discussed above is produced. In particular, the matching procedure provides the reflection coefficient r_d in the form

$$r_d = \frac{e^{-i\pi/2}}{2}\frac{1 - e^{-2i\phi}}{1 + e^{-2i\phi}} = \frac{1}{2}\tan\phi, \tag{III.3.18}$$

where ϕ is the phase delay of the (real) ray in traveling from z_a to z'_a,

$$\phi = k_0 \int_{z_a}^{z'_a} (n^2 - n_0^2 \cosh^2\theta'')^{1/2}\, dz'. \tag{III.3.19}$$

We now observe that r_d diverges when $\phi = (n + 1/2)\pi$. In intuitive terms, this means that a small incident wave can excite a very strong field inside the duct. In other words, a situation for which $\phi = (n + \frac{1}{2})\pi$ corresponds to a *resonance* of the system, which reacts with an infinitely strong field to a finite excitation. The field generated in this way is referred to as a *mode* of the system propagating in the x direction. In particular, Eq. (III.3.19) determines the values of θ'' for which the modes can be excited. Since $k_t \equiv n_0 k_0 \cosh\theta''$ is uniquely related to θ'', the discrete set of allowed values of k_t is also determined. Each mode field is more or less concentrated in the lit region $z_a < z < z'_a$, and the system behaves as a dielectric waveguide propagating a disturbance along the x axis with possible propagation constants k_t.

4 Reflection and Transmission for Arbitrarily Inhomogeneous Media

According to the previous section, reflection occurs only in the presence of turning points. This conclusion is manifestly wrong, since we know that a discontinuity of the refractive index produces a reflected beam in every condition. The discrepancy originates from the fact that the matched asymptotic expansion technique accounts correctly for the field distribution in proximity to a turning point, but leaves outside the analysis of the field in regions where the refractive index changes notably on a scale comparable with the wavelength. Here we will follow an approximate method based on the second-order WKB solution of the wave equation [13]. As will become clear later on, the reflection depends on the field polarization, so that we must approach the problem from a vector point of view.

Drawing on Alexopulos and Uslenghi [13], let us consider the plane wave

$$\mathbf{E}_i = [-\sin(\theta)\cos(\beta)\hat{z} + \sin(\beta)\hat{y} + \cos(\theta)\cos(\beta)\hat{x}]e^{-ik_0(x\sin\theta + z\cos\theta)}, \tag{III.4.1}$$

which propagates in the free half-space $z < 0$ and is incident upon the interface $z = 0$ separating free space from the inhomogeneous half-space $z > 0$, in which the dielectric constant is an unspecified function of the normalized coordinate

$\xi = z/a$, where a represents an appropriate scaling length. In Eq. (III.4.1) β is the polarization angle and θ the incidence angle.

The reflected field in $z < 0$ and that transmitted in the half-space $z > 0$ are, respectively, given by

$$\mathbf{E}_r = [-\sin(\theta)\cos(\beta) r_p \hat{z} + \sin(\beta) r_s \hat{y} - \cos(\theta)\cos(\beta) r_p \hat{x}] e^{-ik_0(x \sin\theta - z\cos\theta)},$$

$$\mathbf{E}_t = \left[-\sin(\theta)\cos(\beta)\frac{g(\xi)}{n^2(\xi)}\hat{z} + \sin(\beta) f(\xi)\hat{y} + i\cos(\beta)\frac{g'(\xi)}{\Lambda n^2(\xi)}\hat{x}\right] e^{-ik_0 x \sin\theta}, \tag{III.4.2}$$

where the prime indicates derivative with respect to ξ, $\Lambda = k_0 a$, and r_p and r_s are the reflection coefficients for polarization parallel ($\beta = 0$) and perpendicular ($\beta = \pi/2$) to the incidence plane. In particular, $g(\xi)$ is a solution of the differential equation

$$\frac{d}{d\xi}\left\{\left[\frac{1}{n^2(\xi)}\right]\frac{dg}{d\xi}\right\} + \Lambda^2\left[1 - \frac{\sin^2\theta}{n^2(\xi)}\right] g = 0. \tag{III.4.3}$$

The continuity of the components of $\mathbf{E}$ and $\mathbf{H}$ tangent to the interface $z = 0$ yields

$$r_p = \left[1 - \frac{i}{\Lambda n^2(0)\cos\theta}\frac{g'(0)}{g(0)}\right] \Bigg/ \left[1 + \frac{i}{\Lambda n^2(0)\cos\theta}\frac{g'(0)}{g(0)}\right]. \tag{III.4.4}$$

If we set

$$\eta = \int^{\xi} n^2(\xi')\, d\xi', \qquad Q(\eta) = \frac{n^2(\xi) - \sin^2\theta}{n^4(\xi)}, \tag{III.4.5}$$

then Eq. (III.4.3) becomes

$$d^2 g/d\eta^2 + \Lambda^2 Q(\eta) g = 0 \tag{III.4.6}$$

This equation can be solved asymptotically for large Λ by expanding g in a manner formally similar to that of Eq. (II.2.5). Thus, if we set

$$g(\eta) \sim C \exp\left[-i\Lambda \sum_{n=0}^{\infty} (i\Lambda)^{-n} S_n(\eta)\right], \tag{III.4.7}$$

and retrace the steps leading to Eq. (II.2.8), we can show that

$$S_0(\eta) = \pm \int^{\eta} Q^{1/2}(\eta')\, d\eta', \tag{III.4.8a}$$

$$S_1(\eta) = (-\tfrac{1}{4}) \ln Q(\eta), \tag{III.4.8b}$$

$$S_2(\eta) = \pm \int^{\eta} \left(\frac{Q''}{8Q^{3/2}} - \frac{5Q'^2}{32 Q^{5/2}}\right) d\eta', \tag{III.4.8c}$$

the primes denoting derivatives with respect to η' (the expressions for S_4 and S_5 can be found in Bender and Orszag [11]).

If no significant reflection occurs inside the inhomogeneous medium, we can take for g the solution of Eq. (III.4.7) associated with a wave traveling from left to right, which amounts to taking the positive determination of S_0, S_2, and so on. The expression for g obtained in this way allows us to evaluate r_p by means of Eq. (III.4.4), which yields

$$\begin{aligned} r_p &= \frac{1 - (i/\Lambda\cos\theta)\,d\ln g/d\eta}{1 + (i/\Lambda\cos\theta)\,d\ln g/d\eta} \sim \frac{\cos\theta - \Sigma_{n=0}^{\infty}(i\Lambda)^{-n}\,dS_n/d\eta}{\cos\theta + \Sigma_{n=0}^{\infty}(i\Lambda)^{-n}\,dS_n/d\eta} \\ &= r_p^{(F)} - (i/\Lambda)r_p^{(1)} - (1/\Lambda^2)r_p^{(2)} + \cdots, \end{aligned} \tag{III.4.9}$$

where

$$r_p^{(F)} = \lim_{\Lambda\to\infty} r_p = \frac{\cos\theta - S_0'}{\cos\theta + S_0'} = \frac{n_0^2\cos\theta - (n_0^2 - \sin^2\theta)^{1/2}}{n_0^2\cos\theta + (n_0^2 - \sin^2\theta)^{1/2}}, \tag{III.4.10}$$

with $n_0 = n(0)$. The other terms can be calculated by applying iteratively the equation

$$r_p^{(q)} = \lim(i\Lambda)^q\left[r_p - \sum_{n=0}^{q-1}(i\Lambda)^{-n}r_p^{(n)}\right]. \tag{III.4.11}$$

The leading term of the asymptotic expansion of $r_p(\Lambda)$ has been indicated by $r_p^{(F)}$ since it coincides with the Fresnel reflection coefficient [see Eq. (III.8.1)]. While $r_p^{(F)}$ depends on the discontinuity of the refractive index at the boundary, $r_p^{(1)}$ is proportional to the first-order derivative of n^2. By continuing the expansion, it can be easily shown that the generic term $r_p^{(m)}$ contains the mth derivative of n^2. The reflection coefficient r_s is obtained by an analogous procedure.

It is important to note that, according to the preceding results, the reflected wave is produced by a discontinuity of a derivative of $n(z)$. This seems to lead to the conclusion that no reflection can occur when $n(z)$ is an analytic function. The fallacy of this conclusion is shown by the following counterexample provided by Epstein, who calculated the magnitude of the reflection coefficient $|r|$ for normal incidence related to a smooth transition between two media with refractive indices n_1 and n_2. He considered a profile $n(z)$ of the form (see Knittl [1], p. 446)

$$n^2(z) = \tfrac{1}{2}(n_1^2 + n_2^2) + \tfrac{1}{2}(n_2^2 - n_1^2)\tanh(z/a), \tag{III.4.12}$$

and obtained

$$|r| = \sinh[\pi^2(a/\lambda)|n_2 - n_1|]/\sinh[\pi^2(a/\lambda)(n_2 + n_1)]. \tag{III.4.13}$$

In particular, we observe that for $a/\lambda \to 0$, $|r| = |n_2 - n_1|/(n_2 + n_1)$ coincides with the amplitude of the Fresnel reflection coefficient related to an abrupt transition [see Eq. (III.8.9)].

The failure of the asymptotic method is not surprising. It parallels the well-known case of the function e^{-kx}, whose asymptotic expansion in the parameter k^{-1} vanishes identically at all orders. The above difficulties can be circumvented by using the characteristic-matrix method, as we will show in Section III.12.4.

5 Exact Solution for the Linearly Increasing Transition Profile

Consider a transition profile of thickness a connecting two homogeneous half-spaces having refractive indices 1 and $n\,(>1)$, respectively, and suppose an s wave is incident on the transition layer at an angle θ. If $n^2(z)$ varies linearly in the transition region $0 \le z \le a$, Eq. (III.4.3) transforms into

$$d^2f/d\xi^2 + \Lambda^2[\cos^2\theta + (n^2 - 1)\xi]f = 0. \qquad \text{(III.5.1)}$$

Under the transformation $\xi \to [\Lambda/(n^2 - 1)]^{2/3}[\cos^2\theta + (n^2 - 1)\xi]$, the above relation reduces to the Airy equation, whose solutions can be written in terms of Bessel functions of order 1/3 [see Eq. (III.3.4)]. Next, applying the continuity of the tangential electric and magnetic field at $z = 0$, $z = a$, we obtain after some algebra the reflection coefficient r_s (see Tyras [14], p. 70):

$$r_s = \frac{F_+^{(1)}(w_0)F_+^{(2)}(w_1) - F_+^{(2)}(w_0)F_+^{(1)}(w_1)}{F_-^{(1)}(w_0)F_+^{(2)}(w_1) - F_-^{(2)}(w_0)F_+^{(1)}(w_1)}. \qquad \text{(III.5.2)}$$

Here $w_0 = (2/3)[\Lambda/(n^2 - 1)]\cos^3\theta$, $w_1 = (2/3)[\Lambda/(n^2 - 1)](n^2 - \sin^2\theta)^{3/2}$, and $F_{+,-}^{(1,2)} = H_{1/3}^{(1,2)} \pm iH_{2/3}^{(1,2)}$. Thus, the reflection coefficient depends on several factors: the variation of $n^2(z)$ between the two homogeneous regions, the thickness of the transition layer, the wavelength, and the incidence angle. As a consequence, it is possible in principle to derive the parameters of the transition layer by measuring r_s for different wavelengths and incidence angles.

6 Stratified Media with Piecewise-Constant Refractive Index Profiles

The simplest example of an inhomogeneous medium is a multilayered region with a piecewise-constant refractive index. In Section III.2 we discussed the extension of the ray optical (RO) method to an inhomogeneous dielectric with a continuous refractive index profile; essential to that analysis was the recourse to matched asymptotic expansions based on the use of Airy functions. In the presence of refractive index discontinuities, this method can still be adopted with some minor changes in the reflection and transmission coefficients. When we encounter a large number of discontinuities, the description of the multiple reflections undergone by a wave passing through the medium becomes so intricate that it requires a systematic study of the dependence of reflection and transmission coefficients on the number, character, and relative positions of the discontinuities of $n(z)$.

A multilayered medium with a piecewise-constant refractive index is particularly suitable for the analysis of the propagation phenomena typical of groups of discontinuities. In particular, layered media with equispaced discontinuity surfaces exhibit the phenomenon of *stopbands*, which can be found in general in media with periodic refractive index [15] profiles, so that no wave can propagate without substantial attenuation at some frequency intervals. The stopbands allow one to use multilayered structures as selective mirrors and filters, which can be easily manufactured by thin-film deposition techniques.

Dielectric multilayers are widely used in electro-optic devices to reduce the reflectivity of a surface, to obtain a bandpass filter, or to enhance the reflectivity at desired wavelengths [16, 17]. They are deposited by evaporation or sputtering techniques on many types of substrates (glasses, polymers, metals, composite materials) [18] (see Fig. III.7). The simplest system consists of *quarter-wave stacks*, in which each layer has an optical thickness $nd = \lambda_0/4$. Structure with layers of slightly different thicknesses are now replacing the quarter-wave stacks, as they permit better control of the transmissivity in a large frequency range. Multilayer coatings are, in general, designed for specified incidences. Most frequently angles of 0 and $\pi/4$ are used.

A quarter-wave stack is usually indicated by specifying the succession of layers in a form such as air HL HL $\cdots$ HL glass or, equivalently, $\mathrm{A(HL)}^m\mathrm{G}$, which emphasizes the fact that the basic period HL (high n, low n) is repeated m times, while the substrate is made of glass and the whole structure is embedded

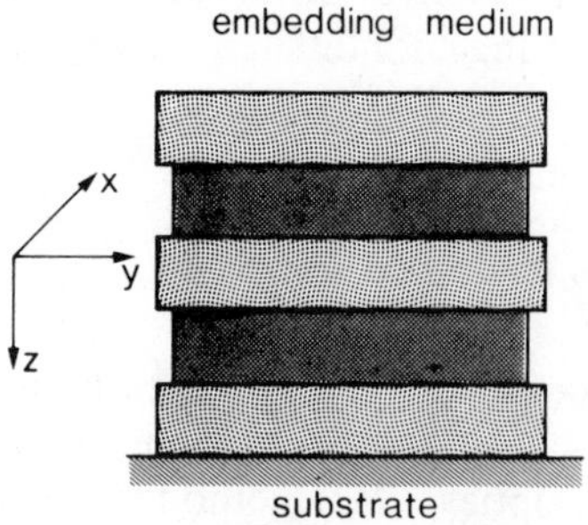

Fig. III.7. Schematic of a multilayer.

in air. More complex structures of the form $A(HL)^m HG$ are used as high-reflectivity mirrors. Sometimes the optical thickness of the basic cell HL is slightly and steadily changed in going from the first cell to the last one adjacent to the substrate. When the refractive indices of the quarter-wave stack $(HL)^m$ are chosen in such a way that $(n_H/n_L)^{2m} = n_S/n_I$, where n_S and n_I are, respectively, the refractive indices of substrate and embedding medium, then the reflectance vanishes at a wavelength λ_0 (*in vacuo*) equal to four times the optical thickness of each layer of the stack. This property is used in *multilayer antireflection* (*AR*) *coatings*, sometimes called *V coatings* because of the typical dependence of r on the frequency. For $m = 1$, and $n_L = 1$, the above condition reduces to $n_H^2 = n_S/n_I$. This means that a small reflectance is obtained by applying to the substrate a coating of approximately a quarter-wavelength optical thickness made up of a material with an index of refraction $\cong (n_S/n_I)^{1/2}$. The most commonly used coating material for single-layer AR coatings of glass embedded in air is MgF_2. In fact, its refractive index 1.37–1.38 (depending on the polarization) is not far from $1.22 \cong 1.5^{1/2}$, which represents the square root of the refractive index of most optical glasses.

7 Electric Network Formalism

As we mentioned in the Introduction, multilayer dielectric coatings are now widely used in optical instruments, a typical example being the dielectric mirrors employed in laser cavities as either output couplers or total reflectors. All these devices belong to the class of plane-stratified media. However, their main feature is a scale of inhomogeneity comparable to the wavelength. As a consequence, they cannot be analyzed by the approach developed before, based on the use of transition functions. A special approach should be used that accounts correctly for the multiple reflection effects due to a sequence of discontinuity surfaces separating the single dielectric layers forming a stack. The problem can be simplified by neglecting the effects due to the finite transverse dimensions. For example, the transmittivity of a multilayer can be calculated without appreciable errors by assuming that the mirror diameter is infinite. In addition, we can assume that the refractive index is constant over the whole thickness of each layer, which implies an abrupt variation of the refractive index when crossing a separation surface. For more general situations see Hinderi and Beckman and Spizzichino, cited in bibliography. To summarize, our model of a multilayer consists of a succession of slabs of unlimited transverse extension separated by perfectly plane-parallel surfaces, each of which is characterized by a constant value of the refractive index (see Fig. III.8). The slabs are numbered progressively from right to left, using the index 1 for the medium farthest from the source of the incident wave. If we call

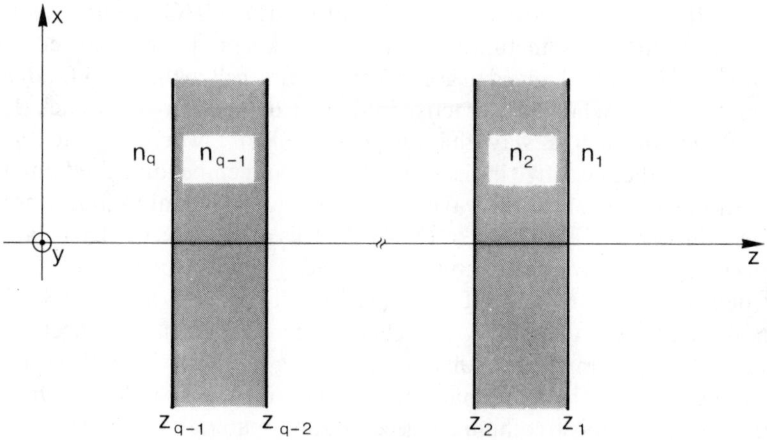

Fig. III.8. Geometry of a multilayer.

z the stratification axis and xz the plane of incidence of a plane wave originating at $z = -\infty$, we can easily show that a generic field component $u(x, z)$ is of the form

$$u(x, z) = e^{-ik_x x} f(z), \tag{III.7.1}$$

where k_x is a constant that depends on the direction of the plane wave illuminating the first slab. It is useful to adopt the electric network formalism (see Chapter II, Felsen and Marcuvitz [3], Chapters 2 and 5), that is, to set

$$\begin{aligned} V_e(z) &= E_x(0, z), \qquad V_h(z) = -E_y(0, z), \\ I_e(z) &= H_y(0, z), \qquad I_h(z) = H_x(0, z). \end{aligned} \tag{III.7.2}$$

Equations (III.7.1,2), with the help of Eqs. (I.1.1–4), ensure that in the generic qth layer characterized by refractive index n_q,

$$-dV_e/dz = i\beta_q Z_q^{(e)} I_e, \tag{III.7.3a}$$

$$-dI_e/dz = i\beta_q V_e/Z_q^{(e)}, \tag{III.7.3b}$$

where

$$Z_q^{(e)} = \zeta_0 \cos\theta_q/\tilde{n}_q. \tag{III.7.4}$$

In the above equation, ζ_0 represents the free-space impedance, $\tilde{n}_q$ the (generally complex) refractive index of the qth medium, and

$$\cos\theta_q = [1 - k_x^2/(\tilde{n}_q^2 k_0^2)]^{1/2}. \tag{III.7.5}$$

For real n_q, θ_q (see Fig. III.9) is the angle between the wave propagation direction in the qth slab and the z axis perpendicular to the faces of the single

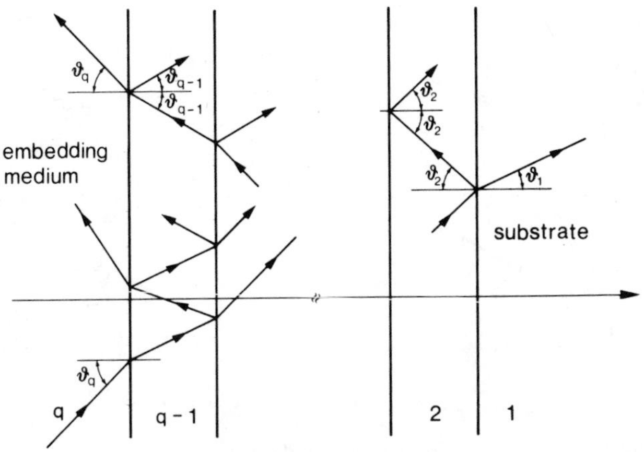

Fig. III.9. Directions of the plane waves interfering in a multilayer structure.

slabs (for lossy media see, e.g., Section III.21). Furthermore, the propagation factor β_q is given by

$$\beta_q = (\tilde{n}_q^2 k_0^2 - k_x^2)^{1/2} = \tilde{n}_q k_0 \cos\theta_q, \tag{III.7.6}$$

so that Eq. (III.7.4) can be written as

$$Z_q^{(e)} = Z_q \cos\theta_q = \beta_q/(\omega\varepsilon_0 \tilde{n}_q^2), \tag{III.7.7}$$

where

$$Z_q = \zeta_0/\tilde{n}_q \tag{III.7.8}$$

is the *characteristic wave impedance* of the qth layer. The index "e" (or TM) indicates a *transverse-magnetic wave*, characterized by a magnetic component orthogonal to the z axis ($H_z = 0$), that is, tangent to the interface between successive slabs (see Fig. III.10a). In this case one has $H_x = 0$ and $E_y = 0$.

A complementary situation is obtained if we assume $E_z = 0$ (*TE waves*, labeled with the index "h"). For these *transverse-electric waves* (Fig. III.10b), we have $H_y = 0$, $E_x = 0$, together with a set of equations analogous to Eqs. (III.7.3) with the substitution $Z_q^{(e)} \to Z_q^{(h)}$, where

$$Z_q^{(h)} = \zeta_0/(\tilde{n}_q \cos\theta_q) = \omega\mu_0/\beta_q. \tag{III.7.9}$$

Usually in optics TM fields are indicated with the symbol "p," denoting that **E** is *parallel* to the plane of incidence, while TE fields are indicated with the symbol "s," denoting an electric field *orthogonal* to the plane of incidence (s stands for *senkrecht*, which in German means orthogonal).

The set of Eqs. (III.7.3) allows us to study the propagation of a TM wave (or a TE wave) by using as a model an electrical transmission line composed of

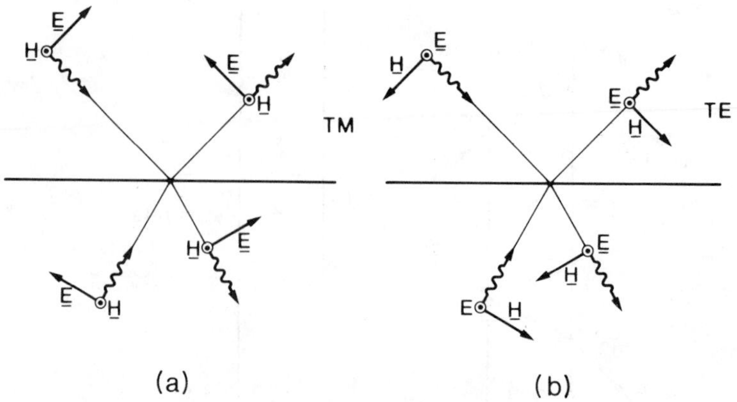

Fig. III.10. (a) Transverse-magnetic and (b) transverse-electric waves.

uniform sections (Fig. III.11) with suitable impedances. For each section, the general solution of Eqs. (III.7.3) is of the form

$$\begin{aligned} V(z) &= V_q^{(+)} e^{-i\beta_q z} + V_q^{(-)} e^{i\beta_q z}, \\ I(z) &= V_q^{(+)} e^{-i\beta_q z}/\hat{Z}_q - V_q^{(-)} e^{i\beta_q z}/\hat{Z}_q, \end{aligned} \tag{III.7.10}$$

where $V_q^{(+)}$ and $V_q^{(-)}$ are constant quantities and the impedance $\hat{Z}_q$ coincides with $Z_q^{(e)}$ for TM waves and with $Z_q^{(h)}$ for TE waves. The quantities $V(z)$ and $I(z)$ [coinciding with $V_e(z)$, $I_e(z)$ for TM waves and with $V_h(z)$, $I_h(z)$ for TE waves] can be formally interpreted as the voltage and current of the electrical line. As is customary in the theory of transmission lines, the current in the upper side is assumed to be positive when flowing along the positive z axis. In order to deal with the stratified structure of the medium, we associate a transmission line with characteristic impedance $\hat{Z}_q$ and length d_q with each homogeneous slab of the dielectric (labeled with the subscript q).

The above analogy is completed by introducing the *local impedance*

$$\vec{Z}(z) = V(z)/I(z), \tag{III.7.11}$$

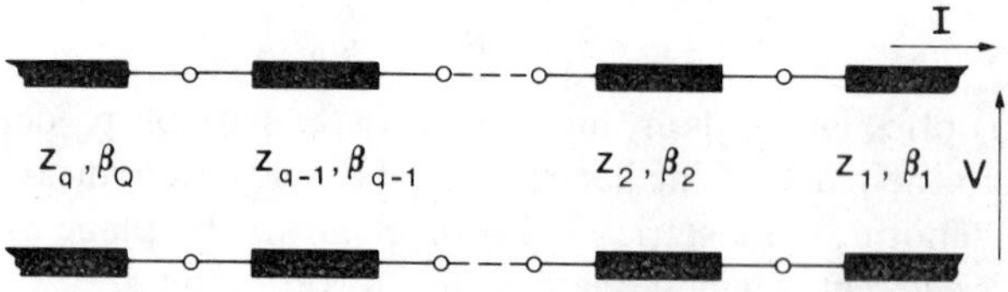

Fig. III.11. Transmission line equivalent to the multilayer of Fig. III.8. Each section has a length equal to the relevant slab thickness.

which can be written, with the help of Eqs. (III.7.10), in the form

$$\vec{Z}(z) = \hat{Z}_q \frac{V_q^{(+)} e^{-i\beta_q z} + V_q^{(-)} e^{i\beta_q z}}{V_q^{(+)} e^{-i\beta_q z} - V_q^{(-)} e^{i\beta_q z}}. \qquad \text{(III.7.12)}$$

Assuming that the section $z = 0$ is contained in the slab we are considering, we have

$$\vec{Z}(0) = \hat{Z}_q \frac{V_q^{(+)} + V_q^{(-)}}{V_q^{(+)} - V_q^{(-)}}, \qquad \text{(III.7.13)}$$

$$\frac{V_q^{(-)}}{V_q^{(+)}} = \frac{\vec{Z}(0) - \hat{Z}_q}{\vec{Z}(0) + \hat{Z}_q}. \qquad \text{(III.7.14)}$$

Equations (III.7.12) and (III.7.13) allow us to write

$$\vec{Z}(z) = \hat{Z}_q \frac{\vec{Z}(0) - i\hat{Z}_q \tan(\beta_q z)}{\hat{Z}_q - i\vec{Z}(0)\tan(\beta_q z)}. \qquad \text{(III.7.15)}$$

The quantity $\vec{Z}(z)$ represents the measured ratio of potential difference and current at section z, provided the part of the line to the left of z has been removed and replaced with a voltage generator (see Fig. III.12). Equation (III.7.15) can be interpreted as the relation giving the local impedance at a generic position z, when it is known at a given point z_1 of the same slab. In fact, the arbitrariness of the origin $z = 0$ allows us to rewrite Eq. (III.7.15) as

$$\vec{Z}(z) = \hat{Z}_q \frac{\vec{Z}(z_1) - i\hat{Z}_q \tan[\beta_q(z - z_1)]}{\hat{Z}_q - i\vec{Z}(z_1)\tan[\beta_q(z - z_1)]}. \qquad \text{(III.7.16)}$$

Let us now consider the *amplitude reflection coefficient* r, defined as the ratio of the backward and forward components of $V(z)$, that is,

$$r(z) = e^{2i\beta_q z} V_q^{(-)}/V_q^{(+)}. \qquad \text{(III.7.17)}$$

Equations (III.7.10) and (III.7.11) allow us to rewrite the above relation in the form

$$r(z) = [\vec{Z}(z) - \hat{Z}_q]/[\vec{Z}(z) + \hat{Z}_q]. \qquad \text{(III.7.18)}$$

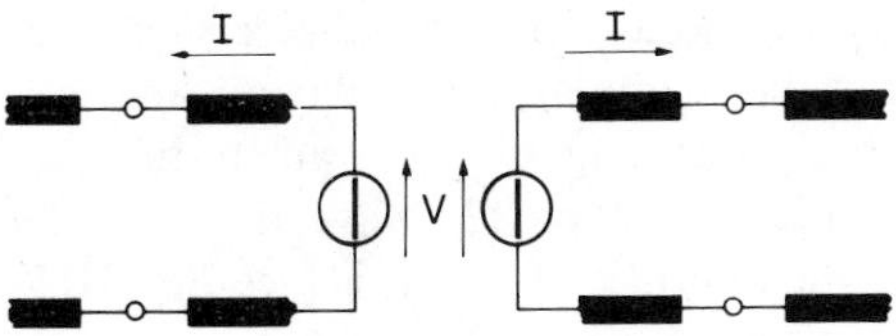

Fig. III.12. Schematic representation of the measurement of the impedances relative to a generic section of the transmission line of Fig. III.11.

We note that while $\vec{Z}(z)$ is a continuous function of z, as are $I(z)$ and $V(z)$, $\hat{Z}_q$ is a piecewise function, so that $r(z)$ is discontinuous (not piecewise) for $z = z_q$, where z_q denotes the left limit of the qth slab.

On each discontinuity surface $z = z_q$, one can also define the *amplitude transmission coefficient* t_q, given by

$$t_q = V(z_q)/[V_{q+1}^{(+)} e^{-i\beta_{q+1} z_q}], \tag{III.7.19}$$

that is, the ratio of $V(z_q)$ and the forward component of $V(z)$ immediately before z_q. We must now mention a distinction between TM and TE waves. If we are interested in the complex amplitude of the *total* electric field $\mathbf{E} = \hat{x}E_x + \hat{z}E_z$ immediately after z_q, it is useful to introduce a new coefficient $t_q^{(\mathrm{tot})}$, obtained from t_q by substituting for $V(z_q)$ and $V_{q+1}^{(+)} e^{-i\beta_{q+1} z_q}$ in Eq. (III.7.19) the corresponding total electric amplitudes. In this way, we easily obtain for TM (p) and TE (s) waves

$$\begin{aligned} t_{q,\mathrm{p}}^{(\mathrm{tot})} &= t_{q,\mathrm{p}} \cos\theta_q / \cos\theta_{q+1} \\ &= t_{q,\mathrm{p}} (n_{q+1}^2 - n_{q+1}^2 \sin^2\theta_q)^{1/2} / (n_{q+1}^2 - n_q^2 \sin^2\theta_q)^{1/2}, \\ t_{q,\mathrm{s}}^{(\mathrm{tot})} &= t_{q,\mathrm{s}}, \end{aligned} \tag{III.7.20}$$

since in this case V represents the total electric field. The formal description of propagation in a succession of slabs can be concluded by observing that the continuity of the tangential components of the field relates $V_{q+1}^{(+)}$ and $V_{q+1}^{(-)}$ to $V_q^{(+)}$ and $V_q^{(-)}$ by the expressions

$$\begin{aligned} V_q^{(+)} e^{-i\beta_q z_q} + V_q^{(-)} e^{i\beta_q z_q} &= V_{q+1}^{(+)} e^{-i\beta_{q+1} z_q} + V_{q+1}^{(-)} e^{i\beta_{q+1} z_q}, \\ \frac{V_q^{(+)} e^{-i\beta_q z_q}}{\hat{Z}_q} - \frac{V_q^{(-)} e^{i\beta_q z_q}}{\hat{Z}_q} &= \frac{V_{q+1}^{(+)} e^{-i\beta_{q+1} z_q}}{\hat{Z}_{q+1}} + \frac{V_{q+1}^{(-)} e^{i\beta_{q+1} z_q}}{\hat{Z}_{q+1}}, \end{aligned} \tag{III.7.21}$$

8 Fresnel Formulas

Let us consider the simple case of two dielectric media with refractive indices n_1 and n_2 separated by the plane interface $z = 0$. According to our convention, n_2 is the refractive index of the medium where the waves enter first. We follow this somewhat unnatural choice since it agrees with the one already adopted for the analysis of multilayers (see Fig. III.8), where the reflection starts, loosely speaking, from the substrate.

The reflection coefficient r_p for a TM wave at the interface (more precisely, immediately before the interface of $2 \to 1$) is [see Eq. (III.7.18)]

$$r_\mathrm{p} = -\frac{Z_2 \cos\theta_2 - Z_1 \cos\theta_1}{Z_2 \cos\theta_2 + Z_1 \cos\theta_1}, \tag{III.8.1}$$

where use has been made of the coincidence of $\vec{Z}(0)$ with $\hat{Z}_1^{(e)}$ [since medium 1 extends from $z = 0$ to $+\infty$, no backward-traveling wave is present in it and Eq. (III.7.12) yields $\vec{Z}(z) = \hat{Z}_1$ for $z > 0$] and of Eq. (III.7.7). Snell's law and Eq. (III.7.8) allow us to rewrite Eq. (III.8.1) and the analogous expression for r_s, t_p, and t_s as

$$r_p = -\frac{\tan(\theta_2 - \theta_1)}{\tan(\theta_1 + \theta_2)}, \qquad r_s = -\frac{\sin(\theta_2 - \theta_1)}{\sin(\theta_1 + \theta_2)},$$
$$t_p = \frac{2 \sin\theta_1 \cos\theta_1}{\sin(\theta_1 + \theta_2)\cos(\theta_1 - \theta_2)}, \qquad t_s = \frac{2 \sin\theta_1 \cos\theta_2}{\sin(\theta_1 + \theta_2)}. \tag{III.8.2}$$

Here θ_1 and θ_2 are coincident, respectively, with the refraction and incidence angles.

Equations (III.8.2) are known as *Fresnel formulas* and hold exactly true only for perfectly plane interfaces. The presence of small rugosities, scratches, or pits produces scattering of the incoming radiation, which affects the actual reflectivity. We observe that r_p vanishes for $\theta_1 + \theta_2 = \pi/2$. The incidence angle θ_B (*Brewster angle*) satisfying this relation turns out to be given, in accordance with Snell's law, by

$$\tan\theta_B = n_1/n_2. \tag{III.8.3}$$

The vanishing of $r_p(\theta_B)$ has many applications in beamsplitter polarizers (see Fig. III.13). In gas laser tubes, the terminating windows are tilted at the Brewster angle in order to enhance the transmittivity of the p component.

Physically, the reflected field is generated by the dipoles produced in the second medium by the refracted field **E**, which are parallel to **E** itself. On the other hand, a dipole does not radiate in a direction coincident with its orientation. As a consequence, when the reflected rays are parallel to the dipoles produced in the second medium, the associated field must vanish. This is exactly what occurs for $\theta = \theta_B$.

While $r_p(\theta_B) = 0$ and $t_p(\theta_B) = 1$, Eqs. (III.8.2,3) imply

$$r_s(\theta_B) = (n_2^2 - n_1^2)/(n_1^2 + n_2^2), \tag{III.8.4a}$$

$$t_s(\theta_B) = 2n_2^2/(n_1^2 + n_2^2). \tag{III.8.4b}$$

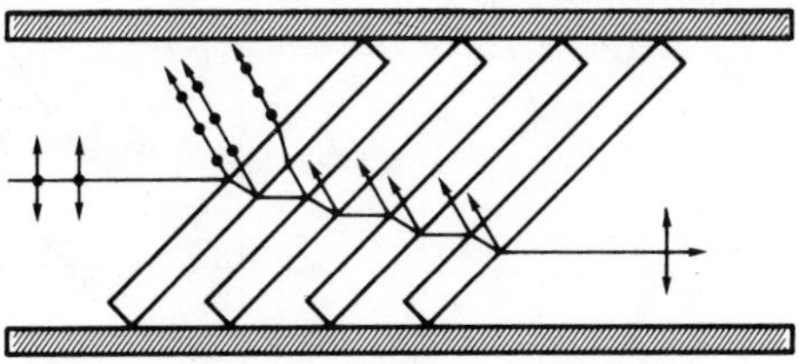

Fig. III.13. Pile of glass plates mounted at the Brewster angle and used as a polarizer.

It is worth stressing that, at any angle, the definitions of r_p, r_s, t_p, and t_s given by Eqs. (III.7.17) and (III.7.19) entail, with the help of Eq. (III.7.10a),

$$t_s - r_s = t_p - r_p = 1. \tag{III.8.5}$$

Let us now relate the reflection and transmission coefficients to the corresponding *optical intensities* (watts per meter squared) I_r and I_t. If we denote by I_i the incident intensity, we can write the following string of relations valid for s-waves

$$I_i = E_i H_i^*/2 = n_2 |E_i|^2/(2\zeta_0), \tag{III.8.6a}$$

$$I_r = I_i |r_s|^2 \equiv I_i R_s, \tag{III.8.6b}$$

$$I_t = I_i |t_s|^2 n_1/n_2 \equiv I_i T_s, \tag{III.8.6c}$$

while for p-waves,

$$I_r = I_i |r_p|^2 \equiv I_i R_p, \tag{III.8.7a}$$

$$I_t = I_i |t_p|^2 (n_1/n_2) |\cos\theta_2/\cos\theta_1|^2 \equiv I_i T_p. \tag{III.8.7b}$$

The coefficients R and T are, respectively, the *reflectance* and *transmittance*. In particular, Eqs. (III.8.2) yield (see Fig. III.14)

$$R_s = \sin^2(\theta_1 - \theta_2)/\sin^2(\theta_1 + \theta_2), \tag{III.8.8a}$$

$$R_p = \tan^2(\theta_1 - \theta_2)/\tan^2(\theta_1 + \theta_2). \tag{III.8.8b}$$

In the limit of normal incidence, the above relations reduce to

$$R_s = R_p = (n_1 - n_2)^2/(n_1 + n_2)^2, \tag{III.8.9}$$

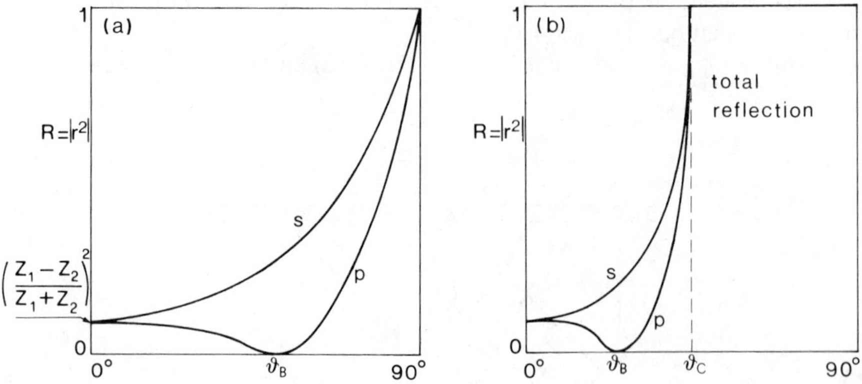

Fig. III.14. Reflectance versus incidence angle for p- and s-waves for a beam traveling (a) from the air to a dielectric and (b) from the dielectric to air.

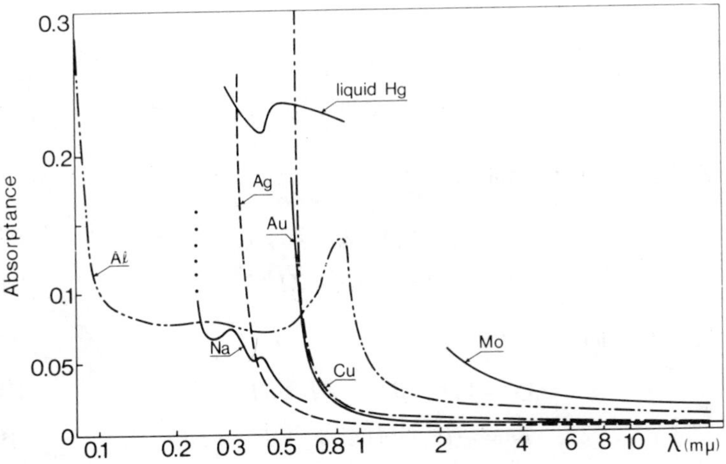

Fig. III.15. Absorptance versus wavelength for some typical metals.

while

$$T_s = T_p = 4n_1n_2/(n_1 + n_2)^2. \tag{III.8.10}$$

Whenever $\tilde{n}_1$ and $\tilde{n}_2$ are complex quantities, that is, for nonnegligible losses, Eqs. (III.8.9) and (III.8.10) generalize to

$$R_s = R_p = |\tilde{n}_1 - \tilde{n}_2|^2/|\tilde{n}_1 + \tilde{n}_2|^2 \qquad (\theta_1 = 0), \tag{III.8.11a}$$

$$T_s = T_p = 4(\operatorname{Re}\tilde{n}_1 \operatorname{Re}\tilde{n}_2 + \operatorname{Im}\tilde{n}_1 \operatorname{Im}\tilde{n}_2)/|\tilde{n}_1 + \tilde{n}_2|^2 \quad (\theta_1 = 0). \tag{III.8.11b}$$

If the refracted power is dissipated in medium 1, which happens in a metal of reasonable thickness, the relative absorption losses are usually described in terms of the *absorptances* A_s and A_p, where $A_s = 1 - R_s$ and $A_p = 1 - R_p$. The absorptances at normal incidence for some typical metals are plotted in Fig. III.15.

9 Characteristic Matrix Formalism

Let us consider a homogeneous dielectric slab of thickness d. Equations (III.7.10) allow us to derive a simple relation between the equivalent voltage and current V_1, I_1 on face 1 and those on face 2. In fact, if we set $z_1 = 0$ and $z_2 = -d$, we obtain

$$\begin{aligned} V_1 &= V^{(+)} + V^{(-)}, \\ I_1 &= V^{(+)}/\hat{Z} - V^{(-)}/\hat{Z}, \end{aligned} \tag{III.9.1}$$

and

$$V_2 = V^{(+)}e^{-i\beta d} + V^{(-)}e^{i\beta d}, \tag{III.9.2a}$$

$$I_2 = V^{(+)}e^{-i\beta d}/\hat{Z} - V^{(-)}e^{i\beta d}/\hat{Z} \tag{III.9.2b}$$

($\hat{Z} = Z_p$ for TM waves and $\hat{Z} = Z_s$ for TE waves), so that

$$V^{(+)} = V_1/2 + \hat{Z}I_1/2, \tag{III.9.3a}$$

$$V^{(-)} = V_1/2 - \hat{Z}I_1/2, \tag{III.9.3b}$$

which in turn yield

$$V_2 = V_1 \cos \beta d - i\hat{Z}I_1 \sin \beta d, \tag{III.9.4a}$$

$$I_2 = -iV_1 \sin(\beta d)/\hat{Z} + I_1 \cos \beta d. \tag{III.9.4b}$$

In matrix form, the above equations read

$$\begin{bmatrix} V_2 \\ I_2 \end{bmatrix} = \begin{bmatrix} \cos \beta d & -i\hat{Z}\sin \beta d \\ -i\sin(\beta d)/\hat{Z} & \cos \beta d \end{bmatrix} \begin{bmatrix} V_1 \\ I_1 \end{bmatrix} \equiv \begin{bmatrix} A & B \\ C & D \end{bmatrix} \begin{bmatrix} V_1 \\ I_1 \end{bmatrix}. \tag{III.9.5}$$

This 2×2 matrix is called the *transmission* or *characteristic matrix* for the slab [19]. It was introduced in optics by Herpin and Muchmore for ordinary media and later extended to anisotropic crystals [20]. It depends on the optical thickness βd and impedance $\hat{Z}$ of the homogeneous layer, and its determinant is equal to unity.

Whenever we deal with a stack of homogeneous slabs, numbered as usual from right to left, we can relate the pair of quantities V_n, I_n for the nth interface to V_1, I_1 by applying $n - 1$ times the product of a vector by a matrix of the kind given in Eq. (III.9.5). More precisely, if we indicate by $\mathbf{M}_{i,i-1}$ the matrix related to the terminal sections of the ith slab, we have

$$\begin{bmatrix} V_n \\ I_n \end{bmatrix} = \mathbf{M}_{n,n-1} \cdot \mathbf{M}_{n-1,n-2} \cdots \mathbf{M}_{2,1} \begin{bmatrix} V_1 \\ I_1 \end{bmatrix} \equiv \mathbf{M}_{n,1} \begin{bmatrix} V_1 \\ I_1 \end{bmatrix}, \tag{III.9.6}$$

where $\mathbf{M}_{n,1}$ represents the matrix product $\mathbf{M}_{n,n-1} \cdots \mathbf{M}_{2,1}$. Note that the product must be performed in the order indicated above, since the matrices do not commute in general among themselves, except for the case of slabs having equal characteristics. In conclusion, we can say that for every multilayer the pairs V, I related to a generic interface can be obtained from each other by making use of the corresponding characteristic matrix.

The discussion above refers to an interface of a multilayer. However, we can remove this condition by considering a generic section of abscissa z. Since we are not restrained from considering this section as an interface between two media, there exists a matrix $\mathbf{M}(z,z')$ such that, in general,

$$\begin{bmatrix} V(z) \\ I(z) \end{bmatrix} = \mathbf{M}(z,z') \cdot \begin{bmatrix} V(z') \\ I(z') \end{bmatrix}. \qquad \text{(III.9.7)}$$

With a further step, we can remove the hypothesis of constant refractive index between two sections where it is discontinuous. In fact, it is possible to approximate a continuous profile with a multilayer formed by slabs of vanishing thickness. Thus, we conclude that the characteristic-matrix formalism applies to the most general plane-stratified medium. In particular, since the determinant of $\mathbf{M}$ is equal to unity for a single homogeneous slab [Eq. (III.9.5)] and the determinant of the product of two matrices is equal to the product of their determinants, then $\det \mathbf{M}(z,z') = 1$ for a generic stratified medium.

In the following, we will first consider the case of a slab with a generic profile $n(z)$, and then successively analyze a periodic stack of slabs that are all equal and characterized by the same matrix $\mathbf{M}$.

9.1 *Equation of Motion of the* $\mathbf{M}$ *Matrix*

For a medium with varying n, we have, with the help of Eq. (III.9.5),

$$\begin{aligned} \mathbf{M}(z+dz,z') &= \mathbf{M}(z+dz,z) \cdot \mathbf{M}(z,z') \\ &= \mathbf{M}(z,z') + i\,dz \begin{bmatrix} 0 & \beta\hat{Z} \\ \beta/\hat{Z} & 0 \end{bmatrix} \cdot \mathbf{M}(z,z'), \end{aligned} \qquad \text{(III.9.8)}$$

so that

$$i \lim_{dz\to 0} \frac{\mathbf{M}(z+dz,z') - \mathbf{M}(z,z')}{dz} = -i\frac{d}{dz}\mathbf{M}(z,z') = \begin{bmatrix} 0 & \beta\hat{Z} \\ \beta/\hat{Z} & 0 \end{bmatrix} \cdot \mathbf{M}(z,z'). \qquad \text{(III.9.9)}$$

Now, if we indicate by $A = M_{1,1}$, $B = M_{1,2}$, $C = M_{2,1}$, and $D = M_{2,2}$ the components of the matrix $\mathbf{M}(z,z')$, we have from Eq. (III.9.9)

$$dA/dz = i\beta\hat{Z}\,C, \qquad \text{(III.9.10a)}$$

$$dB/dz = i\beta\hat{Z}\,D, \qquad \text{(III.9.10b)}$$

$$dC/dz = i\beta A/\hat{Z}, \qquad \text{(III.9.10c)}$$

$$dD/dz = i\beta B/\hat{Z}, \qquad \text{(III.9.10d)}$$

where β and $\hat{Z}$ are both functions of z. Whenever n is a continuous function of z, Eqs. (III.9.10a) and (III.9.10c) yield

$$\hat{Z}\frac{d}{d\phi}\left(\frac{1}{\hat{Z}}\frac{dA}{d\phi}\right) + A = 0, \qquad \text{(III.9.11)}$$

where $\phi = \int \beta\, dz$ is the *phase thickness*. This equation must be solved by imposing the additional conditions $A(z', z') = 1$ and $dA/dz = 0$ for $z = z'$ [the last relation following from $C(z', z') = 0$]. If the medium is lossless, β and $\hat{Z}$ are real quantities and, according to Eq. (III.9.11) and to the initial condition $A = 1$, $A(z, z')$ is a real quantity. As a consequence, according to Eq. (III.9.10c), C is imaginary. Analogous considerations can be repeated for B and D. Therefore, we conclude that, for a lossless medium, the diagonal terms are real, while the off-diagonal ones are imaginary.

We observe that $\beta(z) = k_0[n^2(z) - k_x^2/k_0^2]^{1/2}$, so that Eq. (III.9.11) contains the large parameter k_0^2 and can be rewritten as Eq. (III.4.6) by making the substitutions $d\eta = \hat{Z}(n^2 - k_x^2/k_0^2)^{1/2}\, dz$, $\Lambda = k_0$, $Q(\eta) = 1/\hat{Z}^2$. Then for $n(z)$ slowly varying over a wavelength, we can expand A in the asymptotic series of Eq. (III.4.7). In this way, by imposing the initial conditions, we easily obtain

$$A(z, z') \sim [\hat{Z}(z')/\hat{Z}(z)]^{-1/2} \cos[k_0(S_0 - S_2/k_0^2 + \cdots)]. \qquad \text{(III.9.12)}$$

where

$$S_0 = \int_z^{z'} \left(n^2 - \frac{k_x^2}{k_0^2}\right)^{1/2} dz'', \qquad \text{(III.9.13a)}$$

$$S_2 = \int_z^{z'} \left[\frac{3}{8}\frac{\hat{Z}'^2}{\hat{Z}^2} - \frac{1}{4}\frac{\hat{Z}''}{\hat{Z}}\right]\left(n^2 - \frac{k_x^2}{k_0^2}\right)^{1/2} dz'', \qquad \text{(III.9.13b)}$$

where $\hat{Z}'$ and $\hat{Z}''$ are the first- and second-order derivatives of $\hat{Z}$ with respect to the phase variable $\int (n^2 - k_x^2/k_0^2)^{1/2}\, dz''$. Similar expressions can be obtained for the other components of the **M** matrix.

9.2 *Calculation of the* **M** *Matrix for a Multilayer*

Let us now consider a structure consisting of m layers, characterized by a matrix **M** repeated m times, which leads us to calculate the mth power of **M**. To this end, it is convenient to represent **M** as a combination of two matrices $\mathbf{O}^{(+)}$ and $\mathbf{O}^{(-)}$ such that $\mathbf{O}^{(+)2} = \mathbf{O}^{(+)}$, $\mathbf{O}^{(-)2} = \mathbf{O}^{(-)}$, and $\mathbf{O}^{(+)} \cdot \mathbf{O}^{(-)} = 0$. In order to find this representation, we proceed in three steps: first, we find the two eigenvalues $\gamma_\pm$ of **M**, which we suppose for simplicity to be different; second, we calculate the right and left eigenvectors of **M** corresponding to γ_+ and γ_-; third, we construct $\mathbf{O}^{(+)}$ and $\mathbf{O}^{(-)}$ by using these eigenvectors. Finding the eigenvalues corresponds to solving the equation

$$\det\begin{bmatrix} A - \gamma & B \\ C & D - \gamma \end{bmatrix} = 0, \qquad \text{(III.9.14)}$$

which, with the help of the relation det $\mathbf{M} = 1$, yields

$$\gamma_{\pm} = \tfrac{1}{2}(A + D) \mp i[1 - (A + D)^2/4]^{1/2} = e^{\mp i\delta}, \qquad \text{(III.9.15)}$$

where

$$\delta = \arccos[(A + D)/2]. \qquad \text{(III.9.16)}$$

Next, we evaluate the right eigenvector $\mathbf{V}^{(+)} \equiv (\mathrm{V}^{(+)}, \mathrm{I}^{(+)})$, corresponding to γ_+, by solving the homogeneous system

$$(A - \gamma_+)V^{(+)} + BI^{(+)} = 0, \qquad \text{(III.9.17a)}$$

$$CV^{(+)} + (D - \gamma_+)I^{(+)} = 0, \qquad \text{(III.9.17b)}$$

and the left eigenvector $\tilde{\mathbf{V}}^{(+)} \equiv [\tilde{V}^{(+)}, \tilde{\mathbf{I}}^{(+)}]$ by solving the system

$$(A - \gamma_+)\tilde{V}^{(+)} + C\tilde{I}^{(+)} = 0, \qquad \text{(III.9.18a)}$$

$$B\tilde{V}^{(+)} + (D - \gamma_+)\tilde{I}^{(+)} = 0, \qquad \text{(III.9.18b)}$$

all eigenvectors being determined apart from a multiplicative constant. The same equations obviously hold true with the substitution $\gamma_+, \mathbf{V}^{(+)}, \tilde{\mathbf{V}}^{(+)} \rightarrow \gamma_-, \mathbf{V}^{(-)}, \tilde{\mathbf{V}}^{(-)}$.

The matrix $\mathbf{O}^{(+)}$ ($\mathbf{O}^{(-)}$) is easily constructed by combining the components of $\mathbf{V}^{(+)}$ and $\tilde{\mathbf{V}}^{(+)}$ in the form

$$\mathbf{O}^{(+)} = \frac{\mathbf{V}^{(+)}\tilde{\mathbf{V}}^{(+)}}{\tilde{\mathbf{V}}^{(+)} \cdot \mathbf{V}^{(+)}} \frac{i}{2\sin\delta}\begin{bmatrix} A - e^{i\delta} & B \\ C & D - e^{i\delta} \end{bmatrix} \qquad \text{(III.9.19)}$$

$\mathbf{O}^{(-)}$ being defined in a similar way. In fact, we have

$$\begin{aligned} \mathbf{O}^{(+)2} &= \frac{1}{(\tilde{\mathbf{V}}^{(+)} \cdot \mathbf{V}^{(+)})^2}\mathbf{V}^{(+)}(\tilde{\mathbf{V}}^{(+)} \cdot \mathbf{V}^{(+)})\tilde{\mathbf{V}}^{(+)} \\ &= \frac{1}{\tilde{\mathbf{V}}^{(+)} \cdot \mathbf{V}^{(+)}}\mathbf{V}^{(+)}\tilde{\mathbf{V}}^{(+)} = \mathbf{O}^{(+)}, \qquad \text{(III.9.20)} \\ \mathbf{O}^{(-)2} &= \mathbf{O}^{(-)}, \end{aligned}$$

and

$$\mathbf{O}^{(+)} \cdot \mathbf{O}^{(-)} = \frac{1}{\tilde{\mathbf{V}}^{(+)} \cdot \mathbf{V}^{(+)}} \frac{1}{\tilde{\mathbf{V}}^{(-)} \cdot \mathbf{V}^{(-)}}\mathbf{V}^{(+)}(\tilde{\mathbf{V}}^{(+)} \cdot \mathbf{V}^{(-)})\tilde{\mathbf{V}}^{(-)} = \mathbf{0}. \qquad \text{(III.9.21)}$$

The vanishing of the right-hand side of Eq. (III.9.21) is implied by the relation

$$\tilde{\mathbf{V}}^{(+)} \cdot \mathbf{V}^{(-)} = \frac{1}{\gamma_-}\tilde{\mathbf{V}}^{(+)} \cdot \mathbf{M} \cdot \mathbf{V}^{(-)} = \frac{1}{\gamma_+}\tilde{\mathbf{V}}^{(+)} \cdot \mathbf{M} \cdot \mathbf{V}^{(-)}, \qquad \text{(III.9.22)}$$

which can be satisfied, for $\gamma_+ \neq \gamma_-$, only if $\tilde{\mathbf{V}}^{(+)} \cdot \mathbf{V}^{(-)} = 0$.

We can now express $\mathbf{M}$ as a combination of $\mathbf{O}^{(+)}$ and $\mathbf{O}^{(-)}$, viz.

$$\mathbf{M} = C_+\mathbf{O}^{(+)} + C_-\mathbf{O}^{(-)}. \qquad \text{(III.9.23)}$$

If we multiply $\mathbf{M}$ by $\mathbf{O}^{(+)}$ we have

$$\mathbf{M}\cdot\mathbf{O}^{(+)} = C_+\mathbf{O}^{(+)} = \gamma_+\mathbf{O}^{(+)} \rightarrow C_+ = \gamma_+. \qquad \text{(III.9.24)}$$

Since an analogous relation holds true for C_-, we can finally set

$$\mathbf{M} = \gamma_+\mathbf{O}^{(+)} + \gamma_-\mathbf{O}^{(-)}. \qquad \text{(III.9.25)}$$

Furthermore, with the help of the orthogonality ($\mathbf{O}^{(+)}\cdot\mathbf{O}^{(-)} = 0$) and idempotent ($\mathbf{O}^{(\pm)2} = \mathbf{O}^{(\pm)}$) properties, we easily obtain (see Problem 7 for a different expression for $\mathbf{M}^n$)

$$\mathbf{M}^n = \gamma_+^n\mathbf{O}^{(+)} + \gamma_-^n\mathbf{O}^{(-)}. \qquad \text{(III.9.26)}$$

Example: $\mathbf{M}$ *Matrix of a Multilayer with Alternating Indices.* The above relation is particularly useful for evaluating the matrix of a structure consisting of $2m$ layers with alternating refractive indices n_a and n_b and thicknesses d_a and d_b. In this case, if we indicate by $\mathbf{M}_a$ and $\mathbf{M}_b$ the matrices of the two layers, Eq. (III.9.5) yields, for the matrix $\mathbf{M}_{ba}$ of their combination,

$$\mathbf{M}_{ba} = \mathbf{M}_b\cdot\mathbf{M}_a$$

$$= \begin{bmatrix} \cos\phi_a\cos\phi_b - \dfrac{\hat{Z}_b}{\hat{Z}_a}\sin\phi_a\sin\phi_b & -i\hat{Z}_a\sin\phi_a\cos\phi_b - i\hat{Z}_b\cos\phi_a\sin\phi_b \\ \dfrac{-i}{\hat{Z}_b}\cos\phi_a\sin\phi_b - \dfrac{i}{\hat{Z}_a}\sin\phi_a\cos\phi_b & \dfrac{-\hat{Z}_a}{\hat{Z}_b}\sin\phi_a\sin\phi_b + \cos\phi_a\cos\phi_b \end{bmatrix}, \qquad \text{(III.9.27)}$$

where $\phi_a = \beta_a d_a$ and $\phi_b = \beta_b d_b$ are the phase thicknesses. In order to set $\mathbf{M}_{ba}$ in the form of Eq. (III.9.25), we must calculate δ by means of Eq. (III.9.16) and solve the systems of Eqs. (III.9.17) and (III.9.18).

We now wish to remark that the matrix of any stratified medium is equivalent, at a given wavelength, to that of a two-film combination (Herpin's theorem). Mathematically, this amounts to saying that we can find four quantities $\hat{Z}_a$, $\hat{Z}_b$, ϕ_a, and ϕ_b such that the matrix elements of Eq. (III.9.27) coincide with the assigned values A, B, C, and D.

10 Bloch Waves

Let us continue the analysis of a structure consisting of m identical layers of thickness Λ, having a basic cell characterized by the matrix $\mathbf{M}$. Their periodicity produces the well-known phenomenon of stopbands; that is, a progressive wave propagates only for frequencies in particular intervals, called *passbands*. Outside the passbands the field follows an exponential law similar to that of waves in lossy media. The extension of these stopbands can be studied by means of Floquet's theorem, which will be examined in Section III.17. Their location and width depend on the characteristics of the basic cell of a multilayer (thickness and refractive index of the constituent thin films). From a mathematical point of view, a stopband corresponds to the frequency intervals for which the moduli of the eigenvalues $\gamma_{\pm}$ of the characteristic matrix relevant to the basic cell are different from unity, that is, $|A + D| > 2$.

The eigenvectors $\mathbf{V}^{(+)}$ and $\mathbf{V}^{(-)}$ of $\mathbf{M}$ are also eigenvectors of a generic power $\mathbf{M}^n$. This leads us to discuss the values taken by $\mathbf{V}(z)$ at each interface between two adjacent basic cells, located at $z = z_q = -q\Lambda$ with integer q (the right face of the first basic cell is located at the origin of the z axis). If $\mathbf{V}(0)$ coincides with either $\mathbf{V}^{(+)}$ or $\mathbf{V}^{(-)}$, then

$$\mathbf{V}(z_q) = e^{\pm iq\delta}\mathbf{V}^{(\pm)}. \tag{III.10.1}$$

To complete the description of the field, we must consider the behavior of $\mathbf{V}$ for a generic value of z. To this end, we use *Floquet's theorem* (Section III.17.1), according to which $\mathbf{V}(z)$ can be expressed in the form

$$\mathbf{V}^{(\pm)}(z) = \mathbf{f}^{(\pm)}(z)e^{\mp i\delta z/\Lambda}, \tag{III.10.2}$$

where $\mathbf{f}^{(\pm)}(z)$ is a function with period Λ, which depends on the index profile of the basic cell. In particular, if the basic cell has a plane of symmetry, then $\mathbf{f}^{(+)}(z - \Lambda/2) \propto \mathbf{f}^{(-)}(-z - \Lambda/2)$. The above distribution of $\mathbf{V}^{(\pm)}(z)$ can be considered as the electromagnetic analog of the quantum mechanical electron waves propagating through a crystal with lattice constant Λ. Because of this analogy, $\mathbf{V}^{(+)}(z)$ and $\mathbf{V}^{(-)}(z)$ are called *Bloch waves* [21].

We note that $\mathbf{V}^{(+)}(z)$ and $\mathbf{V}^{(-)}(z)$ represent forward- and backward-propagating waves only when δ is a real quantity, that is [Eq. (III.9.16)],

$$|A + D| \leqq 2. \tag{III.10.3}$$

For a basic cell formed by two slabs, Eqs. (III.9.27) and (III.9.16) yield

$$\cos\delta = \cos\phi_a \cos\phi_b - (1/2)(\hat{Z}_a/\hat{Z}_b + \hat{Z}_b/\hat{Z}_a)\sin\phi_a \sin\phi_b. \tag{III.10.4}$$

In the case $\phi_a = \phi_b \equiv \phi$, Eq. (III.10.4) reduces to

$$\cos\delta = \cos^2\phi - \frac{1}{2}\frac{\hat{Z}_a^2 + \hat{Z}_b^2}{\hat{Z}_a\hat{Z}_b}\sin^2\phi. \tag{III.10.5}$$

Thus, δ is real if

$$\cos^2\phi \geqq (\hat{Z}_a - \hat{Z}_b)^2/(\hat{Z}_a + \hat{Z}_b)^2 \equiv \cos^2\phi_t. \tag{III.10.6}$$

Since $\mathbf{V}^{(+)}$ and $\mathbf{V}^{(-)}$ are defined apart from a multiplicative constant, it is convenient to characterize the Bloch waves through the impedances $Z^{(+)}$ and $Z^{(-)}$, defined by

$$Z^{(+)} = V^{(+)}(z_q)/I^{(+)}(z_q), \qquad Z^{(-)} = V^{(-)}(z_q)/I^{(-)}(z_q). \qquad \text{(III.10.7)}$$

If we take into account Eq. (III.9.17), we immediately obtain

$$Z^{(+)} = -B/(A - e^{-i\delta}), \qquad Z^{(-)} = -B/(A - e^{i\delta}). \qquad \text{(III.10.8)}$$

Note that $I^{(+)}$ and $I^{(-)}$ refer to a common positive direction for the current. Thus, for a wave progressing in a uniform medium from right to left, $I^{(-)}$ and $Z^{(-)}$ are negative.

11 Passbands and Stopbands of Quarter-Wave Stacks

When the characteristic exponent δ is real, the Bloch waves propagate through the multilayer without attenuation and the stack behaves like a transparent dielectric. However, when $\operatorname{Im}\delta \neq 0$, the amplitude of one Bloch wave decays exponentially, while the other one grows exponentially. This means that a wave from outside impinging on the multilayer cannot penetrate it. In this case, the stack acts as a reflector.

Because of these properties, the frequency intervals for which $\operatorname{Im}\delta = 0$ are called *passbands*, while the remaining intervals form the *stopbands*. The passbands are implicitly defined by Eq. (III.10.6) for a multilayer of uniform phase thickness (i.e., $\phi_a = \phi_b$). Each passband extends from $\Omega_m - \Delta\Omega$ to $\Omega_m + \Delta\Omega$, where Ω_m is the central angular frequency (see Fig. III.16). Ω_m and

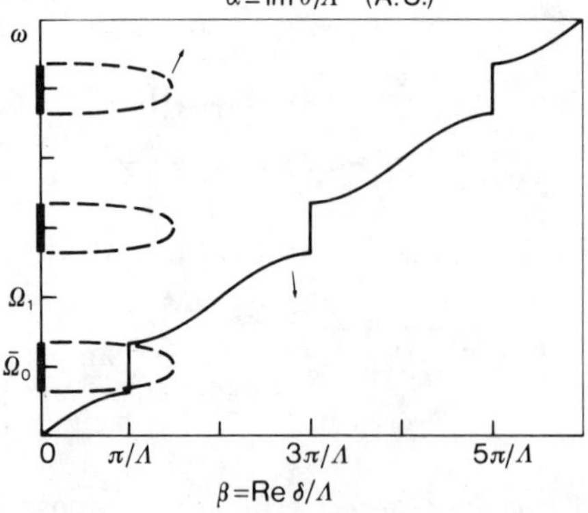

Fig. III.16. Schematic plot of the dispersion curve $\beta = \beta(\omega)$ (continuous line) and the attenuation $\alpha = \alpha(\omega)$ of a periodic structure. The stopbands are indicated with thick bars.

$\Delta\Omega$ can be immediately calculated by noting that for $\omega = \Omega_m$, $\cos^2\phi_a = 1$, while for $\omega = \Omega_m \pm \Delta\Omega$, $\cos^2\phi_a = \cos^2\phi_t$. Accordingly, we obtain, with the help of Eq. (III.10.6),

$$\Omega_m = m\pi c/(n_a d_a \cos\theta_a) \qquad (m \text{ integer}), \tag{III.11.1}$$

and

$$\Delta\Omega = \frac{c}{n_a d_a \cos\theta_a} \arccos|(\hat{Z}_a - \hat{Z}_b)/(\hat{Z}_a + \hat{Z}_b)|. \tag{III.11.2}$$

In particular, for $n_b = n_a$, Eq. (III.11.2) ensures that $2\,\Delta\Omega = \Omega_{m+1} - \Omega_m$, which corresponds to the obvious statement that all frequencies are transmitted through a homogeneous dielectric. The central angular frequencies $\bar{\Omega}_m$ of the stopbands are given by

$$\bar{\Omega}_m = (m + 1/2)\pi c/(n_a d_a \cos\theta_a), \tag{III.11.3}$$

while $\Delta\bar{\Omega}$, the half-width of the stopband, is

$$\Delta\bar{\Omega} = \frac{c}{n_a d_a \cos\theta_a} \arcsin|(\hat{Z}_a - \hat{Z}_b)/(\hat{Z}_a + \hat{Z}_b)|. \tag{III.11.4}$$

In particular,

$$\Delta\bar{\Omega}/\bar{\Omega}_0 = (2/\pi)\arcsin|(\hat{Z}_a - \hat{Z}_b)/(\hat{Z}_a + \hat{Z}_b)|, \tag{III.11.5}$$

while the wavelength λ_0 *in vacuo* corresponding to the center $\bar{\Omega}_0$ of the first stopband is given by

$$n_a d_a \cos\theta_a = n_b d_b \cos\theta_b = \lambda_0/4. \tag{III.11.6}$$

It is worth noting that Ω_m and $\bar{\Omega}_m$ are independent of the polarization, while $\Delta\Omega$ and $\Delta\bar{\Omega}$ are not, except for the case of normal incidence.

Equations (III.11.1) and (III.9.27) imply the reduction of $\mathbf{M}_{ba}$ to the unity matrix at the center of any passband, so that $\gamma_+ = \gamma_- = 1$.

In the same way, at the center of any stopband $\mathbf{M}_{ba}$ reduces to

$$\mathbf{M}_{ba} = -\begin{bmatrix} \hat{Z}_a/\hat{Z}_b & 0 \\ 0 & \hat{Z}_b/\hat{Z}_a \end{bmatrix}, \tag{III.11.7}$$

so that $\gamma_+ = -\hat{Z}_a/\hat{Z}_b$, $\gamma_- = -\hat{Z}_b/\hat{Z}_a$, and $Z^{(+)} = \infty$, $Z^{(-)} = 0$.

For $\phi_a = \phi_b$ we can show by using Eqs. (III.10.8) and (III.9.27) that the impedances of the Bloch waves are given by

$$Z^{(\pm)} = \frac{i(\hat{Z}_a + \hat{Z}_b)\sin\phi_a\cos\phi_a}{\frac{1}{2}\left(\frac{\hat{Z}_a}{\hat{Z}_b} - \frac{\hat{Z}_b}{\hat{Z}_a}\right)\sin^2\phi_a \pm i\left[1 - \left(\cos^2\phi_a - \frac{1}{2}\left(\frac{\hat{Z}_a}{\hat{Z}_b} + \frac{\hat{Z}_b}{\hat{Z}_a}\right)\sin^2\phi_a\right)^2\right]^{1/2}}, \tag{III.11.8}$$

Next, using Eq. (III.10.6), we have $\hat{Z}_a/\hat{Z}_b = (1 + \cos\phi_t)/(1 - \cos\phi_t)$, so that $Z^{(\pm)}$ can be rewritten as

$$Z^{(\pm)} = i\hat{Z}_{\text{cb}} \frac{\sin\phi_t \cos\phi_a}{\cos\phi_t \sin\phi_a \pm i(\sin^2\phi_t - \sin^2\phi_a)^{1/2}}, \tag{III.11.9}$$

where $\hat{Z}_{\text{cb}}$ represents the Bloch impedance $Z^{(+)} = -Z^{(-)}$ at the center of a passband,

$$\hat{Z}_{\text{cb}} = (\hat{Z}_a + \hat{Z}_b)/(2 + \hat{Z}_a/\hat{Z}_b + \hat{Z}_b/\hat{Z}_a)^{1/2}. \tag{III.11.10}$$

According to the above expressions the product $Z^{(+)}Z^{(-)} = -Z_{\text{cb}}^2$ is constant. In addition, inside a stopband $\arg(Z^{(+)}) = \arg(Z^{-}) = \pi/2$. On the other hand, at the center of a stopband $\cos\phi_a = 0$, which in turn implies $Z^{(+)} \to \infty$ and $Z^{(-)} = 0$, as noted before. Inside a passband the amplitude $|Z^{(\pm)}| = Z_{\text{cb}}$ is constant while $\arg(Z^{(+)}) + \arg(Z^{(-)}) = \pi$. In addition, $\arg(Z^{(+)})$ increases monotonically from zero to $\pi/2$ moving from the center to the edge of a passband.

12 Reflection Coefficient of a Multilayer

An important problem is that of determining the relation between backward- and forward-traveling Bloch waves in a stack terminating on a substrate S (supposed indefinite), with impedance $\hat{Z}_{\text{S}}$. To this end, we note that the ratio between $V = V^{(+)} + V^{(-)}$ and $I = I^{(+)} + I^{(-)}$, on the substrate interface, is given by $\hat{Z}_{\text{S}}$, that is, according to Eqs. (III.10.7),

$$\frac{V^{(+)} + V^{(-)}}{V^{(+)}/Z^{(+)} + V^{(-)}/Z^{(-)}} = \hat{Z}_{\text{S}} \tag{III.12.1}$$

Thus, if we indicate by r_{S} the ratio $V^{(-)}/V^{(+)}$, we obtain

$$r_{\text{S}} = \frac{-Z^{(-)}}{Z^{(+)}} \frac{\hat{Z}_{\text{S}} - Z^{(+)}}{\hat{Z}_{\text{S}} - Z^{(-)}}. \tag{III.12.2}$$

At the center of a stopband $Z^{(+)} = \infty$, $Z^{(-)} = 0$ (see Section III.11), and $r_{\text{S}} = 0$, that is, it does not depend on the substrate. In this case, vanishing of the reflected wave does not imply absorption of the incident wave by the load. In fact, in a stopband $V^{(+)}$ represents a *stationary wave* (see once again the analogy with the electron waves in a crystal), and the ordinary picture in terms of fields propagating back and forth loses its meaning. According to Eqs. (III.10.2), the reflection coefficient r_q at the left section of the qth cell is

$$r_{q+1} = r_{\text{S}} e^{-2iq\delta} \qquad (q = 1, 2, \ldots, m). \tag{III.12.3}$$

Particular attention must be paid to the discrepancy between the above definition of reflection coefficient and that given in Section III.3.3. The former refers to Bloch waves, which can be interpreted as forward- and backward-traveling fields only in an approximate sense, because of the failure of this interpretation when approaching the stopband center. The latter refers to traveling waves in a strict sense. Of course, the two definitions approach one another for an almost homogeneous medium.

Let us now evaluate the reflection coefficient r of a structure of the type embedding medium–(HL)^m–substrate at the separation surface between the embedding medium and the stack. We note that the local impedance $\vec{Z}_q$ of the qth section is related to the reflection coefficient r_q and to $Z^{(+)}$ and $Z^{(-)}$ (which, as already noted, represent the local impedances for forward and backward Bloch waves) by the expression

$$\vec{Z}_q = Z^{(+)} \frac{1 + r_q}{1 + r_q Z^{(+)}/Z^{(-)}}. \tag{III.12.4}$$

As a consequence, the input impedance $\vec{Z}_{\text{in}}$ of the stack is given by

$$\vec{Z}_{\text{in}} = Z^{(+)} \frac{1 + e^{-2im\delta} r_{\text{S}}}{1 + e^{-2im\delta} r_{\text{S}} Z^{(+)}/Z^{(-)}}, \tag{III.12.5}$$

which can be expressed in terms of $\hat{Z}_{\text{S}}$, and of the quantities A, B, C, and D characterizing the stack, by using Eq. (III.9.5). The coefficient r is now obtained by applying Eq. (III.7.18) for the case $\vec{Z}(z) = \vec{Z}_{\text{in}}$, $\hat{Z}_q = \hat{Z}_{\text{I}}$, where $\hat{Z}_{\text{I}}$ represents the impedance of the embedding medium, and it reads

$$\begin{aligned} r &= \frac{Z^{(+)}Z^{(-)} - \hat{Z}_{\text{I}} Z^{(-)} + (Z^{(+)}Z^{(-)} - Z^{(+)}\hat{Z}_{\text{I}}) e^{-2im\delta} r_{\text{S}}}{Z^{(+)}Z^{(-)} + \hat{Z}_{\text{I}} Z^{(-)} + Z^{(+)}Z^{(-)} + Z^{(+)}\hat{Z}_{\text{I}}) e^{-2im\delta} r_{\text{S}}} \\ &= \frac{A\hat{Z}_{\text{S}} + B - C\hat{Z}_{\text{S}}\hat{Z}_{\text{I}} - D\hat{Z}_{\text{I}}}{A\hat{Z}_{\text{S}} + B + C\hat{Z}_{\text{S}}\hat{Z}_{\text{I}} + D\hat{Z}_{\text{I}}}. \end{aligned} \tag{III.12.6}$$

If we introduce the set of real parameters X, Y, W, and V, defined by

$$\begin{aligned} X &= \text{Re}(A/\hat{Z}_{\text{I}} + B/\hat{Z}_{\text{I}}\hat{Z}_{\text{S}}), \\ Y &= \text{Im}(B/Z_{\text{I}}\hat{Z}_{\text{S}}), \\ W &= \text{Re}(D/\hat{Z}_{\text{S}}), \\ V &= \text{Im}(C + D/\hat{Z}_{\text{S}}), \end{aligned} \tag{III.12.7}$$

we can easily show that Eq. (III.12.6) yields for the reflectance $R = |r|^2$

$$R = [(X - W)^2 + (Y - V)^2]/[(X + W)^2 + (Y + V)^2] \tag{III.12.8}$$

and for $\arg(r) \equiv \psi$

$$\tan\psi = \frac{(X + W)(Y - V) - (X - W)(Y + V)}{(X + W)(X - W) + (Y + V)(Y - V)}. \qquad \text{(III.12.9)}$$

It is important to notice that the reflectance of a lossless layered system seen from the embedding medium coincides with the reflectance seen from the substrate. This property can be easily proved by exploiting the *reciprocity theorem*, or by using the simple transformation properties of the **M** matrix under inversion of the positive propagation axis (see Problem 8). Accordingly, the reflectance and the transmittance $T = 1 - R$ can be considered characteristic of the stratified system, independent of the side of arrival of the incident wave.

12.1 *Airy Formula*

In some cases it is preferable to calculate the transmittance of a system as a function of the reflectance of the subsystems into which it can be decomposed. For instance, with reference to Fig. III.17, if we know the reflection coefficient of two adjacent interfaces, calculated by assuming as embedding medium the relative slab, supposed to be infinitely thick, we have the *Airy sum formula*, which provides the transmittance T of a spacer layer coated with two multilayer systems,

$$T = 1 - R = \frac{T_1 T_2}{(1 - R_g)^2}\left[1 + \frac{4R_g}{(1 - R_g)^2}\sin^2\frac{1}{2}(\psi_1 + \psi_2 - 2\phi)\right]^{-1}. \qquad \text{(III.12.10)}$$

Here T_1 and T_2 are the transmission coefficients of the two subsystems, $R_g = \sqrt{R_1 R_2}$, ψ_1 and ψ_2 are the phases of r_1 and r_2, and ϕ is the phase thick-

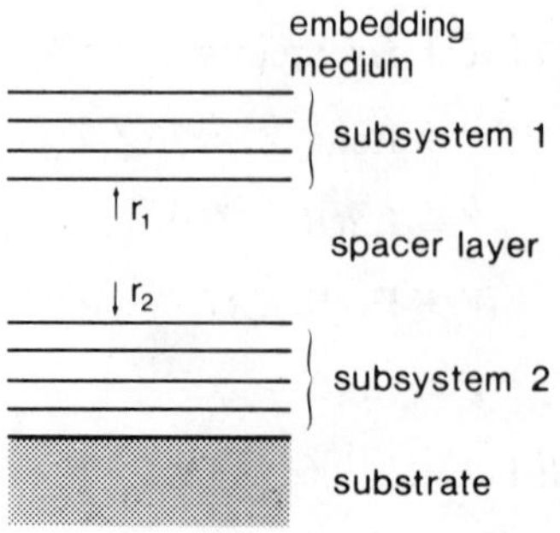

Fig. III.17. Geometry relative to the Airy sum.

ness of the layer between the two selected interfaces. The proof of this formula is left as an exercise for the reader (see, e.g., Stone [22]).

From the structure of Eq. (III.12.10) it can be seen that the transmittance of the whole system can be equal to unity only for $R_1 = R_2$ and $\sin(\psi_1 + \psi_2 - 2\phi)/2 = 0$. This means that a multilayer system symmetrical with respect to a central slab can be made highly transparent by controlling the relative phase thickness in such a way as to satisfy the above condition. The behavior of these structures is strictly related to that of Fabry–Perot interferometers, in which the physical separation between the two symmetric multilayers is changed in such a way as to control the frequencies yielding the transmission peaks (see Section VII.21).

When referring to an interferometer, the Airy formula is usually written as

$$T = T_{\max}\{1 + (2F_R/\pi)^2 \sin^2[\tfrac{1}{2}(\psi_1 + \psi_2 - 2\phi)]\}^{-1}, \qquad \text{(III.12.11)}$$

where

$$T_{\max} = T_1 T_2/[1 - (R_1 R_2)^{1/2}]^2 \qquad \text{(III.12.12)}$$

is the maximum transmissivity (also called *throughput*), while

$$F_R = \pi R_g^{1/2}/(1 - R_g) \qquad \text{(III.12.13)}$$

is the *finesse* of the interferometer (for additional details see Section VII.21).

12.2 *Quarter-Wave Stacks*

Inside the passband of a lossless quarter-wave stack ($\phi_a = \phi_b$) Eq. (III.12.6) simplifies to

$$r = \frac{Z^{(+)}}{Z^{(-)}} \frac{Z_{\mathrm{I}} - Z^{(-)}}{Z_{\mathrm{I}} + Z^{(+)}} \frac{r_{\mathrm{I}} - r_{\mathrm{S}} e^{-2im\delta}}{1 - r_{\mathrm{I}}^* r_{\mathrm{S}} e^{-2im\delta}} \qquad \text{(III.12.14)}$$

where r_{I} is the reflection coefficient relative to the embedding medium (lossless):

$$r_{\mathrm{I}} = -(Z^{(-)}/Z^{(+)})(Z_{\mathrm{I}} - Z^{(+)})/(Z_{\mathrm{I}} - Z^{(-)}), \qquad \text{(III.12.15)}$$

as a consequence of the relation $Z^{(-)} = -Z^{(+)*}$, which is valid for lossless dielectric films. Squaring the modulus of the above expression for r, we obtain

$$R = \frac{|r_{\mathrm{I}}|^2 + |r_{\mathrm{S}}|^2 + 2|r_{\mathrm{I}} r_{\mathrm{S}}| \cos(2m\delta + \phi)}{1 + |r_{\mathrm{I}} r_{\mathrm{S}}|^2 + 2|r_{\mathrm{I}} r_{\mathrm{S}}| \cos(2m\delta + \phi)}, \qquad \text{(III.12.16)}$$

with $\phi = \pi + \arg(r_{\mathrm{S}}) - \arg(r_{\mathrm{I}})$. Lord Rayleigh and others investigated the behavior of a multilayer in the limit $m \to \infty$. In this case, either of two conditions obtains for the reflectance: (1) R tends to 1 or (2) R is bounded

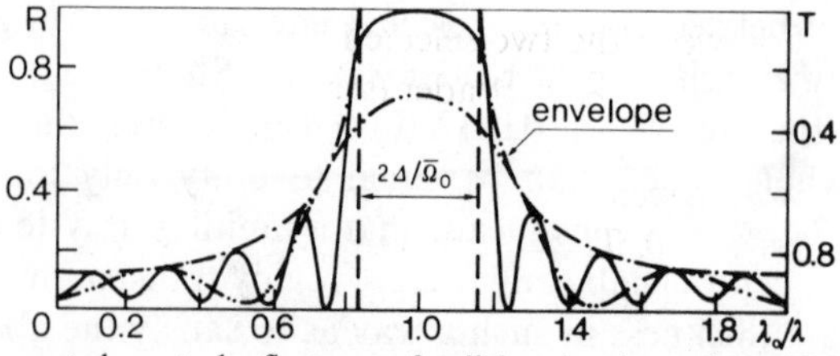

Fig. III.18. Computed spectral reflectance of a dielectric mirror glass–$(HL)^p$. Here $n_S = 1.51$, $n_H = 2.3$ (ZnS), $4n_H d_H = \lambda_0$; $n_L = 1.38$ (MgF_2), and $4n_L d_L = \lambda_0$. (—··—) $p = 2$; (——) $p = 5$. (From Baumeister [22a].)

between the limits $R_{\min}$ and $R_{\max}$, which define an envelope. If we consider R as a function of the continuous parameter m, we see immediately that

$$R_{\min} = \left(\frac{|r_I| - |r_S|}{1 - |r_I r_S|}\right)^2 < R < R_{\max} = \left(\frac{|r_I| + |r_S|}{1 + |r_I r_S|}\right)^2. \qquad \text{(III.12.17)}$$

In Fig. III.18 is plotted the reflectance of a five-double-layer stack versus the frequency normalized with respect to the center of a stopband. We observe that R oscillates between the maximum and minimum envelopes determined by $R_{\max}$ and $R_{\min}$ [see Arndt and Baumeister [23] for a derivation of $R_{\max}$ and $R_{\min}$ from Eq. (III.12.8)].

12.3 *Single-Layer Coatings*

When the coating reduces to a single layer, Eq. (III.12.5) gives the input impedance

$$\hat{Z}_{\text{in}} = \hat{Z}_S \frac{1 + i(\hat{Z}_c/\hat{Z}_S)\tan\phi_c}{1 + i(\hat{Z}_S/\hat{Z}_c)\tan\phi_c}, \qquad \text{(III.12.18)}$$

where the subscript "c" refers to the coating. As a consequence, the reflection coefficient for p-waves reads, with the help of Eq. (III.7.8),

$$r_p = \frac{(\hat{Z}_S - \hat{Z}_I)\hat{Z}_c + i(\hat{Z}_c^2 - \hat{Z}_S\hat{Z}_I)\tan\phi_c}{(\hat{Z}_S + \hat{Z}_I)Z_c + i(\hat{Z}_c^2 + \hat{Z}_S\hat{Z}_I)\tan\phi_c}$$

$$= \frac{\dfrac{\cos\theta_S}{n_S} - \dfrac{\cos\theta_I}{n_I} + i\left(\dfrac{\cos\theta_c}{n_c} - \dfrac{\cos\theta_S\cos\theta_I}{\cos\theta_c}\dfrac{n_c}{n_S n_I}\right)\tan\phi_c}{\dfrac{\cos\theta_S}{n_S} + \dfrac{\cos\theta_I}{n_I} + i\left(\dfrac{\cos\theta_c}{n_c} + \dfrac{\cos\theta_S\cos\theta_I}{\cos\theta_c}\dfrac{n_c}{n_S n_I}\right)\tan\phi_c}. \qquad \text{(III.12.19)}$$

For an assigned direction of the incident beam r_p is a periodic function of the angular frequency ω, $r_p(\omega) = r_p(\omega + \Omega)$, whose period Ω is given by

$$\Omega = \frac{2\pi c}{n_c d_c \cos\theta_c}. \qquad \text{(III.12.20)}$$

The extreme values of $|r_p(\omega)|$ are

$$|(\hat{Z}_S - \hat{Z}_I)/(\hat{Z}_S + \hat{Z}_I)|, \qquad |(\hat{Z}_c^2 - \hat{Z}_S\hat{Z}_I)/(\hat{Z}_c^2 + \hat{Z}_S\hat{Z}_I)|, \qquad \text{(III.12.21)}$$

corresponding to the angular frequencies

$$\omega = [c/(n_c d_c \cos\theta_c)]q\pi, \qquad \omega = [c/(n_c d_c \cos\theta_c)](2q+1)\pi/2, \qquad \text{(III.12.22)}$$

for every integer q. When $\hat{Z}_c = (\hat{Z}_S\hat{Z}_I)^{1/2}$, the minimum of r_p is equal to zero since the coating "transforms" the impedance of the substrate into that of the embedding medium, in full analogy with the quarter-wave transformers used to match two transmission lines. Analogous considerations can be developed for r_s.

12.4 *Reflectance of a Slab with a Slowly Varying Profile*

When n is a slowly varying function of z, we can asymptotically solve Eqs. (III.9.10), which describe the evolution of the **M** matrix components. More precisely, the representation of A given by Eq. (III.9.12) can be easily extended to the other components, yielding

$$\begin{aligned}
A(z,z') &\sim [\hat{Z}(z')/\hat{Z}(z)]^{-1/2}\cos k_0(S_0 - S_2/k_0^2 + \cdots),\\
B(z,z') &\sim -i[\hat{Z}(z')\hat{Z}(z)]^{1/2}\sin k_0(S_0 - S_2'/k_0^2 + \cdots),\\
C(z,z') &\sim -i[\hat{Z}(z)\hat{Z}(z')]^{-1/2}\sin k_0(S_0 - S_2/k_0^2 + \cdots),\\
D(z,z') &\sim [\hat{Z}(z)/\hat{Z}(z')]^{-1/2}\cos k_0(S_0 - S_2'/k_0^2 + \cdots),
\end{aligned} \qquad \text{(III.12.23)}$$

where S_0 and S_2 are defined by Eqs. (III.9.13) and S_2' is obtained by replacing $\hat{Z}$ with $1/\hat{Z}$ in Eq. (III.9.13b). Now, if we consider a lossless medium, inserting Eqs. (III.12.23) into Eq. (III.12.8) and setting $\hat{Z}(z) = \hat{Z}_I$, $\hat{Z}(z') = \hat{Z}_S$ yields

$$R = \tan^2[(S_2 - S_2')/2k_0]. \qquad \text{(III.12.24)}$$

13 Metallic and Dielectric Reflectors

Most reflectors used for ultraviolet, infrared, and visible radiation are fabricated by evaporating a metal on a polished substrate [24]. The spectral absorptances of several metal films are shown in Fig. III.15. Generally, except for rhodium, these absorptances can seldom be attained for practical purposes, due to oxidation and tarnish. This limits the use of silver to second-surface mirrors, even though it has the highest visible and infrared reflectance. For the great majority of applications aluminum is the preferred coating material because of its broad spectral band of high reflectivity and its reasonable durability when properly applied. There is a small dip in the reflectance of Al near 0.825 μm (1.4 eV) arising from a weak interband transition. Frequently, aluminum mirrors are coated with a thin layer of either magnesium fluoride or silicon monoxide either for protection or to enhance the reflectivity in the UV region.

Most metallic reflectors have a reflectivity less than 99%. When higher reflectances are needed, dielectric mirrors must be used. This is the case for total reflectors used in low-gain lasers and for interferometric systems. Dielectric multilayer coatings have very low absorption losses, so that partially transmitting mirrors can be easily fabricated. These are largely used as output couplers for stable optical resonators.

The simplest dielectric mirrors have the structure glass–(HL)^m, which at the center of a stopband presents an impedance [see Eqs. (III.12.2) and (III.12.5) and remember that $e^{i\delta} = -n_H/n_L$, $Z^{(+)} = \infty$, and $Z^{(-)} = 0$]

$$\vec{Z}_{in} = \hat{Z}_S e^{2im\delta} = \hat{Z}_S (n_H/n_L)^{2m}, \qquad \text{(III.13.1)}$$

which, with the help of Eq. (III.7.18), gives

$$r = \frac{\hat{Z}_S (n_H/n_L)^{2m} - \hat{Z}_I}{\hat{Z}_S (n_H/n_L)^{2m} + \hat{Z}_I}. \qquad \text{(III.13.2)}$$

For a given combination of refractive indices, we can increase the reflectance of a substrate by depositing a sufficiently large number of double layers operating at the center of a stopband. In these cases it is customary to express the dielectric mirror reflectance R as a function of the *standing wave ratio* V relative to the field pattern localized in front of the structure [25]. This parameter is widely used in microwaves [21] to characterize the impedance mismatching in a waveguide. In our case V measures the ratio of the maximum and minimum field amplitudes produced by the constructive and destructive interference between the incident and reflected plane waves. In the absence of reflection the field amplitude along the propagation axis is constant. When there is a reflected wave, the two waves interfere to produce a standing-wave pattern and the field is given by

$$E_x(z, t) = \mathrm{Re}[(e^{-ikz} + re^{ikz})e^{i\omega t}E^+], \qquad \text{(III.13.3)}$$

where r is the reflection coefficient. Accordingly, $E_x(z, t)$ oscillates back and forth between maximum values $|E^+|(1 + |r|)$ when z is such that $1 + re^{2ikz} = 1 + |r|$, and minimum values $|E^+|(1 - |r|)$ when $1 + e^{2ikz} = 1 - |r|$. The ratio of the maximum and minimum values at $\lambda = \lambda_0$ reads

$$V = \frac{1 + |r|}{1 - |r|} = \frac{|\vec{Z}_{in}|}{\hat{Z}_I} = \frac{n_I}{n_S}\left(\frac{n_H}{n_L}\right)^{2m}, \qquad \text{(III.13.4)}$$

(for those structures for which $|\vec{Z}_{in}| < Z_I$, $V = \hat{Z}_I/|\vec{Z}_{in}|$), and as a consequence

$$R = \left(\frac{V-1}{V+1}\right)^2$$
$$T = \frac{4V}{(V+1)^2} \qquad \left(\cong \frac{4}{V} \text{ for high-reflectivity mirror}\right). \tag{III.13.5}$$

If we introduce the dimensionless frequency $g = \lambda_0/\lambda$, where λ_0 refers to the center of the first stopband, the frequency difference Δg from the center of the first stopband at $g = 1$ to its edge, [see Eqs. (III.11.5), (III.9.15), and (III.11.7)]

$$\Delta g = (2/\pi)\arcsin[(n_H/n_L - 1)/(n_H/n_L + 1)]. \tag{III.13.6}$$

According to Eq. (III.13.4), V increases with the number of double layers. However, in practice V cannot exceed 10^3–10^4, no matter how many layers have been deposited. In fact, absorption and scattering losses in the thin films increase with m, thus limiting the maximum achievable reflectivity. Analogous considerations apply for the field inside the multilayer (see Problem 13).

The reflectance R of a periodic structure such as glass–$(\mathrm{HL})^m$ can be increased further by adding another H layer so that it now has the form glass–$(\mathrm{HL})^m\mathrm{H}$. In this case $V = [n_H^2/(n_S n_I)](n_H/n_L)^{2m}$.

By depositing one or more stacks on a suitable substrate and centering the stopbands on those wavelengths that are to be rejected, either a *long-wavelength pass* (LWP) or a *short wavelength pass* (SWP) filter can be fabricated (see Fig. III.19). LWP filters with at cutoff at 0.7 μm are used in film projectors and other optical instruments where most of the energy produced by lamps or carbon arcs is infrared radiation. The so-called *heat reflectors* or *cold mirrors* can efficiently protect the optical components against the heavy thermal loads.

Example: Calculation of the Spectral Reflectance of a Dielectric Mirror. Let us calculate the spectral reflectance at normal incidence of a

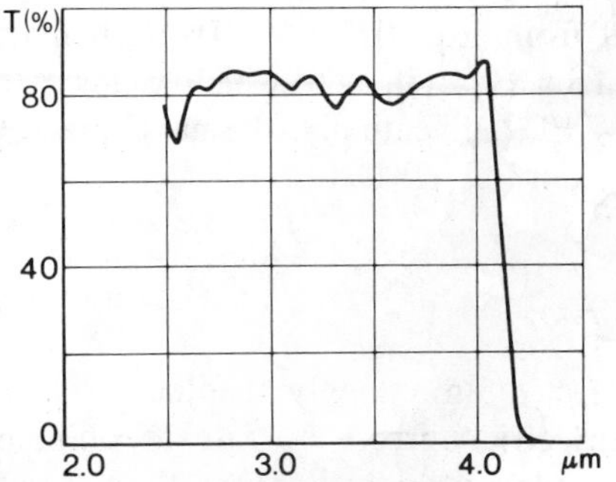

Fig. III.19. Spectral transmittance of a short-wavelength pass (SWP) filter (Optical Coating Laboratory).

dielectric mirror obtained by depositing 22 quarter-wave layers on a glass substrate [glass–$(HL)^{11}$]. We choose a wavelength λ_0 of maximum reflectivity equal to 660 nm. The high- and low-index dielectric materials are TiO_2 ($n_H = 2.3$) and SiO_2 ($n_L = 1.45$), respectively.

With these data, we can immediately calculate the auxiliary quantities $\cos\phi_t = (n_H - n_L)/(n_H + n_L) = 0.226$ and $\hat{Z}_{cb} = 0.548\zeta_0$ [see Eqs. (III.10.6) and (III.11.10)]. It will be convenient in the following to normalize the impedances to $\hat{Z}_{cb}$. For simplicity, we will use the same symbol for normalized and unnormalized quantities.

We now obtain for the dephasing angle δ, with the help of Eqs. (III.10.5) and (III.10.6), $\cos\delta = 1 - 2\sin^2\phi/\sin^2\phi_t = 1 - 2.108\sin^2\phi$, where $\phi = 2\pi nd/\lambda$ is the phase thickness of each layer. Thus, δ will be real for $0 \leqq \phi \leqq 1.34$ and $1.80 \leqq \phi \leqq 4.48$, while it will be of the form $\delta = \pi + i\delta''$ for $1.34 < \phi < 1.80$. Consequently, the stopband centered at $\lambda_0 = 660$ nm ($\phi = \pi/2$) extends from $\lambda_{min} = \pi\lambda_0/(2)(1.80) = 576$ nm to $\lambda_{max} = \pi\lambda_0/(2)(1.34) = 779$ nm. The impedances $Z^{(\pm)}$ will be given by $Z^{(+)} = i\exp(-i\psi)$ and $Z^{(-)} = i\exp(i\psi)$ in the passband, where $\psi = \arctan[(\tan^2\phi_t/\sin^2\phi - 1/\cos^2\phi_t)]^{1/2}$ [see Eq. (III.11.9)], while in the stopband $Z^{(+)} = iZ$ and $Z^{(t)} = i/Z$, with $Z = \sin\phi_t\cos\phi/[\cos\phi_t\sin\phi - (\sin^2\phi - \sin^2\phi_t)^{1/2}]$.

Because of the above results, the relevant quantity r_S [Eq. (III.12.2)] reads

$$
\begin{aligned}
r_S &= -\left(\frac{1 - [2Z_S/(1 - Z_S^2)]\sin\psi}{1 + [2Z_S/(1 + Z_S^2)]\sin\psi}\right)^{1/2} \\
&\quad \times \exp i\left(2\psi - \arctan\frac{\sin\psi}{\cos 2\omega + Z_S}\right) \qquad \text{passband} \\
&= -\left(\frac{1 + Z_S^2/Z^2}{1 + Z_S^2 Z^2}\right)^{1/2} \\
&\quad \times \exp i[\arctan 1/(Z_S Z) - \arctan(Z/Z_S)] \qquad \text{stopband}
\end{aligned}
\tag{III.13.7}
$$

where $Z_S = 1/(0.548 n_S) = 1.2$ for $n_S = 1.52$. The quantity r_I is given by an expression similar to the above in which Z_S is replaced by $Z_I = 1.82$ for $n_I = 1$.

The reflectivity R can be obtained from Eq. (III.12.16). In particular, at the stopband center r is also given by Eq. (III.13.2), which yields $r = 1 - 4.66 \times 10^{-5}$, $R = |r|^2 = 1 - 0.93 \times 10^{-4}$. Eventually, the envelopes of the reflectivity can be easily calculated from Eq. (III.12.17).

13.1 *Operation at Oblique Incidence*

In some cases—for example, in ring lasers or in systems with folded cavities (dye lasers, cavity dumpers, etc.)—dielectric mirrors operate at oblique incidence. In these cases the layers are made a little thicker than the normal

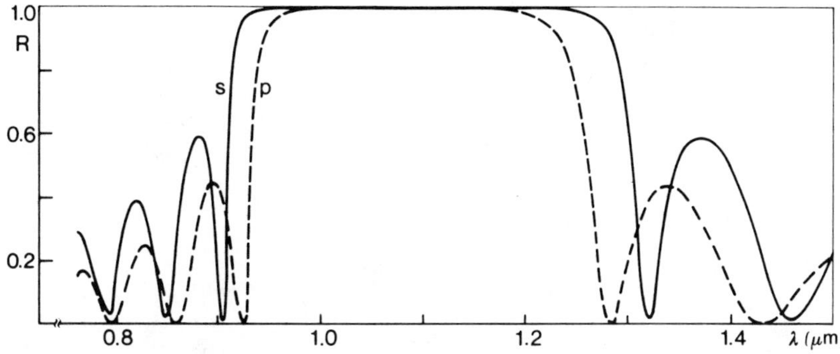

Fig. III. 20. Computed reflectance of a dielectric mirror $(HL)^8HG$ designed for $\lambda_0 = 1.06\ \mu m$ (Nd laser) and working at an incidence angle of 33.5° (Courtesy of C. Misiano, Selenia Industrie Elettroniche).

laser mirrors in order to compensate the factor $\cos\theta$, which occurs in the resonance condition $4nd\cos\theta = \lambda_0$. Because of the different values of $Z^{(h)}$ and $Z^{(e)}$, the reflectance is different for s- and p-polarizations. In these cases, for slabs of equal phase thickness, V_s and V_p at the stopband center are given, respectively, by

$$V_S(\theta) = \left(\frac{n_H^2 - n_I^2\sin^2\theta}{n_L^2 - n_I^2\sin^2\theta}\right)^m \frac{n_I\cos\theta}{(n_S^2 - n_I^2\sin^2\theta)^{1/2}}, \qquad \text{(III.13.8a)}$$

$$V_p(\theta) = \left(\frac{n_H}{n_L}\right)^{4m} \frac{n_I^2}{n_S^2}\frac{1}{V_s(\theta)}, \qquad \text{(III.13.8b)}$$

θ being the incidence angle. Since $V_s > 1$ ($n_H > n_L$) and $V_s > V_p$, $R_s > R_p$. This explains why most ring lasers have s-polarization. Analogously, it follows from Eq. (III.14.4) that $\Delta\bar{\Omega}_s > \Delta\bar{\Omega}_p$. Figure III.20 shows the spectral reflectance of a mirror operating at an incidence angle of 33.5°. The difference between the stopbands for the two polarizations is evident. The difference in the peak reflectivity cannot be observed in this case without expanding the scale.

When a dielectric mirror designed for normal incidence is operated at nonnormal incidence, up to about 30°, there is frequently little degradation in the response. In general, the effect of increasing the incidence angle is a shift in the whole reflectance curve down to slightly shorter wavelengths. This kind of behavior is evidenced by several naturally occurring periodic structures, e.g., in peacock or butterfly wings.

13.2 *Polarizing Beamsplitters*

The difference in reflectance between the p and s components can be used to implement polarizing beamsplitters [26], which are widely used in several electro-optic devices. The most popular version of these polarizers was

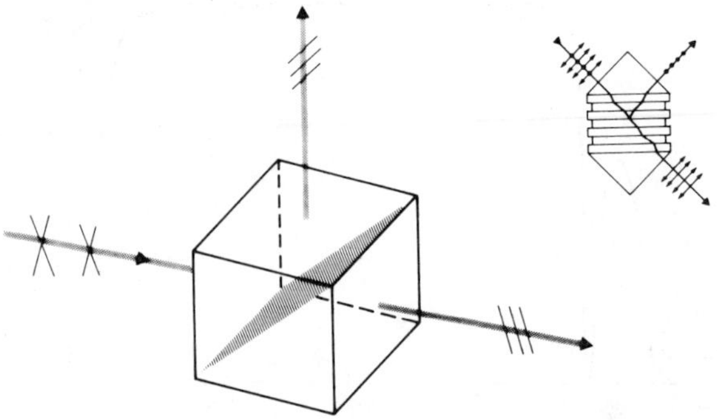

Fig. III.21. Schematic diagram of a MacNeille polarizing beamsplitter. The multilayer sandwich can be considered as the limiting form of the plate polarizer illustrated in Fig. III.13.

patented by MacNeille in 1946 and later developed by Banning. It consists of a multilayer stack deposited on the hypotenuse of a Porro prism, which is subsequently cemented to an identical prism to form a cemented cube (see Fig. III.21).

The multilayer structure adopted is [27, 28] primarily designed to produce the highest reflectance for the s-component and the highest transmittance for the p-component. In addition, a certain tolerance in the acceptance angle is desirable in order to permit the use of diverging beams. As long as the phase thickness of each layer is $\lambda_0/2$, any of the following designs satisfy the above conditions: $(LH)^m$, $(H/2, L, H/2)^m$, and $(L/2, H, L/2)^m$. Generally, these systems are designed to operate at an incidence angle of 45°.

It follows from Eqs. (III.13.8) that to obtain zero reflectance for the p-component at $\theta = \pi/4$, that is, $V_p(\pi/4) = 1$, the relation $V_s(\pi/4) = V_s^2(0)$ must be fulfilled. Using Eq. (III.13.8a) for $\theta = \pi/4$ and putting $n_l = n_S = n_{glass} \equiv n_G$, we obtain

$$[(2n_H^2 - n_G^2)/(2n_L^2 - n_G^2)]^m = (n_H/n_L)^{4m}. \tag{III.13.9}$$

It can easily be shown that this equation is satisfied for any m only if the index of the glass is related to those of the multilayer by

$$n_G^2 = 2n_L^2 n_H^2/(n_L^2 + n_H^2). \tag{III.13.10}$$

The first system designed by Banning was based on a combination of zinc sulfide, which has an index of 2.3, and cryolite, with an index of 1.25. The index of the glass prism satisfying the above relation should be 1.55, a value very close to that of most optical glasses. More complex solutions have been worked out by Schroder and Schlafer [29] in order to improve the physical quality of these devices and to increase the bandwidth.

When the condition (III.13.10) is exactly fulfilled, the transmittance relative to the s-component can be made very small by increasing m, as shown by

$$T_{s'} = 4(n_L/n_H)^{4m} \tag{III.13.11}$$

14 Antireflection (AR) Coatings

At the beginning of this century, the British optical manufacturer Dannis Taylor was occasionally led to the artificial aging of glass surfaces by etching in order to reduce unwanted reflections from lenses. This effect was attributed to the formation of a graded-index region, which produced a smooth transition from glass to air. In the middle 1930s, with the development of vacuum evaporation processes, A. Smakula in Germany and J. Strong in the United States discovered the antireflection potential of evaporated single-dielectric layers. Later, in 1944, Walter Geffcken patented a process of antireflection coating with three dielectric layers. Since then, a host of AR multilayer systems have been investigated [30, 31], and procedures for synthesizing coatings with prescribed characteristics have been developed [32–34]. In the following, we will discuss a few of them.

We saw above that the reflectance of a dielectric can be reduced to zero and the relative transmittance increased to unity by depositing a $\lambda_0/4$ film having a refractive index equal to $\sqrt{n_S n_I}$. Unfortunately, this last condition cannot be exactly satisfied in most practical cases. For example, the most popular coating for optical glass, obtained by depositing a layer of MgF_2, cannot provide a reflectance lower than 1.26% when the embedding medium is air (see Fig. III.22a). In order to reduce the reflection losses exactly to zero it is necessary to use at least a double-layer coating. In fact, if the refractive indices of the two materials that we intend to deposit are assigned, we can obtain zero reflectivity by choosing the phase thicknesses in agreement with the following relations [30]:

$$\tan^2\phi_1 = n_1^2 \frac{(n_I - n_S)(n_I n_S - n_2^2)}{(n_S n_1^2 - n_I n_2^2)(n_S n_I - n_1^2)},$$
$$\tan^2\phi_2 = n_2^2 \frac{(n_I - n_S)(n_I n_S - n_1^2)}{(n_S n_1^2 - n_I n_2^2)(n_S n_I - n_2^2)}. \tag{III.14.1}$$

In particular, when $n_1^2 n_S = n_2^2 n_I$ then $\phi_1 = \phi_2 = \pi/2$. This coating is usually referred to as a *double-quarter* coating. When we look at the variation of R with frequency we notice that such a system is characterized by a typical V-shaped dependence (see Fig. III.22b), which has given rise to the term "V-coating."

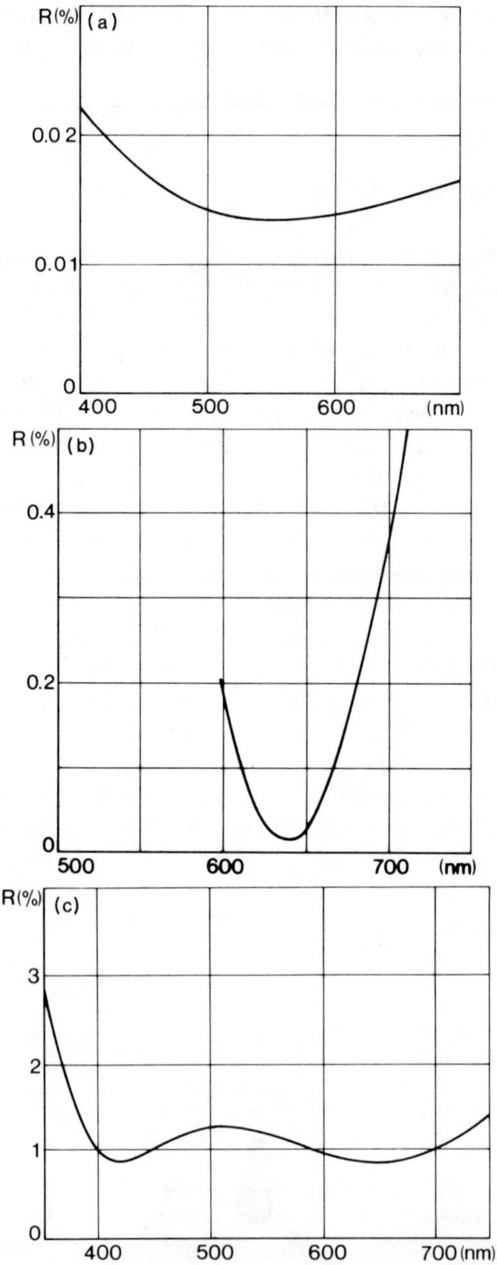

Fig. III.22. Reflectance of a glass substrate ($n_S = 1.51$) with different AR coatings. (a) Single layer of MgF_2 ($n = 1.38$) with optical thickness $n_c d_c = 555/4$ nm; (b) two-layer V-coating: $n_1 = 1.38$ (MgF_2), $4n_1 d_1 = 633$ nm, $n_2 = 2.3$ (TiO_2), $4n_2 d_2 = 633$ nm (He–Ne laser line); (c) two-layer W-type coating: $n_1 = 1.38$ (MgF_2), $4n_1 d_1 = 510$ nm, $n_2 = 1.6$ (CeF_3), $4n_2 d_2 = 1020$ nm. (After Musset and Thelen [30].)

When a low reflectance in a broader frequency interval is desired, it is preferable to employ the so-called *quarter–half*, consisting of a half-wave film in contact with the substrate, and a quarter-wave film. By using Eqs. (III.12.8) and (III.9.8) and setting $R = X/(1 + X)$, it can easily be shown that at a generic frequency for which $\phi = \phi_1 = \phi_2/2$

$$X = \frac{n_S}{4n_I}\left\{\left(\frac{n_I}{n_1} - \frac{n_1}{n_S}\right)^2 + \left[\left(\frac{n_I}{n_S}\right)^2\left(1 + 2\frac{n_2}{n_1}\right)^2 + \left(1 + 2\frac{n_1}{n_2}\right)^2 - \left(\frac{n_I}{n_1}\right)^2\left(5 + 4\frac{n_1}{n_2}\right) - \left(\frac{n_1}{n_S}\right)^2\left(5 + 4\frac{n_2}{n_1}\right)\right]\cos^2\phi + A_4\cos^4\phi + A_6\cos^6\phi\right\}, \tag{III.14.2}$$

where A_4 and A_6 are factors that depend on the refractive indices [30]. For $\phi = \pi/2$ we have

$$R = [(n_I n_S - n_1^2)/(n_I n_S + n_1)]^2. \tag{III.14.3}$$

The spectral reflectance exhibits a typical W shape with two minima. For this reason such a deposit is called a W-coating (see Fig. III.22c), or BBAR.

Other two-layer AR coatings useful for high-index substrates (e.g., $n_{Si} = 3.45$, $n_{Ge} = 4$) are based on the combinations $\phi_1 = \phi_2$ and $n_2 = n_1\sqrt{n_S/n_I}$ (the so-called *group I-type coating*) and $\phi_1 = \phi_2$ and $n_2 = n_S n_I/n_1$ (the so-called *group II-type coating*). The first type produces a single broad minimum with zero reflectance at $\phi = \pi/2$ (see Fig. III. 23). The second has a small maximum R_0 at $\phi = \pi/2$ and two zero points at $\phi = \arctan(R_{unc}/R_0)^{1/4}$, where R_{unc} represents the reflectivity of the uncoated sample.

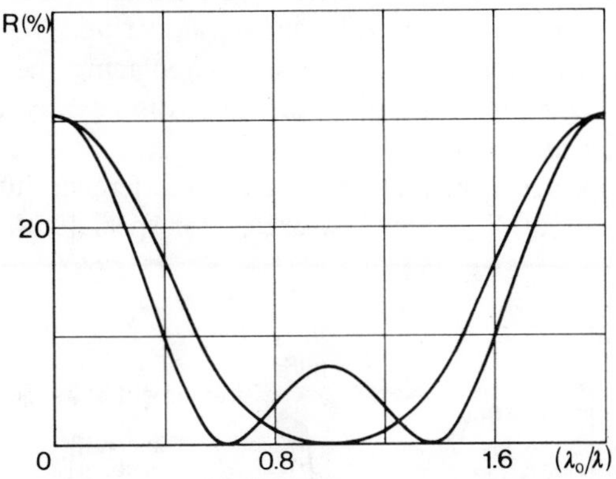

Fig. III.23. Reflectance of two-layer group I (single minimum) and group II (double minimum) coatings on a silicon substrate ($n_S = 3.45$), $n_I = 1$, $n_1 = 1.56$, $n_2 = 2.896$ (single minimum), 2.21 (double minimum) ($n_1 d_1 = n_2 d_2$). (From Musset and Thelen [30].)

Three-layer coatings are also used. In the so-called group I and group II types the phases ϕ_1, ϕ_2, and ϕ_3 are chosen equal while $n_2 = \sqrt{n_S n_I}$, $n_3 = n_1\sqrt{n_S/n_I}$ for the I type and $n_2 = \sqrt{n_S n_I}$, $n_3 = n_S n_I/n_1$ for the II type. These coatings produce, respectively, two and three zero-reflectance points.

Because of the lack of coating materials with a low enough index of refraction, the above systems cannot be used for glass substrates. In this case one can use the combination $n_2 = n_1\sqrt{n_D/n_I}, n_3 = \sqrt{n_D n_S}$, where n_D is the refractive index of an arbitrary very thin dummy layer interposed between layers 3 and 2. This combination of indices is derived by requiring that the transmittances of the two subsystems air–layer 1–layer 2–dummy medium and dummy medium–layer 3–substrate be equal to unity. If we use the two-layer group I-type combination for the first system and the $\lambda_0/4$ condition for the second system, we end up with the above combinations of indices.

The procedure described above can be extended to four-layer coatings synthesized by combining two group II-type coatings. Accordingly, the refractive indices must satisfy the relations $n_1 n_2 = n_I n_D$ and $n_3 n_4 = n_D n_S$. If we require that the reflectivity at λ_0 be zero, we must impose the additional condition $n_1 n_3/n_2 n_4 = \sqrt{n_I/n_S}$.

For more complex structures the reader is referred to Knittl [1], MacLeod [16], Musset and Thelen [30], Cox and Hass [31], and Dobrowolski [35].

15 Interference Filters

Interference filters depend for their action on the interference of the multiple reflected beams within a multilayer system. The simplest form of this filter is the three-layer coating invented by Geffecken in Germany in 1939. It consists of a dielectric *spacer layer*, having refractive index n_{sp}, coated on both sides with a semitransparent metal film. By replacing the metal film with dielectric multilayers it is possible to approximate any desired spectral characteristic of transmission [35].

The typical spectral transmittance of an interference filter is plotted in Fig. III.24. In general, T is very low except for special values of the wave-

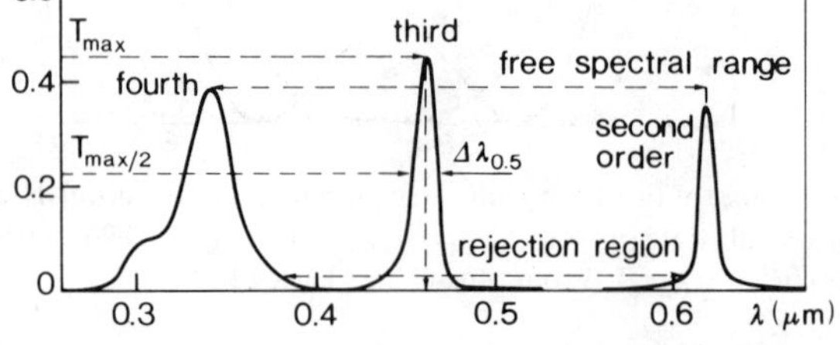

Fig. III.24. Typical spectral transmittance of a Fabry–Perot narrowband filter.

length that form the sequence $\bar{\lambda}_0/q$, $q = 1, 2, \ldots$, with $\bar{\lambda}_0 = 2\cos\theta\, dn_{sp}$, θ being the refraction angle inside the spacer of thickness d.

Accordingly, by changing the incidence angle we can shift the position of the transmission maxima. The quantity $q\pi$ represents the phase thickness of the spacer corresponding to the transmittance peaks and q is called the *spacer order*. For very large q ($> 10^3$), the ratio of the transverse dimension and the thickness of the film is an important parameter that affects the transmissivity. These devices can be best understood by treating them as optical resonators, and the reader is referred to Chapter VII for the pertinent theory. For low values of q the effects of the aperture are negligible and the system can be analyzed as a stratified medium extending indefinitely in the directions perpendicular to the stratification axis. The following analysis will be based on this assumption.

The transmittance spectrum is characterized by the quantities T_{max}, T_{min}, λ_0, $\Delta\lambda_{0.5}$, and the *free spectral range* (FSR). Here T_{max} represents the transmittance at the peak of the passband and T_{min} its value in the rejection region, while $\Delta\lambda_{0.5}$ is the half-width of the transmission line centered at λ_0. The distance between the two transmission maxima adjacent to the principal transmission band is called the free spectral range. It will become apparent later that these quantities are interconnected.

In Section III.12 we reported the Airy sum formula [Eq. (III.12.10)] giving the transmittance of the system. F_1DF_2, where F_1 and F_2 represent the coatings of the two sides of the dielectric spacer D. The films may be metallic or dielectric and, in special cases, might just be simple boundaries between D and the surrounding medium. If R_1 and R_2 indicate the reflectivities of the interfaces F_1–D and F_2–D, Eq. (III.12.10) yields

$$T = T_{max}/(1 + F^2 \sin^2\theta), \tag{III.15.1}$$

where F, the *finesse*, is given by

$$F = 2(R_1 R_2)^{1/4}/(1 - \sqrt{R_1 R_2}), \tag{III.15.2}$$

while

$$\theta = \tfrac{1}{2}(\psi_1 + \psi_2 - 2\phi). \tag{III.15.3}$$

The effect of small absorption losses A is serious in Fabry–Perot filters. For a symmetric structure the maximum transmittance is given by $T_{max} = 1/(1 + A/T)^2$ where $T = T_1 = T_2$. Since T is generally quite small the quantity A/T can be comparable with unity and T_{max} can be of the order of 25%.

If we indicate by λ_0 a wavelength for which $T = T_{max}$, that is,

$$\psi_1(\lambda_0) + \psi_2(\lambda_0) - 2\phi(\lambda_0) = -2\pi q\pi, \tag{III.15.4}$$

then, by taking into account that $\phi(\lambda) = 2\pi d n_{sp} \cos\theta/\lambda$, we can write for $\lambda \cong \lambda_0$

$$\psi_1(\lambda) + \psi_2(\lambda) - 2\phi(\lambda) = -2q\pi + \frac{\partial}{\partial\lambda}(\psi_1 + \psi_2)|_{\lambda_0}(\lambda - \lambda_0) + 2\frac{\phi(\lambda_0)}{\lambda_0}(\lambda - \lambda_0)$$

$$\equiv -2q\pi + \alpha(\lambda - \lambda_0) \qquad \text{(III.15.5)}$$

and

$$T(\lambda) = \frac{T_{\max}}{1 + 4(\lambda - \lambda_0)^2/\Delta\lambda_{0.5}^2}, \qquad \text{(III.15.6)}$$

where

$$\Delta\lambda_{0.5} = \frac{1}{(\partial/\partial\lambda)(\psi_1 + \psi_2) + 2\phi(\lambda_0)/\lambda_0}\frac{4}{F} \qquad \text{(III.15.7)}$$

represents the half-width of the transmission band and $F = \pi R_g^{1/2}/(1 - R_g)$ is the finesse of the interference filter [see Eq. (III.15.2)]. According to Eq. (III.15.6), a Fabry–Perot filter has a *lorentzian line shape*. In particular, the 1/100 width (called the *base width* and indicated by $\Delta\lambda_{0.01}$) is about $10\,\Delta\lambda_{0.5}$. The ratio $\Delta\lambda_{0.01}/\Delta\lambda_{0.5}$ is called the *shape factor* and indicates how square the transmission band is. For a Fabry–Perot filter having a lorentzian shape, this factor is equal to 10, while for a gaussian profile it is equal to 2.57.

For large deviations from λ_0 the response of a Fabry–Perot filter depends on the behavior of the spectral reflectances R_1 and R_2. For a spacer coated with metallic films, the reflectances are quite insensitive to the frequency in the visible and IR ranges. In this case the transmissivity is a periodic function of λ^{-1}, as mentioned before. For $\bar{\lambda}_0$ corresponding to a spacer thickness $\phi = \pi + (\psi_1 + \psi_2)/2$, the peaks of T are situated at $\lambda = \bar{\lambda}_0 \times [q + (\psi_1 + \psi_2)/2\pi]^{-1}$, the index q being the *order* of the transmission maximum, so that the free spectral range, measuring the maximum allowable bandwidth of a beam from which we can select a single narrow line of given order, is $2\bar{\lambda}_0^{-1}$. Usually, the order of interference of the transmitted band is the first or the second and the peak transmissivity is between 20 and 40%. Since the phase dispersion of the reflection coefficient of metallic films is negligible, the half-width is given for $(\psi_1 + \psi_2)/2\pi \ll 1$ by

$$\Delta\lambda_{0.5} = (2/F)(\lambda_0/q\pi) = [(1 - R_g)/R_g^{1/2}](\bar{\lambda}_0/q\pi), \qquad \text{(III.15.8)}$$

while the *rejection ratio* $T_{\min}/T_{\max}$ is related to the half-width through

$$\frac{T_{\min}}{T_{\max}} = \frac{1}{1 + F^2} = \frac{1}{1 + [2\bar{\lambda}_0/(q\pi\Delta\lambda_{0.5})]^2} \qquad \text{(III.15.9)}$$

The last two quantities depend on the order of the spacer and the reflectivity of the coating. Typically, $\Delta\lambda_{0.5}/\lambda_0$ for $q = 1$ is between 1 and 8%.

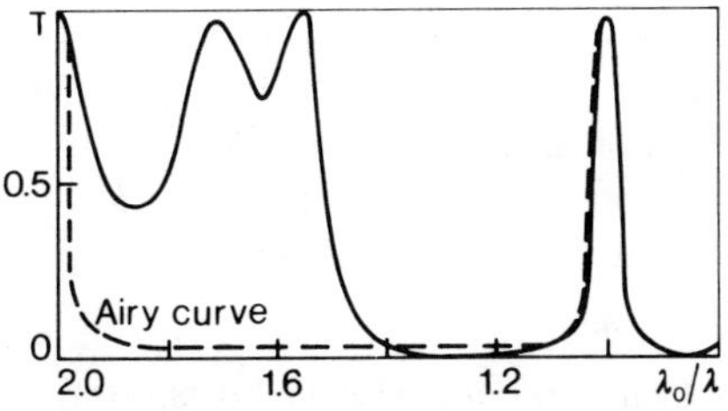

Fig. III.25. Fabry–Perot band shape from Airy sum (---) and computed transmittance of the filter HLHHLH (——). (From Smith [37].)

If we want to reduce the filter linewidth, it is necessary to increase the reflectance of the metal film, which can be done by increasing its thickness. Unfortunately, this implies an increase of the absorption losses, to the detriment of T_{max}. To obtain filters with half-widths less than 1% it is necessary to use dielectric mirrors fabricated by depositing on the two faces of the spacer some suitable multilayers. In this case, due to the resonant behavior of such structures, R_1 and R_2 depend strongly on the frequency [36], so that the resulting transmittance of the filter deviates from the simple Airy profile (see Fig. III.25) on the higher-frequency side of the first-order peak [37]. Despite the loss of periodicity, dielectric multilayers are invariably used in the UV, where metals are almost transparent. Examples of these structures are given by $(HL)^m HH(LH)^m = [(HL)^m H]^2$.

If we integrate two F–P systems in the same multilayer structure [37], we obtain a system of the form $(HL)^m HH(LH)^m C(HL)^m HH(LH)^m$, with C a coupling layer. In this case the coupling of two resonances produces a splitting of the two peaks and the relative transmittance curve presents a small dip at λ_0. If we increase the number of Fabry–Perot systems we can obtain a shape with a more rapid transition from stopband to passband and a flatter top. The shape factors are approximately equal to 3.5, 2, and 1.5 for two, three, and four cavities, respectively. In addition, the half-width and the rejection ratio can be varied independently. They are generally preferred to metallic Fabry–Perot filters because of the higher transmittance and the sharper transition.

16 Anisotropic Stratified Media

Whenever the stratified medium is characterized by a tensorial dielectric constant, the calculation of the field generated by an incident plane wave becomes tremendously involved. The reason for this complexity can be easily understood if we remember that, for each propagation direction, a plane wave can propagate only if the electric field is suitably oriented. In addition, if we fix two of the four parameters $\omega, \mathbf{k}$ on which the normal modes depend, the other two are determined through the dispersion relation $D(\omega, \mathbf{k}) = 0$.

In order to clarify these remarks, let us consider the stratified medium of Fig. III.8, and assume field solutions depending on the transverse coordinates x, y according to the simple law $\exp(-ikx)$. This means that for each slab two different normal waves can exist, with the $k_z(\omega; k_x)$ propagation constant taking either of the two values determined by the dispersion relation.

16.1 *Four-by-Four Characteristic Matrix Formalism*

We saw in Section III.7 that, for an isotropic medium, the equations for the TE and TM components separate. For the reasons mentioned above, this is no longer true for anisotropic media. In fact, for a homogeneous slab, Eqs. (III.7.3) are replaced by [20]

$$\left(\frac{i}{k_0}\right)\frac{d}{dz}\begin{bmatrix} V_\mathrm{e} \\ I_\mathrm{e} \\ V_\mathrm{h} \\ I_\mathrm{h} \end{bmatrix} = \mathbf{\Delta} \cdot \begin{bmatrix} V_\mathrm{e} \\ I_\mathrm{e} \\ V_\mathrm{h} \\ I_\mathrm{h} \end{bmatrix} \equiv \mathbf{\Delta} \cdot \boldsymbol{\Psi}, \tag{III.16.1}$$

where $\boldsymbol{\Psi}$ is a four-component column vector and $\mathbf{\Delta}$ is a matrix that reduces, for a medium characterized by a scalar permeability μ_0, to the expression

$$\Delta = \begin{bmatrix} \Delta_{11} & \zeta\Delta_{12} & \Delta_{13} & 0 \\ \Delta_{21}/\zeta_0 & \Delta_{11} & \Delta_{23}/\zeta_0 & 0 \\ 0 & 0 & 0 & \zeta \\ \Delta_{23}/\zeta_0 & \Delta_{13} & \Delta_{43}/\zeta_0 & 0 \end{bmatrix}, \tag{III.16.2}$$

where

$$\begin{aligned} &\Delta_{11} = -(k_x/k_0)\varepsilon_{xz}/\varepsilon_{zz}, && \Delta_{12} = 1 - (k_x^2/k_0^2)\varepsilon_0/\varepsilon_{zz}, \\ &\Delta_{13} = (k_x/k_0)\varepsilon_{yz}/\varepsilon_{zz}, && \\ &\Delta_{21} = (\varepsilon_{xx}/\varepsilon_0) - \varepsilon_{xz}^2/(\varepsilon_0\varepsilon_{zz}), && \Delta_{23} = -\varepsilon_{xy}/\varepsilon_0 + \varepsilon_{xz}\varepsilon_{yz}/(\varepsilon_0\varepsilon_{zz}), \\ &\Delta_{43} = \varepsilon_{yy}/\varepsilon_0 - \varepsilon_{yz}^2/(\varepsilon_0\varepsilon_{zz}) - k_x^2/k_0^2. && \end{aligned} \tag{III.16.3}$$

The components Δ_{ij} for more general media can be found in Berreman [20]. Let us now indicate by $k_{z(j)}$ the four eigenvalues of $k_0\mathbf{\Delta}$ that can be obtained by solving the quartic polynomial equation in $k_{z(j)}$ resulting from expanding the determinantal equation $\det(k_0\mathbf{\Delta} - k_z) = 0$. From a physical point of view, the $k_{z(j)}$ are the z components of the four normal waves that can propagate in the crystal for assigned k_x and k_y equal to zero. For a uniaxial crystal, two are ordinary waves and the other two are extraordinary ones, so that we have

$$k_z = \begin{cases} (n_o^2 k_0^2 - k_x^2)^{1/2} \equiv k_{z(1)}, \\ -(n_o^2 k_0^2 - k_x^2)^{1/2} \equiv k_{z(2)}, \\ (n_{e1}^2 k_0^2 - k_x^2)^{1/2} \equiv k_{z(3)}, \\ -(n_{e2}^2 k_0^2 - k_x^2)^{1/2} \equiv k_{z(4)}, \end{cases} \qquad \text{(III.16.4)}$$

where n_o is the refractive index of the ordinary wave, while n_{e1} and n_{e2} are the refractive indices of the extraordinary waves, whose k_z is, respectively, positive and negative. In addition, one can easily show that $k_{z(3)}$ and $k_{z(4)}$ satisfy the quadratic equation

$$\begin{aligned} &k_z^2[1 + c_z^2(n_e^2 - n_o^2)/n_o^2] + 2k_z k_x c_x c_z(n_e^2 - n_o^2)/n_o^2 \\ &\quad + k_x^2[1 + c_x^2(n_e^2 - n_o^2)/n_o^2] - n_e^2 k_0^2 = 0, \end{aligned} \qquad \text{(III.16.5)}$$

where c_x and c_z are the direction cosines of the optic axis. In most practical cases, n_e is almost coincident with n_o.

Having obtained the four eigenvalues $k_{z(j)}$, we evaluate the right and left eigenvectors $\boldsymbol{\Psi}_j$ and $\tilde{\boldsymbol{\Psi}}_j$ of $\boldsymbol{\Delta}$, viz.

$$\begin{aligned} k_0 \boldsymbol{\Delta} \cdot \boldsymbol{\Psi}_j &= k_{z(j)} \boldsymbol{\Psi}_j, \\ k_0 \tilde{\boldsymbol{\Psi}}_j \cdot \boldsymbol{\Delta} &= k_{z(j)} \tilde{\boldsymbol{\Psi}}_j, \end{aligned} \qquad \text{(III.16.6)}$$

where each $\boldsymbol{\Psi}_j = (\Psi_{j,1}, \Psi_{j,2}, \Psi_{j,3}, \Psi_{j,4})$ is a four-component column vector. Then, following the standard procedure outlined in Section III.9, we express $k_0 \boldsymbol{\Delta}$ in the form

$$k_0 \boldsymbol{\Delta} = \sum_j \frac{k_{z(j)}}{\tilde{\boldsymbol{\Psi}}_j \cdot \boldsymbol{\Psi}_j} \boldsymbol{\Psi}_j \tilde{\boldsymbol{\Psi}}_j \equiv \sum_j k_{z(j)} \mathbf{O}_j, \qquad \text{(III.16.7)}$$

where the $\mathbf{O}_j$ are projectors, in the sense that $\mathbf{O}_j^n = \mathbf{O}_{j'}$ and $\mathbf{O}_j \cdot \mathbf{O}_{j'} = 0$, for $j \neq j'$. For a homogeneous slab, Eq. (III.16.1) can be integrated by using the representation Eq. (III.16.7) of $\boldsymbol{\Delta}$, which yields

$$\boldsymbol{\Psi}(z) = \left[\sum_{j=1}^{4} e^{-ik_{z(j)}(z-z')} \mathbf{O}_j \right] \cdot \boldsymbol{\Psi}(z') \equiv \mathbf{M}(z - z') \cdot \boldsymbol{\Psi}(z'), \qquad \text{(III.16.8)}$$

where $\mathbf{M}$ is the 4×4 version of the characteristic matrix of the slab, which can be used in a way similar to that followed for the 2×2 matrix.

16.2 *Reflection and Transmission Coefficients of a Multilayer*

Let us consider a multilayer, described by the 4×4 characteristic matrix $\mathbf{M}$, interfaced on the right and left sides, respectively, with a substrate and an incident medium having infinite extension. If a wave represented by the 4-

vector $\boldsymbol{\Psi}^{(\mathrm{i})}$ is incident on interface 2, then a reflected wave $\boldsymbol{\Psi}^{(\mathrm{r})}$ originates in the incident medium and a wave $\boldsymbol{\Psi}^{(\mathrm{t})}$ is transmitted to the substrate. In particular, $\boldsymbol{\Psi}^{(\mathrm{r})}$ is given by a superposition of the two eigenvectors of the incident medium representing waves traveling from right to left. According to the convention established in Eqs. (III.16.4) for uniaxial media, which remains valid for biaxial crystals, the above two waves are labeled with the indices 2 and 4, while those moving from left to right have indices 1 and 3. Thus, we can write $\boldsymbol{\Psi}^{(\mathrm{i})}$, $\boldsymbol{\Psi}^{(\mathrm{r})}$, and $\boldsymbol{\Psi}^{(\mathrm{t})}$ in the form

$$\begin{aligned}\boldsymbol{\Psi}^{(\mathrm{i})} &= A_1^{(\mathrm{I})}\boldsymbol{\Psi}_1^{(\mathrm{I})} + A_3^{(\mathrm{I})}\boldsymbol{\Psi}_3^{(\mathrm{I})},\\ \boldsymbol{\Psi}^{(\mathrm{r})} &= A_2^{(\mathrm{I})}\boldsymbol{\Psi}_2^{(\mathrm{I})} + A_4^{(\mathrm{I})}\boldsymbol{\Psi}_4^{(\mathrm{I})},\\ \boldsymbol{\Psi}^{(\mathrm{t})} &= A_1^{(\mathrm{S})}\boldsymbol{\Psi}_1^{(\mathrm{S})} + A_3^{(\mathrm{S})}\boldsymbol{\Psi}_3^{(\mathrm{S})}.\end{aligned} \tag{III.16.9}$$

where $\boldsymbol{\Psi}^{(\mathrm{I})}$ and $\boldsymbol{\Psi}^{(\mathrm{S})}$ refer, respectively, to the incident and substrate media.

Next, we relate $\boldsymbol{\Psi}^{(\mathrm{t})}$ to $\boldsymbol{\Psi}^{(\mathrm{i})}$ and $\boldsymbol{\Psi}^{(\mathrm{r})}$ by using the matrix $\mathbf{M}$, viz.

$$\boldsymbol{\Psi}^{(\mathrm{i})} + \boldsymbol{\Psi}^{(\mathrm{r})} = \mathbf{M}\cdot\boldsymbol{\Psi}^{(\mathrm{t})}. \tag{III.16.10}$$

Exploiting the orthogonality of the eigenvectors, we easily obtain from the above relations

$$\begin{aligned}T_{11}A_1^{(\mathrm{S})} + T_{13}A_3^{(\mathrm{S})} &= F_1,\\ T_{31}A_1^{(\mathrm{S})} + T_{33}A_3^{(\mathrm{S})} &= F_3,\\ R_{22}A_2^{(\mathrm{I})} + R_{24}A_4^{(\mathrm{I})} &= F_2,\\ R_{42}A_2^{(\mathrm{I})} + R_{44}A_4^{(\mathrm{I})} &= F_4,\end{aligned} \tag{III.16.11}$$

where $T_{ij} = \tilde{\boldsymbol{\Psi}}_i^{(\mathrm{I})}\cdot\mathbf{M}\cdot\boldsymbol{\Psi}_j^{(\mathrm{S})}$, $R_{ij} = \tilde{\boldsymbol{\Psi}}_i^{(\mathrm{S})}\cdot\mathbf{M}^{-1}\cdot\boldsymbol{\Psi}_j^{(\mathrm{I})}$, $F_{1,3} = \tilde{\boldsymbol{\Psi}}_{1,3}^{(\mathrm{I})}\cdot\boldsymbol{\Psi}^{(\mathrm{i})}$, and $F_{2,4} = \tilde{\boldsymbol{\Psi}}_{2,4}^{(\mathrm{S})}\cdot\mathbf{M}^{-1}\cdot\boldsymbol{\Psi}^{(\mathrm{i})}$. Solving the above system allows us to obtain the reflected and transmitted waves.

An interesting situation is that in which the slab has a vanishing thickness and the substrate is directly interfaced with the incident medium. In this case, $\mathbf{M}$ reduces to the unity matrix and the coefficients of the system of Eqs. (III.16.11) are given by $T_{ij} = \tilde{\boldsymbol{\Psi}}_i^{(\mathrm{I})}\cdot\boldsymbol{\Psi}_j^{(\mathrm{S})}$, $R_{ij} = \tilde{\boldsymbol{\Psi}}^{(\mathrm{S})}\cdot\boldsymbol{\Psi}_j^{(\mathrm{I})}$, $F_{1,3} = \tilde{\boldsymbol{\Psi}}_{2,3}^{(\mathrm{I})}\cdot\boldsymbol{\Psi}^{(\mathrm{i})}$, and $F_{2,4} = -\tilde{\boldsymbol{\Psi}}_{2,4}^{(\mathrm{S})}\cdot\boldsymbol{\Psi}^{(\mathrm{i})}$. In particular, when the incident medium is isotropic, we can choose $\boldsymbol{\Psi}_1^{(\mathrm{I})}$ and $\boldsymbol{\Psi}_3^{(\mathrm{I})}$, respectively, coincident with a p-wave ($\boldsymbol{\Psi}_{\mathrm{p}}^{(+)}$) and an s-wave ($\boldsymbol{\Psi}_{\mathrm{s}}^{(+)}$), and this is also possible for $\boldsymbol{\Psi}_2^{(\mathrm{I})}$ $(= \boldsymbol{\Psi}_{\mathrm{p}}^{(-)})$ and $\boldsymbol{\Psi}_4^{(\mathrm{I})}$ $(= \boldsymbol{\Psi}_{\mathrm{s}}^{(-)})$. Thus, for an incident p-wave $\boldsymbol{\Psi}^{(\mathrm{I})} = \boldsymbol{\Psi}_{\mathrm{p}}^{(+)}$,

$$\begin{aligned}\tilde{\boldsymbol{\Psi}}_2^{(\mathrm{S})}\cdot\boldsymbol{\Psi}_{\mathrm{p}}^{(-)}A_{\mathrm{p}}^{(-)} + \tilde{\boldsymbol{\Psi}}_2^{(\mathrm{S})}\cdot\boldsymbol{\Psi}_{\mathrm{s}}^{(-)}A_{\mathrm{s}}^{(-)} &= -\tilde{\boldsymbol{\Psi}}_2^{(\mathrm{S})}\cdot\boldsymbol{\Psi}_{\mathrm{p}}^{(+)},\\ \tilde{\boldsymbol{\Psi}}_4^{(\mathrm{S})}\cdot\boldsymbol{\Psi}_{\mathrm{p}}^{(-)}A_{\mathrm{p}}^{(-)} + \tilde{\boldsymbol{\Psi}}_4^{(\mathrm{S})}\cdot\boldsymbol{\Psi}_{\mathrm{s}}^{(-)}A_{\mathrm{s}}^{(-)} &= -\tilde{\boldsymbol{\Psi}}_4^{(\mathrm{S})}\cdot\boldsymbol{\Psi}_{\mathrm{p}}^{(+)}.\end{aligned} \tag{III.16.12}$$

Since $A_{\mathrm{s}}^{(-)}$ in general does not vanish, the reflected wave is formed by a superposition of p- and s-waves even when the incident field is a pure p-wave. An analogous result can be easily proved for an incident s-wave.

If we define the vectors $\boldsymbol{\Psi}_{\rm p}^{(\pm)}$ and $\boldsymbol{\Psi}_{\rm s}^{(\pm)}$ in such a way that the respective electric fields have unit amplitudes, then the coefficients $A_{\rm p,s}^{(-)}$ of the above system represent the reflection coefficients of the anisotropic medium. We can use the system of Eqs. (III.16.12) together with Eqs. (III.16.5) to obtain the following reflection coefficients for uniaxial crystals, originally derived by Drude for an optic axis normal to the plane of incidence:

$$r_{\rm s} = \frac{n\cos\theta - (n_{\rm e}^2 - n^2\sin^2\theta)^{1/2}}{n\cos\theta + (n_{\rm e}^2 - n^2\sin^2\theta)^{1/2}},$$
$$r_{\rm p} = \frac{n_{\rm o}^2\cos\theta - n(n_{\rm o}^2 - n^2\sin^2\theta)^{1/2}}{n_{\rm o}^2\cos\theta + n(n_{\rm o}^2 - n^2\sin^2\theta)^{1/2}}, \tag{III.16.13}$$

and an optic axis parallel to the plane of incidence and forming an angle α with the normal to the surface:

$$r_{\rm s} = \frac{n\cos\theta - (n_{\rm o}^2 - n^2\sin^2\theta)^{1/2}}{n\cos\theta + (n_{\rm o}^2 - n^2\sin^2\theta)^{1/2}},$$
$$r_{\rm p} = \frac{n_{\rm o}n_{\rm e}\cos\theta - n(n_{\rm e}^2\cos^2\alpha + n_{\rm o}^2\sin^2\alpha - n^2\sin^2\theta)^{1/2}}{n_{\rm o}n_{\rm e}\cos\theta + n(n_{\rm e}^2\cos^2\alpha + n_{\rm o}^2\sin^2\alpha - n^2\sin^2\theta)^{1/2}}, \tag{III.16.14}$$

where $n_{\rm o}$ and $n_{\rm e}$ are the ordinary and extraordinary refractive indices of the substrate, n is the index of the incident medium, and θ is the incidence angle.

17 Propagation through Periodic Media

There are many examples in holography [38], integrated optics [39], and semiconductor lasers of waves propagating through media characterized by index profiles having the form

$$n^2(z) = n_0^2\left[1 + 2\sum_{m=1}^{\infty} X_m \cos\left(\frac{2\pi}{\Lambda}mz\right)\right]. \tag{III.17.1}$$

For example, when the thickness of the material used to record a hologram is larger than the detail of the recorded diffraction pattern, the hologram acquires the properties of a three-dimensional diffraction grating. Then the diffraction must be described in terms of Bragg angle reflections, analogous to x-ray diffraction from crystals. The properties of these systems—that is, the sensitivity of the reconstructed images to the incidence angle and wavelength of the beam used to read the hologram—can be discussed by considering a sinusoidal spatial grating, such as that obtained when the emulsion is exposed to the interference pattern of two generic plane waves. In this case the refractive index profile is represented by Eq. (III.17.1) with the z axis parallel to $\mathbf{k}_1 - \mathbf{k}_2$, the $\mathbf{k}$'s being the wave vectors of the two plane waves.

Advances in microelectronics and thin-film technology have permitted the fabrication of corrugated thin-film waveguides. These have been applied to many optical devices in integrated optics and semiconductor lasers, such as distributed feedback (DFB) lasers, distributed Bragg reflector (DBR) lasers, wavelength filters, beam couplers, and directional couplers. These structures have in common the propagation through a dielectric thin film whose thickness varies periodically, a circumstance that produces, approximately, a corresponding periodic modulation of the transverse propagation constant k_x. Since the longitudinal wave number k_z is related to k_x through $k_z^2 + k_x^2 = k_0^2 n_f^2$, n_f being the index of the film, the periodic change in k_x results in a corresponding change in k_z. Accordingly, the longitudinal propagation occurs as in a medium periodically stratified along z (see Figs. III.26 and III.27).

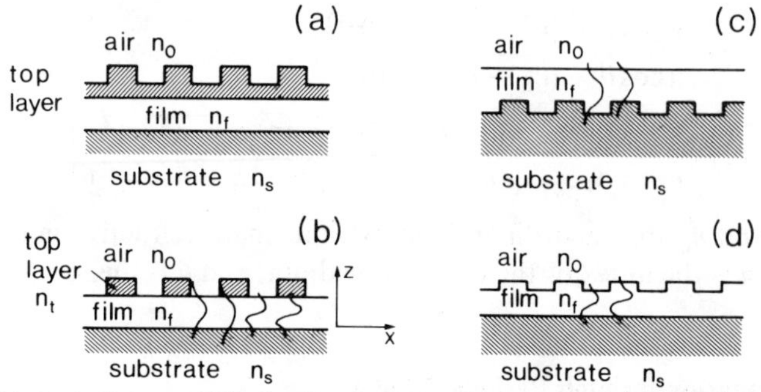

Fig. III.26. Periodic waveguides for distributed feedback lasers. (a) and (b) Periodic variation in thickness of the top dielectric layer. (c) and (d) Periodic variation in thickness of the film. (From Wang [39a] © 1974 IEEE.)

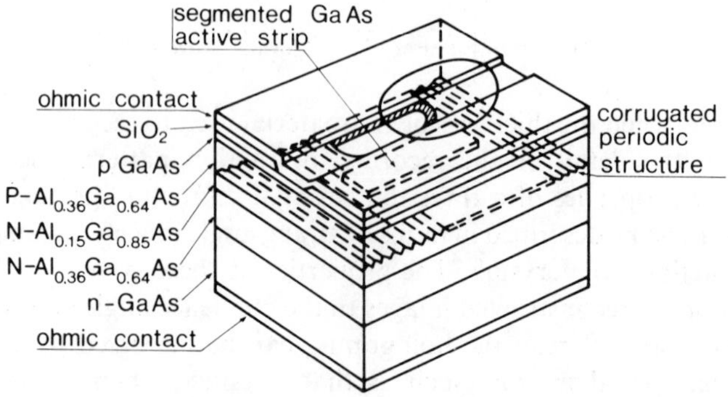

Fig. III.27. Schematic structure of a strip buried heterostructure laser with distributed feedback Bragg reflectors at both ends. (From Tsang *et al.* [39b] © 1979 IEEE.)

A TE wave propagating through the medium described by Eq. (III.17.1) satisfies *Hill's equation* [40],

$$\frac{d^2u}{d\zeta^2} + \left(\theta_0 + 2\sum_{m=1}^{\infty} \theta_m \cos 2m\zeta\right)u = 0, \qquad \text{(III.17.2)}$$

where $\zeta = \pi z/\Lambda$, $\theta_0 = (2\Lambda/\lambda)^2$, and $\theta_m = X_m\theta_0$.

17.1 *Floquet's Theorem*

If we denote by $u_1(\zeta)$ and $u_2(\zeta)$ two linearly independent solutions of Eq. (III.17.2), then in view of the periodicity of the function multiplying u, $u_1(\zeta + \pi)$ and $u_2(\zeta + \pi)$ must also be solutions. This means that we can find four coefficients, say a_{ij}, such that

$$\begin{aligned} u_1(\zeta + \pi) &= a_{11}u_1(\zeta) + a_{12}u_2(\zeta), \\ u_2(\zeta + \pi) &= a_{21}u_1(\zeta) + a_{22}u_2(\zeta). \end{aligned} \qquad \text{(III.17.3)}$$

Let us now look for two particular solutions $V^{(+)}$ and $V^{(-)}$ of Eq. (III.17.2) having the property

$$V^{(\pm)}(\zeta + \pi) = \lambda_\pm V^{(\pm)}(\zeta). \qquad \text{(III.17.4)}$$

If we write $V^{(\pm)} = V_1^{(\pm)}u_1 + V_2^{(\pm)}u_2$ and take into account Eq. (III.17.3), $V^{(\pm)}(\zeta + \pi)$ becomes $\lambda_\pm V^{(\pm)}(\zeta)$ if

$$\begin{aligned} (a_{11} - \lambda_\pm)V_1^{(\pm)} + a_{12}V_2^{(\pm)} &= 0, \\ a_{21}V_1^{(\pm)} + (a_{22} - \lambda_\pm)V_2^{(\pm)} &= 0. \end{aligned} \qquad \text{(III.17.5)}$$

Accordingly, the two eigenvalues λ_+ and λ_- are obtained by imposing the vanishing of the determinant of the above homogeneous system.

If we consider the Wronskian W of $V^{(+)}$ and $V^{(-)}$,

$$W(\zeta) = V^{(+)}(\zeta)\,dV^{(-)}/d\zeta - V^{(-)}(\zeta)\,dV^{(+)}/d\zeta, \qquad \text{(III.17.6)}$$

we can easily show, with the help of Eq. (III.17.4), that $W(\zeta + \pi) = \lambda_+\lambda_- W(\zeta)$. Since the Wronskian of two solutions of the wave equation (III.17.2) is independent of ζ, it follows that

$$\lambda_+\lambda_- = 1. \qquad \text{(III.17.7)}$$

If we denote λ_+ by $e^{-i\delta}$, we will complete the proof of Floquet's theorem by writing the two linearly independent solutions $V^{(\pm)}$ in the form

$$\begin{aligned} V^{(+)}(\zeta) &= e^{-i\delta\zeta/\pi} f^{(+)}(\zeta), \\ V^{(-)}(\zeta) &= e^{i\delta\zeta/\pi} f^{(-)}(\zeta), \end{aligned} \qquad \text{(III.17.8)}$$

where $f^{(\pm)}$ are two periodic functions.

In particular, since in our case $n(\zeta) = n(-\zeta)$, then $f^{(+)}(-\zeta)e^{i\delta\zeta/\pi}$ is a solution of Eq. (III.17.2), linearly independent of $V^{(+)}(\zeta)$ and satisfying the periodicity condition Eq. (III.17.4). As a consequence, it must coincide with $V^{(-)}(\zeta)$. In conclusion, every solution of Eq. (III.17.2) can be expressed by (*Floquet's theorem*)

$$u(\zeta) = Ae^{-i\delta\zeta/\pi}f(\zeta) + Be^{i\delta\zeta/\pi}f(-\zeta), \tag{III.17.9}$$

where the characteristic exponent δ is a suitable constant.

17.2 *Hill's Determinant*

The function $f(\zeta)$ appearing in Eq. (III.17.9) being periodic, with period π, it can be expanded in a Fourier series, that is,

$$u(\zeta) = Ae^{-i\delta\zeta/\pi} \sum_{m=-\infty}^{\infty} f_m e^{2mi\zeta} + Be^{i\delta\zeta/\pi} \sum_{m=-\infty}^{\infty} f_m e^{-2mi\zeta}. \tag{III.17.10}$$

Inserting the right side of Eq. (III.17.10) in Eq. (III.17.2), we obtain a trigonometric series that must vanish identically. Accordingly, we have

$$\frac{1}{\theta_0 - 4n^2}\left[\left(2n - \frac{\delta}{\pi}\right)^2 f_n - \sum_{m=-\infty}^{\infty} \theta_m f_{n-m}\right] \equiv \sum_{m=-\infty}^{+\infty} A_{nm} f_m = 0, \tag{III.17.11}$$

where $\theta_{-m} = \theta_m$. If we indicate by $\Delta(\delta)$ the determinant of the above system (also called *Hill's determinant*), the characteristic exponent δ is determined from the relation

$$\Delta(\delta) = 0. \tag{III.17.12}$$

Hill has proved that the roots of the above equation can be found among those of the equation

$$\sin(\delta/2) = \pm\sqrt{\Delta(0)}\sin[(\pi/2)\sqrt{\theta_0}], \tag{III.17.13}$$

which is an extension of Eq. (III.9.15), established for piecewise-constant periodic media, to systems exhibiting any index profile symmetric with respect to the origin of coordinates.

When the modulus of the right side of Eq. (III.17.13) is less than unity, δ is real. Otherwise, $\delta = (2q + 1)\pi + i\,\mathrm{Im}\,\delta$. In particular, we are often interested in the first Bragg resonance occurring when $q = 0$. In general, we are faced with the problem of calculating the determinant for $\delta = 0$ of the infinite order matrix

$$\mathbf{A} = \begin{bmatrix} \cdots & \cdots & \cdots & \cdots & \cdots \\ \cdots & \dfrac{[(\delta/\pi + 2)^2 - \theta_0]}{(4 - \theta_0)} & \dfrac{-\theta_1}{(4 - \theta_0)} & \dfrac{-\theta_2}{(4 - \theta_0)} & \cdots \\ \cdots & \dfrac{\theta_1}{\theta_0} & \dfrac{(\theta_0 - \delta^2/\pi^2)}{\theta_0} & \dfrac{\theta_1}{\theta_0} & \cdots \\ \cdots & \dfrac{-\theta_2}{(4 - \theta_0)} & \dfrac{-\theta_1}{(4 - \theta_0)} & \dfrac{[(\delta/\pi - 2)^2 - \theta_0]}{(4 - \theta_0)} & \cdots \\ \cdots & \cdots & \cdots & \cdots & \cdots \end{bmatrix}. \tag{III.17.14}$$

For $\theta_0 \simeq 1$ we can approximate Eq. (III.17.13) with

$$\cosh \operatorname{Im} \delta/2 = [\Delta_0(0)]^{1/2} \sin(\theta_0^{1/2}\pi/2), \tag{III.17.15}$$

where $\Delta_0(0)$ is calculated for $\theta_0 = 1$. Accordingly, when $\Delta_0(0)$ is slightly larger than one, $\operatorname{Im} \delta \neq 0$ for $|\theta_0^{1/2} - 1| < 2\{[\Delta_0(0)]^{1/2} - 1)\}/\pi$; that is, the periodic medium behaves as a so-called *Bragg reflector* for $|\lambda - \lambda_0| < \{[\Delta_0(0)]^{1/2} - 1\}\lambda_0 2/\pi$.

17.3 *Coupled-Mode Theory*

After evaluating δ, the coefficients f_m of the Fourier series of $f(\zeta)$ can be calculated by solving the homogeneous system of Eqs. (III.17.11). When $\theta_0 \cong 1, |\theta_1, ..| \ll 1$, it can be shown that all the coefficients f_m can be neglected with the exception of f_0 and f_1; as an example, the reader can examine the expression of $f(\zeta)$ for a piecewise-constant medium given in Problem 14 or can approximate the infinite system with the 3×3 matrix of Eq. (III.17.14). This in conjunction with the relation $\delta = \pi + i \operatorname{Im} \delta$ (valid in proximity with the Bragg resonance) allows us to write Eq. (III.17.10) in the form

$$\begin{aligned} u(\zeta) &\simeq Ae^{-i\zeta}e^{\operatorname{Im}\delta\zeta/\pi}(f_0 + f_1 e^{2i\zeta}) + Be^{i\zeta}e^{-\operatorname{Im}\delta\zeta/\pi}(f_0 + f_1 e^{-2i\zeta}) \\ &= e^{-i\zeta}(Af_0 e^{\operatorname{Im}\delta\zeta/\pi} + Bf_1 e^{-\operatorname{Im}\delta\zeta/\pi}) + e^{i\zeta}(Af_1 e^{\operatorname{Im}\delta\zeta/\pi} + Bf_0 e^{\operatorname{Im}\delta\zeta/\pi}) \\ &\equiv e^{-i\zeta}R(\zeta) + e^{i\zeta}S(\zeta) \end{aligned} \tag{III.17.16}$$

In view of the preceding hypotheses ($\operatorname{Im} \delta \ll 1$) $R(\zeta)$ and $S(\zeta)$ are slowly varying functions of ζ.

While we arrived at Eq. (III.17.16) by exploiting the properties of the Floquet's solutions, we can use it as an ansatz valid when we are not too far from the Bragg resonance. If we insert the last expression of Eq. (III.17.16) into Eq. (III.17.2), neglect the second derivatives of S and R, and retain only θ_0 and

θ_1, we obtain the system of coupled equations

$$dR/d\zeta - i(\theta_0 - 1)R = -i\theta_1 S, \qquad dS/d\zeta + i(\theta_0 - 1)S = i\theta_1 R. \qquad \text{(III.17.17)}$$

This approach is referred to as *coupled-mode theory* and it can be immediately extended to cases in which the medium exhibits loss or gain. An important application will be examined in Chapter VIII.

18 Analytical Properties of the Reflection Coefficient

In a multilayered region with piecewise-constant refractive index, the reflection coefficient $r(z)$ at each section depends on the various propagation constants $\beta_q = (n_q^2 k_0^2 - k_x^2)^{1/2}$ relative to the layers [see Eq. (III.7.17)]. Each β_q, if regarded as a function of the complex variable k_x, has two branch points $k_x = \pm n_q k_0$, where $\partial\beta_q/\partial k_x$ and all the higher derivatives become singular.

The double-valued function $\beta = (\tilde{n}^2 k_0^2 - k_x^2)^{1/2}$ can be made single-valued by providing the complex k_x plane with the branch cuts defined by the relations (see Fig. III.28b)

$$\begin{aligned} &\operatorname{Im}(\tilde{n}^2 k_0^2 - k_x^2) = -2k_0^2 n\kappa - 2\operatorname{Im} k_x \operatorname{Re} k_x = 0, \\ &\operatorname{Re}(\tilde{n}^2 k_0^2 - k_x^2) \leqq 0. \end{aligned} \qquad \text{(III.18.1)}$$

The two curves obeying Eqs. (III.18.1) are portions of the hyperbola running from the branch points toward the real axis in the k_x plane. For $\kappa = 0$, they

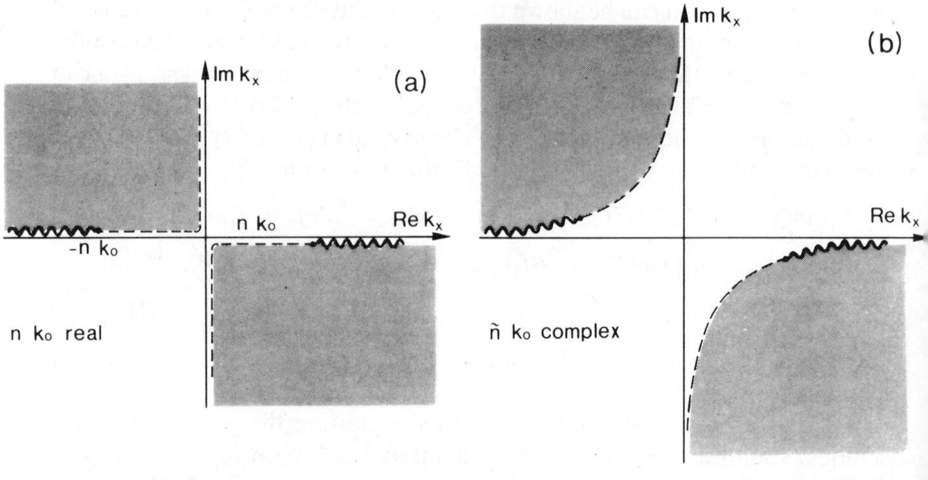

Fig. III.28. Branch cuts in the complex domain of k_x used for a single-valued determination of the reflectivity versus k_x for (a) real and (b) complex refractive indices.

degenerate into the half-lines of Fig. III.28a. Since the above cuts define all the points for which $\tilde{n}^2 k_0^2 - k_x^2$ is a nonpositive real quantity, the function $\mathrm{Re}[(\tilde{n}^2 k_0^2 - k_x^2)^{1/2}]$ will always be positive (or negative) on the remaining part of the plane. This property is particularly useful from a physical point of view. In fact, since the incident plane wave, to which Eq. (III.7.17) refers, travels from left ($z = -\infty$) to right ($z = +\infty$), $\mathrm{Re}\,\beta = \mathrm{Re}[(\tilde{n}^2 k_0^2 - k_x^2)^{1/2}]$ must be positive. Thus, we have established a one-to-one correspondence between forward-traveling waves and values of β defined on the proper Riemann sheet characterized by the above branch cuts. If we consider the complete hyperbola given by Eq. (III.18.1a), we can easily show that the positive sign of $\mathrm{Re}\,\beta$ implies $\mathrm{Im}[(\tilde{n}^2 k_0^2 - k_x^2)^{1/2}] < 0$ in the unshaded region and $\mathrm{Im}[(\tilde{n}^2 k_0^2 - k_x^2)^{1/2}] > 0$ in the shaded one. In more concise terms,

$$\operatorname{sgn} \mathrm{Im}(\tilde{n}^2 k_0^2 - k_x^2)^{1/2} = \operatorname{sgn} \mathrm{Im}(\tilde{n}^2 k_0^2 - k_x^2). \qquad \text{(III.18.2)}$$

When we consider the reflection coefficient $r(k_x)$ at a given section as a function of the transverse component k_x of the incident field wave vector, we note, with the help of Eq. (III.7.18), that it presents *polar singularities* corresponding to the zeros of $(\vec{Z} + \hat{Z})$ and *branch points* coinciding with those of $\vec{Z}$ and $\hat{Z}$ [7]. The branch points of $\vec{Z}(z)$ are conveniently investigated by rewriting Eq. (III.7.16) in the form

$$\vec{Z}(z_2) = \frac{\vec{Z}(z_1) - i\hat{Z}\tan[\beta(z_2 - z_1)]}{1 - i\vec{Z}(z_1)\tan[\beta(z_2 - z_1)]/\hat{Z}}. \qquad \text{(III.18.3)}$$

Since both expressions $\hat{Z}\tan[\beta(z_2 - z_1)]$ and $\tan[\beta(z_2 - z_1)]/\hat{Z}$ are even functions of β, they are regular functions of k_x for $\beta = 0$. As a consequence, the branch points of $\vec{Z}(z_2)$ are only those of $\vec{Z}(z_1)$. By iterating Eq. (III.18.3) to the various layers, we conclude that the branch points of the impedance seen in a given direction do not depend on the local position on the stratification axis. In particular, they coincide with the two branch points of the last medium on the right, if the impedance is oriented to the right, and on the left in the opposite case. As a consequence, $r(k_x)$ relative to a substrate coated with a dielectric multilayer has four branch points at $k_x = \pm n_S k_0$ and $k_x = \pm n_1 k_0$, n_1 being the refractive index of the embedding medium.

In particular, in a limited region of the complex k_x plane $r(k_x)$ takes the rational form

$$r(k_x) = \frac{(k_x^2 - k_1'^2)(k_x^2 - k_2'^2)\cdots}{(k_x^2 - k_1^2)(k_x^2 - k_2^2)\cdots}, \qquad \text{(III.18.4)}$$

which fulfills the obvious condition of being an even function of k_x. The poles of the *scattering coefficient* r of the multilayered structure represent the resonant transverse wave numbers distinguishing all possible *source-free solutions* of Maxwell's equations. From an intuitive point of view, since $|r|$ tends to infinity at a pole, the amplitude of the incident wave can tend to zero and still produce a reflected wave with finite amplitude.

It is to be expected that these poles correspond to vectors $\mathbf{k}$ with complex components, whenever they represent confined waves. In fact, as will become clear when dealing with propagation in optical fibers, the guided modes supported by the dielectric structure decay in the direction *transverse* to the propagation direction. We stress that, for the plane-stratified structure, the modes will propagate parallel to the slab interface and will decay exponentially in the stratification z direction.

For lossless media, the real roots of the equation

$$\vec{Z}_{\text{in}}(k_x) + \hat{Z}_{\text{l}}(k_x) = 0 \tag{III.18.5}$$

provide poles of the reflection coefficients associated with *propagation modes* (along the x axis) of the structure. The remaining complex roots correspond to a discrete set of source-free nonmodal solutions of Maxwell's equations. In a strict sense, they cannot be called "modes," since this definition implies the independence of $|\mathbf{E}(\mathbf{r})|^2$ from x, which holds true only for real values of k_x, as implied by Eq. (III.7.1); they are usually referred to as *leaky waves, damped resonances*, or *radio-active states.* In spite of their physically unacceptable behavior in remote regions, where their magnitude tends to infinity, such waves can be employed as good approximations of actual fields traveling along the x axis while undergoing a progressive energy leakage in the z direction.

The above discussion shows that the poles of $r(k_x)$ occur at complex values $\beta - i\alpha$ of k_x. As we will see in the following, each of these poles refers either to a *surface wave* (the case in which α tends to zero by assuming negative values) or to a *leaky wave* ($\alpha > 0$), which can be guided by the planar layer.

As we will see in Chapter VIII, the surface waves correspond, in the frame of propagation in cylindrical optical fibers (whose symmetry axis coincides with z), to *guided modes*, and the leaky waves to *leaky modes.* The analytical description is different, however, since there we consider waves impinging on the fiber from $z = -\infty$ and propagating in the positive z direction, while if we adopt the present point of view, we would have waves impinging on the fiber orthogonally to the z axis and propagating, as surface or leaky waves, along the z axis itself.

The existence of localized modes near the interfaces of a layered medium and a homogeneous one can be easily understood on the basis of the analogy between thin-films optics and electron band theory of solids. By analogy with the so-called surface states describing impurities localized in proximity to a boundary in solid-state physics, particular waves can be excited in proximity

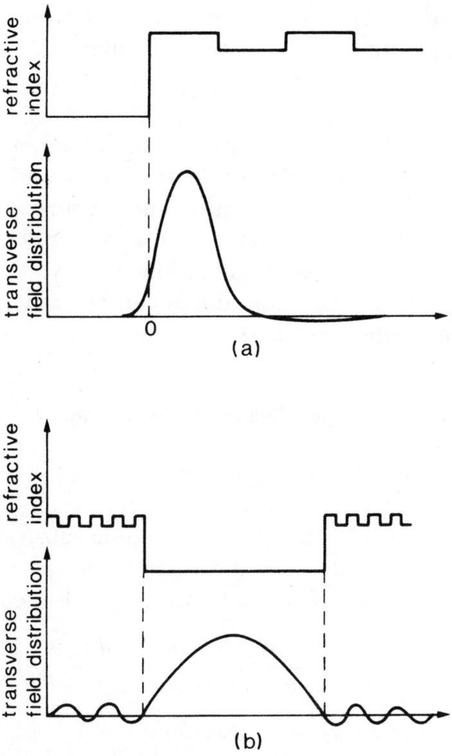

Fig. III.29. (a) Profile of a surface wave propagating perpendicularly to the stratification axis of a Bragg reflector; (b) surface wave in the gap between two Bragg reflectors. (From Yeh *et al.* [41].)

to the interfaces of a multilayer. Their properties can be studied [41] by finding the real roots of Eq. (III.18.5) and obtaining the field distribution by means of the electrical network formalism illustrated before. In Fig. III.29a we show the transverse field distribution for a typical fundamental surface mode guided by a periodic structure.

18.1 *Transverse-Resonance Condition*

The resonance condition described by Eq. (III.18.5) was established for a multilayer interfaced with a homogeneous embedding medium extending to infinity (in our conventions $z = -\infty$). If we now indicate by $\overleftarrow{Z}(z)$ and $\overrightarrow{Z}(z)$ the impedances seen looking to the left and to the right of a generic section of a stratified medium, we can easily prove, by iteratively transforming $\overleftarrow{Z}$ and $\overrightarrow{Z}$

by means of Eq. (III.7.15), that the condition represented by Eq. (III.18.5) holds true for every section of the stratified medium, viz.

$$\overleftarrow{Z}(z,k_x,k_0) + \overrightarrow{Z}(z,k_x,k_0) = 0. \tag{III.18.6}$$

This relation, known as the *transverse-resonance condition*, allows us to find the values k_x and k_0 corresponding to the free oscillations of the system by calculating the impedances $\overleftarrow{Z}$ and $\overrightarrow{Z}$ at the more convenient sections of the multilayer. For example, for the structure of Fig. III.29b, which is symmetrical with respect to the origin of the coordinate z, we have $\overrightarrow{Z}(0,k_x,k_0) = \overleftarrow{Z}(0,k_x,k_0)$, so that the propagation modes can be easily found by looking for the zeros of the function $\overrightarrow{Z}(0,k_x,k_0)$.

19 Propagation of Surface and Leaky Waves through a Thin Film

We mentioned in the Introduction that multilayered structures can support the propagation of particular waves called surface and leaky waves. The simplest example is provided by an ideal metallic substrate ($\kappa_S \gg 1$) coated with a single dielectric layer (*grounded dielectric sheet*) (see Fig. III.30). In this case, the reflection coefficient r_p given by Eq. (III.12.19) reduces to

$$r_p = \frac{-\cos\theta_I + (i\cos\theta_c/n)\tan[\omega n_c d_c(\cos\theta_c)/c]}{\cos\theta_I + (i\cos\theta_c/n)\tan[\omega n_c d_c(\cos\theta_c)/c]}, \tag{III.19.1}$$

where $n = n_c/n_I$ and the subscript c is used for quantities related to the coating. Now, assuming $n_I = 1$ and $\cos\theta_c = \sqrt{1 - k_x^2 c^2/\omega^2 n^2}$ in the above equation, we can immediately show that the poles and zeros of $r_p(k_x)$ are determined by the eigenvalue equation (see Collin [41] for the TE case)

$$(k_0^2 n^2 - k_x^2)^{1/2}\tan[d_c(k_0^2 n^2 - k_x^2)^{1/2}] = \pm i n^2 (k_0^2 - k_x^2)^{1/2}, \tag{III.19.2}$$

the plus and minus signs referring to poles and zeros, respectively. Thus, for each frequency, the reflection coefficient r_p of the structure vanishes or diverges whenever the transverse propagation coefficient k_x satisfies Eq. (III.19.2). For fixed k, n, and d_c, if $k_x^{(p)}$ is a root of Eq. (III.19.2) with the sign $+$, $k_x^{(p)*}$ is a root of the same equation with the sign $-$. In fact, for real n, changing $k_x^{(p)}$ into $k_x^{(p)*}$ implies the change of both functions $(k_0^2 n^2 - k_x^2)^{1/2}$ and

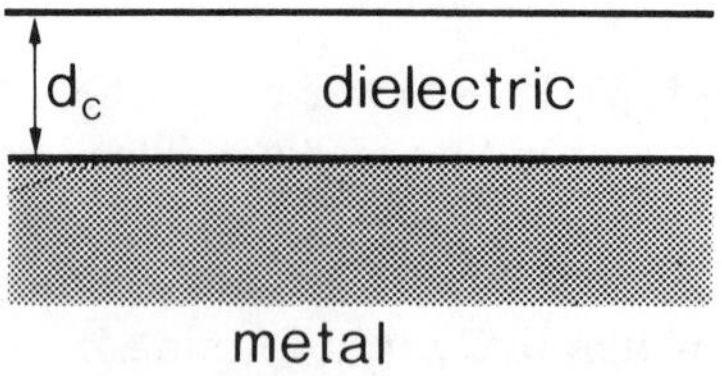

Fig. III.30. Grounded dielectric waveguide.

$(k_0^2 - k_x^2)^{1/2}$ into their complex conjugates. Thus, the zeros of $r_p(k_x)$ coincide with the complex conjugates of the poles.

In order to determine the poles, we write Eq. (III.19.2) in the form

$$u \tan u = n^2 v, \qquad u^2 + v^2 = V^2, \tag{III.19.3}$$

where V (the *normalized frequency* in the theory of optical fibers), v, and u are defined as

$$V = (n^2 - 1)^{1/2} k_0 d_c, \qquad v = i d_c (k_0^2 - k_x^2)^{1/2}, \qquad u = d_c (k_0^2 n^2 - k_x^2)^{1/2}. \tag{III.19.4}$$

Since $\operatorname{sgn} \operatorname{Im}(k_0^2 - k_x^2)^{1/2} = -\operatorname{sgn} \operatorname{Im} k_x$, then for n real and $\operatorname{Re} k_x > 0$ we have $\operatorname{sgn} \operatorname{Re} v = \operatorname{sgn} \operatorname{Im} k_x$.

We note in passing that the incident and reflected magnetic fields are represented in terms of v as

$$H_{yi} \propto e^{-ik_x x - vz/d_c}, \qquad H_{yr} \propto e^{-ik_x x + vz/d_c}. \tag{III.19.5}$$

Equations (III.19.3) admit an infinite number of generally complex roots, each of which can be regarded as a function of V. In particular, for $V \to \infty$, one immediately sees that $u(V) \to (q - \frac{1}{2})\pi$ with integer q. Accordingly, we can label each root by means of the index q corresponding to its asymptotic value, i.e., $u \to u_q$.

When u and v are both real, Eqs. (III.19.3) can be solved graphically as illustrated in Fig. III.31a, where u_q is defined by the intersection of the curve $(u \tan u)/n^2$ with the circle of radius V. If we reduce V, u and v move to the left of the respective real axes, while k_x (see Fig. III.31b) first moves to the left of the point nk (we suppose $n > 1$) and then, after reaching branch point k, goes back to the right in the lower quadrant of the right side of the complex k_x plane. If V becomes smaller than the *critical value* V_m for which the circle becomes tangent to $(u \tan u)/n^2$, the root u_q becomes complex.

In order to determine the real and imaginary parts of $u_q = u'_q + iu''_q$, we square Eqs. (III.19.3) and, after simple algebra, we obtain

$$\frac{[\sin^2(2u') - \sinh^2(2u'')](u'^2 - u''^2) - 4u'u'' \sin(2u') \sinh(2u'')}{n^4[\cos(2u') + \cosh(2u'')]^2} + u'^2 - u''^2 = V^2, \tag{III.19.6a}$$

$$\frac{\sin(2u') \sinh(2u'')}{2 + [\cos(2u') + \cosh(2u'')]^2(n^4 - 1) + 2\cos(2u')\cosh(2u'')} + \frac{u'u''}{u'^2 - u''^2} = 0 \tag{III.19.6b}$$

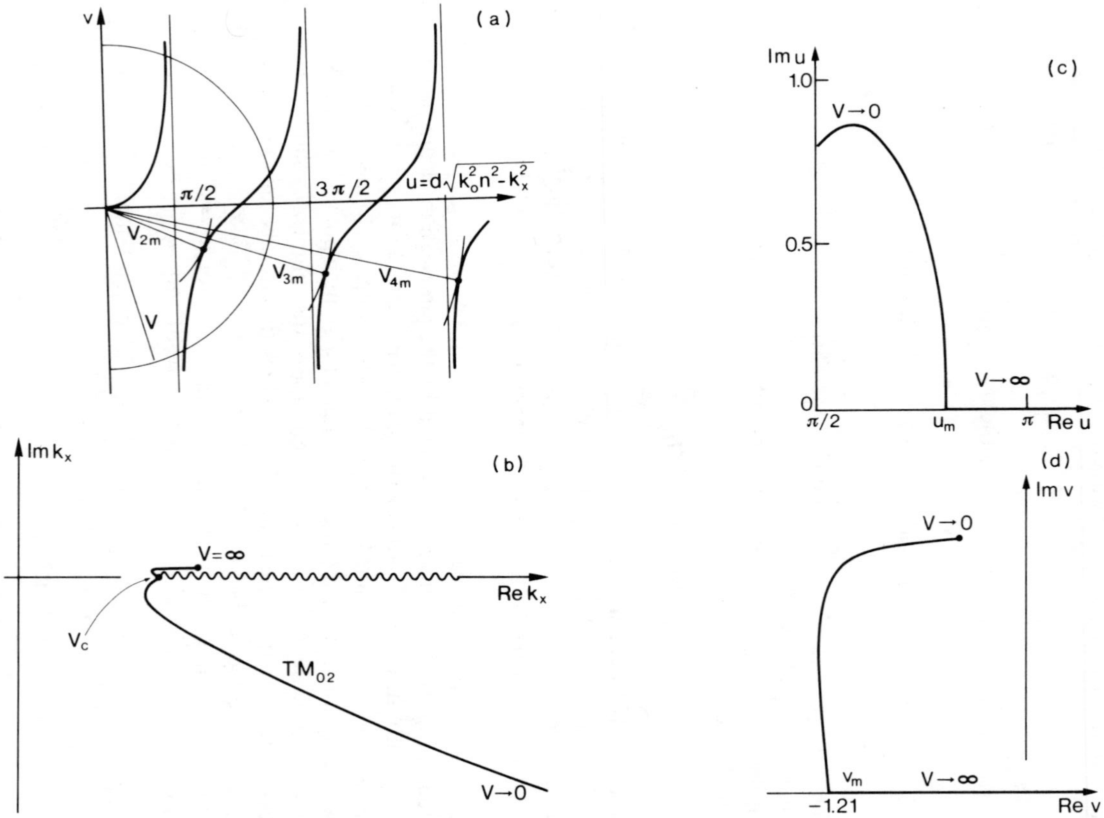

Fig. III.31. (a) Graphical solution of the eigenvalue Eq. (III.19.3). The thick curves represent the equation $v = u\tan(u)/n^2$, while $V^2 = u^2 + v^2$. Loci of the solution point for a TM_{02} mode for a grounded dielectric guide with $n^2 = 1.5$ in the (b) complex k_x plane, (c) u plane, and (*d*) v plane. The wavy line in (b) represents the branch cut of $(k_0^2 - k_x^2)^{1/2}$. $k_x(V_m)$ lies on the lower side of the branch cut at the right side of k_0. The locus of k_x presents a cusp in $k_x(V_m)$ and runs parallel at a vanishing [illegible] by achieving a minimum value less than k_0.

While Eq. (III.19.6b) relates u'' to u', Eq. (III.19.6a) allows us to determine the corresponding value of V. For $u'' = 0$, V is given by

$$u'^2 \tan^2 u' + n^4 u'^2 = n^4 V^2. \qquad \text{(III.19.7)}$$

The minimum value V_m of V (coinciding with the critical V_m introduced above) is attained when $u \equiv u_m$ satisfies the relation

$$\tan^2 u_m + (u_m \tan u_m)/\cos^2 u_m = -n^4. \qquad \text{(III.19.8)}$$

For $V < V_m$, u becomes complex. In Fig. III.31c we have plotted u'' versus u' for $q = 2$. Using Eq. (III.19.3a), it is possible to derive a plot of v'' versus v' (Fig. III.31d). On physical grounds, v'' must be positive if H_{yi} in Eq. (III.19.5) represents a wave impinging on the dielectric and traveling from $z = -\infty$ to $z = \pm\infty$. In particular, Fig. III.31c shows that u_2'' vanishes identically for $2.52 < u_2' < 3\pi/2$, and then rises quite rapidly and reaches a maximum value for $V \cong 0$, which can be easily determined by using Eq. (III.19.3a) together with the condition $v = \pm iu$. This means that $u_q(V = 0)$ is obtained by solving the equation $\tan u_q = \pm in^2$. Thus, if we write $u_q(0) = (q - \frac{3}{2})\pi + iu_q''(0)$, we find that $u_q''(0)$ is independent of the mode index and is given by

$$u''(0) = \ln[(n^2 - 1)^{1/2}/(n^2 + 1)^{1/2}]. \qquad \text{(III.19.9)}$$

The values of β and α ($k_x \equiv \beta - i\alpha$), for V increasing from zero to ∞, are shown in Fig. III.32. When k_x lies in the first quadrant of the complex plane immediately above the branch cut, the wave in the embedding medium travels parallel to the interface and is evanescent perpendicular to it (surface wave) (see Fig. III.33a).

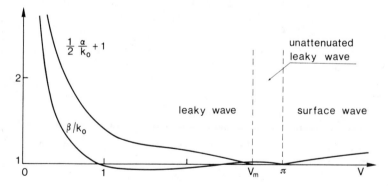

Fig. III.32. Attenuation (α) and propagation (β) constant of a TM_{02} mode propagating through a thin film ($n^2 = 1.5$) deposited on a metallic substrate. When the normalized frequency V is larger than π a surface wave propagates; for $V_m < V < \pi$ the TM_{02} is an unattenuated leaky wave; for $V < V_m$ the attenuation constant α of the leaky wave becomes rapidly comparable with k. For $V \to \infty$, $\beta \to nk_0$ (see Fig. VIII.11).

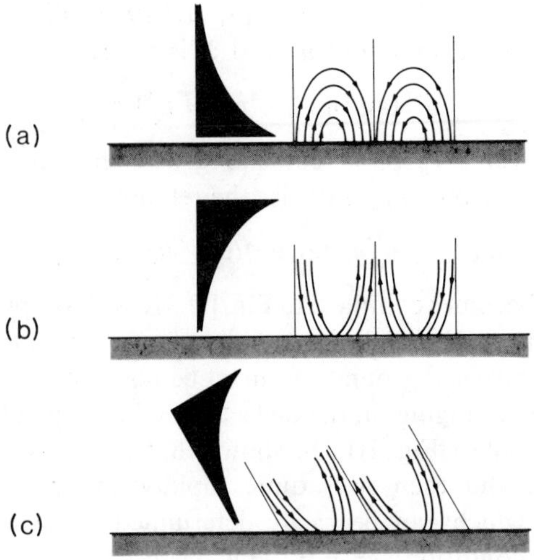

Fig. III.33. Schematic plot of the streamlines of the electric field relative to (a) a surface wave, (b) an unattenuated leaky wave (Im $k_x = 0$), and (c) a radiating leaky wave (Im $k_x < 0$).

A situation of particular relevance is that in which k_x lies in the fourth quadrant. In this case the external field is an inhomogeneous wave moving away from the slab. As a consequence, a fraction of the power carried by the mode propagating inside the slab is gradually lost by the continuous radiative leakage into the external medium. This mechanism is at the origin of the exponential decay of the mode amplitude along the propagation axis parallel to the slab. Incidentally, this leakage effect is used in dielectric antennas, where a wave guided inside a dielectric rod produces a radiation field outside.

If we measured the field created outside by these leaky modes, we would observe the physical paradox of a field tending to infinity at an infinite distance from the slab. This paradox is easily explained by noting that we have considered a slab of infinite extension, so that the mode amplitude must become infinite when we proceed in the opposite direction of propagation. Since the external field at an infinite distance from the slab is generated by these far upstream sources, it must tend to infinity.

Surface and leaky waves, guided by a dielectric slab bounded by media having different refractive indices, have been studied by Hsue and Ternir, cited in the bibliography.

20 Illumination at an Angle Exceeding the Critical One

For $\mu = n/n' < 1$, Snell's law gives a real refraction angle smaller than $\pi/2$. When $\mu > 1$, and the angle of incidence θ is greater than the *critical angle*

$$\theta_c = \arcsin(1/\mu), \tag{III.20.1}$$

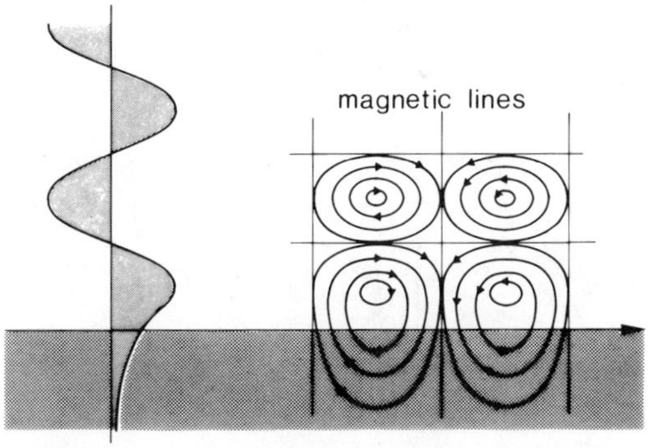

Fig. III.34. On the left is shown the electric field amplitude along the normal to the interface between two media illuminated by a plane wave incident at an angle greater than the critical one. On the right is shown streamlines of the magnetic field.

then $\sin\theta'$ is larger than unity, so that the refraction angle $\theta' = \pi/2 + i\theta'^{('')}$ is complex, with the imaginary component $\theta'^{('')}$ given by

$$\cosh\theta'^{('')} = \mu\sin\theta. \tag{III.20.2}$$

Thus, the refracted field can be associated with a *complex* refracted ray, whose direction $\hat{s}$, is determined from Eqs. (II.11.9) and (II.11.11) and reads

$$\hat{s}' = (\hat{s} + \hat{n}\cos\theta) + i\hat{n}\sinh\theta'^{('')} = \mu\hat{n}\times\hat{s}\times\hat{n} + i\hat{n}\sinh\theta'^{('')}, \tag{III.20.3}$$

where $\hat{s}$ and $\hat{n}$ are, respectively, the direction of the incident ray and the normal to the separation surface (directed into the incident medium). Because of the complex value of $\hat{s}'$, the refracted field is an evanescent wave, i.e.,

$$u(\mathbf{r}) \propto \exp[-ik_0 n(\hat{n}\times\hat{s}\times\hat{n})\cdot\mathbf{r} + n'k_0\hat{n}\cdot\mathbf{r}\sinh\theta'^{('')}], \tag{III.20.4}$$

which decays exponentially in the direction normal to the interface. If we look at the vector component **H** of the TE evanescent field, we note that, as a general property of the evanescent waves discussed in Section II.7, the magnetic lines in the less dense medium start from and terminate on the surface, while in the more dense medium they describe closed orbits or connect two points of the surface. The whole set translates parallel to the surface (see Fig. III.34).

If we apply Eqs. (III.8.2), we obtain, with the help of Eq. (III.20.2),

$$r_s = \frac{\mu\cos\theta + i\sinh\theta'^{('')}}{\mu\cos\theta - i\sinh\theta'^{('')}} = e^{2i\phi_s}, \tag{III.20.5}$$

with (see Fig. III.35)

$$\phi_s = \arctan\left(\frac{\sinh\theta'^{('')}}{\mu\cos\theta}\right) = \arctan[(\mu^2\sin^2\theta - 1)/(\mu^2\cos^2\theta)]^{1/2}, \tag{III.20.6}$$

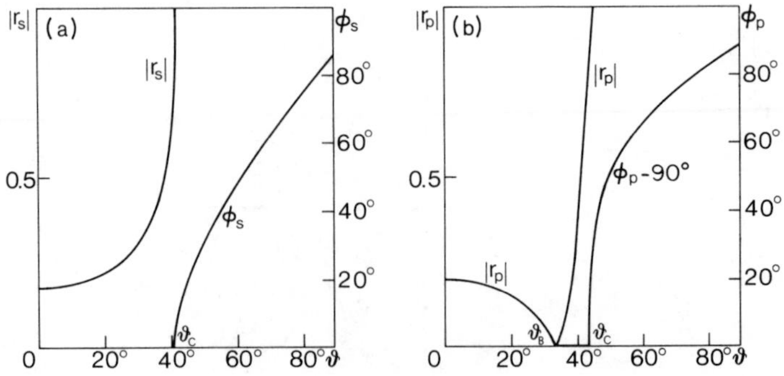

Fig. III.35. Reflection coefficient and phase shift versus angle of incidence for (a) an s-wave and (b) a p-wave; $n_1 = 1.5$, $n_2 = 1.0$. $\phi_s = -90°$ for $\theta < \theta_c$ while $\phi_p = -90°$ for $\theta < \theta_B$ and $0°$ for $\theta_B < \theta < \theta_c$.

and

$$r_p = e^{2i\phi_s + 2i\phi(\theta)} \equiv e^{2i\phi_p}, \tag{III.20.7}$$

with

$$2\phi(\theta) = \pi + 2\arctan\left[\frac{\cos\theta(\mu^2\sin^2\theta - 1)^{1/2}}{\mu\sin^2\theta}\right]. \tag{III.20.8}$$

We observe that $|r_p| = |r_s| = 1$, which is an obvious consequence of energy conservation, since no energy is transported by the refracted evanescent field.

The difference $2\phi(\theta)$ between the phase changes of the TM and TE waves can be used to modify the polarization state of a field. More precisely, $\phi = \pi/2$ for grazing incidence $\theta = \pi/2$ and for incidence at the critical angle, while its maximum value is attained when

$$\sin^2\theta = 2/(1 + \mu^2) \tag{III.20.9}$$

$$\phi_{max} = \arctan[(\mu^2 - 1)/2\mu] + \pi/2. \tag{III.20.10}$$

For example, if we use a glass with refractive index 1.51, the first medium being vacuum, we obtain $\phi_{max} = 45°\,16'$, corresponding to the incidence angle $\theta = 51°\,20'$. Thus, in the above condition, a phase difference of $\cong 90°$ is obtained after a single total reflection, which, e.g., corresponds to shifting from linear to circular polarization. After two total reflections, the polarization state coincides with the initial one. This effect is used in the so-called *Fresnel rhomb*, which is used in place of quarter-wave plates to convert a linear polarization into a circular one.

When we consider a wave passing from a medium of assigned density to a more dense medium, then $\mu < 1$ and Eq. (III.20.1), defining the critical angle, is satisfied by a complex value of θ_c, that is,

$$\theta_c = \pi/2 - i\theta_c'', \tag{III.20.11}$$

with $\cosh\theta_c'' = 1/\mu$. Accordingly, total reflection can be observed only with evanescent waves propagating parallel to the interface. More precisely, we apply the definition of total reflection to the present situation in a formal way, so as to consider as totally reflected the evanescent waves propagating parallel to the interface with an incidence angle $\theta = \pi/2 - i\theta''$, provided $\theta'' < \theta_c''$. In this case, the refraction angle θ' is immediately obtained by means of Snell's law,

$$\theta' = \arcsin(\cosh\theta''/\cosh\theta_c''), \tag{III.20.12}$$

and represents a real quantity corresponding to a real refracted ray. To summarize, we have a dual situation: if $n/n' > 1$, total reflection implies real incident and reflected waves and evanescent refracted wave, while if $n/n' < 1$, total reflection corresponds to evanescent incident and reflected waves and real refracted wave.

20.1 *Frustrated Total Reflection*

In order to control the power transfer between two components of integrated optical devices, it has become customary to interpose a less dense medium, which acts as the dark barrier discussed in Section III.3.3. The optical thickness of the gap separating the two coupled devices is the parameter that controls the coupling. More precisely (see Fig. III.6), a wave impinges on the interface between the incidence medium $n = n_I$ and the gap ($n = n_g < n_I$) at an incidence angle θ larger than the cirtical angle $\theta_c = \arcsin(n_g/n_I)$, while a substrate ($n = n_S > n_g$) interfaces with the gap at a distance d_g from the incidence medium.

If we neglect the presence of the substrate, the perturbation in the gap coincides with a field evanescent in the direction entering the gap itself. Actually, due to the finite distance d_g between incidence medium and substrate, the tail of the evanescent wave can reach the substrate, thus producing a tunneling of power, while the presence of the second interface produces a second evanescent field extending toward the incidence medium. This process continues indefinitely, producing a final field configuration made of two evanescent waves in the gap, a transmitted wave in the substrate, and a reflected wave in the incidence medium. The amplitudes of the above fields can be calculated with the characteristic-matrix formalism, taking care of the complex value of $\theta_g = \pi/2 + i\theta_g''$ in determining the impedance and the phase thickness ($\phi_g = n_g d_g k_0 \cos\theta_g = -in_g d_g k_0 \sinh\theta_g''$). In particular, let us apply

Eq. (III.12.19) to evaluate the reflection coefficient for a p-wave. If we assume for simplicity that the substrate has the same refractive index as the incidence medium, Eq. (III.12.19) yields

$$r_p = \frac{\mu^2 - 1 - (\mu^4 - 1)\sin^2\theta}{\mu^2 + 1 - (\mu^4 + 1)\sin^2\theta - 2i\mu\cos\theta\sinh\theta''_g\coth(d_g n_g k_0 \sinh\theta''_g)}, \tag{III.20.13}$$

with $\mu = n_S/n_g$, where $\sinh\theta''_g = (\mu^2\sin^2\theta - 1)^{1/2}$, as implied by Snell's law. For $d_g \to \infty$, r_p tends to the Fresnel coefficient $\exp(2i\phi_p)$ [see Eq. (III.20.7)], while it vanishes for $d_g \to 0$. In general, $|r_p| < 1$ (*frustrated* total reflection), while for a lossless gap energy conservation entails the relation $|t_p| = 1 - |r_p|$.

21 Reflection and Refraction at a Dielectric–Lossy Medium Interface

The refraction angle for a field impinging from the vacuum on a lossy medium (hereafter, all equations can be easily generalized to the case in which the incidence medium is a homogeneous dielectric without losses different from the vacuum) is a complex quantity θ' which is easily determined by means of Snell's law as

$$\theta' = \theta'^{(')} + i\theta'^{('')} = \arcsin[\sin\theta/(n - i\kappa)], \tag{III.21.1}$$

If we set $\cos\theta' = qe^{-i\gamma}$ with q and γ real quantities and assume that the fields propagate in the plane xz, with z orthogonal to the plane separation surface, the refracted wave vector $\mathbf{k}'$ is given by

$$\mathbf{k}' = k_0[\hat{x}\sin\theta + \hat{z}(n - i\kappa)qe^{-i\gamma}] \equiv \mathbf{k}'^{(')} - i\mathbf{k}'^{('')}, \tag{III.21.2}$$

where use has been made of the constancy of k_x after refraction [see Eq. (III.7.1)], with θ representing the incidence angle and

$$\begin{aligned} \mathbf{k}'^{(')} &= k_0[\hat{x}\sin\theta + \hat{z}q(n\cos\gamma - \kappa\sin\gamma)], \\ \mathbf{k}'^{('')} &= k_0\hat{z}q(\kappa\cos\gamma + n\sin\gamma). \end{aligned} \tag{III.21.3}$$

The above relations show that the *equiphase planes* (normal to $\mathbf{k}'_r$) and the *equiamplitude planes* (normal to $\mathbf{k}'^{('')}$ and thus to $\hat{z}$) are not mutually perpendicular, which is a peculiar difference from the case of evanescent fields and real refractive indices (see Section II.7).

We observe that a complex refractive index implies a complex Brewster angle. In fact, Eq. (III.8.3) yields

$$\tan\theta_B \equiv \tan(\theta'_B - i\theta''_B) = n - i\kappa. \tag{III.21.4}$$

For a good conductor ($\kappa \gg n$), the above equation is satisfied by

$$\theta_B \cong \pi/2 - i\,\mathrm{arctanh}(1/\kappa) \cong \pi/2 - i/\kappa \cong \pi/2. \qquad \text{(III.21.5)}$$

When we are near the metal plasma frequency, κ is no longer much larger than n, so that in this case n cannot be neglected with respect to $i\kappa$ and the Brewster angle deviates sensibly from the grazing direction.

We now wish to determine the main features of reflection at the plane interface between a dielectric and a lossy medium by means of the Fresnel formulas [43, 44]. To this end, we note that Eqs. (III.8.2a) and (III.8.2b) entail, with the help of Snell's law,

$$\frac{r_p - r_s}{r_p + r_s} = \frac{-\sin\theta\tan\theta}{(\tilde{n}^2 - \sin^2\theta)^{1/2}} \cong \frac{-\sin\theta\tan\theta}{\tilde{n}}, \qquad \text{(III.21.6)}$$

where we have neglected $\sin^2\theta$ in comparison with $\tilde{n}^2$. If we set

$$r_s/r_p = \tan(\delta)e^{-2i\phi}, \qquad \text{(III.21.7)}$$

we obtain

$$n = \frac{-\sin\theta\tan\theta\cos(2\delta)}{1 - \sin(2\delta)\cos(2\phi)}, \qquad \text{(III.21.8a)}$$

$$\kappa = \tan(2\delta)\sin(2\phi)n. \qquad \text{(III.21.8b)}$$

Equations (III.21.8) are used to determine n and κ from polarization measurements. In particular, for $\theta = 0$, we have $\delta = \pi/4$, $\phi = 0$, and $r_s/r_p = 1$, which corresponds to the loss of distinction between TE and TM waves for normal incidence.

Let us now look for the *real* incidence angle $\bar{\theta}$, for which $|r_s/r_p|$ attains the maximum value. We have, from Eq. (III.21.8b),

$$\tan\delta = \frac{-\sin(2\phi) + [\sin^2(2\phi) + \kappa^2/n^2]^{1/2}}{\kappa/n}, \qquad \text{(III.21.9)}$$

from which

$$|r_s/r_p|_{\max} = \tan\delta_{\max} = \frac{1 + (1 + \kappa^2/n^2)^{1/2}}{\kappa/n}, \qquad \text{(III.21.10)}$$

for $2\phi = -\pi/2$. The angle $\bar{\theta}$ is now easily evaluated by means of Eqs. (III.21.8a) and (III.21.10) and is given by the relation

$$\sin\bar{\theta}\tan\bar{\theta} = (n^2 + \kappa^2)^{1/2} = |\tilde{n}|. \qquad \text{(III.21.11)}$$

The circumstance that the maximum value ($<\infty$) attained by $|r_s/r_p|$ in the range of real angles of incidence occurs for $\theta = \bar{\theta}$ leads us to consider $\bar{\theta}$ as a

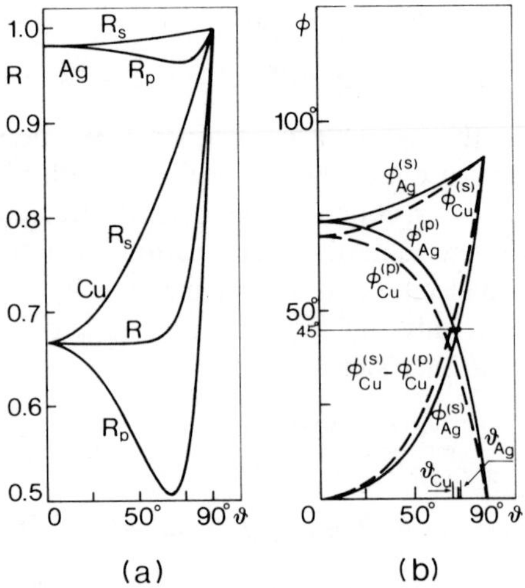

Fig. III.36. (a) Reflectance versus incidence angle for green mercury light ($\lambda = 5450$ Å) for Ag ($n = 0.055, \kappa = 3.32$) and Cu ($n = 0.76, \kappa = 2.42$); $R = (R_p + R_s)/2$ refers to unpolarized light. (b) Phase shifts ϕ of the reflection coefficients $r = R^{1/2}\exp(2i\phi)$ versus the incidence angle for the metals in (a); θ_{Cu} and θ_{Ag} are the pseudo-Brewster angles corresponding to $2\phi_s - 2\phi_p = \pi/2$. (From Abelès [43]. © North-Holland Physics Publishing, Amsterdam, 1972.)

pseudo-Brewster angle. The angle $\bar{\theta}$ is also called by some authors the *principal angle* of incidence, for which the phase shift in reflection is 90° (see Fig. III.36).

21.1 *Reflection and Refraction at Grazing Incidence*

Due to the proximity of θ_B to $\pi/2$, for grazing incidence the intensity of the reflected field depends critically on the polarization of the incident field. In fact, Eq. (III.8.2) yields

$$r_s = \frac{-1 + \cos\theta/\{\tilde{n}^2 - \sin^2\theta\}^{1/2}}{1 + \cos\theta/\{\tilde{n}^2 - \sin^2\theta\}^{1/2}}$$

$$\cong -1 + \frac{2\cos\theta}{(\tilde{n}^2 - 1)^{1/2}}, \qquad \text{(III.21.12)}$$

so that the absorptance $A_s = 1 - R_s$ is given by

$$A_s \cong 4\cos\theta\,\mathrm{Re}[1/(\tilde{n}^2 - 1)^{1/2}] \cong 4n\cos\theta/(n^2 + \kappa^2), \qquad \text{(III.21.13)}$$

the last expression holding true for $|\tilde{n}| \gg 1$. Analogously,

$$A_p \cong 4\cos\theta\,\mathrm{Re}[\tilde{n}^2/(\tilde{n}^2 - 1)^{1/2}] \cong 4n\cos\theta. \qquad \text{(III.21.14)}$$

Therefore, the absorptance of a metal at grazing incidence for a TM (p) wave is $\cong |\tilde{n}|^2$ (10^3–10^4) times the absorptance for a TE (s) wave, this being clearly related to the Brewster angle phenomenon.

21.2 *Reflection and Refraction at Normal Incidence*

For normal incidence, the reflectance $R = R_s = R_p$ and the absorptance $A = A_s = A_p = 1 - R$ is given by [see Eq. (III.8.12a)]

$$A = \frac{4n}{n^2 + \kappa^2 + 1 + 2n}. \qquad \text{(III.21.15)}$$

Both R and A depend critically on the wavelength in the visible region, as shown in Figs. III.14 and III.15.

The amplitude reflection coefficient $r_s(\theta = 0) = r_p(\theta = 0) = r$ is easily obtained by means of Eqs. (III.8.2) and of Snell's law and reads

$$r = R^{1/2}e^{i\phi} = (1 - \tilde{n})/(1 + \tilde{n}). \qquad \text{(III.21.16)}$$

The quantities n and κ, and thus R and ϕ, depend on the angular frequency ω. In particular, it can be shown by means of the Kramers–Kronig dispersion relations that $\phi(\omega)$ is connected with $R(\omega)$ through the principal value of the integral

$$\phi(\omega) = -\frac{\omega}{\pi} \mathrm{P} \int_0^\infty \frac{\ln R(\omega')}{\omega^2 - \omega'^2} d\omega' + \text{const.} \qquad \text{(III.21.17)}$$

The above relation is used in the *Robinson–Price* method for measuring the refractive index spectrum of materials in the region of strong absorption. In this method, one first measures the reflectance $R(\omega)$ over a frequency range as wide as possible, and then calculates the phase $\phi(\omega)$ of the amplitude reflection coefficient by using Eq. (III.21.17). Once $R(\omega)$ and $\phi(\omega)$ are known, it is easy to evaluate $n(\omega)$ and $\kappa(\omega)$ by using Eq. (III.21.16). For a complete description of the methods used for measuring optical constants see Bell [45].

21.3 *Reflection at a Generic Angle*

For a generic angle of incidence θ, the reflectances R_p and R_s can be easily evaluated by means of Eqs. (III.8.2), with the help of Snell's law. If $|\tilde{n}| \gg 1$, we obtain

$$R_p = |r_p|^2 = \frac{1 + (n^2 + \kappa^2)\cos^2\theta - 2n\cos\theta}{1 + (n^2 + \kappa^2)\cos^2\theta + 2n\cos\theta}. \qquad \text{(III.21.18)}$$

An analogous expression for R_s can be obtained by replacing $\cos\theta$ with $1/\cos\theta$ in the above equation.

22 Surface Waves at the Interface between Two Media

We wish to consider the surface waves originating at the separation surface between a dielectric and a medium with complex refractive index, in connection with incident waves propagating at the Brewster angle.

To this end, some preliminary comments about the connection of the Brewster angle with the zeros and poles of $r_p(k_x)$ are in order. In fact, if we deal with complex-index media and extend the reflection coefficient r_p defined by Eq. (III.8.1) to the complex domain, the function $r_p(k_x)$ (remember that "2" refers to the first medium) can take two determinations, so that the Brewster angle may appear to be not uniquely defined. As a first step, let us observe that, if we replace the direction of the incident wave with that of the reflected one in Eq. (III.8.1), r_p transforms into its reciprocal, as immediately seen by replacing $k_z^{(2)}$ with $-k_z^{(2)}$. As a consequence, r_p can be either zero or infinite at the Brewster angle, depending on the determination of $k_z^{(2)}$. In particular, according to our conventions, $|r_p| = 0$ for $\operatorname{Re} k_z^{(2)} > 0$, and $|r_p| = \infty$ for $\operatorname{Re} k^{(2)} < 0$. On the other hand, in many problems we prefer to express r_p as a function of k_x. In this case, r_p is a two-valued function of k_x, each determination being the reciprocal of the other one. In fact [see Eq. (III.8.1) and Snell's law], if we assume for simplicity that medium 2 coincides with the vacuum,

$$r_p(k_x) = \frac{(\tilde{n}^2 - k_x^2/k_0^2)^{1/2} - \tilde{n}^2(1 - k_x^2/k_0^2)^{1/2}}{(\tilde{n}^2 - k_x^2/k_0^2)^{1/2} + \tilde{n}^2(1 - k_x^2/k_0^2)^{1/2}}. \qquad \text{(III.22.1)}$$

The square roots can be made single-valued by introducing branch cuts in the complex k_x plane. In this case, when we look for the Brewster angle, we must bear in mind that it corresponds either to a zero or to a pole of $r_p(k_x)$, depending on the determination chosen for r_p itself. To summarize, corresponding to the values of k_x for which one of the two determinations of r_p vanishes, the other one diverges. Thus, zeros and poles of the two-valued function $r_p(k_x)$ coincide. In physical terms, this is related with the fact that Brewster's angle, corresponding to "no reflection," can obviously be associated with an infinite response to a vanishing excitation by interchanging the roles of incident and reflected fields. Therefore, we can limit ourselves to considering the situation $r_p(k_x) = 0$, directly corresponding to absorption of the radiation by the second medium, which is in general a stratified structure.

Whenever the field in the substrate (last half-space of the stratified medium) escapes in the z direction orthogonal to the plane separating the first and second medium without attenuation (real value of k_z in the substrate), the incident power is completely transferred to $z = +\infty$. In the opposite case the perturbation is confined near the separation plane, the resulting field corresponding to a leaky or to a surface wave (see also Section III.18). More precisely, a nonvanishing imaginary part of k_x yields a leaky wave, while a real k_x gives rise to a surface wave. While the later case corresponds to a field propagating along the separation surface in the x direction without attenuation, a leaky wave loses energy while traveling along the x direction. The distinction between leaky and surface waves is particularly relevant when a beam of finite section impinges on the separation surface, since only the surface wave gives rise to effective energy transport over large distances orthogonal to the stratification axis of the structure.

If we choose the usual coordinate system with the z axis perpendicular to the (plane) interface and the x axis coplanar with the incident ray, the wave vector component k_z of a TM wave propagating at the Brewster angle [Eq. (III.8.3)] in the first medium (2), supposed for simplicity to coincide with the vacuum, is

$$k_z^{(2)} \equiv k_z^{\prime(2)} - ik_z^{\prime\prime(2)} = k_0 \cos\theta_{\mathrm{B}} = \frac{k_0}{(1+\tilde{n}^2)^{1/2}} = \frac{\omega}{c}\left(\frac{\varepsilon_0}{\varepsilon_0+\tilde{\varepsilon}}\right)^{1/2}, \qquad \text{(III.22.2)}$$

$\tilde{\varepsilon} = \varepsilon_0 \tilde{n}^2$ being the dielectric constant of the second medium. After refraction, Snell's law yields

$$k_z^{(1)} \equiv k_z^{\prime(1)} - ik_z^{\prime\prime(1)} = k_0 \frac{\tilde{n}^2}{(1+\tilde{n}^2)^{1/2}} = \frac{\omega}{c}\frac{\tilde{\varepsilon}}{(\varepsilon_0+\tilde{\varepsilon})^{1/2}\varepsilon_0^{1/2}}, \qquad \text{(III.22.3)}$$

while k_x does not vary.

22.1 *Surface Waves in Metals*

In good conductors, well below the plasma frequency (visible and IR region), the extinction coefficient κ is much larger than unity, so that the above relations are well approximated by

$$k_z^{(2)} \cong k_0(n/\kappa^2 + i/\kappa), \qquad k_z^{(1)} \cong k_0(n - i\kappa), \qquad \text{(III.22.4)}$$

where we have retained the leading terms in the smallness parameter $1/\kappa$ for real and imaginary parts. Thus, although in the first medium the planes of constant phase and constant amplitude are orthogonal to each other ($\mathbf{k}^{\prime(2)} \cdot \mathbf{k}^{\prime\prime(2)} = k_x' k_x'' + k_z^{\prime(2)} k_z^{\prime\prime(2)} = 0$), the same property does not hold true in the second medium, this being connected with the presence of a complex refractive index.

Equations (III.22.4) correspond to a field attaining its maximum value on the interface and evanescent from both sides. The attenuation in the direction $\hat{x}$ parallel to the interface is much slower than that in the orthogonal direction.

In fact, it can be easily shown that

$$-k_z''^{(2)}/k_x'' = \kappa^2/n \gg 1, \qquad k_z''^{(1)}/k_x'' = \kappa^4/n \gg 1. \tag{III.22.5}$$

Since the decay length along the surface is much larger than the optical wavelength, the field propagates as a surface wave connected with the natural oscillation modes of the electronic gas of the metal (*surface plasmons*, SP) [10, 46]. On the other hand, the decay length $(k_0\kappa)^{-1}$ into the metal is large enough to ensure that the surface wave characteristics are essentially determined by the bulk dielectric constant of the metal. Nevertheless, they are sensitive to perturbations on the surface, due for example to layers of adsorbed gas.

For $\tilde{\varepsilon} = \varepsilon_L[1 - \omega_p^2/\omega(\omega - i\Gamma)]$ [see Eq. (I.2.47) and the legend of Fig. I.5], Eq. (III.22.2) allows us to write the real part of k_x as

$$k_x' = \begin{cases} \dfrac{\omega}{c}\left(\dfrac{\varepsilon_L}{\varepsilon_0+\varepsilon_L}\right)^{1/2}\left(\dfrac{\omega^2-\omega_p^2}{\omega^2-\varepsilon_L\omega_p^2/(\varepsilon_0+\varepsilon_L)}\right)^{1/2} & \left[\dfrac{\omega}{\omega_p}<\left(\dfrac{\varepsilon_L}{\varepsilon_0+\varepsilon_L}\right)^{1/2}, \quad \dfrac{\omega}{\omega_p}>1\right], \\ 0 & \left[\left(\dfrac{\varepsilon_L}{\varepsilon_0+\varepsilon_L}\right)^{1/2}<\dfrac{\omega}{\omega_p}<1\right], \end{cases} \tag{III.22.6}$$

where we have neglected the collision frequency Γ. Accordingly, the surface wave cannot propagate for ω in the frequency range between $\omega_p\varepsilon_L^{1/2}/(\varepsilon_0 + \varepsilon_L)^{1/2}$ and ω_p. This situation is usually considered to be due to the presence of a *forbidden band* (see Fig. III. 37). For finite values of Γ, the dispersion

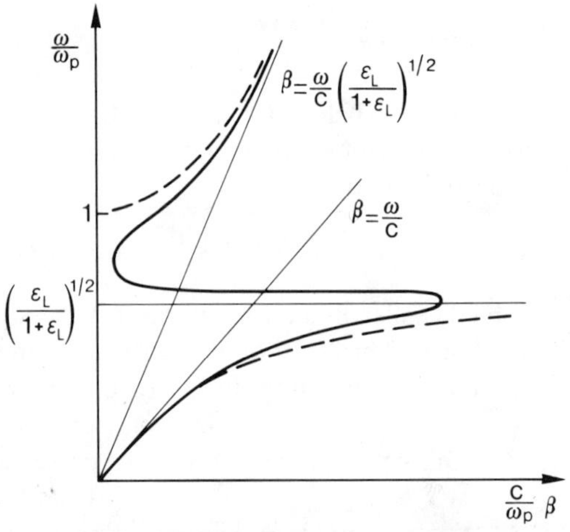

Fig. III.37. Schematic plot of the dispersion curve of the surface waves propagating at the interface of a metal with a dielectric. The dashed curve refers to an ideal metal without damping, whose dielectric function is that of a collisionless plasma. The plasma frequency of the metal is denoted by ω_p.

curve modifies into the continuous line. For $\omega = \omega_p \varepsilon_L^{1/2}/(\varepsilon_0 + \varepsilon_L)^{1/2}$, k_x' does not diverge any more, but it presents a peak and then bends back and reaches a minimum at a value of ω/ω_p that depends on Γ. Thus, the phenomenon of the forbidden band tends to disappear for increasing Γ. We note that, for $\omega > \omega_p$, the medium becomes transparent, so that the fields corresponding to Eq. (III.22.6a) do not represent surface plasmons.

Since the surface electromagnetic wave is confined close to the surface, it will not leak out into a radiative wave unless the surface is subject to perturbations or imperfections. Conversely, it is also impossible to excite a surface wave by shining a light beam directly on a smooth surface. In order to study the surface-wave properties, various excitation and detection methods have been devised, namely linear and nonlinear optical excitation and detection achieved through perturbation of the surface, or use of a prism sitting on top of the surface with a small gap of the order of a wavelength (see Fig. III.6 and the discussion in Section III.3.3). Such a technique is known as *attenuated total reflection* (ATR) and employs the evanescent wave generated at a medium–air interface when light in the medium undergoes total internal reflection. It is the reduction of the reflected wave due to absorption that is called ATR. The first system, proposed by Otto, consists of a prism (P) separated from a thick sample of the medium (M) by a small air (A) or vacuum gap (PAM ATR configuration, indicated in Fig. III.38a). Provided the air gap is sufficiently small, the evanescent field generated in the gap owing to total reflection in the prism can reach the air–medium interface and excite a surface plasmon.

In the prism–medium–air system (PMA, *Kretschmann technique*), an active medium is deposited onto the base of a prism. The surface wave at the medium–air interface is excited by the evanescent wave generated at the prism–medium interface and penetrating the film (Fig. III.38b).

The excitation of a surface wave can be observed as a sharp minimum in the reflected intensity of a TM wave. By measuring the angle at which such a minimum is observed, we can immediately derive the value of the propagation factor k_x' of the excited surface wave, so that, if we change the frequency, we can reconstruct the dispersion diagram schematically represented in Fig. III.37.

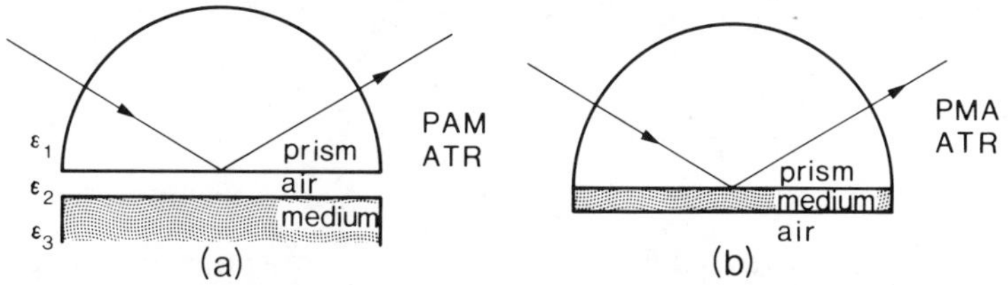

Fig. III.38. Systems used for exciting surface plasmons by exploiting a cylindrical prism. Configurations (a) and (b) are also referred to as the Otto and Kretschmann systems.

23 Impedance Boundary Conditions

The boundary conditions on a metal surface can be determined by taking into account the decay of the field away from the interface. Let us consider a plane TE wave incident from vacuum on a metal. The transmitted field is of the type

$$u'(x,z) = \exp[-ik_0\tilde{n}(z\cos\theta' + x\sin\theta')], \tag{III.23.1}$$

the z axis pointing inside the metal in the direction orthogonal to the plane separation surface and xz being the plane of incidence. Since for a metal $|\tilde{n}| \gg 1$, Snell's law implies a refraction angle $\theta' \cong 0$, so that

$$u'(x,z) \cong e^{-ik_0(nz + x\sin\theta)}e^{-k_0\kappa z}, \tag{III.23.2}$$

according to which the penetration depth of the wave inside the metal is of the order of $(k_0\kappa)^{-1}$. We now observe that u is continuous on the interface, provided it represents the electric component E_y parallel to the interface itself. On the other hand, $\partial u/\partial z$ is proportional to H_x, which is, in turn, continuous. Therefore, the relation

$$\hat{n}\cdot\nabla u \cong iuk_0\tilde{n} \qquad \text{(TE waves)} \tag{III.23.3}$$

where $\hat{n}$ is the normal to the interface pointing outside the metal, holds true *in vacuo* close to the separation surface, as an immediate consequence of Eq. (III.23.2) and of the continuities of E_y and H_x.

The above argument can be applied to TM waves, provided u represents H_y, the relevant continuous quantities in this case being u and $(1/\tilde{n}^2)\,\partial u/\partial z$ (proportional to E_x). Then Eq. (III.23.3) transforms into

$$\hat{n}\cdot\nabla\mathbf{u} = iuk_0/\tilde{n} \qquad \text{(TM waves).} \tag{III.23.4}$$

The above relations connecting u and $\partial u/\partial n$ immediately outside the metal are usually termed the *Leontovich impedence boundary conditions* [47] for imperfectly conducting surfaces. According to Eqs. (III.23.3) and (III.23.4), we can characterize the surfaces by

$$Z_S \equiv (i\zeta_0/k_0u)(\partial u/\partial z), \tag{III.23.5}$$

which yields $Z_{S,s} = \zeta_0\tilde{n}$ for TE waves and $Z_{S,p} = \zeta_0/\tilde{n}$ for TM waves.

Problems

Section 3

1. Show that Eqs. (III.3.9) and (III.3.11), representing the field in the neighborhood or away from a turning point, may be replaced by Langer's

formula which is valid for all z,

$$u(z) \propto S_0^{1/6}(z)[n^2(z) - n_0^2 \sin^2 \theta]^{-1/4} \operatorname{Ai}\left\{\left[\frac{3}{2} k_0 S_0(z)\right]^{2/3}\right\},$$

where

$$S_0(z) = \int_{z_a}^{z} [n^2(z') - n_0^2 \sin^2 \theta]^{1/2} \, dz'.$$

2. Calculate the transmission coefficient t_d of a duct.

Section 8

3. Prove that for an incidence angle of 45° the Fresnel reflection coefficients satisfy the so-called *Abelès condition*, $R_s^2 = R_p$ and $2\psi_s = \psi_p$, ψ_s and ψ_p being the absolute phase changes on reflection for the s and p components.

4. Show that the transmittance of a lossless plate is given by $T_{s,p} = (1 - R_{s,p})/(1 + R_{s,p})$ when the beam undergoes multiple incoherent reflections within the plate. Calculate the *extinction ratio* T_s/T_p as a function of the incidence angle.

5. Consider the pile-of-plates transmission polarizer shown in Fig. III.13. Calculate the total transmittance for p- and s-polarization by assuming incoherent multiple reflections within each plate and none between plates. Show that

$$T_{s,p} = [(1 - R_{s,p})^{2m} e^{-m\alpha d}]/[(1 + R_{s,p}^2 e^{-2\alpha d})^m]$$

where m is the number of plates, $\alpha = 4\pi\kappa/(\lambda \cos \theta')$, and κ is the extinction coefficient (see Section III.17), θ' the refraction angle, and d the plate thickness.

6. Using the formula of Problem 5, show that the *degree of polarization* that can be achieved with the polarizer of Fig. III.13 is equal to

$$\frac{T_p - T_s}{T_p + T_s} = \frac{1 - \cos^{4m}(\theta - \theta')(1 - R_p^2 e^{-2\alpha d})^m/(1 - R_s^2 e^{-2\alpha d})^m}{1 + \cos^{4m}(\theta - \theta')(1 - R_p^2 e^{-2\alpha d})^m/(1 - R_s^2 e^{-2\alpha d})^m},$$

where θ and θ' are the angles of incidence and refraction, respectively.

Section 9

7. Express the characteristic matrix $\mathbf{M}_{ba}$ of a double layer in the form

$$\mathbf{M}_{ba} = a\boldsymbol{\sigma}_0 + b\boldsymbol{\sigma}_1 + c\boldsymbol{\sigma}_2 + d\boldsymbol{\sigma}_3,$$

where

$$\boldsymbol{\sigma}_0 = \begin{bmatrix} 1 & 0 \\ 0 & 1 \end{bmatrix}, \quad \boldsymbol{\sigma}_1 = \begin{bmatrix} 0 & 1 \\ 1 & 0 \end{bmatrix}, \quad \boldsymbol{\sigma}_2 = \begin{bmatrix} 0 & -i \\ i & 0 \end{bmatrix}, \quad \boldsymbol{\sigma}_3 = \begin{bmatrix} 1 & 0 \\ 0 & -1 \end{bmatrix}$$

are the Pauli spin matrices and $a^2 - b^2 - c^2 - d^2 = 1$. Then, putting $A + D = 2\cos\delta$ and using the properties of these matrices, show that

$$\mathbf{M}_{ba}^{m} = \frac{1}{\sin\delta}\begin{bmatrix} A\sin(m\delta) - \sin[(m-1)\delta] & B\sin(m\delta) \\ C\sin(m\delta) & D\sin(m\delta) - \sin[(m-1)\delta] \end{bmatrix}$$

8. Show that

$$\mathbf{M}_b \cdot \mathbf{M}_a - \mathbf{M}_a \cdot \mathbf{M}_b = [-(Z_b/Z_a) + (Z_a/Z_b)] \sin\phi_a \sin\phi_b \boldsymbol{\sigma}_3,$$

where

$$\boldsymbol{\sigma}_3 = \begin{bmatrix} 1 & 0 \\ 0 & -1 \end{bmatrix}.$$

9. Use the **M** matrix of $(\mathrm{HL})^p$ given in Problem 7 to calculate the matrix relative to the symmetrical system $(\mathrm{HL})^p\mathrm{H}$. In addition, calculate the refractive index and the phase thickness of the equivalent single layer (see Problem 10).

10. Show that any thin-film combination is equivalent at one wavelength to a two-film combination (Herpin's theorem). *Hint*: Show that the more general **M** matrix can be put in the form of Eq. (III.9.27).

11. Consider a three-layer combination of the form aba where the outer layers are alike in thickness and index. Show that

$$A = D = \cos 2\phi_{\mathrm{a}} \cos\phi_{\mathrm{b}} - (1/2)(n_{\mathrm{a}}/n_{\mathrm{b}} + n_{\mathrm{b}}/n_{\mathrm{a}}) \sin 2\phi_{\mathrm{a}} \sin\phi_{\mathrm{b}},$$

$$B = (i/n_{\mathrm{a}})[\sin 2\phi_{\mathrm{a}} \cos\phi_{\mathrm{b}} + (1/2)(n_{\mathrm{a}}/n_{\mathrm{b}} + n_{\mathrm{b}}/n_{\mathrm{a}}) \cos 2\phi_{\mathrm{a}} \sin\phi_{\mathrm{b}} + (1/2)(n_{\mathrm{b}}/n_a - n_{\mathrm{a}}/n_{\mathrm{b}}) \sin\phi_{\mathrm{b}}].$$

12. Discuss the variation of the $\mathbf{M}_{ba}$ matrix with the angle of incidence. In particular, calculate $\partial\mathbf{M}/\partial\theta$ for $\theta = 0$. *Hint*: Use Eq. (III.9.27) and notice that $\partial\phi_b/\partial\phi_a = (d_b/d_a)^2 \phi_a/\phi_b$.

Section 10

13. Consider a quarter-wave stack $(\mathrm{HL})^m$. Show that at the *Bragg resonance*, i.e., $\lambda = \lambda_0$, the voltage along the equivalent transmission line is proportional to

$$V(z) = \frac{1}{\sqrt{2}}\left(V_0 - \frac{V'_0}{k_0 n_{\mathrm{H}}}\right)\left(\frac{n_{\mathrm{H}}}{n_{\mathrm{L}}}\right)^{q(z)-q(0)}\left(\frac{n_{\mathrm{H}}}{n_{\mathrm{L}}}\right)^{\varepsilon} \cos\left(\phi + \frac{\pi}{4}\right) + \frac{1}{\sqrt{2}}\left(V_0 + \frac{V'_0}{k_0 n_{\mathrm{H}}}\right)\left(\frac{n_{\mathrm{L}}}{n_{\mathrm{H}}}\right)^{q(z)-q(0)} \cos\left(\frac{\pi}{4} - \phi\right).$$

Here $q(z)$ is a function that takes integer values equal to the index of the double layer containing the point z; ε is equal to 0 or 1 if z lies in a high- or low-index layer, respectively; $\phi \equiv k_0 \int_0^z n(z')\,dz'$. The origin of coordinates coincides with the midplane of a high-index layer $[V_0 = V(-\frac{1}{2}d_H), V_0' = dV/dz]$.

14. With reference to the system considered above, show that the *Bloch* function $V^{(+)}(z)$ is given by $f(z)\exp(-i\delta z/\Lambda)$, where ($\Lambda = d_H + d_L$)

$$e^{-i\delta z/\Lambda} = e^{-i\pi z/\Lambda}(n_L/n_H)^{z/\Lambda},$$

$$f(z) = |\cos[(\pi/4) - \phi]|(n_H/n_L)^{z/\Lambda}e^{i\pi z/\Lambda} \qquad (-d_H/2 < z < (d_H/2) + d_L).$$

In addition, show that $V^{(-)}(z) \propto V^{(+)}(-z)$.

15. Consider a medium consisting of infinitely alternating layers of two different lossless dielectrics a and b. Calculate the phase factor δ for $\lambda \gg d_a, d_b$. Show that for s- and p-polarization

$$\delta^2 + (d_a + d_b)^2(\beta_x^2 + \beta_y^2) = n_a^2 d_a^2 + n_b^2 d_b^2 + (n_a^2 + n_b^2)d_a d_b \qquad \text{(s-wave)}$$

$$\delta^2 + (d_a + d_b)^2 + (n_a/n_b + n_b/n_a - 2)d_a d_b(\beta_x^2 + \beta_y^2)$$
$$= n_a^2 d_a^2 + n_b^2 d_b^2 + (n_a^2 + n_b^2)d_a d_b \qquad \text{(p-wave)}$$

Notice that the field can be assimilated to a plane wave with wave vector $\beta_x, \beta_y, \delta/(d_a + d_b)$. Consequently, the stratified medium behaves at long wavelengths as a homogeneous uniaxially anisotropic medium, whose ordinary and extraordinary refractive indices are, respectively, equal to

$$n_o = \frac{n_a^2 d_a^2 + n_b^2 d_b^2 + (n_a^2 + n_b^2)d_a d_b}{(d_a + d_b)^2},$$

$$n_e = \frac{n_a^2 d_a^2 + n_b^2 d_b^2 + (n_a^2 + n_b^2)d_a d_b}{d_a^2 + d_b^2 + (n_a^2/n_b^2 + n_b^2/n_a^2)d_a d_b}.$$

Hint: Calculate the characteristic matrix by making use of the approximation $\cos\phi = 1 - \phi^2/2$. In addition, put $\cos\theta_{a,b} = (n_{a,b}^2 k_0^2 - \beta_x^2 - \beta_y^2)^{1/2}/(n_{a,b}k_0)$. (See Yariv and Yeh [48].)

16. Show that at the center of a passband ($\delta = 2n\pi$, $\phi = n\pi$) $\partial\delta/\partial\phi = (2 + n_H/n_L + n_L/n_H)^{1/2}$.

Section 11

17. Prove Eq. (III.11.9). *Hint*: Use Eqs. (III.10.8) and (III.9.27) by imposing $\phi_a = \phi_b$ and using the definition of ϕ_t given by Eq. (III.10.6).

18. Show that at the edges of a passband for a multilayer with alternating indexes, the phase of $Z^{(+)}$ tends to $\pi/2$ while its amplitude is given by

$$Z^{(+)} = Z_{cb}\left(1 + \frac{3}{2 + n_a/n_b + n_b/n_a}\right)^{-1/2}.$$

Section 12

19. Show that the input impedance Z_{in} of a multicoated substrate can be expressed in the form

$$\vec{Z}_{\text{in}} = (AZ_S + B)/(CZ_S + D)$$

where the parameters A, B, C, and D refer to the multilayer.

20. Show that the input impedance of the system $(\text{HL})^p$ glass can be expressed by

$$\vec{Z}_{\text{in}} = \frac{Z_g[A \sin p\delta - \sin(p-1)\delta] + B \sin p\delta}{Z_g C \sin p\delta + D \sin p\delta - \sin(p-1)\delta}.$$

Hint: See Problem 7.

21. Prove Eq. (III.12.8) by using Eqs. (III.12.6) and (III.12.7).

22. Use the second law of thermodynamics to prove that the transmittance of a multilayer separating two lossless volumes is the same for either direction of propagation of a plane wave. Notice that the reflectances are necessarily equal only when the diaphragm is lossless.

23. Use Eq. (III.12.16) to show that the reflectance of a single-layer coating is given by

$$R = \frac{R_I + R_S + +2\sqrt{R_I R_S} \cos(2n_c d_c k_0 \cos\theta_c)}{1 + R_I R_S + 2\sqrt{R_I R_S} \cos(2n_c d_c k_0 \cos\theta_c)},$$

where R_I and R_S are the Fresnel reflectances relative to the embedding medium–coating and coating–substrate interfaces, respectively [see Eqs. (III.8.8)].

Section 13

24. Consider a lossless thin film deposited on a metallic substrate having optical constants n_S and κ_S. Show that the reflectance R is given by

$$R = \frac{1 + \sqrt{R_I}(a_1 \cos 2\phi_c + b_1 \sin 2\phi_c) + R_I(a_1^2 + b_1^2)}{R_I + 2\sqrt{R_I}(a_1 \cos 2\phi_c + b_1 \sin 2\phi_c) + a_1^2 + b_1^2},$$

where R_I is the Fresnel reflection coefficient of the thin film–embedding medium interface, $\phi_c = n_c d_c k_0$ is the coating phase thickness, and

$$a_1 = \frac{n_c^2 - n_S^2 - \kappa_S^2}{(n_c + n_S)^2 + \kappa_S^2}, \qquad b_1 = \frac{2n_c \kappa_S}{(n_c + n_S)^2 + \kappa_S^2}.$$

25. Consider a lossless dielectric plate illuminated at normal incidence. Show that the amplitude of the electric field in the vicinity of the entrance

surface is less than that of the incident field by a factor $2/(n + 1)$. By contrast, interference between the light incident on and reflected from the exit surface results in an increase of E at the boundary by the factor $2n/(n + 1)$. Assume that the beam undergoes incoherent reflections between the two faces of the plate. Note that this difference between the values of E explains why optical damage occurs more frequently on the output face of a plate.

26. Consider a metallic substrate coated with a double layer $L_1 L_2$ of equal phase thickness ϕ. Show that the reflectance is given by

$$R = \frac{R_S + R_I - 2\sqrt{R_S R_I}\cos(2\phi - \delta)}{1 + R_S R_I - 2\sqrt{R_S R_I}\cos(2\phi - \delta)},$$

where R_S is the Fresnel coefficient relative to the metal–dielectric 2 interface, δ is the relative phase shift,

$$\delta = \arctan[2n_2\kappa_S/(n^2 - n_S^2 - \kappa_S^2)],$$

R_I is the Fresnel coefficient relative to the embedding medium–dielectric 1 system, and n_S and κ_S are the real and imaginary parts, respectively of the complex refractive index of the metal.

27. Consider a quarter-wave dielectric reflector of the form glass–$(\mathrm{HL})^p$. If we indicate by $n_H - i\kappa_H$ and $n_L - i\kappa_L$ the complex refractive index of the alternate layers, show that for $p \to \infty$ the reflectance tends to

$$R = 1 - 2\pi[(\kappa_H + \kappa_L)/(n_H^2 - n_L^2)]n_I.$$

(See Koppelmann [49].)

Section 14

28. Show that the spectral reflectance of a single-layer AR coating $(n_c^2 = n_I n_S)$ is given by

$$R(\lambda, \theta) = \frac{R_I(\theta) + R_S(\theta_c) + 2\sqrt{R_I(\theta)R_S(\theta_c)}\cos[\pi(\lambda_0/\lambda)\cos\theta_c]}{1 + R_I(\theta)R_S(\theta_c) + 2\sqrt{R_I(\theta)R_S(\theta_c)}\cos[\pi(\lambda_0/\lambda)\cos\theta_c]}.$$

Here θ is the incidence angle, θ_c the refraction angle in the coating, and $\lambda_0 = \frac{1}{4}n_c d_c$.

29. Plot the reflectance of a glass $(n = 1.51)$ coated with MgF_2 $(n = 1.38)$ versus λ and θ for the s- and p-polarizations.

30. Express the reflectance in the form $R = X/(1 + X)$ and calculate X for the double-layer group I and group II types of AR coatings (see Eqs. (18) and (19) in Musset and Thelen [30]).

31. Calculate the reflectance of a three-layer group I-type AR coating of the form $\phi_1 = \phi_2 = \phi_3$, $n_2 = \sqrt{n_S n_I}$, $n_3 = n_1\sqrt{n_S/n_I}$. Show that this coating

produces two zero-reflectance points, while the reflectance in the center ($\phi = \pi/2$) is equal to

$$R_{\max} \simeq (n_S/4n_I)[(n_I/n_1)^2 - n_1^2/n_I n_S]^2.$$

Hint: Use Eq. (III.12.10). See also Musset and Thelen [30], Eq. (25).

32. Calculate the reflectance of a three-layer group II-type AR coating ($\phi_1 = \phi_2 = \phi_3 = \phi$ and $n_2 = \sqrt{n_S n_I}$, $n_3 = n_S n_I/n_1$). Show that this coating produces three zero-reflectance points, one in the center ($\phi = \pi/2$). *Hint*: Use Eq. (III.12.10). See also Musset and Thelen [30], Eq. (28).

Section 15

33. Calculate the equivalent refractive index and phase thickness of the symmetrical system aba. Determine the interval of frequency in which ϕ_{eq} is complex. *Hint*: See Problem 11.

34. Calculate the transmittance of the double half-wave filter $(\text{HL})^m\text{HH}(\text{LH})^m\text{H}(\text{HL})^m\text{HH}(\text{LH})^m$. *Hint*: Use Eqs. (III.12.10), (III.12.8), and (III.12.9) and the results of the preceding problem.

35. Plot the equivalent refractive index and phase thickness of the symmetrical thin-film combination pqp versus $2\phi_p + \phi_q$. (See Epstein [50].)

36. Consider two symmetrical multilayers such that the refractive index of each layer of the first system is the reciprocal of the index of the corresponding layer of the second one. In addition, the phase thicknesses of the corresponding layers of the two systems are equal. Show that the equivalent refractive index of the first system is the reciprocal of that of the second system. (See Thelen [51].)

Section 19

37. Calculate the attenuation constant α of a surface wave propagating adjacent to a lossless thin film deposited on a lossy metal. *Hint*: Calculate the input impedance of the thin-film–metal system for a multilayer vacuum thin film-metal, and find a complex $k_x = \beta - i\alpha$ such that the real and imaginary parts of the impedance satisfy the resonance condition.

Section 20

38. Consider a hollow-core slab waveguide formed by two lossless dielectric plates spaced by a distance $d \gg \lambda$. Calculate the constant of propagation and the coefficient of attenuation α of a radiative mode propagating in this waveguide. Show that α is approximately given by

$$\alpha_{\text{TE}} = \lambda^2/8d^3\sqrt{n^2 - 1}, \qquad \alpha_{\text{TM}} = n^2\lambda^2/8\text{d}^3\sqrt{n^2 - 1},$$

for TE and TM polarization of the mode with respect to the dielectric faces; n is the dielectric refractive index. *Hint*: Represent the mode in the gap between the dielectrics as a superposition of two evanescent plane waves which undergo total reflection at the dielectric faces and transform into one another because of these reflections. (See Marcuse [52].)

39. Consider a lossless dielectric plate coated with a thin layer of a material having a refractive index less than unity (e.g., a metal near ω_p or a dielectric near a narrow and pronounced resonance) and a small extinction coefficient. Calculate the reflectance as a function of the incidence angle and the wavelength. Show that, to a good approximation, the critical angle $\theta_c = \arcsin n$ of the coating coincides with the angle θ_m where the slope of $R(\theta)$ is a maximum. In addition, discuss the effects due to the multiple reflections inside the coating. (See Hunter [53].)

40. Show that the magnetic streamlines relative to the field generated by an s-wave incident at an angle greater than the critical one are described in the more dense medium by the equation

$$\sin(\omega t + kx\sin\theta)\sin(ky\cos\theta) = \text{const.}$$

Section 21

41. Calculate the attenuation constant α of TE and TM modes propagating through a planar waveguide with lossy metallic sidewalls. Assuming a distance d much larger than λ between the two metallic planes, show that

$$\alpha = m\lambda A(\theta)/2d^2,$$

where $\theta = \pi/2 - m\lambda/2d$ is the angle of incidence of the plane waves constituting the mth mode with the walls. Then, using Eqs. (III.21.14) and (III.21.15), prove that

$$\alpha_m^{\text{TE}} = (m^2\lambda^2/d^3)\,\text{Re}\,(1/\tilde{n}), \qquad \alpha_m^{\text{TM}} = (m^2\lambda^2/d^3)\,\text{Re}\,\tilde{n}.$$

(See Garmire *et al.* [54].)

42. Draw a plot of the pseudo-Brewster angle versus $(n^2 + \kappa^2)^{1/2}$.

43. Calculate the complex refractive index of a metal for which you know the reflectivity at normal incidence and the pseudo-Brewster angle $\bar{\theta}$.

References

1. Knittl, Z., "Optics of Thin Films." Wiley, New York, 1976.
2. Hass, G., *J. Opt. Soc. Am.* **72**, 27 (1982).
3. Spiller, E., *Appl. Opt.* **15**, 2333 (1976).

4. Rosenbluth, A. E., and Foresyth, J. M., "Reflecting Properties of X-Ray Multilayer Devices." Inst. Opt., Univ. of Rochester, Rochester, New York, 1982.
5. Leith, E. H., Kozma, A., Upatnieks, J., Marks, J., and Maney, N. *Appl. Opt.* **5**, 1303 (1966).
6. George, N., and Matheus, J. W., *Appl. Phys. Lett.* **9**, 212 (1966).
7. Brekovskikh, L. M., "Waves in Layered Media." Academic Press, New York, 1960.
8. Wait, J. R., "Electromagnetic Waves in Stratified Media." Macmillan, New York, 1962.
9. Tien, P., *Rev. Mod. Phys.* **19**, 361 (1977).
10. Agranovich, V. M., and Mills, D. L., "Surface Polaritons." North-Holland Publ., Amsterdam, 1982.
11. Bender, C. M., and Orszag, S. A., "Advanced Mathematical Methods for Scientists and Engineers." McGraw-Hill, New York, 1978.
12. Jacobsson, R., *Prog. Opt.* **5**, 247–286 (1966).
13. Alexopulos, N. G., and Uslenghi, P. L. E., *J. Opt. Soc. Am.* **71**, 1508 (1981).
14. Tyras, G., "Radiation and Propagation of Electromagnetic Waves." Academic Press, New York, 1969.
15. Brillouin, L., "Wave Propagation in Periodic Structures," 2d Ed. Dover, New York, 1953.
16. MacLeod, H. A., "Thin Film Optical Filters." Am. Elsevier, New York, 1969.
17. Vasicek, A., "Optics of Thin Films." North-Holland Publ., Amsterdam, 1960.
18. Hass, G., ed., "Physics of Thin Films," Vol. 1. Academic Press, New York, 1963.
19. Abelès, F., *Ann. Phys.* (*Paris*) **5**, 596; **5**, 706 (1950).
20. Berreman, D. W., *J. Opt. Soc. Am.* **65**, 502 (1972).
21. Collin, R. E., "Foundations for Microwave Engineering." McGraw-Hill, New York, 1965.
22. Stone, J. M., "Radiation and Optics." McGraw-Hill, New York, 1963.
22a. Baumeister, P. "A Survey of Optical Interference Coatings," Inst. Opt., Univ. of Rochester, Rochester, New York, 1977.
23. Arndt, J., and Baumeister, P., *J. Opt. Soc. Am.* **56**, 1760 (1966).
24. Hass, G., *J. Opt. Soc. Am.* **45**, 945 (1955).
25. Baumeister, R., and Arnon, O., *Appl. Opt.* **16**, 439 (1977).
26. Thelen, A., *J. Opt. Soc. Am.* **61**, 365 (1971).
27. Clapham, P. B., Downs, M. J., and King, R. J., *Appl. Opt.* **8**, 1965 (1969).
28. Bennet, J. M., and Bennet, H. E., *in* "Handbook of Optics" (W. G. Driscoll and W. Vaughan, eds.), p. 10-1–10-100 McGraw-Hill, New York, 1978.
29. Schroder, H., and Schlafer, R., *Z. Naturforsch.* **49**, 576 (1949).
30. Musset, A., and Thelen, A., *Prog. Opt.* **8**, 201–237 (1970).
31. Cox, J. T., and Hass, G., *in* "Physics of Thin Films" (G. Hass and R. E. Thun, eds.), Vol. 2, p. 239–303. Academic Press, New York, 1964.
32. Kard, P., "The Analysis and Synthesis of Multilayers Interference Coatings," pp. 79–91. Valgus, Tallin, Estonia, 1971. (in Russ.)
33. Delano, E., and Pegis, R. J., *Prog. Opt.* **7**, 67–137 (1969).
34. Baumeister, P., Moore, R., and Walsh, K., *J. Opt. Soc. Am.* **67**, 1039 (1977).
35. Dobrowolski, J. A., *in* "Handbook of Optics" (W. G. Driscoll and W. Vaughan, eds.), p. 8-1–8-117. McGraw-Hill, New York, 1978.
36. Seeley, J. S., *J. Opt. Soc. Am.* **54**, 342 (1964).
37. Smith, S. D., *J. Opt. Soc. Am.* **48**, 43 (1958).
38. Kogelnik, H., *Bell Syst. Tech. J.* **48**, 2909 (1969).
39. Tamir, T., ed., "Integrated Optics." Springer-Verlag, Berlin and New York, 1975.
39a. Wang, S., *IEEE J. Quantum Electron.* **QE-10**, 413 (1974).
39b. Tsang, W. T., Logan, R. A., Johnson, L. F. J., Hartman, R. L., and Koszi, L. A., *J. Quantum Electron.* **QE-15**, 1091 (1979).
40. Magnus W., and Winkler, S., "Hill's Equation." Wiley, New York, 1966.

41. Yeh, P., Yariv, A., and Hong, C.-S. *J. Opt. Soc. Am.* **67**, 423 (1977).
42. Collin, R. E., "Field Theory of Guided Waves." McGraw-Hill, New York, 1960.
43. Abelès, F., ed. "Optical Properties of Solids." North-Holland Publ., Amsterdam, 1972.
44. Abelès, F., *in* "Advanced Optical Techniques" (A. C. S. van Heel, ed.), p. 143–188. North-Holland Publ., Amsterdam, 1963.
45. Bell, E. E., *in* "Handbuch der Physik" (S. Flügge, ed.), Vol. XXV-2A, p. 1–57. Springer-Verlag, Berlin and New York, 1967.
46. Burstein, E., and De Martini F., eds., "Polaritons." Pergamon, New York, 1974.
47. Senior, B. A., *Appl. Sci. Res., Sect. B* **8**, 437 (1960).
48. Yariv, A. and Yeh, P., *J. Opt. Soc. Am.* **67**, 438 (1977).
49. Koppelmann, G. *Ann. Phys.* (*Paris*) **5**, 388 (1950).
50. Epstein, I. I., *J. Opt. Soc. Am.* **42**, 806 (1952).
51. Thelen, A., *J. Opt. Soc. Am.* **56**, 1533 (1966).
52. Marcuse, D., *IEEE J. Quantum Electron.* **QE-8**, 661 (1972).
53. Hunter, W. R., *J. Opt. Soc. Am.* **54**, 15 (1964).
54. Garmire, E., McMahon, T. M., and Bass, M., *IEEE J. Quantum Electron.* **QE-16**, 23 (1980).

Bibliography

Agranovich, V. M., and Loudon, R., eds., "Surface Excitations," North-Holland Publishing, Amsterdam, 1984.

Azzam, R. M. A., and Bashara, N. M., "Ellipsometry and Polarized Light." North-Holland Publ., Amsterdam, 1977.

Beckman, P., and Spizzichino, A., "The Scattering of Electromagnetic Waves from Rough Surfaces." Pergamon, Oxford, 1963.

Hsue, C. W., and Tamir, T., *J. Opt. Soc. Am.* **A2**, 923 (1985).

Hunderi, O., *Surf. Sci.* **96**, 1 (1980).

Jacobsen, R. T., *J. Opt. Soc. Am.* **54**, 1170 (1964).

Koch, E. E., ed., "Handbook on Synchrotron Radiation," Vol. 1. North-Holland Publ., Amsterdam, 1983.

Kortum, G. F., "Reflectance Spectroscopy." Springer-Verlag, Berlin and New York, 1969.

Rovard, P., and Bousquet, P., in "Progress in Optics" (E. Wolf, ed.) Vol IV, pp. 145–197. North-Holland Publ., Amsterdam, 1965.

Tamir, T., Wang, H. C., and Oliver, A. A., *IEEE Trans. Microwave Theory Tech.* **MTT-12**, 323 (1964).

Yeh, P., *J. Opt. Soc. Am.* **69**, 742 (1979).

Yeh, P., *Surf. Sci.* **96**, 41 (1980).

Chapter IV

Fundamentals of Diffraction Theory

1 Introduction

1.1 *Historical Account*

Wave theory, as it stands today, is the result of a long and interesting evolution originating from the ideas illustrated in 1690 by Christian Huygens in his celebrated "Traité de la Lumière". As he says in the first chapter, he was impressed by the fact that

> the undulations produced by small movements and corpuscules, should spread to such immense distances; as for example from the Sun or from the Stars to us. For the force of these waves must grow feeble in proportion as they move away from their origin, so that the action of each one in particular will without doubt become incapable of making itself felt to our sight. But one will cease to be astonished by considering how at a great distance from the luminous body an infinitude of waves, though they have issued from different points of this body, unite together in such a way that they sensibly compose one single wave only, which, consequently, ought to have enough force to make itself felt. Thus this infinite number of waves which originate at the same instant from all points of a fixed star, big it may be as the Sun, make practically only one single wave which may well have force enough to produce an impression to our eyes.

This description of wave propagation has become known as the *Huygens principle.*

Thomas Young had the great merit of reintroducing the ideas of wave propagation to explain, following Newton, the corpuscular theory of optical phenomena. In fact, in his three Bakerian lectures read at the Royal Society in 1801, 1802, and 1803, he introduced the *principle of interference*, which, applied to the waves accompanying the corpuscular components of the light, already postulated by Newton, explained the formation of the rings observed when the curved surface of a convex lens is pressed against a flat optical surface.

By virtue of this principle, Young was able to compute for the first time the wavelengths of different colors. Unfortunately, the prevailing corpuscular theory led the scientific community to oppose his ideas.

Later, on October 15, 1815, Augustin Jean Fresnel presented to the French Academy the famous treatise "La Diffraction de la Lumière," in which, developing the ideas of Huygens and Young, he presented a systematic description of the fringes observed on the dark side of an obstacle illuminated by a thin light source. In this way, he was able to show agreement between the measured spacings of the fringes and those calculated by means of the wave theory. Fresnel also put the Huygens principle in a more correct form by stressing the role of the phases of the single contributions. In fact, Huygens had no conception of transverse vibrations, of the principle of interference, or of the existence of the ordered sequence of waves in trains. In July 1819 the French Academy commemorated Fresnel with a special prize, and this represented the final victory of the wave theory over Newton's corpuscular theory.

Another remarkable success of the wave theory of light was recorded in 1835 with the publication in the *Transactions of the Cambridge Philosophical Society* of a fundamental paper by Sir George Biddell Airy, director of the Cambridge observatory, in which he derived his famous expression for the image of a star seen through a well-corrected telescope. The image consists of a bright nucleus, known since then as *Airy's disk*, surrounded by a number of fainter rings, of which only the first is usually bright enough to be visible to the eye.

Successive developments, until Maxwell's publication in 1873 of the "Treatise on Electricity and Magnetism," took advantage of Fresnel's ideas for solving a host of scattering and diffraction problems by using propagation through an elastic medium as a physical model. In particular, in 1861 Clebsch explained the diffraction of a plane wave by a spherical object. Surprisingly, most of these solutions remained valid when the electromagnetic phenomena were interpreted in the light of Maxwell's equations. The solutions found by Clebsch for the sphere are a typical example. The reason for their success lies in the fact that both electromagnetic and elastic fields can, in principle, be described by scalar functions that satisfy the scalar wave equation. Thus, we can see this equation as the unifying principle of many fields occurring in nature, which explains the astonishing phenomenon that a small perturbation originating in a finite volume can propagate throughout physical space and yet be felt at astronomical distances. The fascinating history of optical theories is repeated in the modern attempts to describe the more complex fields that rule the subnuclear world, which is the domain of high-energy physics.

Since the initial success of the Airy formula, the theory of diffraction has enjoyed increasing popularity, providing the fundamental tools for quantitatively assessing the quality of images and measuring the ability of optical systems to provide well-resolved images. This success can be explained with a well-known example. The size of the central diffraction disk of the image of a point object does not depend critically on the correct positioning of the observation plane or on the actual entity of the spherical aberration. For instance, it has been calculated that for a quarter-wavelength of defocusing or one wavelength of deformation of the wave front, the size of the disk remains appreciably constant. This means that the power of the instrument to resolve two points under these conditions is the same as that of an ideal aberrationless system. Surprisingly, the situation changes when we observe extended objects. In fact, a marked loss of contrast in the finest details of the image of an extended object is observed when the above aberrations are present. The explanation for this must be found in the modification of the whole pattern of the diffraction image. In fact, while the size of the central disk remains constant, the aberrations modify the distribution of the total intensity among the bright disk and the concentric rings. Calculations show that, with the above aberrations, the intensity of the rings is increased by 17%, with a consequent equal reduction of the brightness of the central disk.

In order to deal with this complex situation, Duffieux proposed in 1946 that the imaging of sinusoidal intensity patterns be examined as a function of their period. Thus, the optical system becomes known through an *optical transfer function* (OTF) that gives the system response versus the number of lines of the object per unit length. This function can be calculated by suitably manipulating the integrals of diffraction theory, while using the aberration function W_0 of the system (see Section II.15) evaluated with the ray optical (RO) formalism.

More recently, with the development of ideas put forth by Gabor and Toraldo di Francia, optical systems have been characterized by means of the numerable set of object fields that are faithfully reproduced (also for finite pupils and in the presence of aberrations). This approach, based on the solution of Fredholm's integral equations derived from the standard diffraction integrals, has allowed the ideas of information theory to be applied to optical instruments. Thus, quantizing the information in an image and measuring the information capacity of an optical instrument allow one to define in a straightforward way the information capacity of the electronic communication channels needed for image transmission.

1.2 *The Mathematical Apparatus*

Wave optics deals with the differences between the actual behavior of electromagnetic fields and that predicted by ray optics. While ray optical results are based on the approximation that waves propagate along certain trajectories called rays, the electromagnetic fields are actually ruled by the Helmholtz wave equation supplemented with suitable boundary conditions. While the solutions of the electromagnetic boundary-value problems are bounded and continuous in regions without sources, ray optical fields are singular on the caustics and discontinuous across the *shadow boundaries* produced by obstacles obstructing the ray bundles.

Unfortunately, the singularities of ray fields do not disappear when we consider the higher-order contributions of the LK series [Eq. (II.2.5)]. In fact, two successive terms of the series are connected by the recursive relations of Eq. (II.6.2), so that when the first term diverges it becomes impossible to calculate the successive ones. Thus, we must look for different representations of the fields, at least in proximity to these critical regions. The aim of wave optics is to remove these unphysical features of ray fields and to improve the evaluation of fields propagating over very long distances.

If we analyze the available analytic solutions of Maxwell's equations, which imply the presence of caustics or shadow boundaries in the limit $\lambda = 0$, we observe that the field amplitude systematically undergoes spatial oscillation in proximity to the critical regions. In other words, the caustics and shadow boundaries appear to be surrounded by a sort of boundary layer, in a way similar to that in a fluid lapping a surface (see Fig. IV.1). The thickness and the rate of variation of the field within this layer depend on the wave number k.

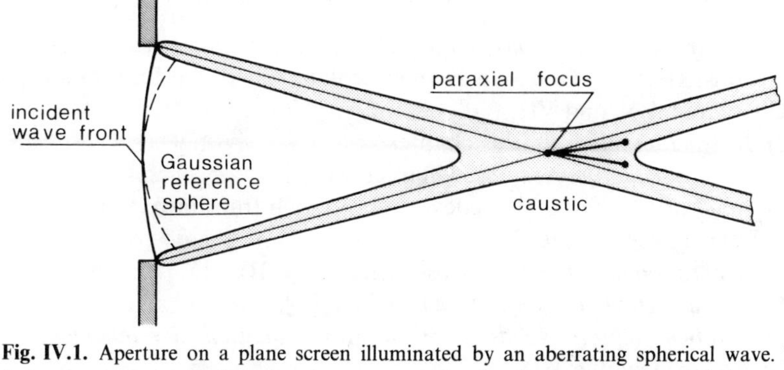

Fig. IV.1. Aperture on a plane screen illuminated by an aberrating spherical wave. The shaded region indicates the boundary layer surrounding the shadow boundary and the caustic.

As $k \to \infty$, the thickness of the layer vanishes and the solution of the wave equation tends to a limiting form [1].

In general, the deviation of actual fields from RO fields increases as we move away from the sources and from the obstacles hindering free propagation. As an example, let us consider an aperture on a metallic screen illuminated by a plane wave. The ray field corresponds to patterns of unaltered size and shape over planes parallel to the aperture. The actual irradiance pattern is known as a *diffraction pattern*, and the similarity between it and the initial illumination is progressively lost. Eventually, a far-field region is reached beyond which the size of the pattern, but not the shape, keeps changing with increasing distance.

A different situation arises in the presence of refractive index inhomogeneities. In this case, we must introduce several waves propagating in many directions, with coupling between these fields provided by the inhomogeneities. Physically, this can be interpreted as a *scattering* process; an incident wave produces a scattered wave propagating in various directions.

Wave theory problems can be roughly grouped into three classes:

(1) *Propagation through inhomogeneous media* (see Chapter III),
(2) *Diffraction* (this chapter),
(3) *Scattering by obstacles* (see Chapter VI).

In particular, diffraction theory (DT) deals mainly with the description of fields in proximity to caustics, foci, and shadow boundaries associated with wave fronts delimited by apertures (or stops). In a strict sense, any obstacle corresponds to a region with a refractive index different from that of the embedding medium, so that diffraction by apertures and scattering by obstacles could be considered examples of propagation through inhomogeneous media. Thus, the above grouping is essentially a matter of convenience.

Wave theory makes use of a number of analytical tools [2, 3]:

(1) *Spectral representations of the fields* (plane, cylindrical, spherical wave expansions; Hermite–Gaussian beams; prolate spheroidal harmonics) (e.g., see Chapters IV, V, and VII),

(2) *Diffraction integrals* (this chapter),

(3) *Integral equations* (e.g., see Chapter VII),

(4) *Integral transforms* (Lebedev–Kontorovich transform, Watson transform) [4] (e.g., see Chapter V),

(5) *Separation of variables* (e.g., see Chapter VIII).

(6) *Wiener–Hopf–Fock functional method* [5],

(7) *WKB asymptotic solutions of the wave equations for inhomogeneous media* (e.g., see Chapter III),

(8) *Variational methods* [6, 7],

(9) *Perturbative solutions of Maxwell's equations* [8] *for weakly inhomogeneous media* (*tenuous media*) (e.g., see Chapter VI).

In many cases the solutions are expressed by complex integrals and series, which can be evaluated either asymptotically or numerically by resorting to:

(1) *Stationary-phase and saddle-point methods* (e.g., see Chapter V),
(2) *Boundary-layer theory*,
(3) *Two-dimensional fast Fourier transform* (*FFT*) *algorithm* [9].

The saddle-point method, which is in general more accurate than the stationary-phase method, consists of deforming the integration path in the complex plane and then evaluating the integral by means of an asymptotic series. It can be shown that this series coincides globally with a modified LK representation of the field, in which the leading term is proportional to a fractional power of k^{-1}.

The anomalies on a caustic or on a shadow boundary can, in principle, be eliminated by using a conveniently "stretched" coordinate system, which allows one to describe with good approximation the rapid variations of the field. This approach is usually referred to as the *boundary-layer theory of diffraction* [1].

We recall that a complete knowledge of the electromagnetic field requires the determination of the scalar components of **E** and **H**. In general, the three components of the field (e.g., **E**) oscillate at the common frequency ω with different phases. As a consequence, the tip of the vector $\mathbf{E}(\mathbf{r}, t)$ at a given $\mathbf{r}$ describes a plane figure having the form of an ellipse as the time t spans a period $T = 2\pi/\omega$. When the ellipse degenerates into a segment, the field is *linearly polarized*. In many cases, relevant changes of polarization occur as $\mathbf{r}$ varies. For example, if we focus a linearly polarized beam by means of a lens, linear polarization is progressively lost when approaching the focus. These processes can be satisfactorily described only by means of vector wave theory (see Section IV.13).

An important feature of the integral representation of the field in a structure is the fact that, in many cases, we can study the resonances of the system by looking at the poles of the integrand. In particular, poles corresponding to real values of the frequency yield oscillation or propagation modes of the system, while complex frequencies identify the leaky modes (see Section III.19.). In addition, the presence of *branch cuts* in the integrand is connected with some special waves (e.g., surface and *lateral waves*). In general, relevant information can be obtained by studying the domain of analyticity of the integrand.

2 Green's Function Formalism

2.1 *Helmholtz–Kirchhoff Integral Theorem*

Let us consider a volume V bounded by a closed surface S (generally multisheeted) and two generic functions $f(\mathbf{r},\mathbf{r}')$ and $u(\mathbf{r})$, where $\mathbf{r}$ and $\mathbf{r}'$ represent points in V (see Fig. IV.2). If we indicate by $\partial/\partial n_0$ the derivative along the outward normal $\hat{n}_0$ to S, we can write, with the help of Gauss's theorem,

$$\iiint_V [f(\mathbf{r},\mathbf{r}')\nabla^2 u(\mathbf{r}) - u(\mathbf{r})\nabla^2 f(\mathbf{r},\mathbf{r}')]\,dV$$

$$= \iiint_V \nabla\cdot[f(\mathbf{r},\mathbf{r}')\nabla u(\mathbf{r}) - u(\mathbf{r})\nabla f(\mathbf{r},\mathbf{r}')]\,dV$$

$$= \oiint_S \left[f(\mathbf{r},\mathbf{r}')\frac{\partial u(\mathbf{r})}{\partial n_0} - u(\mathbf{r})\frac{\partial f(\mathbf{r},\mathbf{r}')}{\partial n_0}\right] dS, \tag{IV.2.1}$$

where the operator ∇ acts on $\mathbf{r}$.

For our purposes, it is convenient to introduce a function $G(\mathbf{r},\mathbf{r}')$ obeying the inhomogeneous wave equation

$$[\nabla^2 + n^2(\mathbf{r})k_0^2]G(\mathbf{r},\mathbf{r}') = -\delta(\mathbf{r}-\mathbf{r}'), \tag{IV.2.2}$$

where $\delta(\mathbf{r})$ denotes the three-dimensional delta function. Any G satisfying a linear differential equation whose right-hand side is that of Eq. (IV.2.2) is said to be a *Green's function* for the operator on the left-hand side of the same

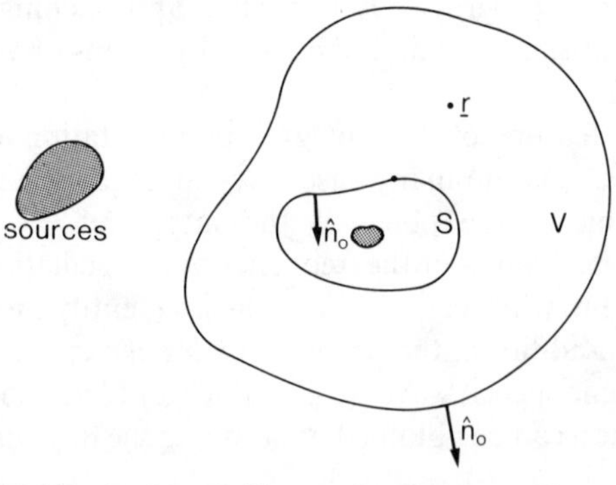

Fig. IV.2. Typical field region bounded by two closed surfaces. (Vectors are underlined in the figures and boldface in the text.)

equation. From Maxwell's equations, it follows that

$$\nabla^2\mathbf{E} + k_0^2 n^2(\mathbf{r})\mathbf{E} + 2\,\nabla[\mathbf{E}\cdot\nabla(\ln n)] = i\omega\mu_0\mathbf{J} - (i/\omega\varepsilon_0)\,\nabla[\nabla\cdot\mathbf{J}/n^2(\mathbf{r})], \tag{IV.2.3}$$

which reduces to Eq. (II.8.1) for $\mathbf{J} = 0$. If we neglect the third term on the left side of Eq. (IV.2.3) (scalar theory approximation), we have

$$\nabla^2 u + k_0^2 n^2(\mathbf{r})u = -h(\mathbf{r}), \tag{IV.2.4}$$

where u denotes any cartesian component of $\mathbf{E}$ and h is the corresponding component of the right side of Eq. (IV.2.3). Since u depends linearly on h, we can apply the superposition property, so that

$$u(\mathbf{r}) = \iiint_{-\infty}^{+\infty} G(\mathbf{r},\mathbf{r}')h(\mathbf{r}')\,d\mathbf{r}' \tag{IV.2.5}$$

where G obeys Eq. (IV.2.2). In the particular case $h(\mathbf{r}') = \delta(\mathbf{r}' - \mathbf{r}_0)$, Eq. (IV.2.5) yields

$$u(\mathbf{r}) = G(\mathbf{r},\mathbf{r}_0), \tag{IV.2.6}$$

so that any $u(\mathbf{r})$ associated with a δ-type source in $\mathbf{r}'$ coincides with a given $G(\mathbf{r},\mathbf{r}')$. Equation (IV.2.2) admits infinite solutions, each of which is determined by the values taken on the surface S. In many cases, by using the formal identity between Eqs. (IV.2.2) and (IV.2.4), it is convenient to select the solution representing the field generated by a source concentrated in $\mathbf{r}'$, without sources external to S. Other kinds of Green's functions are obtained by considering additional sources outside the volume V, so as to satisfy suitable boundary conditions.

For an unbounded homogeneous medium (refractive index independent of $\mathbf{r}$), the appropriate solution of Eq. (IV.2.2) is given by (see Chapter I Problems 5 and 6)

$$G(\mathbf{r},\mathbf{r}') = G(\mathbf{r}',\mathbf{r}) = G(|\mathbf{r}-\mathbf{r}'|) = \frac{\exp(-ik_0 n|\mathbf{r}-\mathbf{r}'|)}{4\pi|\mathbf{r}-\mathbf{r}'|}. \tag{IV.2.7}$$

If we turn our attention to a component u of the field obeying the wave equation

$$[\nabla^2 + k_0^2 n^2(\mathbf{r})]u(\mathbf{r}) = 0, \tag{IV.2.8}$$

we obtain the simple relation

$$G(\mathbf{r},\mathbf{r}')\,\nabla^2 u(\mathbf{r}) - u(\mathbf{r})\,\nabla^2 G(\mathbf{r},\mathbf{r}') = u(\mathbf{r})\delta(\mathbf{r}-\mathbf{r}'), \tag{IV.2.9}$$

as easily seen by multiplying Eqs. (IV.2.8) and (IV.2.2), respectively, by G and u and subtracting the resulting equations. If we set $G \equiv f$ and insert Eq. (IV.2.9) into Eq. (IV.2.1), we obtain the *integral theorem of Helmholtz and Kirchhoff* (also referred to as *Green's theorem*) in the form

$$u(\mathbf{r}) = \oiint_S \left[G(\mathbf{r}',\mathbf{r})\frac{\partial u(\mathbf{r}')}{\partial n_0} - u(\mathbf{r}')\frac{\partial G(\mathbf{r}',\mathbf{r})}{\partial n_0} \right] dS'$$

$$= \oiint_S [G(\mathbf{r}',\mathbf{r})\,\nabla' u(\mathbf{r}') - u(\mathbf{r}')\,\nabla' G(\mathbf{r}',\mathbf{r})] \cdot \hat{n}_0\, dS'$$

$$= \oiint_S \mathbf{v}(\mathbf{r},\mathbf{r}') \cdot \hat{n}_0\, dS', \qquad \text{(IV.2.10)}$$

where

$$\mathbf{v}(\mathbf{r},\mathbf{r}') \equiv G(\mathbf{r}',\mathbf{r})\,\nabla' u(\mathbf{r}') - u(\mathbf{r}')\,\nabla' G(\mathbf{r}',\mathbf{r}) \qquad \text{(IV.2.11)}$$

is known as the *Helmholtz vector field.* In writing Eq. (IV.2.10), we interchanged the roles of $\mathbf{r}$ and $\mathbf{r}'$, so that integrals and derivatives must be performed with respect to $\mathbf{r}'$, as indicated by the symbol ∇'.

It is worth noting that Eq. (IV.2.9) implies the relation

$$\nabla' \cdot \mathbf{v}(\mathbf{r},\mathbf{r}') = 0, \qquad \text{(IV.2.12)}$$

except for $\mathbf{r}' = \mathbf{r}$. Therefore, Gauss's theorem ensures that the choice of the surface S bounding the volume in which the field point $\mathbf{r}$ is located does not influence the value of the integral on the right side of Eq. (IV.2.10), and thus of $u(\mathbf{r})$.

When the Helmholtz–Kirchhoff integral is extended to a surface S characterized by a *surface impedance* Z_S, we can express $\partial u/\partial n_0$ as a function of u [see Eq. (III.23.5)], so that Eq. (IV.2.10) becomes

$$u(\mathbf{r}) = -\oiint_S \left[ikZ_S(\mathbf{r}')\frac{G(\mathbf{r}',\mathbf{r})}{\zeta} + \frac{\partial G(\mathbf{r}',\mathbf{r})}{\partial n_0} \right] u(\mathbf{r}')\, dS', \qquad \text{(IV.2.13)}$$

where $k = k_0 n$ and $\zeta = \zeta_0/n$.

This relation is particularly useful when dealing with diffraction effects due to metallic obstacles having a finite conductivity. In this case, Z_S/ζ_0 coincides with either the *complex refractive* index $n - i\kappa$ or its inverse, according whether the field is polarized normal or parallel to the plane of incidence [Eqs. (III.23.6) and (III.23.7)]. Whenever the region V delimited by the surface S is homogeneous, the Green's function G reduces to the right side of Eq. (IV.2.7) so that Eq. (IV.2.13) yields

$$u(\mathbf{r}) = \frac{i}{2\lambda} \oiint_S \frac{e^{-ik_0 nR}}{R} u(\mathbf{r}') \left[\frac{Z_S(\mathbf{r}')}{\zeta} + \hat{n}_0 \cdot \hat{R}\left(1 - \frac{i}{kR}\right) \right] dS', \qquad \text{(IV.2.14)}$$

where $\hat{R} = (\mathbf{r}' - \mathbf{r})/|\mathbf{r}' - \mathbf{r}|$.

According to Eq. (IV.2.10), the field $u(\mathbf{r})$ is determined, when it is known together with its normal derivative on a closed surface containing the region of interest. This result solves the problem of the field distribution only apparently. In fact, in order to use Eq. (IV.2.10), we must know the Green's function related to the particular refractive index distribution and to the boundary conditions imposed by obstacles, apertures, and so on. From a formal point of view, we can consider u and $\partial u/\partial n_0$ on S as the input of a linear system whose response $u(\mathbf{r})$ is represented by the integral of Eq. (IV.2.10). This suggests that we may compare an optical system to a black box whose input is the set of values of u and $\partial u/\partial n_0$ on S, although, as we will see in the following, u and $\partial u/\partial n_0$ cannot be assigned independently of each other. In this context, the Green's function represents the analog of the impulse response of an electronic device.

A first approach to determining the $G(\mathbf{r}, \mathbf{r}')$ corresponding to a field generated by a point source in $\mathbf{r}'$ hinges on the methods of ray optics. If we put a source in $\mathbf{r}'$, we can derive the trajectories of the rays departing from $\mathbf{r}'$ and the family of wave fronts. In general, the trajectories are curved lines due to the inhomogeneities of the medium. When a discontinuity surface of the refractive index is met, the rays undergo partial reflection and refraction. In some situations the reflected-ray congruence overlaps the original congruence, giving rise to an interference pattern (e.g., see Fig. IV.3). In addition, the refracted rays leave the dielectric if they hit the outside surface with an angle of incidence less than the critical angle, so that we need the Fresnel formulas (see Chapter III) relative to the transmission and reflection coefficients of the waves impinging on the discontinuity surfaces of $n(\mathbf{r})$. Once the trajectories of the multiple ray congruences are determined, we must evaluate the amplitudes

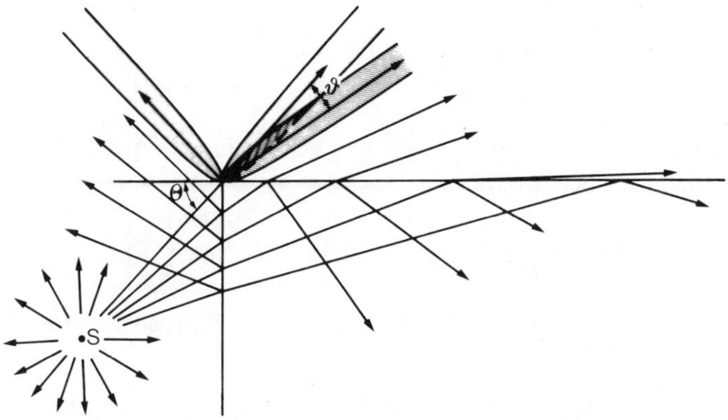

Fig. IV.3. Schematic representation of the field diffracted by a right angle dielectric wedge. The transmitted, reflected, and refracted rays departing from the edge are surrounded by critical regions, where the field differs notably from the geometrical optics one.

$A_m(\mathbf{r})$ of the field, which can be done in principle by using the transport equations [see Eqs. (II.6.4)]. The structure of these equations does not allow us to neglect the higher-order terms of the Luneburg–Kline series $A_m (m > 1)$ when A_0 changes rapidly in space. As an example, we note that in proximity to the edge of the dielectric represented in Fig. IV.3, the rays undergo an abrupt change of direction. In fact, the angular deviation of a ray after crossing the $\pi/2$ dielectric wedge is given by $\theta = \Theta - \arcsin(1 + \sin^2\Theta - n^2)^{1/2}$, where Θ is the angle of incidence of a ray hitting the edge and penetrating from the vacuum in a dielectric with refractive index n, as easily seen by a double application of Snell's law [Eq. (II.11.8)]. Thus, the field in the shaded region undergoes a rapid variation and higher-order amplitudes A_m must be explicitly taken into account.

The Green's functions of inhomogeneous media can also be evaluated by using completely different approaches. This happens, for example, in the presence of an axially symmetric refractive index distribution associated with a discrete set of propagation modes [Eq. (II.12.17)]. As we shall see, the Green's function turns out to be expressed by a suitable series of modes.

In general, the program of constructing the Green's function is developed as a sort of patchwork combining RO, diffraction by canonical obstacles (apertures, wedges, and so on) and smooth objects, and reflection and transmission formulas for discontinuities. In many cases, the evaluation of G is simplified if we consider sources located at infinity, which corresponds to considering fields tending asymptotically to plane waves.

2.2 *Huygens Principle*

Let us consider a field propagating in a *homogeneous* region. Equations (IV.2.10) and (IV.2.7) imply

$$u(\mathbf{r}) = \oiint_S \left[\frac{e^{-ikR}}{4\pi R}\frac{\partial u(\mathbf{r}')}{\partial n_0} - u(\mathbf{r}')\frac{\partial}{\partial n_0}\frac{e^{-ikR}}{4\pi R}\right] dS', \tag{IV.2.15}$$

where $\mathbf{R} = \mathbf{r}' - \mathbf{r}$. Therefore, the field can be considered as the superposition of the elementary perturbations

$$du(\mathbf{r}) = \frac{e^{-ikR}}{4\pi R}\left[\frac{\partial u(\mathbf{r}')}{\partial n_0} + u(\mathbf{r}')\hat{n}_0 \cdot \mathbf{R}\left(ik + \frac{1}{R}\right)\right] dS'. \tag{IV.2.16}$$

Whenever $Rk \gg 1$, as it usually is at optical frequencies, Eq. (IV.2.16) reduces to

$$du(\mathbf{r}) = \frac{e^{-ikR}}{4\pi R}\left[\frac{\partial u(\mathbf{r}')}{\partial n_0} + ik\hat{n}_0 \cdot \mathbf{R}u(\mathbf{r}')\right] dS'. \tag{IV.2.17}$$

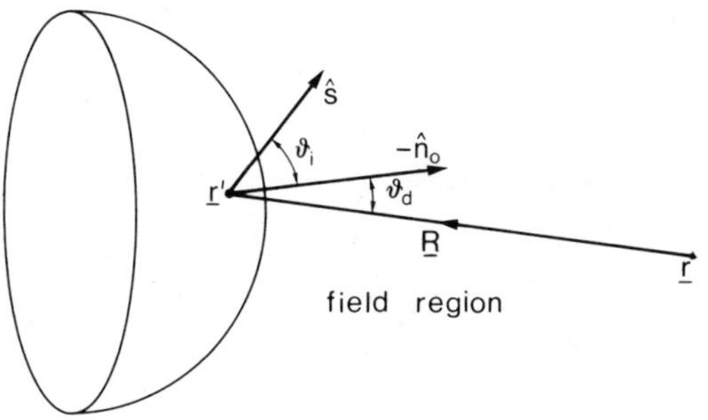

Fig. IV.4. Geometry for Huygens principle.

For u admitting an RO representation $u = A(\mathbf{r})e^{-ikS(\mathbf{r})}$ we can write

$$du(\mathbf{r}) = -iku(\mathbf{r}')\frac{e^{-ikR}}{4\pi R}\frac{\partial(S-R)}{\partial n_0}dS', \tag{IV.2.18}$$

having neglected the slow variation of $A(\mathbf{r})$. Thus, $du(\mathbf{r})$ is proportional to $u(\mathbf{r}')$ and each elementary area dS' radiates a diffraction field whose amplitude decays as $1/R$ and possesses a *radiation pattern* of the form

$$\begin{aligned} -\partial(S-R)/\partial n_0 &= -\hat{n}_0 \cdot \hat{s} + \hat{n}_0 \cdot \hat{R} = \cos\theta_i + \cos\theta_d \\ &= 2\cos[(\theta_i + \theta_d)/2]\cos[(\theta_i - \theta_d)/2], \end{aligned} \tag{IV.2.19}$$

where θ_i and θ_d are, respectively, the angle between the incident ray passing through $\mathbf{r}'$ and the inward normal $-\hat{n}_0$ to S, and the angle between the direction $-\hat{R}$ along which the diffracted field is calculated and $-\hat{n}_0$ (see Fig. IV.4). Accordingly, the diffraction integral [Eq. (IV.2.15)] reduces to

$$u(\mathbf{r}) = \frac{i}{2\lambda}\oint\!\!\oint_S \frac{A(\mathbf{r}')}{R}e^{-ik(R+S)}(\cos\theta_i + \cos\theta_d)\,dS' \tag{IV.2.20}$$

(no confusion should arise between the eikonal S in the exponent and the surface S). In particular, when the surface S coincides with a wave front, $\theta_i = 0$ and $\cos\theta_i + \cos\theta_d$ reduces to the obliquity factor $1 + \cos\theta_d$. The representation of a field as a superposition of many elementary wavelets of the form (IV.2.17) is known as the *Huygens principle*; and it was presented by Rayleigh as follows (article for *Encyclopaedia Britannica* 1889):

> If around the origin of waves an ideal closed surface be drawn, the whole action of the waves in the region beyond may be regarded as due to the motion continually propagated across the various elements of this surface. The wave motion due to any element of the surface is called a "secondary wave," and in estimating the total effect regard must be paid to the phases as well as the amplitudes of the components....

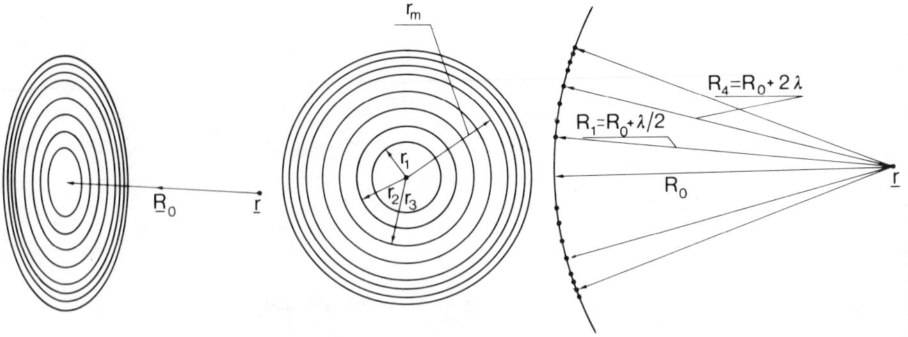

Fig. IV.5. Fresnel rings on a surface seen from the field point **r**.

In practice, it can be shown that only the contributions $du(\mathbf{r})$, arising from a number of well-selected regions on an assigned wave front are needed in order to build up the field at a given point. In order to illustrate this, let us consider the case of a spherical wave front A of curvature radius R divided into elementary rings, called *Huygens* or *Fresnel zones*, by spheres centered at the field point $\mathbf{r}$ (Fig. IV.5). We suppose that the first sphere of radius R_0 is tangent to A, and the successive spheres with radii of $R_m = R_0 + m\lambda/2$ intersect A along a circumference of radius $r_m \cong (m\lambda R_0 R/(R - R_0))^{1/2}$, where $\lambda = 2\pi/k$ is the wavelength. In this way, A is divided into a succession of circular rings of equal area $\cong \pi\lambda|R_0 R/(R - R_0)|$. If we call $u_m(\mathbf{r})$ the field associated with the mth ring, $u(\mathbf{r})$ will be obtained by summing up the u_m. Two successive contributions have approximately equal amplitudes and opposite signs, due to the relation $e^{-ikR_m} = -e^{-ikR_{m+1}}$. Thus, since two successive terms tend to neutralize each other, the whole field is essentially due to the contributions of the low-order terms. This means that only a portion of area $\cong \pi\lambda|R_0 R/(R - R_0)|$ of the wave front A influences the field in $\mathbf{r}$. This intuitive argument will be put into a more rigorous form in connection with the asymptotic evaluation of the diffraction integral. We will also show that the preceding result is valid for more general wave fronts.

In conclusion, we can state approximately that the field $u(\mathbf{r})$ is due to the secondary waves coming from a number of regions A_q of the wave front, each of which is centered around a point $\mathbf{r}_q^*$ (*stationary-phase point*) coinciding with the intersection with the wave front of a ray passing through $\mathbf{r}$ (see Fig. IV.6). For $\lambda \to 0$, each region A_q reduces to the relative stationary point, so that the contributions to the field $u(\mathbf{r})$ become proportional to $u(\mathbf{r}_q^*)$, in agreement with the results of ray optics. For finite values of λ, the contribution to $u(\mathbf{r})$ coming from A_q is still proportional to $u(\mathbf{r}_q^*)$, provided $u(\mathbf{r}')$ does not vary appreciably over an area $\cong \pi\lambda|\mathbf{r} - \mathbf{r}_q^*|$. This provides a correct criterion for the RO application.

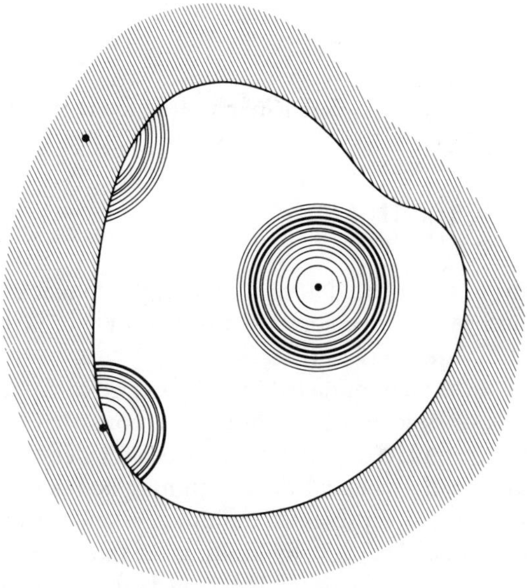

Fig. IV.6. Stationary-phase points of a wave front contributing to the field. The circle families represent the relative Fresnel zones. The aperture has the effect of either neutralizing or diminishing the contributions from the points near the edge.

3 Kirchhoff–Kottler Formulation of the Huygens Principle

In a region with uniform refractive index, the Helmholtz–Kirchhoff integral can be represented in the form of Eq. (IV.2.15). Since this holds true for each cartesian component of the field **E**, we have in a sourceless region bounded by the closed surface S,

$$\mathbf{E}(\mathbf{r}) = \oint\!\!\oint_S \left[\frac{e^{-ikR}}{4\pi R} \frac{\partial \mathbf{E}(\mathbf{r}')}{\partial n_0} - \mathbf{E}(\mathbf{r}') \frac{\partial}{\partial n_0} \frac{e^{-ikR}}{4\pi R} \right] dS'. \qquad \text{(IV.3.1)}$$

This integral can be written in an equivalent form containing only the fundamental quantities **E** and **H**. More precisely, by using Maxwell's equations together with suitable vector identities, it is possible to show that for a closed surface

$$\mathbf{E}(\mathbf{r}) = -\oint\!\!\oint_S [-i\zeta k(\hat{n}_0 \times \mathbf{H}) G(\mathbf{r}, \mathbf{r}') + (\hat{n}_0 \times \mathbf{E}) \times \nabla' G(\mathbf{r}, \mathbf{r}')$$
$$+ (\hat{n}_0 \cdot \mathbf{E}) \nabla' G(\mathbf{r}, \mathbf{r}')] \, dS'. \qquad \text{(IV.3.2)}$$

Analogously, we obtain

$$\mathbf{H}(\mathbf{r}) = -\oint\!\!\!\oint_S \left[\frac{ik}{\zeta}(\hat{n}_0 \times \mathbf{E})G(\mathbf{r},\mathbf{r}') + (\hat{n}_0 \times \mathbf{H}) \times \nabla' G(\mathbf{r},\mathbf{r}') + (\hat{n}_0 \cdot \mathbf{H})\nabla' G(\mathbf{r},\mathbf{r}')\right] dS'. \quad \text{(IV.3.3)}$$

The above relations also hold true for multiply connected domains. For example, the surface S can be composed of two or more closed surfaces all contained in an external surface (Fig. IV.2.)

For calculating the field diffracted by a screen, we can assume, by analogy with Kirchhoff's principle in scalar diffraction theory, that the actual field on the aperture may be replaced by the unperturbed incident field, and similarly that the field immediately behind the screen may be taken as zero. Accordingly, we are tempted to represent the diffracted field by using the integral representations of Eqs. (IV.3.2,3), by limiting the integration to the aperture surface A and using the incident fields $\mathbf{E}_i$ and $\mathbf{H}_i$ in the integrand. Unfortunately, the integrals so obtained give a diffracted field differing from that used in the integrand when the observation point comes closer to the aperture. This inconsistency is removed by adding a contour integral, that is, writing [10]

$$\mathbf{E} = -\iint_{\bar{A}} [-i\zeta k(\hat{n}_0 \times \mathbf{H}_i)G + (\hat{n}_0 \times \mathbf{E}_i) \times \nabla' G + (\hat{n}_0 \cdot \mathbf{E}_i)\nabla' G]\, dS' + i\frac{\zeta}{k}\int_{\partial\bar{A}} \mathbf{H}_i \cdot \hat{l}\,\nabla' G_0\, dl, \quad \text{(IV.3.4)}$$

$$H = -\iint_{\bar{A}} \left[i\frac{k}{\zeta}(\hat{n} \times \mathbf{E}_i)G + (\hat{n}_0 \times \mathbf{H}_i) \times \nabla' G + (\hat{n}_0 \cdot \mathbf{H}_i)\nabla' G\right] dS' - \frac{i}{k\zeta}\int_{\partial\bar{A}} \mathbf{E} \cdot \hat{l}\,\nabla' G\, dl, \quad \text{(IV.3.5)}$$

where $\hat{l}$ is a unit vector tangential to the contour element dl of the aperture edge $\partial\bar{A}$ described in the counterclockwise sense when seen from the field point. According to the above equations, application of the scalar Kirchhoff procedure to the cartesian components of $\mathbf{E}$ and $\mathbf{H}$ is not legitimate. However, it will be shown in Section IV.13 that the line integral can be neglected when the observation point is many wavelengths away from the aperture edge.

4 Sommerfeld Radiation Condition

In the argument following Eq. (IV.2.20) there is a hidden assumption. We have considered the field as composed only of elementary wavelets originating on the wave front. In so doing, we have neglected the wavelets associated with a surface at infinity, which should compose, together with the wave front, the closed surface appearing in the Helmholtz–Kirchhoff theorem. We now wish to show the vanishing of the contributions coming from a surface at infinity, for a rather general class of fields. To this end, we consider a closed surface made up of a finite almost plane region (not necessarily a wave front) and a spherical section A_S of radius $R \gg \lambda$ centered at the field point $\mathbf{r}$ (see Fig. IV.7). The contribution to $u(\mathbf{r})$ originating from A_S reads [Eq. (IV.2.16)]

$$\iint_{A_S} \frac{e^{-ikR}}{4\pi R}\left[\frac{\partial u(\mathbf{r}')}{\partial n_0} + ik\hat{n}_0 \cdot \mathbf{R}u(\mathbf{r}')\right] dS' = \frac{R}{4\pi} e^{-ikR} \iint_{A_S} \left(\frac{\partial u}{\partial R} + iku\right) d\Omega, \qquad \text{(IV.4.1)}$$

where $d\Omega = dS'/(4\pi R^2)$ is the element of solid angle.

If the field sources lie on the side of the plane opposite to that containing A_S, we can safely let the radius of A_S go to infinity, thus being concerned with the asymptotic evaluation of the integrand in Eq. (IV.4.1). The above integral tends to zero for $R \to \infty$, whenever

$$\lim_{R \to \infty} R[(\partial u/\partial R) + iku] = 0, \qquad \text{(IV.4.2)}$$

which is known as *Sommerfeld radiation condition* [11] (in German, *Ausstrahlungbedingung*).

For fields satisfying Eq. (IV.4.2) (*radiation fields*), the Helmholtz–Kirchhoff integral can be calculated on an infinite *open* surface S separating all the sources from the field points. Equation (IV.4.2) is consistent with the asymptotic relation

$$u(\mathbf{r}) \propto f(\hat{R}) \frac{e^{-ikR}}{R}, \qquad R \to \infty, \qquad \text{(IV.4.3)}$$

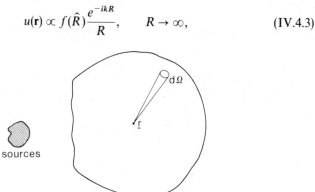

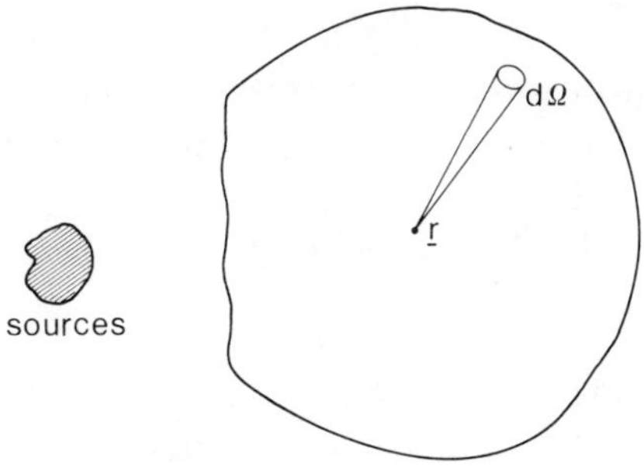

Fig. IV.7. Geometry of the Sommerfeld radiation condition.

which corresponds to an outgoing spherical wave with radiation pattern $f(\hat{R})$. On the other hand, it is often difficult to infer from the integral expression at our disposal whether the field satisfies the Sommerfeld condition. As a practical rule based on the above condition of an outgoing spherical wave, we can state that a field obeys Eq. (IV.4.2) provided that its expression modified by the substitution $k \to k - i\varepsilon$, where ε is a small positive quantity, vanishes for $R \to \infty$.

5 Rayleigh's Form of Diffraction Integrals for Plane Screens

We now wish to turn our attention to the situation in which a plane Π separates the region I containing the sources from the homogeneous region II, where the field must be calculated. In this case, it can be convenient to choose a Green's function G such that either G or $\partial G/\partial n_0$ vanishes on the plane. In so doing, we simplify the Helmholtz–Kirchhoff integral, since one of the two terms in the integrand disappears. The appropriate Green's function can be easily constructed by adding to the virtual source at the field point $\mathbf{r}$ a second virtual source with the same magnitude and with equal or opposite sign at the specular image $\mathbf{r}_S$ of $\mathbf{r}$ with respect to Π. Thus, we have

$$G_{\pm}(\mathbf{r}, \mathbf{r}') = \frac{\exp(-ik|\mathbf{r} - \mathbf{r}'|)}{4\pi|\mathbf{r} - \mathbf{r}'|} \pm \frac{\exp(-ik|\mathbf{r}_S - \mathbf{r}'|)}{4\pi|\mathbf{r}_S - \mathbf{r}'|}. \tag{IV.5.1}$$

It is easy to verify that

$$\partial G_+(\mathbf{r}, \mathbf{r}')/\partial n_0 = G_-(\mathbf{r}, \mathbf{r}') = 0, \tag{IV.5.2}$$

for $\mathbf{r}'$ on Π. If we use G_+, the *diffraction integral* [Eq. (IV.2.10)] assumes the form

$$u(\mathbf{r}) = \iint_{\Pi} G_+(\mathbf{r}, \mathbf{r}') \frac{\partial u(\mathbf{r}')}{\partial n_0} \, dx' \, dy' = -2 \iint_{\Pi} \frac{\partial u(x', y', z')}{\partial z'} \frac{e^{-ikR}}{4\pi R} \, dx' \, dy', \tag{IV.5.3}$$

where the axis z', orthogonal to Π, is directed into region II and $R = [(x - x')^2 + (y - y')^2 + (z - z')^2]^{1/2}$. Analogously, if we use G_-,

$$\begin{aligned} u(\mathbf{r}) &= -\iint_{\Pi} \frac{\partial G_-(\mathbf{r}, \mathbf{r}')}{\partial n_0} u(\mathbf{r}') \, dx' \, dy' \\ &= 2 \iint_{\Pi} \left[u(x', y', z') \left(ik + \frac{1}{R} \right) \frac{e^{-ikR}}{4\pi R} \cos\theta \right] dx' \, dy', \end{aligned} \tag{IV.5.4}$$

where $\theta\,(<\pi/2)$ is the angle between $\mathbf{R} = \mathbf{r}' - \mathbf{r}$ and $\hat{n}_0$. In general, it is useful to replace the expression for u given by Eq. (IV.2.14) with either Eq. (IV.5.3) or Eq. (IV.5.4).

Example: Aperture on a Plane Screen For an RO field different from zero on a plane region (aperture), Eq. (IV.5.4) yields

$$u(\mathbf{r}) = 2 \iint_{-\infty}^{+\infty} \left[P(x', y')A(x', y', z')\left(ik + \frac{1}{R}\right) \times \frac{\exp[-ikR - ikS(x', y', z')]}{4\pi R} \cos\theta \right] dx'\, dy', \qquad \text{(IV.5.5)}$$

where $P(x', y')$ (*pupil function*) takes on the value one if (x', y') belongs to the aperture and vanishes otherwise. It can be shown that the main contribution to the above integral comes from the point P^* for which $|\partial(S - R)/\partial n_0|$ is maximum [see also Eq. (IV.2.18)] and from the edge of the aperture.

According to Eq. (IV.5.5), in the paraxial limit $\cos\theta \cong 1$, the field on the plane $z = \text{const}$ can be compared to the output of a linear system characterized by an *impulse response* $K(x', y'; x, y, z)$ [field on the plane $z = \text{const}$ corresponding to a δ-type field $\delta(x - x')\delta(y - y')$ on the aperture (cf. Section IV.15)] given by

$$K(x', y'; x, y, z) = iP(x', y')\exp\{-ik[(x - x')^2 + (y - y')^2]/(2d) - ikd\}/(\lambda d), \qquad \text{(IV.5.6)}$$

where $d \equiv |z - z'| \gg 1/k$.

6 Babinet's Principle

We observe that the disturbances u_1 and u_2 diffracted by two *complementary screens* (corresponding to a situation in which the second screen is obtained from the first one by interchanging apertures and opaque portions) satisfy the relation (*Babinet's principle*) [12]

$$u_1(\mathbf{r}) + u_2(\mathbf{r}) = u(\mathbf{r}) \qquad \text{(IV.6.1)}$$

where $u(\mathbf{r})$ represents the field in the absence of a screen. Equation (IV.6.1) immediately follows from Eq. (IV.5.4), once the integration is performed over the plane surface containing both screens, provided we assume that the two fields $u_1(\mathbf{r}')$ and $u_2(\mathbf{r}')$ in the apertures coincide with the field $u(\mathbf{r}')$ obtained without a screen. In general, this hypothesis is only approximately verified, since $u_1(\mathbf{r}')$ and $u_2(\mathbf{r}')$ are not equal to $u(\mathbf{r}')$, even if the difference becomes relevant only close to the edges of the apertures. Nevertheless, an exact form of Babinet's principle may be derived for perfectly conducting plane screens (see Chapter I, Born and Wolf [11], p. 559).

7 Diffraction Integrals for Two-Dimensional Fields

In some cases the electromagnetic field can be written in the form

$$u(x, y, z) = \exp[-i(k^2 - \chi^2)^{1/2} y] u(x, z), \tag{IV.7.1}$$

so that u satisfies the *two-dimensional Helmholtz equation*

$$\partial^2 u/\partial x^2 + \partial^2 u/\partial z^2 + \chi^2 u = 0 \tag{IV.7.2}$$

as implied by Eqs. (I.1.12). These fields are usually called *cylindrical waves*.

By retracing the steps leading to Eq. (IV.2.10), it is easy to show that

$$u(\boldsymbol{\rho}) = \oint_C \left[G(\boldsymbol{\rho}', \boldsymbol{\rho}) \frac{\partial u(\boldsymbol{\rho}')}{\partial n_0} - u(\boldsymbol{\rho}') \frac{\partial G(\boldsymbol{\rho}', \boldsymbol{\rho})}{\partial n_0} \right] ds', \tag{IV.7.3}$$

where C is a closed curve containing a plane sourceless field region at a given value $y = \text{const}$, $\boldsymbol{\rho} \equiv \hat{x}x + \hat{z}z$, and the *two-dimensional Green's function* G obeys the equation

$$(\partial^2/\partial x'^2 + \partial^2/\partial z'^2 + \chi^2) G(\boldsymbol{\rho}', \boldsymbol{\rho}) = -\delta(x - x')\delta(z - z'). \tag{IV.7.4}$$

The simplest form of G satisfying the radiation condition at infinity and singular only for $\boldsymbol{\rho} = \boldsymbol{\rho}'$ is given by

$$G(\boldsymbol{\rho}, \boldsymbol{\rho}') = G(\boldsymbol{\rho}', \boldsymbol{\rho}) = (i/4) H_0^{(2)}(\chi R), \qquad R \equiv \rho' - \rho, \tag{IV.7.5}$$

where $H_0^{(2)}$ is the Hankel function of zeroth order and the second kind, which admits the asymptotic representation

$$H_0^{(2)}(\chi R) \cong [2/(\pi \chi R)]^{1/2} e^{-i(\chi R - \pi/4)}, \qquad R\chi \to \infty. \tag{IV.7.6}$$

If the cylindrical wave satisfies the radiation condition of Eq. (IV.4.2), we can apply to our plane geometry the considerations of Section IV.4., so that

$$u(\boldsymbol{\rho}) = \frac{i}{4} \int_{C_\infty} \left[H_0^{(2)}(\chi R) \frac{\partial u}{\partial n_0} - u \frac{\partial H_0^{(2)}(\chi R)}{\partial n_0} \right] ds', \tag{IV.7.7}$$

where C_∞ is an infinite curve separating the field point $\boldsymbol{\rho}$ from the sources. The analog of Eqs. (IV.5.3) and (IV.5.4) reads

$$\begin{aligned} u(\boldsymbol{\rho}) &= -\frac{i}{2} \int_{-\infty}^{+\infty} P(x') H_0^{(2)}(\chi R) \frac{\partial u(x', z')}{\partial z'} dx' \\ &= -\frac{i\chi}{2} \int_{-\infty}^{+\infty} P(x') u(x', 0) \frac{dH_0^{(2)}(\chi R)}{d(\chi R)} \cos\theta \, dx', \end{aligned} \tag{IV.7.8}$$

where $P(x')$ is the analog of the pupil function (see Section IV.5), while $\theta(<\pi/2)$ is the angle between $-\mathbf{R}$ and $\hat{z}'$, that is, the normal to the straight line $z'=0$ separating the field region $z' \geqq 0$ from the source region $z'<0$. For $R \gg 1/\chi$, we can approximate $H_0^{(2)}$ by means of Eq. (IV.7.6), thus obtaining

$$u(\rho) \cong -\left(\frac{\chi}{2\pi}\right)^{1/2} e^{i\pi/4} \int_{-\infty}^{+\infty} P(x')u(x',0)\frac{e^{-i\chi R}}{R^{1/2}} \cos\theta \, dx'. \tag{IV.7.9}$$

8 Plane-Wave Representation of the Field

We wish to express the field in the homogeneous half-space, limited by the plane on which it is assigned, as a superposition of plane waves [13].

As a preliminary step, let us evaluate the bidimensional Fourier transform of $G(\mathbf{r},\mathbf{r}')$ considered as a function of x' and y' for fixed values of z', x, y, and z with $z>z'$. We observe that Eq. (IV.5.3) is equivalent to

$$u(\mathbf{r}) = -2 \iint_{\Pi} G(|\mathbf{r}-\mathbf{r}'|)\frac{\partial u(x',y',z')}{\partial z'} dx'\, dy', \tag{IV.8.1}$$

where Π is a plane at constant z' and the sourceless field-region is on the side $z>z'$. We can now apply Eq. (IV.8.1) to the plane wave

$$\exp(-ik_x x - ik_y y - ik_z z), \tag{IV.8.2}$$

where k_x and k_y are real quantities, provided the relation

$$k_z = (k^2 - k_x^2 - k_y^2)^{1/2} \tag{IV.8.3}$$

(k_z being real positive or imaginary negative) is fulfilled. In fact, Eq. (IV.8.3) implies that Eq. (IV.8.2) represents a radiation field moving in the direction of increasing z, which implies the validity of Eq. (IV.8.1) (see Section IV.4). Therefore, we can write

$$\begin{aligned}\exp(-ik_x x - ik_y y - ik_z z) &= 2ik_z \exp(-ik_z z') \\ &\quad \times \iint_{-\infty}^{+\infty} \exp(-ik_x x' - ik_y y') G\{[(x-x')^2 \\ &\quad + (y-y')^2 + (z-z')^2]^{1/2}\}\, dx'\, dy' \\ &= 2ik_z \exp(-ik_z z') F[G(x',y'); -k_x, -k_y], \end{aligned} \tag{IV.8.4}$$

where F is the Fourier transform of G,

$$F[G(x',y'); k_x, k_y] = \iint_{-\infty}^{+\infty} G(x',y') \exp(ik_x x' + ik_y y')\, dx'\, dy', \tag{IV.8.5}$$

where for notational simplicity we have omitted the variables x, y, z, and z', which are kept constant with respect to the transform operation. Thus, we obtain

$$F[G(x', y'); k_x, k_y] = -\frac{i}{2k_z}\exp[ik_x x + ik_y y - ik_z(z - z')]. \tag{IV.8.6}$$

Since the operations of Fourier transformation with respect to x' and y' and of differentiation with respect to z' are mutually independent, we can also write, for the transform of $\partial G/\partial z'$,

$$F\left[\frac{\partial G(x', y', z')}{\partial z'}; k_x, k_y\right] = \frac{1}{2}\exp[ik_x x + ik_y y - ik_z(z - z')]. \tag{IV.8.7}$$

By inverting the above transform, we have

$$\frac{\partial G}{\partial z'} = \frac{1}{2(2\pi)^2}\iint\limits_{-\infty}^{+\infty} \exp[-ik_x(x' - x) - ik_y(y' - y) + ik_z(z' - z)]\, dk_x\, dk_y, \tag{IV.8.8}$$

which yields, with the help of Eq. (IV.5.4),

$$u(x, y, z) = \frac{1}{(2\pi)^2}\iiiint\limits_{-\infty}^{+\infty} \exp[-ik_x(x' - x) - ik_y(y' - y) + ik_z(z' - z)] \times u(x', y', z')\, dx'\, dy'\, dk_x\, dk_y. \tag{IV.8.9}$$

If we introduce the Fourier transform

$$F[u(x, y, z'; k_x, k_y] = \iint\limits_{-\infty}^{+\infty} u(x, y, z')\exp(ik_x x + ik_y y)\, dx\, dy, \tag{IV.8.10}$$

Eq. (IV.8.9) can be rewritten as

$$\begin{aligned} u(x, y, z) &= \frac{1}{(2\pi)^2}\iint\limits_{-\infty}^{+\infty} \exp[-ik_x x - ik_y y - ik_z(z - z')] \\ &\quad \times F[u(x', y', z'); k_x, k_y]\, dk_x\, dk_y \\ &= \frac{1}{(2\pi)^2}\iint\limits_{-\infty}^{+\infty} \exp(-ik_x x - ik_y y) F[u(x', y', z); k_x, k_y]\, dk_x\, dk_y, \end{aligned} \tag{IV.8.11}$$

where we have defined

$$F[u(x',y',z);k_x,k_y] \equiv \exp[-ik_z(z-z')]F[u(x',y',z');k_x,k_y]. \tag{IV.8.12}$$

Equation (IV.8.11) is the expression of the most general field satisfying the radiation condition in the half-space $z > z'$ as a superposition of plane waves. We note that the existence of evanescent fields is accounted for by means of the imaginary negative values assumed by k_z when $k_x^2 + k_y^2 > k^2$. The corresponding waves are evanescent in the positive z direction and propagate orthogonal to the z axis.

9 Angular Spectrum Representation

It is sometimes convenient to express k_x, k_y, and k_z in the form

$$k_x = k\cos\beta\sin\gamma, \tag{IV.9.1a}$$

$$k_y = k\cos\beta\cos\gamma, \tag{IV.9.1b}$$

$$k_z = k\sin\beta \tag{IV.9.1c}$$

where β ranges from $0 - i\infty$ to $\pi + i\infty$ (*Sommerfeld's path*) and γ from 0 to π (Fig. IV.8). With the above definitions, Eq. (IV.8.11) reads

$$u(x,y,z) = \int_{0-i\infty}^{\pi+i\infty} d\beta \int_0^{\pi} d\gamma\, S(\beta,\gamma)\exp(-ikr\cos\theta), \tag{IV.9.2}$$

where $r = (x^2 + y^2 + z^2)^{1/2}$, θ is the (eventually complex) angle between $\mathbf{r} \equiv (x,y,z)$ and $\mathbf{k} \equiv (k_x,k_y,k_z)$, so that $\cos\theta = \cos\beta\sin\delta\cos(\phi-\gamma) +$

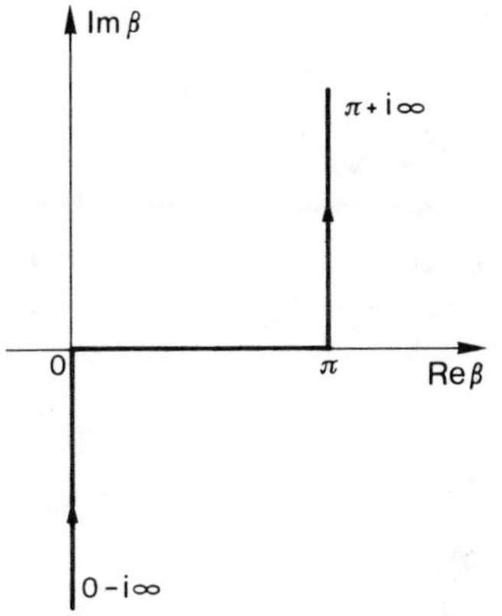

Fig. IV.8. Sommerfeld's integration path.

$\sin\beta\cos\delta$, δ and ϕ being the angles formed by $\mathbf{r}$ with $\hat{z}$ and by $x\hat{x} + y\hat{y}$ with $\hat{x}$, respectively, and

$$S(\beta,\gamma) = e^{ik_z z'} F[u(x',y',z');\, k_x, k_y] \sin(2\beta)/2\lambda^2. \tag{IV.9.3}$$

The function $S(\beta,\gamma)$, referred to as the *angular spectrum*, has the important property of being independent of the reference plane $z = z'$ [see Eq. (IV.8.12)]. For imaginary values of β, it represents the evanescent portion of the radiation. We observe that, for $kr \to \infty$, the integral in Eq. (IV.9.2) can be evaluated by the *stationary-phase method* [see Section V.11 and Eq. (V.11.7)], which yields

$$u(x,y,z) \underset{kr\to\infty}{\propto} -2\pi i \frac{e^{-ikr}}{kr} S(\beta,\gamma)_{\theta=0}. \tag{IV.9.4}$$

If we move on a surface at constant large distance r from the origin [Eq. (IV.9.4)] we observe a field proportional to the value assumed by $S(\beta,\gamma)$ when the direction of $\mathbf{k}(\beta,\gamma)$ and of the radius vector $\mathbf{r}$ coincide, which justifies the definition of angular spectrum for the function S.

In particular, if we consider a field that is rotationally invariant around the z axis, the angular spectrum $S(\beta,\gamma)$ is independent of the angle γ and does not vary when β is replaced by $\pi - \beta$, that is, $S(\beta,\gamma) \equiv S(\beta) = S(\pi - \beta)$. As a consequence, Eq. (IV.9.2) reduces to

$$\begin{aligned} u(r,\delta) &= 2\int_{0-i\infty}^{\pi/2} S(\beta)\,d\beta \int_0^{\pi} \exp\{-ikr[\cos\beta\sin\delta\cos(\phi-\gamma) \\ &\qquad + \sin\beta\cos\delta]\}\,d\gamma \\ &= 2\pi \int_{0-i\infty}^{\pi/2} S(\beta) J_0(kr\cos\beta\sin\delta)\exp(-ikr\sin\beta\cos\delta)\,d\beta. \end{aligned} \tag{IV.9.5}$$

9.1 *Angular Spectrum for Two-Dimensional Waves*

If we refer to the two-dimensional wave given by Eq. (IV.7.1), it is easy to show that Eqs. (IV.8.11), (IV.9.2) and (IV.9.3) are replaced by

$$\begin{aligned} u(x,z) &= \frac{1}{2\pi}\int_{-\infty}^{+\infty} \exp[-ik_x x - ik_z(z-z')] F[u(x',z');\, k_x]\,dk_x \\ &= \int_{0-i\infty}^{\pi+i\infty} \exp[-i\chi\rho\cos(\beta-\phi)] S(\beta)\,d\beta \end{aligned} \tag{IV.9.6}$$

where $x = \rho\cos\phi$, $z = \rho\sin\phi$,

$$F[u(x',z');k_x] = \int_{-\infty}^{+\infty} \exp(ik_x x')u(x',z')\,dx', \tag{IV.9.7}$$

and

$$S(\beta) = \exp(i\chi z'\sin\beta)F[u(x',z');k_x]\frac{\chi}{2\pi}\sin\beta. \tag{IV.9.8}$$

In order to derive Eq. (IV.9.6) we have set $k_x = \chi\cos\beta$, $k_z = \chi\sin\beta$, and, once again, the evanescent (along the z axis) portion of the radiation is accounted for by means of the imaginary values of β.

Example: Gaussian Angular Spectrum. Let us consider

$$S(\beta) = \exp[-(\beta-\theta)^2/\sigma^2]. \tag{IV.9.9}$$

Correspondingly, we have

$$F[u(x',z');k_x] = \frac{2\pi}{\chi\sin\beta}\exp\left[-\frac{(\beta-\theta)^2}{\sigma^2} - i\chi z'\sin\beta\right], \tag{IV.9.10}$$

and

$$u(\rho,\phi) = \int_{0-i\infty}^{\pi+i\infty} \exp\left[-i\chi\rho\cos(\beta-\phi) - \frac{(\beta-\theta)^2}{\sigma^2}\right]d\beta. \tag{IV.9.11}$$

The last integral cannot be calculated in closed form. However, the *steepest-descent method* (see Section V.6) allows us to evaluate it in the limit $\chi\rho \to \infty$.

9.2 *Angular Spectrum in the Complex Domain*

The angular spectrum representation is not restricted to the field in a half-space region. We will see in Section VI.2 [Eq. (VI.2.2)] that the field diffracted by a wedge can be represented by two angular spectrum integrals extended to two Sommerfeld paths displaced by 2π. Consequently, we are led to generalize the above discussion by considering the integral

$$I = \int_\Gamma S(\beta)\exp[-i\chi\rho\cos(\beta-\phi)]\,d\beta, \tag{IV.9.12}$$

Γ being an arbitrary and generally complex path of integration. This expression represents a cylindrical wave as a superposition of plane waves of amplitude $S(\beta)$ and wave vector $\mathbf{k} = [\chi\cos\beta, (k^2-\chi^2)^{1/2}, \chi\sin\beta]$. The function $S(\beta)$ may present poles and branch points, and whenever it is a periodic function with period equal to a multiple of 2π, the integration path Γ can be displaced by an amount equal to this period without altering the value of the integral.

It is possible to determine the integration paths that allow Eq. (IV.9.12) to represent physically possible fields, that is, fields that do not diverge for $\rho \to \infty$. To this end, we set $\beta = \beta' + i\beta''$, so that the ρ-dependent contribution to the exponent of I reads

$$-i\chi\rho\cos(\beta' + i\beta'' - \phi) = -i\chi\rho\cos(\beta' - \phi)\cosh\beta'' - \chi\rho\sin(\beta' - \phi)\sinh\beta''. \tag{IV.9.13}$$

Therefore, the integral I will remain finite provided Γ is made up of complex values of β lying in the shaded regions of Fig. IV.9 defined by the relations

$$\begin{aligned} &\beta'' \le 0, \qquad 2\pi + 2n\pi \ge \beta' - \phi \ge \pi + 2n\pi \qquad (n = 0, \pm 1, \pm 2, \ldots), \\ &\beta'' \ge 0, \qquad \pi + 2n\pi \ge \beta' - \phi \ge 2n\pi \qquad (n = 0, \pm 1, \pm 2, \ldots). \end{aligned} \tag{IV.9.14}$$

In general, an assigned path fulfils the above relations only for a certain range of values of the observation angle ϕ, since the described regions depend on ϕ itself. As an example, Sommerfeld's path $(0 - i\infty, \pi + i\infty)$ satisfies Eqs. (IV.9.14) for $0 \le \phi \le \pi$, and thus can be safely used for the field region $z = \rho\sin\phi \ge 0$.

Integrals of the kind appearing in Eq. (IV.9.12) are often evaluated by means of the method of steepest descent (see Section V.6). To this end, it is necessary to perform the integral along the *steepest-descent path* (SDP), which can be done by means of a continuous deformation of the initially assigned path Γ (for example, Sommerfeld's path) into an SDP. More precisely, we have

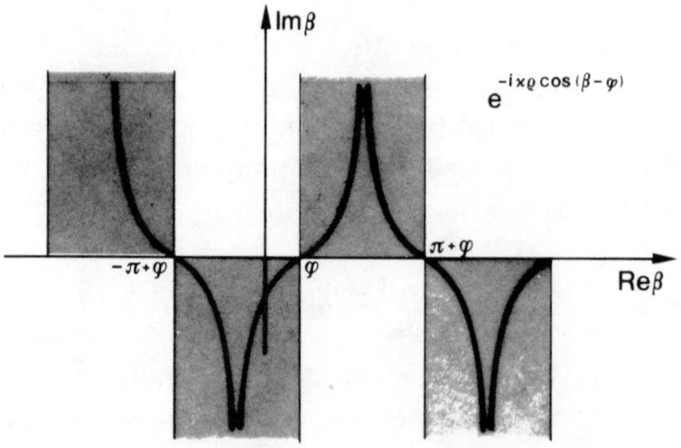

Fig. IV.9. Regions of the complex β plane in which $\exp[-i\chi\rho\cos(\beta - \phi)]$ remains limited as $\chi\rho \to \infty$. The continuous lines represent the steepest-descent path (see Section V.6).

[see Eq. (V.6.4)]

$$I = \left(\int_{\text{SDP}} + \int_{\Gamma_B} \right) S(\beta) \exp[-i(k^2 - \chi^2)^{1/2} y - i\chi\rho\cos(\beta - \phi)]\, d\beta$$

$$+ \sum_q r_q \exp[-i(k^2 - \chi^2)^{1/2} y - i\chi\rho\cos(\beta_q - \phi)] \qquad \text{(IV.9.15)}$$

where Γ_B is a path encircling the branch cuts, if any, of S located between Γ and SDP, and r_q is the qth residue of S, that is (see Fig. V.16),

$$r_q = 2\pi i \lim_{\beta \to \beta_q} S(\beta)(\beta - \beta_q) \qquad \text{(IV.9.16)}$$

β_q being the qth pole of S comprised between Γ and SDP.

In general, the integral extended to an SDP decays as $1/\rho^{1/2}$, as we shall see in the next chapter. Accordingly, it represents a cylindrical wave characterized by a far-field pattern function of ϕ [see Eq. (V.6.8)].

9.3 *Polar Singularities*

Let us suppose that the plane $z = 0$ is partially illuminated by the field

$$u(x, y, z = 0) = U(x)\exp(-ikx\cos\phi_0) \qquad \text{(IV.9.17)}$$

$U(x)$ being the unit step function [$U(x) = 0$ for $x < 0$, $U(x) = 1$ for $x > 0$] and ϕ_0 an angle with a small imaginary component, so that Im cos $\phi_0 < 0$. In the present case, the field is independent of y, which implies that $k = \chi$. Equations (IV.9.7) and (IV.9.8), respectively, yield

$$F[u(x', 0); k_x] = \int_0^\infty \exp[i(k_x - k\cos\phi_0)x']\, dx',$$

$$= \frac{i}{k_x - k\cos\phi_0} = \frac{i}{k}\frac{1}{\cos\beta - \cos\phi_0}, \qquad \text{(IV.9.18)}$$

and

$$S(\beta) = \frac{1}{2\pi i}\frac{\sin\beta}{\cos\phi_0 - \cos\beta}. \qquad \text{(IV.9.19)}$$

In turn, Eqs. (IV.9.12) and (IV.9.15) yield

$$u(\rho, \phi) = \frac{1}{2\pi i}\int_{0 - i\infty}^{\pi + i\infty} \frac{\sin\beta}{\cos\phi_0 - \cos\beta} \exp[-ik\rho\cos(\beta - \phi)]\, d\beta$$

$$= \frac{1}{2\pi i}\int_{\text{SDP}} \frac{\sin\beta}{\cos\phi_0 - \cos\beta} \exp[-ik\rho\cos(\beta - \phi)]\, d\beta$$

$$+ \exp[-ik\rho\cos(\phi_0 - \phi)]U(-\phi + \tilde{\phi}_0), \qquad \text{(IV.9.20)}$$

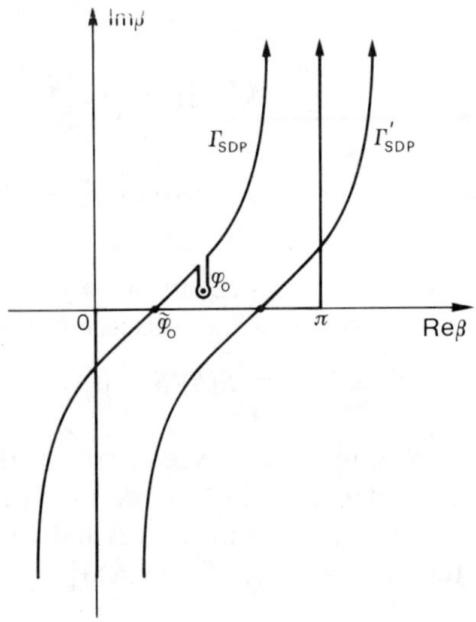

Fig. IV.10. Steepest-descent paths [see Eq. (IV.9.20)] in the presence of a complex pole in ϕ_0.

where the SDP is indicated in Fig. IV.10 with a thick line and $\tilde{\phi}_0$ is the intersection with the real β axis of the SDP passing through the complex point ϕ_0. The last term of the above equation represents a plane wave confined in the angular sector $0 \leqq \phi \leqq \tilde{\phi}_0$, as ensured by the presence of the unit step function $U(x)$. This wave is discontinuous on the boundary $\phi = \tilde{\phi}_0$, which marks the separation between the lit zone (see Section II.12.1.c) and the shadow. The half-line $\phi = \tilde{\phi}_0$ is a typical example of a *shadow boundary* (SB). The present example argues a correspondence between poles of the angular spectrum and the SB of a generic field (see Section V.6.1).

9.4 *Branch-Point Singularities*

We now wish to consider the y-independent field reflected by the plane $z = 0$ separating two lossless dielectric media (see Fig. IV.11) and illuminated by a unitary line source parallel to y and lying along the line $(x = x_S, z = z_S > 0)$. In this case, the incident field $u_i(\boldsymbol{\rho})$ coincides with the Green's function $G(\boldsymbol{\rho}, \boldsymbol{\rho}_S)$ $[\boldsymbol{\rho}_S = (x_S, z_S)]$, so that the two-dimensional analog of Eq. (IV.8.6) allows us to write

$$\begin{aligned} F[U_i(x', 0); k_x] &= (-i/2k_z)\exp(ik_x x_S - ik_z z_S) \\ &= (-i/2k\sin\beta)\exp[-ik(-x_S\cos\beta + z_S\sin\beta)], \quad \text{(IV.9.21)} \end{aligned}$$

where use has been made of Eq. (IV.9.8) and of the relation $k = \chi$.

We now observe that the reflected field propagating with a given value of k_x is obtained by multiplying the corresponding incident wave by the appro-

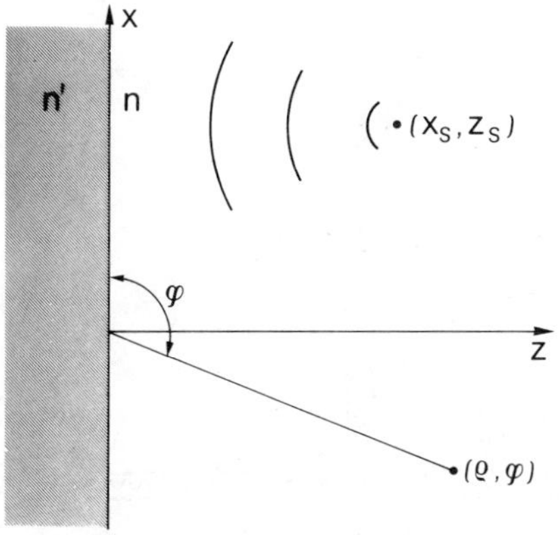

Fig. IV.11. Geometry of reflection by a plane separating two dielectric media.

priate reflection coefficient $r(k_x)$. Therefore, the total reflected field $u_r(x, z)$ reads, with the help of Eq. (IV.9.5),

$$u_r(x,z) = \frac{1}{2\pi}\int_{-\infty}^{+\infty} \exp(-ik_x x - ik_z z) r(k_x) F[u_i(x,0); +k_x]\, dk_x$$

$$= \int_{0-i\infty}^{\pi+i\infty} \exp[-ik\rho\cos(\beta-\phi)] S_i(\beta) r(k_x)\, d\beta, \qquad \text{(IV.9.22)}$$

where $k_x = k\cos\beta$, and S_i refers to the incident field,

$$S_i(\beta) = F[u_i(x,0); k_x](k/2\pi)\sin\beta. \qquad \text{(IV.9.23)}$$

If the line source is an electric current flowing parallel to the y axis, the electric field is parallel to the y axis as well, so that we can set $u(x,z) = E_y(x,z)$, and the reflection coefficient appearing in Eq. (IV.9.22) is that related to a TE wave [see Eq. (III.8.2)],

$$r_s = [\cos\theta - (n^2 - \sin^2\theta)^{1/2}]/[\cos\theta + (n^2 - \sin^2\theta)^{1/2}], \qquad \text{(IV.9.24)}$$

θ being the angle of incidence relative to the plane wave component directed along $\hat{x}k_x + \hat{z}k_z$, and n representing the ratio n_2/n_1 between the refractive index of the half-space $z < 0$ and that of the half-space $z > 0$. Inserting Eqs. (IV.9.23) and (IV.9.24) into Eq. (IV.9.22), we easily obtain, with the help of Eq. (IV.9.21),

$$u_r(\rho,\phi) = \frac{-i}{4\pi}\int_{0-i\infty}^{\pi+i\infty} \frac{\sin\beta - (n^2-\cos^2\beta)^{1/2}}{\sin\beta + (n^2-\cos^2\beta)^{1/2}}$$

$$\times \exp\{-ik[-x_S\cos\beta + z_S\sin\beta + \rho\cos(\beta-\phi)]\}\, d\beta \qquad \text{(IV.9.25)}$$

where use has been made of the relations $\cos\theta = \sin\beta$, $\sin^2\theta = \cos^2\beta$.

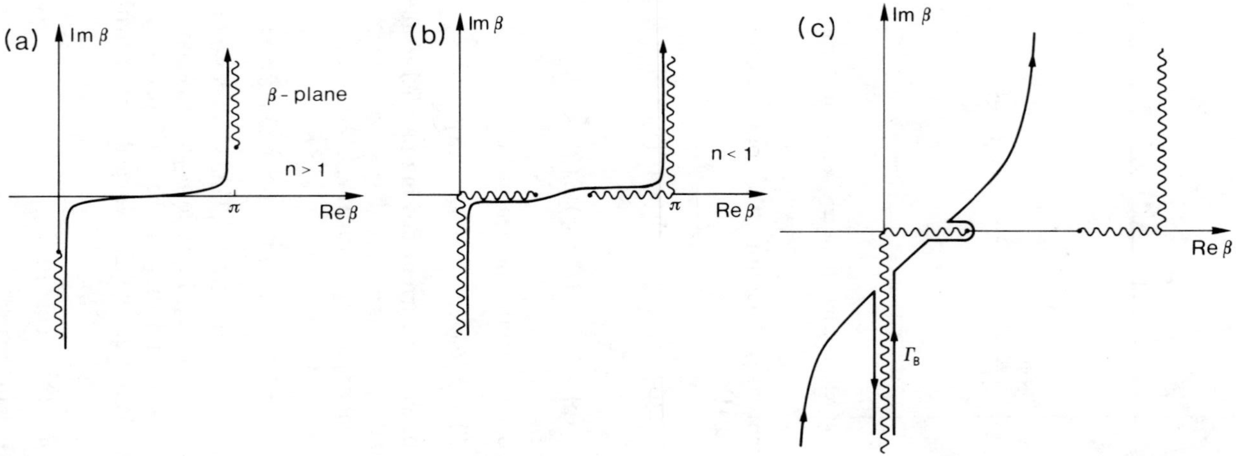

Fig. IV.12. (a) and (b) Branch cuts and Sommerfeld paths relative to the angular spectrum representation of the field reflected by a plane interface between two dielectrics illuminated by a cylindrical wave. (c) Steepest-descent path obtained by modifying the Sommerfeld path in Fig. IV. 8 and partially surrounding (Γ_B) the wavy line representing a branch cut.

The above integrand has a number of branch points defined by the equation $\cos\beta = \pm n$. More precisely, while for $n > 1$ there is one purely imaginary branch point, for $n < 1$ we find two real branch points (Fig. IV.12a,b). Thus, if we wish to shift from Sommerfeld's integration path to the SDP, we must take into account the contributions of the branch cuts indicated in Fig. IV.12 with wavy lines. Figure IV.12c represents a possible modification of Sommerfeld's path that includes the path Γ_B partially encircling the branch cuts (see Section V.6.2).

The two examples discussed above suggest two interesting conclusions. First, whenever the field contains plane waves in some regions not coinciding with the whole space, the angular spectrum $S(\beta)$ possesses a sequence of poles distributed in the β plane. Second, the presence of discontinuities in the medium gives rise to a sequence of branch points and poles of $S(\beta)$, which are associated with the phenomenon of *lateral* and *leaky* waves, as we will see in Section V.7.1.

10 Fresnel and Fraunhofer Diffraction Formulas

10.1 *Fresnel Diffraction Formula*

We now wish to consider the case of a field having a *band-limited* angular spectrum. More precisely, we assume that $F(u; -k_x, -k_y) = 0$ for $k_x^2 + k_y^2 > k_t^2 \ll k^2$. This allows us to replace k_z with $k - \frac{1}{2}(k_x^2 + k_y^2)/k$ in Eq. (IV.8.11), so that

$$u(x, y, z) = \frac{\exp[-ik(z-z')]}{(2\pi)^2} \iint_{-\infty}^{+\infty} \exp\left[-ik_x x - ik_y y + \frac{i}{2k}(k_x^2 + k_y^2)(z-z')\right] \times F[u(x, y, z'); k_x, k_y]\, dk_x\, dk_y$$

$$= \exp[-ik(z-z')]F^{-1}\left\{\exp\left[\frac{i}{2k}(k_x^2+k_y^2)(z-z')\right]; x, y\right\} * u(x, y, z'), \tag{IV.10.1}$$

where the symbol $*$ indicates the convolution operation, and use has been made of the convolution theorem for the inverse Fourier transform (indicated by F^{-1} of the product of two functions. Equation (IV.10.1) can easily be rewritten as

$$u(x, y, z) = \frac{i\exp\{-ik(z - z') - ik(x^2 + y^2)/[2(z - z')]\}}{\lambda(z - z')} \times F\left\{\exp\left[-ik\frac{x'^2 + y'^2}{[2(z - z')]}\right]u(x', y', z'); \frac{kx}{z - z'}, \frac{ky}{z - z'}\right\}. \tag{IV.10.2}$$

Thus, the diffraction integral is equivalent to a Fourier transform of the field on the reference plane $z' = \text{const} < z$ times a suitable phase factor. The practical relevance of this result is related to the possibility of performing a numerical evaluation of the field by means of the FFT algorithm. Equation (IV.10.2) is a form of the *Fresnel diffraction formula.*

The error made in replacing k_z by $k - \frac{1}{2}(k_x^2 + k_y^2)/k$ is negligible only if $|k_z - k + \frac{1}{2}(k_x^2 + k_y^2)/k||z - z'| \ll 2\pi$. This means that Eq. (IV.10.2) can be used only for points whose distance from the integration surface does not exceed z_k, a distance loosely defined by the relation

$$|k_z - k + \tfrac{1}{2}(k_t^2/k)|z_k \cong \tfrac{1}{8}(k_t^4 z_k/k^3) = \alpha, \qquad \text{(IV.10.3)}$$

where α is a small quantity (e.g., 10^{-1}). Consequently, the Fresnel diffraction formula for bond-limited fields can be used only for propagating the field over a finite distance z_k.

Let us now consider a field different from zero only on a finite plane aperture. If we indicate by a the radius of the smallest circumference encircling the aperture and assume that $|z - z'| \gg a$, we can approximate R with $|z - z'| + \frac{1}{2}[(x - x')^2 + (y - y')^2]/|z - z'|$ so that the diffraction integral of Eq. (IV.5.5) reads

$$u(x, y, z) = i\frac{\exp(-ik|z - z'|)}{\lambda|z - z'|}\iint_{-\infty}^{+\infty} P(x', y')u(x', y', z') \times \exp\left\{-i\frac{k}{2|z - z'|}\left[(x - x')^2 + (y - y')^2\right]\right\} dx'\,dy'. \qquad \text{(IV.10.4)}$$

We note immediately that this expression coincides with that provided by Eq. (IV.10.2) even though they were obtained under different assumptions. In fact, we can use the latter Fresnel integral only if $|z - z'| > z_R$, z_R being a distance defined through a relation similar to Eq. (IV.10.3), i.e.,

$$k\left||r - r'| - |z - z'| + \frac{1}{2}\frac{(x - x')^2 + (y - y')^2}{|z - z'|}\right| \cong \tfrac{1}{8}(ka^4/z_R^3) = \alpha. \qquad \text{(IV.10.5)}$$

If $z_R < z_k$, then the Fresnel integral applies, for space- and band-limited fields, at any distance $|z - z'|$. From Eqs. (IV.10.3,5) we can immediately derive a simple criterion for the uniform validity of the Fresnel approximation, that is,

$$ak_t < 8\alpha(k/k_t)^2, \qquad \text{(IV.10.6)}$$

which sets an upper limit to the product "dimension × bandwidth." For a uniformly illuminated slit the plane-wave spectrum is proportional to $\sin(k_t a)/(k_t a)$, that is, the product $k_t a$ is roughly equal to 2π and Eq. (IV.10.6) yields

$$2\pi < 8\alpha(ka/2\pi)^2. \tag{IV.10.7}$$

Consequently, for sufficiently large and uniformly illuminated (in both amplitude and phase) apertures such that $ka \gg 2\pi$, the Fresnel integral can be used uniformly for observation points placed at any distance from the integration plane.

Let us now consider the effect of a nonuniform phase distribution. We will show later on that in the limit of $k \to \infty$, k_t is given by $k[\sin\theta_{max} + 2\pi/(ak)]$, where θ_{max} represents the maximum angle the rays crossing the aperture form with the z axis. Accordingly, Eq. (IV.10.6) reads

$$ak[\sin\theta_{max} + (2\pi/ak)]^3 < 8\alpha. \tag{IV.10.8}$$

so that we can use the Fresnel formula for propagating the field through the whole half-space only under particular conditions. For example, for a beam having a numerical aperture NA $= \sin\theta_{max} = 10^{-2}$ and $\alpha = 0.1$, the aperture width should be smaller than $10^5\,\lambda$ for the Fresnel formula to be applied uniformly. This, in turn, means that for a 1-μm wavelength, the maximum allowable cross section of the illuminated region on the integration plane is about 10 cm. When the numerical aperture increases to 10^{-1}, this last dimension is drastically reduced to only 100 μm.

Finally, we note that the field defined by Eq. (IV.10.4) satisfies the parabolic wave equation introduced in Section II.6 [see Eq. (II.6.12)]. For this reason some authors, especially in the Soviet Union, refer to the Fresnel approximation as the *parabolic approximation.*

10.2 *Fraunhofer Diffraction Formula*

Returning to the general expression provided by Eq. (IV.10.2), note that for $z - z' \gg D^2\pi/\lambda$, where D is a characteristic distance from $(0, 0, z')$ at which $u(x', y', z')$ is considered to vanish, we can neglect the terms of the exponential in the integrand of Eq. (IV.10.4) proportional to $x'^2 + y'^2$, so that

$$u(x, y, z) = \frac{i\exp(-ikR_0)}{\lambda|z - z'|} \iint_{-\infty}^{+\infty} \exp\left(ik\frac{xx' + yy'}{|z - z'|}\right) u(x', y', z')\, dx'\, dy', \tag{IV.10.9}$$

where

$$\begin{aligned} R_0 &= |z - z'| + \tfrac{1}{2}(x^2 + y^2)/|z - z'| \\ &\cong [(z - z')^2 + x^2 + y^2]^{1/2}. \end{aligned} \tag{IV.10.10}$$

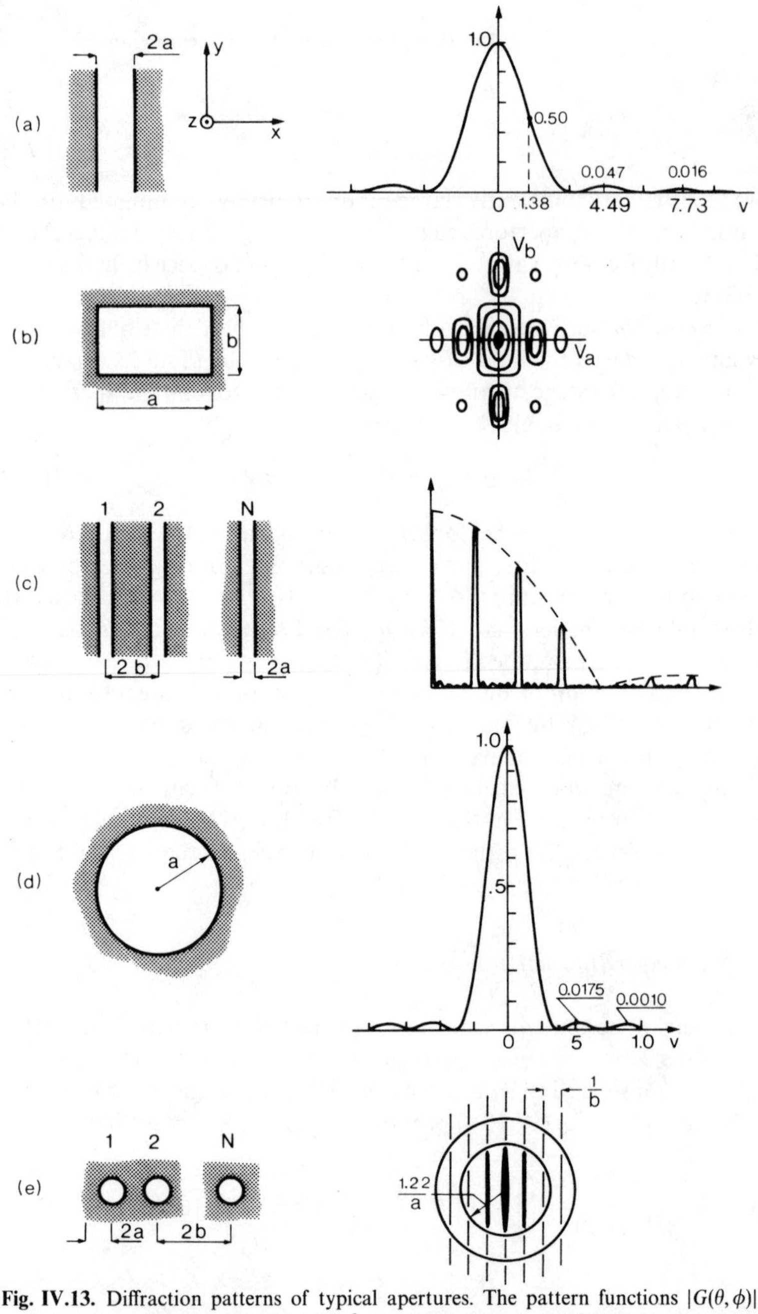

Fig. IV.13. Diffraction patterns of typical apertures. The pattern functions $|G(\theta, \phi)|^2$ are respectively proportional to (a) $[\sin(v)/v]^2$, $v = ka \sin\theta \cos\phi$; (b) $\{[\sin(v_a)/v_a][\sin(v_b)/v_b]\}^2$, $v_a = ka \sin\theta \cos\phi$, $v_b = kb \sin\theta \sin\phi$; (c) $\{[\sin(v_a)/v_a][\sin(Nv_b)/\sin v_b]\}^2$, $v_a = ka \sin\theta \cos\phi$, $v_b = kb \sin\theta \sin\phi$; (d) $(2J_1(v)/v)^2$, $v = ka \sin\theta$; (e) $\{[2J_1(v_a)/v_a][\sin(Nv_b)/\sin v_b]\}^2$, $v_a = ka \sin\theta$, $v_b = kb \sin\theta \cos\phi$.

The above equation, usually referred to as the *Fraunhofer diffraction formula*, allows us to express the far field in terms of the two-dimensional Fourier transform of u on the reference plane, evaluated for $k_x = kx/|z - z'|$ and $k_y = ky/|z - z'|$. In Fig. IV.13 we have plotted the Fraunhofer fields relative to some typical apertures illuminated by plane waves.

The far field can be measured at a finite distance by placing a lens in front of the aperture and observing the field on the focal plane. Accordingly, a lens can be used to replace a field distribution on a plane with its Fourier transform. This property of lenses is widely used in coherent optics for implementing optical correlators and optical matched filters, for recognizing assigned patterns, or for filtering images.

10.3 *Extension to Two-Dimensional Fields*

The preceding considerations can be easily extended to the two-dimensional waves represented by Eq. (IV.7.1). More precisely, it is easy to show that Eqs. (IV.10.4) and (IV.10.9) are replaced by

$$u(x, z) = \frac{\exp(i\pi/4 - i\chi|z - z'|)}{(2\pi/\chi)^{1/2}|z - z'|^{1/2}} \int_{-\infty}^{+\infty} \exp\left[-i\chi\frac{(x - x')^2}{2|z - z'|}\right] u(x', z')\,dx', \tag{IV.10.11}$$

and

$$u(x, z) = \frac{\exp(i\pi/4 - i\chi\rho_0)}{(2\pi/\chi)^{1/2}|z - z'|^{1/2}} \int_{-\infty}^{+\infty} \exp\left(i\chi\frac{xx'}{|z - z'|}\right) u(x', z')\,dx', \tag{IV.10.12}$$

where

$$\rho_0 \cong [(z - z')^2 + x^2]^{1/2}. \tag{IV.10.13}$$

10.4 *Diffraction of Periodic Fields*

Among the very few cases in which diffracted fields can be evaluated analytically, there is the important one related to periodic functions, which we treat here in some detail because of its interesting implications for the theory of gratings [14] (cf. Section VI.10).

Let us consider a cylindrical field $u(x, z)$ with phase independent of the y coordinate [see Eq. (IV.7.1)], expressed on the plane $z' = 0$ by

$$U(x', 0) = f(x') = \sum_{n=-N}^{+N} f_n \exp\left(in2\pi\frac{x'}{d}\right), \tag{IV.10.14}$$

where $f(x') = f(x' + d)$ is a periodic function of x' containing a number of harmonics N such that $N\lambda/d \ll 1$. Then for $kz < 4\alpha(d/N\lambda)^4$ [see Eq. (IV.10.3)] we can apply the Fresnel diffraction formula [Eq. (IV.10.4)], obtaining

$$u(x,z) = \frac{\exp(i\pi/4 - ikz)}{\lambda^{1/2}z^{1/2}} \sum_{n=-N}^{+N} f_n \int_{-\infty}^{+\infty} \exp\left[-ik\frac{(x-x')^2}{2z} + in2\pi\frac{x'}{d}\right] dx'$$

$$= \exp(-ikz) \sum_{n=-N}^{+N} f_n \exp\left[in2\pi\frac{x}{d} + i\left(\frac{n2\pi}{d}\right)^2 \frac{z}{2k}\right]. \qquad \text{(IV.10.15)}$$

It is interesting to note that for the special values $z = z_q = qd^2/\lambda$, $\exp[i(n2\pi/d)^2 z_q/(2k)] = \exp(in^2\pi q) = \exp(in\pi q)$ and

$$u(x, z_q) = \exp(-ikz)qf(x + \tfrac{1}{2}qd), \qquad \text{(IV.10.16)}$$

so that on any plane of observation at distance z_q, the field intensity is the same as on the initial plane. This property (the *Talbot* or *self-imaging effect*), first reported by Talbot in 1836, has found applications in Fourier spectroscopy and interferometry.

11 Field Expansion in Cylindrical Waves

In this section, we wish to show that the most general field satisfying Sommerfeld's radiation condition can be expanded as a superposition of *cylindrical modes*. This representation can be used as an alternative to the plane-wave representation introduced in Section IV.8, and turns out to be particularly useful for fields presenting a rotational symmetry around an axis (say z). Let us start from the representation of the Green's function

$$\frac{\exp(-ik|\mathbf{r} - \mathbf{r}'|)}{4\pi|\mathbf{r} - \mathbf{r}'|} = \frac{-i}{4\pi} \sum_{m=-\infty}^{+\infty} \exp[im(\phi - \phi')] \int_0^\infty \frac{J_m(\chi\rho)J_m(\chi\rho')}{(k^2 - \chi^2)^{1/2}}$$

$$\times \exp[-i(k^2 - \chi^2)^{1/2}|z - z'|]\chi\, d\chi, \qquad \text{(IV.11.1)}$$

which is obtained from the *Sommerfeld–Ott representation* of G and *Graf's addition theorem* for Bessel functions (see Problem 8). In Eq. (IV.11.1) J_m is the Bessel function of first kind and mth order, and use is made of the cylindrical coordinates ρ, ϕ, z.

If we insert the above series into Eq. (IV.5.4) and choose a plane perpendicular to the z axis as the integration domain, the field in $\mathbf{r}$ is

$$u(x,y,z) = u(\rho,\phi,z) = \frac{1}{2\pi}\int_0^\infty \rho'\, d\rho' \int_0^{2\pi} u(\rho',\phi',z') \sum_{m=-\infty}^{+\infty} \exp[im(\phi - \phi')]\, d\phi'$$

$$\times \int_0^\infty J_m(\chi\rho)J_m(\chi\rho')\exp[-i(k^2 - \chi^2)^{1/2}|z - z'|]\chi\, d\chi. \qquad \text{(IV.11.2)}$$

On the other hand, the field on the reference plane can be expanded in a

Fourier series with respect to ϕ',

$$u(\rho', \phi', z') = \sum_{m=-\infty}^{+\infty} C_m(\rho', z') e^{im\phi'} \qquad \text{(IV.11.3)}$$

(m ranges over integer values, and u is a single-valued quantity), so that

$$\begin{aligned}\int_0^\infty \rho' \, d\rho' \int_0^{2\pi} u(\rho', \phi', z') J_m(\chi\rho') e^{im(\phi - \phi')} \, d\phi' \\ = 2\pi e^{im\phi} \int_0^\infty C_m(\rho', z') J_m(\chi\rho') \rho' \, d\rho' \\ \equiv 2\pi e^{im\phi} H_m[C_m(\rho', z'); \chi], \qquad \text{(IV.11.4)}\end{aligned}$$

where $H_m[f(\rho'); \chi]$ is the *Hankel transform* of order m and argument χ (see Sneddon [15] and Appendix E). The coefficients $C_m(\rho, z)$ of the field Fourier expansion (in ϕ) on the plane of coordinate z are now obtained by means of Eqs. (IV.11.2,3,4):

$$\begin{aligned}C_m(\rho, z) &= \int_0^\infty H_m[C_m(\rho', z'); \chi] \exp[-i(k^2 - \chi^2)^{1/2}|z - z'|] J_m(\chi\rho) \chi \, d\chi \\ &= H_m[H_m\{C_m(\rho', z'); \chi\} \exp[-i(k^2 - \chi^2)^{1/2}|z - z'|]; \rho] \\ &= C_m(\rho, z') * H_m\{\exp[-i(k^2 - \chi^2)^{1/2}|z - z'|]; \rho\}. \qquad \text{(IV.11.5)}\end{aligned}$$

This result was obtained by using the convolution property of the Hankel transform and its self-reciprocity, that is, $H_m\{H_m\{f\}\} = f$. In particular, it is easily verified that, for $z = z'$, $C_m(\rho, z) = C_m(\rho, z')$. In conclusion, we can write the field in the form

$$u(\rho, \phi, z) = \hat{T} \sum_{m=-\infty}^{+\infty} e^{im\phi} C_m(\rho, z') = \sum_{m=-\infty}^{+\infty} C_m(\rho, z) e^{im\phi}, \qquad \text{(IV.11.6)}$$

$\hat{T}$ being the linear operator that transforms $C_m(\rho, z')$ into $C_m(\rho, z)$ according to Eq. (IV.11.5).

If we isolate the generic mth term of the above series, we can regard it as a superposition of *cylindrical modes* of the form

$$H_m[C_m(\rho, z); \chi] J_m(\chi\rho) \exp[im\phi - i(k^2 - \chi^2)^{1/2}|z - z'|] \chi \, d\chi. \qquad \text{(IV.11.7)}$$

Since plane and cylindrical wave expansions refer to the same class of fields, it is possible to interchange the two representations. While the plane-wave representation is based on three continuous parameters k_x, k_y, and k_z, related by $k_x^2 + k_y^2 + k_z^2 = k^2$, the cylindrical one needs a discrete parameter m correlated with the angular behavior $\exp(im\phi)$ and a continuous parameter χ defining the z dependence $\exp[-i(k^2 - \chi^2)^{1/2}|z - z'|]$. The radial behavior $J_m(\chi\rho)$ depends on both m and χ.

11.1 *Fresnel Approximation*

Whenever the spectrum $A_m(\chi)$ of each cylindrical mode vanishes for $\chi \gg \chi_{\max}$, with $\chi_{\max} \ll k$, we can simplify Eq. (IV.11.5) by approximating the propagation factor by

$$\exp[-i(k^2 - \chi^2)^{1/2}|z - z'|] \cong \exp[-ik|z - z'| + i\chi^2|z - z'|/(2k)], \tag{IV.11.8}$$

so that we obtain, after some algebra,

$$C_m(\rho, z) = \frac{-ik}{|z - z'|}\exp\left(-ik|z - z'| - ik\frac{\rho^2}{2|z - z'|}\right) \times H_m\left\{C_m(\rho', z')\exp\left(-ik\frac{\rho'^2}{2|z - z'|}\right); \frac{k\rho}{|z - z'|}\right\}, \tag{IV.11.9}$$

having used the integral identity

$$\int_0^\infty \exp\left(i\chi^2\frac{|z - z'|}{2k}\right)J_m(\chi\rho)J_m(\chi\rho')\chi\, d\chi = \frac{-ik}{|z - z'|}J_m\left(k\frac{\rho\rho'}{|z - z'|}\right)\exp\left(-ik\frac{\rho^2 + \rho'^2}{2|z - z'|}\right). \tag{IV.11.10}$$

Equation (IV.11.9) can be considered equivalent to the Fresnel formula (IV.10.2) established for cartesian coordinates.

11.2 *Expansion of a Plane Wave in Cylindrical Waves*

As already observed, any field having a plane-wave representation can also be expressed in terms of cylindrical modes. This is achieved in a natural way by developing a generic plane wave into cylindrical modes. To this end, we note that a plane wave can be expressed on the reference plane $z = z'$ by

$$u(\rho', \phi', z') = \exp[-ik\rho'\cos(\phi' - \phi'_0)\sin\theta], \tag{IV.11.11}$$

where θ is the angle between the wave vector $\mathbf{k}_0$ and the z axis. If we develop the right-hand side of Eq. (IV.11.11) in cylindrical coordinates, we have

$$u(\rho', \phi', z') = \sum_{m=-\infty}^{+\infty}(-i)^m J_m(k\rho'\sin\theta)\exp[im(\phi' - \phi'_0)], \tag{IV.11.12}$$

which yields [see Eq. (IV.11.3)]

$$C_m(\rho', z') = (-i)^m e^{-im\phi'_0}J_m(k\rho'\sin\theta). \tag{IV.11.13}$$

Thus, the Hankel transform operation yields

$$H_m[C_m(\rho', z'); \chi] = (-i)^m e^{-im\phi_0'} \int_0^\infty J_m(k\rho' \sin\theta) J_m(\chi\rho')\rho'\, d\rho'$$
$$= (-i)^m e^{-im\phi_0'} \chi^{-1} \delta(\chi - k \sin\theta), \qquad \text{(IV.11.14)}$$

so that, finally, Eq. (IV.11.5.) becomes

$$C_m(\rho, z) = C_m(\rho, z') \exp(-ik_0|z - z'| \cos\theta). \qquad \text{(IV.11.15)}$$

As a result of the above discussion, the expansion in cylindrical modes of the plane wave associated with the field distribution of Eq. (IV.11.11) on the reference plane reads

$$u(\rho, \phi, z) = \exp(-ik|z - z'| \cos\theta) \sum_{m=+\infty}^{-\infty} (-i)^m J_m(k\rho \sin\theta) e^{im(\phi - \phi_0')}. \qquad \text{(IV.11.16)}$$

11.3 *Aperture-Limited Rotationally Invariant Field*

We now wish to give an example of the Fresnel approximation in cylindrical coordinates. More precisely, we consider a field different from zero on a disk of radius a on the plane $z = z'$ and independent of ϕ. This rotational symmetry implies the vanishing of C_m for $m \neq 0$, so that Eq. (IV.11.6) reduces to

$$u(\rho, z) = C_0(\rho, z). \qquad \text{(IV.11.17)}$$

Let us suppose that the field has, on the reference plane, the RO representation

$$u(\rho', z') = A(\rho') e^{-ikS(\rho')} = C_0(\rho', z') \qquad \text{(IV.11.18)}$$

for $\rho' \leqq a$, while $u(\rho', z') = 0$ for $\rho' > 0$. Equations (IV.11.5,17,18) allow us to write

$$u(\rho, z) = a^2 \int_0^\infty \chi\, d\chi \int_0^1 x A(ax) \exp[-ikS(ax) - i(k^2 - \chi^2)^{1/2}|z - z'|]$$
$$\times J_0(\chi a x) J_0(\chi\rho)\, dx, \qquad \text{(IV.11.19)}$$

which, in the Fresnel approximation, reduces to (see Eq. (IV.11.9))

$$u(\rho, z) = -\imath \mathrm{NA} ka \exp\left\{\left(-ik|z - z'| - ik \frac{\rho^2}{2|z - z'|}\right)\right.$$
$$\left. \times \int_0^1 A(ax) J_0 \frac{(kax\rho)}{|z - z'|} \exp\left[-ikaNA\frac{x^2}{2} - ikS(ax)\right] x\, dx\right\}, \qquad \text{(IV.11.20)}$$

where $NA = a/|z - z'|$ is the *numerical aperture* of the disk as seen from the distance $|z - z'|$.

For large distances, $\exp(-ika\text{NA}x^2/2) \cong 1$, so that Eq. (IV.11.20) simplifies to

$$u(\rho, z) = -i\text{NA}ka \exp\left\{\left(-ik|z - z'| - ik\frac{\theta^2|z - z'|}{2}\right)\right.$$
$$\left. \times \int_0^1 A(ax)J_0(ka\theta x)\exp[-ikS(ax)]\, x\, dx\right\}, \qquad \text{(IV.11.21)}$$

where $\theta = \rho/|z - z'|$ (*Fraunhofer diffraction formula for rotationally invariant fields*).

11.3.a *Boivin's Series Expansions*

If we set $A(ax)\exp(-ikS - ika\text{NA}x^2/2) = f(x^2)$, we may follow Boivin [16] by introducing the set of functions $A_p(\xi)$ defined by

$$\frac{1}{2p + 1} A_p(\xi) = \int_0^1 x^{2p+1} J_0(\xi x)\, dx, \qquad \text{(IV.11.22)}$$

coincident with the Hankel transforms of zeroth order of the function $g(x) = 2(p + 1)x^{2p}$ for $x \leqq 1$, $g(x) = 0$ otherwise. Then, by expanding $f(x^2)$ in a power series we can represent the field as a series of the form

$$u(\rho, z) = C \sum_{p=0}^{\infty} \frac{f^{(p)}(0)}{(p + 1)!} A_{p+1}\left(ka\frac{\rho}{|z - z'|}\right), \qquad \text{(IV.11.23)}$$

C standing for the factors multiplying the integral of Eq. (IV.11.20). The A_p are easily evaluated by using the recurrence relation [16]

$$A_p(\xi) + [\xi^2/4p(p + 1)]A_{p+1}(\xi) = J_0(\xi) + [\xi J_1(\xi)/2p], \qquad \text{(IV.11.24)}$$

together with

$$A_1 = 2J_1(\xi)/\xi. \qquad \text{(IV.11.25)}$$

Boivin has obtained the following series expansion [16]

$$u(\rho, z) = C \sum_{p=0}^{\infty} (-1)^p f^{(p)}(1) 2^p \frac{J_{p+1}(\xi)}{\xi^{p+1}}, \qquad \text{(IV.11.26)}$$

with $\xi = ka\rho/|z - z'|$. In particular, for $\exp\{i(\omega_1 x^2 + \omega_2 x^4)\}$ one obtains $f^{(p)}(1) = (-1)^p f^{(0)}(1)\{(i\alpha)^p - (i\alpha)^{p-2} i\beta(p)_2/2 \cdots\}$ with $(p)_n = \Gamma(p-1)/\Gamma(p-n)$, $\alpha = -\omega_1 - 2\omega_2$ and $\beta = -2\omega_2$. In particular, it can be shown that $\alpha = 0$ corresponds to the marginal focus, $\alpha = \beta$ to the paraxial focus, and $\alpha = \beta/2$ indicates the circle of least confusion, that is the minimal image appearing on a screen perpendicular to the optic axis.

12 Cylindrical Waves of Complex Order and Watson Transformation

Series expansions in terms of cylindrical waves are often only intermediate steps in the search for simple analytic representations of the field. In fact, in many cases these series converge so slowly that it is necessary to retain a large number of terms to represent the field with fairly good accuracy. A typical example of this is the representation of a plane wave scattered by a cylindrical obstacle. In spite of the simple behavior of the field in the geometrical optics limit, it is not easy to conjecture the presence of a dark region by inspecting the relative Fourier–Bessel series. A way to circumvent this difficulty was suggested by Watson, who transformed the original expansion into a new one that converges more rapidly to the ray optics limit as $\lambda \to 0$. In the following, we will illustrate the main steps leading to this series transformation by using a particular example, while deferring to Section VI.5 the application of the Watson transformation to dielectric cylinders.

In the case of diffraction of a plane wave by a circular cylinder, heuristic considerations lead us to presume that the field behaves like an evanescent wave on the portion of the cylindrical surface lying in the shadow region, that is,

$$u(a, \phi) \propto \exp[-\nu(\phi - \phi_{SB})], \qquad \text{(IV.12.1)}$$

where ϕ_{SB} is the angle relevant to the shadow boundary and ν is a complex coefficient with $\operatorname{Re} \nu > 0$. From a formal point of view, the above solution corresponds to a cylindrical wave "rotating" around the cylinder with a complex propagation constant ν/a, a being the cylinder radius. Thus, the field in the lit region is a combination of cylindrical wave components of real order m, while in the dark region components of complex order ν come into play.

Before showing that Eq. (IV.12.1) is correct for $\lambda \to 0$, we must modify the expansion Eq. (IV.11.3) in order to include the contributions from waves having a generic form $J_\nu(\chi\rho)\exp(i\nu\phi)$, with complex ν. To this end, we observe that for a field independent of z, Eq. (IV.11.6) reduces to

$$u(\rho, \phi) = \sum_{m=-\infty}^{\infty} A_m J_m(k\rho) e^{im\phi}$$

$$= \frac{1}{2} \sum_{m=-\infty}^{\infty} A_m H_m^{(1)}(k\rho) e^{im\phi} + \frac{1}{2} \sum_{m=-\infty}^{\infty} A_m H_m^{(2)}(k\rho) e^{im\phi}, \qquad \text{(IV.12.2)}$$

where $H_m^{(1),(2)}$ are the Hankel functions of the first (index 1) and second (index 2) kind. They can be defined for every complex index ν and, for fixed real ν and $|x| \to \infty$, they exhibit the asymptotic behavior

$$H_\nu^{(2)*}(x) = H_\nu^{(1)}(x) \underset{|x|\to\infty}{\sim} \left(\frac{2}{\pi x}\right)^{1/2} \exp\left(ix - i\nu\frac{\pi}{2} - i\frac{\pi}{4}\right), \qquad \text{(IV.12.3)}$$

so that $H_\nu^{(2)}$ represents a wave traveling to infinity, while the opposite applies to $H_\nu^{(1)}$. Thus, if we are interested only in the field component traveling to infinity, we can express $u(\rho, \phi)$ in the simple form

$$u(\rho, \phi) = \sum_{m=-\infty}^{+\infty} A_m H_m^{(2)}(k\rho) e^{im\phi}. \qquad \text{(IV.12.4)}$$

If $u(\rho, \phi)$ is known on a circle of radius a, we can recast the above series in the form

$$u(\rho, \phi) = \sum_{m=-\infty}^{+\infty} \langle u(a, \phi')e^{-im\phi'}\rangle_{\phi'} \frac{H_m^{(2)}(k\rho)}{H_m^{(2)}(\beta)} e^{im\phi}, \qquad \text{(IV.12.5)}$$

where $\beta = ka$ is the so-called *size parameter* and

$$\langle u(a, \phi')e^{-im\phi'}\rangle_{\phi'} = \frac{1}{2\pi}\int_0^{2\pi} u(a, \phi')e^{-im\phi'}\, d\phi'. \qquad \text{(IV.12.6)}$$

In writing Eq. (IV.12.5), we have taken into account the fact that, for given $\rho = a$, Eq. (IV.12.4) is the Fourier series expansion of the periodic function $f(\phi) = f(\phi + 2\pi) = u(\rho, \phi)$.

When both β and $k\rho$ are very large, we can approximate Eq. (IV.12.5) by replacing the Hankel functions with the asymptotic expansion of Eq. (IV.12.3), thus obtaining

$$\begin{aligned} u(\rho, \phi) &\sim \left(\frac{a}{\rho}\right)^{1/2} e^{-ik(\rho - a)} \sum_{m=-\infty}^{+\infty} \langle u(a, \phi')e^{-im\phi'}\rangle_{\phi'} e^{im\phi} \\ &= \left(\frac{a}{\rho}\right)^{1/2} e^{-ik(\rho - a)} u(a, \phi), \end{aligned} \qquad \text{(IV.12.7)}$$

which is just the geometrical optics expression of the field. The asymptotic approximation of $H_\nu^{(2)}$ expressed by Eq. (IV.12.3) fails when the contribution of the terms of large index m cannot be neglected. In fact, this asymptotic expansion loses its validity when $|x|$ is of the order of or smaller than $|\nu|$. In these cases, for constant x/ν and $|\nu| \to \infty$, $H_\nu^{(2)}$ tends asymptotically to (see Watson [4], p. 71)

$$H_\nu^{(2)}(x) \sim \left(\frac{2}{\pi\nu}\right)^{1/2}\left[-e^{i2\pi\nu}\left(\frac{ex}{2\nu}\right)^\nu + i\left(\frac{2\nu}{ex}\right)^\nu\right], \qquad \pi > \arg\nu > 0 \qquad \text{(IV.12.8)}$$

{an analogous expression can be obtained for $H_\nu^{(1)}$ by using the relation $H_\nu^{(1)}(x) = \exp(-\nu\pi i)H_\nu^{(2)}[\exp(-\pi i)x]$}. Accordingly,

$$H_\nu^{(2)}(k\rho)/H_\nu^{(2)}(\beta) \underset{|\nu|\to\infty}{\sim} (a/\rho)^\nu, \qquad \pi/2 > \arg\nu > 0, \qquad \text{(IV.12.9a)}$$

$$H_\nu^{(2)}(k\rho)/H_\nu^{(2)}(\beta) \underset{|\nu|\to\infty}{\sim} (\rho/a)^\nu, \qquad \pi > \arg\nu > \pi/2, \qquad \text{(IV.12.9b)}$$

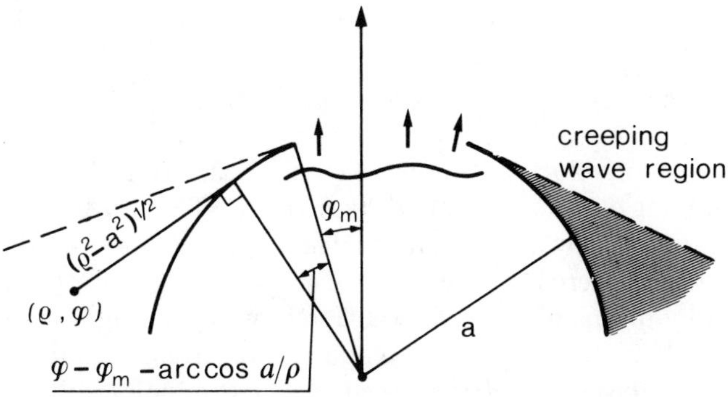

Fig. IV.14. Geometry related to the creeping waves excited in the shadow region through an aperture on a metallic cylinder.

which, in particular, imply that the higher-order terms of the series Eq. (IV.12.5) decrease as $(a/\rho)^m$, in contrast to the behavior shown by Eq. (IV.12.7). The drastic attenuation of the higher-order terms implies a sensitive smoothing of the far-zone pattern with respect to the geometrical optics behavior.

To be more specific, let us consider a metallic circular cylinder having radius a, with an aperture extending from $\phi = -\phi_m$ to $\phi = \phi_m$ (see Fig. IV.14), and assume that the sources are located inside the cylinder and produce a field $u(a, \phi)$ on the aperture. This field will propagate outside the cylinder, in the RO approximation, as a wave confined to the sector $|\phi| < |\phi_m|$, dropping abruptly to zero outside. However, due to the progressive attenuation of the amplitude of the higher harmonics, the passage from the lit to the dark sector will be less abrupt the more the observation point moves into the far zone.

These qualitative considerations can be put in quantitative form by trans-

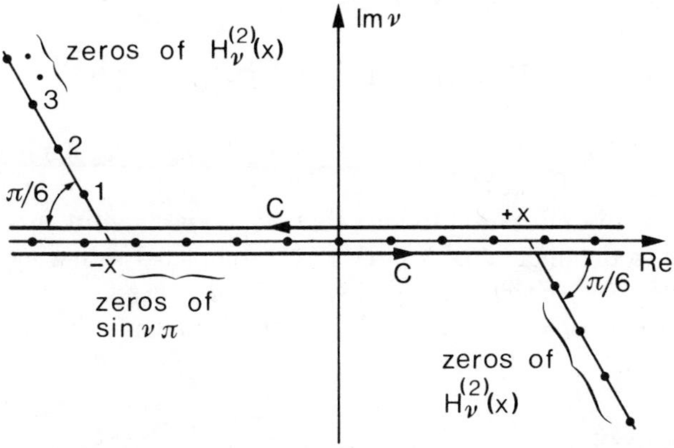

Fig. IV.15. Distributions of the zeros of $H_\nu(x)$ and $\sin(\nu\pi)$. The integration path C includes the zeros of $\sin(\nu\pi)$ and leaves outside the zeros of $H_\nu(x)$.

forming the series of Eq. (IV.12.5) into the contour integral (see Fig. IV.15)

$$u(\rho, \phi) = -\frac{i}{2}\oint_C \frac{e^{i\nu(\phi-\pi)}}{\sin \nu\pi}\langle u(a, \phi')e^{-i\nu\phi'}\rangle_{\phi'} \frac{H_\nu^{(2)}(k\rho)}{H_\nu^{(2)}(\beta)}\, d\nu, \tag{IV.12.10}$$

where C is a counterclockwise-oriented contour enclosing all of the poles of $\sin \nu\pi$ and leaving outside those of $H_\nu^{(2)}(k\rho)/H_\nu^{(2)}(\beta)$. The above integral can be immediately justified by noting that its integrand has inside C a series of poles coinciding with the zeros of $\sin \nu\pi$, that is, at $\nu = m$. Therefore, by calculating the integral by means of the residue theorem we again obtain the series of Eq. (IV.12.5). If we now take into account the recurrence relation $H_{-\nu}^{(2)}(x) = e^{-i\nu\pi}H_\nu^{(2)}(x)$ and the fact that $H_\nu^{(2)}$ has neither zeros nor poles with $\text{Im}\, \nu = 0$, we can reduce the contour C to $-\infty + i\varepsilon$, $i\varepsilon + \infty$, with $\varepsilon > 0$, i.e.,

$$u(\rho, \phi) = i\int_{-\infty+i\varepsilon}^{\infty+i\varepsilon} \frac{\langle u(a, \phi')\cos \nu(\pi - \phi + \phi')\rangle_{\phi'}}{\sin \nu\pi} \frac{H_\nu^{(2)}(k\rho)}{H_\nu^{(2)}(\beta)}\, d\nu. \tag{IV.12.11}$$

If we take into account Eqs. (IV.12.9), we observe that the integrand of Eq. (IV.12.11) tends to zero as $|\nu| \to \infty$ for $|\pi - \phi + \phi'| < \pi$ and $\pi > \arg \nu > 0$. Consequently, we can complete the integration path with a semicircle in the upper half-plane. Since $H_\nu^{(2)}(x)$ can be represented as a combination of Bessel functions J_ν and $J_{-\nu}$,

$$H_\nu^{(2)}(x) = i\{[J_{-\nu}(x) - J_\nu(x)e^{i\nu\pi}]/\sin \nu\pi\} \tag{IV.12.12}$$

and $J_\nu(x)$ can be expressed by the series

$$J_\nu(x) = \sum_{m=-\infty}^{+\infty} (-1)^m \left(\frac{x}{2}\right)^{\nu+2m} [m!\,\Gamma(\nu + 1 + m]^{-1}, \tag{IV.12.13}$$

$\Gamma(z)$ being the Gauss gamma function, we have

$$H_\nu^{(2)}(x) = \frac{i}{\sin \nu\pi}\left\{\left(\frac{2}{x}\right)^\nu \sum_{m=0}^{+\infty}(-1)^m\left(\frac{x}{2}\right)^{2m} [m!\,\Gamma(1 + m - \nu)]^{-1} \right.$$

$$\left. -\left(\frac{x}{2}\right)^\nu e^{i\nu\pi} \sum_{m=0}^{+\infty}(-1)^m\left(\frac{x}{2}\right)^{2m} [m!\,\Gamma(1 + m + \nu)]^{-1}\right\}. \tag{IV.12.14}$$

Now, observing that $\Gamma^{-1}(\nu)$ is an entire function of ν, i.e., is regular in the whole complex ν plane, and that $H_\nu^{(2)}(x)$ tends to a finite limit for $\nu = m$ (m being a positive or negative integer), it follows that $H_\nu^{(2)}(x)$ is an entire function of ν.

Returning to the integral of Eq. (IV.12.11) completed with a semicircle in the upper half-plane, the *integrand presents only polar singularities corresponding to the zeros* ν_n ($n = 1, \ldots, \infty$) of $H_\nu^{(2)}(\beta)$. Consequently, by applying the residue theorem, we finally obtain

$$u(\rho, \phi) = -2\pi \sum_{n=1}^{+\infty} \frac{\langle u(a, \phi')\cos[\nu_n(\pi - \phi + \phi')]\rangle_{\phi'}}{\sin(\nu_n\pi)} \frac{H_{\nu_n}^{(2)}(k\rho)}{\partial H_\nu^{(2)}(\beta)/\partial\nu|_{\nu=\nu_n}}. \tag{IV.12.15}$$

Thus, as a result of a rather involved sequence of transformations, the initial series of Eq. (IV.12.5) has been replaced by a new series containing functions of the generally complex index ν_n. This procedure was initially proposed by G. N. Watson in 1918 for improving the convergence of the spherical wave representation of the field scattered by a spherical obstacle in the dark zone. In so doing, he was able to show that only the first term of this new series can be retained, thus explaining the exponential decay of the field radiated by a transmitter beyond the line of sight and into the region of geometric shadow of the earth. In more recent years (1958) T. Regge has rediscovered this technique for solving the problem of the scattering of the Schrödinger wave function of a particle by a central potential. In the latter case, the index m represents, apart from Planck's constant $\hbar$, the quantum mechanical momentum of the particle. The recourse to complex values of ν can thus be rephrased by saying that in the dark regions the particles are characterized by a *complex angular momentum* [17].

It now remains to examine the distributions of these zeros and the expressions for the terms of the *Watson series* of Eq. (IV.12.15).

12.1 *Zeros of $H_\nu^{(2)}(\beta)$ in the Upper Half-Plane*

The zeros of $H_\nu^{(2)}(\beta)$ ($H_\nu^{(1)}, H_\nu^{(2)\prime}$, and $H_\nu^{(1)\prime}$) have been studied by several authors [18–21]. In particular, Schöbe [19] derived the following expansion for $H_\nu^{(2)}(\beta)$, valid when $|\nu + \beta| = O(x^{1/3})$, $|\beta| \gg 1$:

$$\begin{aligned} H_\nu^{(2)}(\beta) &\sim 2e^{i\pi/3}\left(\frac{2}{\beta}\right)^{1/3} \mathrm{Ai}(-\xi) \sum_{n=0}^{\infty} (-1)^n \left(\frac{2}{\beta}\right)^{2n/3} P_n(\xi) \\ &\quad - e^{i\pi/3}\, \mathrm{Ai}'(-\xi) \sum_{n=1}^{\infty} (-1)^n \left(\frac{2}{\beta}\right)^{2n/3} Q_n(\xi) \\ &\underset{|\beta|\to\infty}{\sim} 2e^{i\pi/3}\left(\frac{2}{\beta}\right)^{1/3} \mathrm{Ai}(-\xi), \end{aligned} \tag{IV.12.16}$$

where $\mathrm{Ai}(z)$ and $\mathrm{Ai}'(z)$ are the Airy function and its derivative, respectively, and

$$\xi \equiv -e^{i\pi/3}(2/\beta)^{1/3}(\nu + \beta), \tag{IV.12.17}$$

while $P_0(\xi) = 1$ and the other first four factors P_n and Q_n are given by

$$\begin{aligned} P_1(\xi) &= e^{-i\pi/3}\frac{\xi}{15}, & Q_1(\xi) &= -e^{i\pi/3}\frac{\xi^2}{60}, \\ P_2(\xi) &= e^{i\pi/3}\left(\frac{\xi^5}{7200} - \frac{13\xi^2}{1260}\right), & Q_2(\xi) &= -\frac{\xi^3}{420} + \frac{1}{140}. \end{aligned} \tag{IV.12.18}$$

Starting from the above expansion of Eq. (IV.12.16) and adapting a formula of

Streifer and Kodis [18], we can show that the low-index zeros ν_n of $H_\nu^{(2)}(\beta)$ are given with good accuracy by the formula [see also Eq. (VI.6.3) and Nussenzveig [22], Eq. (A8)]

$$\nu_n(\beta) = -\beta - e^{-i\pi/3}(\beta/2)^{1/3}(x_n - \delta_n) \underset{|\beta| \gg 1}{\sim} -\beta - e^{-i\pi/3}(\beta/2)^{1/3} x_n, \tag{IV.12.19}$$

where x_n is the nth zero of $\mathrm{Ai}(-x)$, and

$$\delta_n(\beta) = -e^{-i\pi/3}\frac{x_n^2}{60}\left(\frac{2}{\beta}\right)^2 - e^{i\pi/3}\left(\frac{x_n^3}{400} - \frac{1}{140}\right)\left(\frac{2}{\beta}\right)^4 - \left(\frac{281 x_n^4}{4{,}536{,}000} - \frac{29 x_n}{12{,}600}\right)\left(\frac{2}{\beta}\right)^6. \tag{IV.12.20}$$

The first five values of x_n are given by

$$x_1 = 2.338, \quad x_2 = 4.088, \quad x_3 = 5.521, \quad x_4 = 6.787, \quad x_5 = 7.944, \tag{IV.12.21}$$

while for large values of n (less than β), x_n is given by the asymptotic formula

$$x_n \sim \{(3/2)\pi[n + (3/4)]\}^{2/3}. \tag{IV.12.22}$$

On the other hand, for $n \to \infty$ (see [21])

$$\mathrm{Re}\,\nu_n(\beta) \sim -\frac{\pi^2}{2}\left(n - \frac{1}{4}\right)\left\{\ln\left[\frac{2\pi(n - 1/4)}{e\beta}\right]\right\}^{-2}, \tag{IV.12.23a}$$

$$\mathrm{Im}\,\nu_n(\beta) \sim \pi\left(n - \frac{1}{4}\right)\left\{\ln\left[\frac{2\pi(n - 1/4)}{e\beta}\right]\right\}^{-1}. \tag{IV.12.23b}$$

Finally, the distribution of the zeros ν_n is illustrated schematically in Fig. IV.15, which shows clearly that $\mathrm{Im}\,\nu_n$ is an increasing function of n. In addition, for x sufficiently large, the first zeros are aligned on a line forming an angle of 60° with the real axis.

For the first zeros, we can replace $H_\nu^{(2)}$ with the Airy function $\mathrm{Ai}(-\xi)$, as shown in Eq. (IV.12.16). Accordingly, we can write

$$\left.\frac{\partial H_\nu^{(2)}(\beta)}{\partial \nu}\right|_{\nu=\nu_n} \sim -2e^{i\pi/3}\left(\frac{2}{\beta}\right)^{1/3} \mathrm{Ai}'(-\xi)\left.\frac{\partial \xi}{\partial \nu}\right|_{\nu=\nu_n} = 2e^{i2\pi/3}\left(\frac{2}{\beta}\right)^{2/3} \mathrm{Ai}'(-x_n). \tag{IV.12.24}$$

For the high-order zeros ($|\nu_n| \gg \beta$) we can use Eq. (IV.12.8) to write

$$\left.\frac{\partial H_\nu^{(2)}(\beta)}{\partial \nu}\right|_{\nu=\nu_n} \sim -2\left(\frac{2}{\pi\nu_n}\right)^{1/2} e^{i2\pi\nu_n}\left(\frac{e\beta}{2\nu_n}\right)^{\nu_n}\left[\ln\left(-\frac{\beta}{2\nu_n}\right) - \frac{1}{2\nu_n}\right], \tag{IV.12.25}$$

so that, taking into account Eq. (IV.12.23a),

$$\left|\frac{H_{\nu_n}^{(2)}(k\rho)}{\partial H_\nu^{(2)}(\beta)/\partial\nu|_{\nu=\nu_n}}\right| \underset{n\to\infty}{\sim} \frac{1}{2}\left(\frac{\rho}{a}\right)^{\mathrm{Re}(\nu_n)}\left|\ln\left(-\frac{\beta}{2\nu_n}\right)\right|^{-1} \underset{n\to\infty}{\to} 0 \qquad \text{(IV.12.26)}$$

12.2 *Creeping Waves*

In view of the rapid decay of the terms of the Watson series for $n \to \infty$, confirmed by Eq. (IV.12.26), we can consider only the low-order terms for which $|\nu_n + \beta| = O(\beta^{1/3})$. Consequently, Eq. (IV.12.24) applies and we can write

$$u(\rho,\phi) \sim \pi e^{i\pi/3}\left(\frac{\beta}{2}\right)^{2/3} \sum_{n=1}^{\infty} \frac{\langle u(a,\phi')\cos[\nu_n(\pi-\phi+\phi')]\rangle_{\phi'} H_{\nu_n}^{(2)}(k\rho)}{\sin(\nu_n\pi)\,\mathrm{Ai}'(-x_n)}. \qquad \text{(IV.12.27)}$$

If we wish to consider the field for $a < \rho < \infty$ we need an expression for $H_{\nu_n}^{(2)}(k\rho)$ that is valid uniformly for $k\rho$ either very large or comparable to $|\nu_n|$. Debye and, later, Watson derived asymptotic expressions valid for x and $|\nu|$ both very large and $|x + \nu| > O(x^{1/3})$. Using their results, we can write (see Watson [4], p. 262)

$$H_\nu^{(2)}(x) \sim [2/\pi(x^2+\nu^2)^{1/2}]^{1/2} \exp(-i\{(x^2-\nu^2)^{1/2} + \nu[\pi - \arccos(\nu/x)] - \pi/4\}), \qquad \pi > \arg\nu > \pi/2. \qquad \text{(IV.12.28)}$$

If, in view of Eq. (IV.12.19), we make the approximations $(k^2\rho^2 - \nu_n^2)^{1/2} \cong k(\rho^2 - a^2)^{1/2}$ and $\pi - \arccos(\nu_n/k\rho) \cong \arccos(a/\rho)$, we have

$$H_{\nu_n}^{(2)}(k\rho) \sim \left[\frac{2}{k(\rho^2-a^2)^{1/2}}\right]^{1/2} \times \exp\left\{-i\left[k(\rho^2-a^2)^{1/2} + \nu_n \arccos\left(\frac{a}{\rho}\right) - \frac{\pi}{4}\right]\right\}. \qquad \text{(IV.12.29)}$$

On the other hand, since $\mathrm{Im}(\nu_n) \gg 1$ [see Eq. (IV.12.23b)], we can put $\cos[\nu_n(\pi - \phi + \phi')]/\sin(\pi\nu_n) \cong e^{i\nu_n[\pi - |\pi+\phi'-\phi|]}$, and

$$\frac{\langle u(a,\phi')\cos[\nu_n(\pi-\phi+\phi'])\rangle_{\phi'}}{\sin(\nu_n\pi)} \underset{\mathrm{Im}\,\nu_n\to\infty}{\sim} \frac{u(a,\phi_m)}{2\pi\nu_n} e^{i\nu_n(\phi-\phi_m)}. \qquad \text{(IV.12.30)}$$

Consequently, the Watson expansion reduces to

$$u(\rho, \phi) \sim -i \frac{e^{i\pi/2}}{4\pi^{1/2} 2^{1/6}} \frac{1}{\beta^{1/6}} \sum_{n=1}^{\infty} \frac{1}{\mathrm{Ai}'(-x_n)} \left(\frac{a^2}{\rho^2 - a^2} \right)^{1/4}$$

$$\times \exp\left\{ -ik \left[\left(\rho^2 - a^2 \right)^{1/2} + a \left(\phi - \phi_m - \arccos\left(\frac{a}{\rho} \right) \right) \right] \right.$$

$$\left. - e^{-i\pi/3} \left[\phi - \phi_m - \arccos\left(\frac{a}{\rho} \right) \right] \left(\frac{\beta}{2} \right)^{1/3} x_n \right\}. \qquad \text{(IV.12.31)}$$

Each term of this expansion can be interpreted as the field amplitude along a ray departing from the aperture edge, traveling along the cylinder $\rho = a$ at an angle equal to $\phi - \phi_m - \arccos(a/\rho)$, and then following the tangent to the cylinder (Fig. V.14) until it reaches the observation point (ρ, ϕ). The factor $a^{1/2}/(\rho^2 - a^2)^{1/4}$ accounts for the geometric attenuation along the straight portion of the trajectory. The phase factor contains two terms, the first of which (in square brackets) measures the delay accumulated by the ray in describing the curved trajectory from the edge to the field point, while the second accounts for the deviations from the law of geometric optics due to the curvature of the circular part of the trajectory. It is the presence of this factor that marks the particular features of these waves; they undergo an exponential attenuation along the cylindrical surface, which increases with order n according to the coefficient x_n [cf. Eq. (IV.12.21)]. These waves are known in the literature as *creeping waves*. Their properties will be discussed again in Section VI.5 in connection with the problem of scattering by a dielectric cylinder.

It is noteworthy that Eq. (IV.12.31) holds true for $|k\rho - \beta| \gg \beta^{1/3}$. To calculate the field closer to the cylinder, we can represent $H_{\nu_n}^{(2)}(k\rho)$ with the asymptotic formula of Eq. (IV.12.16). In addition, for β not very large, the zeros ν_n are given by Eqs. (IV.12.19) in conjunction with Eq. (IV.12.20). This situation arises when the size of the obstacle is comparable to the wavelength (for additional details see Section VI.5).

13 Field Patterns in the Neighborhood of a Focus

Imaging systems are designed with the aim of conveying a finite conical ray congruence, radiated by a point source placed on the object plane, toward a focal point on the image plane (see Chapter II). In most cases, the field relative to the region between the source and the exit pupil can be calculated by RO methods, that is, evaluating the trajectories of the rays propagating through the sequence of refracting surfaces, and then calculating the amplitude A along each ray. However, when we try to extend this approach to the image space, downstream from the exit pupil, we are faced with the unphysical result of a field vanishing abruptly across the shadow boundary surface formed by the envelope of the rays passing through the edge of the exit pupil. In order to eliminate this discontinuity connected with the use of ray optics, it is necessary to resort to the diffraction integral representation. In particular, the field on the exit aperture can be assumed to coincide with that existing in the absence of the aperture itself; this approximation, known as *Kirchhoff's principle*, is equivalent to the assumption that a finite exit pupil does not perturb the field on the pupil plane. Since presumably the actual perturbation is significant only near the pupil edge, we expect the error related to the application of Kirchhoff's principle to be negligible, provided the aperture is sufficiently large.

The exact analysis (see Chapter VI) of the effects produced by some simple apertures (e.g., half-plane, slit) confirms the validity of Kirchhoff's hypothesis for calculating the field near the shadow boundaries; the error becomes relevant only for field points in either the lit or the dark regions. Luckily, the field in these zones is represented with good accuracy by ray optics.

The above considerations explain why Kirchhoff's principle is generally accepted without criticism in the optical range, while many attempts have been made to go beyond it in radio-wave physics. In this frame, the weak field in dark regions must be evaluated correctly, for example, when one is concerned with the radiation on the back side of a reflector antenna. In other cases, one is interested is calculating the exact field on the aperture by deriving it in a self-consistent way from the expression for the field radiated from a generic point of the aperture itself. Whenever both RO and DT in the limit of Kirchhoff's principle give rise to unreliable results, an alternative approach is offered by the *geometric theory of diffraction* (GTD), which will be presented in later chapters.

The assumption of a scalar description of the field may be considered implicit in the above discussion. However, when the numerical aperture of the beam entering or leaving the lens is quite large, it is necessary to account for the vector character of **E** and **H**. This occurs, for example, in microscope imaging, where the aperture of the beam entering the objective can be very large. The analysis shows considerable departures from the behavior predicted by paraxial theory, even in the absence of aberrations. In particular, for rotationally symmetric lenses, the focal spot obtained when the entering beam is linearly polarized is not radially symmetric, a fact that affects the resolving power of the instrument. In addition, the electric field has both a transverse and an axial component.

Since the behavior in the focal region of an optical system of small numerical aperture is discussed in a number of textbooks (see, e.g., Chapter I,

Born and Wolf [11]), we will concentrate in the following discussion on the main features of the vector field, deriving the scalar results as a special case of the vector *Luneburg–Debye integral* [23, 24].

13.1 *Electromagnetic Point Sources*

We wish to mention some properties of point sources radiating vector fields. While a point source originating a scalar field is proportional to a three-dimensional delta function appearing as a driving term of the wave equation [see Eq. (IV.2.2)], in the vector case we must imagine the radiation as a suitable combination of electric and magnetic elementary multipoles. The simplest situation deals with the presence of an electric dipole **p** and a magnetic dipole **m** located in $\mathbf{r}_S \equiv (x_0, y_0, z_0)$. When the source is located in a homogeneous medium the field radiated by **p** and **m** is given by (see Chapter I, Papas [17], pp. 90–93),

$$\mathbf{E}(\mathbf{r}) = \frac{1}{\varepsilon}[(\mathbf{p}\cdot\nabla)\nabla G + k^2\mathbf{p}G] - i\omega\mu_0\mathbf{m}\times\nabla G \underset{kR\to\infty}{\to} \frac{e^{-ikR}}{R}\mathbf{E}'(\hat{n}), \tag{IV.13.1}$$

where the *constant-amplitude* field $\mathbf{E}'$ [see Eq. (II.8.3)] is

$$\mathbf{E}'(\hat{n}) = (-\omega^2\mu_0/4\pi)[\hat{n}\times(\hat{n}\times\mathbf{p}) + (n/c)\mathbf{m}\times\hat{n}], \tag{IV.13.2}$$

with $\hat{n} = (\mathbf{r} - \mathbf{r}_S)/|\mathbf{r} - \mathbf{r}_S| \equiv \mathbf{R}/R$ and $n = \sqrt{\varepsilon/\varepsilon_0}$. The Green's function $G(\mathbf{r}, \mathbf{r}_S) = G(R)$ coincides with that given in Eq. (IV.2.7) for $\mathbf{r}' = \mathbf{r}_S$.

In many applications, point sources are obtained by illuminating a small circular aperture (*pinhole*) of diameter about 10–100 μm, placed on the focal plane of a laser beam focused by a microscope objective (see Fig. IV.16). In this case, if we neglect the edge contribution of Kottler's diffraction formula [Eqs. (IV.3.4,5)], the aperture can be approximately compared to a pair of electric and magnetic dipoles proportional, respectively, to the electric and magnetic fields integrated over the aperture itself.

For optical systems used in connection with synchrotron radiation, the source is represented by an electronic current at a distance R from the observer, radiating a field whose Fourier transform is given (apart from an

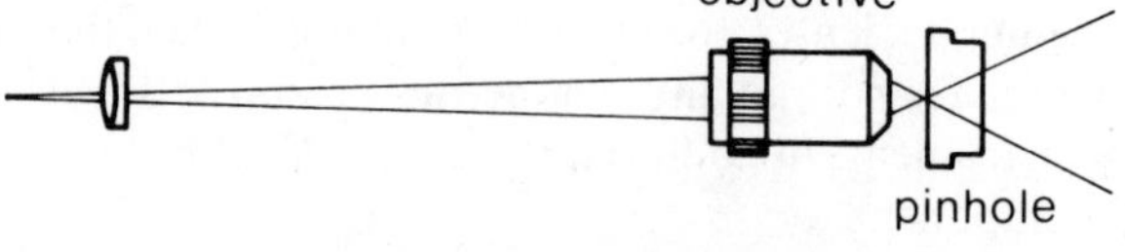

Fig. IV.16. Combination of divergent lens, microscope objective, and pinhole used in conjunction with a laser to obtain a spherical wave of large aperture.

inessential phase factor) by

$$\mathbf{E}(\omega) = \frac{-e\zeta_0}{8\pi^2 R}\int_{-\infty}^{+\infty} \exp\left\{-i\omega\left[t' - \hat{n}\cdot\frac{\mathbf{r}_e(t')}{c}\right]\right\}\frac{\hat{n}\times[(\hat{n}-\boldsymbol{\beta})\times\dot{\boldsymbol{\beta}}]}{(1-\boldsymbol{\beta}\cdot\hat{n})^2}\,dt' \tag{IV.13.3}$$

($\hat{n}$ being the direction from the small finite region, swept by the electron, to the field point), where $\mathbf{r}_e(t')$ represents the electron trajectory, $\boldsymbol{\beta} = (1/c)d\mathbf{r}_e/dt'$, $\dot{\boldsymbol{\beta}} = d\boldsymbol{\beta}/dt'$, and $-e\zeta_0$ represents the electron charge times the free-space impedance. In other cases, the source coincides with an electric dipole $d\mathbf{p}$ induced by a field $\mathbf{E}_i$ incident on an elementary volume dV of a medium with dielectric constant ε, that is,

$$d\mathbf{p}(\mathbf{r}') = [\varepsilon(\mathbf{r}') - \varepsilon_0]E_i(\mathbf{r}')\,dV, \tag{IV.13.4}$$

dV being centered around $\mathbf{r}'$. When the incident field is produced by a thermal source, as in the case of the condenser of a microscope illuminated by a lamp, we must account for the fluctuating phase of $\mathbf{E}_i$, which affects the phase of $\mathbf{p}$ and, ultimately, of the imaged field.

13.2 *Transport of Vector Fields from the Object Space to the Exit Pupil*

Let us now look for the most relevant transport properties of vector fields from the object to the image space, by considering, for simplicity, the case of a source placed on the optic axis of a centered system. If we use spherical coordinates r', θ', ϕ centered on the source $\mathbf{r}_S$ for the object space, with $\theta' = 0$ representing the optic axis pointing in the direction $z = +\infty$, and an analogous system r, θ, ϕ for the image space, with the optic axis pointing in the direction $z = -\infty$, a simple relation connects the field amplitudes in the object and image spaces, relative to a generic ray departing from the source along the direction θ', ϕ and arriving at the exit pupil along the direction θ, ϕ in the absence of reflection losses at the discontinuity surfaces met during its trajectory. This relation reads

$$|\mathbf{E}'(\theta,\phi)| = |\mathbf{E}'(\theta',\phi)|(d\Omega'/d\Omega)^{1/2} = |\mathbf{E}'(\theta',\phi)|[(\sin\theta'\,d\theta')/\sin\theta\,d\theta)]^{1/2}$$
$$\equiv |\mathbf{E}'(\theta',\phi)|g(\theta) \tag{IV.13.5}$$

(where $d\Omega = \sin\theta\,d\theta\,d\phi$ and $d\Omega' = \sin\theta'\,d\theta'\,d\phi$ are the solid angles of the ray pencil in the image and object spaces, respectively), and expresses the conservation of the power flow along the ray pencil.

For a lens designed well enough to satisfy *Abbe's sine condition* (see Chapter I, Born and Wolf [11], p. 168), we can put $\sin\theta'/\sin\theta = M$, where M

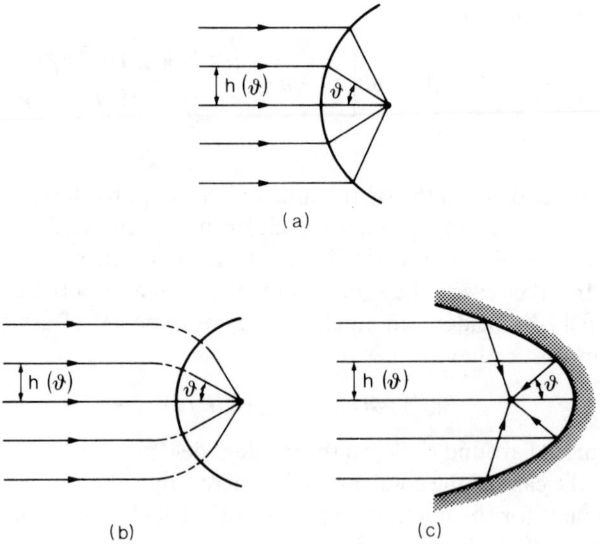

Fig. IV.17. Height $h(\theta)$ of a ray measured with respect to the optic axis versus the angle θ for (a) aplanatic, (b) uniform, and (c) parabolic focusing systems.

is the linear magnification, so that $g(\theta = M(\cos\theta/\cos\,\theta')^{1/2}$. If the source is located at infinity, Eq. (IV.3.5) becomes

$$|\mathbf{E}'(\theta,\phi)| = |\mathbf{E}(h\cos\phi, h\sin\phi)|[(h\,dh)/(\sin\theta\,d\theta)]^{1/2}$$
$$\equiv |\mathbf{E}(h\cos\phi, h\sin\phi)|\,fg_0(\theta), \qquad \text{(IV.13.6)}$$

where $h(\theta)$ is the height of the ray in the object space, and f is a constant that plays the role of a focal length. For a system satisfying the sine condition, $h/\sin\theta = f$, $g_0(\theta) = (\cos\theta)^{1/2}$. This approximation amounts to assuming that the incident plane wave is completely unaffected until it hits a spherical surface, which refracts it directly into a converging wave front having the same curvature radius as the sphere (see Fig. IV.17).

As another example, if we assume that the plane wave is gradually converted into a spherical wave in such a way that equal radial distances from the optic axis are converted into equal angles on the sphere, the so-called *uniform convergence projection*, then $h(\theta) = f\theta$ and $g_0(\theta) = (\theta/\sin\theta)^{1/2}$. This projection can represent more accurately than the aplanatic one the behavior of optical systems consisting of a series of thin lenses, the entire length of the optical system being long compared to the final focal length. For a plane wave focused by a parabolic mirror of focal length f, it can easily be shown that $h(\theta) = 2f(1 - \cos\theta)$, so that in this case $g_0(\theta) = 2/(1 + \cos\theta)$.

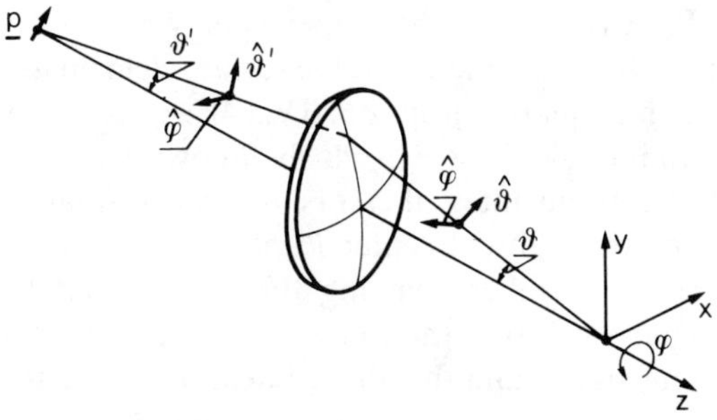

Fig. IV.18. Mutual orientation of the spherical coordinate systems relative to the source and the image formed by a lens.

Equation (IV.13.5) can be recast in vector form (see Fig. IV.18)

$$\mathbf{E}'(\theta, \phi) = g(\theta)(\hat{\phi}\hat{\phi} + \hat{\theta}\hat{\theta}') \cdot \mathbf{E}'(\theta', \phi), \tag{IV.13.7}$$

which reduces, for a source located at infinity ($R \to \infty$, $\hat{n} \to \hat{z}$), to

$$\mathbf{E}'(\theta, \phi) = fg_0(\theta)[\hat{\phi}\hat{\phi} + \hat{\theta}(\hat{z} \times \hat{\phi})] \cdot \mathbf{E}(h\cos\phi, h\sin\phi). \tag{IV.13.8}$$

When $\mathbf{E}'(\theta', \phi)$ is radiated by the dipoles $\mathbf{p}$ and $\mathbf{m}$ [see Eq. (IV.13.2)], Eq. (IV.13.7) reads *in vacuo*

$$\mathbf{E}'(\theta, \phi) = (\omega^2\mu_0/4\pi)g(\theta)\{[\mathbf{p} - \hat{n} \times (\mathbf{m}/c)] \cdot \hat{\phi}\hat{\phi} + (\mathbf{m}/c + \hat{n} \times \mathbf{p}) \cdot \hat{\phi}\hat{\theta}\}. \tag{IV.13.9}$$

Analogous considerations hold true for the magnetic field, which is approximately related to the electric vector by [see Eq. (II.8.12)] $\zeta_0 \mathbf{H}'(\theta, \phi) = \hat{s} \times \mathbf{E}'(\theta, \phi)$, $\hat{s}$ being the propagation direction of the image field.

13.3 *Luneburg–Debye Diffraction Integral*

Whenever **E** and **H** can be represented by RO fields the Kottler–Kirchhoff diffraction formula (IV.3.4) reduces to

$$\mathbf{E}(\mathbf{r}) \cong \frac{i}{\lambda}\iint_{\bar{A}} e^{-ik(R+S)}\mathbf{E}'(\hat{n}_0)\,d\Omega + \frac{i}{\lambda}\iint_{\bar{A}} e^{-ik(R+S)}(\hat{R} - \hat{n}_0) \times (\mathbf{E}' \times \hat{n}_0)\,d\Omega + \frac{\zeta_0}{4\pi}\oint_{\partial\bar{A}} \mathbf{H}'(\hat{n}_0) \cdot \hat{l}e^{-ik(R+S)}\frac{\hat{R}}{R^2}\,dl, \tag{IV.13.10}$$

where $\mathbf{E}'(\hat{n}_0) = \mathbf{E}_0 R$ is a vector that depends only on the ray direction $-\hat{n}_0$ and $d\Omega = d\bar{A}/R^2$ is the solid angle that the surface element $d\bar{A}$ subtends at the focus. In deriving this equation from Eq. (IV.3.4), we made use of the relations $\nabla' G = -ik\hat{R}G$ and $\mathbf{H}'\zeta = \mathbf{E}' \times \hat{n}_0$, which follows from Eq. (II.8.12). For $R/\lambda \to \infty$, the last contour integral can be neglected. Analogously, when the wave front passing through the center E' of the exit pupil $\bar{A}$ is slightly different from a reference sphere passing through E' and having its center in the gaussian image $\bar{x}, \bar{y}, 0$ of the source (see Section II.15.4), the difference $\hat{R} - \hat{n}_0$ becomes negligible and the above integral reduces to

$$\mathbf{E}(\mathbf{r}) = \frac{i}{\lambda} \iint_{\bar{A}} e^{-ik(R+S)} \mathbf{E}'(\hat{n}_0)\, d\Omega. \qquad \text{(IV.13.11)}$$

Let us now choose a cartesian system with the z axis parallel to the optic axis and the plane $z = 0$ coinciding with the gaussian image of the object plane $z = z_0$ (<0). The coordinates of the source are x_0, y_0, z_0. We can rely on Hamilton's mixed characteristic $W(x_0, y_0, z_0; p, q)$ relative to the source position x_0, y_0, z_0 and to the direction cosines p and q of the ray in the image space (see Chapter II) for determining the quantity $R + S$ appearing in the phase factor of Eq. (IV.3.11). If we indicate by ξ, η, ζ the coordinates of a wave front point Q and with x, y, z those of the field point P, we have (see Fig. IV.19)

$$\begin{aligned} S + R = S &- (p\xi + q\eta + r\zeta) + (p\xi + q\eta + r\zeta) \\ &+ [(\xi - x)^2 + (\eta - y)^2 + (\zeta - z)^2]^{1/2}, \end{aligned} \qquad \text{(IV.13.12)}$$

where $p = -n_{0x}, q = -n_{0y}$, and $r = -n_{0z}$ are the direction cosines of the ray passing through ξ, η, ζ. Since the wave front is generated by a point source in (x_0, y_0, z_0), the eikonal $S(\xi, \eta, \zeta)$ coincides with the point characteristic

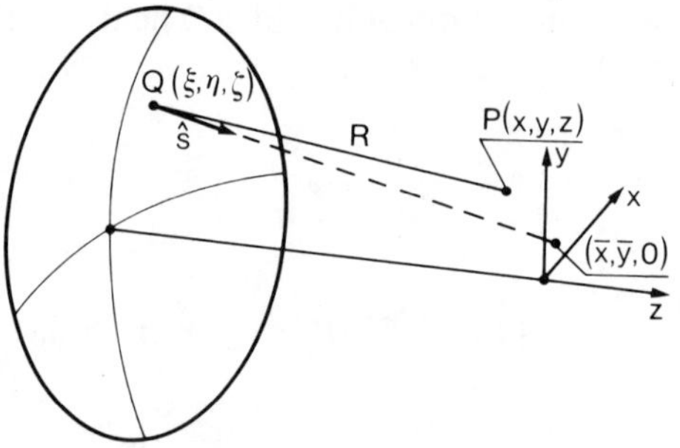

Fig. IV.19. Schematic representation of field point $P(x, y, z)$, wave front $Q(\xi, \eta, \zeta)$, and gaussian image $(\bar{x}, \bar{y}, 0)$.

$V(x_0, y_0, z_0; \xi, \eta, \zeta)$, so that

$$S - (p\xi + q\eta + r\zeta) = V(x_0, y_0, z_0; \xi, \eta, \zeta) - (p\xi + q\eta + r\zeta)$$
$$= W(z_0, z_1; x_0, y_0, p, q), \qquad \text{(IV.13.13)}$$

where W is independent of the particular choice of the wave front (we recall that W represents the optical length between the object point and the base point of the perpendicular dropped from the point $x = y = z = 0$ onto the ray).

For field points P (x, y, z) close to the gaussian image $(\bar{x}, \bar{y}, 0)$ of the source, we have (see Fig. IV.19)

$$[(\xi - x)^2 + (\eta - y)^2 + (\zeta - z)^2]^{1/2}$$
$$\cong [(\xi - \bar{x})^2 + (\eta - \bar{y})^2 + \zeta^2]^{1/2} + p(x - \bar{x}) + q(y - \bar{y}) + rz$$
$$+ \frac{1 - p^2}{2R}(x - \bar{x})^2 + \frac{1 - q^2}{2R}(y - \bar{y})^2 + \frac{1 - r^2}{2R}z^2$$
$$\cong p(x - \xi) + q(y - \eta) + r(z - \zeta)$$
$$+ \frac{1 - p^2}{2R}(x - \bar{x})^2 + \frac{1 - q^2}{2R}(y - \bar{y})^2 + \frac{1 - r^2}{2R}z^2, \qquad \text{(IV.13.14)}$$

which yields, with the help of Eqs. (IV.13.12) and (IV.13.13)

$$S(\xi, \eta, \zeta) + R(\xi, \eta, \zeta; x, y, z) = W(z_0, z_1; x_0, y_0, p, q) + px + qy + rz$$
$$+ \frac{1-p^2}{2R}(x - \bar{x})^2 + \frac{1-q^2}{2R}(y - \bar{y})^2 + \frac{1-r^2}{2R}z^2. \qquad \text{(IV.13.15)}$$

After introducing the aberration function W_0 [see Eq. (II.15.31)], we can write

$$W(z_0, z_1; x_0, y_0, p, q) = V(x_0, y_0, z_0; \bar{x}, \bar{y}, 0) - p\bar{x} - p\bar{y} + W_0(x_0, y_0, z_0; p, q, r), \qquad \text{(IV.13.16)}$$

which allows us to cast Eq. (IV.13.15) in the form

$$S + R = V_0 + p(x - \bar{x}) + q(y - \bar{y}) + rz + W_0$$
$$+ \frac{p^2 + q^2}{2R}z^2 + \frac{1 - p^2}{2R}(x - \bar{x})^2 + \frac{1 - q^2}{2R}(y - \bar{y})^2, \qquad \text{(IV.13.17)}$$

where $V_0 = V(x_0, y_0, z_0; \bar{x}, \bar{y}, 0)$.

It is worth noting that

$$\frac{k}{2R}[(1 - p^2)(x - \bar{x})^2 + (1 - q^2)(y - \bar{y})^2] \leqq \frac{\pi\rho_{\max}^2}{\lambda R} = \pi N_{\max}, \qquad \text{(IV.13.18)}$$

where ρ_{max} is the radius of the area in which the focal image is practically contained and N_{max} the relative *Fresnel number*, equal to *the number of Fresnel rings contained in the circle* $\pi\rho^2_{max}$ *seen from the exit pupil.* Since it can be shown that N_{max} is of the order of the reciprocal of the Fresnel number N_e relative to the exit pupil seen from the focus, the term on the left in Eq. (IV.13.18) can be neglected when evaluating the exponent of Eq. (IV.13.11), provided $N_e \gg 1$. Analogously, we have $k(p^2 + q^2)z^2/(2R) < \bar{u}^2_{max}/N_e$, where $\bar{u}_{max} = k|z_{max}|\mathrm{NA}^2$ [see Eq. (IV.13.21)], and $|z_{max}|$ is the maximum distance of interest from the focus. Since $\bar{u}_{max}$ is of the order of 10^2, the above term is also negligible. However, when N_e becomes comparable to unity, the presence of the term in z^2 induces a displacement of the focal point with respect to the geometric one [see Eq. (IV.13.37)]. For a field point so close to the gaussian image as to make negligible the last three terms of Eq. (IV.13.17), Eq. (IV.13.11) reduces to the *Luneburg–Debye* integral

$$\mathbf{E}(\mathbf{r}) = \frac{ie^{-ikV_0}}{\lambda}\iint_{\bar{A}} \mathbf{E}'(\hat{n}_0)\exp[-ikW_0 - ik(p(x-\bar{x}) + q(y-\bar{y}) + rz)]\,d\Omega, \tag{IV.13.19}$$

the equivalent expression for $\mathbf{H}'$ being obtained by means of the substitution $\mathbf{E}' \to \mathbf{E}' \times \hat{n}_0/\zeta_0$.

If the field point is sufficiently close to the coordinate origin, we can replace the solid angle $d\Omega$ subtended by the wave front element at the focus $(\bar{x}, \bar{y}, 0)$ with that subtended at the origin $(0, 0, 0)$. In this case, if we use a coordinate system such that the direction $z = -\infty$ corresponds to $\theta = 0$, the above integral reads

$$\mathbf{E}(\mathbf{r}) = \frac{ie^{-ikV_0}}{\lambda}\iint_{2\pi} P(\theta,\phi)\mathbf{E}'(\theta,\phi)\exp[-ikW_0 + ik\rho\sin(\theta)\cos(\phi-\psi)] \times \exp[-ik\cos(\theta)z]\sin(\theta)\,d\theta\,d\phi, \tag{IV.13.20}$$

where the pupil function P coincides with unity on the aperture and vanishes outside (see Section IV.5) and $x - \bar{x} = \rho\cos\psi$, $y - \bar{y} = \rho\sin\psi$, $p = -\sin\theta\cos\phi$, $q = -\sin\theta\sin\phi$, r (direction cosine, not spherical coordinate) $= \cos\theta$.

It is now useful to introduce the *optical coordinates* v, $\bar{u}$ defined by [see Eq. (II.15.28)]

$$v = k\rho NA, \qquad \bar{u} = kzNA^2, \tag{IV.13.21}$$

where $\mathrm{NA} = \sin\theta_{max}$ represents the *numerical aperture*, θ_{max} indicating the half-aperture on the image space (in Chapter VIII the coordinate v will be shown to play an important role in the description of electromagnetic propagation in fibers; in that context it is known as the *normalized frequency* and indicated with V). In this way, the Luneburg–Debye integral can be rewritten in a form independent of z and ρ, i.e.,

$$\mathbf{E}(\mathbf{r}) = \frac{i\exp(-ikV_0)}{\lambda}\iint_{2\pi} P(\theta,\phi)\mathbf{E}'(\theta,\phi)$$

$$\times \exp\left[-ikW_0 + iv\frac{\sin(\theta)\cos(\phi-\psi)}{NA} - i\bar{u}\frac{\cos(\theta)}{NA^2}\right]\sin(\theta)\,d\theta\,d\phi, \tag{IV.13.22}$$

which, for NA sufficiently small, reduces to

$$\mathbf{E}(\mathbf{r}) = \frac{i\exp(-ikV_0 - i\bar{u}/NA^2)}{\lambda}NA^2\iint P(\Theta,\phi)\mathbf{E}'(\Theta,\phi)$$

$$\times \exp\left[-ikW_0 + iv\Theta\cos(\phi-\psi) + i\bar{u}\frac{\Theta^2}{2}\right]\Theta\,d\Theta\,d\phi,$$

$$P(\Theta,\phi) = \begin{cases} 0 & \text{for} \quad \Theta > 1, \\ 1 & \text{for} \quad \Theta < 1, \end{cases} \tag{IV.13.23}$$

where $\Theta \equiv \theta/\text{NA}$. For an aberration-free lens ($W_0 = 0$), the field on the focal plane $\Pi(\bar{u} = 0)$ can be rewritten as

$$\mathbf{E}(x,y,0) = \frac{ie^{-ikV_0}}{\lambda f_1^2}\iint_{-\infty}^{\infty} P(x',y')\mathbf{E}'(x',y',-f_1)\exp\left(ik\frac{xx'+yy'}{f_1}\right)dx'\,dy'$$

$$= \frac{ie^{-ikV_0}}{\lambda f_1^2}F\left[P(x',y')\mathbf{E}'(x',y',-f_1);\,k\frac{x}{f_1},k\frac{y}{f_1}\right], \tag{IV.13.24}$$

which expresses the field on the focal plane in the form of a Fourier transform. We observe that Eq. (IV.13.24) readily follows from Eq. (IV.13.22) if we use the relation $\sin(\theta)\,d\theta\,d\phi \cong dx'\,dy'/f_1^2$, where the focal distance f_1 coincides with the distance between exit pupil and focus.

In the above paraxial limit, the field $\mathbf{E}'_2$ on the entrance plane of a second lens whose front focal plane coincides with Π itself (see Fig. IV.20) is in turn a Fourier transform of the distribution on Π, so as to reproduce the distribution $\mathbf{E}'$ on the exit pupil of the first lens. However, if the field on Π is multiplied by a suitable function $t(x, y)$, inserting on Π a transparency that modifies amplitude and phase distributions, the similarity of $\mathbf{E}'_2$ and $\mathbf{E}'$ is lost. More precisely, $\mathbf{E}'_2$

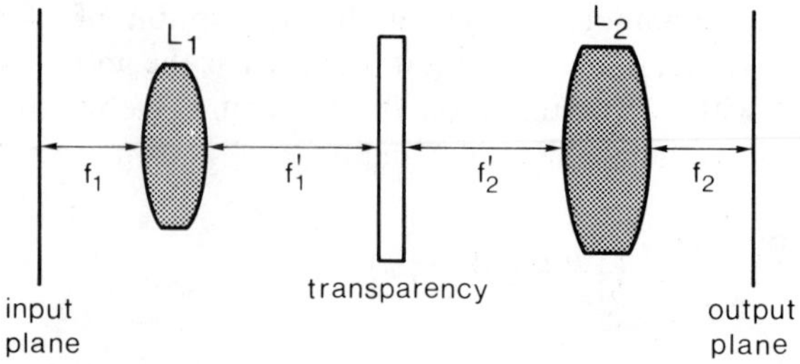

Fig. IV.20. Schematic diagram suggesting the inclusion of a transparency in an afocal system. The image of a point x', y' on the input plane is described by the function $K(x/M - x', y/M - y')$ [see Eq. (IV.13.25)]. For this configuration, $M = f_2/f_1$.

is proportional to the convolution of $\mathbf{E}'$ with a Fourier transform of t and reads

$$\mathbf{E}_2'(x'', y'') = \iint_{-\infty}^{+\infty} K\left(\frac{x''}{M} - x', \frac{y''}{M} - y'\right)\mathbf{E}'(x', y')\, dx'\, dy', \tag{IV.13.25}$$

where $M = f_2/f_1$ (f_2 being the focal distance of the second lens) (Fig. IV.20) is the magnification factor (see Chapter II) of the afocal system, and

$$K(\xi, \eta) = F[t(x, y); k\xi/f_2, k\eta/f_2] \tag{IV.13.26}$$

Therefore, each transparency corresponds to a particular linear integral transform of the field $\mathbf{E}'$. This property is used to transform fields $\mathbf{E}'$ of uniform intensity and nonuniform phase into fields $\mathbf{E}''$ with nonuniform intensity, thus visualizing phase variations, as originally suggested by Zernike (phase-contrast method, Schlieren, shadowgraph method). Other methods for improving the quality of images and for optical correlation applications have reached a high degree of development, giving rise to the new branch of coherent optics [25, 26]. (For additional details see Section IV.15.)

13.4 *Focusing of Linearly Polarized Plane Waves*

We wish to apply some of the above results to evaluate the behavior of the field $\mathbf{E}(\mathbf{r})$ present in the focal region of an aplanatic lens (see Chapter II) in connection with an electromagnetic source emitting a linearly polarized field $\mathbf{E}_i$ [24, 27, 28]. To this end, we insert the right side of Eq. (IV.13.8) into Eq. (IV.13.22), thus obtaining for a circular pupil with half-aperture $\theta_{\max}$

$$\begin{aligned}\mathbf{E}(\mathbf{r}) = \frac{if}{\lambda}\int_0^{2\pi} d\phi \int_0^{\theta_{\max}} \sin\theta\, g_0(\theta) \\ \times \exp\left[-ikW_0 + iv\frac{\sin\theta\cos(\phi - \psi)}{NA} - i\bar{u}\frac{\cos\theta}{NA^2}\right] \\ \times [\hat{\phi}\hat{\phi} + \hat{\theta}(\hat{z} \times \hat{\phi})] \cdot \mathbf{E}_i\, d\theta,\end{aligned} \tag{IV.13.27}$$

having omitted the inessential phase factor $\exp(-ikV_0)$. For $\mathbf{E}_i$ parallel to the x axis, we have

$$[\hat{\phi}\hat{\phi} + \hat{\theta}(\hat{z} \times \hat{\phi})] \cdot \mathbf{E}_i = \{[\cos\theta + (1 - \cos\theta)\sin^2\phi]\hat{x} + (1 - \cos\theta)\sin\phi\cos\phi\,\hat{y} + \sin\theta\cos\phi\,\hat{z}\}E_i. \tag{IV.13.28}$$

After inserting Eq. (IV.13.28) into Eq. (IV.13.27) and assuming that the incident field amplitude depends only on the height h, i.e., $E_i = E_i(f\sin\theta)$, we can show that the field $\mathbf{E}$ obeys the relation

$$\mathbf{E} = iA\{[I_0 + I_2\cos(2\psi)]\hat{x} + I_2\sin(2\psi)\hat{y} + 2iI_1\cos\psi\,\hat{z}\}, \tag{IV.13.29}$$

where $A = \pi f E_i(0)/\lambda$ and

$$I_i(\bar{u}, v; \theta_{max}) = \int_0^{\theta_{max}} g_0(\theta)f(\theta)b_i(\theta)J_i\left(\frac{v\sin\theta}{\sin\theta_{max}}\right)\exp\left(-i\bar{u}\frac{\cos\theta}{\sin^2\theta_{max}} - ikW_0\right)d\theta, \tag{IV.13.30}$$

with $f(\theta) = E_i(\theta)/E_i(0)$, $b_0 = \sin\theta(1+\cos\theta)$, $b_1 = \sin^2\theta$, $b_2 = \sin\theta(1-\cos\theta)$, and W_0 supposed to depend on θ only. In particular, for $u = 0$, that is, on the focal plane $z = 0$, and $W_0 = 0$ (aberration-free lens), the three integrals I_0, I_1, and I_2 are real and E_z is 90° out of phase with respect to E_x and E_y. Therefore, the field is elliptically polarized on a plane perpendicular to the focal one. In addition, along the y axis ($\psi = \pi/2$) (i.e., perpendicular to the polarization direction of the incident beam) the field on the focal plane is linearly polarized.

On the axis of the system, $v = 0$, and as a consequence $I_1 = I_2 = 0$, due to the vanishing of $J_1(0)$ and $J_2(0)$. Thus, the electric vector at each point on the revolution axis in the image space is linearly polarized in the same direction as the incident beam.

The electric w_e and magnetic w_m time-averaged energy densities corresponding to Eqs. (IV.13.29) read

$$w_e(\bar{u}, v, \psi) = (\varepsilon_0/4)|A|^2[|I_0|^2 + 4|I_1|^2\cos^2\psi + |I_2|^2 + 2\cos(2\psi)\,\mathrm{Re}(I_0 I_2^*)] = w_m(\bar{u}, v, \psi + \pi/2). \tag{IV.13.31}$$

Boivin *et al.* [27] calculated I_0, I_1, and I_2 for $f(\theta) = 1$ (uniformly illuminated exit pupil), $g_0(\theta) = (\sin\theta)^{1/2}$ (aplanatic lens), and $\theta_{max} = 45°$. They found that the electric energy distribution on the focal plane is not rotationally symmetrical. The energy contours for $v > 4$ are approximately elliptical, with their major axes in the direction of the electric vector of the incident wave. The energy density vanishes only along the meridional line $\psi = \pi/2$ (y axis), while it presents a succession of minima different from zero along $\psi = 0$ (x axis).

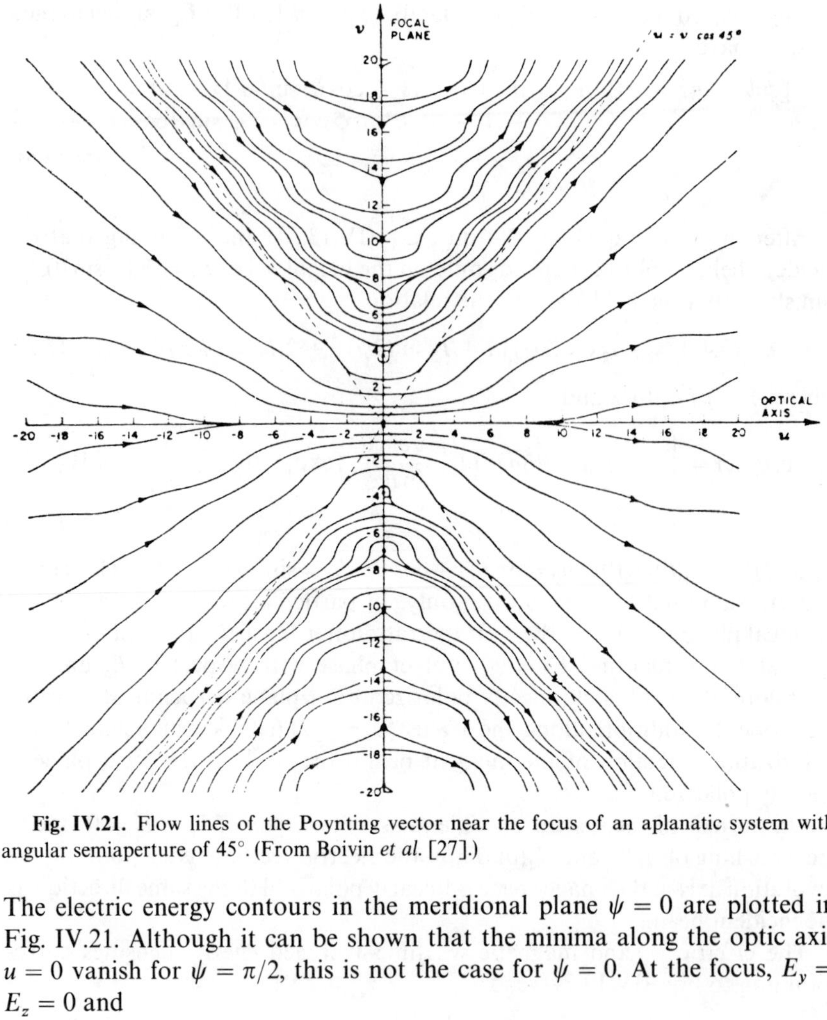

Fig. IV.21. Flow lines of the Poynting vector near the focus of an aplanatic system with angular semiaperture of 45°. (From Boivin *et al.* [27].)

The electric energy contours in the meridional plane $\psi = 0$ are plotted in Fig. IV.21. Although it can be shown that the minima along the optic axis $u = 0$ vanish for $\psi = \pi/2$, this is not the case for $\psi = 0$. At the focus, $E_y = E_z = 0$ and

$$E_x^{\max} = iAI_0(0, 0; \theta_{\max}) = (i\pi f/\lambda)E_i\left\{\frac{2}{3}[1-(\cos\theta_{\max})^{3/2}] + \frac{2}{5}[1-(\cos\theta_{\max})^{5/2}]\right\} \tag{IV.13.32}$$

As a further result, E_z attains a maximum value on the focal plane (for $\theta_{\max} = 45°$) of about $0.28E_x^{\max}$ at the points $v \cong 2.25$ (about a half-wavelength from the axis) in the azimuth 0, π. This means that, for large apertures, the longitudinal components attain values comparable with the transverse ones. In Fig. IV.21 the flow lines of the time-averaged Poynting vector (see Section I.8) for the illumination conditions considered above are represented. In particular, close to the focal plane the Poynting vector streamlines present a series of vortices with centers on the focal plane itself.

13.5 *Focusing of Small-Aperture Beams*

For θ_{max} sufficiently small ($\theta_{max} \leqq 5°$), the integrals I_1 and I_2 become negligible, so that the field $\mathbf{E} \cong iAI_0\hat{x}$ is linearly polarized and the scalar theory emerges as an accurate approximation. More precisely, if for completeness we remove the hypothesis of independence of W_0 from ϕ, Eq. (IV.13.29) holds true with $I_1 = I_2 = 0$ and I_0 replaced by

$$I(\bar{u}, v, \psi; \theta_{max}) = \frac{\exp(-i\bar{u}/\theta_{max}^2)}{\pi}\theta_{max}^2 \int_0^{2\pi} d\phi \times \int_0^1 \Theta f(\Theta, \phi) \exp\left[iv\Theta\cos(\phi - \psi) - ikW_0 + i\bar{u}\frac{\Theta^2}{2}\right] d\Theta \tag{IV.13.33}$$

where $\Theta \equiv \theta/\theta_{max}$, $f(\Theta, \phi) \equiv E_i(\Theta, \phi)/E_i(0, 0)$, allowing for the possible lack of rotational symmetry of the incident field amplitude. A relevant result is that I turns out to be independent of ψ if f and W_0 do not depend on ϕ, which means that an initial rotational symmetry of the field is preserved in the focal region; this is *not* the case for larger values of θ_{max}, as implied by the presence of the angular factors multiplying I_1 and I_2 in Eq. (IV.13.29).

For small Fresnel numbers ($N_e \leqq 1$), Eq. (IV.13.33) loses its validity. Li and Wolf (1984) have shown that a rotationally symmetrical field can be still represented by an integral similar to Eq. (IV.13.33) by replacing the optical coordinates $\bar{u}$ and v with the new quantities $\bar{u}_N = \bar{u}/\{1 + \bar{u}/(2\pi N_e)\}$, $v_N = v/\{1 + \bar{u}/(2\pi N_e)\}$.

As we saw in Section II.15, for small circular apertures the aberration function reduces to the *Seidel primary aberrations*, that is,

$$kW_0 = kW^{(4)}(\theta, \phi) = \frac{BR_4^{(0)}(\theta)}{24} + y_0^2\frac{CR_2^{(2)}(\theta)}{2}\cos(2\phi) - y_0^2\frac{DR_2^{(0)}(\theta)}{4} + y_0^3 ER_1^{(1)}(\theta)\cos\phi + y_0\frac{FR_3^{(1)}(\theta)}{3}\cos\phi, \tag{IV.13.34}$$

where we have oriented the x, y axes so that $x_0 = 0$. Here the functions $R_n^{(m)}(\theta)$ are the Zernike polynomials (see Chapter I, Born and Wolf [11]) normalized in such a way that $R_n^{(m)}(1) = 1$. As a consequence, the coefficients B, C, D, E, and F measure the phase deviation of the marginal rays with respect to the principal ray. In particular, the above five coefficients refer in order to the *spherical aberration*, the *astigmatism*, the *field curvature*, the *distortion*, and the *coma*.

The integral $I(\bar{u}, v, \psi, \theta_{max})$ has been calculated by Nijboer for different types of small primary aberrations. For spherical aberrations, the focal distribution of a two-dimensional field can be expressed by means of a function studied by Pearcey (Section V.5). The reader is referred to Chapter IX

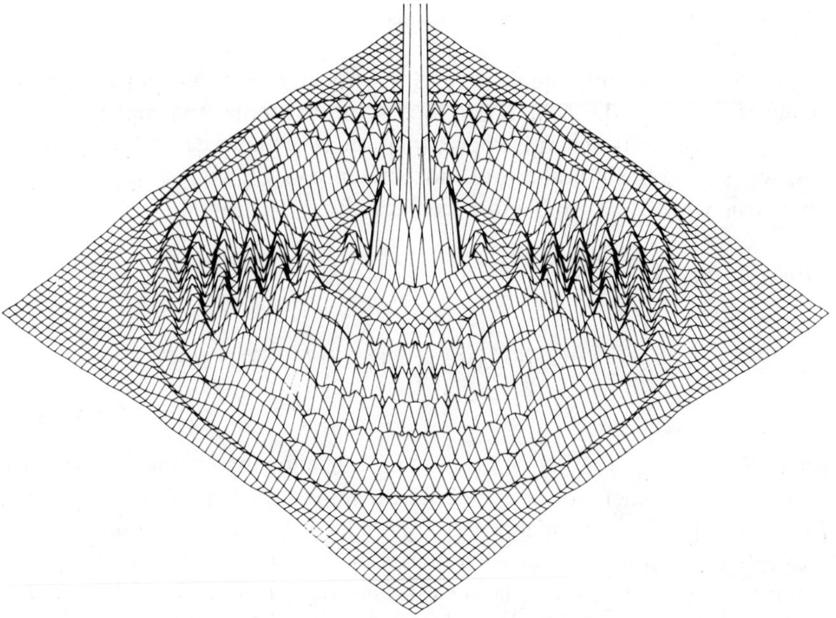

Fig. IV.22. Diffraction pattern of a ring-shaped aperture with internal diameter 0.8 times the external one, illuminated by a plane wave, at a distance corresponding to a Fresnel number of 15. The three-dimensional plot was obtained by using an improved fast Fourier transform algorithm (From Luchini [28a].) © North-Holland, Amsterdam, 1984.

of Born and Wolf for an accurate description of the analytical methods and the special functions used to calculate the integral. Here we wish to stress that I can be evaluated under some conditions with the asymptotic formulas discussed in Sections V.2 and V.8. Figure IV.22 shows the intensity distributions in proximity to the focal plane computed by means of the FFT algorithm.

13.6 *Field Distribution along the Optic Axis for an Aperture-Limited Gaussian Beam*

On the optic axis $\rho = v = 0$, for a gaussian illumination $f(\theta) = \exp[-\theta^2/(2\delta^2)]$ truncated at $\theta = \theta_{\max}$ and $W_0 = 0$, it can be shown that

$$I(\bar{u}, 0; \theta_{\max}) = |I| \exp[-i\bar{u}/\theta_{\max}^2 - i\Phi(\bar{u}) - i\pi/2] \qquad \text{(IV.13.35a)}$$

where

$$|I|^2 = 4\theta_{\max}^4 \frac{\exp(-\theta_{\max}^2/\delta^2) + 1 - 2\cos(\bar{u}/2)\exp[-\theta_{\max}^2/(2\delta^2)]}{\theta_{\max}^4/\delta^4 + \bar{u}^2} \qquad \text{(IV.13.35b)}$$

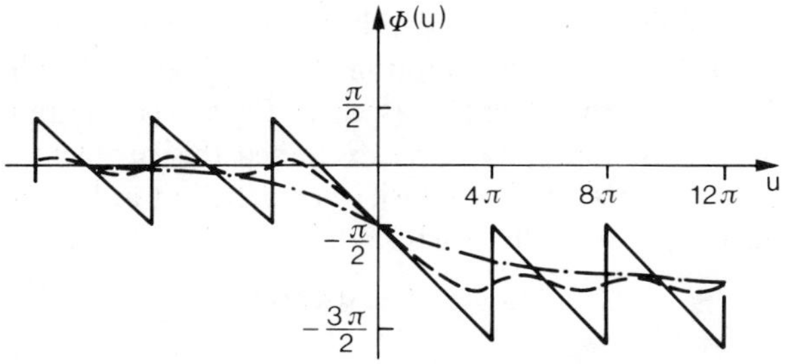

Fig. IV.23. Phase anomaly in proximity to the focus for a gaussian beam. (———) Uniform wave; (-----) $\theta_m^2 = 2\delta^2$; (— - — - —) $\theta_m^2 = 6\delta^2$, where θ_m represents the beam aperture delimited by the circular aperture, while $\delta 2^{1/2}$ is the effective aperture of the Gaussian beam.

and

$$\Phi(\bar{u}) = -\pi/2 + \arctan\left[\frac{\sin(\bar{u}/2)}{\exp[\theta_{\max}^2/(2\delta^2)] - \cos(\bar{u}/2)}\right] - \arctan(\bar{u}\delta^2/\theta_{\max}^2). \tag{IV.13.35c}$$

The behavior of the phase factor $\Phi(\bar{u})$ close to the focal point $\bar{u} = 0$ is connected with the Gouy phase anomaly of the field, which was mentioned at the end of Section II.9. We showed in Fig. IV.23 the function $\Phi(u)$ for different values of the parameter $\theta_{\max}^2/\delta^2$. For $\delta \gg \theta_{\max}$ (uniform illumination), we observe a series of sawtooth oscillations, of amplitude $\pi/2$, about the average values of 0 ($u < 0$) and $-\pi$ ($u > 0$). In the opposite situation of $\theta_{\max} \gg \delta$ the oscillations tend to disappear and the transition becomes smooth. This leads us to consider the phase oscillations as due to the aperture edge illumination. In general, they become negligible when the field on the exit pupil is sufficiently tapered. In any case, the increment of π undergone by $-\Phi(\bar{u})$ across the focus corresponds to the jump of π obtained in Section II.9 in the RO frame.

The intensity $\propto |I|^2$ given by Eq. (IV.13.35b) reduces to $|I(\bar{u} = 0)|^2[\sin(\bar{u}/4)/(\bar{u}/4)]^2$ for $\delta/\theta_{\max} \to \infty$ (uniform illumination), and to $|I(\bar{u} = 0)|^2/(1 + \bar{u}^2\delta^4/\theta_{\max}^4)$ for $\delta/\theta_{\max} \to 0$ (aperture large with respect to the gaussian beam spot). Therefore, if we regard a loss of 20% as the intensity tolerance for sharp focusing, the *normalized focal depth* $\bar{u}_{\max}$ is $\cong 3.2$ for the uniform case and $\cong 0.5\theta_{\max}^2/\delta^2$ for a gaussian illumination slightly obstructed by the exit pupil. If we indicate with $N_e = a^2/(\lambda f)$ the Fresnel number of the circular aperture of radius a seen from the focus, we have, for a uniformly illuminated pupil, a focal depth $|z_{\max}|$ given by

$$|z_{\max}| = \bar{u}_{\max}/(k\text{NA}^2) = \bar{u}_{\max}\lambda/(2\pi\sin^2\theta_{\max})$$
$$= \bar{u}_{\max}\lambda/(2\pi a^2/f^2) \cong 0.51\ f/N_a. \tag{IV.13.36}$$

When N_e is small, the focal depth becomes comparable to f, thus contradicting our tacit initial assumption of a focal region localized far away from the wave front. Li and Wolf [29] (see also Erkkila (1981) and Li and Wolf (1984) cited in the bibliography) have examined this problem carefully and have shown that the maximum intensity is reached at a distance z from the gaussian image, approximately equal to

$$z = -f/(1 + 0.82N_e^2). \tag{IV.13.37}$$

13.7 *Field Distribution on the Focal Plane for an Aperture-Limited Gaussian Beam*

For an aberration-free optical system ($W_0 = 0$) and an illumination truncated at $\theta = \theta_{max}$, under the assumption of small θ_{max} underlying the above results, the image on the focal plane ($\bar{u} = 0$) is given by

$$I(0, v; \theta_{max}) = 2\theta_{max}^2 \int_0^1 f(\Theta) J_0(v\Theta)\Theta \, d\Theta \qquad \text{(independent of } \psi\text{)}, \tag{IV.13.38}$$

which is easily obtained from Eq. (IV.13.30). For a gaussian illumination (see Federov *et al.* (1984) cited in bibliography)

$$f(\Theta) = \exp[-\Theta^2\theta_{max}^2/(2\delta^2)],$$

we can use the expansion (10.2.37) of Chapter II, Abramowitz and Stegun [4] to write

$$\begin{aligned}
\exp\left(\frac{-\Theta^2\theta_{max}^2}{2\delta^2}\right) &= \exp\left(\frac{-\theta_{max}^2}{4\delta^2}\right)\exp\left[-\left(\frac{\theta_{max}^2}{4\delta^2}\right)(2\Theta^2 - 1)\right] \\
&= \exp\left(\frac{-\theta_{max}^2}{4\delta^2}\right)\left(\frac{2\pi\delta^2}{\theta_{max}^2}\right)^{1/2} \\
&\quad \times \sum_{n=0}^{\infty} (2n + 1)(-1)^n I_{n+1/2}\left[\frac{\theta_{max}^2}{4\delta^2}\right] R_{2n}^{(0)}(\Theta),
\end{aligned} \tag{IV.13.39}$$

where $I_{n+1/2}$ are the modified Bessel functions of order $n + 1/2$, and we have made use of the equality $R_{2n}^{(0)}(\Theta) = P_n(2\Theta^2 - 1)$ between Zernike and Legendre polynomials. If we introduce Eq. (IV.13.39) into Eq. (IV.13.38), we obtain, with the help of Problem 27,

$$\frac{I(0, v; \theta_{max})}{I(0, 0; \theta_{max})} \equiv K(v) = 2 \sum_{n=0}^{\infty} (2n + 1) \frac{I_{n+1/2}[\theta_{max}^2/(4\delta^2)]}{I_{1/2}[\theta_{max}^2/(4\delta^2)]} \frac{J_{2n+1}(v)}{v}, \tag{IV.13.40}$$

where $K(v)$ represents the *impulse response*, which will be discussed in Section IV.15. In particular, for $\theta_{max}/\delta = 0$, we have

$$K(v) = 2J_1(v)/v, \tag{IV.13.41}$$

whose square gives the well-known *Airy pattern*. Therefore, the intensity distribution on the focal plane relative to a perfect optical system uniformly enlightened presents a central spot with a normalized radius $v_0 = 3.8$, coincident with the first zero of $J_1(v)$, and containing $\cong 81\%$ of the total focused power. This spot is called the *Airy disk*.

13.8 *Resolving Power*

If we consider the diffraction images of two plane waves forming an angle θ, the normalized distance V between the centers of the two relative Airy disks can be shown to fulfill, in the paraxial limit, the relation $V = \pi D\theta/\lambda_0$, D representing the diameter of the exit pupil. It is customary to assume as a resolution limit the separation v at which the center of one Airy disk falls on the first dark ring of the other (*Rayleigh's criterion of resolution*). This gives for the angular resolution

$$\theta_{min} \cong 1.22\lambda/D. \tag{IV.13.42}$$

[It is worth noting that a generic optical coordinate v is given by $v = 2\pi(N_e N_v)^{1/2}$, where N_e is the Fresnel number of the exit pupil defined in Section IV.13.6 and N_v is that relative to a disk on the focal plane of normalized radius v seen from the exit pupil. Accordingly, the Fresnel number of the Airy disk is approximately furnished by $N_{Airy} = 0.36/N_e$.] An analogous expression can be derived for a microscope by adapting the above procedure to the integral representation [Eq. (IV.15.6)] of the image of an object at finite distance provided by an isoplanatic optical system. By leaving the derivation to the reader, who can consult several textbooks, we give the expression of the resolution limit on the object plane of a microscope

$$d_{min} = 0.61\lambda_0/(n \sin\alpha), \tag{IV.13.43}$$

where α is the semiangle of the cone of rays collected by the lens from the axial object point, λ_0 the wavelength *in vacuo*, and n the refractive index of the medium in which the objective is immersed (air or oil).

Implicit in the Rayleigh criterion is the assumption that the intensity pattern of the image of two points is the superposition of the two respective intensity patterns. This implies that the phases of the two objects are supposed to be uncorrelated. While for telescopes this assumption is generally satisfied, the same is in generally not true in microscopy. This problem has been thoroughly analyzed by taking into account the partial coherence of the object illumination, which has led to different resolution criteria [30].

14 Reduction of Diffraction Integrals to Line Integrals

Whenever the field is limited by an aperture, it can be useful, both for computational reasons and for obtaining an intuitive picture of the diffracted wave, to substitute for the (two-dimensional) diffraction integral a suitable line integral over the contour delimiting the aperture. To this end, we can exploit the properties of the Helmholtz vector [Eq. (IV.2.11)], considered as a function of $\mathbf{r}'$ for an assigned determination of the field and a given field point $\mathbf{r}$,

$$\mathbf{v}(\mathbf{r}') = G(\mathbf{r}',\mathbf{r})\,\nabla' u(\mathbf{r}') - u(\mathbf{r}')\,\nabla' G(\mathbf{r}',\mathbf{r}), \tag{IV.14.1}$$

which satisfies the relation $\nabla' \cdot \mathbf{v} = 0$ [Eq. (IV.2.12)] in a domain that does not contain $\mathbf{r}$ and the singularities (sources) of the field. This allows us to introduce a *vector potential* $\mathbf{w}(\mathbf{r}')$ such that

$$\mathbf{v}(\mathbf{r}') = \nabla' \times \mathbf{w}(\mathbf{r}') + \nabla' f(\mathbf{r}'), \tag{IV.14.2}$$

where f is a function satisfying the Poisson equation $\nabla^2 f(\mathbf{r}') = 0$ in a domain in which $\mathbf{v}$ is regular, so that the term in f accounts for the singularities of $\mathbf{v}$.

Of course, $\mathbf{w}$ is not uniquely defined. In fact, if we add to a given $\mathbf{w}$ the gradient of a generic scalar function $h(\mathbf{r}')$, the curl of the new potential $\mathbf{w}' = \mathbf{w} + \nabla' h$ is still equal to $\nabla' \times \mathbf{w}$. As will become clear in the following, this ambiguity is not prejudicial to using $\mathbf{w}$ to represent the field diffracted by an aperture. The diffraction integral [Eq. (IV.2.10)] can now be rewritten by using Eq. (IV.14.2) in the form

$$u(\mathbf{r}) = \iint_A \mathbf{v}(\mathbf{r}') \cdot \hat{n}_0\, dA = \iint_A [\nabla' \times \mathbf{w}(\mathbf{r}')] \cdot \hat{n}_0\, dA = \oint_{\partial A} \mathbf{w}(\mathbf{r}') \cdot d\mathbf{l} + u'(\mathbf{r}), \tag{IV.14.3}$$

where we have supposed a field vanishing on a given surface, apart from the aperture A. In Eq. (IV.14.3), ∂A denotes the boundary of A oriented in such a way that a point moving in the positive direction appears to move in the counterclockwise direction when observed from the outward normal (with respect to $\mathbf{r}$) $\hat{n}_0$, and $u'(\mathbf{r})$ depends on the singularities of $w(\mathbf{r}')$ on A. Thus, the field $u(\mathbf{r})$ can be considered as the superposition of a wave originating from the edge of the aperture and a wave that depends on the interior field only.

14.1 *Historical Digression*

The line integral representation of the field suggests an intuitive representation of the effects due to the edges of an aperture. As we know from experience, the edges of an illuminated aperture shine when observed from the shadow region. This fact was already analyzed by Newton, who explained it in terms of repulsion of light corpuscles by the edges ("Opticks," Book 3, observation I, Figs. 1 and 2). Later, Thomas Young formulated a wave theory according to which a diffracted wave is formed by reflection of the incident wave on the line elements of the diffracting edge. On the other hand, Fresnel explained the diffraction effect on the basis of the Huygens principle; if the field point is so far away from the geometric shadow that a large number of Fresnel rings (see Section IV.2.2.) are complete, the illumination, mainly determined by the first zone, is the same as if there were no obstruction at all. If, on the contrary, the field point is well immersed in the geometric shadow, the lowest-order rings are missing. As a consequence, the sum of the contributions due to the partially illuminated rings is approximately zero, each contribution being neutralized by the halves of its immediate neighbors, which are of the opposite sign. In the transition region, an oscillating intensity must be expected due to the superposition of the contributions of different rings.

In 1896 Arnold Sommerfeld [31] obtained the rigorous electromagnetic solution of the half-plane diffraction problem. Using his result, it can be shown that the total field splits into a geometric optical wave and a diffracted wave originating from the edge. Subsequently (1917), Rubinowicz recast the (scalar) diffraction integral for a generic aperture illuminated by a spherical wave in the form of a line integral plus a geometric optical field. Miyamoto and Wolf [32] finally proved that a similar splitting of the diffracted field is possible even for an arbitrary incident wave.

Parallel to this development, J. B. Keller successfully generalized the concept of ray by including those diffracted by the edges of an aperture. In order to emphasize the geometric character of his approach, Keller called it the *geometric theory of diffraction* (GTD), and derived it from a generalized Fermat's principle valid for rays reaching the observer from the edges (cf. Chapter VI). The modern approach to diffraction theory is strongly influenced by GTD. In fact, this method is able, in principle, to go beyond the scalar theory and to remove *Kirchhoff's approximation*, according to which the actual field on the aperture is approximated by the one produced by the same sources in the absence of the screen. In addition, GTD makes it possible to account for the different possible shapes and electrical characteristics of the wedges limiting the aperture. Finally, it also applies to smooth obstacles lit at near-grazing incidence, a case in which surface waves are excited.

14.2 *Vector Potential for Spherical Waves*

Let us consider the Helmholtz vector field $\mathbf{v}(\mathbf{r}')$ and set the origin of coordinates at the field point P. It is easy to prove that [33]

$$\mathbf{v}(\mathbf{r}') = \nabla' \times \mathbf{w}(\mathbf{r}') + u(0)\, \nabla'\left(\frac{1}{4\pi r'}\right), \qquad \text{(IV.14.4)}$$

where

$$\mathbf{w}(\mathbf{r}') = -\int_{0^+}^{1} t\mathbf{r}' \times \mathbf{v}(t\mathbf{r}')\,dt. \tag{IV.14.5}$$

In fact, if we calculate the curl of **w** we obtain

$$\begin{aligned}\nabla' \times \mathbf{w} &= -\int_{0^+}^{1} t\,\nabla' \times [\mathbf{r}' \times \mathbf{v}(t\mathbf{r}')]\,dt \\ &= \int_{0^+}^{1} t\left[2\mathbf{v}(t\mathbf{r}') + t\frac{d}{dt}\mathbf{v}(t\mathbf{r}')\right]dt = t^2\mathbf{v}(t\mathbf{r}')\Big|_{0^+}^{1} \\ &= \mathbf{v}(\mathbf{r}') - 0^{+2}\mathbf{v}(\mathbf{r}'0^+) = \mathbf{v}(\mathbf{r}') - u(0)\,\nabla'\left(\frac{1}{4\pi r'}\right),\end{aligned} \tag{IV.14.6}$$

where use has been made of Eq. (IV.14.1) and of the vector identity

$$\begin{aligned}\nabla \times [\mathbf{r} \times \mathbf{v}(t\mathbf{r})] &= [\mathbf{v}(t\mathbf{r}) \cdot \nabla]\mathbf{r} - \mathbf{v}(t\mathbf{r})\,\nabla \cdot \mathbf{r} - (\mathbf{r} \cdot \nabla)\mathbf{v}(t\mathbf{r}) + \mathbf{r}[\nabla \cdot \mathbf{v}(t\mathbf{r})] \\ &= \mathbf{v}(t\mathbf{r}) - 3\mathbf{v}(t\mathbf{r}) - t\frac{d}{dt}\mathbf{v}(t\mathbf{r}) = -2\mathbf{v}(t\mathbf{r}) - t\frac{d}{dt}\mathbf{v}(t\mathbf{r}).\end{aligned} \tag{IV.14.7}$$

Now, if we apply Eq. (IV.14.1), i.e.,

$$\mathbf{v}(\mathbf{r}') = -u(\mathbf{r}')\,\nabla'[e^{-ikr'}/(4\pi r')] + [e^{-ikr'}/(4\pi r')]\,\nabla' u(\mathbf{r}'), \tag{IV.14.8}$$

we can easily show, by inserting the above expression into Eq. (IV.14.5), that

$$\mathbf{w}(\mathbf{r}') = -\frac{\hat{r}'}{4\pi} \times \int_0^1 e^{-iktr'}\,\nabla u(t\mathbf{r}')\,dt. \tag{IV.14.9}$$

When the field is created by a spherical wave originating from $\mathbf{R}_s$, then $u(t\mathbf{r}') = \exp(-iks)/(4\pi s)$, where $\mathbf{s} = t\mathbf{r}' - \mathbf{R}_s$ and $s = |\mathbf{s}|$ is a function of t. Since $\hat{r}' \times \hat{s} = (\mathbf{R}_s \times \mathbf{r}')/(sr')$, Eq. (IV.14.9) transforms into

$$\mathbf{w}(\mathbf{r}') = -\frac{\mathbf{R}_s \times \mathbf{r}'}{(4\pi r')^2}\int_0^1 e^{-iktr'}\frac{r'}{s}\left(\frac{d}{ds}\frac{e^{-iks}}{s}\right)dt. \tag{IV.14.10}$$

Next it can be shown that the above integrand is a total differential (see Chapter I, Born and Wolf [11], p. 452),

$$e^{-iktr'}\frac{r'}{s}\frac{d}{ds}\frac{e^{-iks}}{s} = \frac{d}{dt}\left[\frac{e^{-ik(tr'+s)}}{s^2(1+\cos\theta)}\right], \tag{IV.14.11}$$

θ being the angle formed by **s** and **r**′ (see Fig. IV.24). As a consequence,

$$\mathbf{w}(\mathbf{r}) = -\frac{\mathbf{r}' \times \mathbf{s}'}{r's' + \mathbf{r}' \cdot \mathbf{s}'}\frac{e^{-ik(r'+s')}}{4\pi r's'} - \frac{\mathbf{r}' \times \mathbf{R}_s}{r'R_s - \mathbf{r}' \cdot \mathbf{R}_s}\frac{u(0)}{4\pi r'}, \tag{IV.14.12}$$

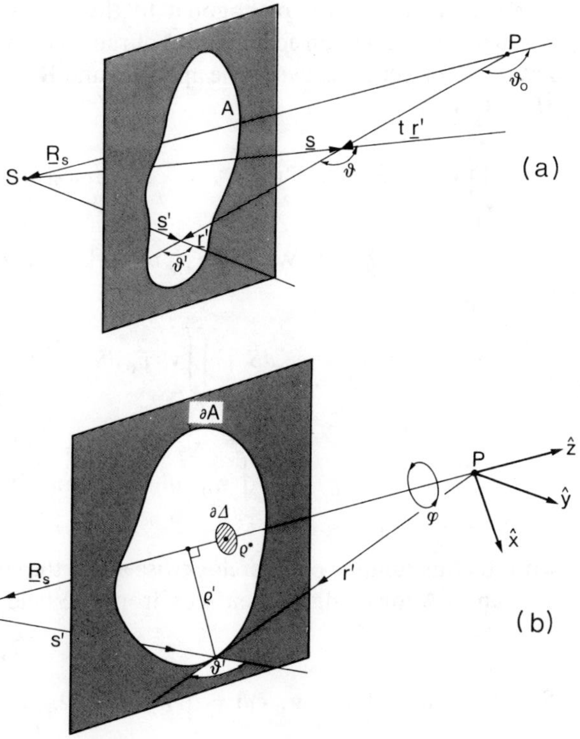

Fig. IV.24. Geometry related to the transformation of surface diffraction integrals into line integrals.

where $\mathbf{s}' = \mathbf{r}' - \mathbf{R}_s$. The vector potential becomes singular when $\mathbf{r}'$ lies on the vector $\mathbf{R}_s$ connecting the source with the field point P. Thus, $\mathbf{w}$ is singular only when $\mathbf{r}$ lies in the zone reached by the rays departing from the source and passing through the aperture.

Now, making use of simple vector algebra, we can show that

$$\nabla' \times \left(\frac{\mathbf{r}' \times \mathbf{R}_s}{r'R_s - \mathbf{r}' \cdot \mathbf{R}_s} \frac{1}{4\pi r'}\right) = -\nabla'\left(\frac{1}{4\pi r'}\right), \qquad \text{(IV.14.13)}$$

so that Eq. (IV.14.4) is superseded by $\mathbf{v}(\mathbf{r}') = \nabla' \times \mathbf{w}_0$, where $\mathbf{w}_0 = -(\mathbf{r}' \times \mathbf{s}') \times e^{-ik(r'+s')}\,[4\pi r's'(r's' + \mathbf{r}' \cdot \mathbf{s}')]^{-1}$. If we adopt cylindrical coordinates with z axis parallel to $-\mathbf{R}_s$, we have (see Fig. IV. 24b)

$$\mathbf{v}(\mathbf{r}') = \nabla' \times [w_0(\rho, z)\hat{\phi}] = -(\partial w_0/\partial z)\hat{\rho} - (1/\rho)(\partial/\partial\rho)(\rho w_0)\hat{n}_0, \qquad \text{(IV.14.14)}$$

where $\hat{\phi} = \hat{z} \times \hat{\rho}$ is a unit vector orthogonal to the z axis and to $\hat{\rho}$ and $\hat{n}_0 = -\hat{R}_s$. Furthermore, let us consider a disk $\bar{\Delta}$ of radius ρ^* perpendicular to $\mathbf{R}_s$, centered at the intersection between the aperture and $\mathbf{R}_s$, so that, with the help of Eq. (IV.14.14).

$$\iint_{\bar{\Delta}} \mathbf{v} \cdot \hat{n}_0 \, dS = 2\pi\rho^* w_0(\rho^*, z) - 2\pi \lim_{\rho \to 0} w_0(\rho, z) \qquad \text{(IV.14.15)}$$

which tends to zero as $\rho^* \to 0$. With these remarks, we can write [see Eq. (IV.14.4)]

$$\iint_{\bar{A}} \mathbf{v} \cdot \hat{n}_0 \, dS = \iint_{\bar{A}-\bar{\Delta}} \mathbf{v} \cdot \hat{n}_0 \, dS + \iint_{\bar{\Delta}} \mathbf{v} \cdot \hat{n}_0 \, dS$$

$$= \oint_{\partial \bar{A}} \mathbf{w}_0 \cdot d\mathbf{l} - \oint_{\partial \bar{\Delta}} \mathbf{w}_0 \cdot d\mathbf{l} + \iint_{\bar{\Delta}} \mathbf{v} \cdot \hat{n}_0 \, dS, \qquad \text{(IV.14.16)}$$

the curvilinear integrals running counterclockwise along the closed curves $\partial \bar{A}$ (aperture edge) and $\partial \bar{\Delta}$ (disk edge) when seen from the field point. Letting $\rho^* \to 0$,

$$\iint_{\bar{A}} \mathbf{v} \cdot \hat{n}_0 \, dS = \oint_{\partial \bar{A}} \mathbf{w}_0 \cdot d\mathbf{l} - \lim_{\rho^* \to 0} \oint_{\partial \bar{A}} \mathbf{w}_0 \cdot d\mathbf{l} = \oint_{\partial \bar{A}} \mathbf{w}_0 \cdot d\mathbf{l} + 2\pi \lim_{\rho^* \to 0} \rho^* w_0(\rho^*, z), \qquad \text{(IV.14.17)}$$

which follows from the definition of $\mathbf{w}_0$ implicit in Eq. (IV.14.14).

On the other hand, $|\mathbf{r}' \times \mathbf{s}'| = R_s \rho^*$ for $\mathbf{r}'$ connecting ρ with the dish edge, so that

$$2\pi\rho^* w_0 = \frac{1}{2R_s} \frac{|\mathbf{r}' \times \mathbf{s}'|^2}{r's' + \mathbf{r}'\mathbf{s}'} \frac{e^{-ik(r'+s')}}{4\pi r's'} \underset{\rho^* \to 0}{\to} \frac{1}{4\pi R_s} e^{-ikR_s}. \qquad \text{(IV.14.18)}$$

Finally we have, by restoring a spherical coordinate system,

$$u(P) \equiv u(\mathbf{r}) = \iint_{\bar{A}} \mathbf{v}(\mathbf{r}') \cdot \hat{n}_0 \, dS$$

$$= -\frac{1}{(4\pi)^2 R_s} \int_0^{2\pi} (1 - \cos\theta') e^{-ik(r'+s')} \, d\phi + C \frac{e^{-ikR_s}}{4\pi R_s} \equiv u_b(P) + u_g(P), \qquad \text{(IV.14.19)}$$

where ϕ is the angular coordinate corresponding to a counterclockwise rotation along the edge of the aperture as seen by an observer in P and θ' is the angle formed by $\mathbf{r}'$ and $\mathbf{s}'$, where $\mathbf{r}'$ is the running vector connecting the field point to the aperture edge. The factor C is equal to one in the geometrically illuminated zone and zero in the shadow region, thus giving rise to the geometric optics (GO) field u_g. We observe that Eq. (IV.14.19) is obtained from Eq. (IV.14.17) and

$$\frac{(\mathbf{r}' \times \mathbf{s}') \times d\mathbf{l}'}{r's'(r's' + \mathbf{r}' \cdot \mathbf{s}')} = -\frac{R_s \rho'^2 \, d\phi}{(r's')^2(1 + \cos\theta')} = -\frac{1 - \cos\theta'}{R_s} d\phi, \qquad \text{(IV.14.20)}$$

all primed quantities referring to the aperture edge. The field $u_b(P)$ represents the contribution of the wave diffracted by the boundary.

Whenever the spherical wave reduces to a plane one of the form $\exp(-i\mathbf{k} \cdot \mathbf{r}')$, Eq. (IV.14.19) can be shown to reduce to

$$u(\mathbf{r}) = Ce^{-i\mathbf{k} \cdot \mathbf{r}} - \frac{1}{4\pi} \int_0^{2\pi} (1 - \cos\theta')e^{-ikr'(1 + \cos\theta')} \, d\phi \equiv u_g + u_b. \qquad \text{(IV.14.21)}$$

From a physical point of view the field in $\mathbf{r}$ is a superposition of the field u_g in the absence of the screen and of a *boundary-diffracted wave* (BDW) u_b departing from the edge of the aperture. The first contribution is absent in the shadow region, so that the whole field coincides with u_b.

The above expressions are rigorously valid for spherical waves. However, it can be shown [32] that they can be used for ray optical fields in the limit of vanishing wavelength. In Section V.10 we will use the BDW representation of the field diffracted by an aperture to obtain an asymptotic representation in the limit of small wavelength.

15 Coherent and Incoherent Imagery

An ideal imaging system can be mathematically described as a mapping of the points of the object plane Π_o, located within the *object domain* Σ_o, into those of the plane Π_i, within the *image domain* Σ_i. In the presence of aberrations and for a finite wavelength and limited pupil, a unit point source located at (x_0, y_0) produces a field distribution $K(x, y; x_0, y_0)$ called an *impulse response*, which differs from a delta function $\delta^{(2)}(x - \bar{x}, y - \bar{y})$ centered on the gaussian image of the object having coordinates $(\bar{x}, \bar{y})$. As a consequence, aberrations and diffraction destroy the one-to-one correspondence between Σ_o and Σ_i. When the aberrations are eliminated by using a composite lens, accurately designed, and reducing the aperture of the instrument pupil, the impulse response depends only on diffraction effects, in which case the system is said to be *diffraction-limited.*

The departure of K from a delta function introduces an amount of uncertainty in the reconstruction of an object through its image. This is indicated by the fact that two point sources are seen through an optical instrument as clearly separate only if their distance is larger than a quantity W roughly coincident with the dimension of the region on Π_i where K is substantially different from zero, divided by the magnification. As we will see in the following, the parameter W, which measures the smallest dimension resolved by an instrument, decreases linearly with wavelength. Thus, higher frequencies correspond to better resolving powers. This explains the popularity of synchrotron radiation sources, which, with their ability to provide high fluxes of UV, vacuum UV, and soft x ray radiation, have permitted the implementation of imaging systems capable of resolutions better than 0.1 μm, a characteristic exploited in microelectronics for the photolithographic production of very large scale integrated (VLSI) circuits with a gap between contiguous elements of only 0.2 μm.

An image is seen by detectors (e.g., the human eye, photographic emulsion, mosaic of microscopic solid-state detectors) that respond only to the intensity. In addition, the phases of the point sources forming the "object" are in some cases spatially uncorrelated. In this situation, a lens establishes a relation only between the *intensity patterns* on two conjugate planes. The term *incoherent imagery* indicates the difference from the imaging of a phase-correlated object (*coherent imagery*). In real life, we often deal with object fields that are *partially correlated*. For example, microscopes are mostly used under conditions where the illumination is not completely incoherent. This situation requires an accurate analysis of the transformation occurring to the *correlation functions* [34] (see Section I.8).

The analysis of correlation functions has become a subject of modern radiometry, whose tremendous evolution in the past twenty years was based on the requirements for accurate radiometric measurements in the space programs. While classical radiometry was mainly concerned with the average spectral radiant energy density, a new dimension has been opened up by the measurement of *first-* and *second-order degrees of coherence* (Section I.8), connected with the growth of systems involving lasers. We are now at a stage where radiometry includes *quantum coherence theory*. This is based on a quantum statistical description of the radiation field, initiated in 1963 by Glauber [35] and Sudarshan [36]. Glauber introduced in quantum electrodynamics the so-called *coherent states* of the field, tending for "vanishing" Planck's constant (corresponding to a large photon number in the field region) to the classical sinusoidal oscillations of field vectors of assigned amplitude and phase, namely $E(\mathbf{r}, t) = E_0 \exp(-i\mathbf{k} \cdot \mathbf{r}) \exp(i\omega t)$. A useful analytical tool for the statistical characterization of the quantized field is furnished by the *P representation*, corresponding in the classical limit to a probability density function for the complex amplitude $\alpha = |E_0| \exp(i\phi)$. A detailed investigation of quantum coherence theory goes beyond the scope of this book. The interested reader can resort to a number of monographs and review papers (see, e.g., Pike [37], Arecchi and De Giorgio [38], and Crosignani *et al.* [39]).

15.1 *Impulse Response and Point Spread Function*

Let us consider a unit point source in (x_0, y_0, z_0) producing a spherical wave front transformed by a composite lens into a wave converging toward the paraxial image point $(\bar{x}, \bar{y}, z)$. Using the Luneburg–Debye integral of Eq. (IV.13.19), we can express the impulse response $K(x, y; x_0, y_0)$ on Π_i as an integral extended to the wave front of the converging wave, viz.

$$K(x, y; x_0, y_0) = \exp(-ikV_0) \iint P(p, q) A(p, q)$$
$$\times \exp[i(v_x p + v_y q) - ikW_0(p, q; x_0, y_0)]\, dp\, dq, \qquad \text{(IV.15.1)}$$

where $v_x = k(x - \bar{x})\text{NA}$ and $v_y = k(y - \bar{y})\text{NA}$ are the optical coordinates of the point (x, y) referred to the paraxial image $(\bar{x}, \bar{y})$, NA is the numerical aperture of the lens in the image space, P is the pupil function, and p and q are proportional to the optical direction cosines of the normal to the converging wave front in the image space in such a way that $p^2 + q^2 = 1$ on the largest circle contained in the exit pupil. The aberration function W_0 measures the distance of the actual wave front at the exit pupil from the gaussian sphere (see Section II.15.4), chosen in such a way as to make W_0 vanish for rays passing through the paraxial image. $V_0 = V(x_0, y_0; \bar{x}, \bar{y})$ represents the optical distance of the object point from its gaussian image. In the paraxial limit, V_0 is given approximately by

$$V_0(x_0, y_0; \bar{x}, \bar{y}) = V_0(0, 0; 0, 0) + (x_0^2 + y_0^2)/(2d_0) + (\bar{x}^2 + \bar{y}^2)/(2d)$$
$$= V_0(0, 0; 0, 0) + [1/(2k\bar{u})](v_{0x}^2 + v_{0y}^2 + v_x^2 + v_y^2) \qquad \text{(IV.15.2)}$$

d_0 and d being the distances of the object and the image from the respective principal planes. The coordinates (v_{0x}, v_{0y}), (v_x, v_y), and $\bar{u} = \bar{u}_0$ are, respectively, the optical coordinates of the points (x_0, y_0) and (x, y) and of the principal planes. The function $A(p, q)$ stands for

$$A(p, q) = R\Omega |U(p, q)|, \qquad \text{(IV.15.3)}$$

where R indicates the radius of the reference sphere, Ω the solid angle of the instrument in the image space, and $|U|$ the field amplitude on the sphere. Since $R|U|$ coincides with the analogous quantity in the image space times the transmission amplitude $T^{1/2}(p, q)$ of the instrument, we have

$$A(p, q) = [\Omega/(4\pi)]\, T^{1/2}(p, q), \qquad \text{(IV.15.4)}$$

in view of the fact that $R|U|$ in the object space is equal to $1/(4\pi)$ for a unit point source.

For an optically centered system, the aberration function W_0 depends only on the three rotational invariants $\xi = x_0^2 + y_0^2$, $\eta = p^2 + q^2$, and $\zeta = x_0 p + y_0 q$. Concerning the dependence of W_0 on the object point, a few definitions are in order. According to Welford, [40] "a generical optical system with any or no symmetry is said to be isoplanatic if the aberrations are stationary for small displacements of the object point." This means that there is no asymmetry of the image in a sufficiently small neighborhood of the field center. Consequently, circular coma must be absent. It can also be shown that, for axisymmetric systems, the absence of coma entails the absence of spherical aberrations on the optic axis. The deviation from isoplanatism can be envisaged as a nonlinear behavior of optical systems, in analogy to what occurs to nonlinear electrical networks whose responses depend on the level of a bias signal. When the instrument is isoplanatic, W_0 obeys the relation

$$W_0(p, q; x_0, y_0) = \bar{W}_0(p - \bar{p}; q - \bar{q}), \qquad \text{(IV.15.5)}$$

where we indicate by $\bar{p}$ and $\bar{q}$ the direction cosines of the principal ray. Then, inserting $\bar{W}_0$ into the integral of Eq. (IV.15.1) and using Eq. (IV.15.2), we obtain for a *clear pupil* ($T = 1$)

$$\begin{aligned} K(x, y; x_0, y_0) &= \exp\left(-ik\frac{x_0^2 + y_0^2}{2d_0} - ik\frac{\bar{x}^2 + \bar{y}^2}{2d} + iv_x\bar{p} + iv_y\bar{q}\right) \\ &\quad \times \frac{\Omega}{4\pi}\iint_\Omega P(p, q)\exp[iv_x p + iv_y q - ik\bar{W}_0(p, q)]\,dp\,dq \\ &\equiv e^{-i\phi}\bar{K}(x - \bar{x}; y - \bar{y}), \end{aligned} \qquad \text{(IV.15.6)}$$

where $\phi = (v_{0x}^2 + v_{0y}^2 + v_x^2 + v_y^2)/(2\bar{u}) \pm (v_x v_{0x} + v_y v_{0y})/v_{max}$, having made use of the coincidence of the optical coordinates of two conjugate points, apart from a plus or minus sign (see Section II.15). Here v_{max} represents the optical coordinate corresponding to the radius of the largest circle contained in the exit pupil.

Physically, we are interested in the intensity of the field, so that it is convenient to define a *point-spread function* $t(x, y; x_0, y_0)$ given by

$$t(x, y; x_0, y_0) = |K(x, y; x_0, y_0)/K(\bar{x}, \bar{y}; x_0, y_0)|^2. \qquad \text{(IV.15.7)}$$

When the system is isoplanatic, the impulse response becomes stationary, i.e., depends only on the distance of the observation point from the gaussian image of the source [$K \equiv \bar{K}(x - \bar{x},\ y - \bar{y})$]. In particular, for diffraction-limited instruments with clear circular and square pupils, $\bar{K}$ is, respectively,

$$\bar{K} \propto 2J_1(v)/v, \qquad \bar{K} \propto (\sin v_x/v_x)(\sin v_y/v_y), \tag{IV.15.8a}$$

with

$$v_x = k\mathrm{NA}(x + \mathrm{M}x_0), \qquad v_y = k\mathrm{NA}(y + My_0), \qquad v = (v_x^2 + v_y^2)^{1/2}. \tag{IV.15.8b}$$

We note that $\bar{K}$ scales down linearly with the wavelength, as already observed. However, for systems with anything approaching a diffraction-limited aberration correction, the frequency increase enhances the phase factor $k\bar{W}_0$, greatly affecting the form of the impulse response.

The integral of Eq. (IV.15.6) giving the impulse response for finite aberrations can be calculated for circular pupils by expressing W_0 in terms of Zernike polynomials [e.g., see Eq. (IV.13.34)]. In particular, for a simple defocusing, the integral can be expressed in terms of Lommel's functions (see Problem 24). For a square pupil with defocusing and spherical aberration, $\bar{K}$ is written in terms of the function $I(u, v)$ calculated by Pearcey and discussed in Section V.5. For small aberrations, we observe a decrease of intensity in the central spot, while more light appears in the outer rings. The dimension of the central spot does not change appreciably. On the basis of this observation, K. Strehl proposed in 1902 that the aberrations be measured by means of the ratio of the maximum value of the intensity in the central core of a point-diffraction image in the actual system to the corresponding value in an aberrationless system of the same aperture and focal length. This ratio $\bar{V}$, referred to as the *Strehl intensity ratio*, gives a measure of the fraction of light contained in the central spot. The Strehl ratio can be easily calculated starting from Eq. (IV.15.6), setting $x = y = \bar{x} = \bar{y} = 0$, subtracting from $\bar{W}_0$ its average value over the aperture, and approximating the phase factor $\exp[-ik(\bar{W}_0 - \langle\bar{W}_0\rangle)]$ by $1 - ik(\bar{W}_0 - \langle\bar{W}_0\rangle)$ in view of the smallness of $k(\bar{W}_0 - \langle\bar{W}_0\rangle)$, viz.

$$\bar{V} = 1 - k^2\left[\iint_\Omega \bar{W}_0^2\,dp\,dq - \left(\iint_\Omega \bar{W}_0\,dp\,dq\right)^2\right]. \tag{IV.15.9}$$

Photographs showing the light distribution for various aberrations can be found in the beautiful "Atlas of Optical Phenomena" (see Chapter II, Cagnet *et al.* [31]).

15.2 *Coherent Imaging of Extended Sources*

At this stage, we can apply the superposition principle to calculate the *image field* $i(x, y, z)$ corresponding to the *extended object field* $o(x_0, y_0, z_0)$, obtaining

$$i(x, y, z) = \iint_{\Sigma_0} K(x, y, z; x_0, y_0, z_0) o(x_0, y_0, z_0)\, dx_0\, dy_0. \qquad \text{(IV.15.10)}$$

Note that we have indicated explicitly the dependence of i, o, and K on z and z_0, in order to remark that the above linear relation between the fields at z_0 and z also applies when the plane $z = \text{const}$ does not coincide with the image plane $z_0 = \text{const}$. Obviously, in this case the function K is no longer given by the Debye–Luneburg integral. In general, since $i(\mathbf{r})$ is a solution of the Helmholtz scalar equation, $K(\mathbf{r}; \mathbf{r}_0)$ must satisfy the same equation with respect to $\mathbf{r}$.

If we are interested in the intensity, we obtain from Eq. (IV.15.10)

$$|i(x, y, z)|^2 = \iint_{\Sigma_0} dx_0\, dy_0 \iint_{\Sigma_0} dx_0'\, dy_0'\, K(x, y, z; x_0, y_0, z_0) K^*(x, y, z; x_0', y_0', z_0') \times o(x_0, y_0, z_0) o^*(x_0', y_0', z_0'). \qquad \text{(IV.15.11)}$$

15.3 *Propagation of the Mutual Intensity; Van Cittert–Zernike Theorem*

If, in particular, the amplitude and phase of $o(x_0, y_0, z_0)$ are partially known, then $o(x_0, y_0, z_0)$ and $i(x, y, z)$ can be treated as stochastic processes described by the set of their moments (see Section 1.8). In this case, relevant information is associated with the *mutual intensity* $\langle i(x, y, z) i^*(x', y', z)\rangle$ relative to the plane $z = \text{const}$, which is obtained by means of Eq. (IV.15.10) in terms of the analogous quantity for the plane $z_0 = \text{const}$ and reads [cf. Eq. (I.8.12)]

$$\begin{aligned}\langle i(x, y, z) i^*(x', y', z)\rangle &= \iint_{\Sigma_0} dx_0\, dy_0 \iint_{\Sigma_0} dx_0'\, dy_0'\, K(x, y, z; x_0, y_0, z_0) \\ &\quad \times K^*(x', y', z; x_0', y_0', z_0) \langle o(x_0, y_0, z_0) o^*(x_0', y_0', z_0)\rangle \\ &\equiv \iint_{\Sigma_0} dx_0\, dy_0\, K(x, y, z; x_0, y_0, z_0) F(x_0, y_0, z_0; x', y', z)\end{aligned} \qquad \text{(IV.15.12)}$$

Now if we fix x', y', the left-hand side of the above relation becomes a function of x, y, z expressed by an integral similar to that of Eq. (IV.15.10).

Since the function defined by this last equation represents a field satisfying the Helmholtz equation, we arrive at the conclusion that

$$(\partial^2/\partial x^2 + \partial^2/\partial y^2 + \partial^2/\partial z^2 + k^2)\langle i(x, y, z)i^*(x', y', z)\rangle = 0. \qquad \text{(IV.15.13)}$$

Because of the symmetry of the mutual intensity with respect to the coordinates **r** and **r**′, an analogous equation holds true for the dependence of $\langle i(x, y, z)i^*(x', y', z)\rangle$ on x', y', and z.

A relevant situation is that in which the mutual intensity $\langle o(x_0, y_0, z_0)o^*(x'_0, y'_0, z_0)\rangle$ is different from zero in a neighborhood of (x_0, y_0, z_0) so small that in it $K(x, y, z; x'_0, y'_0, z_0)$ changes very little. In this case, we can approximate Eq. (IV.15.12) by

$$\langle i(x, y, z)i^*(x', y', z)\rangle = \iint_{\Sigma_0} dx_0\, dy_0\, K(x, y, z;\, x_0, y_0, z_0) \times K^*(x', y', z;\, x_0, y_0, z_0) I_0(x_0, y_0, z_0), \qquad \text{(IV.15.14)}$$

with

$$I_0(x_0, y_0, z_0) = \iint_{\Sigma_0} \langle o(x_0, y_0, z_0)o^*(x'_0, y'_0, z_0)\rangle\, dx'_0\, dy'_0. \qquad \text{(IV.15.15)}$$

The quantity I_0 can be interpreted with some caution as a sort of intensity. To this end, we notice that the correlation length of the mutual intensity depends on the distance of the real or virtual sources from the object plane. A reduction of the correlation distance implies sources very close to Π_0. This has several consequences: the amplitude $o(x, y, z)$ must become very large, the field cannot be considered locally as a plane wave, so that the squared modulus of $o(x, y, z)$ no longer represents the Poynting vector. Whenever these facts do not imply difficulties, we can interpret I_0 as an intensity.

As an example, let us consider an optical system consisting of a circular aperture on the object plane. In this case, K is given by Eq. (IV.5.6), so that, apart from an inessential phase factor,

$$\langle i(x, y, z)i^*(x', y', z)\rangle = \frac{1}{(\lambda d)^2} \iint_{\Sigma_0} \exp\left\{-ik\frac{[x_0(x - x') + y_0(y - y')]}{d}\right\} \times I_0(x_0, y_0, z_0)\, dx_0\, dy_0. \qquad \text{(IV.15.16)}$$

The above relation shows that *the correlation function of the field at a distance d from an aperture illuminated incoherently coincides with the Fourier transform of the aperture illumination*, a result initially recognized by Van Cittert and Zernike and forming the basis of the celebrated theorem bearing their name (see, e.g., Chapter I, Born and Wolf [11]).

15.4 *Evaluation of Optical Systems by Means of the Optical Transfer Function (OTF)*

If we are interested in the intensity distribution $I(x, y) \propto \langle i(x, y) i^*(x, y) \rangle$, Eqs. (IV.15.14,7) give

$$I(x, y) = \iint_{\Sigma_0} I_0(x_0, y_0) t(x, y; x_0, y_0)\, dx_0\, dy_0, \qquad \text{(IV.15.17)}$$

which becomes a convolution integral for isoplanatic systems

$$I(x, y) = \frac{1}{M^2} \iint_{\Sigma_0'} I_0\left(\frac{-x_0}{M}, \frac{-y_0}{M}\right) \bar{t}(x - x_0, y - y_0)\, dx_0\, dy_0, \qquad \text{(IV.15.18)}$$

where $\Sigma_0' = M^2 \Sigma_0$. By Fourier-transforming, the above relation becomes

$$I(\alpha, \beta) \propto T(\alpha, \beta) I_0(\alpha, \beta), \qquad \text{(IV.15.19)}$$

where $T(\alpha, \beta)$ is the *optical transfer function* (OTF) of the system (e.g., see Fig. IV.25) and is proportional to the Fourier transform of the point-spread

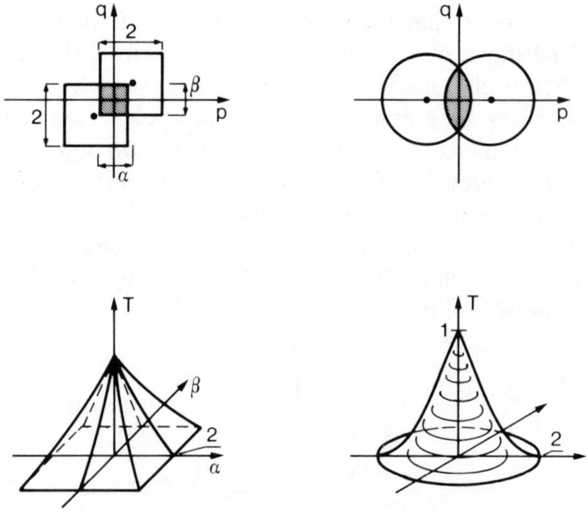

Fig. IV.25. Optical transfer functions related to the incoherent illumination of square (left) and round (right) pupils.

function $\bar{t}$ of an isoplanatic system

$$T(\alpha, \beta) = \frac{\iint_{-\infty}^{+\infty} \bar{t}(v_x, v_y) \exp(i\alpha v_x + i\beta v_y)\, dv_x\, dv_y}{\iint_{-\infty}^{+\infty} \bar{t}(v_x, v_y)\, dv_x\, dv_y}$$

$$\times \frac{\begin{array}{c}\iint_{-\infty}^{+\infty} P\left(p + \frac{\alpha}{2}, q + \frac{\beta}{2}\right) P\left(p - \frac{\alpha}{2}, q - \frac{\beta}{2}\right) \\ \exp\left[-ik\bar{W}_0\left(p + \frac{\alpha}{2}, q + \frac{\beta}{2}\right) + ik\bar{W}_0\left(p - \frac{\alpha}{2}, q - \frac{\beta}{2}\right)\right] dp\, dq\end{array}}{\iint_{\Omega} P(p, q)\, dp\, dq}$$

(IV.15.20)

by virtue of convolution, Parseval's theorems, and Eq. (IV.15.7). The denominator is simply the area of the aperture. For an aberration-free system, the integral in the numerator is related to the area of overlap of two pupil functions P displaced by the quantities α and β with respect to the x and y coordinates. It can be shown that the presence of aberrations reduces the OTF, which is always equal to unity for $\alpha = \beta = 0$. Being the Fourier transform of a real function, the real part of T is an even function of α and β, while the imaginary part is an odd function. The modulus of T is called the *modulation transfer function* (MTF). It is worth noting that the OTF can also be defined for photographic emulsions, television cameras, and other electro-optic devices, a fact that is particularly appreciated when we must assemble a complex electro-optic imaging system. If we can characterize each component by means of its OTF, the design of the complex electro-optic imaging system can be accomplished in a way similar to that followed for a cascade of electronic amplifiers.

For a circular pupil, the OTF is easily evaluated in the absence of aberrations and reads (see Fig. IV.25, right side)

$$T(\omega) = \begin{cases} (2/\pi)[\arccos(\omega/2) - \omega(1 - \omega^2/4)^{1/2}], & 0 \leqq \omega \leqq 2, \\ 0 & \text{otherwise,} \end{cases}$$

(IV.15.21)

where $\omega = (\alpha^2 + \beta^2)^{1/2}$. Then, for a circular pupil, the spatial frequency ω is limited to the interval (0, 2). The dimensionless frequency ω, conjugate of the optical coordinate v, is related to the *spatial frequency* f expressed in cycles per unit length by the relation $f = k\omega\mathrm{NA}$. The expression just given for the OTF indicates that the diffraction of light by the optical system considered sets an upper limit to the ability of the system to resolve a bar target with a normalized spatial frequency greater than 2.

The presence of aberrations notably modifies the OTF, as can be seen from Fig. IV.26 taken from Black and Linfoot [41]. The dashed curve of Fig. IV.26a

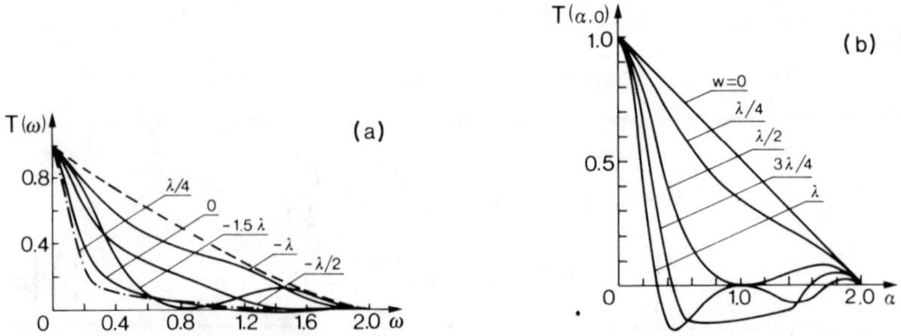

Fig. IV.26. (a) Optical transfer functions relative to a round pupil for a total retardation of 1λ due to spherical aberration. The different curves refer to receiving planes displaced by $\lambda/4$, 0, -1.5λ, $-\lambda$, and -0.5λ from the paraxial focus (from Black and Linfoot [41]). (b) cross section of the OTF for a square pupil and focusing errors of $\lambda/4$, $\lambda/2$, $3\lambda/4$, and λ.

represents the OTF $T(\omega)$ of the aberrationless lens. The other curves refer to a retardation of 1λ due to spherical aberration for different positions of the receiving plane with respect to the gaussian focus. It is noteworthy that the OTF at best approaches the ideal one for a defocusing of $-\lambda$. A significant feature of these curves is that, for larger defocusing, the OTF can actually become negative (see Fig. IV.26b), which implies a phase reversal in the image. For full details on the Fourier methods in the imaging analysis the reader can consult Goodman [25], Gaskill [26], Duffieux [42], Linfoot [43], and Murata (1966) cited in the bibliography.

15.5 *Degrees of Freedom of an Image*

Let us consider an aberrationless system **K** with square pupil and forming an erect image. In this case, Eqs. (IV.15.8,10) yield

$$i(x, y) \propto \iint_{-a}^{+a} \frac{\sin(x - x_0)}{x - x_0} \frac{\sin(y - y_0)}{y - y_0} \times \exp\left(i\frac{x_0^2 + y_0^2 + x^2 + y^2}{2\bar{u}} - i\frac{xx_0 + yy_0}{v_{\max}} \right) o(x_0, y_0)\, dx_0\, dy_0 \tag{IV.15.22}$$

where, for notational simplicity, we have indicated the optical coordinates with the corresponding ordinary coordinate. Here, $2a$ is the side dimension of the object field expressed in optical coordinates. We now wish to inquire whether **K** can faithfully reproduce some objects [44]. This leads us to look for solutions of the above equations in which $i(x, y)$ is proportional to $o(x, y)$. In this way, we obtain a Fredholm integral equation with generally complex symmetric kernel. Since the kernel factorizes in the product of two functions of x and y, respectively, we can express a generic eigenfunction in the form $o(x, y) = \Phi_n(x)\Phi_m(y)$, where Φ_n is an eigenfunction of eigenvalue γ_n of the integral equation

$$\gamma_n\Phi_n(\xi) = \int_{-1}^{+1} \frac{\sin a(\xi - \xi')}{\xi - \xi'} \exp\left(-ia^2\frac{\xi^2}{2\bar{u}} - ia^2\frac{\xi\xi'}{v_{\max}}\right)\Phi_n(\xi')\,d\xi'. \tag{IV.15.23}$$

In particular, for small object fields, we have $\bar{u} \gg a^2$ and $v_{\max} \gg a^2$, so that we can neglect the phase factor of the above kernel. The corresponding equation admits a complete set of eigenfunctions, which are scaled versions of the angular prolate spheroidal functions [45] [cf. Eq. (VII.16.7)]. It has been pointed out [44, 45] that the relative eigenvalues have a modulus almost constant and equal to π for $n < 2a/\pi$, while for $n > 2a/\pi$, γ_n drops abruptly to zero. This property has a dramatic physical consequence. If we expand the generic field o by using the functions Φ_n,

$$o(x_0, y_0) = \sum_{n,m=0}^{\infty} O_{n,m}\Phi_n(x_0)\Phi_m(y_0), \tag{IV.15.24}$$

the image $i(x, y)$ will be given by

$$i(x, y) = \sum_{n,m=0}^{\infty} \gamma_n\gamma_m O_{nm}\Phi_n(x)\Phi_m(y) \simeq \sum_{n,m=0}^{2a/\pi} \gamma_n\gamma_m O_{nm}\Phi_n(x)\Phi_m(y). \tag{IV.15.25}$$

Therefore, the system **K** transmits only $(2a/\pi)^2$ terms, so that all information contained in the other terms is lost. The expansion of Eq. (IV.15.24) can be interpreted as a decomposition of $o(x_0, y_0)$ into its components in the infinite-dimension Hilbert space having as basis the functions Φ_n. Then, each term of the series represents a degree of freedom of the object. Because of Eq. (IV.15.25), the number N of *degrees of freedom of the image* is not greater than $(2a/\pi)^2$. By transforming a in physical coordinates, we have

$$N = (2a/\pi)^2 = U/\lambda^2 = S_{\text{obj}}\Omega_{\text{obj}}/\lambda^2 = S_{\text{im}}\Omega_{\text{im}}/\lambda^2, \tag{IV.15.26}$$

S being the extension of the object (image) field and Ω the solid angle of the instrument aperture [corresponding to the entrance (exit) pupil seen from the object (image) plane]. The above expression implies that the transmission capacity of an optical instrument increases quadratically with frequency, whenever the system is aberrationless.

The quantity N can be also shown to correspond to the number of samples of the object to be taken at regular intervals, which are sufficient to define the image. Consequently, if the image must be "read" by a mosaic of N_{d} microscopic detectors, the étendue U [see Section II.15.3 and Eq. (II.15.29)] of the optical instrument used for imaging the object must be larger than $\lambda^2 N_{\text{d}}$.

The considerations of this section can be shown to be qualitatively true for any shape of the object and the pupils.

For further details the reader is referred to Gori and Guattari [46] and Gori and Ronchi [47].

Problems

Section 10

1. Let a plane diffracting screen contain a rectangular aperture, of dimensions a by b, illuminated by a spherical wave radiated by a point source in S. Show that the field along the normal from S to the screen is given by

$$u(\mathbf{r}) = u_g(\mathbf{r})(-\tfrac{1}{2}i)[F(\xi_2) - F(\xi_1)][F(\eta_2) - F(\eta_1)],$$

where u_g is the field in the absence of the screen, $F(x)$ is the Fresnel complex integral defined by Eq. (V.3.5), and

$$\xi_{2,1} = x_{2,1}[2(d_S + d_P)/(\lambda d_S d_P)]^{1/2}, \qquad \eta_{2,1} = y_{2,1}[2(d_S + d_P)/(\lambda d_S d_P)]^{1/2}.$$

where d_S and d_P are the distances of S and P from the aperture plane, while $x_{1,2}$ and $y_{1,2}$ are the coordinates of the aperture vertices in a cartesian system with center on the normal from S to the screen.

2. With reference to Problem 1, show that the diffraction pattern on a plane parallel to the screen remains unchanged when the screen is moved to a new position such that the new distance d'_S from the source is equal to the old distance d_P from the observation plane.

3. Consider a plane aperture illuminated normally by a plane wave. Let O be the normal projection of the field point on the screen and ρ, ϕ a polar coordinate system with center in O. In this system the aperture edge is represented by $\phi = \phi(\rho)$, where $\phi(\rho)$ is generally a multivalued function of ρ, while ρ is comprised between ρ_{max} and ρ_{min}. Show that the Fresnel diffraction integral can be rewritten as a one-dimensional integral in the variable ρ, viz.

$$u(P) \propto \int_{\rho_{min}}^{\rho_{max}} f(\rho) \exp\left(\frac{-ik\rho^2}{2|z - z'|}\right) \rho \, d\rho,$$

where $f(\rho)$ represents the length of the circumference arc of radius ρ and center in O contained in the aperture. Discuss the analytic expression of $f(\rho)$ for a circular and a triangular aperture (see also Chapter V, Problem 1).

4. Consider a circular aperture of radius a illuminated by a field proportional to $\exp(-2b\rho^2)$. Show that the far-field $G(\theta)$ along a direction

forming an angle θ with the normal to the screen plane is given by the Neumann series [see also Section IV.11.3a and Eqs. (IV.11.23) and (IV.11.24)]

$$G(\theta) \propto \sum_{s=0}^{\infty} G_{2s+1} \frac{J_{2s+1}(v)}{v},$$

where $v = ka \sin \theta$ and

$$G_1 = \frac{1}{2ba^2}(1 - e^{-2a^2b}), \qquad G_3 = -\frac{1}{a^2b}\left[\left(1 - \frac{1}{a^2b}\right) + \left(1 + \frac{1}{a^2b}\right)e^{-2a^2b}\right],$$

$$\frac{G_{2s+1}}{2s+1} = \frac{G_{2s-1}}{a^2b} + \frac{G_{2s-3}}{2s-3},$$

(Chapter II, Cornbleet [18] p. 215).

5. Consider a circular aperture with an amplitude illumination proportional to $f(x) = 0.076 - 0.0441(1 - x^2) + 0.528(1 - x^2)^2 + 0.44$, with $x = \rho/a$, where a is the aperture radius. Plot the far field and show that the side lobes are less than 4×10^{-3} the peak value (relative to $\theta = 0$), while the beamwidth is almost coincident with that of a uniformly illuminated aperture. This distribution is an example of tapered illuminations used to reduce the side-lobe level in order to improve the resolution of optical instruments (especially microscopes and telescopes) and reflector antennas. In optics the illumination tapering is obtained by placing a transparency on the apertures or on the focal planes. This process is known as *apodization.* (see Cornbleet [48] for the properties of the above illumination and the article by Jacquinot and Roizen-Dossier [49] for a review of investigations on the apodization.)

6. Consider a circular aperture with a disk obstruction in the center. Discuss the properties of the far field with particular attention to the dependence of the main lobe beamwidth and the height of the first side lobe on the obstruction ratio. In addition, discuss the class of illuminations that make it possible to reducing the beamwidth to arbitrarily small values (see Toraldo di Francia [50]).

Section 11

7. Show that the free-space Green's function G can be represented by the *Sommerfeld–Ott* integral

$$G(\mathbf{r}, \mathbf{r}') = \frac{\exp(-ik|\mathbf{r} - \mathbf{r}'|)}{4\pi|\mathbf{r} - \mathbf{r}'|}$$

$$= -\frac{i}{4\pi}\int_0^{\infty} \frac{J_0(\chi d)\exp[-i(k^2 - \chi^2)^{1/2}|z - z'|]}{(k^2 - \chi^2)^{1/2}} \chi \, d\chi,$$

where $d = [\rho^2 + \rho'^2 - 2\rho\rho'\cos(\phi - \phi')]^{1/2}$ and $\rho, z,$ and ϕ are the cylindrical coordinates of the point $\mathbf{r}$. *Hint*: Solve the inhomogeneous wave equation by using cylindrical coordinates.

8. Using the Sommerfeld–Ott integral representation, discussed in the preceding problem, and the expansion

$$J_0[\chi\sqrt{\rho^2 + \rho'^2 - 2\rho\rho'\cos(\phi - \phi')}] = \sum_{m=-\infty}^{+\infty} e^{im(\phi-\phi')} J_m(\chi\rho) J_m(\chi\rho'),$$

following from Graf's addition theorem for Bessel functions, prove Eq. (IV.11.1)

9. Show that when the z components E_z and H_z of a generic field depend on z as $e^{-i\chi z}$, then, in a free-space region,

$$\mathbf{H}_t = -i(\omega^2/c^2 - \chi^2)^{-1}[\chi \nabla_t H_z + \omega/(\zeta_0 c)\hat{z} \times \nabla_t E_z],$$

$$\mathbf{E}_t = i(\omega^2/c^2 - \chi^2)^{-1}(-\chi \nabla_t E_z + \zeta_0 \omega/c\hat{z} \times \nabla_t H_z),$$

where $\mathbf{E}_t = \hat{\rho}E_\rho + \hat{\phi}E_\phi$ and $\mathbf{H}_t = \hat{\rho}H_\rho + \hat{\phi}H_\phi$ are the transverse components of $\mathbf{E}$ and $\mathbf{H}$ with reference to a cylindrical coordinate system. Analogously, $\nabla_t = \hat{\rho}\,\partial/\partial\rho + \hat{\phi}\rho^{-1}\,\partial/\partial\phi$.

10. Show that for $\chi > k_0$, a cylindrical wave in an empty region extending to $\rho \to \infty$, is represented by a function of the form

$$\exp(-i\chi z + im\phi)K_m[(\chi^2 - k_0^2)^{1/2}\rho],$$

where $K_m(x)$ is a modified Bessel function of the second type, decaying at infinity as e^{-x} (see Chapter VIII).

11. Solve the scalar wave equation in spherical coordinates and show that a field can be expanded in a homogeneous region in a series of *spherical* modes (see Chapter II, example of Section II.12.3 and Section VI.12):

$$u(r, \theta, \phi) = \sum_{n=0}^{\infty} \sum_{m=-n}^{n} [U_{nm}^{(\text{out})} h_n^{(2)}(k_0 r) + U_{nm}^{(\text{in})} h_n^{(1)}(k_0 r)]\, Y_n^m(\theta, \phi),$$

where $h_n^{(2)}$ and $h_n^{(1)} = h_n^{(2)*}$ are spherical Hankel functions of the second and first kind, respectively. *Hint*: See Chapter 6.

12. Show that an electromagnetic field in a free-space region can be represented in the form

$$\mathbf{E} = \nabla \times \nabla \times (\mathbf{r}v) - i\omega\mu_0 \nabla \times (\mathbf{r}u),$$

$$\mathbf{H} = \nabla \times \nabla \times (\mathbf{r}u) - i\omega\varepsilon_0 \nabla \times (\mathbf{r}v),$$

where u and v are two solutions of the homogeneous wave equation, known as *Debye potentials* (see Chapter I, Papas [17]).

13. Using the Debye potentials and the asymptotic expressions for the spherical Hankel functions, prove that the far field radiated by a distribution of sources located in a finite volume can be expressed by

$$\mathbf{E} = -\frac{e^{-ikr}}{r} \sum_{n=0}^{\infty} \sum_{m=-n}^{n} E'_{nm}\hat{r} \times (\hat{r} \times \nabla) Y_n^m + E''_{nm}(\hat{r} \times \nabla) Y_n^m.$$

Hint: Represent $\mathbf{E}$ by means of the Debye potentials discussed in the preceding problem, then expand u and v in spherical harmonics (see Chapter I, Papas [17], Eq. (168), see also Eqs. (II.12.43) and (VI.12.26)).

14. Consider the electric dipole $\mathbf{J}(\mathbf{r}) = \delta(x)\delta(y)\delta(z)\hat{z}$ oriented parallel to the z axis. Using the integral relation

$$\mathbf{A}(\mathbf{r}) = \mu_0 \iiint \mathbf{J}(\mathbf{r}') G(|\mathbf{r} - \mathbf{r}'|)\, dV'$$

connecting the vector potential $\mathbf{A}$ to the current distribution $\mathbf{J}$, and expressing the electric and magnetic fields as functions of $\mathbf{A}$, i.e., $\mathbf{H} = (1/\mu)\nabla \times \mathbf{A}$, expand $\mathbf{H}$ into cylindrical waves by means of Eq. (IV.11.1).

Section 12

15. The field scattered by a sphere can be represented as a superposition of spherical harmonics. From a mathematical point of view, the field is given by a series of spherical Hankel functions multiplied by the spherical harmonic $Y_n^m(\theta, \phi)$. For a scalar problem, the scattered wave function can be calculated by imposing to the total wave function $u(r, \theta, \phi) = u_i(r, \theta, \phi) + u_d(r, \theta, \phi)$ to satisfy the boundary conditions on the surface of the sphere. For a plane incident wave propagating along the z axis and a total field vanishing on the sphere, expand the scattered field in spherical harmonics, viz.

$$u_d(r, \theta, \phi) = \sum_{n=0} C_n h_n^{(2)}(k_0 r) Y_n^m.$$

Section 13

16. Consider an isotropic point source placed in the focus of a rotationally symmetric paraboloid having a finite aperture. Calculate the far field radiated along a direction forming an angle ψ with the reflector axis, by excluding the contribution of the rays leaving the source without hitting the paraboloid. *Hint*: Compute the field by using the diffraction integral for the field distribution on the paraboloid aperture and calculate the latter by propagating the spherical wave radiated by the source by using the laws of geometric optics. Next, using the relation between the height h of a ray with respect to the axis and the angle θ formed by the ray with the axis [$h(\theta) = 2f(1 - \cos\theta)$; see

Fig. IV.17], the far field can be derived from the Fraunhofer integral

$$u(v) \propto \int_0^1 (1 + \cos\theta) J_0(vx) x \, dx,$$

where $x = h/h_{max}$ and $v = kh_{max} \sin\psi$.

17. Consider a metallic half-plane (knife edge) with the edge coincident with the x axis of a cartesian coordinate system whose z axis coincides with the optic axis of an aberration-free thin lens L_1. Assume that the screen lies on the back focal plane of L_1, which in turn coincides with the front focal plane of a second aberration-free thin lens L_2. Show that the field $i(x, y)$ on the output plane of L_2 is related to the object amplitude $u_0(x, y)$ on the input plane of L_1 by the integral relation

$$i(x, y) = \frac{e^{-ik_0(f_1+f_2)}}{M}\left[o\left(\frac{x}{M}, \frac{y}{M}\right) - \frac{i}{\pi}\int_{-\infty}^{+\infty} \frac{o(x', y)}{(x/M) - x'} dx'\right],$$

where $M = f_2/f_1$ is the linear magnification of the afocal system and f_i is the focal length of the lens L_i. This transformation is used in the *Schlieren method* of analysis of convective motion patterns in wind tunnels.

18. Consider a grating characterized by a transmittance function $t(x, y) = \Sigma_{n=-\infty}^{\infty} T_n \exp(inx2\pi/d)$, placed in the back focal plane of a lens illuminated by a plane wave directed along the optic axis. Show that the far field radiated by the grating is proportional to

$$u(x, y) \propto \iint_{-\infty}^{\infty} \exp[-ik_0 W_0(x', y')] K\left(\frac{x}{M} - x', \frac{y}{M} - y'\right) dx' \, dy',$$

where W_0 is the aberration function of the lens and $M = d/f$, with d the distance of the observation plane from the focal plane and f the lens focal length. The function K is given by

$$K(\xi, \eta) \propto \iint_{-\infty}^{+\infty} t(x'', y'') \exp\left(-ik_0 \frac{x''\xi + y''\eta}{f}\right) dx'' \, dy''$$

$$= \delta(\eta) \sum_{n=-\infty}^{+\infty} T_n \delta\left(\frac{2\pi}{d} n - \frac{k_0}{f}\xi\right).$$

This system is known as *Ronchi's interferometer* and is used for measuring the aberrations of an optical system (see Ronchi [51].)

19. Consider the Ronchi interferometer described in the preceding problem. By supposing that the grating spacing d is such that only two consecutive

orders, say n and $n + 1$, of the grating far field contribute to the intensity measured on a plane at a distance d, show that the diffracted field intensity is proportional to (see Problem 18 for the definitions of the quantities)

$$|u^2(x, y)| \propto \left| 1 + \frac{T_{n+1}}{T_n} \exp\left\{ -ik_0 \left[W_0\left(\frac{x}{M} - \frac{\lambda}{d} f(n+1), \frac{y}{M} \right) - W_0\left(\frac{x}{M} - \frac{\lambda}{d} f(n), \frac{y}{M} \right) \right] \right\} \right|.$$

20. Discuss the form of the fringes produced by a Ronchi interferometer for the primary Seidel aberrations. *Hint*: Use the expression for the intensity derived in Problem 19 together with the approximation

$$W_0[\xi - (n+1)\alpha, \eta] - W_0(\xi - n\alpha, \eta) \cong -\alpha \frac{\partial}{\partial \xi} W_0(\xi - n\alpha, \eta)$$

(see Toraldo di Francia [52]).

21. Consider the far field projected by an aberration-free converging lens illuminated by a plane wave. Calculate the Fraunhofer field $G(v)$ along the direction forming an angle ψ with the optic axis, by following two distinct procedures. First, express G as the Hankel transform of the lens output plane illumination, viz.

$$G(v) \propto \int_0^1 J_0(vx) e^{i\bar{u}x^2} x\, dx,$$

where $v = k_0 a \sin\psi, \bar{u} = k_0 a^2/(2f)$, a and f being the radius of the lens aperture and the focal length, respectively. Alternatively, calculate the field on the lens focal plane and then compute the Hankel transform of the last one. Since the field on the focal plane is the Airy pattern $J_1(v')/v'$, whose Hankel transform is proportional to $\mathrm{circ}(\psi f/a)$, the far field obtained in this way vanishes abruptly outside the shadow boundary. This result contrasts with the general findings of diffraction theory. Consequently, it is necessary to evaluate the far field by using the above integral. Show that the discrepancy is due to the fact that the Airy formula correctly describes the field only in proximity to the optic axis (paraxial approximation).

22. Show that the field projected at a great distance by an objective illuminated by a spherical wave, whose image is at a distance d from the exit pupil, is proportional to the field calculated in proximity to a focus on a plane distant $u = k_0 a^2/(2d)$ from it, a being the pupil diameter. Discuss the variations of the far field produced by either a displacement of the source with respect to the objective or a change of the diameter of the pupil. For what combination of values of a, d, and λ is the central Airy spot replaced by a dark one? Use as a guide the isophotes reported in Chapter I, Born and Wolf [11], p. 440.

Section 13.5

23. Show that the field amplitude in proximity to the focus of an aberration-free spherical wave of small aperture, can be represented by (see the Appendix, Sections C and E)

$$u(\bar{u}, v) = 2\int_0^1 J_0(v\rho)e^{i\bar{u}\rho^2/2}\rho\, d\rho = \frac{2}{\bar{u}}e^{i\bar{u}/2}U_1 + i\frac{2}{\bar{u}}e^{-i\bar{u}/2}U_2$$

$$= i\frac{2}{\bar{u}}e^{-iv^2/(2\bar{u})} - i\frac{2}{\bar{u}}e^{i\bar{u}/2}V_0 - \frac{2}{\bar{u}}e^{i\bar{u}/2}V_1,$$

where U_n and V_n are Lommel's functions, defined by Eqs. (C.21) (see Dekanosidze [53]).

24. Show that along the shadow boundary ($u = v$) of an aberration-free spherical wave the field intensity $I(v, v)$ is given by

$$I(v, v) = \frac{1 - 2J_0(v)\cos v + J_0^2(v)}{v^2} I(0, 0).$$

25. Show that the power flowing through a disk (the so-called *cumulative point-spread function*) of normalized radius v_0 placed on the focal plane of an aberration-free spherical wave is given by the Rayleigh formula

$$\int_0^{v_0} I(0, v)v\, dv = 2I(0, 0)[1 - J_0^2(v_0) - J_1^2(v_0)].$$

26. Show that the image of a slightly aberrated spherical wave is given approximately by [see Eqs. (IV.13.33) and (IV.13.34)]

$$I(0, v) = \pi\int_0^1 d\theta\, \theta J_0(v\theta)\left[1 - i\frac{B}{24}R_4^{(0)}(\theta) + i\frac{D}{4}R_2^{(0)}(\theta)\right]$$

$$+ i\int_0^{2\pi} d\phi \int_0^1 d\theta\, \theta \exp[iv\theta\cos(\phi - \psi)]$$

$$\times \left[-\frac{C}{2}R_2^{(2)}(\theta)\cos 2\phi + ER_1^{(1)}(\theta)\cos\phi + \frac{F}{3}R_3^{(1)}(\theta)\cos\phi\right]$$

for $y_0 = 1$.

27. Using the relation (see Chapter I, Born and Wolf [11], p. 772)

$$\int_0^1 R_n^{(m)}(\theta)J_m(v\theta)\theta\, d\theta = (-1)^{(n-m)/2}\frac{J_{n+1}(v)}{v},$$

calculate the integrals of Problem 26.

28. It is convenient to express the intensity $I(u,v)$ as a fraction of the intensity I^* that would be obtained at the gaussian image point if no aberrations were present, viz.

$$i(u,v) = \frac{I'(u,v)}{I^*} = \frac{1}{\pi^2}\left|\int_0^1 \theta\,d\theta \int_0^2 \phi \exp[iv\theta\cos(\phi - \psi) - ikW_0 + iu\theta^2/2]\right|^2.$$

Show that in the limit of small aberrations the normalized intensity $I'(0,0)$ calculated at the gaussian image point is proportional to the mean square deformation of the wave front, viz.

$$I'(u,v) = 1 - k^2\langle W_0^2 - \langle W_0\rangle^2\rangle,$$

where

$$\langle W_0^2 - \langle W_0\rangle^2\rangle = \frac{1}{\pi}\int_0^1 \theta\,d\theta \int_0^{2\pi} d\phi(W_0^2 - \langle W_0\rangle^2).$$

Section 14

29. Consider an aperture on a plane screen illuminated by a gaussian beam. Express the diffracted field by means of the boundary diffraction wave formalism. Calculate the vector potential W by treating the gaussian beam as a spherical wave emanating from a complex point related to the size and position of the waist and to the beam direction by Eqs. (V.7.9). In particular, consider a circular aperture illuminated perpendicular to the screen and calculate the field along the axis (see Otis [54]).

References

1. Babic, V. M., and Kirpicnikova, N. Y., "The Boundary Layer Method in Diffraction Problems." Springer-Verlag, Berlin and New York, 1979.
2. Baker, B. B., and Copson, E. T., "The Mathematical Theory of Huygens' Principle," Oxford Univ. Press (Clarendon), London and New York, 1950.
3. Bowman, S. I., Senior, T. B. A., and Uslenghi, P. L. E., "Electromagnetic and Acoustic Scattering by Simple Shapes." North-Holland Publ., Amsterdam, 1969.
4. Watson, G. N., "Theory of Bessel Functions." Cambridge Univ. Press, London and New York, 1962.
5. Vaynshteyn, L. A., "The Theory of Diffraction and the Factorization Method." Golem Press, Boulder, Colorado, 1969.
6. Mittra, R., and Lee, S. W., "Analytical Techniques in the Theory of Guided Waves." Macmillan, New York, 1971.
7. Borgnis, F. E., and Papas, C. H., *in* "Handbuch der Physik" (S. Flügge, ed.), Vol. XVI, p. 285–422. Springer-Verlag, Berlin and New York, 1958.
8. van Bladel, J., "Electromagnetic Fields." McGraw-Hill, New York, 1964.
9. Nussbaumer, H. J., "Fast Fourier Transform and Convolution Algorythm." Springer-Verlag, Berlin and New York, 1981.

10. Stratton, J. A., "Electromagnetic Theory." McGraw-Hill, New York, 1941.
11. Müller, C., "Grundprobleme der Matematischen Theorie Elektromagnetischer Schwingungen." Springer-Verlag, Berlin and New York, 1957.
12. Smythe, W. R., "Static and Dynamic Electricity," 2nd Ed. McGraw-Hill, New York, 1941.
13. Clemmow, P. C., "The Plane Wave Spectrum Representation of Electromagnetic Fields." Pergamon, Oxford, 1966.
14. Denisyuk, Y. N., Ramishvihi, N. M., and Chavchanidze, V. V., *Opt. Spectrosc.* (*Engl. Transl.*) **30**, 603 (1971); see also Lohmann, A. W., and Silva, D. A., *J. Opt. Soc. Am.* **61**, 687 (1971); Silva, D. A., *Appl. Opt.* **11**, 2613 (1972).
15. Sneddon, I. N., "The Use of Integral Transforms." McGraw-Hill, New York, 1972.
16. Boivin, A., "Théorie at Calcul des Figures de Diffraction de Révolution." Presses Université Laval, Quebec, 1964.
17. Newton, R. G., "The Complex j-Plane." Benjamin, New York, 1964.
18. Streifer, W., and Kodis, R. D., *Q. Appl. Math.* **21**, 285 (1964).
19. Schöbe, W., *Acta Math.* **92**, 265 (1954).
20. Bremmer, A., "Terrestrial Radio Waves." Elsevier, Amsterdam, 1949; see also Franz, W., *Z. Naturforsch., A* **9A**, 705 (1954).
21. Magnus, W., and Kotin, L., *Numer. Math.* **2**, 228 (1960); see also Keller, J. B., Rubinow, S. I., and Goldstein, M., *J. Math. Phys.* **4**, 829 (1963).
22. Nussenzveig, H. M., *J. Math. Phys.* **10**, 82 (1969); **10**, 125 (1969).
23. Wolf, E., *Proc. R. Soc. London, Ser. A* **253**, 349 (1959).
24. Richards, B., and Wolf, E., *Proc. R. Soc. London, Ser. A* **253**, 358 (1959).
25. Goodman, J. W., "Introduction to Fourier Optics." McGraw-Hill, New York, 1968.
26. Gaskill, J. D., "Linear Systems, Fourier Transforms and Optics." Wiley, New York, 1978.
27. Boivin, A., Dow, J., and Wolf, E., *J. Opt. Soc. Am.*, **57**, 1171 (1977).
28. Linfoot, E. H., "Recent Advances in Optics." Oxford Univ. Press, London and New York, 1955.
28a. Luchini, P., *Comput. Phys. Commun.* **31**, 303 (1984).
29. Li, Y., and Wolf, E., *Opt. Commun.* **39**, 211 (1981).
30. Thompson, B. J., *Prog. Opt.* **7**, 169–230 (1969).
31. Sommerfeld, A., "Optics." Academic Press, New York, 1954.
32. Rubinowicz, A., *Prog. Opt.* **4**, 199–240 (1965).
33. Gordon, W. B., *J. Math. Phys.* **16**, 448 (1975).
34. Beran, M. J., and Parrent, G. B., "Theory of Partial Coherence." Prentice-Hall, Englewood Cliffs, New Jersey, 1964; Glauber, R. J., *in* "Quantum Optics and Electronics" (C. De Witt, A. Blandin, and C. Cohen-Tannoudji, eds.), p. 63–185. Gordon & Breach, New York, 1965; Mandel, L., and Wolf, E., *Rev. Mod. Phys.* **37**, 231 (1965); Klauder, J. R., and Sudarshan, E. C. G., "Fundamentals of Quantum Optics." Benjamin, New York, 1968.
35. Glauber, R. J., *Phys. Rev.* **130**, 2529 (1963); *Phys. Rev.* **131**, 2766 (1963).
36. Sudarshan, E. C. G., *Phys. Rev. Lett.* **10**, 277 (1963).
37. Pike, E. R., *in* "Quantum Optics" (S. M. Kay and A. Maitland, eds.), p. 127. Academic Press, New York, 1970.
38. Arecchi, F. T., and De Giorgio, V., *in* "Laser Handbook" (F. T. Arecchi and E. O. Schulz-Dubois, eds.), Vol. 1, p. 191–264. North-Holland Publ., Amsterdam, 1972.
39. Crosignani, B., Di Porto, P., and Bertolotti, M., "Statistical Properties of Scattered Light." Academic Press, New York, 1975.
40. Welford, W. T., *Prog. Opt.* **13**, 267–292 (1976).
41. Black, G., and Linfoot, E. H., *Proc. R. Soc. London, Ser. A* **239**, 522 (1957).
42. Duffieux, P. M., "The Fourier Transform and its Application to Optics." Wiley, New York, 1983.

43. Linfoot, E. H., "Fourier Methods in Optical Image Evaluation." Focal Press, London, 1960.
44. Toraldo di Francia, G., *J. Opt. Soc. Am.* **59**, 799 (1969).
45. Landau, H. J., and Pollack, H. O., *Bell Syst. Tech. J.* **41**, 1295 (1962).
46. Gori, F., and Guattari, G., *Opt. Commun.* **7**, 163 (1973).
47. Gori, F., and Ronchi, L., *J. Opt. Soc. Am.* **71**, 150 (1981).
48. Cornbleet, S., *Electron. Lett.* **2**, 79 (1966).
49. Jacquinot, P., and Roizen-Dossier, B., *Prog. Opt.* **3**, 29–186 (1964).
50. Toraldo di Francia, G., *Nuovo Cimento Suppl.* **9**, 426 (1952).
51. Ronchi, V., *Appl. Opt.* **3**, 437 (1964).
52. Toraldo di Francia, G., "Optical Image Evaluation." Nat. Bur. Stand., Washington, D. C., 1954.
53. Dekanosidze, E. N., "Tables of Lommel's Functions of Two Variables." Pergamon, Oxford, 1960.
54. Otis, G., *J. Opt. Soc. Am.* **64**, 1545 (1974).

Bibliography

Cowley, J. M., "Diffraction Physics." North-Holland Publ., Amsterdam, 1984.
Erkkila, J. H., *J. Opt. Soc. Am.* **71**, 197 (1981).
Federov, V. B., and Mityakov, V. G., *Opt. Spectros.* (*Engl. Transl.*) **56**, 537 (1984).
Li, Y. and Wolf, E., *J. Opt. Soc. Am.*, **1A**, 801 (1984).
Murata, K., in "Progress in Optics" (E. Wolf, ed.) Vol. V, 199–245. North-Holland Publ., Amsterdam, 1966.
Northover, F. H., "Applied Diffraction Theory." Amer. Elsevier, New York, 1971.
Yu, F. T. S., "Optics and Information Theory." Wiley, New York, 1976.

Chapter V

Asymptotic Evaluation of Diffraction Integrals

1 Introduction

In spite of their apparent simplicity, the Helmholtz–Kirchhoff and Huygens–Fresnel integrals can be evaluated analytically only in a limited number of cases; in general, it is necessary to resort to numerical methods.

All numerical methods are limited by the fact that a finite set of samples of the field on the reference plane can be used. This introduces some errors, which become particularly large when the integrands are rapidly varying functions. Unfortunately, most of the fields diffracted by obstacles have a far-field pattern with strong and narrow peaks. Consequently, in all these cases the numerical approach becomes lengthy and time-consuming, because many samples are needed.

As an alternative to the above approaches, techniques based on the asymptotic evaluation of the integrals have become very popular [1]. There are several reasons for this success: the simplicity of the expressions, the high degree of accuracy achievable by retaining a suitable number of terms of the asymptotic series, the possibility of separating the field regions in domains where a particular behavior of the field is expected, and, most important, the possibility of improving the representation of the field on the reference aperture.

Of all the advantages just described, the last one deserves some additional comments. Usually, the most important problem is that of calculating the correct field distribution on a reference surface, since, when this is known, we can calculate the field in the whole space. In general, this problem cannot be solved exactly, so that in most cases we use as the field distribution that calculated either in the absence of the obstacles if the aperture is large compared to the wavelength (*Kirchhoff approximation*), or in the absence of the aperture in other cases (*Bethe approximation*). When using the Kirchhoff approximation, the field is set equal to zero on the dark side of a screen and equal to the unperturbed field on the aperture itself. This distribution can be further improved by an iterative procedure consisting of calculating the field on the aperture at each step by using the field calculated at the step before. To be practical, this strategy needs a simple expression for the radiated field, which can be provided by its asymptotic representation. This approach has been followed with some ad hoc approximations to calculate, for example, the field diffracted by a slit. This and similar cases have led to a new trend in which the asymptotic representations are, in a sense, the constitutive blocks of electromagnetic fields.

All the asymptotic methods that we will discuss can be considered as modifications of the Wentzel–Kramers–Brillouin (WKB) and *stationary-phase* (SP) methods. While the former is more directly applicable to differential equations, the latter directly concerns integrals containing rapidly oscillating phase factors. In some cases the stationary-phase method is replaced by the *steepest-descent* (SD) method, with which one can correctly account for the complex localization of the stationary points of the phase factor.

In the simplest case the diffraction integral reduces to a Fourier transform of the form

$$I(k) = \int_a^b g(s)e^{-iks}\,ds. \tag{V.1.1}$$

If $g(s)$ is a continuous function together with its derivative of order less than N, we can determine the behavior of I for large k by applying the technique of integration by parts, whose repeated application gives

$$I = \frac{1}{-ik}ge^{-iks}\bigg|_a^b - \frac{1}{(-ik)^2}\frac{dg}{ds}e^{-iks}\bigg|_a^b + \cdots + \frac{(-1)^{N-1}}{(-ik)^N}\frac{d^{N-1}g}{ds^{N-1}}e^{-iks}\bigg|_a^b$$
$$+ \frac{(-1)^N}{(-ik)^N}\int_a^b \frac{d^Ng}{ds^N}e^{-iks}\,ds. \tag{V.1.2}$$

We can estimate the last integral on the right side of the above finite expansion by resorting to the *Riemann–Lebesgue theorem*, which states that if $\int_a^b |d^Ng/ds^N|\,ds < \infty$, then

$$\lim_{k\to\infty}\int_a^b \frac{d^Ng}{ds^N}e^{-iks}\,ds = 0. \tag{V.1.3}$$

Consequently, the integration-by-parts procedure allows the evaluation of a Fourier transform up to $O(k^{-N})$, with N the highest integer for which the integral of $|d^Ng/ds^N|$ is finite. This introduces strong limitations, as can be immediately appreciated by applying the above procedure to the function $g(s) = \sqrt{s}$. In such a case, if $a = 0$ and $b > 0$, we have $N = 1$ and the best we can do is to write $I(k)$ in the form

$$I(k) = \frac{1}{-ik} s^{1/2} e^{-iks}\Big|_0^b + \frac{1}{ik}\frac{1}{\sqrt{2}}\int_0^b \frac{e^{-iks}}{\sqrt{s}}\,ds. \tag{V.1.4}$$

In the following, we will show that $I(k)$ can be represented as an asymptotic series in the large parameter k, in which the terms of order greater than $1/k$ can be obtained by a generalization of the integration-by-parts procedure.

In several cases we encounter diffraction integrals of the form

$$I(k) = \int_{-\infty}^{+\infty} e^{-ikas^2} g(s)\,ds. \tag{V.1.5}$$

As $k \to \infty$ we can obtain an accurate estimate of this integral by assuming that $g(s)$ is constant in the neighborhood of the origin, thus getting

$$I(k) = \frac{g(0)}{\sqrt{k|a|}}\int_{-\infty}^{+\infty} e^{\mp ix^2}\,dx = \frac{g(0)\sqrt{\pi}}{\sqrt{|a|k}}\begin{cases} e^{-i\pi/4}, & a > 0 \\ e^{i\pi/4}, & a < 0 \end{cases} \tag{V.1.6}$$

which is immediately recognized as the formula provided by the stationary-phase method.

The angular spectrum integrals can take several forms, such as those of Eqs. (IV.9.2, 5, 6). In general, they can be written as

$$I(k) = \int_{\Gamma} e^{-ik\rho\cos(\beta - \phi)} S(\beta)\,d\beta, \tag{V.1.7}$$

where $S(\beta)$ is a particular function of β characterized by polar singularities and branch cuts. In order to calculate $I(k)$ for large k it is generally useful to modify the integration path Γ according to the scheme of Eq. (IV.9.15). In this way, $I(k)$ is given by the contributions of the poles $S(\beta)$ plus an integral similar to that above with Γ replaced by the steepest-descent path (SDP) plus the path Γ_B encircling the branch cuts of $S(\beta)$, if any. The SDP is defined as the path passing through a point of the complex β-plane where the phase $k\rho\cos(\beta - \phi)$ is stationary and departing from it by following the trajectory of steepest descent of the imaginary part of $\cos(\beta - \phi)$. Along the SDP the phase factor can be written as $\exp[-ik\rho\cos(\beta - \phi)] \propto \exp(-u^2)$, where u is a real function of s along the SDP. Thus, by means of the change of variables $s \to u$, the integral can be reduced to the form of Eq. (V.1.6).

Near a caustic or a focus the WKB, SP, and SD methods give singular fields. A remedy for these singularities has been found by replacing the integral with *comparison integrals*, the most renowned being the *Airy function*, which is also known as the *rainbow integral* since it was introduced by Airy to explain rainbow formation. By multiplying these comparison integrals by asymptotic series it is possible to obtain a complete representation of the field that is valid in proximity to and away from the critical regions. This approach, which closely resemble Langer's method discussed in Section III.3, is called *uniform asymptotic representation theory* (UAT) [2–6].

2 Stationary-Phase Method

The integrand of diffraction integrals contains a phase factor, which can be made to oscillate almost everywhere in the integration interval as rapidly as we like by letting $k \to \infty$. This is true with the exception of values of the integration variable that correspond to the zeros of the phase derivative. Heuristically, if we divide the integration interval into two domains, one formed by the neighborhoods of the stationary points of the phase, we can expect that, for k sufficiently large, the contribution from the domain containing the points for which the phase factor oscillates will rapidly become negligible compared to that of the other domain. In fact, if we indicate the integrand by $g(s)\exp[-ikh(s)]$, the phase undergoes a 2π variation for an increment $\Delta s = 2\pi/[h'(s)k] = O(1/k)$. By letting $k \to \infty$ for $h' \neq 0$ the last increment Δs can be made so small that the corresponding variation of $g(s)$ over the interval $s, s + \Delta s$ is negligible and the relevant integral vanishes. Accordingly, the integral reduces to the contributions of the neighborhoods of the points for which $h' = 0$ (*stationary points*). On the other hand, near these points the phase can be approximated by $k(h_0 + \frac{1}{2}h_0'' s^2)$ and the relative integral explicitly calculated (*stationary-phase method*).

Stokes [7] is credited with being the first (1856) to take advantage of this reasoning to find an approximate expression for the Airy integral. The general procedure was formulated by Lord Kelvin [8] (1887) and was subsequently placed on rigorous ground by G. N. Watson [9] (1918). Here, we adopt a rigorous if simplified approach developed by Erdelyi [10, 11] and based on the use of successive integrations by parts [12]. We will first discuss the case of an integral with stationary points, if any, coinciding with the end points of the integration interval. Then we will extend the analysis to an integral with any number of stationary points distributed all over the interval.

In order to treat separately the contributions of each critical point, we will use a mathematical device introduced by van der Corput [12, 13], which consists of the introduction of a class of functions, called *neutralizers*, that are equal to unity at an extreme of an interval and zero at the other extreme. In addition, the derivatives of these functions of any order vanish at both end points.

2.1 *Asymptotic Expansion of Fourier-Type Integrals with Monotonic Phase*

As a preliminary step we will consider the asymptotic expansion of the Fourier-type integral [14, 15]

$$I(k) = \int_0^a e^{-ikh(s)} v(s) g(s)\, ds, \tag{V.2.1}$$

where $h(s)$ is a monotonic function of s in the interval $(0, a)$ and

$$v(s) = \frac{\int_s^a \exp[-1/y + 1/(y-a)]\, dy}{\int_0^a \exp[-1/y + 1/(y-a)]\, dy} \tag{V.2.2}$$

is a function introduced by van der Corput, who called it a *neutralizer* [13], such that

$$v(0) = 1, \qquad v(a) = 0, \tag{V.2.3}$$

while the derivatives tend to vanish when either extreme is approached from the interior of the interval $(0, a)$. In addition, we assume that h and g can be expressed for s belonging to the interval $(0, a)$ as

$$h(s) = h_0 + s^\rho u(s), \tag{V.2.4a}$$

$$g(s) = s^\gamma v(s), \tag{V.2.4b}$$

where u and v are analytic functions of s, different from 0 for $s = 0$. In addition, $0 > \operatorname{Re}\gamma > -1$ and $\rho > 0$. Since $s^\rho |u(s)|$ is an increasing function of s, we can introduce the next variable x, defined by

$$x^\rho = s^\rho |u(s)|, \tag{V.2.5}$$

with the branch chosen so that x is an increasing function of s. By replacing s with x in Eq. (V.2.1), we obtain

$$I(k) = \exp(-ikh_0) \int_0^{x(a)} \exp(-ik\varepsilon x^\rho) x^\gamma f(x) v(x)\, dx, \tag{V.2.6}$$

with $x > 0$, $\varepsilon = \operatorname{sgn} u(0)$, and

$$f(x) = (g(s)/|h(s) - h_0|^{\gamma/\rho})\, ds/dx. \tag{V.2.7}$$

If $f(x)$ can be differentiated N times, the integral of Eq. (V.2.6) can be expanded into a polynomial in $1/k$ by successive integrations by parts, as originally pointed out by Erdelyi (see Bleistein and Handelsman [14], Section 3.4),

$$I(k)\exp(ikh_0) = -\sum_{n=0}^{N-1} h^{(-n-1)}(x; k) \frac{d^n}{dx^n} [v(x) f(x)] \Bigg|_0^{x(a)} + R_N, \tag{V.2.8}$$

where

$$R_N = (-1)^N \int_0^{x(a)} h^{(-N)}(x;k) \frac{d^N}{dx^N}[v(x)f(x)]\,dx \tag{V.2.9}$$

and

$$\begin{aligned} h^{(-n-1)}(x;k) &= \int_x^{x+\infty(-i\varepsilon)^{1/\rho}} dx_n \int_{x_n}^{x_n+\infty(-i\varepsilon)^{1/\rho}} dx_{n-1} \cdots \int_{x_1}^{x_1+\infty(-i\varepsilon)^{1/\rho}} dx_0\, x_0^\gamma \exp(-ik\varepsilon x_0^\rho) \\ &= -\frac{1}{n!}\int_x^{x+\infty(-i\varepsilon)^{1/\rho}} (x_0 - x)^n x_0^\gamma \exp(-ik\varepsilon x_0^\rho)\,dx_0. \end{aligned} \tag{V.2.10}$$

The last identity is easily proved by interchanging the order of integration. In particular,

$$\begin{aligned} h^{(-n-1)}(0,k) &= -\frac{1}{n!}\int_0^{\infty(-i\varepsilon)^{1/\rho}} x_0^{n+\gamma} \exp(-ik\varepsilon x_0^\rho)\,dx_0 \\ &= \frac{1}{n!}\frac{\Gamma[(n+\gamma+1)/\rho]}{\rho k^{(n+\gamma+1)/\rho}} \exp\left(-i\frac{\pi}{2}\varepsilon\frac{n+\gamma+1}{\rho}\right). \end{aligned} \tag{V.2.11}$$

$\Gamma(x)$ being the Gauss gamma function. Because of the presence of the neutralizer, the contribution of the end point $x = x(a)$ vanishes identically and Eq. (V.2.8) reduces, with the help of Eq. (V.2.11), to

$$\begin{aligned} &I(k)\exp(ikh_0) \\ &= \frac{1}{\rho}\sum_{n=0}^{N-1}\frac{1}{n!}\frac{\Gamma[(n+\gamma+1)/\rho]\exp[-i(\pi/2)\varepsilon(n+\gamma+1)/\rho]}{k^{(n+\gamma+1)/\rho}}\frac{d^n f(x)}{dx^n}\bigg|_{x=0} + R_N. \end{aligned} \tag{V.2.12}$$

It can be shown [see Bleistein and Handelsman [14], p. 99, Eq. (3.5.46)] that

$$|h^{(-n-1)}(x;k)| < \frac{x^\gamma}{n!}\int_0^\infty \sigma^n \exp(-k\sigma^\rho)\,d\sigma = \frac{x^\gamma\Gamma[(n+1)/\rho]}{n!\,\rho k^{(n+1)/\rho}}, \tag{V.2.13}$$

so that

$$\lim_{k\to\infty} k^{(N+\gamma+1)/\rho} R_N = \frac{1}{\rho N!}\Gamma\left(\frac{N+\gamma+1}{\rho}\right)\exp\left(-i\frac{\pi}{2}\varepsilon\frac{N+\gamma+1}{\rho}\right)\frac{d^N}{dx^N}f\bigg|_{x'=0}, \tag{V.2.14}$$

that is,

$$R_N = O(k^{-(N+\gamma+1)/\rho}). \tag{V.2.15}$$

If f can be indefinitely differentiated, we can choose N as large as we like. Consequently, we can formally replace N with ∞ in Eq. (V.2.12) by omitting R_N and replacing the sign of equality with that of asymptotic equality (i.e., $=$ becomes $\sim$).

Now, if we introduce the function

$$a(s) = (dx/ds)^{-1}, \tag{V.2.16}$$

we have

$$(d^n/dx^n)f(x) = [a(s)\,d/ds]^n a(s)g(s)|h(s) - h_0|^{-\gamma/\rho}, \tag{V.2.17}$$

which, when inserted in Eq. (V.2.12), finally yields

$$I(k) \sim \exp(-ikh_0)\frac{1}{\rho}\sum_{n=0}^{\infty}\frac{1}{n!}\frac{1}{k^{(n+\gamma+1)/\rho}}\exp\left(-i\frac{\pi}{2}\varepsilon\frac{n+\gamma+1}{\rho}\right)$$
$$\times\,\Gamma\left(\frac{n+\gamma+1}{\rho}\right)\left(a\frac{d}{ds}\right)^n(ag|h - h_0|^{-\gamma/\rho})|_{s=0}. \tag{V.2.18}$$

The last expansion can be easily extended to the integral of Eq. (V.2.1) with the integrand deprived of the neutralizer by exploiting the following identity

$$I(k) = \int_0^a e^{-ikh(s)}g(s)\,ds = \int_0^a e^{-ikh(s)}v(s)g(s)\,ds + \int_0^a e^{-ikh(s)}[1 - v(s)]g(s)\,ds$$
$$\equiv I_0(k) + I_a(k). \tag{V.2.19}$$

In particular, $I_a(k)$ can be rewritten as

$$I_a(k) = \exp(-ikh_a)\int_{x(0)}^{0}\exp(-ik\varepsilon_a x^{\rho_a})x^{\gamma_a}f_a(x)v(x)\,dx, \tag{V.2.20}$$

with $x(0) < 0$, $\varepsilon_a = \operatorname{sgn} u(a)$, and the exponents γ_a, ρ_a, and h_a defined by

$$h(s) = h_a + (s - a)^{\rho_a}u_a(s), \tag{V.2.21a}$$

$$g(s) = (s - a)^{\gamma_a}v_a(s), \tag{V.2.21b}$$

while

$$x = (s - a)|u_a(s)|^{1/\rho_a}, \tag{V.2.21c}$$

$$f_a(x) = [g(s)/|h(s) - h_a|^{\gamma_a/\rho_a}](ds/dx). \tag{V.2.21d}$$

Now we can replace x with $-x$ in Eq. (V.2.20) and use the expansion of Eq. (V.2.18) to write

$$I_a(k) = \exp(-ikh_a)(-1)^{\gamma_a}\int_0^{-x(0)}\exp[-ik(-1)^{\rho_a}x^{\rho_a}\varepsilon_a]x^{\gamma_a}f_a(-x)v(x)\,dx$$
$$\sim \exp(-ikh_a)\frac{(-1)^{\gamma_a}}{\rho_a}\sum_{n=0}^{\infty}\frac{(-1)^n\Gamma[(n+\gamma_a+1)/\rho_a]}{n!\,k^{(n+\gamma_a+1)/\rho_a}}$$
$$\times\exp\left[-i\frac{\pi}{2}\varepsilon_a(-1)^{\rho_a}\frac{n+\gamma_a+1}{\rho_a}\right]\frac{d^n}{dx^n}f_a(x)\bigg|_{x=0}, \tag{V.2.22}$$

where $(-1)^{\gamma_a} = e^{i\pi\gamma_a}$ and $(-1)^{\rho_a} = e^{i\pi\rho_a}$.

2.2 *Generalization of the SP Method to Diffraction Integrals Having a Multiplicity of Stationary and Singular Points*

Let us consider one-dimensional diffraction integrals of the form

$$u(\mathbf{r}) = \left(\frac{k}{2\pi}\right)^{1/2} e^{i\pi/4} \int_a^b u(x', 0) \frac{e^{-ikR}}{R^{1/2}} \hat{z} \cdot \hat{R}\, dx'. \tag{V.2.23}$$

If $u(x', z')$ is the ray optical (RO) field $A\exp(-ikS)$, then $u(\mathbf{r})$ will be represented by

$$I(k) = \left(\frac{k}{2\pi}\right)^{1/2} e^{i\pi/4} \int_a^b g(s) e^{-ikh(s)}\, ds, \tag{V.2.24}$$

where s stands for x', $g(x') = A(x')R^{-1/2}\hat{z} \cdot \hat{R}$, and $h(x') = R(x') + S(x')$. If h is a variation-limited function of s and g and its derivatives have a finite number of discontinuities, we can divide the interval (a, b) into a finite set of subintervals separated by the *critical points* of the integrand (i.e., discontinuity points of h and g and their derivatives, end points a and b, and *stationary points* of h). Let the critical points be represented by

$$a = a_0 < a_1 < a_2 < \cdots < a_m = b. \tag{V.2.25}$$

In each interval $h(s)$ is a monotonic function of s, continuous with all its derivatives. The points that are roots of $h'(s) = 0$ are referred to as *stationary points* of $h(s)$. Now we can use for each interval the asymptotic expansions of Eqs. (V.2.18,22) to evaluate the contributions I_i^+ and I_i^- from the right and left sides of each critical point a_i, viz.

$$I(k) = \sum_{i=0}^{m} (I_i^+ + I_i^-). \tag{V.2.26}$$

Notice that at the end points I_a^- and I_b^+ are missing.

In general we have, by using either Eq. (V.2.22) or Eq. (V.2.18),

$$I_{i(\rho,\gamma)}^{\pm} \sim k^{(1/2)-(\gamma+1)/\rho} \frac{1}{(2\pi)^{1/2}\rho} \exp(-ikh_i + i\pi/4)$$
$$\times \sum_{n=0}^{\infty} C_n^{\pm} \frac{\Gamma[(n+\gamma+1)/\rho]}{n!\,k^{n/\rho}} \left(a\frac{d}{ds}\right)^n \left[\frac{a(s)g(s)}{|h - h_i|^{\gamma/\rho}}\right]\Bigg|_{s=a_i}, \tag{V.2.27}$$

where ρ and γ are the indices of the critical point a_i approached either from the left $(-)$ or from the right $(+)$, while

$$C_n^+ = \exp[-i(\pi/2)\varepsilon(n+\gamma+1)/\rho], \tag{V.2.28a}$$

$$C_n^- = (-1)^{n+\gamma} \exp[-i(\pi/2)\varepsilon(-1)^{\rho}(n+\gamma+1)/\rho]. \tag{V.2.28b}$$

In particular, when h and g are continuous in a_i, then the function $a(s)$ defined by Eq. (V.2.16) is also continuous and

$$I_{i(\rho,\gamma)} = I^+_{i(\rho,\gamma)} + I^-_{i(\rho,\gamma)} \sim k^{1/2-(1+\gamma)/\rho} \frac{1}{(2\pi)^{1/2}\rho} \exp\left(-ikh_i + i\frac{\pi}{4}\right)$$
$$\times \sum_{n=0}^{\infty} \left[1 + (-1)^{n+\gamma} \exp\left\{-i\left(\frac{\pi}{2}\right)\varepsilon[(-1)^\rho - 1]\frac{n+\gamma+1}{\rho}\right\}\right]$$
$$\times \exp\left[-i\left(\frac{\pi}{2}\right)\varepsilon\frac{n+\gamma+1}{\rho}\right] \frac{\Gamma[(n+\gamma+1)/\rho]}{n!\,k^{n/\rho}}$$
$$\times \left[a(s)\frac{d}{ds}\right]^n \left[\frac{a(s)g(s)}{|h - h_i|^{\gamma/\rho}}\right]\Bigg|_{s=a_i} \qquad \text{(V.2.29)}$$

2.3 *Stationary-Point Contributions*

When $h(s)$ is continuous in a_i and its derivatives of order less than ρ vanish, a_i is called a *stationary point of order* ρ. In particular, for $\rho = 2$ and $\gamma = 0$, Eq. (V.2.29) yields

$$I_{(2,0)} \sim \frac{\exp[i(1-\varepsilon)\pi/4 - ikh]}{(2\pi)^{1/2}} \sum_{n=0}^{\infty} \frac{\Gamma(n+1/2)}{(2n)!\,k^n}$$
$$\times \exp\left(-i\frac{\pi}{2}\varepsilon n\right)\left(a\frac{d}{ds}\right)^{2n} [a(s)g(s)]\Bigg|_{s=a_i}$$
$$= \exp\left[i(1-\varepsilon)\frac{\pi}{4}\right] 2^{-(1/2)} \exp(-ikh)\left\{ag + \frac{\exp[-i(\pi/2)\varepsilon]}{4k}\right.$$
$$\left.\times [(aa'^2 + a^2a'')g + 3a^2a'g' + a^3g''] + O(k^{-2})\right\}, \qquad \text{(V.2.30)}$$

where we have omitted the argument a_i for notational simplicity and indicated with a prime the derivatives with respect to s. For the successive terms the reader should consult Dingle [16].

Now, it can be easily shown that

$$a(a_i) = [2/\varepsilon h''(a_i)]^{1/2}, \qquad \text{(V.2.31a)}$$

$$a'(a_i) = -\tfrac{1}{3}(h'''/h'')a(a_i), \qquad \text{(V.2.31b)}$$

$$a''(a_i) = [\tfrac{1}{4}(h^{iv}/h'') + (11/36)(h'''/h'')^2]a(a_i). \qquad \text{(V.2.31c)}$$

Substituting the above expressions back into Eq. (V.2.30), we finally obtain

$$I_{(2,0)} \sim \frac{\exp[i(1-\varepsilon)\pi/4 - ikh]}{(\varepsilon h'')^{1/2}}\left\{g + \frac{\exp[-i(\pi/2)\varepsilon]}{2kh''\varepsilon}\right.$$
$$\left.\times\left[\left[\frac{5}{12}\left(\frac{h'''}{h''}\right)^2 + \frac{1}{4}\frac{h^{iv}}{h''}\right]g - \frac{h'''}{h''}g' + g''\right] + O(k^{-2})\right\}, \qquad \text{(V.2.32)}$$

all quantities being calculated for $s = a_i$. Notice that the leading term of the asymptotic series is independent of k [see the Appendix, Eq. (F1)].

Proceeding in a similar way, we obtain for $I_{(3,0)}$

$$I_{(3,0)} \sim \frac{k^{1/6}\exp(i\pi/4 - i\varepsilon\pi/6 - ikh)}{3(2\pi)^{1/2}}\sum_{n=0}^{\infty}\frac{\Gamma[(n+1)/3]}{n!\,k^{n/3}}\exp\left(-i\frac{\pi}{6}\varepsilon n\right)$$
$$\times\left\{1 + (-1)^n\exp\left[i\frac{\pi}{3}\varepsilon(n+1)\right]\right\}\left(a\frac{d}{ds}\right)^n[a(s)g(s)]\bigg|_{s=a_i}. \qquad \text{(V.2.33)}$$

In this case

$$a(a_i) = [6/\varepsilon h'''(a_i)]^{1/3}, \qquad \text{(V.2.34)}$$

so that

$$I_{(3,0)} \sim \frac{k^{1/6}\exp(i\pi/4 - i\varepsilon\pi/6 - ikh)}{3(2\pi)^{1/2}}$$
$$\times\,\Gamma\left(\frac{1}{3}\right)\left[1 + \exp\left(i\varepsilon\frac{\pi}{3}\right)\right]\left(\frac{6}{\varepsilon h'''}\right)^{1/3} g + O(k^{-1/6}). \qquad \text{(V.2.35)}$$

2.4 *End Point Contributions*

At the end points a and b, when g does not vanish, we have $\gamma = 0$, $\rho = 1$, and [see the Appendix, Eq. (F1)].

$$I_{a(1,0)} \sim \frac{\exp(-i\pi/4 - ikh_a)}{(2\pi k)^{1/2}}\sum_{n=0}^{\infty}\frac{1}{(ik)^n}\left(\frac{1}{h'}\frac{d}{ds}\right)^n\frac{g}{h'}\bigg|_{s=a}$$
$$= \frac{\exp(-i\pi/4 - ikh_a)}{(2\pi k)^{1/2}}\left[\frac{g_a}{h'_a} + \frac{1}{ik}\frac{h'_a g'_a - g_a h''_a}{h'^3_a} + O(k^{-2})\right], \qquad \text{(V.2.36a)}$$

$$I_{b(1,0)} \sim -\frac{\exp(-i\pi/4 - ikh_b)}{(2\pi k)^{1/2}}\left[\frac{g_b}{h'_b} + \frac{1}{ik}\frac{h'_b g'_b - g_b h''_b}{h'^3_b} + O(k^{-2})\right]. \qquad \text{(V.2.36b)}$$

When g vanishes at an end point, say a, then $\gamma = 1$ and we have

$$I_{a(1,1)} \sim \frac{\exp(-i3\pi/4 - ikh_a)}{(2\pi)^{1/2}k^{3/2}} \sum_{n=0}^{\infty} \frac{n+1}{(ik)^n}\left(\frac{1}{h'}\frac{d}{ds}\right)^n\left[\frac{g}{h'(h-h_a)}\right]$$

$$= \frac{\exp(-i3\pi/4 - ikh_a)}{(2\pi)^{1/2}k^{3/2}}\left[\frac{g'_a}{h'^2_a} + \frac{2}{ik}\frac{g''_a h'_a - 3g'_a h''_a}{2h'^4_a} + O(k^{-2})\right]. \quad \text{(V.2.37)}$$

It is noteworthy that we cannot neglect the contribution of the end points when the field vanishes on them.

2.5 *Point of Discontinuity Contributions*

For a point of discontinuity of g or h with $\rho = 1$ and $\gamma = 0$, the first two terms of the asymptotic expansion are given by

$$I_{(1,0)} = \frac{\exp(-i\pi/4 - ikh^+)}{(2\pi k)^{1/2}}\left(\frac{g^+}{h'^+} + \frac{1}{ik}\frac{h'^+g'^+ - g^+h''^+}{h'^{+3}}\right)$$

$$- \frac{\exp(-i\pi/4 - ikh^-)}{(2\pi k)^{1/2}}\left[\frac{g^-}{h'^-} + \frac{1}{ik}\frac{h'^-g'^- - g^-h''^-}{(h'^-)^3}\right] + O(k^{-5/2}).$$

(V.2.38)

2.6 *Asymptotic Expression of the Diffraction Integrals*

If we collect the leading terms of the above integrals we can write for $\rho = 1$,

$$I(k) = \sum_i{}' \frac{e^{-ikh_i}}{(h''_i)^{1/2}} g_i + \left(\frac{e^{-ikh_a}}{h'_a} g_a - \frac{e^{-ikh_b}}{h'_b} g_b\right)\frac{e^{-i\pi/4}}{(2\pi k)^{1/2}}$$

$$+ \frac{e^{-i\pi/4}}{(2\pi k)^{1/2}} \sum_i{}'' \left(\frac{e^{-ikh_i^+}}{h'^+_i} g_i^+ - \frac{e^{-ikh_i^-}}{h'^-_i} g_i^-\right) + O(k^{-1}), \quad \text{(V.2.39)}$$

where the summation Σ' is extended to the stationary points, while Σ'' refers to the points of discontinuity.

If we take into account the definition of $h = S + R$ [cf. Eqs. (V.2.23,24)] we immediately verify that the stationary points correspond to the roots of the equation

$$h' = (\nabla' S + \nabla' R) \cdot \hat{x} = (\hat{s}' + \hat{R}) \cdot \hat{x} = 0. \quad \text{(V.2.40)}$$

Since $\hat{s}$ and $-\hat{R}$ are both oriented toward $\hat{x}$, this relation can be satisfied only for $\hat{s}' = -\hat{R}$. Consequently,

$$h'' = S'' + (z - z')^2/R^3 = [(1/\rho_c) + (1/R)](\hat{z} \cdot \hat{R})^2, \quad \text{(V.2.41)}$$

where we have indicated with ρ_c the radius of curvature of the wave front at $x_i = a_i$, so that

$$g_i/(h_i'')^{1/2} = [\rho_c/(\rho_c + R)]^{1/2} A(a_i) \tag{V.2.42}$$

and

$$\sum_i{}' \frac{\exp(-ikh_i)}{(h_i'')^{1/2}} g_i = \sum_i{}' \exp\{-ik[S(a_i) + R(a_i)]\} A(a_i)\left(\frac{\rho_c}{\rho_c + R}\right)^{1/2}. \tag{V.2.43}$$

We immediately recognize in Eq. (V.2.43) the expression for the field given by the leading term of the LK series.

Now, if we indicate with ϕ_a' the angle formed by the ray hitting end point a with $-\hat{x}$ and with ϕ_a the angle formed by $-\hat{R}_a$ with $-\hat{x}$, measured so that it is larger than π (see Fig. V.1a), we have

$$h_a' = \cos\phi_a + \cos\phi_a'. \tag{V.2.44}$$

Consequently, the contribution of end point a can be written as

$$I_a \sim \exp\{-ik[S(a) + R(a)]\} \frac{A(a)}{R_a^{1/2}} \frac{-2\sin\phi_a}{\cos\phi_a + \cos\phi_a'} \frac{\exp(-i\pi/4)}{(8\pi k)^{1/2}}. \tag{V.2.45}$$

By analogous considerations for b and the points of discontinuity of h and g, we finally obtain (see Fig. V.1b for the definition of the angles)

$$\begin{aligned} I(k) \sim{} & \sum_i{}' \exp\{-ik[S(a_i) + R(a_i)]\} A(a_i)\left(\frac{\rho_c}{\rho_c + R}\right)^{1/2} \\ & + \frac{\exp(-i\pi/4)}{(8\pi \mathrm{k})^{1/2}} \exp\{-ik[S(a) + R(a)]\} \frac{A(a)}{R(a)^{1/2}} \frac{-2\sin\phi_a}{\cos\phi_a + \cos\phi_a'} \\ & + \frac{\exp(-i\pi/4)}{(8\pi k)^{1/2}} \exp\{-ik[S(b) + R(b)]\} \frac{A(b)}{R(b)^{1/2}} \frac{-2\sin\phi_b}{\cos\phi_b + \cos\phi_b'} \\ & + \sum_i{}'' \frac{\exp(-i\pi/4)}{(8\pi k)^{1/2}} \left[-\exp[-ik(S_i^+ + R_i)] \frac{A_i^+}{R_i^{1/2}} \frac{-2\sin\phi_i'}{\cos\phi_i + \cos\phi_i'^+} \right. \\ & \left. + \exp[-ik(S_i^- + R_i)] \frac{A_i^-}{R_i^{1/2}} \frac{-2\sin\phi_i'}{\cos\phi_i + \cos\phi_i'^-} \right] \equiv u_g + u_b + u_\Delta, \end{aligned} \tag{V.2.46}$$

where u_g is the contribution of the stationary points, and u_b and u_Δ are those of the end points (boundary) and the discontinuities, respectively. The term u_g coincides with the geometric optics (GO) representation of the field, while the remaining terms account for the diffraction effects due to the finite extension of the wave front and the discontinuities of the phase and the amplitude.

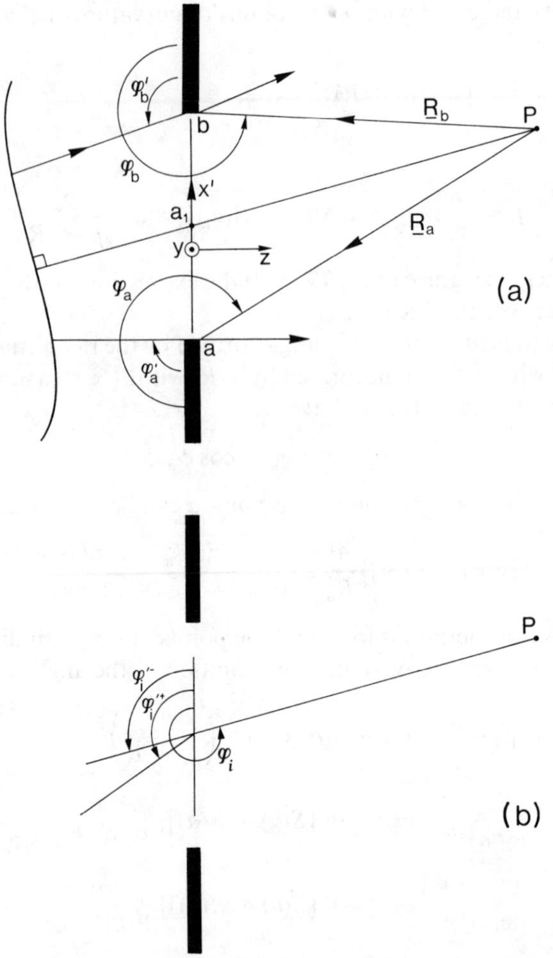

Fig. V.1. Geometry relative to the contribution to the diffraction integral from the aperture edges, the stationary points (a), and the points of discontinuity of the phase (b).

It is now worth rewriting u_b as a function of the incident field u_i, i.e.,

$$u_b = D_1(\phi_a, \phi_a')u_i(a)e^{-ikR(a)}R(a)^{-1/2} + D_1(\phi_b, \phi_b')u_i(b)e^{-ikR(b)}R(b)^{-1/2}, \tag{V.2.47}$$

where we have introduced the *diffraction coefficient* D_1, defined as (see Keller [1], Eq. (27))

$$D_1(\phi, \phi') = [e^{-i\pi/4}/(8\pi k)^{1/2}](-2\sin\phi)/(\cos\phi + \cos\phi'). \tag{V.2.48}$$

According to Eq. (V.2.48), the aperture edge is completely characterized by these diffraction coefficients. On the other hand, Eq. (V.2.47) is based on the Kirchhoff approximation, since u_i is defined as the incident field calculated in absence of the screen. We will see in the following chapter that it is possible to account for the modifications due to the screen itself by retaining Eq. (V.2.47) as valid with the unperturbed incident field and by replacing the above expression for the diffraction coefficients with others suggested by Sommerfeld's solution of the canonical problem of the diffraction of a plane wave by a half-plane.

2.7 *Removal of the SP Singularities by Recourse to Comparison Integrals*

The asymptotic expressions discussed above become singular when either $h''(s_i) = 0$ or a stationary point comes close to an end point. In order to avoid these singularities and obtain an asymptotically correct representation of a diffraction integral, we can replace it with a *comparison integral*, which exhibits the same singularities if expressed with the series of the preceding section. This integral is chosen in such a way as to coincide with a known special function, such as the Fresnel complex integral $F(x)$, the Airy function $\mathrm{Ai}(x)$, or the parabolic cylinder function $D_{-1/2}(e^{i\pi/4}x)$. When we come close to one of the conditions for which the usual expansions diverge, we express the diffraction integral as the product of the comparison integral times an asymptotic series that tends to a constant when the conditions of singularities are satisfied. In most cases we can omit the explicit calculation of this series, since the comparison integral represents the field with sufficient accuracy until we are far enough from the critical regions that the usual expansions are valid. In other words, generally the comparison integral representation transforms gradually and without discontinuity into the LK series. For all these reasons the representation based on comparison integrals is called *uniform* and the relative approach *uniform asymptotic theory*. In the following sections we will analyze some cases of particular interest.

3 Shadow Boundaries: Stationary Point near End Point

The above representation of the end point contribution (Section V.2.6) becomes singular when $\phi_a - \phi'_a = \pi$, that is, when the diffracted ray is parallel to that incident on edge a. From a geometric point of view the prolongation of the incident ray separates the lit region from the dark one, so that it delimits the shadow projected by the aperture. For this property the surface formed by all the rays touching the aperture edge is called the shadow boundary (SB). In the case of a slit illuminated by a cylindrical wave, the shadow boundary reduces to a couple of half-planes. When we limit the analysis of a cylindrical field to the distribution on a plane perpendicular to the relative axis, it is customary to apply the term shadow boundary to the two half-lines obtained as the cross section of the SB surface.

We have frequently called the attention of the reader to the singular behavior of an RO field in proximity to an SB. This behavior parallels the singularity of the edge contribution to the diffracted field, due to the coincidence of an end point of the integral with a stationary point. In principle, this anomaly can be removed by a more accurate asymptotic evaluation of the integral.

As a preliminary step we note that the leading term of the asymptotic series representing the contribution of a stationary point α is given by

$$I_\alpha \sim \frac{\exp(-i\pi/4 - ikh_\alpha)}{\lambda^{1/2}} g(\alpha) \int_{-\infty}^{\infty} \exp(-iks^2 u_\alpha^0)\, ds, \tag{V.3.1}$$

where $u_\alpha^0 = u_\alpha(0) = h''(\alpha)/2$ and $s_\alpha = 0$. When there is an end point in proximity to α the integral becomes

$$\begin{aligned} I_\alpha &\sim \left(\frac{k}{2\pi}\right)^{1/2} \exp(i\pi/4 - ikh_\alpha) g(\alpha) \int_{s_a}^{\infty} \exp(-iks^2 u_\alpha^0)\, ds \\ &= \left(\frac{k}{2\pi}\right)^{1/2} \exp(-ikh_\alpha) \frac{g(\alpha)}{(ku_\alpha^0)^{1/2}} F(\xi) = u_{\mathrm{i}}(P) F(\xi), \end{aligned} \tag{V.3.2}$$

where $u_{\mathrm{i}}(P)$ represents the incident field at field point P and $F(\xi)$ is a complex Fresnel integral accounting for the transition from the lit to the dark region. ξ is the so-called *detour parameter*, defined by [17]

$$\xi \equiv \sqrt{k(h_\alpha - h_a)} = \sqrt{ku_\alpha^0}\, s_a. \tag{V.3.3}$$

We recall that kh represents the phase of the field at P produced by the wavelets departing from the aperture point $P'(s)$. Consequently, $k(h - h_a) = \xi^2$ is the phase difference between the field produced by the ray passing undisturbed and that associated with the ray diffracted by the aperture edge. Speaking in terms of rays, $h_\alpha - h_a$ represents the difference between the optical path of the ray reaching P without deviations and that of the ray diffracted by the edge. With reference to Fig. V.2 we have

$$\xi = \sqrt{k_0([AP] - [CP])} \simeq -(2k\rho)^{1/2} \cos(\psi/2), \tag{V.3.4}$$

with $n\rho = [AP]$, where we have made use of the fact that the ray in C is almost parallel to the ray in A. The sign of the square root is chosen positive in the shadow region and negative otherwise. In Fig. V.3 we schematically indicate the transition region surrounding the shadow boundary for a straight edge

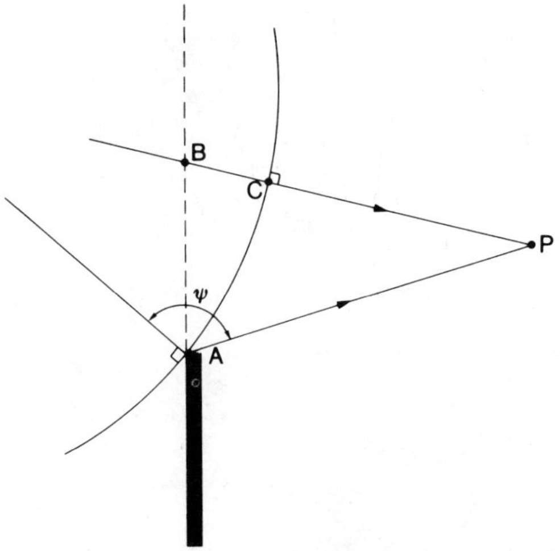

Fig. V.2. Detour parameter $\xi = [k_0(AP - CP)]^{1/2}$ relative to the stationary-phase (C) and edge (A) contributions to the field in P.

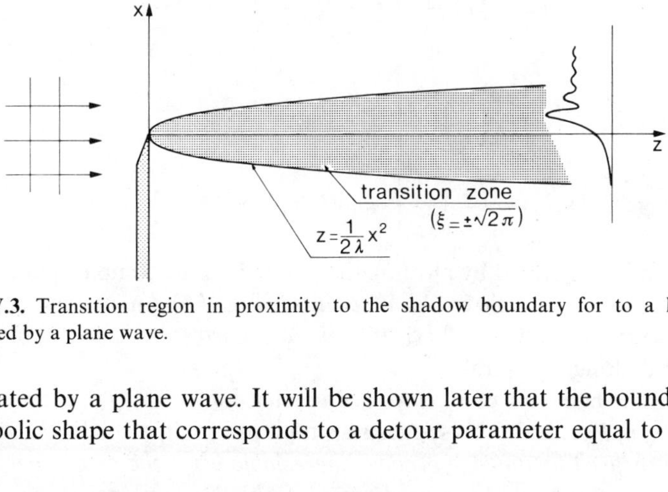

Fig. V.3. Transition region in proximity to the shadow boundary for to a half-plane illuminated by a plane wave.

illuminated by a plane wave. It will be shown later that the boundary has a parabolic shape that corresponds to a detour parameter equal to $\pm\sqrt{2\pi}$.

3.1 *Properties of the Transition Function $F(x)$*

The transition function describing the passage through the shadow boundary coincides with the *complex Fresnel integral*

$$F(x) = \frac{1}{\sqrt{\pi}} e^{i\pi/4} \int_x^\infty e^{-it^2}\, dt. \tag{V.3.5}$$

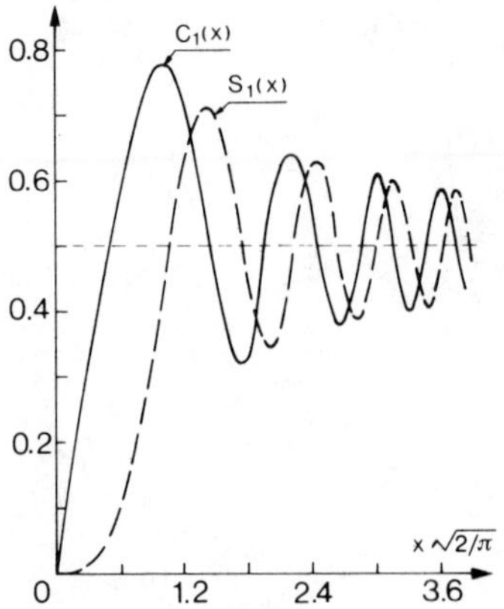

Fig. V.4. Fresnel integrals.

This is an entire transcendental function, which can be represented by means of the Fresnel integrals $C_1(x)$ and $S_1(x)$ (see Fig. V.4)

$$F(x) = \tfrac{1}{2} - \tfrac{1}{2}[C_1(x) - iS_1(x)](1 + i), \tag{V.3.6}$$

$$|F(x)| = \tfrac{1}{2}[1 + 2C_1^2 + 2S_1^2 - 2C_1 - 2S_1]^{1/2}, \tag{V.3.7}$$

where

$$C_1(x) = \left(\frac{2}{\pi}\right)^{1/2} \int_0^x \cos t^2\, dt, \tag{V.3.8}$$

$$S_1(x) = \left(\frac{2}{\pi}\right)^{1/2} \int_0^x \sin t^2\, dt. \tag{V.3.9}$$

The curve generated by plotting the points F in the complex plane for all x from $-\infty$ to $+\infty$ is called a *Cornu spiral* (see Fig. V.5). An interesting property of this curve is that $ds = |dx|$, that is, $|dx|$ corresponds to the arc length measured along the spiral.

From the behavior of $F(x)$ we deduce that for $x < -\sqrt{\pi}$, $F(x) \simeq 1$, while for $x > \sqrt{\pi}$, $F(x) \simeq 0$. This suggests a simple approximate rule: *the diffraction effects produced by an edge become appreciable when the detour parameter is between $-\sqrt{\pi}$ and $\sqrt{\pi}$. In terms of optical paths, this corresponds to a difference less than $\lambda/2$.*

As an example, let us consider a point source S and a field point P with a straight edge in between (see Fig. V.6) (neglect the fact that the source is pointwise). In practice, the field will coincide with the geometric optics field when

$$(h^2/2)[(1/d_1) + (1/d_2)] \gg \lambda/2. \tag{V.3.10}$$

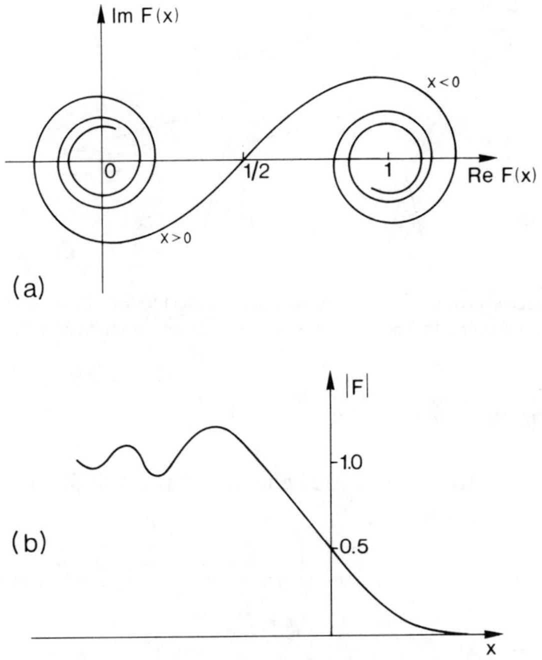

Fig. V.5. (a) Cornu spiral and (b) amplitude of F.

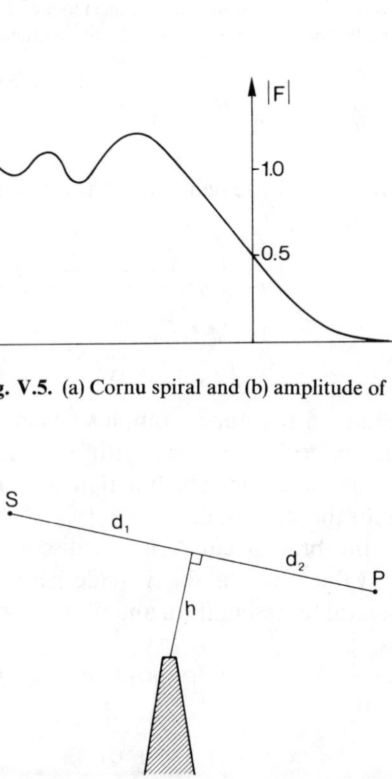

Fig. V.6. Geometry for the calculation of the detour parameter for a source S and a field point P in the presence of a diffracting obstacle represented by the hatched region.

If we consider the Fresnel ellipse having foci in S and P and semiaxis $a = \lambda/4 + SP/2$ (see Fig. V.7), the field will not be affected by the obstacle if its edge lies outside the ellipse. This result has been generalized to more complex fields by Kravtsov and Orlov [18].

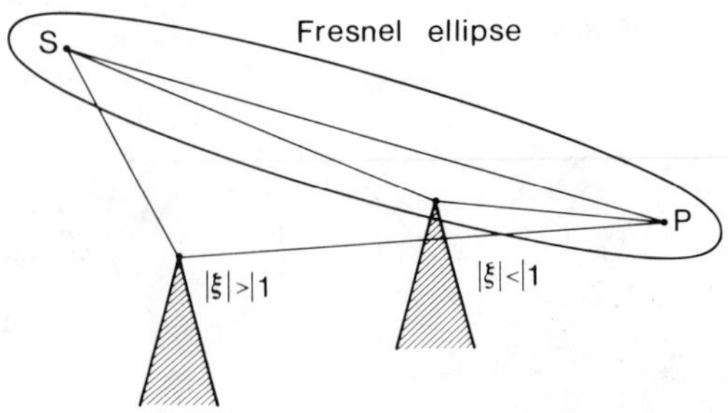

Fig. V.7. Obstacles having different positions with respect to the Fresnel ellipse relative to the source S and the field point P. The field in P is affected only by the obstacles penetrating the Fresnel ellipse.

3.2 *Asymptotic Behavior of* $F(x)$

In order to study the asymptotic behavior of the diffracted field it is useful to introduce the function

$$G(x) = -xe^{-i\pi/4 - ix^2}\frac{1}{2\pi i}\int \frac{e^{-t^2}}{t^2 + ix^2}\,dt - \operatorname{res}\left(\frac{xe^{-1\pi/4}}{t^2 + ix^2}\right)$$

$$= -xe^{-i\pi/4 - ix^2}\frac{1}{2\pi i}\int_{-\infty}^{+\infty} \frac{e^{-t^2}}{t^2 + ix^2}\,dt, \qquad \text{(V.3.11)}$$

where res( ) is the residue in the upper complex t plane (i.e., $\operatorname{Im} t > 0$). The first integration path connects $x = -\infty$ with $x = \infty$ by leaving on the right side the poles of the integrand. The function G is analytic on the whole complex plane except for the straight line $x = \pm|x|e^{i\pi/4}$, which is a *branch cut*. On the two sides of the branch cut G has a discontinuity equal to one (difference between the values taken at the two sides); this can be easily verified by using the above integral representation and observing that the jump is due to the variation of res( ).

After these remarks, we notice that for positive real x, $G(x)$ reduces to $F(x)$. Then, in general, we have

$$F(x) = U_F(-x) + G(x), \qquad \text{(V.3.12)}$$

$U_F(-x)$ being equal to zero on the lower side of the branch cut [i.e., $\operatorname{Im}(xe^{-i\pi/4}) < 0$] and one otherwise. The function U_F has been introduced to compensate the discontinuity of G.

The function $G(x)$ admits the following asymptotic representation, as can be shown by integration by parts of the last integral of Eq. (V.3.11):

$$G(x) \sim \hat{F}(x)\left[1 + \sum_{n=1} \frac{1 \cdot 3 \cdot 5 \cdots (2n-1)}{(-2)^n x^{2n} i^n}\right], \qquad \text{(V.3.13)}$$

where

$$\hat{F}(x) = (1/2\sqrt{\pi})(1/x)e^{-i(x^2+\pi/4)}. \tag{V.3.14}$$

Since this expansion is valid for every value of the phase of x, we finally have

$$F(x) \sim U_F(-x) + \hat{F}(x) + O(1/x^2), \qquad |x| \gg 1. \tag{V.3.15}$$

3.3 *Asymptotic Representation of the Field*

From the above representation of F [see Eq. (V.3.15)] it follows that

$$I_\alpha \sim u_i[U_F(-\xi) + \hat{F}(\xi)] \tag{V.3.16}$$

for $|\xi|$ sufficiently large. Accordingly, when a stationary point approaches an end point, we can neglect the isolated contribution from the latter, and multiply the incident optical field u_i by the transition function F, so that

$$\begin{aligned} u(P) = u_i U_F(-\xi) + u_i G(\xi) &= u_i U_F[(2k\rho)^{1/2}\cos(\psi/2)] \\ &+ u_i G[-(2k\rho)^{1/2}\cos(\psi/2)] \equiv u_g + u_d, \end{aligned} \tag{V.3.17}$$

where $u_d \equiv u_i G$.

3.4 *F(x) at small arguments*

For x small, the function F can be approximated by the first terms of its power series expansion,

$$F(x) \simeq \frac{1}{2} - \frac{e^{i\pi/4}}{\sqrt{\pi}} x \left\{1 - \frac{i}{1!3}x^2 - \frac{x^4}{2!5} + i\frac{x^6}{3!7} + \cdots\right\} \tag{V.3.18}$$

In particular, $F(0) = 1/2$, $F(1.6) \simeq -0.16$, $F(-1.6) \simeq 1.16$.

4 Caustics of Cylindrical Fields: Two Adjacent Stationary Points

When the field point approaches a caustic, then two or more rays having almost equal directions pass through P. Mathematically, this means (see Fig. V.8) that $h(s)$ has two or more stationary points quite close to each other. To be specific, let us assume that we have two such points, say s_1 and s_2. Since $h'(s)$ vanishes at s_1 and s_2 there must be a point between them, say s_0 where $h''(s_0) = 0$. If we now put the origin of our coordinates at s_0, then the leading term of the asymptotic approximation to the diffraction integral can

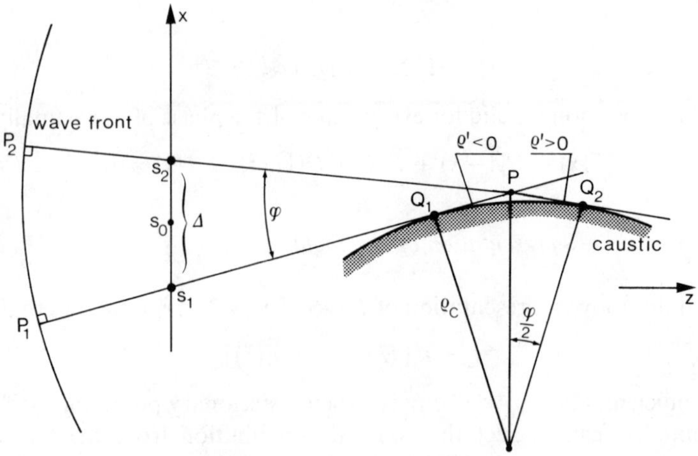

Fig. V.8. Geometry for the calculation of the field in proximity to a caustic.

be obtained by writing

$$I_\alpha \sim \frac{\exp(i\pi/4 - ikh_\alpha)}{\sqrt{\lambda}} g_\alpha \int_{-\infty}^{\infty} \exp[-ik(u_\alpha^1 s + \tfrac{1}{6}u_\alpha^3 s^3)]\, ds, \qquad \text{(V.4.1)}$$

where we indicate with α the point $s_0 = 0$ and with u_α^1 and u_α^3, respectively, the first and third derivatives of h at s_0. It should be noted that s_0 is not a stationary point and that accordingly the derivative u_α^1 is different from zero. When s_1 is very close to s_2, we have

$$u_\alpha^1 + \tfrac{1}{2}u_\alpha^3 s_1^2 = 0, \qquad \text{(V.4.2a)}$$

$$u_\alpha^3 s_1 = h''(s_1) \simeq -h''(s_2), \qquad \text{(V.4.2b)}$$

so that

$$u_\alpha^1 = -\tfrac{1}{2}h''(s_1)s_1 = \tfrac{1}{4}h_1''\,\Delta, \qquad \text{(V.4.3a)}$$

$$u_\alpha^3 = h''(s_1)/s_1 = -2h_1''/\Delta, \qquad \text{(V.4.3b)}$$

$$s_1 = -\tfrac{1}{2}\Delta, \qquad \text{(V.4.3c)}$$

Δ being the distance between the two stationary points. The integral on the rightside of Eq. (V.4.1) is the Airy function Ai (cf. Section III.3 and Fig. V.9), so that we have

$$I_\alpha \sim \frac{\exp(i\pi/4 - ikh_\alpha)}{\sqrt{\lambda}} g_\alpha 2\pi\left(\frac{2}{ku_\alpha^3}\right)^{1/3} \mathrm{Ai}\left[ku_\alpha^1\left(\frac{2}{ku_\alpha^3}\right)^{1/3}\right]$$

$$= (2\pi k)^{1/2}\exp(i\pi/4 - ikh_\alpha)g_\alpha\left(-\frac{\Delta}{kh_1''}\right)^{1/3} \mathrm{Ai}\left[-\frac{1}{4}(k^2 h_1''^2 \Delta^4)^{1/3}\right].$$

(V.4.4)

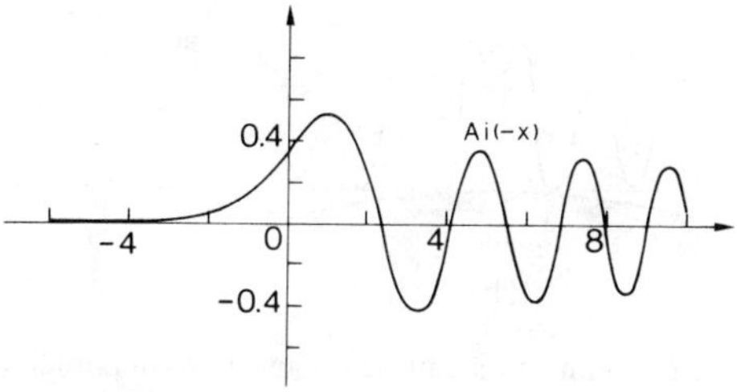

Fig. V.9. Plot of the Airy function Ai(x) for $-10 \leq x \leq 6$. The function Ai(x) is oscillatory for negative x and decays exponentially as $x \to \infty$.

On the other hand, we know that [see Section V.2.6, Eq. (V.2.41)]

$$h_1'' = [(R + \rho)/R\rho](\hat{z} \cdot \hat{R})^2 \simeq -(\rho'/R^2)(\hat{z} \cdot \hat{R})^2, \tag{V.4.5}$$

$$\Delta^2 = (R^2 \sin^2 \phi)/(\hat{z} \cdot \hat{R})^2, \tag{V.4.6}$$

where the signed quantity $\rho' = |\rho| - R$ measures the distance of P from the caustic measured along the ray (ρ' is positive when P lies between the center of curvature Q and the aperture, and is negative otherwise) and ϕ is the angle formed by the two rays passing through P, so that, using Eqs. (V.4.5) and (V.4.6), we arrive at the relation

$$\left(-\frac{\Delta}{kh_1''}\right)^{1/3} = [R/(\hat{z} \cdot \hat{R})][\sin \phi/(k\rho')]^{1/3}. \tag{V.4.7}$$

Since we can show with simple geometric arguments that ϕ, for small values, is related to ρ' and to the curvature radius ρ_c of the caustic through the expression (see Fig. V.8)

$$\phi \simeq 2|\rho'|/\rho_c, \tag{V.4.8}$$

we finally obtain, by replacing g_α with $AR^{-1/2}$ and putting $\hat{z} \cdot \hat{R} = 1$,

$$I_\alpha(\rho') \sim (2\pi k)^{1/2} \exp(i\pi/4 - ikh_\alpha) AR^{1/2} \left(\frac{2}{k\rho_c}\right)^{1/3} \mathrm{Ai}\left[-\left(\frac{k^2\rho'^6}{4\rho_c^4}\right)^{1/3}\right]. \tag{V.4.9}$$

Since $AR^{1/2}$ is constant along a ray, the right side of Eq. (V.4.9) is independent of the position of the reference plane $z = 0$. Figure V.10c shows a schematic plot of $I(\rho')$. On the caustic we have

$$I_\alpha(0) \sim 0.355(2\pi k)^{1/2} AR^{1/2} \left(\frac{2}{k\rho_c}\right)^{1/3} \exp(i\pi/4 - ikh_\alpha), \tag{V.4.10}$$

while at some distance from it we can replace the Airy function with its

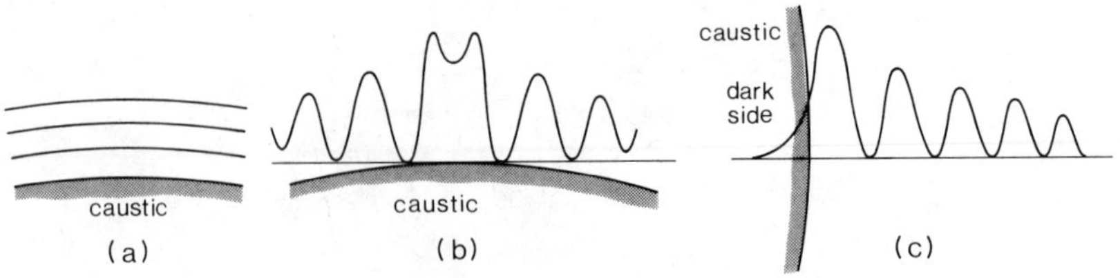

Fig. V.10. Fringes in proximity to a caustic (a) and field amplitude along a ray (b) and a normal to a caustic (c).

asymptotic expression [see Eq. (III.3.6)] and

$$I_\alpha(\rho') = A\left(\frac{R}{|\rho'|}\right)^{1/2} \exp(-ikh_\alpha)\{\exp[ik\rho'^3/(3\rho_c^2)] - i\exp[-ik\rho'^3/(3\rho_c^2)]\}$$
$$= u_g(P, Q_1) + u_g(P, Q_2), \qquad \text{(V.4.11)}$$

where $u_g(P, Q_i) = A(R/|\rho'|)^{1/2}\exp[-ik(h_\alpha - \rho'^3/(3\rho_c^2)]$ represents the field along a ray passing through P and Q_i; ρ' is positive if Q_i lies outside the segment PP_i, P_i being the intersection with the wave front. The two ray fields interfere, thus producing a typical oscillating pattern (see Fig. V.10.b). It is noteworthy that each field undergoes a shift of $\pi/2$ in passing the caustic tangentially. In addition, its effective wave number varies as $k(\rho') = k(1 - \rho'^2/\rho_c^2)$. This permits us to say that the wave "slows down" in proximity to the caustic.

Let us now return to Eq. (V.4.9) to observe that $\rho'^2 = 2\rho_c\rho$, where ρ is the normal distance of P from the caustic. Therefore, we can rewrite Eq. (V.4.9) in the form

$$I_\alpha(\rho) \sim (2\pi k)^{1/2} A R^{1/2}\left(\frac{2}{k\rho_c}\right)^{1/3} \exp\left(\frac{i\pi}{4} - ikh_\alpha\right)\mathrm{Ai}\left[-2\left(\frac{k}{2\rho_c^2}\right)^{2/3}\rho_c\rho\right]. \qquad \text{(V.4.12)}$$

In this expression ρ can assume positive and negative values according to whether P lies in the lit or the shadow region. A plot of the field distribution along a normal to a caustic is shown in Fig. V.10c.

5 Field in Proximity to a Two-Dimensional Cusp: A Model for the Impulse Response in the Presence of Defocusing and Third-Order Aberration

We saw in the preceding section that the field structure near a simple caustic, produced by an advancing cylindrical wave that possesses any nonzero amount of cylindrical aberration, depends on the curvature of the caustic, the nature of the variation of the field in a direction perpendicular to the caustic being expressible in terms of the Airy integral. When we are near the cusp of a caustic (see Figs. II.15,16) the field pattern becomes much more complex due to the interference of three or more rays. This situation can be described by saying that three stationary points of the function $h(s)$ appearing in the diffraction integral tend to coalesce. Accordingly, we can study the field by analyzing the comparison integral

$$I(k) = \frac{\exp(i\pi/4)}{\sqrt{\lambda}} \int_a^b \exp[-i(as + bs^2 + cs^4)]\, ds. \tag{V.5.1}$$

Note that the absence of a term in s^3 does not limit the generality of this expression, since it can always be eliminated by a simple translation of the s coordinate.

In order to find a physical model for the above integral let us consider a slit of width $w = |b - a|$ illuminated by a uniform field having phase $-kS(s) = ks^2/(2f) - (B/4)(s/w)^4$, where f is the distance of the focus from the aperture and B is a measure of the spherical aberration [see Sections II.15 and IV.13 and Eq. (IV.13.34)]. The diffracted field is then

$$I(k) = \frac{\exp(i\pi/4)}{\sqrt{\lambda}} \int_a^b \exp\left\{i\left[\frac{kx}{2z}s + \frac{k}{2}\left(\frac{1}{f} - \frac{1}{z}\right)s^2 + \frac{Bs^4}{4w^4}\right]\right\} ds. \tag{V.5.2}$$

Near the focus ($x \simeq 0, z \simeq f$) the three stationary points of the phase factor tend to coalesce into the focus itself, which implies that the caustic of this field has a cusp that coincides with the paraxial focus. On the other hand, we can immediately obtain the equation of the caustic by imposing the coalescence of two stationary points. If we analyze the cubic equation $h'(s) = 0$, which for $z \simeq f$ reads

$$s^3 + [(\bar{u}w^2 s)/B] + [(w^3 v)/(2B)] = 0 \tag{V.5.3}$$

where $v = k\text{NA}x$ and $\bar{u} = k(\text{NA})^2(z - f)$ are the optical coordinates of the field point, we find that the condition for obtaining a double root is given by

$$\bar{u}^3 = -(27/16)Bv^2. \tag{V.5.4}$$

If we finally neglect the contributions of the end points, we can rewrite Eq. (V.5.2) in the form [19]

$$I(k) \propto \int_{-\infty}^{+\infty} \exp[-i(Vt + Ut^2 + t^4)]\, dt \equiv I(\text{V}, U), \tag{V.5.5}$$

where U and V are nondimensional coordinates connected with $\bar{u}$ and v by the relations

$$U = \bar{u}/B^{1/2}, \qquad V = v/(2^{1/2}B^{1/4}). \tag{V.5.6}$$

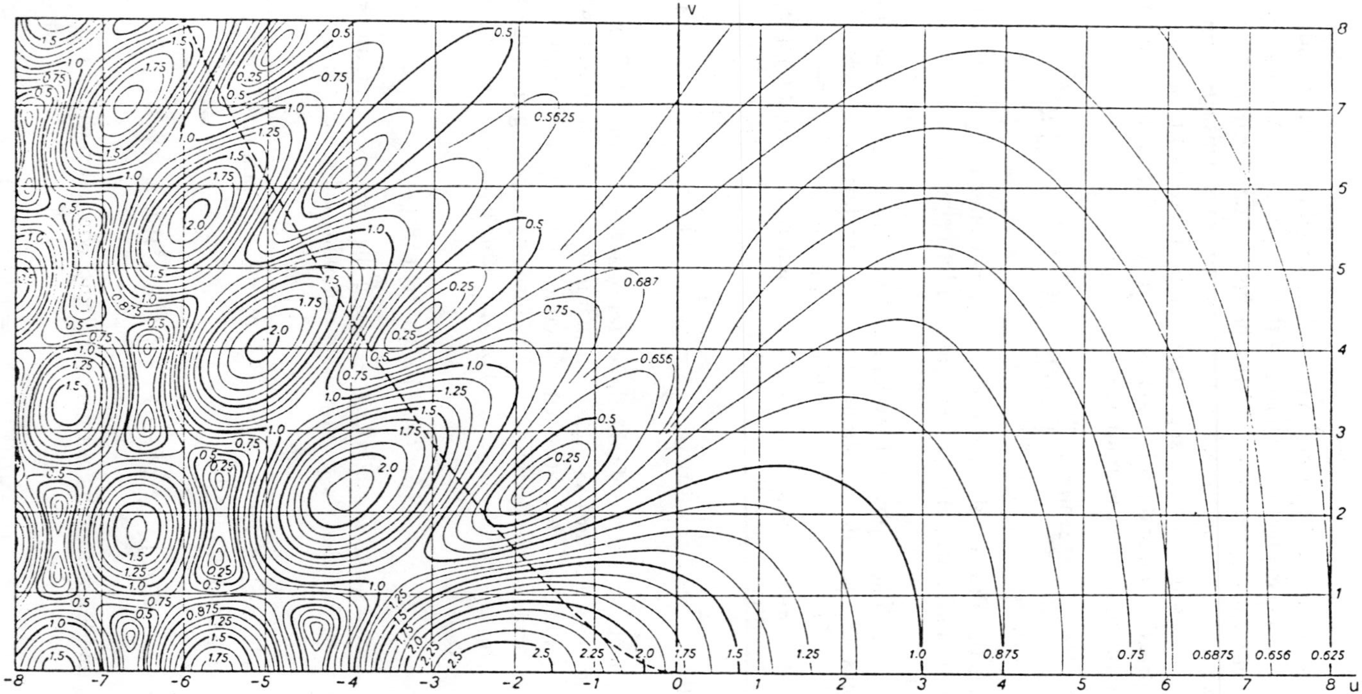

Fig. V.11. Contours of modulus of the cusp function $I(V, U) = \int_{-\infty}^{+\infty} \exp[-i(Vt + Ut^2 + t^4)]\, dt$ calculated by Pearcey. (From Pearcey [19].)

The indefinite integral $I(V, U)$ has been evaluated by Pearcey, and diagrams of the modulus of I are shown in Fig. V.11. Note that a principal focus is formed on the U axis at $U \cong -1.8$. Whereas on the left of the cusp there is a system of three rays lying between the two branches of the caustic, which by interfering give the complex system of maxima and minima seen in Fig. V.11, on the right there is only one family of rays so that I decays monotonically.

Along the U axis $I(U,0)$ can be expressed by means of the *parabolic cylinder function* $D_{-1/2}$ (see Chapter II, Abramowitz and and Stegun [4], p. 687),

$$I(U,0) = 2\int_0^{\infty} \exp[-i(Ut^2 + t^4)]\,dt \propto D_{-1/2}(\exp(i\pi/4)2^{-1/2}U). \tag{V.5.7}$$

Recalling that $D_{-1/2}(x)$ satisfies the differential equation

$$(d^2/dx^2)D_{-1/2}(x) - \tfrac{1}{4}x^2 D_{-1/2}(x) = 0, \tag{V.5.8}$$

we immediately see that

$$(d^2/du^2)I(U,0) + \tfrac{1}{16}U^2 I(U,0) = 0. \tag{V.5.9}$$

This equation can be compared with that of Airy [see Eq. (III.3.2)]; both have a turning point at the origin, but of different order. While the Airy function Ai decreases exponentially for $x \to \infty$, in the present case we have for $U \gg 1$,

$$I(U,0) \sim (A_{\pm}/|U|^{1/2})\sin(U^2/8 + \phi) \tag{V.5.10}$$

6 Steepest-Descent Method

In several cases the stationary-phase method cannot be used because of the slow convergence of the relative asymptotic series representation of a diffraction integral. An important case where this limit is particularly evident is that in which the distribution of the field on the reference plane is gaussian. Let us consider the following diffraction integral:

$$\begin{aligned} I(k) &= \frac{\exp(i\pi/4)}{\sqrt{\lambda}} \int_{-\infty}^{\infty} \exp(-ikas^2 - bs^2 - ikcs)\,ds \\ &= \frac{\exp\{-k^2c^2/[4(b + ika)]\}}{(2a - 2ib/k)^{1/2}}, \end{aligned} \tag{V.6.1}$$

where a and b are real quantities. If we put $g(s) = \exp(-bs^2)$ and $h(s) = as^2 + cs$ and use Eq. (V.2.32), we obtain

$$I(k) \sim \frac{1}{\sqrt{2a}} \exp[(c^2/4)(ik/a - b/a^2)]. \tag{V.6.2}$$

The last expression differs from the exact expression for the integral given by Eq. (V.6.1), especially when c^2k^2 and b are comparable with ka. Physically, $I(k)$ represents the field radiated by a gaussian distribution, viz. $u(x,0) \propto \exp(-bx^2)$ at a distance $z = 1/(2a)$, along the direction $\theta = \arcsin c$. While the exact expression [Eq. (V.6.1)] for the field predicts for $a \to 0$ $(z \to \infty)$ a far field proportional to $\exp(-k^2\theta^2/4b)$, Eq. (V.6.2) gives a field proportional to $\exp(-bx^2)$. This means that in this case the SP method is unable to predict the spreading of the beam during propagation.

A way to improve the SP method consists of introducing a complex function $h(s) = as^2 + cs - ibs^2/k$. The prolongation of the phase function into the complex plane presents some problems concerning the solutions of the equation $h'(s) = 0$, which will, in general, have complex roots. If these roots are not real there will be no point on the real axis where the real and imaginary parts of h' vanish simultaneously. This difficulty can be circumvented by modifying the integration path so that it passes through the complex roots s_s of $h' = 0$. In proximity to s_s, $\operatorname{Re} h(s)$ and $\operatorname{Im} h(s)$ represent quadric surfaces with a saddle point in s_s, as a consequence of the assumption that $h(s)$ is analytic. Accordingly, the stationary points of $h(s)$ are called *saddle points* [14] (see Figs. V.12,13).

When $h(s)$ is analytic it satisfies the *Cauchy–Riemann differential equations*

$$\partial \operatorname{Im} h/\partial s'' = \partial \operatorname{Re} h/\partial s', \qquad \partial \operatorname{Im} h/\partial s' = -\partial \operatorname{Re} h/\partial s'' \tag{V.6.3}$$

where $s = s' + is''$. From a geometric point of view these equations imply that the families of curves $\operatorname{Im} h = \text{const}$ and $\operatorname{Re} h = \text{const}$ are mutually orthogonal. Consequently, the curve $\operatorname{Re} h = \text{const}$ is tangent everywhere to the gradient of $\operatorname{Im} h$. The curve $\operatorname{Re} h = \text{const}$ is the path along which $\operatorname{Im} h$ varies in the most rapid way (see Figs. V.14,15). As a consequence of this property $\operatorname{Re} h = \text{const}$ is called the *steepest-descent* (or *ascent*) *path.* In the following we will indicate with SDP a *steepest-descent path for* Im *h*, *passing through a saddle point*.

These preliminary considerations lead us to the conclusion of deforming the real axis into the SDP passing through the saddle point of $h(s)$, supposed for the time being to be unique, and such that $\operatorname{Im} h(s_s) > \operatorname{Im} h(s)$. We put

$$I(k) = \frac{e^{i\pi/4}}{\lambda^{1/2}}\left[\left(\int_{\text{SDP}} + \int_{\Gamma_{\text{B}}}\right) g e^{-ikh(s)}\, ds + \sum_q r_q e^{-ikh(s_q)}\right]$$
$$\equiv I_{\text{SDP}} + I_{\Gamma_{\text{B}}} + \frac{e^{i\pi/4}}{\lambda^{1/2}} \sum_q r_q e^{-ikh(s_q)}, \tag{V.6.4}$$

where Γ_{B} represents an ensemble of paths encircling the branch cuts of $g(s)$, if

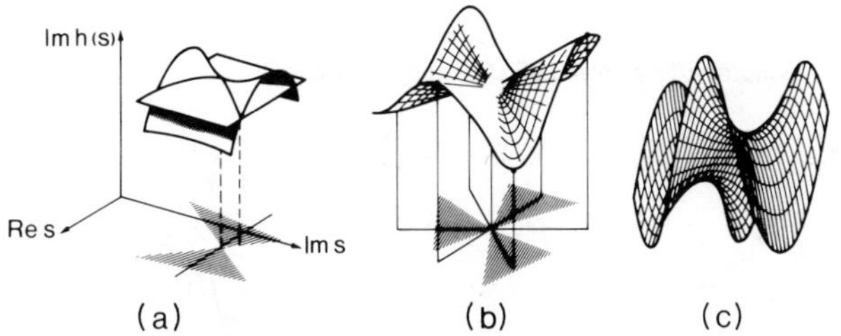

Fig. V.12. Examples of surfaces near a simple saddle point (a), a second-order saddle point (also called *monkey saddle*) (b), and two saddle points (c). The hatched regions represent the projections of the respective valleys on the xy plane.

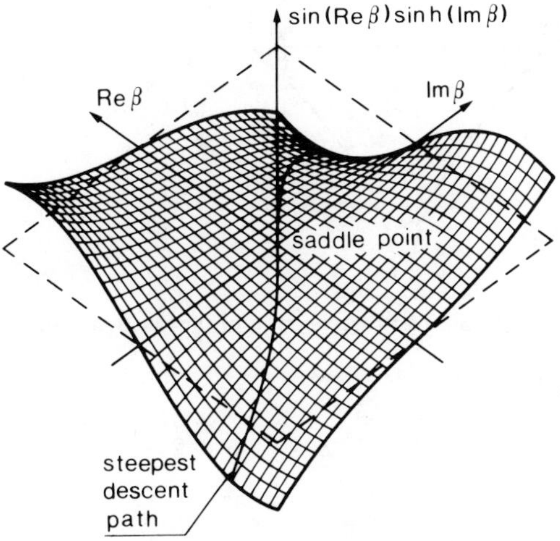

Fig. V.13. Tridimensional plot of the function $\sin(\mathrm{Re}\,\beta)\sinh(\mathrm{Im}\,\beta) = \mathrm{Im}\sin(\beta)$ (courtesy of P. Luchini).

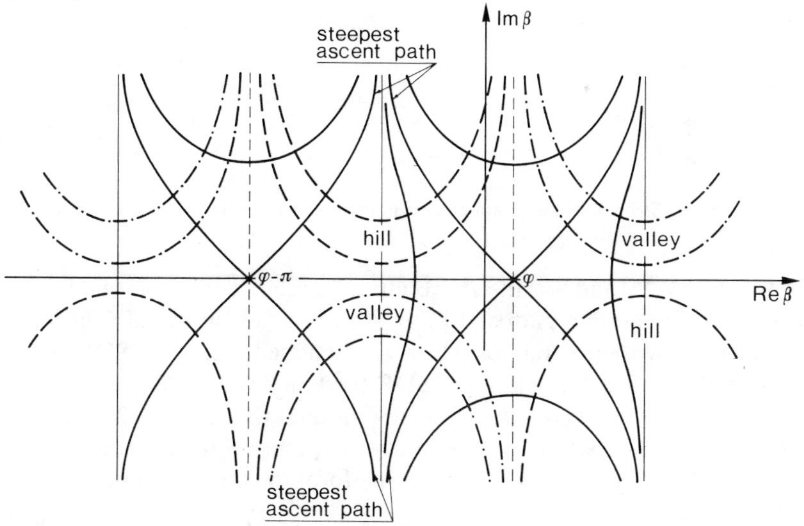

Fig. V.14. Hills and valleys of the function $\mathrm{Im}\cos(\beta - \phi)$. (———) $\mathrm{Re}\cos(\beta - \phi) = \text{const}$; (-----) $\mathrm{Im}\cos(\beta - \phi) = \text{const} > 0$; (—·—·—) $\mathrm{Im}\cos(\beta - \phi) = \text{const} < 0$.

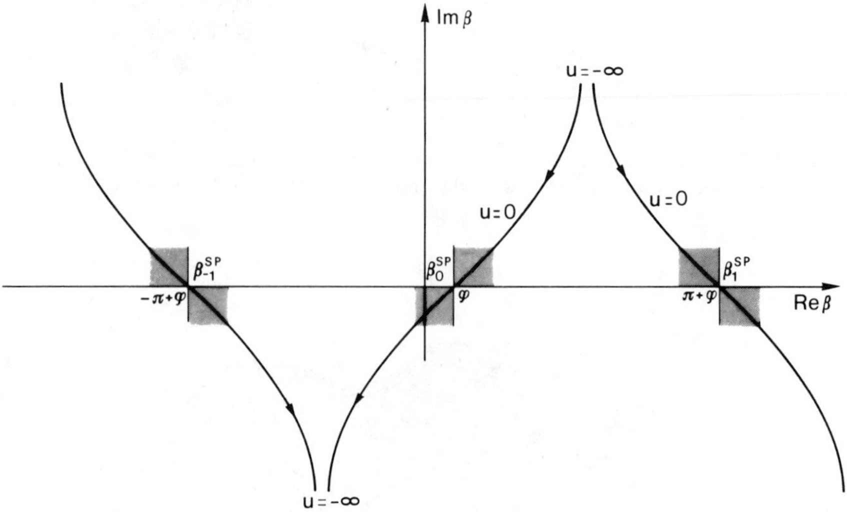

Fig. V.15. Steepest-descent paths of the function $\cos(\beta - \phi)$.

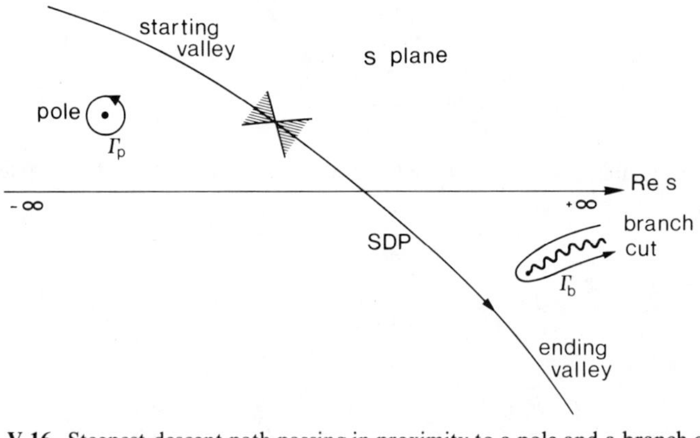

Fig. V.16. Steepest-descent path passing in proximity to a pole and a branch cut.

any (see Fig. V.16) and r_q is the qth residue of g relative to a pole of g between the original integration path and SDP + Γ_B. The advantage of splitting the integral into two parts, one of which is calculated along the SDP, is connected with the fact that along an SDP the function $h(s)$ can be written as $h(s) = \operatorname{Re} h(s_s) + i \operatorname{Im} h(s)$. Therefore, by noting that along the SDP the function $\operatorname{Im} h$ has a single maximum at $s = s_s$ and decreases monotonically when departing from it, it is possible to perform the transformation

$$s \to u = \pm\sqrt{\operatorname{Im} h(s_s) - \operatorname{Im} h(s)} \tag{V.6.5}$$

for $s \in$ SDP. Consequently, we can transform the integral in Eq. (V.6.4) calculated along SDP into

$$I_{\text{SDP}} = e^{-ikh(s_s)} \frac{e^{i\pi/4}}{\sqrt{\lambda}} \int_{-\infty}^{\infty} g[s(u)] e^{-ku^2} \frac{ds}{du} du. \qquad \text{(V.6.6)}$$

As $k \to \infty$, the factor e^{-ku^2} decreases rapidly when we depart from $u = 0$ and the main contribution to the integral comes from the values of g in proximity to the saddle point. Thus,

$$I_{\text{SDP}} = e^{-ikh(s_s)} g(s_s)/[h''(s_s)]^{1/2}. \qquad \text{(V.6.7)}$$

In particular, for the angular spectrum integral of Eq. (IV.9.12) we obtain

$$I_{\text{SDP}} = \int_{\text{SDP}} S(\beta) e^{-ik\rho\cos(\beta - \phi)}\, d\beta = S(\phi) \left(\frac{2\pi}{k\rho}\right)^{1/2} e^{i\pi/4 - ik\rho}. \qquad \text{(V.6.8)}$$

The last expression justifies the interpretation of the angular spectrum integral extended to a steepest-descent path as a cylindrical wave with angular pattern $S(\phi)$ (cf. Section IV.9.2).

6.1 *Pole near a Saddle Point*

There are cases in which a saddle point lies very close to a pole of the integrand. This happens especially when we consider angular spectrum representations of a field. In these cases, we can consider the particular form taken by the integral

$$I(k) = \frac{e^{i\pi/4}}{\sqrt{\lambda}} \int_{-\infty}^{\infty} \frac{f(s)}{s - s_p} e^{-ks^2}\, ds, \qquad \text{(V.6.9)}$$

where s_p represents a pole close to the saddle point $s_s = 0$. If $s_p \simeq 0$ and $f(s)$ is almost constant in the interval $|s| < k^{-1/2}$, we can rewrite the above integral as

$$\begin{aligned} I(k) &= \frac{\exp(i\pi/4)}{\sqrt{\lambda}} f(0) \int_0^{\infty} e^{-ks^2} \left(\frac{1}{s - s_p} - \frac{1}{s + s_p} \right) ds \\ &= \frac{\exp(i\pi/4)}{\sqrt{\lambda}} f(0) s_p \int_{-\infty}^{\infty} \frac{\exp(-ks^2)}{s^2 - s_p^2}\, ds \\ &= -i(2\pi k)^{1/2} \exp\left(\frac{i\pi}{4} - ks_p^2\right) f(0) G\left[\exp\left(\frac{i\pi}{4}\right) k^{1/2} s_p\right], \end{aligned} \qquad \text{(V.6.10)}$$

where $G(x)$ is defined by Eq. (V.3.11). In particular, the integral

$$I(k) = \int_{\text{SDP}} \frac{f(\beta)}{\beta - \beta_p} \exp[-ik\rho\cos(\beta - \phi)]\, d\beta, \qquad \text{(V.6.11)}$$

calculated along the steepest descent path passing through $\beta = \phi$, can be put in the form of Eq. (V.6.9) by making the substitution $i\rho\cos(\beta - \phi) = s^2 + i\rho$ and taking into account that for $\beta \simeq \phi, \beta - \phi \simeq e^{i\pi/4}s(2/\rho)^{1/2}$. In so doing [see the Appendix, Eq. (F2)]

$$I(k) \simeq \exp(-ik\rho)\int_{-\infty}^{+\infty} \frac{f[\beta(s)]\exp(-ks^2)}{s - (\beta_p - \phi)(1 - i)\sqrt{\rho/2}}\,ds$$

$$\simeq -2\pi i \exp[-ik\rho\cos(\beta_p - \phi)]\, G\left[(\beta_p - \phi)\left(\frac{k\rho}{2}\right)^{1/2}\right] f(\phi). \qquad \text{(V.6.12)}$$

6.2 *Branch Point near a Saddle Point*

If we consider the integral

$$I(k) = \frac{e^{i\pi/4}}{\sqrt{\lambda}}\int_{-\infty}^{\infty} \frac{e^{-iks}}{\sqrt{a^2 + s^2}}\,ds = \frac{e^{i\pi/4}}{\sqrt{a}}e^{-ka}, \qquad \text{(V.6.13)}$$

we notice that the terms of its asymptotic expansion in powers of $1/k$ vanish identically because $\lim_{k\to\infty} k^n I(k) = 0$ for every positive integer n. A way to get rid of such a difficulty is suggested by the presence of the factor $\sqrt{a^2 + s^2}$, which presents two branch points at $s_b = \pm ia$. While in the preceding section we considered the polar singularities of the integrand, it is now necessary to take a further step by accounting for branch-point singularities located near the integration path.

The problem we address now is that of producing an accurate representation of the following class of integrals [see the Appendix Eqs. (F3) and (F4)]:

$$I_{\pm}(k) = \frac{e^{i\pi/4}}{\sqrt{\lambda}}\int_{\Gamma} (s - s_b)^{\pm 1/2} g(s) e^{-ikh(s)}\,ds, \qquad \text{(V.6.14)}$$

Γ being a generic path in the complex s plane (see Fig. V.16). The square root in the integrand is supposed to be single-valued, at least for s running along Γ. This means that a branch cut is supposed to depart from s_b and to run parallel to the real axis in the direction $s = -\infty$.

If we transform the integration path by including a portion Γ_b encircling the branch point and the relative branch cut, we can consider only the contribution to $I(k)$ from Γ_b only. To this end, we notice that for Γ_b very close to the half-line $s = s_b - x$, $x > 0$, we can put

$$(s - s_b)^{1/2} = iw, \qquad \text{(V.6.15)}$$

where $w > 0$ for s on the upper side of Γ_b and $w < 0$ for the lower one.

Accordingly, letting w run from $-\infty$ to ∞, s will describe the curve Γ_b. Thus, if we change the integration variable from s to w, the path Γ_b changes to the interval $-\infty$, $+\infty$. In particular, we have for $I_-(k)$ [see Eq. (V.6.14)],

$$I_-(k) = 2i\frac{\exp(i\pi/4)}{\lambda^{1/2}}\int_{-\infty}^{\infty} g[s(w)]\exp\{-ikh[s(w)]\}\,dw. \qquad \text{(V.6.16)}$$

The function $h[s(w)]$ has a saddle point at $w = 0$, which contributes to $I_-(k)$ with

$$I_{-s_b}(k) \sim -2\frac{\exp[-ikh(s_b)]}{[2h'(s_b)]^{1/2}}\left[g(s_b) + \frac{1}{ik}\frac{1}{8h'}\left(3\frac{h''}{h'}g - 2g''\right)\right] + O\left(\frac{1}{k^2}\right), \qquad \text{(V.6.17)}$$

where the functions with unspecified arguments are calculated for $s = s_b$.

Proceeding in an analogous way, we obtain for $I_+(k)$

$$I_+(k) = -2i\frac{\exp(i\pi/4)}{\sqrt{\lambda}}\int_{-\infty}^{\infty} w^2 g[s(w)]\exp\{-ikh[s(w)]\}\,dw$$
$$\sim -\frac{1}{\sqrt{2}}\frac{\exp[-ikh(s_b)]}{[h'(s_b)]^{3/2}}\frac{g(s_b)}{ik}. \qquad \text{(V.6.18)}$$

Here, in contrast to Eq. (V.6.17), the leading term contains the factor k^{-1}, so that the contribution of the branch point tends to disappear as $k \to \infty$. It is noteworthy that both of these asymptotic contributions tend to infinity when the branch point comes close to a stationary point, as evidenced by the presence of the factor h' in the denominator.

If we return to the integral considered at the beginning, we observe that it can be rewritten in the form

$$I = \frac{\exp(i\pi/4)}{\sqrt{\lambda}}\int_{\Gamma_a}\frac{1}{\sqrt{s+ia}}g(s)\exp[-ikh(s)]\,ds, \qquad \text{(V.6.19)}$$

where $g(s) = (s - ia)^{-1/2}$ and $h(s) = s$. Then, using Eq. (V.6.17), we immediately obtain the right side of Eq. (V.6.13).

7 Diffraction Effects at a Plane Interface between Two Dielectrics

We saw in Section IV.9 that the field reflected by the plane interface between two dielectrics illuminated by a TE cylindrical wave is represented by a Sommerfeld-type integral [see Eq. (IV.9.25)] characterized by the presence of branch points in the integrand. Consequently, by using the angular spectrum we can express the integral in the form (see Problem 13)

$$u_r(\rho, \phi) = \frac{i}{4\pi} \int_{0-i\infty}^{\pi+i\infty} \frac{\cos 2\beta + 2(n^2 - \cos^2\beta)^{1/2} \sin\beta - n^2}{1 - n^2}$$
$$\times \exp\{-ik_1[-x_s \cos\beta + z_s \sin\beta$$
$$+ \rho\cos(\beta - \phi)]\}\, d\beta \equiv I_{\mathrm{SDP}} + I_{\Gamma_b}, \qquad \text{(V.7.1)}$$

where k_1 refers to medium 1 containing the source and $n = n_2/n_1$, with n_2 referring to medium 2; I_{SDP} represents the contribution of the saddle point and I_{Γ_b} that of the path Γ_b encircling the portion of the branch cut relative to $(n^2 - \cos^2\beta)^{1/2}$, comprised between the branch point

$$\beta_b = q\pi - \arccos n$$

and the intersection with the SDP (see Fig. IV.12c). The coordinates x_s and z_s are those of the source, while $x = \rho\cos\phi$ and $z = \rho\sin\phi$ are those of the field point (see Fig. IV.11).

To calculate the contribution I_{SDP}, we notice that the saddle point β_s is a root of the equation

$$\tan\beta_s = (z + z_s)/(x - x_s), \qquad \text{(V.7.2)}$$

which lends itself to a geometric interpretation of β_s. In fact, β_s represents the angle formed by the x axis and the vector connecting the specular image of the source located at x_s, $-z_s$ with the observation point.

Now, by applying Eq. (V.6.8) we easily obtain (see Problem 14)

$$I_{\mathrm{SDP}} \sim -[\exp(-i\pi/4 - ik_1\bar{\rho})/(8\pi k_1\bar{\rho})^{1/2}]r_s(\beta_s), \qquad \text{(V.7.3)}$$

where $\bar{\rho} = [(x - x_s)^2 + (z + z_s)^2]^{1/2}$ is the distance of the field point from the image source and r_s is the Fresnel coefficient. Accordingly, I_{SDP} represents a congruence of rays departing from the source, reaching the interface, and being reflected in accordance with Snell's and Fresnel's laws.

7.1 *Lateral Waves*

When the medium containing the source and the observation point is more dense that the other medium, then $n < 1$. Consequently, in this case the branch points are real and the steepest-descent path intersects the branch cut only if $\beta_s > \pi - \arccos n$ or $\arccos n > \beta_s$. Let us assume that the first case applies.

Then we have

$$I_{\Gamma_b} = \frac{\exp(i\pi/4)}{\lambda_1^{1/2}} \int_{\Gamma_b} (\cos\beta - n)^{1/2} g(\cos\beta) \exp[-ik_1 h(\cos\beta)]\, d(\cos\beta), \tag{V.7.4}$$

where

$$h(\cos\beta) = -x_s\cos\beta + z_s\sin\beta + \rho\cos(\beta - \phi), \tag{V.7.5}$$

$$g(\cos\beta) = -[\lambda_1^{1/2}/(2\pi)]\exp(i\pi/4)(n + \cos\beta)^{1/2}/(1 - n^2).$$

Next, using Eq. (V.6.18) with $s_b = n$, we obtain

$$I_{\Gamma_b} \sim \frac{\exp(i\pi/4)}{(2\pi)^{1/2}} n^{1/2}(1 - n^2)^{-1/4} \times \frac{\exp\{-ik_1[(x - x_s)n + (z + z_s)(1 - n^2)^{1/2}]\}}{\{k_1[-(z + z_s)n + (x - x_s)(1 - n^2)^{1/2}]\}^{3/2}}. \tag{V.7.6}$$

Now, if we rewrite the phase in the form

$$nk_1[(x - x_s) - (z + z_s)n(1 - n^2)^{-1/2}] + k_1(z_s + z)(1 - n^2)^{-1/2},$$

it can be interpreted as arising from a ray that propagates from the source to the interface at the critical angle $\theta_c = \arcsin n$, is refracted parallel to the interface in the second medium, and leaves the interface by refraction at the angle θ_c to reach the observation point. If we indicate with L_1, L_2, and L_3 the distances traveled over these three paths (see Fig. V.17), we can write Eq. (V.7.6) in the form

$$I_{\Gamma_b} \sim \frac{\exp(i\pi/4)}{(2\pi)^{1/2}} \frac{n^{1/2}}{(1 - n^2)} \frac{\exp[-ik_1(L_1 + nL_2 + L_3)]}{(k_1 L_2)^{3/2}}. \tag{V.7.7}$$

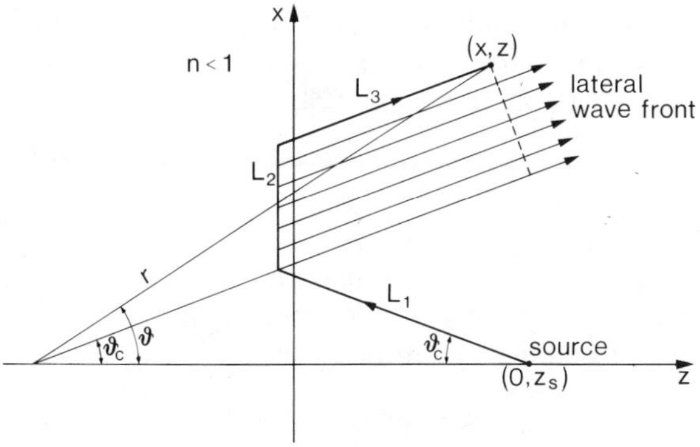

Fig. V.17. Geometry related to the excitation of a lateral wave.

Because of the sideways nature of the propagation of this wave, it has been given the name *lateral wave.* Its amplitude decays with the $\frac{3}{2}$ power of the distance traveled parallel to the interface. The fact that it attenuates more rapidly than a cylindrical wave (i.e., $L^{-1/2}$) is due to leakage of energy along the lateral path (see Chapter III, Brekovskikh [7]).

The discussion above was based on the assumption that the two media were lossless. With slight changes it can be shown that Eq. (V.7.6) remains valid for lossy media. In particular, if the source medium is lossy while the other one is not, when the field point is displaced parallel to the interface, I_{Γ_b} decreases algebraically (i.e., as a power) with the distance x, while I_{SDP} decays exponentially. As a consequence of this behavior, the lateral wave becomes predominant with respect to the GO contribution. Roughly speaking, we can imagine that in these cases the rays are deviated on a path running along the interface in order to avoid the strong losses of the bulk medium, although these waves must not be confused with the surface waves discussed in Sections III.21,22. In fact, while the amplitude of the surface waves decays exponentially in the direction normal to the interface, this is not the case with the lateral waves, which penetrate deeply in the two media.

Lateral waves also exist when the interface is illuminated by point sources. In these cases the main features of these waves remain unaltered, the only difference arising from the presence of the additional factor $\rho^{-1/2} = [(x - x_s)^2 + (y - y_s)^2]^{-1/4}$ in Eq. (V.7.6) (see Chapter II in [3], p. 514), viz.

$$I_{\Gamma_b} \propto \exp[-i(k_1 L_1 + k_2 L_2 + k_1 L_3)]/[\rho^{1/2}(k_1 L_2)^{3/2}]. \qquad \text{(V.7.8)}$$

The presence of this additional factor can be justified heuristically by assuming that the total power carried by the lateral wave is constant.

Finally, it is important to remark that the above asymptotic expressions are valid only when the saddle point is not too close to the branch point. As β_s tends to β_c it is easy to show that $L_2 \to 0$ and Eq. (V.7.7) becomes singular. In this case a more accurate evaluation of the integrals can be obtained by using a transition function found by Fock based on the parabolic cylinder function of order $\frac{1}{3}$ (see Chapter III, Brekovskikh [7]).

7.2 *Goos–Haenchen Effect*

Since the integrals I_{SDP} and I_{Γ_b} discussed above are analytic functions of the coordinates x_s, z_s of the source, we can still use Eqs. (V.7.3,6) in the case in which x_s and z_s are complex quantities:

$$x_s = \operatorname{Re} x_s - ib \sin\psi, \qquad \text{(V.7.9a)}$$

$$z_s = \operatorname{Re} z_s + ib \cos\psi, \qquad \text{(V.7.9b)}$$

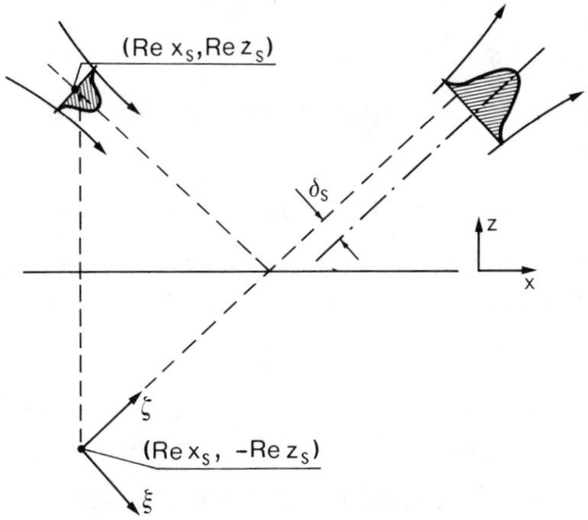

Fig. V.18. Reflection of a Gaussian beam, showing the displacement δ_S of the reflected beam accounting for the Goos–Haenchen effect.

where b and ψ are two generic real quantities. If we introduce two new cartesian coordinates ξ, ζ (see Fig. V.18) with origin in Re x_s, $-$Re z_s and rotated by the angle ψ with respect to x, z, we easily see that $\xi_s = 0$ and $\zeta_s = -ib$. Consequently, we can write the distance $\bar{\rho}$ of the field point from the image source in the form

$$\bar{\rho} = [\xi^2 + (\zeta + ib)^2]^{1/2} \cong \zeta + ib + \tfrac{1}{2}\xi^2/(\zeta + ib), \qquad \text{(V.7.10)}$$

for points close to the ζ-axis. Inserting this expression for $\bar{\rho}$ into Eq. (V.7.3) yields a field with a complex eikonal (see Section II.7)

$$I_{\text{SDP}} \sim -\frac{\exp(-i\pi/4)\lambda_1^{1/2}}{4\pi}\frac{1}{(\zeta + ib)^{1/2}}$$
$$\times \exp\left[-ik_1\left(\zeta + ib + \frac{1}{2}\frac{\xi^2}{\zeta^2 + b^2}\zeta\right) - \frac{1}{2}\frac{k_1 b}{\zeta^2 + b^2}\xi^2\right] r_s(\beta_s). \qquad \text{(V.7.11)}$$

That is, the field has a gaussian distribution along the direction ξ (see Section II.7.1) with width $w^2 = 2(\zeta^2 + b^2)/(k_1 b)$. However, we cannot neglect the factor $r_s(\beta_s)$, which is in general a function of the observation point, as displayed in Eq. (V.7.2).

Of particular relevance is the case in which the gaussian beam undergoes total reflection. In that case the reflection coefficient has a unitary modulus and is characterized by a phase $2\phi_s$ defined by Eq. (III.20.6). If we consider slight deviations of the observation point from the ζ axis (justified by the narrowness of the gaussian beam) we can put

$$r(\beta_s) = \exp[2i\phi_s(\beta_s)] = \exp[2i\phi_s(\bar{\beta}_s) + 2i(d\phi_s/d\beta_s)\,d\beta_s], \qquad \text{(V.7.12)}$$

where $d\beta_s$ is related to the displacements of $P(\rho, \phi)$ through Eq. (V.7.2), which can be written as

$$\bar{\beta}_s + d\beta_s = \tfrac{1}{2}\pi + \psi + \arctan[\xi/(\zeta + ib)] \cong \tfrac{1}{2}\pi + \psi + \xi/(\zeta + ib), \qquad \text{(V.7.13)}$$

ψ being the angle formed by the axial direction of the gaussian beam (ζ axis) with the z axis. On the other hand, from Eq. (III.20.6) we obtain, after the substitution $\theta = \bar{\beta}_s - \frac{1}{2}\pi = \psi$,

$$\frac{d\phi_s}{d\beta_s} = -\frac{\cos\bar{\beta}_s}{(\cos^2\beta_s - n^2)^{1/2}} = \frac{\sin\psi}{(\sin^2\psi - \sin^2\theta_c)^{1/2}} \equiv \phi_s', \qquad \text{(V.7.14)}$$

where $\theta_c = \arcsin(n)$ is the critical angle of the interface.

After inserting Eqs. (V.7.13,14) into Eq. (V.7.12), we can write Eq. (V.7.11) in the form

$$I_{\mathrm{SDP}} \sim -\frac{\exp(-i\pi/4)\lambda_1^{1/2}}{4\pi}\frac{1}{(\zeta + ib)^{1/2}}\exp\left\{-ik_1\left[\zeta + ib + \frac{1}{2}\frac{(\xi - \delta_s)^2}{\zeta + ib}\right.\right.$$
$$\left.\left. -\frac{1}{2}\frac{\delta_s^2}{\zeta + ib} - 2\frac{\phi_s(\psi)}{k_1}\right]\right\}, \qquad \text{(V.7.15)}$$

where $\delta_s = 2\phi_s'/k_1$. This means that in case of total reflection the dispersion of the phase of the Fresnel reflectivity factor leads to a shift of the beam in the direction perpendicular to its axis, given for an s-wave by (see Problem 19)

$$\delta_s = (2/k_1)[\sin\psi/(\sin^2\psi - \sin^2\theta_c)^{1/2}]. \qquad \text{(V.7.16)}$$

Equation (V.7.15) includes only the leading term of the asymptotic expansion of the integral. It can be shown that the next term of this expansion becomes comparable with the leading term for $\psi \to \frac{1}{2}\pi$. In particular, while the right-hand side of Eq. (V.7.16) gives a finite shift for grazing incidence, which is physically unacceptable, δ_s actually tends to zero for $\psi \to \frac{1}{2}\pi$ (see Chapter III, Brekovskikh[7]).

Because of the small factor k_1^{-1}, the shift δ_s can be appreciated only when the angle of incidence is near the critical angle. However, for ψ too close to θ_c, the saddle point β_s approaches the branch point and the asymptotic contributions of I_{SDP} and I_{Γ_b} cannot be separated any longer. In this case it is necessary to make use of a suitable transition function.

The lateral shift of an optical beam was observed by Goos and Haenchen [20] in 1947, thus extending to electromagnetic waves [21] the early observations of shifts of ultra-acoustic waves. More recently [22], a displacement by several wavelengths of a beam impinging on a planar multilayered structure consisting of four slabs was reported. The large shift is a consequence of the dependence of the reflection coefficient on the angle of incidence. In fact, as we noticed in Section III.18, $r(k_x)$ in general contains some poles in the complex k_x plane. The reflected field is given by an integral similar to that of Eq. (V.7.1) in which the integrand is replaced by

$$u_r(\rho, \phi) \propto \int_{0-i\infty}^{\pi+i\infty} \frac{f(\beta)}{(\cos\beta - \cos\beta_1)} \times \exp\{-ik_1[-x_s\cos\beta + z_s\sin\beta + \cos(\beta - \phi)]\}\, d\beta, \qquad \text{(V.7.17)}$$

where $f(\beta) = r(\beta)(\cos\beta - \cos\beta_1)$ and β_1 represents the pole of the reflection coefficient closest to the steepest-descent path. Asymptotic evaluation of this integral shows that the reflected beam is shifted laterally with respect to a specularly reflected one when the complex pole β_1 comes very close to the SDP. Since β_1 is associated with a leaky mode of the multilayer structure (see Section III.18), the large lateral shift can be interpreted [23] as due to the excitation by the incident wave of a leaky wave that propagates for some distance parallel to the plane faces of the structure and then is reradiated into the space containing the source. By a careful choice of the parameters of the multilayer it is possible to produce a very large shift, which can be used for making optical couplers used in integrated optics.

8 Asymptotic Evaluation of the Diffraction Integrals in Cylindrical Coordinates

Here we wish to evaluate the relevant integrals $I(k)$ appearing in connection with the use of cylindrical coordinates, which are typically of the form

$$I(k) = \frac{k}{L}\int_0^a f(s) J_m\left(ks\frac{\rho}{L}\right) \exp\left[-ik\frac{s^2+\rho^2}{2L} - ikT(s)\right] s\, ds \qquad \text{(V.8.1)}$$

[e.g., see Eq. (IV.11.4)], where $f(s)$ and $T(s)$ depend on the values assumed by the field on the reference plane $z = z'$ and reduce, respectively, to the amplitude $A(s)$ times the phase factor $-i\exp(-ikL)$ and to the eikonal $S(s)$ for a rotationally symmetric reference field vanishing outside a circular aperture of radius a. In fact, in this case the right sides of Eqs. (V.8.1) and (IV.11.19) coincide, provided $J_m = J_0$, and $|z - z'| = L$.

If we express J_m as (see Chapter II, Abramowitz and Stegun [4], p. 364)

$$J_m(\zeta) = [2/(\pi\zeta)]^{1/2}\tfrac{1}{2}[P(m,\zeta) + iQ(m,\zeta)]\exp\{i[\zeta - (m/2 + 1/4)\pi]\} + [2/(\pi\zeta)]^{1/2}\tfrac{1}{2}[P(m,\zeta) - iQ(m,\zeta)]\exp\{-i[\zeta - (m/2 + 1/4)\pi]\}, \qquad \text{(V.8.2)}$$

where P and Q are two slowly varying functions of ζ that reduce, respectively, to 1 and 0 in the limit of large ζ, we can rewrite $I(k)$ in the form (see Southwell (1978), in bibliography)

$$I(k) = \frac{\exp(i\pi/4)}{\lambda^{1/2}} \int_0^a g_1(s) \exp[-ikh_1(s)]\, ds + \frac{\exp(i\pi/4)}{\lambda^{1/2}} \int_0^a g_2(s) \exp[-ikh_2(s)]\, ds, \tag{V.8.3}$$

where, for the rotationally invariant case,

$$g_1(s) = [-i/(L\rho)^{1/2}]e^{-ikL} A(s)s^{1/2}[P(0, ks\rho/L) + iQ(0, ks\rho/L)], \tag{V.8.4a}$$

$$h_1(s) = (s - \rho)^2/(2L) + S(s), \tag{V.8.4b}$$

$$g_2(s) = -e^{-2ikL} g_1^*(s), \tag{V.8.4c}$$

$$h_2(s) = (s + \rho)^2/(2L) + S(s). \tag{V.8.4d}$$

The leading terms of the integrals can be evaluated by using Eqs. (V.2.32,36), which yields, with the help of the relation $s^{1/2}(P + iQ) \to s^{1/2}$ (as $k \to \infty$),

$$\begin{aligned} I(k)\exp(ikL + i\pi/2) \sim{} & \sum_p A(s_p)\left[\frac{s_p}{\rho(1 + LS''(s_p))}\right]^{1/2} \exp[-ikh_1(s_p)] \\ & + \sum_q A(s_q)\left[\frac{s_q}{\rho(1 + LS''(s_q))}\right]^{1/2} \exp[-ikh_2(s_q)] \\ & - \frac{L^{1/2}\exp(-i\pi/4)}{(2\pi k\rho)^{1/2}} A(a)\left\{\left[\frac{a}{a - \rho + LS'(a)}\right]^{1/2}\right. \\ & \times \exp[-ikh_1(a)] + \left[\frac{a}{a + \rho + LS'(a)}\right]^{1/2} \\ & \left.\times \exp[-ikh_2(a)]\right\}, \end{aligned} \tag{V.8.5}$$

where s_p and s_q are, respectively, the stationary-phase points of h_1 and h_2, and the third term represents the contribution of the end point $s = a$. No contribution of the other end point $s = 0$ is present because of the vanishing of $s^{1/2}$ $(P + iQ)$ for $s = 0$. The above relations will be used in Section VII.18 to evaluate the field in unstable resonators.

Equation (V.8.5) shows that the field amplitude undergoes a geometric attenuation, as the distance L from the aperture increases, essentially given by

$$\left\{\frac{s_p}{\rho[1 + LS''(s_p)]}\right\}^{1/2}, \tag{V.8.6}$$

which differs from the expression found for cylindrical fields [see Eq. (V.2.43)] by the additional presence of the factor $(s_p/\rho)^{1/2}$. This is due to the fact that our present field has wave fronts with two finite curvature radii, which is not the case for cylindrical wave fronts.

When the field point lies on the optic axis ($\rho = 0$), it is convenient to deal directly with the integral of Eq. (V.8.1), which reduces to

$$I(k) = -i\exp(-ikL)\left(\frac{k}{L}\right)\int_0^a A(s)\exp\left\{-ik\left[\frac{s^2}{2L} + S(s)\right]\right\}s\,ds, \qquad \text{(V.8.7)}$$

where we have considered a rotationally invariant wave. If $A(s) = 1$ and $S(s) = -s^2/(2f) + Bs^4/(4w^4k)$ (see Section V.5) with $w = 2a$, the above integral transforms into

$$\begin{aligned} I(k) &= -i\exp(-ikL)\frac{k}{2L}\int_0^{a^2}\exp\left\{-ik\left[\left(\frac{1}{2L} - \frac{1}{2f}\right)x + \frac{Bx^2}{4w^4k}\right]\right\}dx \\ &= -i\exp(-ikL)\frac{k}{2L}\exp\left[ik^2\left(\frac{1}{L} - \frac{1}{f}\right)^2\frac{4a^4}{B}\right] \\ &\quad\times\int_0^{a^2}\exp\left[-i\frac{B}{4w^4}\left[x + \left(\frac{1}{L} - \frac{1}{f}\right)\frac{4w^4k}{B}\right]^2\right\}dx \\ &= -8\pi N\left(\frac{\pi}{B}\right)^{1/2}\exp\left[i\frac{\pi}{4} - ikL + ik^2\left(\frac{1}{L} - \frac{1}{f}\right)^2\frac{4a^4}{B}\right] \\ &\quad\times\left\{F\left[\frac{8a^2k}{B^{1/2}}\left(\frac{1}{L} - \frac{1}{f}\right)\right] - F\left[\frac{B^{1/2}}{8} + \frac{8a^2k}{B^{1/2}}\left(\frac{1}{L} - \frac{1}{f}\right)\right]\right\}, \end{aligned} \qquad \text{(V.8.8)}$$

where $N = a^2/(L\lambda)$ is the Fresnel number of the aperture, $F(\zeta)$ is the complex Fresnel integral [Eq. (V.3.5)], and we have assumed for simplicity that $B > 0$.

This result allows us to infer some relevant features of the field. If the distance of the field point from the focus is such that

$$|1/L - 1/f|8a^2k > 10B^{1/2}, \qquad \text{(V.8.9)}$$

we can approximate the Fresnel integral with its asymptotic value [Eq. (V.3.15)], thus obtaining an expression coinciding with the GO field. Conversely, the segment of the optic axis in which the GO approximation fails has a length $|L - f|$ given by

$$|L - f| \cong (B^{1/2}/1.6\pi NA^2)\lambda. \qquad \text{(V.8.10)}$$

For $B = 0$, the field is immediately calculated by performing the first integral in Eq. (V.8.8), and its amplitude is

$$|u(\rho = 0, L)| = \left|\frac{\sin[k(1/(4L) - 1/(4f))a^2]}{L[1/(2L) - 1/(2f)]}\right|. \qquad \text{(V.8.11)}$$

In particular, when the aperture is illuminated by a plane wave ($1/f = 0$), the above expression reduces to

$$|u(\rho = 0, L)| = 2|\sin[ka^2/(4L)]|. \tag{V.8.12}$$

Accordingly, the field intensity oscillates, vanishing at the distances $L_q = a^2/(2q\lambda)$ and going to zero as $1/L^2$ for $L \to \infty$. In practice, this means that the field diffracted by the aperture behaves as a spherical wave only when the field point is at a distance L much larger than a^2/λ. Thus, if we consider the Fresnel number N relative to the aperture seen from the field point, the field vanishes on the axis when $N(L) = 2q$ and behaves as a spherical wave only for $N \ll 1$.

We can also consider the case of complex f, that is,

$$1/f = (1/f_r) + i/(kw^2), \tag{V.8.13}$$

which corresponds to considering an aperture illuminated by a gaussian field and affected by spherical aberration. The field on the axis is no longer an oscillating function of L and can generally be evaluated by resorting to Fresnel integrals of complex argument.

Similar oscillations can be observed along the optic axis when approaching the focus of a wave affected by third- and fifth-order spherical aberration, as shown in Fig. V.19 (from Focke [24]), which refers to a Tessar lens.

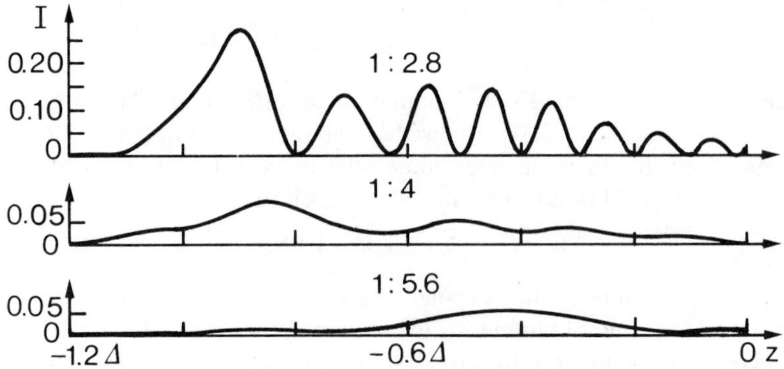

Fig. V.19. Field intensity on the optic axis in proximity to the focus of a Tessar lens, calculated by Focke [24a] for three apertures (1:2.8, 1:4, 1:5.6). The adimensional parameter Δ corresponds to the optical coordinate $\bar{u}$, defined by Eq. (IV.13.21), relative to the distance of the focus of the marginal ray from the paraxial focus: $\Delta = \frac{1}{2}k_0 \mathrm{NA}^2 z_{\mathrm{sph}}$, z_{sph} being the longitudinal spherical aberration relative to a marginal ray of height ρ_0. The field distribution was calculated by assuming a dependence of z_{sph} on ρ_0 of the form $z_{\mathrm{sph}} = -\frac{1}{2}f^2(I\rho_0^2 + I^*\rho_0^4)$, accounting for primary and secondary spherical aberration. The other data were: $f = 52.5$ mm, $\rho_0 = 9.15$ mm, $z_{\mathrm{sph}} = -0.42$ mm, and $\Delta = -75$ at $\lambda = 546$ nm.

9 Asymptotic Series Derived from Comparison Integrals: Chester–Friedman–Ursell (CFU) Method

The expressions obtained in Section V.4 are the leading terms of asymptotic series that can be obtained by following a method developed in 1956 by Chester *et al.* [25]. It is based on the change of variable $s \to u$ defined by the implicit relation

$$-ih(s) = \tfrac{1}{3}u^3 - \zeta u - ih[s(0)] \tag{V.9.1}$$

where ζ is a parameter controlling the relative position of the two saddle points of h. The transformation can be shown to be uniformly regular near $s = 0$. If we now expand $g\,ds/du$ in the power series

$$g(s)\frac{ds}{du} = \sum_m p_m(u^2 - \zeta)^m + \sum_m q_m u(u^2 - \zeta)^m, \tag{V.9.2}$$

and insert the above series into the diffraction integral of Eq. (V.2.24), we obtain

$$I_\alpha(k) \sim \frac{\exp(i\pi/4)}{\sqrt{\lambda}}\left\{\sum_m p_m \int_{-i\infty}^{i\infty} (u^2 - \zeta)^m \exp\left[k\left(\frac{u^3}{3} - \zeta u\right)\right] du \right.$$
$$\left. + \sum_m q_m \int_{-i\infty}^{i\infty} u(u^2 - \zeta)^m \exp\left[k\left(\frac{u^3}{3} - \zeta u\right)\right] du\right\} e^{-ikh_\alpha}. \tag{V.9.3}$$

Let us now introduce the functions

$$F_m(\zeta, k) \equiv \frac{1}{2\pi i}\int_{-i\infty}^{i\infty} (u^2 - \zeta)^m \exp\left[k\left(\frac{u^3}{3} - \zeta u\right)\right] du, \tag{V.9.4a}$$

$$G_m(\zeta, k) \equiv \frac{1}{2\pi i}\int_{-i\infty}^{i\infty} u(u^2 - \zeta)^m \exp\left[k\left(\frac{u^3}{3} - \zeta u\right)\right] du, \tag{V.9.4b}$$

which are related to each other by the recurrence relations

$$F_m = -(2/k)(m-1)G_{m-2}, \tag{V.9.5a}$$

$$G_m = -(1/k)(2m-1)F_{m-1} - (2/k)(m-1)\zeta F_{m-2}, \tag{V.9.5b}$$

which can be easily proved by integration by parts of the integrals of Eq. (V.9.4). In particular,

$$F_0 = k^{-1/3}\,\mathrm{Ai}(\zeta k^{2/3}), \qquad G_0 = -k^{-2/3}\,\mathrm{Ai}'(\zeta k^{2/3}). \tag{V.9.6}$$

If we replace the functions in the series (V.9.3), we obtain

$$I_\alpha(k) \sim 2\pi i\frac{\exp(i\pi/4 - ikh_\alpha)}{\sqrt{\lambda}}\left[\frac{g_0(\zeta)}{k^{1/3}}\mathrm{Ai}(\zeta k^{2/3}) + i\frac{g_1(\zeta)}{k^{2/3}}\mathrm{Ai}'(\zeta k^{2/3})\right], \tag{V.9.7}$$

where

$$g_0(\zeta) = p_0 - \frac{q_1 + 2\zeta q_2}{k} + 4\frac{p_3}{k^2} + \cdots, \tag{V.9.8a}$$

$$g_1(\zeta) = -q_0 + 2\frac{p_2}{k} + \cdots. \tag{V.9.8b}$$

10 Asymptotic Evaluation of the Field Diffracted from an Aperture

As we saw in Section IV.14, the disturbance at a point P reached by a spherical wave originating in S and diffracted by an aperture $\bar{A}$ can be represented by the sum of the GO contribution u_g and of the field u_b originating at the boundary of the aperture (boundary diffraction wave, BDW).

From a mathematical point of view, the introduction of the BDW allows us to replace a two-dimensional diffraction integral with a line integral. This transformation reduces the complexity of the asymptotic evaluation of the field. In fact, we learned before how to obtain an asymptotic series representation of one-dimensional finite integrals exhibiting discontinuities of the integrands. Thus, what we have to do is to apply the above formulas to [see Eq. (IV.14.19)]

$$u_b = -\frac{1}{(4\pi)^2 R_s}\int_0^{2\pi} (1 - \cos\theta') e^{-ik(r'+s')}\, d\phi = u_{b,I} + u_{b,II}, \tag{V.10.1}$$

where $u_{b,I}$ collects the contributions of the phase-stationary points [viz. $\partial(r' + s')/\partial\phi = 0$], and $u_{b,II}$ is due to the points of discontinuity of the derivatives of $\theta'(\phi)$ and $r'(\phi) + s'(\phi)$. The latter are localized on the vertices of the aperture edge.

In particular, the phase-stationary points are characterized by a very simple geometric relation. In fact, if we indicate with s the curvilinear coordinate of the aperture edge and with $\hat{e}$ the unit vector tangent to it, we have (Fig. V.20)

$$(\partial/\partial\phi)(r' + s') = 0 \rightarrow (\partial/\partial s)(r' + s') = \hat{e}\cdot\nabla(r' + s') = \hat{e}\cdot(\hat{r}' + \hat{s}') = 0. \tag{V.10.2}$$

As a consequence, at a stationary point the direction of the incident ray forms with the tangent to the edge an angle β equal to that formed by $\hat{e}$ with the ray diffracted along $\hat{r}'$. This property leads us to say that all the possible *rays diffracted by an edge point Q_e form a half-cone with axis parallel to the edge tangent and aperture β equal to the angle formed by the incident ray with*

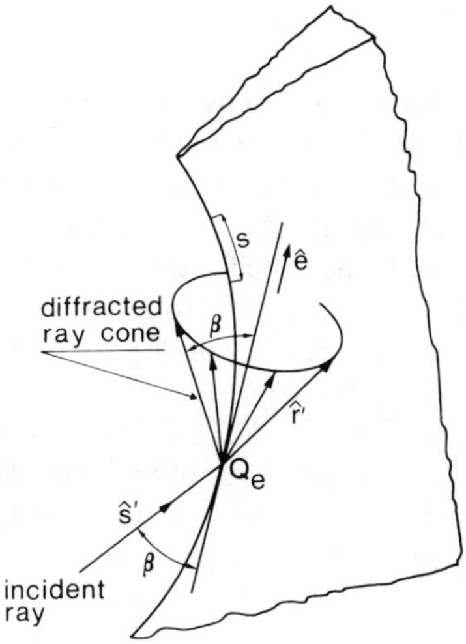

Fig. V.20. Cone of rays diffracted by an edge point Q_e.

the edge. In view of this geometric property, we can locally conceive of the edge diffraction phenomenon as the formation of a conical ray congruence. Thus, the diffraction of the spherical wave by the aperture can be imagined as due to the formation of these conical congruences along the aperture edge. The boundary diffraction wave at the field point P will then be given by the contributions of the diffracted rays departing from various points of the edge and passing through P (see Fig. V.21). This argument is more or less equivalent to the explanation of diffraction proposed by T. Young, according to which the incident ray undergoes a reflection process at the edge, which produces a sort of conical spoke pattern.

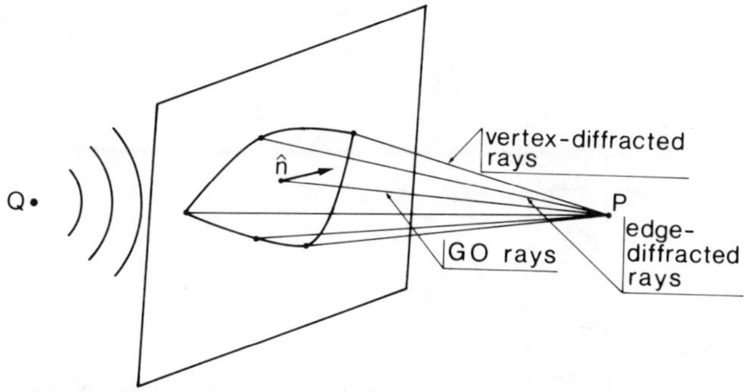

Fig. V.21. Schematic representation of the field in P as due to GO rays, vertex-diffracted rays, and edge-diffracted rays.

10.1 *Caustic of the Diffracted Rays*

The ray congruence formed by the rays diffracted by an edge is on the whole similar to the congruences studied in Chapter II. We notice that the edge itself is a caustic of the diffracted rays. In fact, by definition, a caustic is the locus of the points where the cross section of an elementary ray tube reduces to a segment, and this is what happens on the edge. Thus, a special feature of these congruences is the fact that one of the caustic surfaces reduces to a curved line. The shape and the position of the second caustic surface will depend on the geometry of the diffraction problem (see Fig. V.22) [26].

Let us consider, for example, a straight edge illuminated by a point source in S. It is easy to convince ourselves, by observing the directions of the diffracted rays, that the second caustic (Fig. V.23) is a circle passing through the source, having its center on the edge and lying in a plane perpendicular to it.

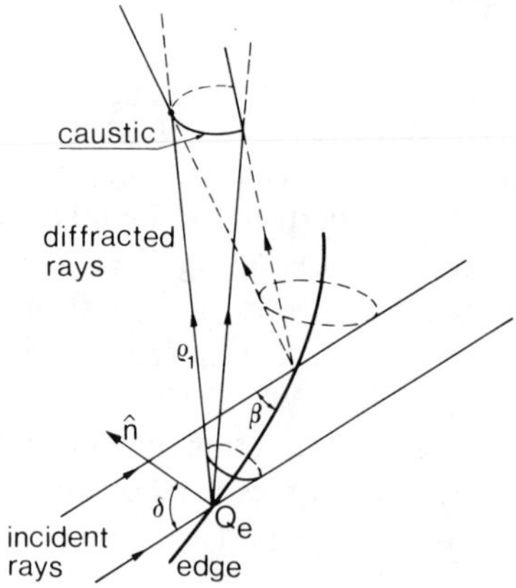

Fig. V.22. Geometry of the diffraction of a ray bundle from an edge. (From Keller [1].)

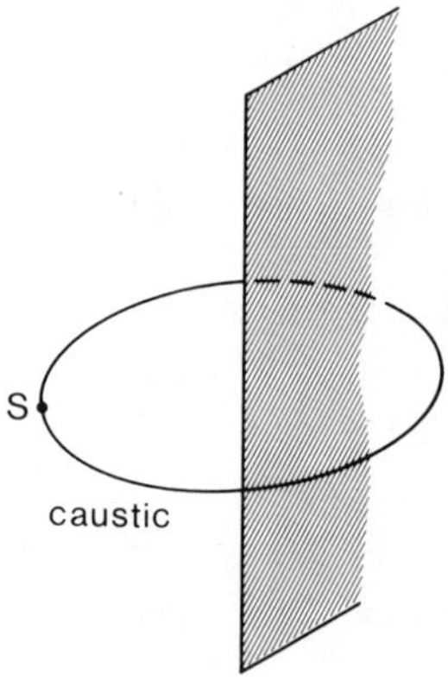

Fig. V.23. Circular caustic of the rays diffracted from a straight edge illuminated by a point source.

Consequently, on each diffracted ray there will be two foci, one (F_1) coincident with the point of intersection with the edge, and the second (F_2) at the same distance from F_1 as F_1 is from the source. From a geometric point of view, the presence of the edge transforms the original ray congruence having its center in S into a congruence with a circle and a straight line as caustics.

For a curved edge, the other caustic is obtained by calculating the envelope of the diffracted-ray congruence, which is represented parametrically by (see Appendix I in Keller [26])

$$[\mathbf{r} - \mathbf{r}_e(s)] \cdot \hat{e}(s) = |\mathbf{r} - \mathbf{r}_e(s)| \cos \beta(s), \tag{V.10.3}$$

where s is the edge arc length and $\hat{e}$ the unit vector tangent to the edge at s, and $\mathbf{r}_e$ represents the edge point Q_e. Differentiating this equation with respect to s yields a second equation, which, together with the above one, defines the position vectors $\mathbf{r}_c$ of the caustic, i.e.,

$$(\mathbf{r}_c - \mathbf{r}_e) \cdot \frac{\hat{n}}{\rho} - 1 + \dot{\beta}|\mathbf{r}_c - \mathbf{r}_e| \sin \beta + \frac{(\mathbf{r}_c - \mathbf{r}_e) \cdot \hat{e} \cos \beta}{|\mathbf{r}_c - \mathbf{r}_e|} = 0, \tag{V.10.4}$$

the dot indicating derivative with respect to s, ρ the curvature radius of the edge, and $\hat{n}$ the principal normal to the edge (see Section II.4.2) directed away from the curvature center. Now, using Eq. (V.10.3), we can replace $(\mathbf{r}_c - \mathbf{r}_e) \cdot \hat{e}$ with $|\mathbf{r}_c - \mathbf{r}_e| \cos \beta$, so that Eq. (V.10.4) becomes

$$|\mathbf{r}_c - \mathbf{r}_e| (\cos \delta + \rho \dot{\beta} \sin \beta) = \rho \sin^2 \beta, \tag{V.10.5}$$

with $\cos \delta = (\mathbf{r}_c - \mathbf{r}_e) \cdot \hat{n}/|\mathbf{r}_c - \mathbf{r}_e|$. Thus, the signed distance $\rho_1 = -|\mathbf{r}_c - \mathbf{r}_e|$ from the edge to the caustic is

$$\rho_1 = -\rho \sin^2 \beta/(\rho \dot{\beta} \sin \beta + \cos \delta). \tag{V.10.6}$$

According to our initial positions, ρ_1 should be negative. However, according to the last expression, ρ_1 can be either positive or negative. When $\rho_1 > 0$, this means that the caustic is *virtual* and lies on the opposite side of the diffracted ray. Consequently, we should start with an expression for the diffracted rays similar to that of Eq. (V.10.3) with the exception that the minus sign in front of $|\mathbf{r} - \mathbf{r}_e|$ is replaced with a plus.

For a straight edge, we have

$$\rho_1 = -\sin \beta/\dot{\beta} \qquad (>0). \tag{V.10.7}$$

In this case, ρ_1 is the same for all the rays departing from the same edge point Q_e. For a point source S, it can immediately be verified that ρ_1 coincides with the distance of Q_e from S, so that the caustic is a circumference passing through S, lying in the plane perpendicular to the edge and having its center on the edge (see Fig. V.23).

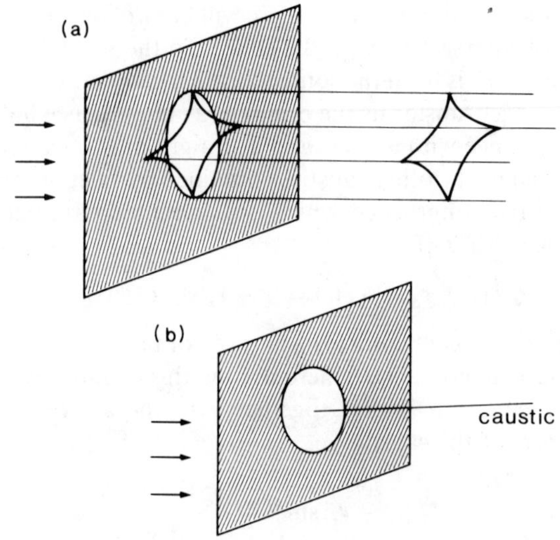

Fig. V.24. Cylindrical caustics formed by the rays diffracted from an aperture on a plane screen illuminated by a plane wave. The sections of the cylinder are the evolutes of the aperture edges. In particular, (a) shows a generalized aperture and (b) whose circular aperture the caustic reduces to the normal passing through the center.

When $\dot{\beta} = 0$,

$$\rho_1 = -\rho \sin^2 \beta / \cos \delta, \tag{V.10.8}$$

and ρ_1 will be positive for $\cos \delta < 0$, that is, whenever the component of the diffracted rays along $\hat{n}$ is negative. In particular, for a plane aperture illuminated perpendicularly $\beta = \pi/2$, so that $\rho = -\rho_1 \cos \delta$; that is, the projection of the curvature center perpendicular to the aperture coincides with the curvature center of the edge. Accordingly, *the caustic is a cylinder perpendicular to the aperture, its cross section coinciding with the edge evolute* (see Fig. V.24). For a circular aperture, the caustic reduces to a line passing through the center. The cylinder continuing behind the aperture represents the virtual caustic of the diffracted rays forming an angle larger than $\pi/2$ with the normal $\hat{n}$ to the edge. The presence of a real caustic is, in general, associated with a strong diffracted field. We saw in Section V.8 an example with the field diffracted by a circular aperture.

We observe that caustics exist in finite regions only when the aperture is not made of straight segments. Thus, relevant enforcements or reductions of the field can be noticed only in the presence of curved apertures. In a certain sense, they have a focusing effect on the diffracted rays.

Finally, we should stress that the above considerations about caustics are independent of the way in which the diffracted rays were generated. If the rays hitting the edge are not produced by a single source, the conclusions of the present section remain valid on the whole.

10.2 *Edge-Diffracted Field*

In order to evaluate $u_{b,I}$, we can use the formulas of Section V.2 by putting $h(s) = r'(s) + s'(s)$ and $g(s) = \lambda^{1/2}\exp(-i\pi/4)(1 - \cos\theta')/[(4\pi)^2 R_S]$. In this way, we obtain

$$u_{b,I} = \frac{\lambda^{1/2}\exp(-i\varepsilon\pi/4)}{(4\pi)^2 R_S}(1 - \cos\theta'_S)\exp[-ik(r'_S + s'_S)]\left|\frac{\partial^2(r' + s')}{\partial\phi^2}\right|^{-1/2}, \tag{V.10.9}$$

where $\varepsilon = \operatorname{sgn}[\partial^2(r' + s')/\partial\phi^2]$ and r'_S, s'_S, and θ'_S refer to a stationary point. Now, the derivative $\partial^2(r' + s')/\partial\phi^2$ can be calculated as follows. First, we note that

$$ds'/ds = \cos\beta, \tag{V.10.10}$$

so that

$$d^2s'/ds^2 = -\dot{\beta}\sin\beta. \tag{V.10.11}$$

On the other hand, we have for the vector $\mathbf{r}' = \mathbf{r}_e(s) - \mathbf{r}(P)$,

$$\mathbf{r}'(s + ds) = \mathbf{r}'(s) + \hat{e}(s)\,ds + \hat{n}(s)\,ds^2/[2\rho(s)], \tag{V.10.12}$$

$\hat{n}$ and ρ being the normal and the curvature radius of the edge. Consequently,

$$\begin{aligned} r'(s + ds) &= r'(s) + \hat{r}' \cdot \hat{e}\,ds + \frac{1}{2}\left[\frac{\hat{r}' \cdot \hat{n}}{\rho} + \frac{1 - (\hat{r}' \cdot \hat{e})^2}{r'}\right]ds^2 \\ &= r'(s) - \sin(\beta)\,ds + \frac{1}{2}\left[\frac{-\cos(\delta)}{\rho} + \frac{\sin^2(\beta)}{r'}\right]ds^2, \end{aligned} \tag{V.10.13}$$

where $\cos\delta$ coincides with the quantity introduced in Eq. (V.10.5). In conclusion, Eqs. (V.10.11,13) furnish

$$\frac{d^2}{ds^2}(r' + s') = -\dot{\beta}\sin\beta - \frac{\cos\delta}{\rho} + \frac{\sin^2\beta}{r'} = \left(\frac{1}{r'} + \frac{1}{\rho_1}\right)\sin^2\beta, \tag{V.10.14}$$

where we have replaced $\cos\delta$ with $-\rho\dot{\beta}\sin\beta - \rho\sin^2\beta/\rho_1$ [see Eq. (V.10.6)]. In view of Eq. (IV.14.20),

$$ds/d\phi = [(1 - \cos\theta')/R_S][r's'(1 + \cos\theta')]/[\hat{e}\cdot(\hat{r}' \times \hat{s}')]. \tag{V.10.15}$$

In order to calculate $\cos\theta'$ and $\hat{e}\cdot(\hat{r}' \times \hat{s}')$, we decompose $\hat{r}'$ and $\hat{s}'$ along $\hat{e}$ and the plane perpendicular to $\hat{e}$. If we indicate with ϕ_e and ϕ'_e the angles formed by the projections of $-\hat{r}'$ and $\hat{s}'$ with the tangent $\hat{t}$ to the screen section

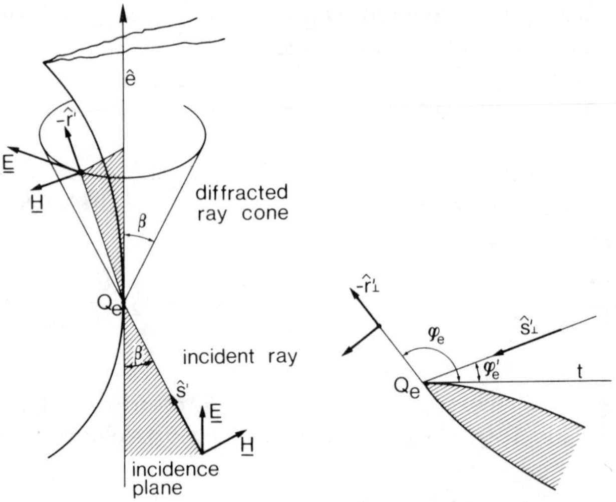

Fig. V.25. Geometry for an edge-diffracted field.

(see Fig. V.25), we can show that

$$1 + \cos\theta' = 1 + \hat{r}' \cdot \hat{s}' = 1 - \cos^2\beta + \sin^2\beta\cos(\phi_e - \phi'_e)$$

$$= 2\sin^2\beta\cos^2\left(\frac{\phi_e - \phi'_e}{2}\right), \tag{V.10.16}$$

$$\hat{e} \cdot (\hat{r}' \times \hat{s}') = -\sin^2\beta\sin(\phi_e - \phi'_e), \tag{V.10.17}$$

so that

$$ds/d\phi = -[(1 - \cos\theta')/R_S]\cot[(\phi_e - \phi'_e)/2]r's'. \tag{V.10.18}$$

As a consequence, combining Eq. (V.10.14) with Eq. (V.10.18), we obtain

$$\frac{d^2}{d\phi^2}(r' + s') = \frac{ds}{d\phi}\frac{d(r' + s')}{ds}\frac{d}{ds}\frac{ds}{d\phi} + \left(\frac{ds}{d\phi}\right)^2\frac{d^2(r' + s')}{ds^2}$$

$$= \sin^2\beta\left(\frac{1}{\rho_1} + \frac{1}{r'}\right)\left(\frac{1 - \cos\theta'}{R_S}\right)^2\cot^2\left(\frac{\phi_e - \phi'_e}{2}\right)r'^2s'^2, \tag{V.10.19}$$

since $d(r' + s')/ds = 0$ corresponding to a stationary point.

Finally, Eqs. (V.10.9,19) yield

$$u_{b,I} = \lambda^{1/2}e^{-i\pi/4}\frac{e^{-ik(r'+s')}}{(4\pi)^2s'}\tan\left(\frac{\phi_e - \phi'_e}{2}\right)\frac{1}{\sin\beta}\left[\frac{\rho_1}{r'(r' + \rho_1)}\right]^{1/2}$$

$$= D_K(\phi_e, \phi'_e; \beta)u_i(Q_e)e^{-ikr'}\left[\frac{\rho_1}{r'(\rho_1 + r')}\right]^{1/2}, \tag{V.10.20}$$

where $u_i(Q_e) = \exp(-iks')/(4\pi s')$ is the amplitude of the field incident on the edge point Q_e, while D_K is a diffraction coefficient similar to D_1, introduced in Section V.2.6, and coincident with Eq. (29) in Keller *et al.* [27], viz.

$$D_K(\phi_e, \phi'_e; \beta) = [e^{-i\pi/4}/(8\pi k)^{1/2}](1/\sin\beta)\tan[(\phi_e - \phi'_e)/2]. \qquad (V.10.21)$$

Myamoto and Wolf (see Chapter IV, Rubinowicz [32]) have shown that the field diffracted by an aperture illuminated by a generic RO field can be represented in the limit of vanishing wavelength by means of a BDW. Therefore, if we retrace the above steps, we easily arrive at the conclusion that Eqs. (V.10.20,21) remain valid for an RO incident field u_i. Keller *et al.* [27] have obtained Eq. (V.10.20) by asymptotically expanding the diffraction integral, and their coefficient coincides with D_K, the subscript K indicating that the diffracted field was calculated by using the complete Helmholtz–Kirchhoff integral formula. We note that D_K differs from D_1 of Eq. (V.2.48), except for $\phi_e - \phi'_e \cong \pi$. This discrepancy will be discussed in Section VI.2.

10.3 *Corner-Diffracted Field*

Corresponding to a corner, the derivative of $r'(\phi) + s'(\phi)$ is discontinuous. Therefore, we can use Eq. (V.2.38) to calculate the contribution $u_{b,II}$ of a corner, which yields

$$u_{b,II} = -\left(\frac{2\pi}{k}\right)^{1/2} e^{-i\pi/2} \frac{1}{(2\pi k)^{1/2}} \frac{e^{-ik(r'+s')}}{(4\pi)^2 R}(1 - \cos\theta')\,\delta\frac{1}{d(r'+s')/d\phi}, \qquad (V.10.22)$$

where $\delta(\cdots)$ indicates the discontinuity of the function on the right.

Now, since

$$\frac{d}{d\phi}(r' + s') = \frac{1 - \cos\theta'}{R} r's' \frac{\cos[(\phi' - \phi)/2]}{\sin[(\phi' - \phi)/2]}(\hat{r}' + \hat{s}')\cdot\hat{e}$$
$$= \frac{1 - \cos\theta'}{R} r's' \frac{1 + \cos\theta'}{(\hat{r}' \times \hat{s}')\cdot\hat{e}}(\hat{r}' + \hat{s}')\cdot\hat{e}, \qquad (V.10.23)$$

by simple algebra we get

$$u_{b,II} = -\frac{i}{k}u_i(Q_e)\frac{e^{-ikr'}}{4\pi r'}\delta\frac{(\hat{r}' \times \hat{s}')\cdot\hat{e}}{(\hat{r}' + \hat{s}')\cdot\hat{e}}$$
$$= -\frac{i}{k}u_i(Q_e)\frac{e^{-ikr'}}{4\pi r'}\frac{(\hat{s}' - \hat{r}')\cdot(\hat{e}_- \times \hat{e}_+)}{[(\hat{s}' + \hat{r}')\cdot\hat{e}_+][(\hat{s}' + \hat{r}')\cdot\hat{e}_-]}$$
$$= C_K u_i(Q_e)\frac{e^{-ikr'}}{r'}, \qquad (V.10.24)$$

where $\hat{e}_+$ and $\hat{e}_-$ stand for the unit vectors tangent to the edge on the two sides of the corner. Thus, a corner produces a spherical wave characterized

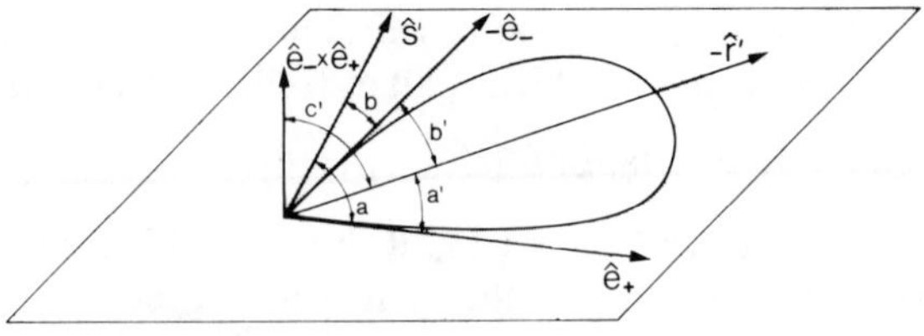

Fig. V.26. Geometry for a corner diffraction coefficient.

by a *corner diffraction coefficient*

$$C_K = -\frac{i}{4\pi k}\frac{(\hat{s}' - \hat{r}')\cdot(\hat{e}_- \times \hat{e}_+)}{[(\hat{s}' + \hat{r}')\cdot\hat{e}_+][(\hat{s}' + \hat{r}')\cdot\hat{e}_-]}, \qquad \text{(V.10.25)}$$

which can be expressed in terms of the angles of Fig. V.26.

In conclusion, when we transform the Helmholtz–Kirchhoff integral into a line integral by means of the Maggi–Rubinowicz formula, an aperture illuminated by a spherical wave gives a field that is the superposition of the geometric optics contribution (u_g), of the stationary edge points ($u_{b,I}$), and of the corners (if any) ($u_{b,II}$)

$$u(P) = u_g + u_{b,I} + u_{b,II} = C\frac{e^{-ikR}}{4\pi R} + \sum{}' D_K u_i(Q_e)e^{-ikr'}\left[\frac{\rho_1}{r'(\rho_1 + r')}\right]^{1/2}$$
$$+ \sum{}'' C_K u_i(Q_e)\frac{e^{-ikr'}}{r'}, \qquad \text{(V.10.26)}$$

where $C = 1$ in the lit region.

The rays contributing to $u_{b,I}$ form a cone around the edge tangent with aperture equal to the angle β formed by the incident ray with the edge itself. A corner point acts as a point source having a directivity pattern C_K becoming singular on the cones $(\hat{s}' + \hat{r}')\cdot\hat{e}_+ = 0$ and $(\hat{s}' + \hat{r}')\cdot\hat{e}_- = 0$, which are analogous to the shadow boundaries.

What we have shown for a spherical-wave illumination holds true for more general ray fields, which can be proved by directly evaluating the Helmholtz–Kirchhoff integral [27] by means of well-known formulas [24, 28].

11 Asymptotic Approximations to Plane-Wave Representation of the Field

We saw in Chapters I and IV that, under rather general conditions, a time-harmonic field $\mathbf{E}(\mathbf{r})$ propagating in an isotropic or anisotropic medium can be represented in a homogeneous half-space $z > 0$ by its plane-wave spectrum

$$\mathbf{E}(\mathbf{r}) = \iint_{-\infty}^{+\infty} \mathbf{E}_0(\mathbf{k})\exp[-i\Phi(\mathbf{k}) - i(k_x x + k_y y + k_z z)]\,dk_x\,dk_y, \qquad \text{(V.11.1)}$$

where $k_z = k_z(k_x, k_y)$ is a function of k_x and k_y derived from the dispersion equation $D(\omega, \mathbf{k}) = 0$ discussed in Chapter I, and $\mathbf{E}_0$ and Φ, respectively, represent the amplitude and the phase of the normal wave propagating along the wave vector $\mathbf{k}$. For k_x and k_y inside a region D_{H} having in general the form of an ellipse, k_z is real and positive while, outside D_{H}, $k_z = -i|k_z|$ is imaginary.

In most cases, the above integral is impossible to calculate analytically, and we must evaluate it asymptotically by resorting to stationary-point formulas [24, 29–33]. In particular, if $\mathbf{E}_0$ is a well-behaved function of k_x and k_y, we can put

$$\mathbf{E}(\mathbf{r}) \sim 2\pi i \sum_n \varepsilon_n \frac{\exp(-i\mathbf{k}_n \cdot \mathbf{r} - i\Phi_n)}{|\Delta|^{1/2}} \mathbf{E}_0(\mathbf{k}_n) + \mathbf{E}_{\mathrm{II}}(\mathbf{r}), \tag{V.11.2}$$

where the summation extends to the stationary vectors $\mathbf{k}_n$,

$$x + (\partial k_z/\partial k_x)z + \partial\Phi/\partial k_x = 0, \tag{V.11.3a}$$

$$y + (\partial k_z/\partial k_y)z + \partial\Phi/\partial k_y = 0 \tag{V.11.3b}$$

[the partial derivative with respect to $k_x(k_y)$ being taken at constant $k_y(k_x)$], while Δ is defined by

$$\Delta = \left(z\frac{\partial^2 k_z}{\partial k_x^2} + \frac{\partial^2 \Phi}{\partial k_x^2}\right)\left(z\frac{\partial^2 k_z}{\partial k_y^2} + \frac{\partial^2 \Phi}{\partial k_y^2}\right) - \left(\frac{\partial^2 k_z}{\partial k_x\, \partial k_y} z + \frac{\partial^2 \Phi}{\partial k_x \partial k_y}\right)^2, \tag{V.11.4}$$

$\Phi_n = \Phi(\mathbf{k}_n)$, and

$$\varepsilon = \begin{cases} +1, & \text{if} \quad \Delta > 0 \quad \text{and} \quad z\,\partial^2 k_z/\partial k_x^2 + \partial^2\Phi/\partial k_x^2 > 0 \\ -1, & \text{if} \quad \Delta > 0 \quad \text{and} \quad z\,\partial^2 k_z/\partial k_x^2 + \partial^2\Phi/\partial k_x^2 < 0 \\ \;\; i, & \text{if} \quad \Delta < 0. \end{cases} \tag{V.11.5}$$

$\mathbf{E}_{\mathrm{II}}(\mathbf{r})$ represents the contributions of the singular points of the phase $\Phi(\mathbf{k}) + \mathbf{k} \cdot \mathbf{r}$.

11.1 *Anisotropic Media*

The saddle point $\mathbf{k}_n$ can be determined for $\Phi = \text{const}$ by plotting in every direction $\hat{k}$ the two wave numbers k_1 and k_2 of the waves propagating in that direction, from the origin of an orthogonal coordinate system. In this way, we obtain a two-sheeted wave vector surface, one sheet of which corresponds to k_1, the other to k_2 (see Problem I.14). Then, according to Eqs. (V.11.3), the normal to the wave vector surface relative to the point $\mathbf{k}_n$ is parallel to $\mathbf{r}$.

In crystal optics, it is customary to represent the dependence of the phase velocities v_1 and v_2 on the directions of the vector $\hat{k}$ or of the relative Poynting vector by using, respectively, the *normal* and *ray surfaces* (see Section I.4.1). The distance of a point on these surfaces from the center is proportional to either v_1 or v_2. They differ in that the normal surface plots $v_{1,2}$ along the direction $\hat{k}$, while the ray surface plots these velocities along the direction of the Poynting vector associated with the vector $\mathbf{k}$. It can be shown that there exists a simple geometric connection between the normal surface and the ray surface (see Chapter IV, Sommerfeld [31]): *the ray surface is the envelope of the planes passing through the end point of the vector* $\hat{k}v_{1,2}$ *and perpendicular to* $\hat{k}$.

When the medium is isotropic and $\Phi = 0$, Eq. (V.11.4) simplifies to

$$\Delta = z^2k^2/k_z^4, \tag{V.11.6}$$

while $\mathbf{k}_n$ is parallel to $\mathbf{r}$, so that Eq. (V.11.2) yields

$$\mathbf{E}(\mathbf{r}) = 2\pi i(ke^{-ikr}/r)(z/r)\mathbf{E}_0(k\hat{r}). \tag{V.11.7}$$

This expression can be used to evaluate asymptotically the angular spectrum integral by putting $d\gamma\, d\beta = -dk_x\, dk_y/(k^2 \sin\beta)$.

12 Willis Formulas

The calculation of the diffraction integrals can be based on the series expansion of the nonoscillating part of the integrand corresponding to the end points of the integration interval. In fact, following an approach due to Willis and discussed by Barakat [34], we consider the following series expansion

$$\begin{aligned}\int_0^\infty g(x)G(k,x)e^{-\alpha x}\,dx &= \sum_{n=0}^\infty \frac{g^{(n)}(0)}{n!}\int_0^\infty G(k,x)x^n e^{-\alpha x}\,dx \\ &\equiv \sum_{n=0}^\infty \frac{(-1)^n g^{(n)}(0)}{n!}\Phi^{(n)}(\alpha),\end{aligned} \tag{V.12.1}$$

where we have assumed that $g(x)$ can be expanded in a Taylor series around $x = 0$. If we let α approach zero, we obtain

$$\int_0^\infty g(x)G(k,x)\,dx = \sum_{n=0}^\infty \frac{(-1)^n g^{(n)}(0)}{n!}\Phi^{(n)}(0). \tag{V.12.2}$$

For example, for $G(k,x) = \sin(kx)/x$ we have

$$\begin{aligned}\Phi(\alpha) &= \int_0^\infty \frac{\sin(kx)}{x}e^{-\alpha x}\,dx = \frac{\pi}{2} - \frac{\alpha}{k} + \frac{\alpha^3}{3k^3} - \frac{\alpha^5}{5k^5} + \cdots \\ &= \Phi(0) + \Phi^{(1)}(0)\alpha + \tfrac{1}{2}\Phi^{(2)}(0)\alpha^2 + \cdots,\end{aligned} \tag{V.12.3}$$

so that

$$\int_0^\infty g(x)\frac{\sin(kx)}{x}\,dx \cong \left(\frac{\pi}{2}\right)g(0) + \frac{g^{(1)}(0)}{k} - \frac{g^{(3)}(0)}{3k^3} + \cdots. \tag{V.12.4}$$

Analogous expression are given in the Appendix, Eq. (F5).

Problems

Section 2

1. Calculate the field diffracted by an aperture on a plane screen illuminated normally by a plane wave. Represent the diffraction integral in the form indicated in Problem IV.3 and replace ρ by $x = \rho^2$. Show that the contributions come from the end points only. Since $f(\rho)$ vanishes for $\rho = \rho_{\max}$, use Eq. (V.2.37). In addition, discuss the case in which either end point coincides with a vertex of the aperture, by noting that the indices of the relative diffraction integral written by using x as a variable are $\rho = 1$ and $\gamma = 1/2$.

2. Using the integral representation of the Bessel function,

$$J_\nu(\nu \sec \beta) = \frac{1}{\pi} \int_0^\infty \cos \nu(\theta - \sec \beta \sin \theta)\, d\theta$$

$$- \frac{\sin \nu\pi}{\pi} \int_0^\infty e^{-\nu(t + \sec \beta \sinh t)}\, dt,$$

for ν real and $\beta \neq 0, \pi$, prove the Debye formula

$$J_\nu(\nu \sec \beta) = \left(\frac{2}{\pi\nu \tan \beta}\right)^{1/2} \cos\left[\nu(\beta - \tan \beta) + \frac{\pi}{4}\right] + O\left(\frac{1}{\nu}\right)$$

as $\nu \to \infty$. *Hint*: Apply the SP method by observing that

$$0 < \int_0^\infty e^{-\nu(t + \sec \beta \sinh t)}\, dt < \int_0^\infty e^{-\nu t}\, dt = \frac{1}{\nu},$$

while the function $f(\theta) = \theta - \sec \beta \sin \theta$ is stationary for $\theta = \beta$ and $f''(\beta) = \tan \beta$.

3. Show that

$$J_\nu(\nu) = (1/\pi)\Gamma(4/3)(6/\nu)^{1/3} 3^{1/6}/2 + O(\nu^{-2/3}).$$

Hint: Use the SP method by observing that $J_\nu(\nu) = J_\nu(\nu \sec \beta)$ with $\beta = 0$. In this case the phase $f(\theta) = \theta - \sec \beta \sin \theta$ introduced in the above problem is stationary at the end point $\theta = 0$ of the integration interval in the integral representation of the Bessel function, while $f''(0) = 0$ and $f'''(0) = 1$. Accordingly, use Eq. (V.2.18) with $\rho = 3$ and $\gamma = 0$.

Section 6

4. Study the hills, valleys, and paths of steepest descent and ascent of the function $\cos z$ (see Fig. V.14) present in the integrand of the angular representation of a field. Show that the steepest-descent paths are

represented by

$$\operatorname{Re} z = n\pi + (-1)^n gd(\operatorname{Im} z),$$

where $n = 0, \pm 1, \pm 2, \ldots$ is an integer and $gd(x)$ is the so-called Gudermannian function defined by (see Chapter II, Abramowitz and Stegun [4], Eq. (4.3.117))

$$\mathrm{gd}(x) = 2 \arctan e^x - \tfrac{1}{2}\pi$$

5. Study the hills, valleys, and paths of steepest descent and ascent of the function $\operatorname{Re}(z - z^3/3)$ present in the integral representation of the Airy function.

6. Study the hills, valleys, and paths of steepest descent and ascent of the function $\cos z + b(z - \frac{1}{2}\pi)$, $b > 1$ and $b < 1$, present in the integral representation of the Bessel and Hankel functions.

7. Find the leading term of the asymptotic representation of the Hankel function $H_n^{(2)}(x)$ by using the integral representation obtained by noting that the Fourier transform of the two-dimensional Green's function [Eq. (IV.7.5)] is given by Eq. (IV.8.6) with $k_y = 0$.

8. Show that the function $\cos z + z - \frac{1}{2}\pi$ has saddle points in $z_s = \frac{1}{2}\pi + 2\pi n$ of order two (monkey saddles), while the steepest-descent paths have directions $\theta = -\pi/6, -5\pi/6, +\pi/2$ and are defined by the equation $\sin(z' - z_s)\cosh z'' = z' - z_s$.

9. With reference to the last two problems, show that the leading term of the asymptotic expansion of $H_n^{(2)}(n)$ for $n \to \infty$ reads

$$H_n^{(2)}(n) \sim -[\Gamma(1/3)/(\pi n^{1/3})](4/27)^{1/6} e^{-2\pi i/3}.$$

10. Consider the Airy function $\mathrm{Ai}(x)$ defined by the integrals

$$\mathrm{Ai}(x) = \frac{1}{\pi}\int_0^\infty \cos\left(\frac{t^3}{3} + xt\right) dt = \frac{1}{2\pi i}\int_{-i\infty}^{+i\infty} e^{-x\tau + \tau^3/3}\, d\tau.$$

Then, introducing the variable transformation $\tau = x^{1/2} z$, where we choose the principal branch of the square root, we obtain

$$\mathrm{Ai}(x) = \frac{x^{1/2}}{2\pi i}\int_{-i\infty}^{+i\infty} \exp\left[x^{3/2}\left(z - \frac{z^3}{3}\right)\right] dz \qquad (x > 0),$$

$$\mathrm{Ai}(x) = \frac{|x|^{1/2}}{2\pi}\int_{-\infty}^{+\infty} \exp\left[-i|x|^{3/2}\left(z - \frac{z^3}{3}\right)\right] dz \qquad (x < 0),$$

while, introducing the integration path L_0 in the complex z plane (see Fig. V.27), we have

$$\mathrm{Ai}(x) = \frac{x^{1/2}}{2\pi i}\int_{L_0} \exp\left[x^{3/2}\left(z - \frac{z^3}{3}\right)\right] dz \qquad (x > 0).$$

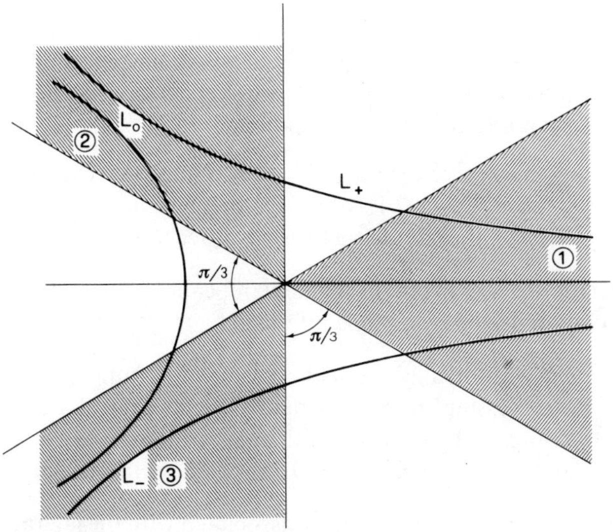

Fig. V.27. Paths for evaluating the Airy functions Ai and Bi.

Finally, the stationary-phase method yields

$$\mathrm{Ai}(x) \sim \sin(2|x|^{3/2}/3 + \pi/4)/(\pi^{1/2}|x|^{1/4}) \qquad (x < 0).$$

Show the above relations.

11. With reference to Problem 10, show that, for $x > 0$, replacing L_0 with the steepest-descent path passing through the saddle point $z = -1$ yields

$$\mathrm{Ai}(x) = \frac{x^{1/2}\exp\left(-2\dfrac{x^{3/2}}{3}\right)}{2\pi i}\int_0^\infty \exp(-sx^{3/2})\left(\frac{1}{z_+^2 - 1} - \frac{1}{z_-^2 - 1}\right) ds,$$

where $z_+(s)$ and $z_-(s) = z_+^*(s)$ are solutions of the equation $s = \frac{1}{3}z^3 - z - \frac{2}{3} = \frac{1}{3}(z - 2)(z + 1)^2$. Then, taking into account the relation

$$1/(z_+^2 - 1) - 1/(z_-^2 - 1) = (2i/s^{1/2}) - (15/72)is^{1/2} + \cdots,$$

show that

$$\mathrm{Ai}(x) \sim \exp\left(-2\frac{x^{3/2}}{3}\right)\Bigg/(2\pi^{1/2}x^{1/4}) \qquad (x > 0).$$

12. The Airy function $\mathrm{Bi}(x)$ is defined by an integral similar to that used for $\mathrm{Ai}(x)$ apart from a change of the integration path L_0 to $L_+ - L_-$ (see Fig. V.27). Show that, for $x > 0$, $\mathrm{Bi}(x)$ tends asymptotically to

$$\mathrm{Bi}(x) \sim \exp(2x^{3/2}/3)/(\pi^{1/2}x^{1/4}).$$

Section 7

13. Calculate the field diffracted by a plane interface between two dielectrics illuminated by a cylindrical p-wave. *Hint*: Use Eq. (V.7.1) by modifying the integrand because of the p-polarization of the wave.

14. Calculate the successive term of the asymptotic expansion of the integral I_{SDP} defined by Eq. (V.7.1), whose leading term is given by Eq. (V.7.3). *Hint*: Use Eq. (V.2.32) to evaluate the second-order term.

15. Calculate the lateral shift of a gaussian beam having a p-polarization, when it is totally reflected by a plane interface.

16. Evaluate asymptotically the field diffracted by a multilayer supporting leaky waves by expressing the Sommerfeld-type integral of Eq. (V.7.17) by means of the complex Fresnel integral, in accordance with the uniform asymptotic formula of Eq. (V.6.12) (see Tamir and Bertoni [23]).

17. Compare the field on the interface associated with the lateral wave with the GO contribution. Show that, for an incidence angle of the illuminating gaussian beam close to the critical angle, the lateral wave becomes comparable to the GO field.

18. Calculate the field transmitted by a plane interface between two media illuminated by a Gaussian beam. Show that it is possible to split the field into a lateral wave and a GO contribution.

19. Show that the Goos–Haenchen lateral shift of a beam reflected by a plane interface between two dielectrics can be described as due to the displacement of the reflection plane slightly inside the second medium. Calculate the distance of this virtual plane of reflection from the interface as a function of the incidence angle.

20. Consider two dielectric media, say 1 and 2, separated by a cylindrical surface of radius a. A collimated gaussian beam illuminates the interface at an incidence angle (referred to the beam axis) greater than the critical one. Calculate the far field transmitted into the second medium for a p- and s-polarized wave as a function of the incidence angle. In addition, calculate the power lost in the reflection by the incident beam due to the partial transmission. *Hint*: Calculate the field on the interface by using the Fresnel transmission coefficient relative to rays directed as the beam axis. Then evaluate the Fraunhofer diffraction integral asymptotically by using the saddle-point method in order to account correctly for the gaussian distribution of the illumination. (See Unger [35].)

Section 9

21. Consider the diffraction integral

$$I(k) = \int_{-\infty}^{+\infty} \exp(-bs^2)\exp\left[-ik\left(\zeta s + \frac{s^3}{3}\right)\right] ds.$$

Expand $I(k)$ asymptotically by using the representation of Eq. (V.9.7). *Hint*: Put $u = is$ and use the expansion [cf. Eq. (V.9.2)]

$$\exp(-bs^2) = \sum_{n=0}^{\infty} \frac{b^n}{n!} u^{2n} = \sum_{m=0}^{\infty} (u^2 - \zeta)^m \sum_{q=0}^{m} \zeta^q b^{m+q} \frac{1}{m!\,q!}$$

to derive g_0 and g_1 from Eqs. (IV.9.2,8).

References

1. Keller, J. B., *J. Opt. Soc. Am.* **52**, 116 (1962).
2. Ludwig, D., *Commun. Pure Appl. Math.* **19**, 215 (1966).
3. Lewis, R. M., and Boersma, J., *J. Math. Phys.* **10**, 2291 (1969).
4. Ufimtsev, P. Y., transl. by Air Force Syst. Command, Foreign Tech. Div., Doc. ID No. FTD-HC-23-259-71 (1971).
5. Ufimtsev, P. Y., *Proc. IEEE* **63**, 1734 (1975).
6. Kouyoumjan, R. G., and Pathak, P. H., *Proc. IEEE* **62**, 1448 (1974).
7. Stokes, G. G., *Cambridge Philos. Trans.* **9**, 166 (1856).
8. Kelvin, Lord (Thomson, W.), *Philos. Mag.* **23**, 252 (1887).
9. Watson, G. N., *Proc. Cambridge Philos. Soc.* **19**, 49 (1918).
10. Erdelyi, A., *SIAM J. Appl. Math.* **3**, 17 (1955).
11. Erdelyi, A., "Asymptotic Expansions." Dover, New York, 1956.
12. van der Corput, J. G., *Proc. Anst. Akad. Wet.* **51**, 650 (1948).
13. van der Corput, J. G., "Asymptotic Expansions I–III," Tech. Rep. Dep. Math., Univ. of California, Berkeley, 1954–1955.
14. Bleistein, N., and Handelsman, R. A., "Asymptotic Expansions of Integrals." Holt, New York, 1975.
15. Sirovich, L., "Techniques of Asymptotic Analysis." Springer-Verlag, Berlin and New York, 1971.
16. Dingle, R. B., "Asymptotic Expansions: Their Derivation and Interpretation." Academic Press, New York, 1973.
17. Lee, S.-W., and Deschamps, G., *IEEE Trans. Antennas Propag.* **AP-24**, 25 (1976).
18. Kravtsov, Yu. A., and Orlov, Y. I., *Sov. Phys.—Usp. (Engl. Transl.)* **23**, 750 (1980).
19. Pearcey, T., *Philos. Mag.* **37**, 311 (1946).
20. Goos, F., and Haenchen, H. H., *Ann. Phys. (Leipzig)* **1**, 333 (1947).
21. Lotsch, H. K., *J. Opt. Soc. Am.* **58**, 551 (1968).
22. Midwinter, J. E., and Zernike, F., *Appl. Phys. Lett.* **16**, 198 (1970).
23. Tamir, T., and Bertoni, H. L., *J. Opt. Soc. Am.* **61**, 1397 (1974).
24. Focke, J., *Ber. Saech. Akad. Wiss. Leipzig* **101**(3), 1 (1954).

24a. Focke, J., *Opt. Acta* **3**, 110 (1956).

25. Chester, C., Friedman, B., and Ursell, F., *Proc. Cambridge Philos. Soc.* **53**, 599 (1957).
26. Keller, J. B., *J. Appl. Phys.* **28**, 426 (1957).
27. Keller, J. B., Lewis, R. M., and Seckler, B. D., *J. Opt. Soc. Am.* **28**, 570 (1957).
28. Hansen, R. C., ed. "Geometric Theory of Diffraction." IEEE Press, New York, 1981.
29. Braun, G., *Acta Phys. Austriaca* **10**, 8 (1956).
30. Jones, D. S., and Kline, M., *J. Math. Phys.* **37**, 1 (1958).
31. Chako, N., *J. Inst. Math. Its Appl.* **1**, 372 (1965).
32. Lalor, E., *J. Opt. Soc. Am.* **58**, 1235 (1968).
33. Sherman, G. C., Stamnes, J. J., and Lalor, E., *J. Math. Phys.* **17**, 760 (1976).
34. Barakat, R., *in* "The Computer in Optical Research" (B. R. Frieden, ed.), p. 35–80. Springer-Verlag, Berlin and New York, 1980.
35. Unger, H. G., "Planar Optical Waveguides and Fibers," Sect. 2.8. Oxford Univ. (Clarendon), London and New York, 1977.

Bibliography

Bleistein, I. N., "Mathematical Methods for Wave Phenomena", Academic Press, Orlando, Florida, 1984.
Southwell, W. H., *Opt. Lett.* **3**, 100 (1978).

Chapter VI

Aperture Diffraction and Scattering from Metallic and Dielectric Obstacles

1 Introduction

In most practical cases diffraction problems [1, 2] (see also Chapter IV, Bowman *et al.* [3]) do not admit closed analytic solutions. Therefore, the most important objective pursued since the last century has been the construction of approximate solutions. The *Kirchhoff principle* has been the starting point for many approximations, especially for diffraction from apertures. As we mentioned before, this principle amounts to approximating the field on the aperture with that created by the same sources when the screen delimiting the aperture has been removed. Physically, this is tantamount to assuming that the screen does not perturb the field on the aperture itself. In fact, this perturbation is significant only in proximity to the aperture edge. Hence, if the aperture is sufficiently large the error due to the deviation from the actual field near the edges becomes negligible. For small apertures, however, the Kirchhoff approximation introduces errors that are too large. In order to quantify the order of magnitude of these errors we must know the extension of the zone near the edges where the above approximation is not valid. To answer this question implies knowledge of the exact solution, which is known only in very few cases and for special geometries.

While diffraction by metallic obstacles can now be considered well described by current approximate theories [geometric theory of diffraction (GTD), uniform asymptotic representation theory (UAT)], the effects of dielectric bodies have not yet been completely clarified. This is because no analytic solution has yet been found for the basic problem of diffraction of a plane wave by a dielectric wedge.

The main achievement of the last decades is the development of asymptotic techniques (see Chapter V, Keller [1, 26], Kouyoumjian and Pathak [6],

Keller *et al.* [27], and Hansen [28]) or constructing the diffracted fields in such a way as to take into account the modifications produced by obstacles and apertures. Based on an initial program of simple tests, the theory has gone so far as to construct field distributions approaching the correct ones in the limit $k \to \infty$ for obstacles of any form and complexity.

The asymptotic construction of diffracted fields will be illustrated by using as canonical examples the wedge with finite surface impedance, the circular cylinder, and the sphere. The wedge is of interest because (1) it provides the coefficients of the *diffraction matrix* $\mathbf{D}$, which is the basic tool for obtaining the field diffracted by a curved wedge delimiting a physical aperture, and (2) the diffraction field also includes two surface waves as a consequence of the finite surface impedance of the two wedge faces. The characteristics of these surface waves will be further illustrated by analyzing a metallic circular cylinder illuminated by plane waves. When the series representing the diffracted field is obtained by applying the Watson transform illustrated in Chapter IV, the field diffracted in the dark region takes the form of a surface wave. Finally, the sphere is used to establish some general theorems for the field scattered by a finite obstacle and to extend the analysis of the surface waves excited by the rays tangent to the scatterer and traveling along the geodesics of its surface.

A peculiarity of all diffraction fields, is their sensitivity to the wavelength. This property finds a great many applications in optical instruments designed for analyzing the spectral composition of light beams. An important example is the diffraction grating formed by combining a large number of wedges. The field diffracted by each edge combines with all the other fields, thus producing a pattern of very narrow lines corresponding to the respective diffraction orders of the grating. The orientation of each line changes with the frequency, so that a detector pointing in a specified direction, and having its front surface delimited by a slit, will measure the intensity of the radiation that is diffracted in the assigned order and has a frequency in a small interval depending on the slit aperture. This is essentially the working principle of modern grating spectrometers, which have almost completely replaced those using a prism as the dispersive element.

2 Diffraction from a Wedge

We consider as a preliminary case a wedge-shaped region, $\rho > 0$, $2\Phi \geq \phi \geq 0$ (see Fig. VI.1), illuminated by an s-polarized plane wave (i.e., having the magnetic field parallel to the edge) propagating along the direction $\phi = \phi'$ perpendicular to the edge of the obstacle, which is supposed to be a perfect conductor:

$$u_i(\rho, \phi) = \exp[ik\rho\cos(\phi - \phi')]. \qquad \text{(VI.2.1)}$$

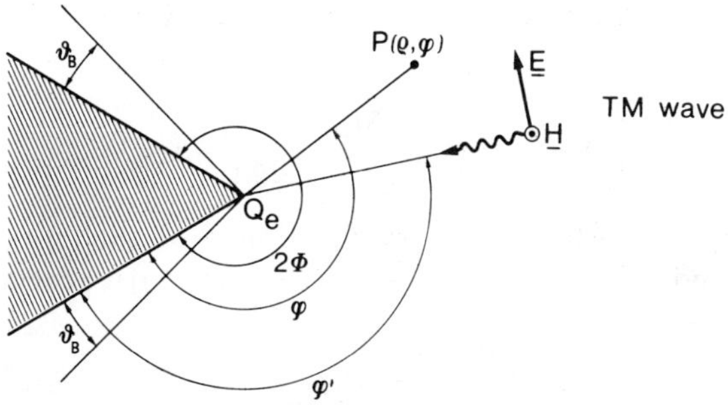

Fig. VI.1. Geometry of a wedge-shaped region.

where the plus sign in the exponent is due to the fact that we think the wave propagates in the direction of the half-line $\phi = \phi'$.

Following Sommerfeld (see Chapter IV [31]), we represent the total field (incident plus reflected plus diffracted) by the integral

$$u(\rho, \phi) = \int_\Gamma e^{-ik\rho\cos\beta} S(\beta + \phi)\, d\beta, \tag{VI.2.2}$$

extended to the complex path Γ indicated in Fig. VI.2. The angular spectrum S has simple polar singularities and is determined by imposing the vanishing of $\partial u/\partial\phi$ on the wedge faces,

$$\int_\Gamma e^{-ik\rho\cos\beta} \frac{\partial S(\beta + \phi)}{\partial\phi}\bigg|_{\phi=0,2\Phi} d\beta = 0. \tag{IV.2.3}$$

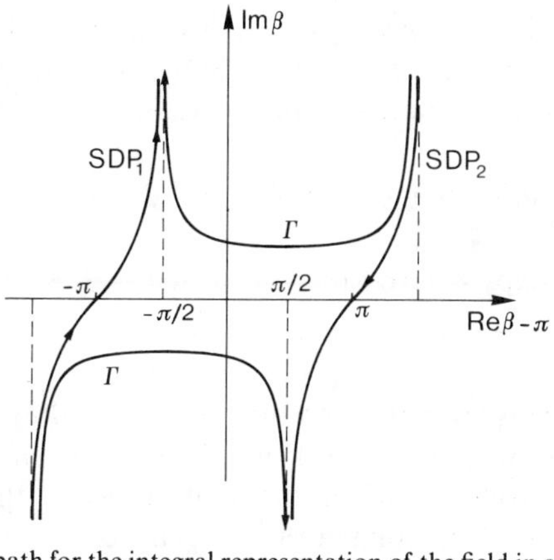

Fig. VI.2. Complex path for the integral representation of the field in a wedge-shaped region [see Eq. (VI.2.2)].

Since $\partial S/\partial\phi = \partial S/\partial\beta$, we can integrate this integral by parts, thus obtaining

$$\int_\Gamma e^{-ik\rho\cos\beta}\sin\beta\begin{Bmatrix} S(\beta) \\ S(\beta+2\Phi)\end{Bmatrix} d\beta = 0. \tag{VI.2.4}$$

This equation is satisfied by every function $S(\beta)$ such that $S(\beta+\pi) = -S(-\beta+\pi)$ and $S(\beta+2\Phi+\pi) = -S(-\beta+2\Phi+\pi)$ for every β. A possible solution is given by

$$S(-\pi) = \frac{i}{2\pi N}\frac{\sin(\alpha/N)}{\cos(\alpha/N) - \cos(\phi'/N)}, \qquad N = \frac{2\Phi}{\pi}. \tag{VI.2.5}$$

We observe that $F(\alpha)$ has an infinity of simple poles located at $\alpha = \beta_q$ with

$$\beta_q = (-1)^q(\phi' - \Phi) + (2q+1)\Phi + \pi, \qquad q = 0, \pm 1, \pm 2, \ldots. \tag{VI.2.6}$$

If we modify the integration path Γ by replacing it with the pair of steepest-descent paths SDP_1 and SDP_2, relative to the factor $\exp(-ik\rho\cos\beta)$ passing through the saddle points 0 and 2π, respectively, we obtain:

$$\begin{aligned} u(\rho,\phi) = {} & \left(\int_{SDP_1} d\beta + \int_{SDP_2} d\beta\right)\exp(-ik\rho\cos\beta)S(\beta+\phi) \\ & + \sum_q r_q \exp[-ik\rho\cos(\beta_q - \phi)], \end{aligned} \tag{VI.2.7}$$

where the r_q represent the residues of the poles of $S(\beta+\phi)$ between SDP_1 and SDP_2:

$$r_q = \lim_{\beta\to\beta_q-\phi} 2\pi i S(\beta+\phi)(\beta+\phi-\beta_q) = 1. \tag{VI.2.8}$$

Where ϕ' is real, the poles β_q are also real, as implied by Eq. (VI.2.6). Therefore, the β_q between SDP_1 and SDP_2 satisfy the inequality $2\pi \geq \beta_q - \phi \geq 0$. As a consequence, we can rewrite Eq. (VI.2.7) as

$$\begin{aligned} u(\rho,\phi) = {} & \int_{SDP_2} d\beta[S(\beta+\phi) - S(\beta+\phi-2\pi)]\exp(-ik\rho\cos\beta) \\ & + \sum_q \exp[-ik\rho\cos(\phi-\beta_q)][U(\phi-\beta_q+2\pi) - U(\phi-\beta_q)], \end{aligned} \tag{VI.2.9}$$

$U(x)$ being the Heavyside unit step function. According to Eq. (VI.2.9), the field comprises some plane waves in the sectors $\max(\beta_q - 2\pi, 0) \leq \phi \leq \min(\beta_q, 2\Phi)$. On the other hand, in the limit of $k \to \infty$ the field must reduce to the superposition of collimated ray congruences, each of them produced by the reflections undergone by the incident wave. It is easy to verify that the ray optical field coincides with the summation of Eq. (VI.2.9).

Since the above expression for $u(\rho, \phi)$ satisfies the boundary conditions and reduces to the correct limit as $k \to \infty$, it correctly represents the field for finite values of k as well.

For $k\rho \to \infty$, the two integrals along SDP_1 and SDP_2 can be replaced by the relative saddle-point contributions, so that:

$$u(\rho, \phi) \sim D_s(\phi, \phi') \frac{\exp(-ik\rho)}{\rho^{1/2}} + \sum_q \exp[-ik\rho\cos(\phi - \phi')][U(\phi - \beta_q + 2\pi) - U(\phi - \beta_q)], \tag{VI.2.10}$$

where

$$D_s(\phi, \phi') = -\frac{e^{-i\pi/4}\sin(\pi/N)}{N(2\pi k)^{1/2}} \times \left\{\frac{1}{\cos(\pi/N) - \cos[(\phi - \phi')/N]} + \frac{1}{\cos(\pi/N) - \cos[(\phi + \phi')/N]}\right\} \tag{VI.2.11}$$

is the diffraction coefficient of the wedge illuminated by an s-wave. For a p-wave (i.e., polarized parallel to the edge) the coefficient D_p can be obtained from D_s by changing to minus the plus sign before the second term on the right side of Eq. (VI.2.11).

When a pole comes close to the saddle points, a more accurate representation of the field can be obtained by using the transition function discussed in Section V.6.

2.1 *Wedges with Surface Impedances*

In the optical range the deviation of a real metal from a perfect conductor becomes quite pronounced. This can be accounted for by representing the surface of a metal with a surface impedance, as already seen at the end of Chapter III. Of course, this representation loses its value when the frequency approaches the plasma frequency too closely. Once the real surface has been replaced with an impedance surface, many diffraction problems can be solved analytically, as shown by the case of an impedance wedge. The same approach can be used to analyze the diffraction from metallic bodies with dielectric coatings. In this case also, some insight into the characteristics of the diffraction phenomenon can be gained by solving for an impedance surface.

In particular, obstacles with imaginary surface impedances share the feature of supporting surface waves. We saw in Chapter III that the Brewster angle of these surfaces is given by $\operatorname{cotan}\theta_B = Z_s/\zeta$. For a good metallic conductor $\theta_B \simeq \pi/2 - i/k$, so that a surface wave can be supported by this surface, which extends for a finite depth into the medium. When we consider the diffraction of a wave from an impedance surface presenting wedges, apertures, or other irregularities, we must take into account that these waves can modify the intensity and the pattern of the total diffracted field.

In order to assess the influence of a finite conductivity on the diffracted field, it is instructive to extend the above analysis to the case of a wedge characterized by a surface impedance Z_s, corresponding to a Brewster angle θ_B. In this case, an s-wave with the magnetic field parallel to the wedge axis z satisfies the boundary condition

$$(1/\rho)(\partial u/\partial\phi) \mp ik\cos\theta_B u = 0 \qquad \text{for} \quad \phi = 0, 2\Phi. \tag{VI.2.12}$$

Here $u(\rho, \phi) = H(\rho, \phi)$, while the minus and plus signs apply to the upper and lower faces of the wedge, respectively (see Fig. VI.1). Maliuzhinets [3] has obtained an angular spectrum representation of u in the form of a generalized Sommerfeld integral similar to that of Eq. (VI.2.2),

$$u(\rho, \phi) = \int_\Gamma e^{-ik\rho\cos\beta} S(\beta + \phi)\frac{g(\beta + \phi)}{g(\pi + \phi')}\, d\beta, \tag{IV.2.13}$$

where $S(\beta)$ is the function defined by Eq. (VI.2.5), while $g(\beta + \phi)$ accounts for the finite surface impedance,

$$g(\alpha + \pi) = \frac{M(\alpha - \theta_B)M(\alpha + \theta_B - 2\Phi)}{M(\alpha - \pi + \theta_B)M(\alpha - \theta_B - 2\Phi + \pi)} \times \frac{\cos[(1/2N)(\alpha + \theta_B - \pi/2)]\sin[(1/2N)(\alpha - \theta_B + \pi/2)]}{\sin(\alpha/N)}. \tag{VI.2.14}$$

The function $M(\alpha)$, introduced by Maliuzhinets, is defined by the infinite product

$$M(\alpha) = \prod_{n,m=1}^{\infty}\left[1 - \left(\frac{\alpha/\pi}{N(2n - 1) + m - 1/2}\right)^2\right]^{(-1)^{m+1}} \tag{VI.2.15}$$

The integral in Eq. (VI.2.13) can be evaluated asymptotically as $k\rho \to \infty$ by following the steps leading to Eq. (VI.2.10). In so doing we obtain:

$$\begin{aligned} u(\rho, \phi) \sim{}& D_s(\phi, \phi')\frac{\exp(-ik\rho)}{\rho^{1/2}} + S_+\exp[-ik\rho\sin(\phi + \theta_B)] \\ &+ S_-\exp[-ik\rho\sin(-\phi + 2\Phi + \theta_B)] \\ &+ \sum_q r_q \exp[-ik\rho\cos(\phi - \beta_q)]. \end{aligned} \tag{VI.2.16}$$

This field differs from that of Eq. (VI.2.10) in three respects. First, the diffraction coefficient D_s depends on the Brewster angle. Second, two surface waves of amplitude S_+ and S_-, respectively, can be present in the angular sectors adjacent to the faces, defined by

$$\begin{aligned} 2\Phi \geq \phi \geq 2\Phi - \operatorname{arccosh}(-1/\theta''_B) + \pi/2 - \theta'_B, \\ 0 \leq \phi \leq \operatorname{arccosh}(-1/\theta''_B) - \pi/2 + \theta'_B. \end{aligned} \tag{VI.2.17}$$

These waves exist only if $\theta'_B + \operatorname{arccosh}(-1/\theta''_B) > \pi/2$. Third, the amplitudes of the waves reflected by the wedge and propagating along $\phi = \beta_q$ differ from unity. These amplitudes can be calculated by using the Fresnel formulas relative to the wedge faces. It can be shown that in the present case D_s reads:

$$\begin{aligned} D_s(\phi, \phi') = -\frac{e^{-i\pi/4}\sin(\pi/N)}{N(2\pi k)^{1/2}}\left[\frac{1}{\cos(\pi/N) - \cos[(\phi - \phi')/N]}\frac{g(\phi)}{g(\phi' + \pi)}\right. \\ \left. + \frac{1}{\cos(\pi/N) - \cos[(\phi + \phi')/N]}\frac{g(\phi + 2\pi)}{g(\phi' + \pi)}\right]. \end{aligned} \tag{VI.2.18}$$

On the other hand, the amplitude r_q of the qth reflected wave is given by

$$r_q = g(\beta_q)/g(\phi' + \pi), \tag{VI.2.19}$$

as can be seen easily by using Eq. (VI.2.8) and noting that the residue of $S(\beta + \phi)g)\beta + \phi)$ at $\phi + \beta = \beta_q$ is given by $g(\beta_q)$. According to Eq. (VI.2.19), the beam diffracted in proximity to the directions $\beta_q \pm \pi$ coincides with that diffracted by a perfect conductor ($\theta_B = \pi/2$) multiplied by the reflection coefficient r_q. Since r_q is independent of the frequency it can be calculated very simply by using the Fresnel formulas (see Chapter III) for a medium of refractive index $\tilde{n} = \tan\theta_B$.

The amplitudes $S_\pm$ of the two surface waves coincide with the residues of the integrand of Eq. (VI.2.13) relative to the two poles of $g(\beta + \phi)$ between SDP_1 and SDP_2. Using Eqs. (VI.2.14,15), we can show that these two poles occur at $\beta_\pm + \phi - \Phi = \pm(3\pi/2 + \Phi - \theta_B) + \pi$.

2.2 *Edge Diffraction Matrix*

The above results have been extended to a curved metallic edge (Fig. VI.3) illuminated by rays forming an angle β with the tangent $\hat{e}$ to the edge at the diffraction point $Q_{\hat{e}}$ (cf. Section V.10 and Fig. V.25). If we consider the projection of the incident and diffracted rays on a plane perpendicular to the edge at Q_e, the position of the diffracted rays, forming a conical surface, is given by the angle ϕ, while the direction of the incident ray is defined by ϕ'

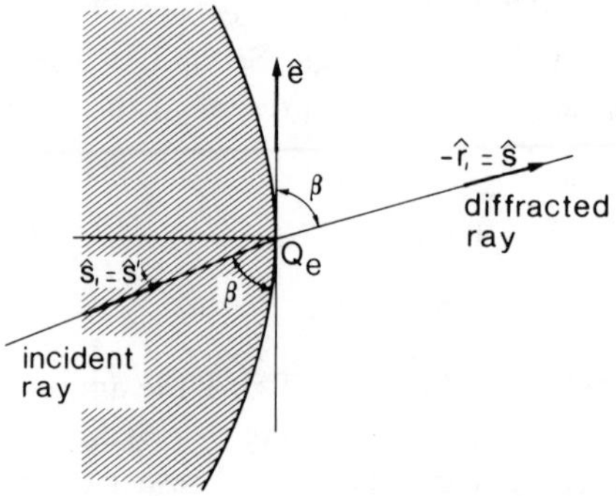

Fig. VI.3. Edge diffraction geometry.

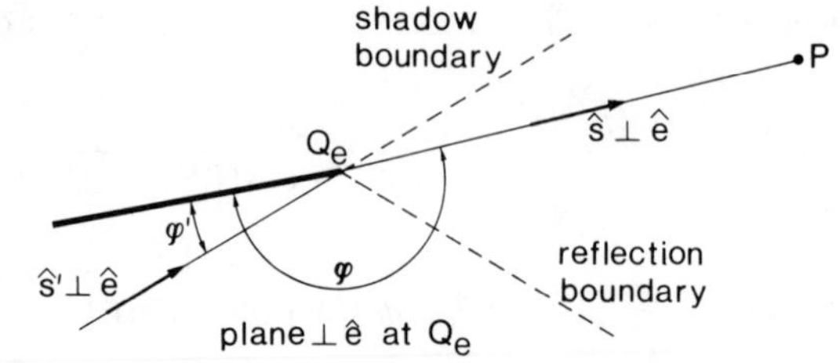

Fig. VI.4. Diffraction by a half-plane.

(Fig. VI.4). The electric component of the edge-diffracted ray can be expressed in the form

$$\mathbf{E}_d(\mathbf{r}) = [\rho_1/[r(\rho_1 + r)]]^{1/2} e^{-ikr} \mathbf{D}(\phi, \phi'; \beta) \cdot \mathbf{E}_i(Q_e), \qquad \text{(VI.2.20)}$$

where $\mathbf{r}$ coincides with r' of Section V.10 while ρ_1 maintains the meaning of Eq. (V.10.20). The diffraction matrix $\mathbf{D}$ has been derived by Kouyoumjian and Pathak [4] and put in the form

$$\mathbf{D}(\phi, \phi'; \beta) = \hat{\beta}_d \hat{\beta}_i D_p + \hat{\phi}\hat{\phi}' D_s, \qquad \text{(VI.2.21)}$$

where $\hat{\beta}_d = \hat{\phi} \times \hat{r}$, $\hat{\beta}_i = \hat{\phi}' \times \hat{s}'$, and D_p and D_s, which are a generalization of the diffraction coefficient of Eq. (VI.2.11) for $\beta \neq \pi/2$, are given by:

$$D_{p,s} = \frac{e^{-i\pi/4} \sin(\pi/N)}{N(2\pi k)^{1/2} \sin\beta} \times \left[-\frac{1}{\cos(\pi/N) - \cos[(\phi - \phi')/N]} \pm \frac{1}{\cos(\pi/N) - \cos[(\phi + \phi')/N]} \right], \qquad \text{(VI.2.22)}$$

with $N = 2\Phi/\pi$. In particular, for a half-plane $N = 2$ and the above equation becomes

$$D_{p,s} = \frac{e^{-i\pi/4}}{(8\pi k)^{1/2} \sin\beta} \left[\frac{1}{\cos[(\phi - \phi')/2]} \mp \frac{1}{\cos[(\phi + \phi')/2]} \right]. \qquad \text{(VI.2.23)}$$

By comparing these diffraction coefficients with D_1 [see Eq. (V.2.48)] and D_k [see Eq. (V.1021)], we observe that they have in common the factor multiplying the term in square brackets. In addition, they become singular for $\phi = \pi + \phi'$, that is, when we consider rays lying on the plane defined by the edge tangent in Q_e and the incident ray. From the point of view of ray optics this plane separates the illuminated region from that in the shadow, and it is consequently called the *shadow boundary* (SB). While D_1 and $D_{p,s}$ are singular for $\phi + \phi' = \pi$, D_k remains finite along these directions. It is easy to show that these directions coincide with those of the rays reflected in the ray optical limit; the half-plane passing through the edge and containing these rays is accordingly called the *reflection boundary* (RB). In conclusion, we note that all these coefficients are almost coincident for directions close to the shadow boundary, while they differ notably for directions departing from the SB. This leads us to affirm that the calculation carried out by representing the field scalarly and applying the Kirchhoff approximation agrees closely with the exact electromagnetic solution whenever we consider rays diffracted in the forward direction and deviating slightly from the shadow boundary. In practice, this means that the Kirchhoff approximation fails at points deeply immersed in either the dark or the lit regions.

In proximity to the SB the field expressed by Eq. (VI.2.20) becomes singular. As we showed in Section V.3, in this case we can use the complex Fresnel integral F and the detour parameter ξ to construct an expression for the total field formally similar to that of Eq. (VI.2.20), with the coefficients $D_{p,s}$ replaced by

$$D = -4\pi F(\xi)\pi^{1/2}\xi e^{i\xi^2}\frac{\sin(\pi/N)}{N(2\pi k)^{1/2}\sin\beta}\frac{1}{\cos(\pi/\mathrm{N}) - \cos[(\phi - \phi')/N]}, \tag{VI.2.24}$$

where the detour parameter is given by

$$\xi^2 = 2k\frac{r(\rho_e^i + r)\rho_1^i\rho_2^i}{\rho_e^i(\rho_1^i + r)(\rho_2^i + r)}\sin^2\beta\cos^2\left(\frac{\phi - \phi'}{2}\right), \tag{VI.2.25}$$

with $\rho_{1,2}^i$ the principal radii of curvature of the incident wave front (cf. Section II.9) and ρ_e^i the curvature radius in the edge-fixed plane of incidence.

It is straightforward to show that for a plane incident wave ξ reduces to $(2kr)^{1/2}\sin\beta\cos[\frac{1}{2}(\phi - \phi')]$, so that, using this expression for the detour factor in Eq. (VI.2.24), we obtain for a straight edge ($\rho_e^i = \infty$) the field expression of Eq. (V.3.2).

2.3 *Diffraction from a Dielectric Wedge*

As already mentioned in the introductory section, no analytic relation is available for the scattering from a dielectric wedge. By limiting the analysis to the zone, the problem can be solved by using two Sommerfeld integrals

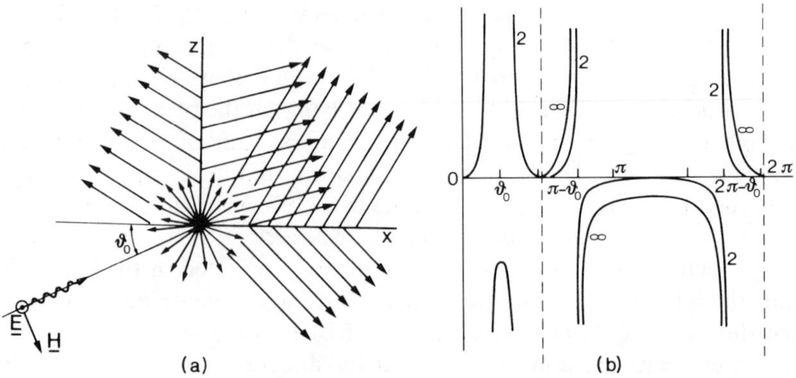

Fig. VI.5. Diffraction of a plane wave from a right angle dielectric wedge for $n = 2, \infty$. Note the vanishing of the diffracted field along the faces of the wedge. (From Joo *et al.* [5].)

containing as unknowns the field angular spectra and imposing matching of the fields on the two sides of the wedge planes. This approach was used [5] to obtain the pattern shown in Fig. VI.5. Here we give only a short summary of the theory.

The ray optical (RO) field is used as the starting distribution. The edge-diffracted field is then represented as a combination of multipolar fields departing from the edge. Then, imposing the vanishing of the analytic continuation of the outside (inside) field prolonged inside (outside) the dielectric, the amplitudes of the multipolar field are obtained by solving a system of linear equations.

3 Diffraction from a Slit

The formalism developed in the preceding section can be conveniently applied to the calculation of the field diffracted by a slit of width $2a$ and infinite length (see Fig. VI.6.). For simplicity, we assume a plane incident wave normal to the edges. As a first approximation we take the field on the aperture coincident with the incident field (Kirchhoff approximation). Then the field point P at a finite distance is reached by two different rays departing from the two edges and by a geometric optics ray, if any. The contribution of the diffracted rays can be calculated by using Eq. (VI.2.21) with the diffraction matrix $\mathbf{D}$ relevant to the edges delimiting the aperture:

$$\mathbf{E}_{\mathrm{d}}(P) = \left[\frac{\exp(-ik_0\rho_1)}{\sqrt{\rho_1}}\mathbf{D}\left(\phi_1, \frac{\pi}{2}; \frac{\pi}{2}\right) + \frac{\exp(-ik_0\rho_2)}{\sqrt{\rho_2}}\mathbf{D}\left(\phi_2, \frac{\pi}{2}; \frac{\pi}{2}\right)\right] \cdot \mathbf{E}_{\mathrm{i}}. \tag{VI.3.1}$$

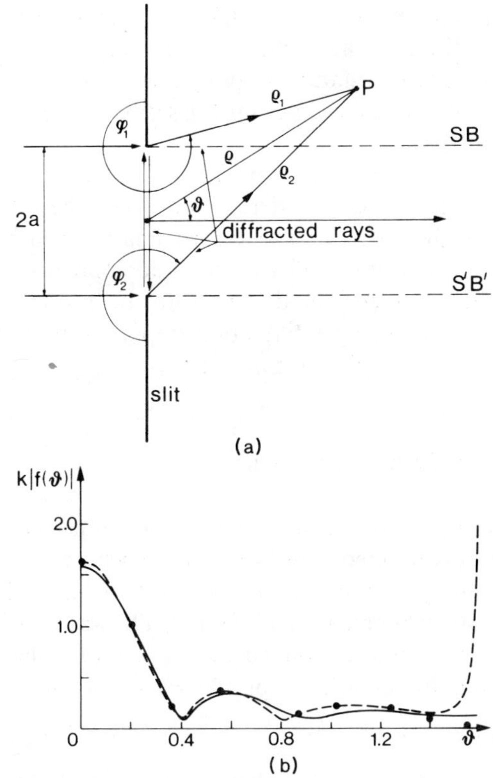

Fig. VI.6. Diffraction by a slit of width $2a$ (a) and relative far-field pattern (b) for a p-wave and $ka = 8$. The solid curve results from single diffraction, while the dashed curve includes the effects of multiple diffraction. The dots represent the exact solution. (From Keller [5a].)

When P is in the dark region, the geometric optics field vanishes. Consequently, the far field is given for every diffraction angle $\theta \neq 0$ by the limiting form of Eq. (VI.3.1):

$$\mathbf{E}_{\mathrm{d}}(\theta) = \frac{\exp(-ik_0\rho)}{\sqrt{\rho}}\left[\mathbf{D}\left(\frac{3}{2}\pi + \theta, \frac{\pi}{2}; \frac{\pi}{2}\right) + \mathbf{D}\left(\frac{3}{2}\pi - \theta, \frac{\pi}{2}; \frac{\pi}{2}\right)\right] \cdot \mathbf{E}_{\mathrm{i}}, \tag{VI.3.2}$$

where the diffraction matrix $\mathbf{D}$ depends on the aperture angle of the wedges delimiting the slit. Since the pattern is a continuous function of θ, this expression is also valid for the forward direction $\theta = 0$.

At a finite distance from the slit the separation in diffracted and geometric optics rays does not hold uniformly. In fact, if we express the field by means of a diffraction integral, we learned in the preceding chapter that the contributions of the end points of the integral can be treated separately from those arising from the stationary-phase points only if they are sufficiently spaced. In particular, when a stationary point approaches an end point (i.e., in proximity to a shadow boundary) we must describe the field by means of the complex Fresnel integral.

Another point that deserves comment is the use of the Kirchhoff approximation, which can be considered valid when the slit width is much larger than the field wavelength. This approximation can be improved by taking into account the multiple diffractions undergone by the rays departing from each edge and diffracted from the opposite one. For moderate values of ka ($\gtrsim 4$) we can consider a single diffraction step, thus obtaining the pattern shown in Fig. VI.6b, from Keller (Chap. V [26]).

4 Diffraction from a Dielectric Cylinder

For illustrative purposes we start with a two-dimensional problem by considering a multiconnected homogeneous dielectric cylinder extending indefinitely along the y axis and a y-independent field. In this case we can separate the field into a combination of TE and TM waves, using the cylinder axis as a reference. If we indicate the field component parallel to the cylinder axis with $u(x, z)$ ($= E_y$ for TE and H_y for TM waves), we have:

$$\partial^2 u/\partial x^2 + \partial^2 u/\partial z^2 + k_0^2 u = -k_0^2(\tilde{n}^2 - 1)P(x, z)u, \qquad \text{(VI.4.1)}$$

$P(x, z)$ being a function equal to 1 when (x, z) lies inside the cylinder cross section and to 0 otherwise, and $\tilde{n}$ the refractive index, assumed independent of the transverse coordinates.

When we move across the boundary C, u and its gradient ∇u are continuous for a TE wave, while for a TM wave the derivative along the tangent to C and the normal derivative divided by the square of the local refractive index are continuous.

We intend to derive the field created inside and outside the dielectric by an incident plane wave u_i. To get u we rewrite Eq. (VI.4.1) in integral form by interpreting the term on the right as a source. In this way we express u as

$$u(x, z) = u_i(x, z) + k_0^2(\tilde{n}^2 - 1)\iint_{S_d} G(x - x'; z - z')u(x', z')\,dx'\,dz', \qquad \text{(VI.4.2)}$$

where S_d represents the dielectric cross section and G is the two-dimensional Green's function written for $x = k_0$ [see Eq. (IV.7.5)]. An iteration technique for dielectric scattering problems represented by the above integral equation has been developed [6], in which the first-order solution is obtained by approximating the total field in the dielectric body by the incident field, and by successively calculating the scattered field through the equivalent volume currents in the dielectric region, assumed to radiate in unbounded free space [see Section VI.8, Eq. (VI.8.2)].

If we set $\varepsilon = k_0^2(n^2 - 1)$ and indicate with G^* the linear convolution operator defined by

$$G^*f \equiv \iint_{S_d} G(x - x', z - z') f(x', z')\, dx'\, dz', \tag{VI.4.3}$$

Eq. (VI.4.2) can be written in the operational form

$$u = u_i + \varepsilon G^* u. \tag{VI.4.4}$$

This is a *Fredholm integral equation* (see Chapter VII, Tricomi [26]) defined on the domain S_d and having as kernel the function

$$G = (i/4) H_0^{(2)}[k_0 \sqrt{(x - x')^2 + (z - z')^2}],$$

which is singular for $x = x'$, $z = z'$, or $k_0 = 0$.

The solution of Eq. (VI.4.2) is generally uniquely determined, except for some discrete values of k_0 corresponding to the modes of oscillation of the dielectric cylinder. For those special values of k_0 Eq. (VI.4.2) also admits a solution for $u_i = 0$.

When we consider dielectric obstacles having an index slightly different from the surrounding medium, i.e., $n \simeq 1$, we can put

$$u = u_i + \varepsilon G^* u_i. \tag{VI.4.5}$$

This approximation is known as the *Rayleigh–Gans* formula. Sometimes it is also named after Born, who made extensive use of it in solving problems of scattering of particles by central potentials. When ε is not very small $G^* u_i$ in Eq. (VI.4.12) can be replaced by $G^* u_g$, where u_g is the geometric optics field.

4.1 *Diffraction from a Dielectric Circular Cylinder*

We now wish to show how the integral equation (VI.4.2) can be solved for a circular cylinder illuminated by a p-wave (i.e., polarized parallel to the axis of the cylinder). In this case it is convenient to expand the fields in cylindrical waves (see Section IV.11) by assuming, with no loss of generality, that the incident plane wave travels along the direction $\phi' = 0$ normal to the y axis. The corresponding results can be extended to the case of oblique incidence by replacing the wavenumber outside the cylinder with $k_0 \sin\theta$ and inside with $k_0(\tilde{n}^2 - \cos^2\theta)^{1/2}$, θ being the angle formed with the y axis and considering HE(EH) fields instead of TE(TM) (see Chapter VIII).

In the present case the field inside the dielectric can be represented by (see Section IV.11)

$$u(\rho', \phi') = \sum_{m=-\infty}^{+\infty} C_m J_m(k_0 \tilde{n} \rho') e^{im\phi'} \qquad (\rho' < a), \tag{VI.4.6}$$

where $\tilde{n}$ is the generally complex refractive index.

Using Graf's addition theorem for cylindrical functions (see Problem IV.8), we now expand the kernel G of the integral equation in a series of cylindrical functions,

$$\frac{i}{4} H_0^{(2)}(k_0 \sqrt{\rho^2 + \rho'^2 - 2\rho\rho' \cos(\phi - \phi')})$$
$$= \frac{i}{4} \sum_{m=-\infty}^{+\infty} H_m^{(2)}(k_0 \rho) J_m(k_0 \rho') e^{im(\phi - \phi')}, \tag{VI.4.7}$$

for $\rho > \rho'$. After inserting the right-hand sides of Eqs. (VI.4.7,6) into Eq. (VI.4.2) we obtain

$$u(\rho, \phi) = \sum_{m=-\infty}^{+\infty} (-i)^m J_m(k_0 \rho) e^{im\phi} + i\frac{\pi}{2} k_0^2 (\tilde{n}^2 - 1) \sum_{m=-\infty}^{+\infty} c_m H_m^{(2)}(k_0 \rho) e^{im\phi}$$
$$\times \int_0^a J_m(k_0 \rho') J_m(k_0 \tilde{n} \rho') \rho' \, d\rho' \equiv \sum_{m=-\infty}^{+\infty} (-i)^m e^{im\phi}$$
$$\times [J_m(k_0 \rho) + r_m^h H_m^{(2)}(k_0 \rho)], \tag{VI.4.8}$$

where the series $\sum_m (-i)^m e^{im\phi} J_m$ represents the incident plane wave [see Eq. (IV.11.12)] and

$$r_m^h = i^{m+1} \frac{\pi}{2} (\tilde{n}^2 - 1) c_m \int_0^\beta J_m(\tilde{n} x) J_m(x) x \, dx. \tag{VI.4.9}$$

Taking into account the relation

$$\frac{d}{dx}\left\{x\left[J_m(\tilde{n}x)\frac{d}{dx}J_m(x) - J_m(x)\frac{d}{dx}J_m(\tilde{n}x)\right]\right\} = (\tilde{n}^2 - 1) J_m(x) J_m(\tilde{n}x) x, \tag{VI.4.10}$$

it is not difficult to deduce

$$r_m^h = -i^{m+1} (\pi/2) C_m \beta [J_m(\beta) J'_m(\alpha) \tilde{n} - J'_m(\beta) J_m(\alpha)], \tag{VI.4.11}$$

where $\beta \equiv k_0 a$ is the *size parameter* and $\alpha = \tilde{n}\beta$.

According to Eq. (VI.4.8), the field outside the dielectric is given by a superposition of cylindrical waves weighed by the factors r_m^h, which can be considered as the reflection coefficients of the cylinder relative to the mth partial wave. The r_m^h can be easily calculated by imposing the continuity of u when moving across the boundary. In this way we obtain

$$(-i)^m[J_m(\beta) + r_m^h H_m^{(2)}(\beta)] = C_m J_m(\alpha), \qquad \text{(VI.4.12)}$$

so that expressing C_m as a function of r_m^h and taking into account the relation $2i/\pi\beta = J'_m(\beta)H_m^{(2)}(\beta) - J_m(\beta)H_m^{(2)\prime}(\beta)$, we get

$$r_m^h = -\frac{J_m(\beta)}{H_m^{(2)}(\beta)} \frac{\ln' J_m(\beta) - \tilde{n} \ln' J_m(\alpha)}{\ln' H_m^{(2)}(\beta) - \tilde{n} \ln' J_m(\alpha)} = -\frac{1}{2}(1 - S_m^h), \qquad \text{(VI.4.13)}$$

where ln′ denotes the logarithmic derivative, while

$$S_m^h = -\frac{H_m^{(1)}(\beta)}{H_m^{(2)}(\beta)} \frac{\ln' H_m^{(1)}(\beta) - \tilde{n} \ln' J_m(\alpha)}{\ln' H_m^{(2)}(\beta) - \tilde{n} \ln' J_m(\alpha)} \equiv S(m, \beta). \qquad \text{(VI.4.14)}$$

In particular, for $\tilde{n}$ real the function $S(m, \beta)$ has unit modulus. In addition, it worth noting that the above equation defines S as a function of the generally complex parameter m. This extension is instrumental for applying the Watson transform to the partial wave representation of the scattered field (see Chapter IV, Newton [17]).

The above analysis can be immediately extended to the case of an incident s-wave, where we obtain:

$$r_m^e = -\frac{J_m(\beta)}{H_m^{(2)}(\beta)} \frac{\ln' J_m(\beta) - \ln' J_m(\alpha)}{\tilde{n} \ln' H_m^{(2)}(\beta) - \ln' J_m(\alpha)} = -\frac{1}{2}(1 - S_m^e), \qquad \text{(VI.4.15)}$$

where

$$S_m^e = -\frac{H_m^{(1)}(\beta)}{H_m^{(2)}(\beta)} \frac{\ln' J_m(\alpha) - \ln' H_m^{(1)}(\beta)}{\ln' J_m(\alpha) - \tilde{n} \ln' H_m^{(2)}(\beta)}. \qquad \text{(VI.4.16)}$$

The limiting case of a metallic cylinder is obtained by letting $\tilde{n} \to \infty$ and the expressions for r_m simplify to

$$r_m^h = -J_m(\beta)/H_m^{(2)}(\beta), \qquad r_m^e = -J'_m(\beta)/H_m^{(2)\prime}(\beta), \qquad \text{(VI.4.17)}$$

where we have replaced $J_m(\tilde{n}\beta)$ with its asymptotic expression.

Accordingly, the reflection coefficient relative to the mth cylindrical multipole depends critically on the polarization. In fact, when $J_m(\beta)$ is a maximum, r_p is very large and r_s is vanishing. By contrast, when r_p is vanishing r_s is large. This property can be used in making a polarizer, by disposing many thin wires parallel and equally spaced. In this way the polarization sensitivity of each wire is enhanced by the periodic arrangement employed. These polarizers find applications in the far-infrared region.

5 S Matrix and Watson–Regge Representation

We can rewrite Eq. (VI.4.8) in the equivalent form

$$\begin{aligned} u(\rho, \phi) &= \frac{1}{2} \sum_{m=-\infty}^{\infty} (-i)^m e^{im\phi} H_m^{(1)}(k_0\rho) + \frac{1}{2} \sum_{m=-\infty}^{+\infty} (-i)^m e^{im\phi} S_m^h H_m^{(2)}(k_0\rho) \\ &\equiv e^{-ik_0 z} U(-z) + [u_d(\rho, \phi) + e^{-ik_0 z} U(z)] \\ &\equiv e^{-ik_0 z} U(-z) + \bar{u}_d(\rho, \phi), \end{aligned} \tag{VI.5.1}$$

where we have assumed that the z axis is coincident with the direction of propagation of the incident wave. The field $\bar{u}_d$ can be thought of as obtained from the action of an operator $\hat{S}$ on the outgoing plane wave $e^{-ik_0 z} U(z)$, i.e.,

$$\bar{u}_d(\rho, \phi) = \frac{1}{2} \sum_{m=-\infty}^{+\infty} (-i)^m e^{im\phi} H_m^{(2)}(k_0\rho) S_m^h \equiv \hat{S} e^{-ik_0 z} U(z). \tag{VI.5.2}$$

In the absence of an obstacle the $\hat{S}$ operator reduces to unity.

If we consider the set of cylindrical modes $(-i)^m e^{im\phi} H_m^{(2)}$ as a basis of representation of the $\hat{S}$ operator, the S_m can be interpreted as the diagonal matrix elements of $\hat{S}(\beta)$. Accordingly, the set of coefficients S_m form the so-called *S matrix*, which diagonalizes in the basis $H_m^{(2)} e^{im\phi}$. It is noteworthy that for lossless scatterers, $|S_m| = 1$. In addition, in the lossless case

$$\hat{S}(\beta) \cdot \hat{S}^\dagger(-\beta) = 1, \tag{VI.5.3}$$

where $\hat{S}^\dagger$ is the transposed conjugate (adjoint) of the operator $\hat{S}$. The above property is referred to by saying that the S matrix is a *unitary operator*.

When considered as a function of the generally complex parameter β, the singularities of the S matrix provide a great deal of information on the parameters and the radial structure of the obstacle. The analysis of this section can be extended to circular cylinders in which the refractive index depends on the radial coordinate ρ. In that case, the complex values of β for which $\hat{S}$ becomes infinite, the so-called *resonances* of the scatterer, can be used to infer the profile of $n(\rho)$.

The resonances corresponding to real values of β represent *bound states* of the electromagnetic field, which appears as if trapped by the obstacle. These resonances correspond to the eigenfunctions of Eq. (VI.4.4), if any. Complex resonances β_q "represent" a field captured temporarily by the obstacle and released after a time that depends inversely on the imaginary part $\operatorname{Im} \beta_q$.

5.1 *Debye Expansion*

Debye [7] showed that the S matrix can be expanded in a power series, i.e.,

$$S_q(\beta) = \frac{H_q^{(1)}(\beta)}{H_q^{(2)}(\beta)} \left[R_q + T_q^2 \frac{H_q^{(2)}(\alpha)}{H_q^{(1)}(\alpha)} \sum_{p=1}^{\infty} \rho_q^{p-1} \right] \equiv \sum_{p=0}^{\infty} S_q^{(p)}(\beta), \tag{VI.5.4}$$

where

$$R_q = -\frac{\ln' H_q^{(2)}(\beta) - \tilde{n} \ln' H_q^{(2)}(\alpha)}{\ln' H_q^{(2)}(\beta) - \tilde{n} \ln' H_q^{(1)}(\alpha)}, \tag{VI.5.5a}$$

$$T_q^2 = \tilde{n}[\ln' H_q^{(2)}(\alpha) - \ln' H_q^{(1)}(\alpha)][\ln' H_q^{(2)}(\beta) - \ln' H_q^{(1)}(\beta)] \\ \times [\ln' H_q^{(2)}(\beta) - \tilde{n} \ln' H_q^{(1)}(\alpha)]^{-2}, \tag{VI.5.5b}$$

$$\rho_q = -\frac{H_q^{(2)}(\alpha) \ln' H_q^{(2)}(\beta) - \tilde{n} \ln' H_q^{(2)}(\alpha)}{H_q^{(1)}(\alpha) \ln' H_q^{(2)}(\beta) - \tilde{n} \ln' H_q^{(1)}(\alpha)}. \tag{VI.5.5c}$$

Here R_q represents the reflection coefficient for the qth partial wave associated with direct reflection from the surface, the T_q term corresponds to transmission into the sphere and transmission to the outside, and ρ_q^{p-1} accounts for $p-1$ internal reflections at the surface. In particular, for $\alpha,\beta \gg 1$, $\ln' H_q^{(2)} \simeq -i$, so that $R_q \simeq -(\tilde{n}-1)/(\tilde{n}+1)$, $T_q^2 \simeq 4\tilde{n}/(\tilde{n}+1)^2$, and $\rho_q \simeq e^{i\pi q}(\tilde{n}-1)/(\tilde{n}+1)$. Then, for $\tilde{n}$ close to 1, $|\rho_q| \ll 1$ and the series converges rapidly.

Inserting the above expansion into Eq. (VI.5.2) yields

$$\bar{u}_d = \bar{u}_d^{(0)} + \cdots + \bar{u}_d^{(p)} + \cdots, \tag{VI.5.6a}$$

where

$$\bar{u}_d^{(0)} = \frac{1}{2} \sum_{q=-\infty}^{\infty} (-i)^q \frac{H_q^{(1)}(\beta)}{H_q^{(2)}(\beta)} R_q e^{iq\phi} H_q^{(2)}(k_0 \rho) \tag{VI.5.6b}$$

and

$$\bar{u}_d^{(p)} = \frac{1}{2} \sum_{q=-\infty}^{+\infty} (-i)^q \frac{T_q^2 H_q^{(1)}(\beta) H_q^{(2)}(\alpha)}{H_q^{(2)}(\beta) \mathrm{H}_q^{(1)}(\alpha)} \rho_q^{p-1} e^{iq\phi} H_q^{(2)}(k_0 \rho). \tag{VI.5.6c}$$

The pth component of the qth partial wave corresponds to transmission into the sphere (T_q) followed by a bouncing back and forth between $\rho = a$ and $\rho = 0$ p times with p internal reflections at the surface (ρ_q^{p-1}) and a final transmission to the outside (T_q). For β large and $\tilde{n} \simeq 1$ the contributions $\bar{u}_d^{(p)}$ can be neglected and $\bar{u}_d$ reduces to $\bar{u}_d^{(0)}$. In Fig. VI.7a we have indicated the dark regions relevant to $p = 0, 1$.

The rate of convergence of the Debye series is determined by the damping produced at each internal reflection. If the cylinder is not perfectly transparent there is an additional damping of successive terms due to absorption. In the particularly important case of water, $n = 1.33$ and more than 98.5% of the total intensity goes into the first three refracted rays, corresponding to the first three terms of the Debye expansion.

5.2 *Watson–Regge Representation*

The scattered field $\bar{u}_d^{(p)}$, corresponding to the pth term of the Debye expansion, is represented by the series of partial waves indicated in Eq. (VI.5.6c). This series can be transformed into an integral by retracing the steps

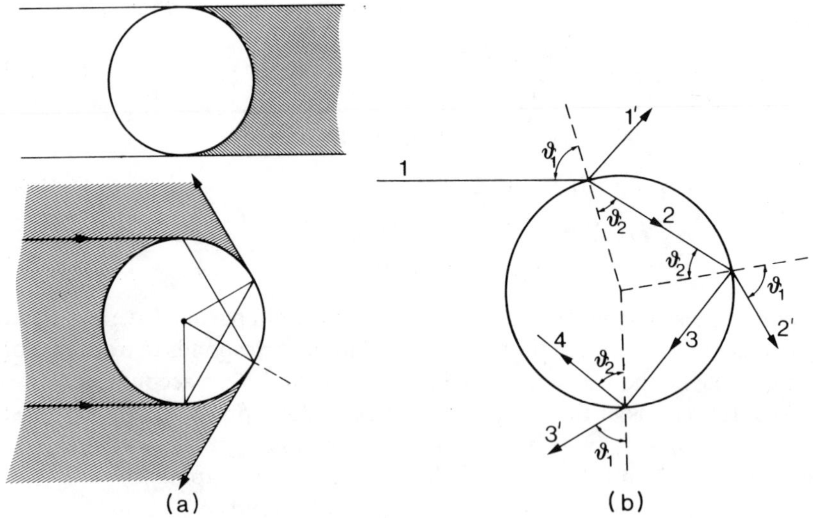

Fig. VI.7. (a) Shadow regions relative to zeroth-order (top) and first-order (bottom) fields and (b) multiple reflected rays resulting from the Debye expansion of the S matrix of a circular cylinder.

leading to Eq. (IV.12.10) and replacing the coefficient $\langle u(a,\phi')e^{-i\nu\phi'}\rangle$ with $\frac{1}{2}e^{i3\pi\nu/2}S_\nu^{(p)}H_\nu^{(2)}$, which yields

$$\bar{u}_{\mathrm{d}}^{(p)}(\rho,\phi) = -\frac{i}{4}\oint_C \frac{e^{i\nu(\phi+\pi/2)}}{\sin\nu\pi} S_\nu^{(p)}(\beta)H_\nu^{(2)}(k_0\rho)\,d\nu. \tag{VI.5.7}$$

Now, taking into account the relation $H_{-\nu}^{(2)} = e^{-i\pi\nu}H_\nu^{(2)}$ and recalling Eq. (VI.5.4), we have $S_{-\nu}^{(p)} = S_\nu^{(p)}\exp[i(1-p)2\pi\nu]$, so that we can reduce the contour C [defined in Eq. (IV.12.10)] to the line running parallel to the real ν axis at a small distance $\varepsilon > 0$, and write accordingly

$$\bar{u}_{\mathrm{d}}^{(p)}(\rho,\phi) = -\frac{i}{2}\int_{-\infty+i\varepsilon}^{\infty+i\varepsilon} e^{i[(1/2)-p]\pi\nu}\frac{\cos[\nu(\phi+p\pi)]}{\sin(\nu\pi)} S_\nu^{(p)}(\beta)H_\nu^{(2)}(k_0\rho)\,d\nu. \tag{VI.5.8}$$

In order to apply the Watson transform the locations of the poles of the function $S_\nu^{(p)}$, given by

$$S_\nu^{(0)} = \frac{H_\nu^{(1)}(\beta)}{H_\nu^{(2)}(\beta)}R_\nu,$$

$$S_\nu^{(p)} = \frac{H_\nu^{(1)}(\beta)H_\nu^{(2)}(\alpha)}{H_\nu^{(2)}(\beta)H_\nu^{(1)}(\alpha)}T_\nu^2\rho_\nu^{p-1}, \qquad p \geq 1 \tag{VI.5.9}$$

have to be known. If we take into account the definitions of the functions R_q,

T_q^2, and ρ_q, we can immediately see that the poles of $S_\nu^{(p)}$ are the roots of the equation (see Chapter IV, Nussenzveig [22])

$$\ln' H_\nu^{(2)}(\beta) = \tilde{n} \ln' H_\nu^{(1)}(\alpha), \qquad \text{(VI.5.10)}$$

which reduce to the zeros of $H_\nu^{(2)}(\beta)$ in the metallic boundary case (see Section IV.12).

We note that the poles of $S_\nu^{(p)}$ in the complex ν plane are the same for all terms, their order being $p + 1$ for the pth term. The poles are close to the zeros of $H_\nu^{(2)}(\beta)$ or to those of $H_\nu^{(1)}(\alpha)$. In particular, it can be shown that (cf. Section IV.12) the first type of poles are given by [cf. Eq. (IV.12.19)]

$$\nu_n \simeq -\beta - e^{-i\pi/3}(\beta/2)^{1/3} x_n - i/(\tilde{n}^2 - 1)^{1/2}, \qquad \text{(VI.5.11)}$$

x_n being the nth zero of $Ai(-x)$. The last term on the right-hand side accounts for the departure of the dielectric boundary conditions from the metallic ones.

Similarly, the second type of poles are given in the upper half-plane by

$$\nu_n' \simeq \alpha + e^{i\pi/3}(\alpha/2)^{1/3} x_n + \tilde{n}/(\tilde{n}^2 - 1)^{1/2}. \qquad \text{(VI.5.12)}$$

In view of the distribution of these poles on the complex ν plane (see Chapter IV, Nussenzveig [22]), we introduce the variables η_1, η_2, and η defined by (see Fig. VI.8)

$$\eta_1 = \delta \ln\left|\frac{2\nu}{e\alpha}\right|, \qquad \eta_2 = \delta \ln\left|\frac{2\nu}{e\beta}\right|, \qquad \eta = \delta \ln\left|\frac{2\nu}{ek_0\rho}\right|, \qquad \text{(VI.5.13)}$$

where $\delta = \arg(\nu) - (\pi/2)\operatorname{sgn}\arg(\nu)$. Taking into account the distributions

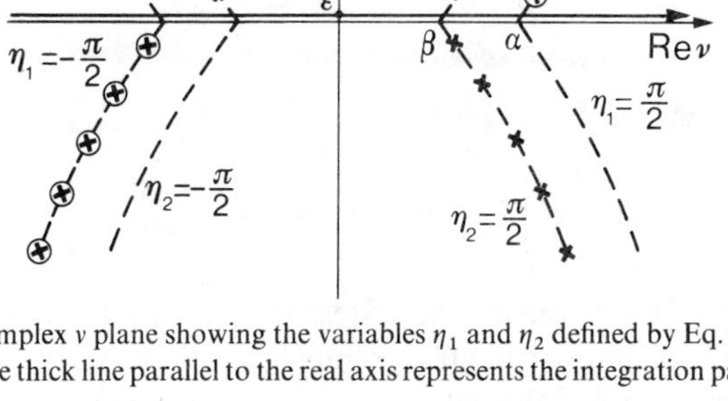

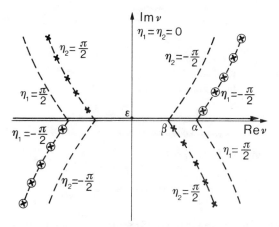

Fig. VI.8. Complex ν plane showing the variables η_1 and η_2 defined by Eq. (VI.5.13); x, poles ν_n: ⊗, poles ν_n'. The thick line parallel to the real axis represents the integration path of the integral of Eq. (VI.5.14).

of the zeros x_n for $n \to \infty$ [see Eqs. (IV.12.22,23)], it is easy to show that the poles ν_n are aligned asymptotically along the curve $\eta_2 = \pi/2$, while the ν'_n are aligned along $\eta_1 = -\pi/2$.

5.3 *First Term of the Debye Expansion*

For $p = 0$ Eq. (VI.5.8) reduces to

$$\bar{u}_{\rm d}^{(0)}(\rho, \phi) = \frac{i}{2}\int_{-\infty + i\varepsilon}^{\infty + i\varepsilon} e^{i\pi\nu/2} \frac{\cos \nu\phi}{\sin \nu\pi} \frac{g_\nu(\beta, \alpha, \rho)}{H_\nu^{(2)}(\beta)}\, d\nu$$
$$- \frac{i}{2}\int_{-\infty + i\varepsilon}^{\infty + i\varepsilon} e^{i\pi\nu/2} \frac{\cos \nu\phi}{\sin \nu\pi} H_\nu^{(1)}(k_0\rho)\, d\nu, \qquad \text{(VI.5.14)}$$

with

$$g_\nu(\alpha, \beta, \rho) = H_\nu^{(1)}(k_0\rho)H_\nu^{(2)}(\beta) + R_\nu(\alpha, \beta)H_\nu^{(2)}(k_0\rho)H_\nu^{(1)}(\beta). \quad \text{(VI.5.15)}$$

Since the last integral of Eq. (VI.5.14) represents the field $-U(-z)e^{-ik_0z}$, in the following we will drop it by replacing $\bar{u}_{\rm d}^{(0)}$ with $u^{(0)} \equiv \bar{u}_{\rm d}^{(0)} + U(-z)e^{-ik_0z}$. In addition, $R_\nu \to -1$ for $|\nu| \to \infty$ in all regions except for the one between $\eta_2 = \pi/2$ and $\eta_2 = -\pi/2$, where $R_\nu \to 0$ like $\beta^2(\tilde{n}^2 - 1)/4\nu^2$. Consequently, in view of the asymptotic representations (IV.12.8) of $H_\nu^{(1,2)}$,the integrand of the first integral of Eq. (VI.5.14) tends to zero in the first quadrant, so that the path of integration $(i\varepsilon, \infty + i\varepsilon)$ may be shifted to the positive imaginary axis $(i\varepsilon, i\infty)$ by sweeping across the poles υ'_n, thus giving

$$u^{(0)} = \frac{i}{2}\left(\int_{-\infty + i\varepsilon}^{i\varepsilon} + \int_{i\varepsilon}^{i\infty}\right) e^{i(\pi/2)\nu} \frac{\cos \nu\phi}{\sin \nu\pi} \frac{g_\nu}{H_\nu^{(2)}(\beta)}\, d\nu + \pi \sum_n e^{i(\pi/2)\nu'_n} \frac{\cos \nu'_n\phi}{\sin \nu'_n\pi} r'_n, \qquad \text{(VI.5.16)}$$

where r'_n is the residue of $g_\nu/H_\nu^{(2)}$ relative to the pole located at $\nu = \nu'_n$.

We can now use the representation

$$\frac{1}{\sin \nu\pi} = -2ie^{i\pi\nu} \sum_{m=0}^{\infty} e^{i2\pi m\nu} \qquad \text{(VI.5.17)}$$

to write the integral in Eq. (VI.5.16) as

$$\sum_{m=0}^{\infty}\left(\int_{-\infty + i\varepsilon}^{i\varepsilon} + \int_{i\varepsilon}^{i\infty}\right) e^{i(3/2)\pi\nu + i2\pi m\nu} \cos(\nu\phi) \frac{g_\nu}{H_\nu^{(2)}(\beta)}\, d\nu. \qquad \text{(VI.5.18)}$$

The calculation of the above integrals can be further simplified by exploiting the vanishing of the integrands for $\nu \to \infty$ in the second quadrant for $m \geq 1$. Consequently, the path $(i\varepsilon, i\infty)$ may be shifted to $(i\varepsilon, -\infty + i\varepsilon)$ for

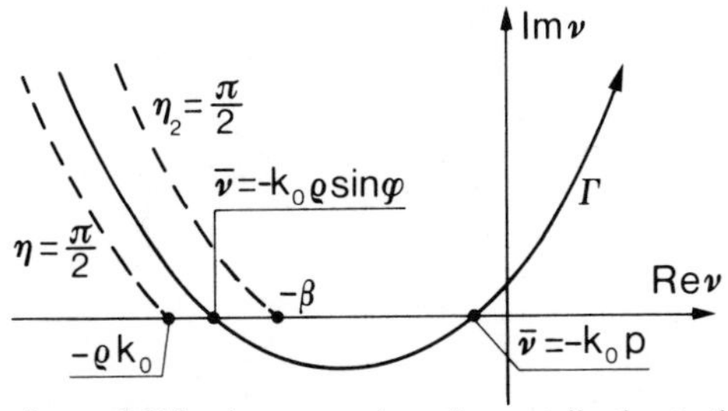

Fig. VI.9. Integration path Γ for the geometric optics contribution to the field diffracted by a cylinder.

the terms with $m \geq 1$ by sweeping across the poles v_m, thus giving the following residue series:

$$u^{(0)}(\rho, \phi) = u_g(\rho, \phi) + \pi \sum_n \sum_{m=1}^{\infty} e^{i[(3/2)\pi + 2\pi m]v_n} \cos(v_n \phi) r_n$$

$$+ \pi \sum_n \sum_{m=0}^{\infty} e^{i[(3/2)\pi + 2\pi m]v'_n} \cos(v'_n \phi) r'_n, \qquad \text{(VI.5.19)}$$

r_n being the residue of $g_v/H_v^{(2)}$ at v_n, while u_g stands for the term $m = 0$ of the series of Eq. (VI.5.18). It can be shown (see Chapter IV, Nussenzveig [22], Sect. 3D) that for refractive index $n > 1$, the residues r'_n are negligibly small.

The integrand of the integral representing u_g tends to zero at infinity for $\eta > \phi$, and for $\phi < \pi/2$ the path $(-\infty + i\varepsilon)$ may be shifted to $\eta = \pi/2$, so that:

$$u_g(\rho, \phi) = \int_\Gamma e^{i(3/2)\pi v} \cos(v\phi) \frac{R_v H_v^{(2)}(k_0\rho) H_v^{(1)}(\beta)}{H_v^{(1)}(\beta)} dv, \qquad \text{(VI.5.20)}$$

where Γ is a path going around the poles of R_v and crossing the real axis twice, first between 0 and $-\beta$, and then between $-\beta$ and $-k_0\rho$. The integrand of Eq. (VI.5.20) has a saddle point in each of these intervals (see Fig. VI.9)

$$v_s^{(1)} = -k_0 p, \qquad v_s^{(2)} = -k_0 \rho \sin \phi. \qquad \text{(VI.5.21)}$$

The saddle points $v_s^{(1)(2)}$ correspond to rays hitting the cylinder with impact parameters p (not to be confused with the index p) and $\rho \sin \phi$. In particular, p is the impact parameter of the incident ray that is geometrically reflected from the cylinder and passes through the field point P (see Fig. VI.10) [in particular, for $\rho \to \infty$, $p = a\cos(\phi/2)$], while $\rho \sin \phi$ is the impact parameter of the ray that reaches P without being deviated by the obstacle.

If we deform Γ by letting it pass through these saddle points, we can asymptotically evaluate the integral in Eq. (VI.5.20). Accordingly, $u_g(\rho, \phi)$ can be easily obtained from geometric optics by multiplying the ray reflected by the cylinder by the Fresnel reflection coefficient.

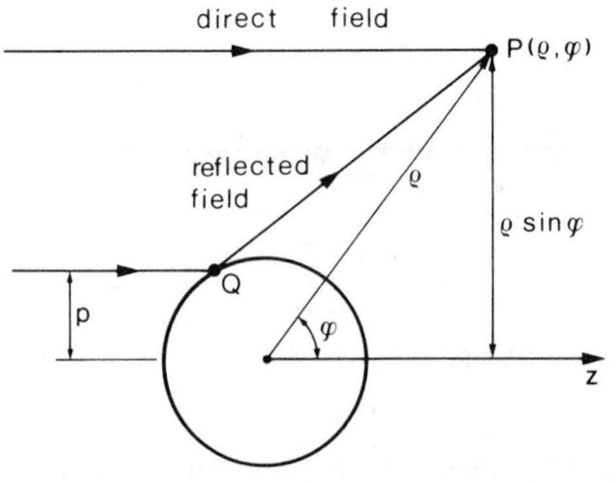

Fig. VI.10. Geometric optics interpretation of the saddle-point contribution to the field $u_g^{(0)}$; p represents the impact parameter of the ray reaching the field point P after reflection at Q.

When we keep ϕ constant and reduce ρ the two saddle points get closer and closer and coincide when P lies on the shadow boundary. When the field point penetrates deeply into the shadow region, the contribution of u_g vanishes.

For $p \neq 0$ an approach similar to that illustrated above can be followed (see Chapter IV, Nussenzveig [22]).

6 Surface Diffraction Waves

The expressions for the scattering amplitudes derived in the last section allow us to write the total field surrounding a metallic circular cylinder, illuminated normal to the axis by a p-wave of unit amplitude, as a sum of partial waves:

$$u(\rho, \phi) = \sum_{m=-\infty}^{+\infty} (-i)^m e^{im\phi} \left[J_m(k_0\rho) - \frac{J_m(\beta)}{H_m^{(2)}(\beta)} H_m^{(2)}(k_0\rho) \right], \quad \text{(VI.6.1)}$$

where $\beta \equiv k_0 a$. Experience with numerical computations indicates that the minimum number of terms that must be kept in the partial-wave expansion of the diffraction field in order to get a good approximation is close to β. When $\beta \gg 1$ the numerous partial waves produce a very irregular diffraction pattern, which depends critically on the frequency through the factor β. In the limit of $\beta \to \infty$ we can use the geometric optics approximation, which leads to dark and lit regions separated by two shadow boundaries. The lit region is spanned by incident and geometrically reflected rays, which can be used as leading terms of the LK series representation of the field for finite k_0. However, the field in the dark region cannot be obtained by the standard ray optical approach since it vanishes for $k_0 \to \infty$. An approximate representation can be obtained in this case by using the Watson transformation discussed in the preceding section. It consists of replacing the partial waves of Eq. (VI.6.1) with waves having complex indices ν_n, the contributions of the poles v_n' being absent in the case of a metallic body. Such waves represent modes "creeping" along the curved surface of the impenetrable cylinder, which is the source of the name *creeping waves* given to these fields. In particular, for a unit-amplitude normally incident p-wave, the formulas of Sections IV.12 and VI.5 give

$$u(\rho, \phi) = \frac{\exp(-ik_0 s)}{(2\pi k_0 s)^{1/2}} \sum_{n=1}^{\infty} D_{n,p}^2 \sum_{m=1}^{\infty} \left(\exp\left\{ i\nu_n \left[\phi - \frac{\pi}{2} - \arccos\left(\frac{a}{\rho}\right) + 2\pi m \right] \right\} + \exp\left\{ -i\nu_n \left[\phi - \frac{3\pi}{2} - \arccos\left(\frac{a}{\rho} - 2\pi m\right] \right\} \right), \tag{VI.6.2}$$

where $s = (\rho^2 - a^2)^{1/2}$ is the distance from Q_2 to P (see Figs. VI.11,12). The

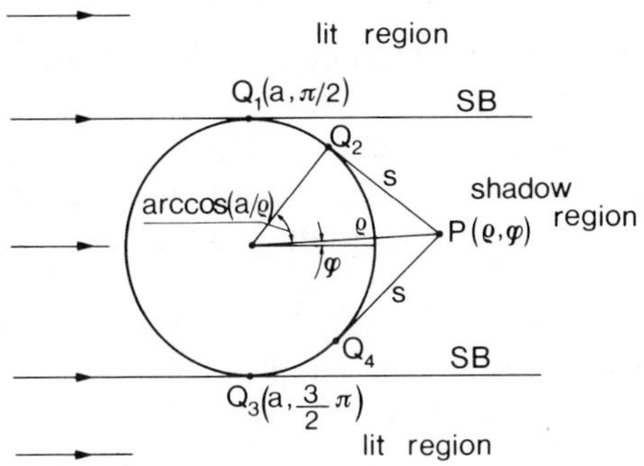

Fig. VI.11. Geometry relative to the creeping waves propagating in the shadow region of a metallic cylinder illuminated by a plane wave.

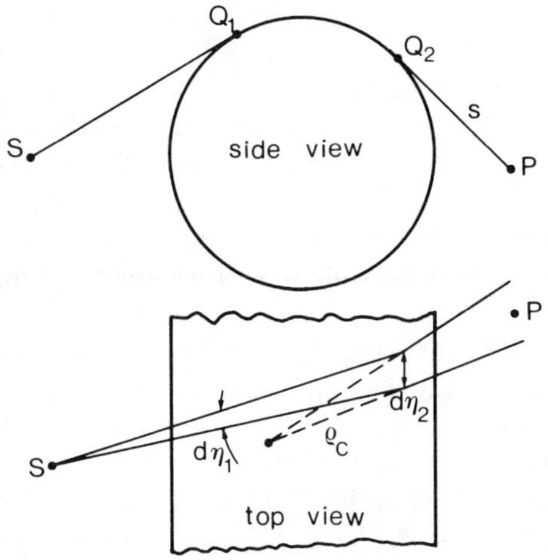

Fig. VI.12. Geometry of the surface diffraction waves relative to a cylinder illuminated by a ray congruence originating from S.

coefficients ν_n are defined by Eq. (IV.12.19) [see also Eq. (VI.5.11) with $\tilde{n} \to \infty$] and read

$$\nu_n = -\beta - \beta^{1/3} e^{-i\pi/3} \begin{cases} 1.856 & (n = 1), \\ 3.245 & (n = 2), \\ 4.381 & (n = 3), \\ 5.387 & (n = 4). \end{cases} \tag{VI.6.3}$$

The quantities $D_{n,p}$ play the roles of *surface diffraction coefficients* and are given by the residues r_n discussed in the preceding section,

$$D_{n,p}^2 = r_n = (\beta/2)^{1/3} e^{-i\pi/12}/[\mathrm{Ai}'^2(-x_n)], \tag{VI.6.4}$$

where the x_n are the zeros of the Airy function $\mathrm{Ai}(-x)$, and Ai' is the respective derivative.

According to the above representation, the field is produced from the rays tangent to the cylinder at $Q_1(\phi = \pi/2)$ and $Q_3(\phi = 3\pi/2)$, which travel along the surface of the cylinder through a distance $a[\phi - \pi/2 - \arccos(a/\rho) + 2\pi m]$ and $a[3\pi/2 - \phi - \arccos(a/\rho) + 2\pi m]$ and encircle the cylinder m times with a complex propagation constant k_n given by

$$k_n \equiv -(\nu_n/a) = k_0 - i\alpha_{n,p}, \tag{VI.6.5}$$

where $\alpha_{n,p} = (1/a)x_n(\beta/2)^{1/3}e^{i\pi/6}$ is the complex attenuation coefficient.

7 Generalized Fermat Principle and Geometric Theory of Diffraction

It is noteworthy that the trajectories of the creeping waves are the shortest paths connecting the source with the field point without penetrating the cylinder. This fact leads us to state a *generalized Fermat principle* for surface-diffracted rays in analogy with rays propagating in a three-dimensional space or rays diffracted by an edge. These three classes of rays describe trajectories that are stationary with respect to small variations subject to suitable constraints. The edge-diffracted rays are constrained to touch the edge at least in one point, while the surface-diffracted rays must be tangent to a smooth obstacle.

This generalized Fermat principle was originally formulated by J. B. Keller (Chapter V [26]) to obtain asymptotic expressions for the field diffracted by very general obstacles. Its importance can be immediately appreciated by looking at the extension of the formulas derived above to a circular cylinder illuminated by a generic ray optical field. In this case we can still assume that the field in the dark region is produced by creeping waves that describe on the cylinder curves which satisfy the Fermat principle. Consequently, each ray of the incident congruence, which is tangent to the cylinder, describes a geodesic of the cylinder, that is, a helix. The creeping wave on the cylinder is fully characterized by the family of helical trajectories formed by the surface rays along which the field propagates with a complex propagation constant derived from Eq. (VI.6.5):

$$k_n = k_0 - id_{n,p} = k_0 - i(x_n/\rho_g)(k_0\rho_g/2)^{1/3}e^{i\pi/6}, \qquad \text{(VI.7.1)}$$

where ρ_g represents the curvature radius of the surface ray. For a helical trajectory ρ_g is constant, as is k_n.

The above considerations can be extended to a generic smooth surface, in which case the field is given by the superposition of infinite contributions of the form

$$u(P) = \frac{u(Q_1)}{(2\pi k_0)^{1/2}}\left[\frac{\rho_c}{s(\rho_c + s)}\right]^{1/2}\left(\frac{d\eta_1}{d\eta_2}\right)^{1/2} D_{n,p}(Q_1)D_{n,p}(Q_2)$$
$$\times \exp\left[-ik_0 s - ik_0 t - \int_0^t \alpha_{n,p}\,dt'\right]. \qquad \text{(VI.7.2)}$$

Here the factor $[\rho_c/s(\rho_c + s)]^{1/2}$ is due to the divergence of the diffracted rays, which leave the surface at Q_2 and present a focus at a distance ρ_c from Q_2, while $d\eta_1/d\eta_2$ represents the ratio of the distance between two nearby rays at Q_1 and Q_2, so that $(d\eta_1/d\eta_2)^{1/2}$ accounts for the divergence of the surface rays. The diffraction coefficient $D_{n,p}^2$ has been split into the product $D_{n,p}(Q_1)D_{n,p}(Q_2)$ since the curvature radius (and consequently β) of the trjectory in Q_1 is in general different from that in Q_2. Finally, s represents the distance of P from Q_2, while t is the arc length of the trajectory between Q_1 and Q_2.

As a final step, the above expressions can be extended to s-polarized waves by replacing the diffraction coefficient $D_{n,p}$ with $D_{n,s}$, defined by

$$D_{n,s}^2 = -[(\beta/2)^{1/3}e^{-i\pi/12}]/[x_n' \,\mathrm{Ai}^2(-x_n')], \qquad \text{(VI.7.3)}$$

and v_n with $\bar{v}_n$, defined by [cf. Eq. (VI.6.3)]:

$$\bar{v}_n = -\beta - \beta^{1/2}e^{-i\pi/3}\begin{cases} 0.809 & (n = 1), \\ 2.578 & (n = 2), \\ 3.826 & (n = 3), \\ 4.892 & (n = 4), \end{cases} \qquad \text{(VI.7.4)}$$

the x_n' being the zeros of $\mathrm{Ai}'(-x)$.

The above scalar diffraction formulas can be combined in a single vector equation providing the electric field $\mathbf{E}_d(P)$ diffracted by a smooth obstacle illuminated by the incident field $\mathbf{E}_i$, viz. [4]

$$\mathbf{E}_d(P) = [\rho/s(\rho + s)]^{1/2}(d\eta_1/d\eta_2)^{1/2}e^{-ik_0 s}(\hat{n}_2\hat{n}_1 F + \hat{b}_2\hat{b}_1 G)\cdot\mathbf{E}_i(Q_1), \qquad \text{(VI.7.5)}$$

in which $\hat{n}$ and $\hat{b}$ are the principal normal and the binormal, respectively, to the surface ray trajectory at Q_1 and Q_2, while

$$\lambda^{-1/2}F = D_{n,s}(Q_1)D_{n,s}(Q_2)\exp\left(-\int_0^t \alpha_{n,s}\,dt'\right)\exp\left(-ik_0t - i\frac{\pi}{12}\right), \qquad \text{(VI.7.6)}$$

G being obtained from the above equation by changing s to p.

In conclusion, the field diffracted by a smooth surface can be represented as a superposition of modes, each obtained by multiplying the incident field by the diffraction matrix $F\hat{n}_2\hat{n}_1 + G\hat{b}_2\hat{b}_1$, in a manner similar to that used for diffraction by an edge.

The geometric theory of diffraction is the name given to the procedures involving rays diffracted by any sort of obstacle (see Fig. VI.13). The name GTD was proposed by J. B. Keller, who developed the basic GTD. The current GTD has evolved into a sophisticated tool for analysis and design that makes systematic use of diffraction matrices and uniform representations for the special regions (e.g., shadow boundaries, caustics) examined in the preceding chapter. In a sense, the GTD can be considered the electromagnetic counterpart of the Feynman diagrams used in quantum field theories. In fact, they share the common feature of providing rules for separating complicated field configurations into a superposition of standard fields represented,

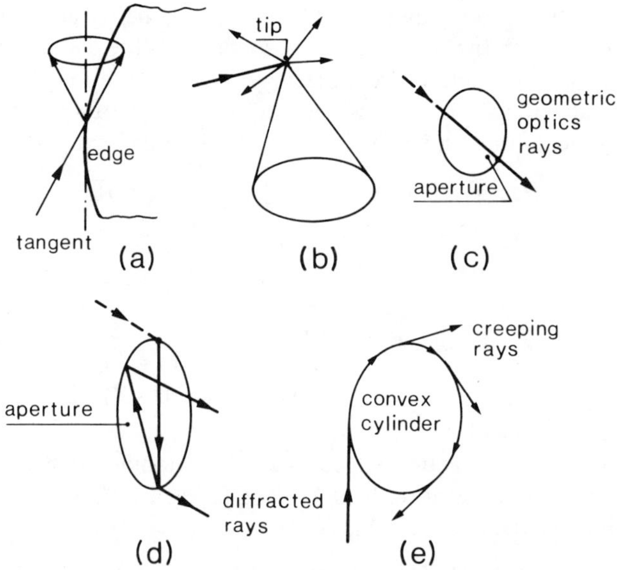

Fig. VI.13. Rays diffracted from obstacles. Note the presence in (d) of doubly diffracted rays and in (c) of transmitted rays.

respectively, by diffracted rays (GTD) and graphs (Feynman diagrams). The interested reader can find some of the applications and developments of GTD in the book edited by Hansen (see Chapter V, Hansen [28]).

8 Scattering from a Dielectric Body

In considering scattering from a dielectric body two different approaches can be developed, based respectively on the solution of a surface or a volume integral equation [8]. As a starting point we consider the integral representation of Eq. (IV.3.1) for the field in a homogeneous region as a function of the field on a surface in order to obtain an integral representation of the field outside and inside the scatterer. Then, letting the field point tend to the surface and taking into account the continuity of the normal component of **D** and that of the tangential components of **E** and **H**, after some suitable manipulations of Eqs. (IV.3.2,3) we arrive at the following relations involving the fields on the *outside* of the surface S of the scatterer:

$$\hat{n}_0 \times \mathbf{E}_i(\mathbf{r}) = \hat{n}_0 \times \oint\!\!\!\oint_S \Big\{ i\omega\mu(\hat{n}'_0 \times \mathbf{H})(G_1 + G_2) - (\hat{n}'_0 \times \mathbf{E}) \times \nabla'(G_1 + G_2) - (\hat{n}'_0 \cdot \mathbf{E})\, \nabla'\Big(G_1 + \frac{\varepsilon_1}{\varepsilon_2} G_2\Big)\Big\}\, dS'_0, \tag{VI.8.1a}$$

$$\hat{n}_0 \times \mathbf{H}_i(\mathbf{r}) = -\hat{n}_0 \times \oint\!\!\!\oint_S \Big\{ i\omega\varepsilon(\hat{n}'_0 \times \mathbf{E})\Big(G_1 + \frac{\varepsilon_2}{\varepsilon_1} G_2\Big) + (\hat{n}'_0 \times \mathbf{H}) \times \nabla'(G_1 + G_2) + \hat{n}'_0 \cdot \mathbf{H}\, \nabla(G_1 + G_2)\Big\}\, dS', \tag{VI.8.1b}$$

where the subscripts 1 and 2 refer, respectively, to the outside and inside regions of the scatterer, $G_{1,2} = G(n_{1,2}\mathbf{r}) \cdot \mathbf{E}$ and **H** represent the field on the outside of S, and $\hat{n}_0$ is the outward normal. Considering **E** and **H** as the unknown functions, Eqs. (VI.8.1) constitute a system of Fredholm integral equations, in which the forcing term is represented by the incident field. As a rule, these equations ensure a unique solution except for special values (generally complex) of k_0, for which they also admit solutions for vanishing incident fields. These special values of k_0 correspond to the resonances of the dielectric body, an example of which will be encountered in Section VI.13 when we discuss the resonances of a dielectric sphere. If we divide the surface into a number of sufficiently small elements that the different functions entering the integral are practically constant over each of them, the system of

Eqs. (VI.8.1) can be transformed into a matrix equation. A systematic application of this reasoning is provided by the *method of moments* (MOM) (see Harrington [9]). A wealth of improvements and extensions has been produced since the introduction of the MOM by Harrington. Although this is not the place to present all these techniques, we will mention the *singularity expansion technique* (SEM), which allows the treatment of scattering problems in the time domain [10].

An alternative approach to the problem at hand makes use of the following integral representation of the field inside the scatterer, which can be easily derived by treating the quantity $(\varepsilon_2 - \varepsilon_1)\mathbf{E}$ as a source term for Maxwell's equations:

$$\mathbf{E}(\mathbf{r}) = \mathbf{E}_i(\mathbf{r}) + \frac{\varepsilon_2 - \varepsilon_1}{\varepsilon_1} \iiint_V [\omega^2 \mu \varepsilon_1 G_1 \mathbf{E} + \nabla' \cdot \mathbf{E} \nabla' \mathbf{G}_1] \, dV. \tag{VI.8.2}$$

9 Physical Optics Approximation for a Perfect Conductor

When the wavelength of the radiation incident on a metallic obstacle is short compared to a typical radius of curvature of the scatterer surface, the surface can be divided approximately into an illuminated ($S_{\rm ill}$) and a shadow ($S_{\rm sh}$) region. Reasoning from the exact integral equation for the induced surface current, Fock (see Chapter IV in [3]) estimated that the transition region between these two areas has a width of the order of $[2/(k\rho_{\rm g})]^{1/3}\rho_{\rm g}$, where $\rho_{\rm g}$ is the local curvature radius of the surface.

Under the above assumptions we can calculate the diffracted field by resorting to the so-called *physical optics* (PO) approximation. This approach consists of two steps. First, it is assumed that at every point of the illuminated portion $S_{\rm ill}$ of the scatterer, the incident field is reflected as though an infinite plane wave were incident on the infinite tangent plane; the field on the shadow portion of the surface is assumed to be zero. This corresponds to assuming that $\mathbf{E}_{\rm d} + \mathbf{E}_{\rm i} = 2\hat{n}_0(\hat{n}_0 \cdot \mathbf{E}_{\rm i})$ and $\mathbf{H}_{\rm d} + \mathbf{H}_{\rm i} = -2\hat{n}_0 \times (\hat{\mathbf{n}}_0 \times \mathbf{H}_{\rm i})$ on $S_{\rm ill}$, where $\hat{n}_0$ is the inward-pointing normal to S. Second, the scattered field is obtained by an integration over $S_{\rm ill}$.

In agreement with this procedure and using Eq. (IV.3.3), we easily obtain the following expression for the scattered field:

$$\mathbf{H}_{\rm d}(\mathbf{r}) = \iint_{S_{\rm ill}} [(\mathbf{H}_{\rm i} + \mathbf{H}_{\rm d}) \times \hat{n}_0] \times \nabla' G \, dS'. \tag{VI.9.1}$$

In the far zone $\nabla' G \cong ikG\hat{n}$, where $\hat{n} = \hat{r}$, so that

$$\mathbf{H}_d(\mathbf{r}) \cong 2ik \iint_{S_{ill}} (\mathbf{H}_i \times \hat{n}_0) \times \hat{n}_0 G \, dS'$$

$$\cong 2ikG(r)\hat{n} \times \iint_{S_{ill}} \hat{n}_0 \times \mathbf{H}_i e^{ik\hat{n}\cdot\mathbf{r}'} \, dS'$$

$$\equiv \frac{e^{-ikr}}{\zeta r} \hat{n} \times \mathbf{E}'_d(\hat{n}). \qquad \text{(VI.9.2)}$$

As $kr \to \infty$, $\mathbf{H}_d \cong \zeta^{-1} \hat{n} \times \mathbf{E}_d$ and the above equation yields

$$\mathbf{E}_d(\mathbf{r}) \cong (e^{-ikr}/r)\mathbf{E}'_d(\hat{r}). \qquad \text{(VI.9.3)}$$

When the incident field is a plane wave, $\mathbf{H}_i = \mathbf{H}_0 e^{-k\hat{n}_i \cdot \mathbf{r}}$, $\mathbf{E}'_d(\hat{n})$ reduces to

$$\mathbf{E}'_d(\hat{n}) = -\frac{i}{\lambda} \zeta \mathbf{H}_0 \times \iint_{S_{ill}} \hat{n}_0(\mathbf{r}') e^{ik(\hat{n} - \hat{n}_i)\cdot\mathbf{r}'} \, dS'$$

$$= \frac{i}{\lambda} \iint_{S_{ill}} e^{ik(\hat{n} - \hat{n}_i)\cdot\mathbf{r}'} (\hat{n}_i \hat{n}_0 - \hat{n}_i \cdot \hat{n}_0) \cdot \mathbf{E}_0 \, dS' \equiv \mathbf{S} \cdot \mathbf{E}_0 \qquad \text{(VI.9.4)}$$

where $\mathbf{S}$ is the scattering matrix, which will be discussed in Section VI.11. According to the last equation $\mathbf{E}'_d$ is proportional to the Fourier transform of the vector $\hat{n}_0(r')$ relative to the illuminated portion of the metallic obstacle. This transform can be evaluated asymptotically by using the stationary-phase method illustrated in Chapter V. It is easy to show that the stationary points, if any, satisfy the relation $\hat{n}_0 \cdot (\hat{n} + \hat{n}_i) = 0$; that is, they coincide with the reflection points of the incident rays scattered along the direction $\hat{n}$. Analogously, we have already presented a discussion leading to the conclusion that the terms of the asymptotic series arising from each stationary point coincide with those appearing in the Luneburg–Kline series. Therefore, at this point we can omit the detailed discussion of these contributions to the scattered field and refer the reader to the methods illustrated in Chapter II. What cannot be predicted by ray optics is the contribution of the boundary of the domain of integration S_{ill}. To calculate the diffraction effects due to the rim of the illuminated region S_{ill}, we must deal directly with the diffraction integral. To simplify the calculation of the leading terms of these diffraction fields, we can use the GTD.

The integral representation of the vector diffraction amplitude $\mathbf{E}'_d$ simplifies notably in the forward direction. In particular, for $\hat{n} = \hat{n}_i \equiv \hat{z}$, the integral of

Eq. (VI.9.4) reduces to

$$\mathbf{E}'_{\mathrm{d}}(\hat{z}) = -i\frac{\zeta}{\lambda}\mathbf{H}_0 \times \iint\limits_{S_{\mathrm{ill}}} \hat{n}_0 \, dS = -i\frac{\zeta}{\lambda}\mathbf{H}_0 \times \left(\hat{x}\iint\limits_{S_{\mathrm{ill}}} dS_x + \hat{y}\iint\limits_{S_{\mathrm{ill}}} dS_y + \hat{z}\iint\limits_{S_{\mathrm{ill}}} dS_z\right), \tag{VI.9.5}$$

where dS_x, dS_y, and dS_z are the projections of the oriented surface $d\mathbf{S}$ with the normal $\hat{n}_0$ on the coordinate planes yz, xz, and xy. For simple surfaces it is easy to show that $\iint_{S_{\mathrm{ill}}} dS_x = \iint_{S_{\mathrm{ill}}} dS_y = 0$. Accordingly, $\mathbf{E}'_{\mathrm{d}}(\hat{z})$ reduces to $\hat{z} \times \mathbf{H}_0 S_{\mathrm{ill},z}$; that is, $\mathbf{E}'_{\mathrm{d}}$ is proportional to the incident electric field times the projection of the illuminated surface along the direction of the incident wave.

When $\hat{n}$ is close to but not coincident with $\hat{n}_{\mathrm{i}}$, we can identify $\hat{n} - \hat{n}_{\mathrm{i}}$ with the component of $\hat{n}$ normal to $\hat{n}_{\mathrm{i}}$ and approximate Eq. (VI.9.4) with

$$\mathbf{E}'_{\mathrm{d}}(\hat{n}) = -i\frac{\zeta}{\lambda}(\mathbf{H}_0 \times \hat{n}_{\mathrm{i}}) \iint\limits_{S_{\mathrm{ill}}} dx'\,dy' \exp[ik\mathbf{n}_\perp \cdot (x'\hat{x} + y'\hat{y})], \tag{VI.9.6}$$

where $x'\hat{x} + y'\hat{y} = \mathbf{r}' - (\mathbf{r}' \cdot \hat{n}_{\mathrm{i}})\hat{n}_{\mathrm{i}}$. The above integral is the Fourier transform of a plane aperture lying on the plane x, y, obtained by the projection along $\hat{n}_{\mathrm{i}}$ of the illuminated portion of the obstacle. Physically, this result allows us to affirm that the field diffracted in the forward direction is approximately coincident with that diffracted by an aperture obtained by squeezing the obstacle along the direction of illumination.

Example: Field Diffracted by a Sphere. As an example let us consider a sphere of radius a and center O illuminated by a plane wave directed along the z axis and polarized in the y direction. In this case the vector scattering amplitude $\mathbf{E}'_{\mathrm{d}}$ for $\hat{n} \simeq \hat{n}_{\mathrm{i}}$ is given by

$$\mathbf{E}'_{\mathrm{d}} = -\frac{i}{\lambda}\mathbf{E}_0 \int_0^a \rho\,\mathbf{d}\rho \int_0^{2\pi} d\phi \exp(ik|\mathbf{n}_\perp|\rho\cos\phi) = -ika\frac{J_1(ka|\mathbf{n}_\perp|)}{ka|\mathbf{n}_\perp|}\mathbf{E}_0. \tag{VI.9.7}$$

This result shows that the forward-diffracted field coincides with that diffracted by a disk of radius coincident with that of the sphere.

10 Electromagnetic Theory of Diffraction from Perfectly Conducting and Dielectric Gratings

In the history of physics the diffraction grating stands out as one of the most important instruments. Until 1891, when Michelson introduced the interferometer bearing his name, the grating with known spacing was the only means of accurately measuring the characteristic wavelengths in atomic spectra.

The grating seems to have been invented by the American astronomer David Rittenhouse in about 1785 and rediscovered some years later by Joseph von Fraunhofer, who in 1819 published his original investigations of the diffraction by gratings. The earliest devices consisted of a grid of fine wire or thread wound about and extending between two parallel screws, which served as spacers. These devices were multiple-slit assemblies, which modulated a wave front in amplitude by confronting it with alternate opaque and transparent regions.

More common forms of gratings are made by ruling fine grooves with a diamond point either on a plane glass surface to produce a transmission grating or on a polished metal mirror to produce a reflection grating. In the past, reflecting gratings were ruled on *speculum metal*, a very hard alloy of copper and tin. Today, they are ruled on an evaporated layer of aluminum, which gives better reflection in the U.V.

Great progress in the manufacture of ruled gratings was due to Harry A. Rowland, who in 1882 built a machine for ruling gratings with an uncorrected periodic error much less than 1/300,000 of an inch. In more recent years considerable progress has been made in the construction of gratings as a consequence of the work of Strong and Babcock, and later of Harrison [11] and Stroke [12] at the Massachusetts Institute of Technology (MIT). With the MIT machine controlled interferometrically, Harrison has been able to construct remarkable gratings 26 cm long with 360 cm^2 of useful surface.

The manufacture of ruled gratings is extremely difficult and time-consuming. In actuality, most gratings are plastic *replicas* of ruled gratings (masters). Replicas and masters behave differently when illuminated by intense light radiation. In particular, gratings used in optical cavities of high-power lasers must be masters to withstand the elevated intensities present in the cavities. For a review of the history and manufacture of gratings the reader is referred to Stroke [12].

Because of the severe mechanical problems of ruling gratings many alternative methods of production have been considered, starting with Michelson, who suggested in 1927 that gratings be produced by photographic techniques. In 1962 Denisyuk [13] mentioned the possibility of obtaining gratings by recording interference fringes. With modern technologies periodic surface corrugations can be generated by the so-called *holographic method*, in which a grating relief pattern is produced by interferometric exposure and development of a photoresist. Successively, the grating is transferred to the substrate by ion-beam milling or chemical etching. The first holographic gratings for spectroscopic use were made in 1967 at the Optical Laboratory of the Göttingen Observatory by Schmahl and Rudolph [14] and in 1968 by Labeyrie and Flamand [15] in Paris. Since then holographic gratings have been fabricated for the visible, UV, and soft x ray regions. Today, it is possible to fabricate gratings more than 600 mm wide, with a line density of more than 10,000 grooves per millimeter, that are completely free of ghosts and exhibit a very low amount of stray light. Holographic gratings can be formed on curved surfaces and the grating frequency can, in principle, be varied over the surface, thus providing the possibility of focusing the diffracted beams. The reflectivity of holographic gratings is lower than that of ruled gratings and is strongly dependent on the polarization of the incident wave.

Before ending this introduction, it is worth noting that a few natural systems also have the dimensions and the high degree of order needed for an optical grating. For example, in liquid crystals molecules are stacked with sufficient regularity to act as diffraction gratings. Because the spacing between molecules depends on the temperature, so does the spectrum of the diffracted field, which explains their use as temperature sensors. A preeminent diffraction grating is the opal, which is made of silicon dioxide and little water spheres, which are closely packed in a three-dimensional array with a spacing of about 0.25 μm. The transparent matrix that fills the space between the spheres has a similar composition but a slightly different index of refraction. The three-dimensional grating structure, resolved for the first time in 1964 with the electron microscope, makes the stone to appear colored when illuminated by white light; the color change as the observer's eye or the stone is moved.

10.1 *Use and Applications of Gratings*

When illuminated by a plane wave of assigned wavelength, a reflection (or transmission) grating produces several beams, which can be labeled with an integer m $(0, \pm 1, \pm 2)$ that represents the *order of diffraction.* If we indicate with θ the incidence angle and d the period of the grating grooves, we will show later on that the mth diffracted beam forms a reflection (transmission) angle θ_m, which satisfies the relation

$$\sin \theta_m - \sin \theta = m(\lambda/d), \qquad \text{(VI.10.1)}$$

known as the grating equation.

For $m = 0$, $\theta = \theta_0$; that is, the zeroth order corresponds to the specular reflection found in ordinary mirrors. For m sufficiently large the above relation can be satisfied only by a complex angle $\theta_m = i\theta''_m + \text{sgn}(m)\pi/2$. This means that the grating produces evanescent waves. When the incident beam travels almost parallel to the grating surface ($\theta = \pi/2$) the diffracted rays deviate from the grazing direction by pointing toward the region of incidence or refraction according to the sign of m.

If we consider a dielectric waveguide, a simple means of drawing some power from a propagating mode consists of corrugating the surface. In this case, the mode acts as a beam incident on the grating at the grating angle and scattered partially outside as a result of the corrugations. These devices are used as input–output couplers for selectively exciting the modes of optical waveguides. Integrated narrowband filters, light deflectors, and phase-matching elements have been made with corrugated surfaces. For an analysis of these devices the reader should consult the review contained in Tamir [16]. Additional applications include the use of corrugations for making Bragg reflectors in semiconductor lasers.

The main application of gratings is in spectroscopy, where they have replaced the dispersive prisms because of their much larger resolving power. Replacing a prism in a spectrometer by a grating gives at least ten times as much flux since the slits can be set at a much greater width to give the same resolving power. In addition, it is possible to utilize a greater dispersive area with a grating, which means a greater resolving power. However, grating spectrometers have the disadvantage compared to prism spectrometers that spectra from unwanted orders of interference must be eliminated.

Gratings used in spectrometers are almost always of the reflection type, where the source and the observation point lie on the same side of the grating. The grooves may be of any shape. However, it is advantageous to give them a saw-tooth profile with the sides optically flat. This profile, introduced by Wood, has the advantage of enhancing the intensity of a given order, thus improving the efficiency of a spectrometer.

A parameter widely used in the characterization of gratings is the *efficiency*, defined as the flux diffracted in a given order per unit incident monochromatic flux. The efficiency depends on the groove profile and on the line density. Figure VI.14 shows the efficiency of a typical blazed grating having 1200 grooves per millimeter and a *blaze angle* $\alpha \doteq 17°27'$ for polarization parallel (p) and perpendicular (s) to the grooves.

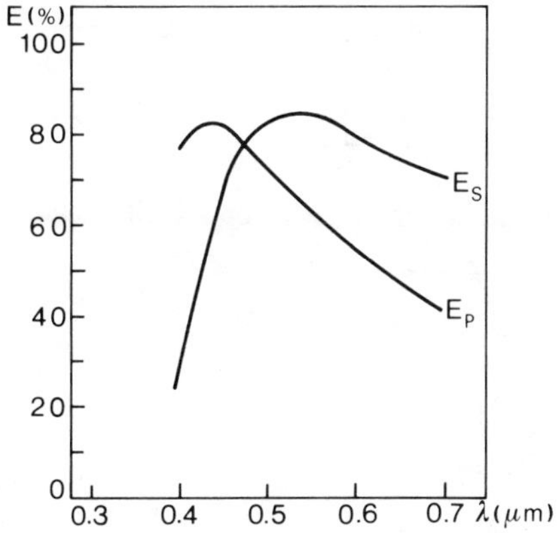

Fig. VI.14. Typical efficiency curve versus λ for a blazed grating with $\alpha = 17°27'$ and 1200 grooves per millimeter (from Jobin–Yvon catalog).

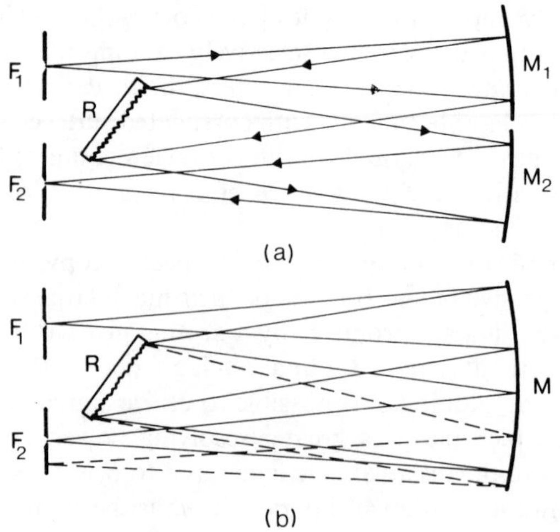

Fig. VI.15. (a) Czerny–Turner and (b) Ebert–Fastie mountings used in scanning spectrometers. The input radiation is analyzed by observing the intensity variations at the exit slit with a photomultiplier, and rotating the plane grating. A common feature of these devices is that the gratings operate with constant deviation angle α between incident and diffracted beams.

Gratings used in monochromators operate with constant deviation between incident and diffracted beams (see Fig. VI.15). In this case $\theta + \theta_m = \alpha$ and Eq. (VI.10.1) reduces to

$$\sin[\theta - (1/2)\alpha] = -[(m\lambda)/(2d)][1/\cos(\alpha/2)]. \qquad \text{(VI.10.2)}$$

In particular, when the angular deviation α is equal to zero, the grating

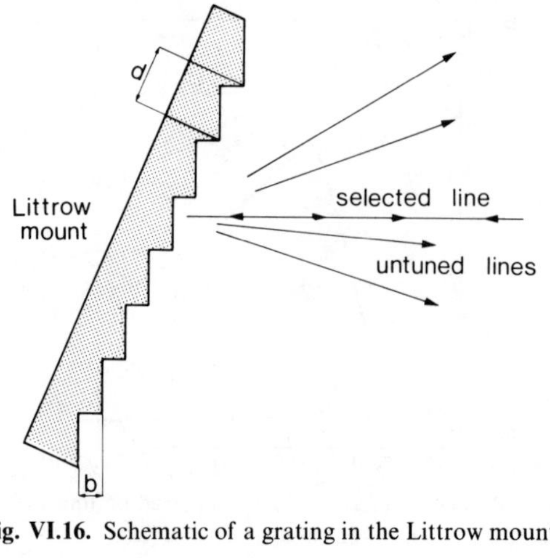

Fig. VI.16. Schematic of a grating in the Littrow mount.

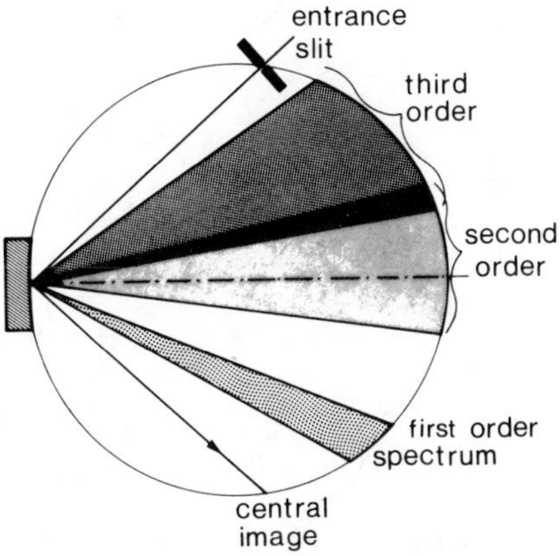

Fig. VI.17. Concave grating with an indication of the Rowland circle. The angular sectors represent the different orders of diffracted fields produced by a beam of given bandwidth illuminating the entrance slit. Due to the focusing property of the grating, the spectrum of the illuminated slit is focused on the Rowland circle.

operates in the so-called *Littrow mount* (see Fig. VI.16). According to the last equation, the wavelength transmitted by a monochromator varies in proportion to the sine of the rotation angle θ of the grating. To achieve linear time scanning of the wavelength, the grating is rotated by a sine-drive mechanism.

A grating can, in principle, be used to examine spectra of every frequency. In fact, Eq. (VI.10.2) can be satisfied for every value of λ/d by choosing a suitable combination of incidence angle θ and order m of operation. However, as a general rule the intensity of the diffracted beam diminishes drastically with an increase of the order m. In addition, for a given order the efficiency is sufficiently large only for $\lambda/(2d\cos(\alpha/2))$ comprised in a small interval (see Fig. VI.14). If we increase m we can use the grating at a wavelength λ/m. From the electromagnetic point of view, the grating should exhibit the same efficiency, apart from the effects produced by the reduced reflectivity of the material it is made of.

In the UV region concave gratings [17] are generally used. The basic principle of the concave grating is such that the diffracted images of the source are sharply focused on a circle, called the *Rowland circle*, equal in diameter to the radius of curvature of the grating, provided the surface of the grating is tangential to the Rowland circle, the ruled lines are at right angles to the Rowland plane, and the illuminated entrance slit is on the Rowland circle and parallel to the ruled lines of the grating. Figure VI.17 shows the basic optical layout of a spectrograph. There are two basic types of spectrographs: the normal incidence spectrograph, suitable for studies from 300 to 2000 Å, and the *grazing incidence spectrograph*, used for studies below 300 Å. When the angle of incidence is less than approximately 10° the radiation is considered to

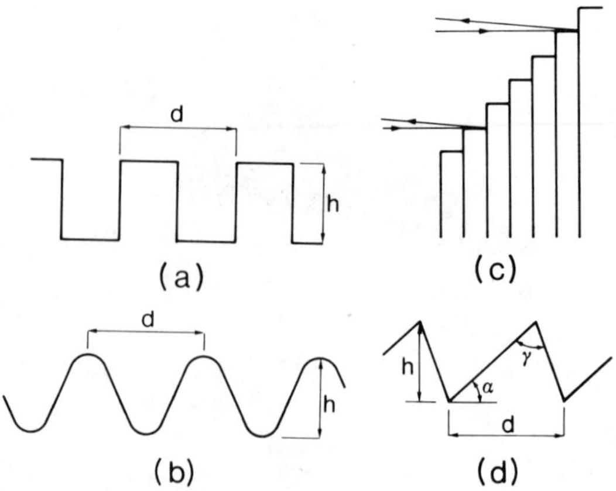

Fig. VI.18. Profiles of (a) lamellar, (b) sinusoidal, (c) echelon, and (d) echelette gratings. For (d), the apex angle is about 90°.

be directed at normal incidence to the grating. To disperse wavelengths shorter than 300 Å grazing incidence mounts are used in order to compensate the decrease in reflectance of the grating material by exploiting the total reflection experienced at extreme grazing incidence.

Different types of grating profiles are shown in Fig. VI.18.

10.2 *Field Diffracted by a Grating*

Let us consider an infinitely wide, periodic cylindrical surface $z = f(x)$ (the grating surface), whose generatrices are parallel to the y axis. $f(x)$ is a periodic function with period d $[f(x + d) = f(x)]$, oscillating between f_{max} and f_{min}, and represents the surface separating a homogeneous and isotropic material of index n_1 lying in region 1 $[z > f(x)]$ from a homogeneous and isotropic medium of index n_2 lying in region 2 $[z < f(x)]$. We assume that the normal $\hat{z}$ to the grating is oriented toward the region from which the incident beam is coming. Notice that this convention is opposite to that adopted in the analysis of multilayer structures. The reason for the present choice is that we want to analyze reflection gratings, which are the ones most frequently adopted in spectroscopy. Medium 2 is characterized by a complex refractive index $\tilde{n} = n - i\kappa$, which becomes purely imaginary for a perfectly conducting metal. The influence of the finite conductivity of the grating material is negligible above 4 μm, but may be fundamental below 1 μm, with 1 to 4 μm being a transition region. In the UV (0.1–0.2 μm) dielectric coatings are used in place of less efficient aluminum gratings. In addition, dielectric-coated gratings are used as wavelength selectors in dye lasers to enhance the efficiency. The grating is considered to be illuminated from region 1 by a monochromatic plane wave at incidence angle θ. The electric vector of the incident wave can be either parallel (p-polarization) or perpendicular (s-polarization) to the grooves (y axis).

Let $u_i(r) = \exp(-i\mathbf{k}_i \cdot \mathbf{r})$ be a generic scalar component of unit amplitude of the incident beam, and $u_d(\mathbf{r})$ be the corresponding component of the diffracted field. If we displace the origin of coordinates of the quantity d (grating period) along the x axis, the incident and diffracted fields become in the new system

$$u_{i(new)}(x, y, z) = u_{i(old)}(x + d, y, z) = u_{i(old)}(x, y, z)\exp(-ik_{ix}d), \qquad \text{(VI.10.3a)}$$

$$u_{d(new)}(x, y, z) = u_{d(old)}(x + d, y, z). \qquad \text{(VI.10.3b)}$$

Here "new" and "old" refer to the field expressions in the new and old reference frames, respectively. Since in the new system the grating profile is identical to the old one $[f_{(new)}(x) = f_{(old)}(x + d) = f_{(old)}(x)]$, while the new and old incident fields differ by the phase factor $\exp(-ik_{ix}d)$, then

$$u_{d(new)}(x, y, z) = u_{d(old)}(x, y, z)\exp(-ik_{ix}d). \qquad \text{(VI.10.3c)}$$

As a consequence, combining Eq. (VI.10.3b) with Eq. (VI.10.3c), we obtain

$$u_d(x + d, y, z)\exp[ik_{ix}(d + x)] = u_d(x, y, z)\exp(ik_{ix}x). \qquad \text{(VI.10.4)}$$

This means that $u_d\exp(ik_{ix}x)$ is a periodic function of x having period d.

Before going on to a discussion of a Fourier series representation of u_d we pause to comment on the boundary conditions imposed by the grating. In analogy with the plane interface case discussed in Section III.6 we set $u_d = u - u_i$ in medium 1 and $u_d = u$ in medium 2. Next, if we assume that the plane of incidence is normal to the grooves (i.e., $k_{iy} = 0$), a condition met in gratings used in monochromators, then we can choose $u = E_y$ for p-waves and $u = H_y$ for s-waves.

Now, with these conventions it is easy to prove that the continuity of the tangent components yields the following boundary conditions for u:

$$u_i + u_{d1} = u_{d2} \qquad \text{p- and s-waves}$$

$$-i(k_{ix}f' - k_{iz})u_i + \frac{\partial u_{d1}}{\partial x}f' - \frac{\partial u_{d1}}{\partial z} = \begin{cases} \dfrac{\partial u_{d2}}{\partial x}f' - \dfrac{\partial u_{d2}}{\partial z}, & \text{p-wave} \\[2ex] \left(\dfrac{\partial u_{d2}}{\partial x}f' - \dfrac{\partial u_{d2}}{\partial z}\right)\dfrac{n_2^2}{n_1^2}, & \text{s-wave} \end{cases} \qquad \text{(VI.10.5)}$$

for $\mathbf{r} = [x, y, f(x)]$ belonging to the grating surface, where we have set $f' = df/dx$. In particular, if medium 2 is a perfect conductor, the above conditions simplify to

$$u_{\mathrm{i}} + u_{\mathrm{d1}} = 0, \qquad \text{p-wave}$$

$$(\partial u_{\mathrm{d1}}/\partial x)f' - \partial u_{\mathrm{d1}}/\partial z = i(k_{ix}f' - k_{iz})u_{\mathrm{i}} \qquad \text{s-wave} \tag{VI.10.6}$$

Let us now return to the initial discussion of the representation of $u_{\mathrm{d}}\exp(ik_{ix}x)$, by observing that $u_{\mathrm{i}}\exp(ik_{ix}x)$ and, consequently, $u\exp(ik_{ix}x)$ are also periodic functions of x. If we repeat the same considerations for coordinate y, we can easily show that $u\exp(ik_{ix}x + ik_{iy}y)$ is constant with respect to y. At this stage we observe that if u is a p-wave, the total field is continuous together with its gradient in correspondence with the grating surface, so that we can expand the field in a Fourier series in x,

$$u(x, y, z) = \exp(-ik_{ix}x - ik_{iy}y) \sum_{m=-\infty}^{+\infty} \exp(-i\,2\pi\, mx)V_m(z), \tag{VI.10.7}$$

where the $V_m(z)$ are continuous functions to be determined. In addition, plugging the above series in the wave equation yields

$$\sum_{m=-\infty}^{+\infty} \left[\frac{d^2 V_m}{dz^2} + \beta_m^2(x, z)V_m\right]\exp\left(-i\frac{2\pi}{d}mx\right) = 0, \tag{VI.10.8}$$

where

$$\beta_m^2(x, z) = k_0^2 n^2(x, z) - [k_{ix} + (2\pi/d)m]^2 - k_{iy}^2. \tag{VI.10.9}$$

If we multiply each equation by $\exp[i(2\pi/d)qx]$, with q an integer, and integrate over a period d, we obtain

$$\frac{d^2 V_q}{dz^2} + k_0^2 V_q(z)g_0(z) = -k_0^2 \sum_{m\neq q} V_m(z)g_{q-m}(z), \tag{VI.10.10}$$

where

$$k_0^2 g_{q-m}(z) = \frac{1}{d}\int_0^d \beta_m^2(x, z)\exp\left[i\frac{2\pi}{d}(q - m)x\right]dx. \tag{VI.10.11}$$

In particular, for a sinusoidal profile $z = (1/2)h\cos(2\pi x/d)$,

$$g_l(z) = g_{-l}(z) = [(n_2^2 - n_1^2)/\pi l]\sin[(2\pi/d)lx) + n_1^2\delta_l. \tag{VI.10.12}$$

We can now distinguish three regions, A, B, and C, defined respectively by $z > f_{\max}$ (region A), $f_{\max}(z) > z > f_{\min}(z)$ (region B), and $f_{\min}(z) > z$ (region C). In regions A and C the functions $g_l(z)$ vanish identically for $l \neq 0$, so that we can solve for the V_m by putting

$$V_m(z) = \begin{matrix} R_{m1}\exp(-i\beta_{m1}z) + \delta_m\exp(ik_{iz}z), & z > f_{\max}, \\ T_{m2}\exp(i\beta_{m2}z), & z < f_{\min}, \end{matrix} \tag{VI.10.13}$$

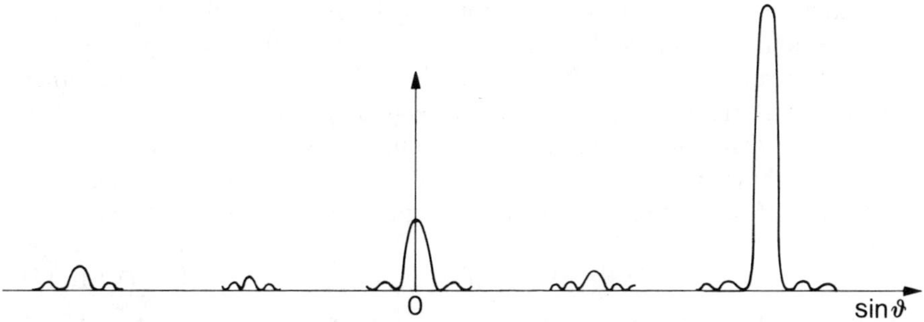

Fig. VI.19. Schematic representation of the far-field pattern of a grating. The scattering angle θ is calculated with respect to the outward normal to the grating surface and the sign of θ is chosen by following the convention adopted for reflection from a surface. The broadening of the single orders is due to the finite size of the grating.

where Re $\beta_m > 0$ and Im $\beta_m > 0$. The generally complex coefficients R_{m1} and T_{m2} can be determined by matching the above solutions with those related to region B. The techniques used to solve the set of coupled differential equations are somewhat involved and by no means routine. Now, inserting the above expressions for V_m into the Fourier series expansion [Eq. (VI.10.7), we see that the field diffracted by a grating is made of an infinite discrete set of plane waves corresponding to the different orders of diffraction. The finite size of the grating produces a broadening of each order, as shown schematically in Fig. VI.19.

10.3 *Rayleigh Representation for a Metallic Grating*

In 1897 Rayleigh put forward a method for calculating the scattering from periodic surfaces based on the assumption that *the field representation found for the region in front of the grating* (region A, according to the above convention) *should also be valid in proximity to the grating surface* (*Rayleigh hypothesis*). Accordingly, on the boundary we obtain for a p-wave with $k_{iy} = 0$

$$\sum_{m=-\infty}^{+\infty} R_m \exp\left[-i\beta_m f(x) - i\frac{2\pi}{d} mx\right] = -\exp[-ik_{iz} f(z)]. \qquad \text{(VI.10.14)}$$

The Rayleigh hypothesis was unquestioned until about 30 years ago, when Deriugin (1952) and Lippmann (1953) argued that the Rayleigh hypothesis seemed unrealistic, because between the corrugations both incoming and outgoing secondary waves as well as the corresponding exponentially growing and evanescent waves are expected to exist (see the Chapter by M. Cadilhac in Petit [18]). Millar showed in 1969 that a necessary and sufficient condition for the Rayleigh hypothesis to hold is that the relative series be an analytic function of x and z. A careful analysis of the domain of analyticity has led several authors to establish upper limits for the groove height. In particular, Petit, Cadilhac, and Millar proved that for the sinusoidal profile $z = (1/2)h\cos(2\pi x/d)$ the hypothesis is valid for $h/d \leq 0.\ 1426$. On the other hand, for a triangular profile the Rayleigh hypothesis is *never* satisfied. Additional theorems useful for establishing the validity of the Rayleigh hypothesis have been proved by van den Berg and Fokkema.

In spite of the failure of the Rayleigh hypothesis, we can still use the set of plane waves

$$u_m(x, z) = \exp(-i\beta_m z - i\gamma_m x), \tag{VI.10.15}$$

where $\gamma_m = k_{ix} + (2\pi/d)m$, to represent the field. In fact, Yasuura has shown that the above set of outgoing and evanescent plane waves is *complete* and can be used to calculate the coefficients by solving the finite system:

$$\sum_{m=-N}^{N} R_m^N \int_0^d u_m[x, f(x)]u_m^*[x, f(x)](1 + f'^2)^{1/2}\, dx$$
$$= -\int_0^d u_i[x, f(x)]u_m^*[x, f(x)](1 + f'^2)^{1/2}\, dx, \tag{VI.10.16}$$

for $n=0, \pm 1, \ldots, \pm N$. Ikuno and Yasuura proved in 1973 that $\lim_{N\to\infty} R_m^N = R_m$. According to this method, generally referred to as the *least-squares approximation method* (LSAM), the reflection coefficient R_m corresponding to the mth diffracted wave is obtained by solving the system of Eq. (VI.10.16) by choosing a value N that is sufficiently high. The coefficients of the R_m's must be calculated by integrating the product of the incident wave times the mth Rayleigh wave u_m^* along the grating profile. A variation of the LSAM consists of replacing $u_m^*[x, f(x)]$ $(1 + f'^2)$ with $u_m^*(x, 0)$. The method is called the *Fourier series method* (FSM).

10.4 *Integral Method*

Let us now consider a different approach that does not have the limitations of the Rayleigh approach (see the chapter by D. Maystre in Petit [18]). In general, we have [cf. Eq. (VI.9.1)]

$$\mathbf{H}(\mathbf{r}) - \mathbf{H}_i(\mathbf{r}) = -\oiint_{\text{obstacle}} (\hat{n}_0 \times \mathbf{H}) \times \nabla' G_0(\mathbf{r} - \mathbf{r}')\, dS', \tag{VI.10.17}$$

where $\hat{n}_0$ is the normal to the grating pointing toward medium 2. For $\mathbf{r} \in S_{\text{obst}}$ the above relation becomes on inhomogeneous integral equation in the unknown $\mathbf{H}_{\text{obst}}(\mathbf{r})$.

In the case of a grating the integral equation becomes two-dimensional when we replace G with the two-dimensional Green's function of Eq. (IV.7.5).

The surface integral becomes a line integral extended to the grating profile. In addition, exploiting the periodicity of $\mathbf{H}\exp(ik_{ix}x)$ along the x axis, we can limit the integral to a single period $(0, d)$, thus obtaining

$$\mathbf{H}(x,z) - \mathbf{H}_{\mathrm{i}}(x,z) = -\int_0^d (\hat{n}_0 \times \mathbf{H}) \times \nabla' G_{\mathrm{per}}[x - x', z - f(x')] \times (1 + f'^2)^{1/2}\,dx', \tag{VI.10.18}$$

where

$$G_{\mathrm{per}}(x - x', z - z') \equiv \sum_{n=-\infty}^{+\infty} \exp(-ik_{ix}\,dn) G^{(2)}(x - x' - nd, z - z'). \tag{VI.10.19}$$

Since $G_{\mathrm{per}}e^{ik_{ix}x}$ is periodic with respect to x with period d and satisfies the inhomogeneous wave equation

$$\left(\frac{\partial^2}{\partial x^2} + \frac{\partial^2}{\partial z^2} + k_0^2\right) G_{\mathrm{per}} = -\sum_{m=-\infty}^{+\infty} \exp(-ik_{ix}md)\,\delta(x - md)\,\delta(z) \tag{VI.10.20}$$

and the radiation condition for $z \to \infty$, it can be expressed as

$$G_{\mathrm{per}}(x,z) = \frac{i}{2d} \sum_{m=-\infty}^{+\infty} \frac{1}{\beta_m} \exp[-i(\gamma_m x + \beta_m z)]. \tag{VI.10.21}$$

For an s-wave the vector $\mathbf{H} = \hat{y}u$ is parallel to the grooves, so that the integral equation (VI.10.18) reduces to

$$u^{\mathrm{i}}(x,z) - u(x,z) = \int_0^d u(x', z')\left(f' \frac{\partial}{\partial x'} - \frac{\partial}{\partial z'}\right) G_{\mathrm{per}}(x - x', z - z')\,dx', \tag{VI.10.22}$$

where $z = f(x)$ and $z' = f(x')$. This integral equation can be solved by reducing it to a matrix by the method of moments.

Since the Green's function G_{per} is singular at $\mathbf{r} = \mathbf{r}'$, it is important to accurately compute the contributions from this singularity. The interested reader can consult Zaki and Neurenther [19]. The integral method can be extended to dielectric gratings for p- and s-polarizations.

10.5 *Resolving Power*

We reported in Section VI.10.1 the general grating formula, which holds true for every sort of grating (independent of the profile). The sign of the diffraction angles has been chosen to give $\theta_0 > 0$ for specular (or zeroth-order) reflection. The amplitudes of the different orders are given by the coefficients R_m, depending on the grating profile, polarization, wavelength, and incidence angle. These coefficients can be calculated by using either the plane-wave expansion methods (say LSAM or FM) or the integral approach, discussed in the last subsection. In general, gratings are used as dispersive elements. Therefore, the most important parameters are those related to their ability to discriminate waves of different wavelengths, say λ and $\lambda + d\lambda$. This feature depends on the spacing d of the grating grooves, on the length and the order m of the diffracted beam, we are considering, and on the *size* of the whole grating. Concerning the parameters d/λ and m, we observe that for a fixed angle of incidence, the grating formula gives the *dispersion equation*

$$d\theta_m/d\lambda = m/(d\cos\theta_m). \qquad \text{(VI.10.23)}$$

When the grating is used in low order, $m = \pm 1, \pm 2, \ldots$, then high angular dispersion is obtained only with a finely ruled grating, hence with small d. On the other hand, it is also possible to obtain large angular dispersion with a coarsely ruled grating by using diffracted beams of high order and large incidence angles. In some cases for assigned θ, λ, and d, there exists an order, say m_L, that corresponds to a beam diffracted in almost the same direction as the incident one, so that the grating appears as a retroreflector for order m. Such a situation is called the *Littrow condition* and the grating is said to be in *Littrow mount*. Now, from the grating equation we immediately derive the order m_L by imposing $\theta_m \sim -\theta$, thus obtaining

$$m_L \simeq -2(d/\lambda)\sin\theta, \qquad \text{(VI.10.24)}$$

where m_L is a negative integer. When the Littrow condition is satisfied, the angular dispersion reduces to

$$d\theta_{mL}/d\lambda \simeq -(2\tan\theta)/\lambda. \qquad \text{(VI.10.25)}$$

As a consequence, the dispersion is independent of the number of grooves for a Littrow-mounted grating. In addition, by choosing an incidence angle sufficiently close to 90°, the dispersion can be made very high.

In principle, every grating could be used in the Littrow configuration with large incidence angles. From a practical point of view this requires a sufficiently high amplitude of the diffracted beam. To this end *echelette* gratings are used which have an asymmetric triangular profile characterized by the blaze angle $\alpha = \theta$ approximately coincident with the incidence angle θ (see Fig. VI.18d). When the wavelength satisfies the Littrow condition, the efficiency $E_{m_L} = |R_{m_L}|^2$ is equal to unity for an incident s-wave. Generally, echelettes in $m = -1$ Littrow mounts are used. The relative efficiency E_{-1} is a function of the incidence angle or, equivalently, of the ratio $\lambda/d = 2\sin\theta$, which ranges from 0 to 2. In Fig. VI.14 a typical efficiency curve is plotted against the wavelength for s- and p-polarizations.

Up to now we have neglected the finite size of the grating, which allowed us to represent the diffracted field as a discrete set of plane waves. The grating width W can be taken into account by assuming that each plane wave is limited by an aperture corresponding to the projection of the grating surface along the direction of the mth order. When we do this, the far-zone diffracted field takes the form:

$$u_d(x, y, z) = \exp(-ik_{iy}y) \sum_{\substack{m=-\infty \\ m\neq 0}}^{+\infty} R_m \exp(-i\gamma_m x - i\beta_m z)$$
$$\times \operatorname{sinc}[k_0 W \cos^2\theta_d(\tan\theta_m - \tan\theta_d)] \qquad \text{(VI.10.26)}$$

where $\theta_d = \arctan(x/z)$ represents the scattering angle measured with respect to the grating normal. The difference from Eq. (VI.10.16) is represented by the presence of the factor sinc( ), which accounts for the finite width W in the x direction. For simplicity we have neglected the analogous factor accounting for the finite size along y. In particular, for $\theta_d \sim \theta_m$, the intensity $i(\theta_d)$ of the pattern relative to the mth order reads

$$i_m(\theta_d) \propto |R_m|^2 \operatorname{sinc}^2[k_0 W \cos\theta_m(\theta_d - \theta_m)]. \qquad \text{(VI.10.27)}$$

Accordingly, the mth diffraction order presents an angular dispersion $d\theta_m \sim 2\pi/(k_0 W \cos\theta_m)$. Then, if we take into account the dispersion equation (VI.10.23), we conclude that the grating can be used to measure the wavelength of monochromatic radiation with an accuracy $d\lambda$ of the order of

$$d\lambda \simeq (d\cos\theta_m/m)\, d\theta_m = \lambda/(mN), \qquad \text{(VI.10.28)}$$

N being the total number of grooves. By dividing λ by $d\lambda$ we obtain the *resolving power* $\mathscr{R}$ of the grating,

$$\mathscr{R} = mN. \qquad \text{(VI.10.29)}$$

Thus, $\mathscr{R}$ is equal to the product of the order of diffraction times the number of grooves *illuminated*. On the other hand, from the grating equation (VI.10.2) for constant-deviation spectrometers we can derive m as a function of θ, obtaining

$$\mathscr{R} = (2W/\lambda)\cos(\alpha/2)\sin[\theta - (\alpha/2)] \leq 2W/\lambda \equiv \mathscr{R}_{\max}. \qquad \text{(VI.10.30)}$$

We observe that the maximum resolving power is proportional to the width of the illuminated portion of the grating and independent of the number of grooves. We can get close to $\mathscr{R}_{\max}$ by reducing the deviation angle and illuminating the grating at grazing incidence. This means that the grating should be used in the Littrow configuration ($\alpha = 0$) at grazing incidence. For a blazed grating it can be proved that the maximum energy goes into the wavelength diffracted in the same direction as the beam reflected specularly from the facets. Therefore, the resolving power can approach $\mathscr{R}_{\max}$ for blaze angles sufficiently small. With these configurations resolving powers of 1 million have been achieved. At 5000 Å this represents a resolution of 600 MHz, a bandwidth inadequate for resolving the spectra of laser beams or studying hyperfine structures. In these cases it is essential to combine a grating spectrometer with a Fabry–Perot interferometer, a device that will be illustrated at the end of the following chapter.

10.6 *Grating Efficiency and Wood's Anomalies*

We have seen that the complex amplitudes R_m of the partial waves diffracted by a grating depend in a very complicated way on the polarization, incidence angle, wavelength, and profile. When gratings are used in spectrometers and laser cavities to achieve line selections or frequency tuning (see Section VII.20) we are interested only in the propagating orders. In other cases, which are much less relevant here, it is important to know the field in proximity to the grooves. Although the former case seems much less complicated, by requiring the calculation of terms $E_n = |R_n|^2$ giving the *efficiency* of the few propagating modes, the analytic difficulties are, in principle, not less than those in the latter case. In fact, the efficiencies of these propagating orders must be calculated by including in the truncated expansions of the field a certain number of modes both propagating and evanescent. However, there are cases in which some predictions about the efficiencies can be made. A case of particular interest is that of the blazed grating operating in the Littrow configuration with incidence angle coincident with the blaze angle or its complementary angle, in which case the efficiency is equal to unity for s-polarization.

In 1902 Wood discovered a rapid variation in the efficiency of the various diffracted spectral orders in certain narrow frequency bands for s-polarized radiation. They appear as unexpected narrow bright and dark bands in the spectrum of an optical reflection grating illuminated by a light source with a slowly varying spectral distribution. In 1907 Rayleigh explained these anomalies by observing that the scattered field is singular at wavelengths for which one of the spectral orders emerges from the grating at the grazing angle. He observed that the wavelengths λ_R, which have come to be called the *Rayleigh wavelengths*, for which there exists an order m such that $\beta = 0$, correspond to Wood anomalies. In view of Eq. (VI.10.9), the possible wavelength $m\lambda_R$ can take the values

$$\lambda_R = d(1 - \cos\theta)/n \qquad (n \text{ integer}). \qquad \text{(VI.10.31)}$$

U. Fano was the first to suggest that anomalies could be associated with the excitation of surface waves along the grating. A different approach was developed by researchers working on surface plasmons (see the end of Chapter III).

Starting from the plasmon resonances of a plane surface, they included the groove profile as a perturbation in the calculation of the probability of excitation of surface plasmons. An accurate presentation of these methods can be found in Petit [18] or in the paper by Hessel and Oliner [20], in which the anomalies are analyzed by representing the grating by a plane surface with a periodic surface impedance.

11 Scattering from Finite Bodies

We are concerned in this section with collisions in which a plane wave propagating *in vacuo* interacts with an obstacle. The introduction of the obstruction in the path of the incident wave produces scattering. At a distance from the obstacle that is large relative both to the wavelength and to its size, the field consists of a plane wave and a spherical one, representing the scattered field. If we indicate with $\mathbf{k}_0$ the wave vector of the incident beam and with $\mathbf{k}'_0$ that relative to the scattered field observed along the direction $\hat{k}'$, we choose the plane parallel to $\mathbf{k}_0$ and $\mathbf{k}'_0$ as the common plane of reference for the two beams, referred to in the following as the *scattering plane*. Accordingly, the polarization of the incident beam and the relative Jones vectors (see Section I.3) are referred to axes $\hat{r}$ and $\hat{l}$ perpendicular and parallel, respectively, to the scattering plane and perpendicular to $\mathbf{k}_0$. The letters r and l were originally suggested by Chandrasekhar and stand for the last letters of the words perpendicular and parallel. The vectors $\hat{r}'$ and $\hat{l}'$ for the scattered wave are perpendicular to $\mathbf{k}_0$ and to the scattering plane.

Since the scattered field behaves asymptotically as a spherical wave, it is convenient to indicate explicitly the radial dependence factor $e^{-ik_0 r}/r$, where r represents the distance from the origin of coordinates located in proximity to the scatterer, by expressing it in the form

$$\begin{bmatrix} E_{dl} \\ E_{dr} \end{bmatrix} = \begin{bmatrix} S_2 & S_3 \\ S_4 & S_1 \end{bmatrix} \cdot \begin{bmatrix} E_{il} \\ E_{ir} \end{bmatrix} \frac{e^{-ik_0 r}}{r}, \qquad \text{(VI.11.1)}$$

where E_{il} and E_{ir} represent the amplitudes of the two components of the incident wave varying in space as $\exp(-ik_0 r)$. It is important to refer the ray vector $\mathbf{r}$ to the same origin O used for the spherical coordinate system adopted for the scattered field. Whenever possible, it is preferable to take the polar axis of these spherical coordinates coincident with the direction of $\mathbf{k}_0$. $\mathbf{S}$ is the scattering matrix of the obstacle, which depends on the way in which it is oriented with respect to $\mathbf{k}_0$ and $\mathbf{k}'_0$. It differs from the matrix used by van de Hulst by the omission of a factor ik_0.

For assigned $\mathbf{k}_0$ and $\mathbf{k}'_0$ there are in general three other positions in which the scattering matrix may be expressed in terms of the original matrix components. From the reciprocity theorem (see Problem I.8) it follows that the scattering process remains unchanged under the transformation $\mathbf{k}_0 \to -\mathbf{k}'_0$, $\mathbf{k}'_0 \to -\mathbf{k}_0$ [21]. This transformation corresponds to considering an incident beam traveling in the opposite direction to the scattered beam considered before. Under this transformation $\mathbf{S}$ becomes

$$\mathbf{S} = \begin{bmatrix} S_2 & -S_4 \\ -S_3 & S_1 \end{bmatrix}. \tag{VI.11.2}$$

In addition, it can be shown (see van de Hulst, Chapter I in [12]) that mirroring of the obstacle with respect to the scattering plane transforms $\mathbf{S}$ into

$$\mathbf{S} = \begin{bmatrix} S_2 & -S_3 \\ -S_4 & S_1 \end{bmatrix}. \tag{VI.11.3}$$

Combining these two transformations, we obtain a mirroring with respect to the plane normal to that of scattering and bisecting of the angle formed by $\mathbf{k}_0$ and $\mathbf{k}'_0$. In this case we have

$$\mathbf{S} = \begin{bmatrix} S_2 & S_4 \\ S_3 & S_1 \end{bmatrix}. \tag{VI.11.4}$$

In particular, for a spherical obstacle $\mathbf{S}$ must remain unchanged under the above transformations. Consequently, S_3 and S_4 must vanish.

11.1 *Extinction Cross Section and the Optical Theorem*

An important parameter measuring the global effects produced by an obstacle is the *extinction cross section* σ_{ext}, defined as the ratio between the total power lost by scattering (P_{sc}) and dissipation (P_{dis}) into the obstacle and the Poynting vector amplitude of the incident plane wave,

$$\sigma_{\text{ext}} = 2\zeta_0(P_{\text{sc}} + P_{\text{dis}})/E_{\text{i}}^2. \tag{VI.11.5}$$

The scattered power can be calculated by integrating the Poynting vector of the scattered wave over a sphere of very large radius,

$$P_{\text{sc}} = \frac{1}{2\zeta_0}\oiint_{4\pi} |E_{\text{d}}|^2 r^2 \, d\Omega = \frac{1}{2\zeta_0}\oiint_{4\pi} |\mathbf{S}\cdot\mathbf{E}_{\text{i}}|^2 \, d\Omega, \tag{VI.11.6}$$

where use has been made of Eq. (VI.11.1). The total power (incident, dissipated, and carried away from the scattered wave) crossing the above sphere must be equal to zero,

$$\begin{aligned}
\mathrm{Re}\oiint_{4\pi}[(\mathbf{E}_\mathrm{i} + \mathbf{E}_\mathrm{d}) \times (\mathbf{H}_\mathrm{i}^* + \mathbf{H}_\mathrm{d}^*)] \cdot \hat{r}\, d\Omega
&= \frac{2P_\mathrm{sc}}{r^2} + \oiint_{4\pi}(\mathbf{E}_i \times \mathbf{H}_\mathrm{i}^*) \cdot \hat{r}\, d\Omega \\
&\quad + \mathrm{Re}\oiint_{4\pi}(\mathbf{E}_i \times \mathbf{H}_\mathrm{d}^* + \mathbf{E}_\mathrm{d} \times \mathbf{H}_\mathrm{i}^*) \cdot \hat{r}\, d\Omega \\
&= -\frac{2P_\mathrm{dis}}{r^2},
\end{aligned} \tag{VI.11.7}$$

so that, since

$$\oiint_{4\pi}(\mathbf{E}_\mathrm{i} \times \mathbf{H}_\mathrm{i}^*) \cdot \hat{r}\, d\Omega = (\mathbf{E}_\mathrm{i} \times \mathbf{H}_\mathrm{i}^*) \cdot \oiint_{4\pi}\hat{\mathbf{r}}\, d\Omega = 0, \tag{VI.11.8}$$

we have

$$P_\mathrm{sc} + P_\mathrm{dis} = -2\,\mathrm{Re}\lim_{r\to\infty} r^2 \oiint_{4\pi}(\mathbf{E}_\mathrm{i} \times \mathbf{H}_\mathrm{d}^* + \mathbf{E}_\mathrm{d} \times \mathbf{H}_\mathrm{i}^*) \cdot \hat{r}\, d\Omega. \tag{VI.11.9}$$

On the other hand, since $\mathbf{H}_\mathrm{d} = -\zeta_0^{-1}\mathbf{E}_\mathrm{d} \times \hat{r}$ as $r \to \infty$, then

$$\begin{aligned}
\zeta_0 \lim_{r\to\infty}(\mathbf{E}_\mathrm{i} \times \mathbf{H}_\mathrm{d}^* + \mathbf{E}_\mathrm{d} \times \mathbf{H}_\mathrm{i}^*) \cdot \mathbf{r} &= -\lim_{r\to\infty}[\mathbf{E}_\mathrm{i} \times (\mathbf{E}_\mathrm{d}^* \times \hat{r}) + \mathbf{E}_\mathrm{d} \times (\mathbf{E}_\mathrm{i}^* \times \hat{k})] \cdot \mathbf{r} \\
&= \exp(-i\mathbf{k}_0 \cdot \mathbf{r} + ik_0 r)\mathbf{E}_\mathrm{i} \cdot \mathbf{S}^* \cdot \mathbf{E}_\mathrm{i}^* \\
&\quad + \exp(i\mathbf{k}_0 \cdot \mathbf{r} - ik_0 r)\hat{k} \cdot \hat{r}\mathbf{E}_\mathrm{i}^* \cdot \mathbf{S} \cdot \mathbf{E}_\mathrm{i} \\
&\quad - \exp(i\mathbf{k}_0 \cdot \mathbf{r} - ik_0 r)\mathbf{E}_\mathrm{i}^* \cdot \hat{r}\hat{k} \cdot \mathbf{S} \cdot \mathbf{E}_\mathrm{i}
\end{aligned} \tag{VI.11.10}$$

Finally, by applying the stationary-phase method we obtain

$$\begin{aligned}
\lim_{r\to\infty} r \oiint_{4\pi} \exp(i\mathbf{k}_0 \cdot \mathbf{r} - ik_0 r) f(\hat{k}, \hat{r})\, d\Omega &= -i\frac{2\pi}{k_0}[f(\hat{k}, \hat{k}) \\
&\quad - \exp(-2ik_0 r) f(\hat{k}, -\hat{k})],
\end{aligned} \tag{VI.11.11}$$

so that inserting the right-hand side of Eq. (VI.11.10) into Eq. (VI.11.9) and using Eq. (VI.11.11) yields

$$P_{sc} + P_{dis} = -(\lambda/2\zeta_0)\mathrm{Re}[i\mathbf{E}_i \cdot \mathbf{S}^*(\hat{k},\hat{k}) \cdot \mathbf{E}_i^* - ie^{2ik_0 r}\mathbf{E}_i \cdot \mathbf{S}^*(\hat{k},-\hat{k}) \cdot \mathbf{E}_i^*$$
$$-i\mathbf{E}_i^* \cdot \mathbf{S}(\hat{k},\hat{k}) \cdot \mathbf{E}_i - i\mathbf{E}_i^* \cdot \mathbf{S}(\hat{k},-\hat{k}) \cdot \mathbf{E}_i e^{-2ik_0 r}]$$
$$= -(\lambda/\zeta_0)\,\mathrm{Im}[\mathbf{E}_i^* \cdot \mathbf{S}(\hat{k},\hat{k}) \cdot \mathbf{E}_i], \qquad \text{(VI.11.12)}$$

and

$$\sigma_{ext} = -2\lambda\{\mathrm{Im}[\mathbf{E}_i^* \cdot \mathbf{S}(\hat{k},\hat{k}) \cdot \mathbf{E}_i]\}/|\mathbf{E}_i|^2. \qquad \text{(VI.11.13)}$$

According to the last expression the extinction cross section depends on the values taken by the scattering matrix in the forward direction. In particular, for spherical particles the two diagonal elements are equal for forward scattering so that Eq. (VI.11.13) simplifies into (*optical theorem*)

$$\sigma_{ext}^{(sphere)} = -2\lambda\,\mathrm{Im}\,S(\hat{k},\hat{k}). \qquad \text{(VI.11.14)}$$

12 Spherical Harmonics Representation of the Scattered Field

As a preliminary step let us consider a scalar function $f(r)$ that is solution of the Helmholtz wave equation in an assigned region, namely

$$\left(\frac{1}{r^2}\frac{\partial}{\partial r}r^2\frac{\partial}{\partial r} + \frac{\hat{D}^2}{r^2} + k^2\right)f(r,\theta,\phi) = 0, \qquad \text{(VI.12.1)}$$

where [cf. Eq. (II.12.40)]

$$\hat{D}^2 = \frac{1}{\sin\theta}\frac{\partial}{\partial\theta}\left(\sin\theta\frac{\partial}{\partial\theta}\right) + \frac{1}{\sin^2\theta}\frac{\partial^2}{\partial\phi^2}, \qquad \text{(VI.12.2)}$$

is known as *Beltrami's operator* for the sphere. In particular, $\hat{D}^2$ admits an infinite set of eigenfunctions Y_n^m, known as *spherical harmonics*,

$$\hat{D}^2 Y_n^m(\theta,\phi) = -n(n+1)Y_n^m, \qquad \text{(VI.12.3)}$$

with

$$Y_n^m(\theta,\phi) = \left[\frac{(2n+1)(n-m)!}{(n+m)!}\right]^{1/2} P_n^m(\cos\theta)e^{im\phi}, \qquad \text{(VI.12.4)}$$

$P_n^m(\cos\theta)$ being the *associated Legendre function of the first kind*, defined

$$P_n^m(\cos\theta) = \sin^m\theta(d/d\cos\theta)^m P_n(\cos\theta), \qquad 0 \le m \le n \qquad \text{(VI.12.5)}$$

with the *Legendre polynomial* $P_n(\cos\theta)$ defined by Rodrigues' formula

$$P_n(\cos\theta) = [(-1)^n/(2^n n!)](d/d\cos\theta)^n \sin^{2n}\theta. \qquad \text{(VI.12.6)}$$

The Legendre polynomials have a parity of n, i.e., $P_n(-x) = (-1)^n P_n(x)$, and satisfy the recurrence relations

$$nP_{n-1} = n \cos\theta P_n + \sin^2\theta P'_n, \tag{VI.12.7}$$

the prime denoting a derivative with respect to $\cos\theta$

Since the spherical harmonics form an orthogonal complete set on the surface of a sphere, viz.

$$\frac{1}{4\pi} \oiint_{4\pi} Y_n^{m*} Y_{n'}^{m'} \, d\Omega = \delta_{nn'}\delta_{mm'}, \tag{VI.12.8}$$

the scalar wave function $f(r, \theta, \phi)$ can be expanded in the series

$$f(r, \theta, \phi) = \sum_{n=0}^{\infty} \sum_{m=-n}^{n} R_{nm}(r) Y_n^m(\theta, \phi), \tag{VI.12.9}$$

where

$$R_{nm}(r) = \frac{1}{4\pi} \int_0^{2\pi} d\phi \int_0^{\pi} d\theta \sin\theta \, Y_n^{m*}(\theta, \phi) f(r, \theta, \phi). \tag{VI.12.10}$$

Example: Scalar Plane Wave. In particular, for a scalar wave $\exp(-ikr\cos\theta)$ propagating along the polar axis z, we have from Eq. (VI.12.10)

$$\begin{aligned} R_{nm}(r) &= (2n+1)^{1/2} P_n[-i(d/dx)](\sin x/x)\, \delta_{m0} \\ &= (2n+1)^{1/2}(-i)^n [\psi_n(kr)/(kr)] \delta_{m0}, \end{aligned} \tag{VI.12.11}$$

where $x = kr$ and $P_n(-id/dx)$ is a differential operator obtained by replacing the argument w of $P_n(w)$ with the differential operator $-id/dw$. The function $\psi_n(kr)$, defined by

$$\psi_n(x) = i^n x P_n[-i(d/dx)] \sin x/x, \tag{VI.12.12}$$

is the *Riccati–Bessel function*, which is regular at $x = 0$ and for $x \to \infty$ tends asymptotically to $\sin(x - n\pi/2)$. By inserting the right-hand side of Eq. (VI.12.11) into Eq. (VI.12.9), we obtain *Bauer's formula*.

$$e^{-ikr\cos\theta} = \frac{1}{kr} \sum_{n=0}^{\infty} (-i)^n (2n+1) \psi_n(kr) P_n(\cos\theta). \tag{VI.12.13}$$

12.1 *Radial Wave Equation*

Since each term of the series on the right side of Eq. (VI.12.9) must satisfy Eq. (VI.12.1), then in view of Eq. (VI.12.2) the function $f_{nm}(kr) = (2kr/\pi)^{1/2} R_{nm}(r)$ satisfies the *Bessel equation* of half-integer order $n + 1/2$,

$$d^2 f_{nm}/dx^2 + (1/x)(df_{nm}/dx) + [1 - (n + 1/2)^2/x^2] f_{nm} = 0. \tag{VI.12.14}$$

Accordingly, f_{nm} is independent of the index m, which will be dropped from now on. If we introduce the *spherical Hankel functions* h_n, defined by

$$h_n^{(1)}(x) = h_n^{(2)*}(x) = \left(\frac{\pi}{2x}\right)^{1/2} H_{n+1/2}^{(1)}(x) = \frac{e^{ix}}{x} \sum_{q=0}^{n} \frac{i^{q-n-1}(n+q)!}{q!(n-q)!} \frac{1}{(2x)^q} \tag{VI.12.15}$$

we can, in general, put

$$R_n(r) = A_n h_n^{(1)}(kr) + B_n h_n^{(2)}(kr) = [\pi/(2kr)]^{1/2} f_n(kr). \tag{VI.12.16}$$

Starting from the above functions we can form the *spherical Bessel function* $j_n(x) = \operatorname{Re} h_n^{(1)}$. The j_n is regular at $x = 0$ and is related to the Riccati–Bessel function ψ_n by the simple relation $\psi_n = x j_n$. Analogously, we can form the *Riccati–Hankel functions* $\zeta_n^{(1),(2)}$ by putting $\zeta_n^{(1)} = x h_n^{(1)}$.

In view of Eq. (VI.12.13), a plane wave can be expanded in a series of ingoing and outgoing waves according to the relation

$$\begin{aligned} e^{-ik\cos\theta} &= \frac{1}{2} \sum_{n=0}^{\infty} (-i)^n (2n+1) h_n^{(1)}(kr) P_n(\cos\theta) \\ &+ \frac{1}{2} \sum_{n=0}^{\infty} (-i)^n (2n+1) h_n^{(2)}(kr) P_n(\cos\theta). \end{aligned} \tag{VI.12.17}$$

12.2 *Vector Spherical Harmonics*

We can construct a set of *vector spherical harmonics* $\mathbf{Y}_n^m$ by applying the differential operator[22]

$$\hat{\mathbf{L}} = \mathbf{r} \times \nabla = \hat{\phi}(\partial/\partial\theta) - \hat{\theta}(1/\sin\theta)(\partial/\partial\phi), \tag{VI.12.18}$$

to the scalar harmonics Y_n^m. Before discussing the properties of the functions $\mathbf{Y}_n^m$, we note that the scalar components $\hat{L}_x$, $\hat{L}_y$, and $\hat{L}_z$ of $\hat{\mathbf{L}}$ can be written conveniently in the combinations

$$\hat{L}_+ = \hat{L}_-^* = \hat{L}_x + i\hat{L}_y = e^{i\phi}[i(\partial/\partial\theta) - \cot\theta(\partial/\partial\phi)], \qquad \hat{L}_z = \partial/\partial\phi. \tag{VI.12.19}$$

In view of the above expressions the combination $\hat{L}_+ \hat{L}_- + \hat{L}_z^2$ coincides with the operator $\hat{D}^2$ defined by Eq. (VI.12.2). In addition, the following commutation properties of $\hat{\mathbf{L}}$, $\hat{D}^2$, and ∇^2 can be easily proved:

$$\hat{D}^2\hat{\mathbf{L}} = \hat{\mathbf{L}}\hat{D}^2, \qquad \hat{\mathbf{L}} \times \hat{\mathbf{L}} = -\hat{\mathbf{L}}, \qquad \hat{L}_j\nabla^2 = \nabla^2\hat{L}_j, \tag{VI.12.20}$$

together with the relations

$$\hat{L}_z Y_n^m = im Y_n^m, \qquad \hat{L}_+ Y_n^m = i[(n-m)(n+m+1)]^{1/2} Y_n^{m+1}. \tag{VI.12.21}$$

Then the scalar spherical harmonics Y_n^m are the eigenfunctions of $\hat{L}_z$ and $\hat{D}^2$. Readers familiar with quantum mechanics will notice the coincidence of these operators with the angular momentum operators.

We can now construct the vector spherical harmonics by applying the operator $\hat{\mathbf{L}}$ to Y_n^m:

$$\mathbf{Y}_n^m \equiv \hat{\mathbf{L}} Y_n^m = i(m/\sin\theta) Y_n^m \hat{\theta} + (m \cot\theta Y_n^m + e^{i\phi} Y_n^{m+1})\hat{\phi}. \tag{VI.12.22}$$

Since it is evident from the definition of $\mathbf{L}$ that $\mathbf{r} \cdot \hat{\mathbf{L}} = 0$, then $\mathbf{r} \cdot \mathbf{Y}_n^m = 0$.

The spherical harmonics can be used to represent vector fields tangent to a sphere. In particular, we can construct a set of *electric multipole fields*

$$\mathbf{B}_{nm}^{(e)}(\mathbf{r}) = R_n(kr)\mathbf{Y}_n^m(\theta, \phi), \qquad \mathbf{E}_{nm}^{(e)}(\mathbf{r}) = -[(i/(\omega\mu_0\varepsilon)] \nabla \times \mathbf{B}_{nm}^{(e)}. \tag{VI.12.23}$$

Now we observe that the radial component of $\mathbf{E}_{nm}^{(e)}$, reads

$$\begin{aligned} \mathbf{r} \cdot \mathbf{E}_{nm}^{(e)} &= (1/k)\mathbf{r} \cdot [\nabla \times (R_m \mathbf{r} \times \nabla Y_n^m)] = -(R_n/k)\mathbf{r} \cdot [\nabla \times (\mathbf{r} \times \nabla Y_n^m)] \\ &= [n(n+1)/k] R_n Y_n^m, \end{aligned} \tag{VI.12.24}$$

where we have used the vector identities (A.8) and (A.10). Analogously, we can consider multipole fields with transverse electric fields

$$\mathbf{E}_{nm}^{(h)} = R_n(kr)\mathbf{Y}_n^m, \qquad \mathbf{B}_{nm}^{(h)} = (i/\omega) \nabla \times \mathbf{E}_{nm}^{(h)}. \tag{VI.12.25}$$

Finally, the more general electromagnetic field can be represented as a superposition of TE (h) and TM (e) spherical multipolar fields,

$$\mathbf{E}(\mathbf{r}) = \sum_{n=1}^{\infty} \sum_{m=-n}^{+n} [a_{nm}\mathbf{E}_{nm}^{(e)}(\mathbf{r}) + b_{nm}\mathbf{E}_{nm}^{(h)}(\mathbf{r})], \tag{VI.12.26}$$

where the coefficients a_{nm} can be calculated by multiplying the above equation scalarly by $\mathbf{r}$ and using Eq. (VI.12.24), together with the orthogonality of the $\mathbf{Y}_n^m$. Relying on the orthogonality of the multipolar fields and proceeding in a way similar to that followed for the expansion of a scalar plane wave, Bauer's formula for a vector plane wave is easily obtained in the form

$$\hat{x}e^{-ikz} = \frac{1}{2}\sum_{n=1}^{\infty} (-i)^n \left\{ j_n(\mathbf{Y}_n^1 + \mathbf{Y}_n^{-1}) + \frac{1}{k} \nabla \times [j_n(\mathbf{Y}_n^1 - \mathbf{Y}_n^{-1})] \right\}. \tag{VI.12.27}$$

13 Scattering from Spherical Particles

The scattering from spherical particles is the best understood example of scattering from finite bodies. The physical interest in this problem was originally motivated by the studies carried out by Lord Rayleigh to explain the blue color displayed by the sky and aerosols.

In the nineteenth century, as described by Kerker [23], this problem was the object of accurate experimental analysis by Brücke (1853) and M. G. Govi (1860), who observed the blue color in alcoholic suspensions of mastic and in smokes of tobacco and alcohol. In almost the same period, Tyndall (1869), superintendent of the Royal Institution, was led to investigate these phenomena by the interest manifested in them by John Herschel, the son of the great astronomer Wilhelm Herschel. For his experiments, Tyndall used aerosols prepared by condensation of the products of gaseous reactions. The experiments of Govi and Tyndall showed that (1) when the particles were small they scattered blue light, and (2) the light scattered at right angles to the incident was completely linearly polarized.

In 1871 Lord Rayleigh published his first paper on scattering of light by small particles in the earth's atmosphere, in which he conceived of a sphere as the simplest model for such scatterers [24]. With his research Rayleigh was able to show by approximate considerations that the total power of the light scattered by particles that are small compared to the wavelength is proportional to $1/\lambda^4$.

In his initial investigations Rayleigh considered scatterers whose refractive indices differ little from those of the embedding medium. This assumption, currently referred to as the Rayleigh–Gans approximation (see Section VI.4), was instrumental in replacing the scatterer internal fields by the corresponding incident values.

When the dimension of the scatterer is comparable to the wavelength, the particle can be treated as a series of multipoles (dipole, quadrupole, octupole, etc.) and the scattered field can be expanded into a power series in k_0 (*Rayleigh series*), viz. $\mathbf{E}_{\rm d} = \Sigma_n k_0^n \mathbf{E}^{(n)}$. Each term $\mathbf{E}^{(n)}$ of the series can be derived from the term $\mathbf{E}^{(n-1)}$ by expressing the scattered field as a surface integral extended to the scatterer boundary. For an accurate description of this method the reader should consult Noble [25] and Kleinman [26].

The scattering of an electromagnetic wave by a sphere was solved in 1881 by H. Lamb by means of an analysis closely related to methods envisaged by A. Clebsch in 1861 for using separation of variables to solve the class of boundary-value problems in which a wave propagating in an elastic medium impinges on a spherical surface.

In 1908 Gustav Mie, in an attempt to explain the brillant colors displayed by colloidal metal suspensions, discussed with great accuracy the scattering from a sphere having a complex refractive medium and embedded in a lossy medium, by making calculations that involved summing the first several partial waves in which the field was expanded [27].

One year after Mie's paper, P. Debye published his investigations [28] on the light pressure on a conducting sphere. Later contributions came from T. J. I'A. Bromwich (1899) and his collaborators at Cambridge University.

In particular, Bromwich's student White proposed to decompose the Watson solution into a contour integral, containing the reflected wave and a residue series, an approach that was illustrated in Section VI.5 for the cylinder case. Other important contributions were made by the Dutch physicists B. van der Pol and H. Bremmer, by V. A. Fock in Soviet Union, who obtained an integral representation of the field in the penumbra region, and by H. C. van de Hulst (see Chapter I, van de Hulst [12]).

13.1 *Partial Wave Expansion of the Scattered Field*

Let us consider a dielectric sphere of radius a and refractive index $\tilde{n}$, illuminated by a plane wave traveling along the positive direction of the z axis and linearly polarized parallel to the x axis. According to Bauer's formula [Eq. (VI.12.27)], the incident plane wave can be decomposed in the region outside the sphere into an infinite superposition of transverse electric ($\mathbf{E}^{(\mathrm{h})}_{n,\pm 1}$) and transverse magnetic ($\mathbf{E}^{(\mathrm{e})}_{n,\pm 1}$) modes. In addition, each angular mode can be split into a wave traveling toward infinity and another wave approaching the center of the sphere, the last decomposition arising from the splitting of the radial function $2j_n = h^{(1)}_n + h^{(2)}_n$. Each partial wave contributing to the incident field gives rise to reflected and transmitted waves having the same angular dependence. That is, an incident mode (say $\mathbf{E}^{(\mathrm{h})}_{n,1}$) produces a reflected wave $r^h_n \mathbf{Y}^1_n h^{(2)}_n(k_0 r)$ and a transmitted one $t^n_{n,1} \mathbf{Y}^1_n \; j_n(\tilde{n} k_0 r)$.

The reflection and transmission coefficients can be determined by imposing the continuity of the electric and magnetic components on the sphere's surface. In particular, the continuity of $\mathbf{E}^{(\mathrm{h})}_{q,n}$ entails

$$j_q(\beta) + r^{\mathrm{h}}_q h^{(2)}_q(\beta) = t^{\mathrm{h}}_q j_q(\alpha), \qquad \text{(VI.13.1)}$$

where $\alpha = \tilde{n}\beta$. Analogously, imposing the continuity of the tangent component of $\mathbf{B}^{(\mathrm{h})}_{qm} = i\omega^{-1} \nabla \times \mathbf{E}^{(\mathrm{h})}_{qm}$ gives

$$j'_q(\beta) + r^{\mathrm{h}}_q h^{(2)\prime}_q(\beta) = \tilde{n} t^{\mathrm{h}}_q j'_q(\alpha). \qquad \text{(VI.13.2)}$$

Then, solving with respect to the reflection coefficient r^{h}_q (also called the *Mie coefficient*) yields [cf. Eq. (VI.4.13)]

$$r^{\mathrm{h}}_q = -\frac{\ln'\psi_q(\beta) - \tilde{n}\ln'\psi_q(\alpha)}{\ln'\zeta^{(2)}_q(\beta) - \tilde{n}\ln'\psi_q(\alpha)} \frac{\psi_q(\beta)}{\zeta^{(2)}_q(\beta)} \equiv -\frac{1}{2}(1 - S^{(\mathrm{h})}_q(\beta)), \qquad \text{(VI.13.3)}$$

where ψ_q and $\zeta^{(2)}_q$ are the Riccati–Bessel and Riccati–Hankel functions, respectively. The functions $S^{\mathrm{h}}_q(\beta)$ are given by

$$S^{\mathrm{h}}_q = -\frac{\ln'\zeta^{(1)}_q(\beta) - \tilde{n}\ln'\psi_q(\alpha)}{\ln'\zeta^{(2)}_q(\beta) - \tilde{n}\ln'\psi_q(\alpha)} \frac{\zeta^{(1)}_q(\beta)}{\zeta^{(2)}_q(\beta)} \equiv \exp(-i2\alpha^{\mathrm{h}}_q), \qquad \text{(VI.13.4)}$$

where α_q^{h} is the *phase shift of the qth partial wave*, which can be expressed by

$$\tan \alpha_q^{\mathrm{h}} = \frac{\ln' j_q(\beta) - \tilde{n} \ln' j_q(\alpha)}{\ln' n_q(\beta) - \tilde{n} \ln' j_q(\alpha)} \frac{j_q(\beta)}{n_q(\beta)}, \tag{VI.13.5}$$

$n_q(\beta)$ being the *spherical Neumann function* ($h_q^{(2)} = j_q - in_q$). In particular, for n real S_q^{h} has unit modulus and the angle α_q^{h} is real.

The reflection coefficient r_q^{e} for the electric case can be easily obtained from the above formulas by replacing n with its reciprocal $1/n$.

In particular, for a metallic sphere the reflection coefficients reduce to [cf. Eq. (VI.4.17)]

$$r_q^{\mathrm{h}} = -j_q(\beta)/h_q^{(2)}(\beta) \qquad r_q^{\mathrm{e}} = -j_q'(\beta)/h_q^{(2)\prime}(\beta). \tag{VI.13.6}$$

13.2 *Scattering Amplitudes S and Efficiency Factors Q*

The field scattered by a sphere illuminated by a linearly polarized plane wave is represented by a series of partial waves that, for $r \to \infty$, reduce to

$$\mathbf{E}_{\mathrm{d}}(\mathbf{r}) \underset{r\to\infty}{\to} \frac{e^{-ik_0 r}}{2k_0 r} \sum_{q=1}^{\infty} (-i)^q \left[r_q^{\mathrm{h}}(\mathbf{Y}_q^1 + \mathbf{Y}_q^{-1}) + \frac{r_q^{\mathrm{e}}}{k_0} \nabla \times (\mathbf{Y}_q^1 - \mathbf{Y}_q^{-1}) \right]$$

$$\equiv \frac{e^{-ik_0 r}}{r} [\cos\phi S_2(\theta)\hat{\theta} + \sin\phi S_1(\theta)\hat{\phi}], \tag{VI.13.7}$$

where S_1 and S_2 are the *scattering amplitudes* for magnetic and electric polarization, respectively,

$$S_1 = \frac{i}{k_0} \sum_{q=1}^{\infty} \frac{2q+1}{q(q+1)} [r_q^{\mathrm{e}} \pi_q(\cos\theta) + r_q^{\mathrm{h}} \tau_q(\cos\theta)], \tag{VI.13.8a}$$

$$S_2 = \frac{i}{k_0} \sum_{q=1}^{\infty} \frac{2q+1}{q(q+1)} [r_q^{\mathrm{h}} \pi_q(\cos\theta) + r_q^{\mathrm{e}} \tau_q(\cos\theta)], \tag{VI.13.8b}$$

π_q and τ_q being functions expressible in terms of the first and second derivatives of the ordinary Legendre functions,

$$\pi_q(\cos\theta) = P_q^1(\cos\theta)/\sin\theta, \qquad \tau_q(\cos\theta) = (d/d\theta)P_q^1(\cos\theta). \tag{VI.13.9}$$

If we consider that the unit vectors $\hat{\theta}$ and $\hat{\phi}$ coincide with the vectors $\hat{l}'$ and $\hat{r}'$ relative to the scattering plane defined in Section VI.11, and we compare the expression for the far field given by Eq. (VI.13.7) with that of Eq. (VI.11.1), we conclude that S_1 and S_2 coincide with the elements of the scattering matrix **S**. The corresponding *intensities* $i_j = |S_j|^2$ $(j = 1, 2)$ together with the *phase difference* $\delta = \arg S_1 - \arg S_2$ completely characterize the scattering.

Once the components of the scattering matrix are determined, we can derive the extinction cross section σ_{ext} from Eq. (VI.11.14), as implied by the optical theorem,

$$\frac{\sigma_{ext}}{\pi a^2} \equiv Q_{exp}(\tilde{n}, \beta) = -\frac{4}{k_0 a^2} \operatorname{Im} S_1(0) = -\frac{2}{\beta^2} \sum_{q=1}^{\infty} (2q+1) \operatorname{Re}(r_q^{e} + r_q^{h}), \tag{VI.13.10}$$

where use has been made of the relations $\pi_n(1) = \tau_n(1) = (1/2)n(n+1)$. We have indicated with $Q_{ext}(\tilde{n}, \beta)$ the *efficiency factor*, which measures the extinction cross section with respect to the geometric cross section of the sphere. Mie has also obtained a simple expression for the total scattered power by integrating the Poynting vector of the scattered field over the sphere of infinite radius, i.e.,

$$\frac{\sigma_{sc}}{\pi a^2} = Q_{sc}(\tilde{n}, \beta) = \frac{1}{\beta^2} \int_0^{\pi} [i_1(\theta) + i_2(\theta)] \sin\theta \, d\theta = \frac{2}{\beta^2} \sum_{q=1}^{\infty} (2q+1)(2|r_q^{h}|^2 + |r_q^{e}|^2), \tag{VI.13.11}$$

where $\sigma_{sc} = 2\zeta P_{sc}/E_i^2$. The difference between these two cross sections gives a measure of the power dissipated into the sphere. In particular, for a lossless sphere $\sigma_{sc} = \sigma_{ext}$, so that the two series of Eqs. (VI.13.10,11) must be coincident. In fact, when the refractive index n is real, the angles α_q^{h} and α_q^{e} are real and $-\operatorname{Re}(r_q^{h} + r_q^{e}) = |r_q^{h}|^2 + |r_q^{e}|^2 = \sin^2 \alpha_q^{h} + \sin^2 \alpha_q^{e}$. A comprehensive tabulation of scattering and extinction cross section can be found in Wickramasinghe [29].

Another quantity that is useful for characterizing the scattering from a sphere is the total momentum $\mathbf{p}_{sc}$ carried by the scattered field per unit incident wave,

$$\mathbf{p}_{sc} = \hbar \int_0^{\pi} [i_1(\theta) + i_2(\theta)] \mathbf{k}_0'(\theta) \sin\theta \, d\theta \equiv \mathbf{p}_i \langle \cos\theta \rangle_\theta \beta^2 Q_{sc}, \tag{VI.13.12}$$

where $\mathbf{p}_i = \hbar k_0 \hat{z}$. $\langle \cos\theta \rangle_\theta$ is the so-called *asymmetry factor*, which characterizes the relative importance of forward and backward scattering. In particular, Debye derived the important relation

$$\beta^2 \langle \cos\theta \rangle_\theta Q_{sc} = 4 \sum_{q=1}^{\infty} \frac{q(q+2)}{q+1} \operatorname{Re}(r_q^{h} r_{q+1}^{h*} + r_q^{e} r_{q+1}^{e*}) + 4 \sum_{q=1}^{\infty} \frac{2q+1}{q(q+1)} \operatorname{Re}(r_q^{h} r_q^{h*}). \tag{VI.13.13}$$

Since the momentum removed from the original beam is proportional to σ_{ext} while the part proportional to σ_{sc} is partially replaced by the forward component of momentum of the scattered light, which, as shown above, is

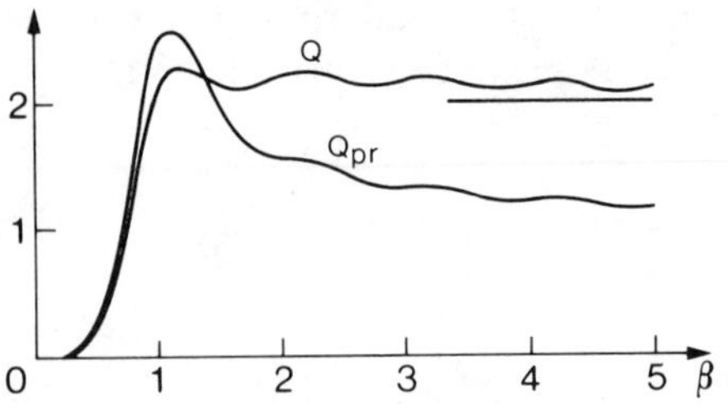

Fig. VI.20. Extinction (Q) and pressure (Q_{pr}) efficiencies versus size parameter β for a metallic sphere.

proportional to $\langle\cos\theta\rangle_\theta\sigma_{sc}$, then the part of the forward momentum that is removed from the incident beam and not replaced by the forward momentum of the scattered light is proportional to

$$\sigma_{pr} = \sigma_{ext} - \langle\cos\theta\rangle_\theta\sigma_{sc} = \sigma_{abs} + (1 - \langle\cos\theta\rangle_\theta)\sigma_{sc} \equiv \pi a^2 Q_{pr}, \qquad \text{(VI.13.14)}$$

Q_{pr} being the radiation pressure efficiency (see Fig. VI.20).

As a consequence of the transfer of momentum to the scattering particle, a force F is exerted on the particle in the direction of propagation of the incident wave,

$$F = (S/c)\sigma_{pr}, \qquad \text{(VI.13.15)}$$

S being the Poynting vector of the wave. This phenomenon is well known in astrophysics, where it gives rise to the *radiation pressure*, discovered and measured in 1899 by P. N. Lebedev in Moscow [30]. This is the most important external force acting on atoms and interstellar dust [31].

When the frequency of the incident light coincides with an absorption line of the scatterer, σ_{abs} can become very large, thus enhancing the magnitude of the radiation pressure. In recent years it has been exploited for freezing the motion of a gas of particles by achieving equivalent temperatures of the order of a few degrees Kelvin (see Minogin and Letokhov [32] and Chapter I, Shen [5]).

The asymmetry factor and the efficiency Q_{pr} have been investigated in detail by Irvine [33]. His accurate numerical calculations show a regular series of sharp optical resonances. Ashkin and Driedzic have accurately measured the variation of radiation pressure with wavelength [34]. In particular, by using a dye laser beam illuminating almost uniformly, they have observed a sequence of narrow resonance peaks for drops made from highly transparent silicon oil which confirm the numerical results of Irvine. They have also shown that on partially illuminating the drops with a focused laser beam, which completely misses the edges of the sphere, the sharp resonances disappear.

13.3 *Stokes Parameters of the Scattered Field*

The scattering matrix defined by Eq. (VI.11.1) is equivalent to the Jones matrix **A** introduced in the first chapter [see Eq. (I.3.15)]. Consequently, we can rely on the formalism illustrated in Section I.3 for calculating the Stokes parameters of the scattered field. In particular, for a spherical scatterer the Jones matrix has only the components S_1 and S_2 different from zero, so that the matrix **F** defined by Eq. (I.3.18) takes the very simple form

$$\mathbf{F}_{\text{sphere}} = \begin{bmatrix} i_2 & 0 & 0 & 0 \\ 0 & i_1 & 0 & 0 \\ 0 & 0 & S_{21} & -D_{21} \\ 0 & 0 & D_{21} & S_{21} \end{bmatrix}, \qquad \text{(VI.13.16)}$$

in which $i_j = |S_j|^2$ are the intensities, $S_{21} = \text{Re}(S_1 S_2^*)$, and $D_{21} = \text{Im}(S_1 S_2^*)$.

Consequently, the degree of polarization $m(\theta)$ of the light scattered along the direction θ is given by [see Eq. (I.3.14)]

$$m(\theta) = \frac{\begin{array}{c}[(i_1 + i_2)^2(s_0^2 + s_1^2) + 2(i_2^2 - i_1^2)s_0 s_1 + 2i_1 i_2(s_1^2 - s_0^2) \\ + 4(S_{21}^2 + D_{21}^2)(s_2^2 + s_3^2)]^{1/2}\end{array}}{(i_1 + i_2)s_0 + (i_2 - i_1)s_1}, \qquad \text{(VI.13.17)}$$

where the s_i are the Strokes parameters of the incident light.

13.4 *Small Spheres*

For $\alpha, \beta < 0.8$ only the first two electric modes and the first magnetic mode are excited and the scattering amplitude S_1 is given to a good approximation by

$$S_1(\theta) \cong \frac{1}{k_0}\frac{n^2 - 1}{n^2 + 2}\beta^3\left[1 + \left(\frac{3}{5}\frac{n^2 - 2}{n^2 - 2} + \frac{n^2 + 2}{30}\cos\theta + \frac{1}{6}\frac{n^2 + 2}{2n^2 + 3}\cos\theta\right)\beta^2 - i\frac{2}{3}\frac{n^2 - 1}{n^2 + 2}\beta^3\right], \qquad \text{(VI.13.18)}$$

an analogous expression holding true for S_2.

Next, by entering the above expressions for S_1 and S_2 into Eq. (VI.13.11), the scattering efficiency

$$Q_{\text{sc}} = \frac{8}{3}\beta^4\left(\frac{n^2 - 1}{n^2 + 2}\right)^2\left(1 + \frac{6}{5}\frac{n^2 - 1}{n^2 + 2}\beta^2\right) \qquad \text{(VI.13.19)}$$

can be obtained.

In particular, for $\beta \ll 1$, the scattering pattern reduces to that obtained by Rayleigh in his celebrated work of 1871:

$$i_1(\theta) = \frac{i_2(\theta)}{\cos^2\theta} = \frac{\beta^6}{k_0^2}\left(\frac{n^2-1}{n^2+2}\right)^2. \qquad \text{(VI.13.20)}$$

Then, entering the above intensities into Eq. (VI.13.17) gives

$$m(\theta) = \sin^2\theta/(1+\cos^2\theta). \qquad \text{(VI.13.21)}$$

While for small spheres the light scattered through a right angle is linearly polarized, when the terms of order higher than β^3 cannot be neglected in the expressions for S_1 and S_2, $|S_2|$ goes through a minimum that is almost zero for an angle $\theta_{\min}$ given by (see Chapter I, van de Hulst [12], p. 146)

$$\cos\theta_{\min} = -\beta^2[(n^2-1)(n^2+2)]/[15(2n^2+3)]. \qquad \text{(VI.13.22)}$$

If n is a known quantity, the position of the angle $\theta_{\min}$ for which the light is linearly polarized can be used to measure β, that is, the size of the particles.

Another parameter that marks the deviation from the Rayleigh scattering described by Eq. (VI.13.20) is the ratio between the forward and backward scattered intensities,

$$|S_1(0)|^2/|S_1(\pi)|^2 = 1 + (4/15)\beta^2[(n^2+4)(n^2+2)]/(2n^2+3). \qquad \text{(VI.13.23)}$$

For transparent particles the forward scattering is always stronger than the backscattering as long as $n\beta < 0.8$.

13.5 *Oscillations of a Dielectric Sphere*

The poles of the scattering matrix represent the free modes of oscillation of a dielectric sphere. They can be easily found for real n by looking for those values of $\beta = k_0 a = (a/c)(\omega' + i\omega'')$ for which the denominator of Eq. (VI.13.3) vanishes, i.e.,

$$\begin{aligned} \zeta_q^{(2)\prime}(\beta)\psi_q(\alpha) - n\psi_q'(\alpha)\zeta_q^{(2)}(\beta) &= 0, \\ n\zeta_q^{(2)\prime}(\beta)\psi_q(\alpha) - \psi_q'(\alpha)\zeta_q^{(2)} &= 0, \end{aligned} \qquad \text{(VI.13.24)}$$

for h- and e-mode vibrations, respectively. In the limiting case of $n \to \infty$, the above equations reduce to $h_q^{(2)}(\beta) = 0$ and $h_q^{(2)\prime}(\beta) = 0$. For $q = 1$, this gives (see Chapter I, van de Hulst [12], Sect. 10.51) $\beta = \pm 0.86 + i0.5$ (e modes) and $\beta = i$ (h mode). Consequently, the scattering by spheres is hardly influenced at all by the existence of these modes and those with larger q, which do not give rise to typical resonance phenomena. In contrast, an important class of resonances corresponds to waves that are kept inside the sphere because of the total reflection at its boundary. The damping of these modes is very small for β

and n sufficiently large, so that they represent the *free modes of vibration of a dielectric sphere*. These were originally found by Debye in 1909 while looking for parameters α close to $q + 1/2$.

By approximating $\zeta_q^{(2)\prime}(\beta)$ with $-(q/\beta)\zeta_q^{(2)}(\beta)$, it can be shown that the imaginary parts ω'' of these resonances are given by (see Chapter VII, Vaynshteyn [2])

$$(\omega'/\omega'')_{q,m} = [q + (1/2)](n^2 - 1)^{1/2} \exp(2T_{q,m})\begin{cases} 1/n & \text{(magnetic modes),} \\ n & \text{(electric modes),} \end{cases}$$

where (VI.13.25)

$$T_{q,m} = \left(q + \frac{1}{2}\right)\left(\operatorname{arccosh} n - \sqrt{1 - \frac{1}{n^2}}\right) + \left[\frac{q + (1/2)}{2}\right]^{1/3} \sqrt{1 - \frac{1}{n^2}}\, x_m + \frac{1}{n^2}, \tag{VI.13.26}$$

$x_m(<0)$ being the mth root of the equation

$$j_q\{q + 1/2 - [(q + 1/2)/2]^{1/3} x_m\} = 0.$$

Then the quality factor $Q_{qm}(=\omega'/(2\omega''))$ (cf. Section VII.3) increases with q and decreases with growing m.

The resonance frequencies ω'_{qm} are given by

$$\omega'_{qm} = \left(\frac{c}{an}\right)\{q + \tfrac{1}{2} - [(q + \tfrac{1}{2})\tfrac{1}{2}]^{1/3} x_m\}. \tag{VI.13.26}$$

In the next chapter we will encounter optical devices that oscillate at some discrete frequencies given by expressions [see, e.g., Eq. (VII.11.5)] similar to the one above. More precisely, ω'_{res} is inversely dependent on the size of the oscillator and proportional to an integer index (q in the present case, the longitudinal mode number for the Fabry–Perot resonator) plus an additional quantity (x_m in the present case, the phase of the transverse gaussian modes for the Fabry–Perot resonator).

13.6 *Van de Hulst Theory for Very Large Spheres*

In the limit of $\beta \to \infty$ we can analyze the scattering of a plane wave by a dielectric sphere in the framework of geometric optics. The link between this asymptotic approach and the Mie series is provided by the *localization principle*, which states that a partial wave of order q corresponds for $\beta \to \infty$ to a family of rays passing the origin of the sphere at a distance $(q + 1/2)/k_0$. Accordingly, the modes of index q hit the sphere with an incidence angle $\theta_{1,q} = \arcsin[(q + 1/2)/\beta]$, while the remaining ones pass along without being either reflected or refracted. The last modes form a complete plane wave front missing a central disk of radius equal to that of the sphere. This incomplete plane wave front gives rise in the far region to a field composed of a plane wave from which must be subtracted the Fraunhofer diffraction pattern of a circular aperture. The rays hitting the sphere may emerge after some internal reflections and thus, together with the light directly reflected from the outer surface and the Fraunhofer pattern, contribute to the total scattering (the energy that does not emerge is lost by absorption inside the sphere). In conclusion, the scattered field can be decomposed into patterns due to (1) reflection and refraction from the sphere, and (2) diffraction of the wave front passing outside the scatterer.

The splitting of the scattered field described above has a counterpart in the form of the Mie coefficients. In fact, each of them consists of two terms: one equal to $-1/2$, independent of the nature of the scatterer, and another, $(1/2)e^{-i2\alpha_q}$, dependent on it through the phase α_q. The former term gives rise to the Fraunhofer pattern, while the latter one is associated with the scattering by reflection and refraction.

The different contributions to the scattered intensity optically interfere, and this gives rise to rapidly oscillating intensities with respect to the scattering direction.

In the following we will give only an outline of the theory, the details being obtainable from the book by van de Hulst. In particular, we will concentrate on the sum containing the phase shifts, assuming that the remaining terms proportional to the constant coefficient 1/2 give the diffraction pattern of a disk.

In view of the localization principle, we can associate the qth partial wave with a ray incident at angle $\theta_{1,q}$. By using the asymptotic expressions for the cylindrical functions, first Debye and later van de Hulst were able to obtain the following limiting expressions:

$$\exp(-i2\alpha_q^h) \sim \exp[i2(\beta f_q - \alpha f'_q)]\frac{1 - ir_p\exp(i2\alpha f'_q)}{1 + ir_p\exp(-i2\alpha f'_q)}$$

$$= -i\exp(i2\beta f_q)\sum_{p=0}^{\infty} \varepsilon_p^h[i\exp(-i2\alpha f'_q)]^p, \qquad \text{(VI.13.27)}$$

with the coefficients

$$\varepsilon_p^{\mathrm{h}} = \begin{cases} r_p & p = 0 \\ (1 - (r^{\mathrm{h}})^2)(-r_p)^{p-1} & p = 1, 2, \ldots \end{cases} \qquad \text{(VI.13.28)}$$

where $r_p^h(\theta_1)$ is the Fresnel coefficient of an h-wave incident at angle $\theta_{1,q}$ while

$$f_q = \cos\theta_{1,q} + (\theta_{1,q} - \pi/2)\sin\theta_{1,q}, \qquad f'_q = \cos\theta_{2,q} + (\theta_{2,q} - \pi/2)\sin\theta_{2,q}, \qquad \text{(VI.13.29)}$$

$\theta_{2,q}$ being the refraction angle. An analogous expression can be obtained for

$\exp(-i2\alpha_q^e)$ by replacing r^h with r^e. Next, using the asymptotic expressions for π_q and τ_q for large q and finite θ,

$$\begin{aligned}\pi_q &\sim [1/(q+1)][2/(\pi q \sin^3\theta)]^{1/2}\sin[(q+1/2)\theta - \pi/4],\\ \tau_q &\sim [(1/(q+1)][(2q)/(\pi\sin\theta)]^{1/2}\cos[(q+1/2)\theta - \pi/4],\end{aligned} \tag{VI.13.30}$$

we obtain for $S_1(\theta)$ (see [22], section 3.5.2)

$$\begin{aligned}S_1(\theta) &= \sum_{p=0}^{\infty}\sum_{q=1}^{\beta}\sum_{t=-1,1}\varepsilon_p^h(\sin\theta_{1q})^{1/2}e^{i\xi_q},\\ \xi_q &= 2\beta(npf'_q - f_q) - \frac{\pi}{2}p - \frac{\pi}{4}t + t\beta\theta\sin\theta_{1q} + 2\pi m,\end{aligned} \tag{VI.13.31}$$

where m is an arbitrary integer. At this point the summation over q is replaced by an integral, since it contains many relatively slowly varying terms, and hence

$$S_1(\theta) \sim a\left(\frac{a}{\lambda\sin\theta}\right)^{1/2}\sum_{\substack{p=0\\ t=-1,1}}^{\infty}\int_0^1 d\sin\theta_1(\sin\theta_1)^{1/2}\varepsilon_p^h e^{i\xi}. \tag{VI.13.32}$$

The integral is now evaluated by the stationary-phase method. This means that for large β most of the contributions come from that value $\bar{\theta}_1$ of θ_1 where $d\xi/d\theta_1 = 0$,

$$\Theta + t\theta - 2\pi m = 0, \tag{VI.13.33}$$

with

$$\Theta \equiv 2p(\bar{\theta}_2 - \tfrac{1}{2}\pi) + \pi - 2\bar{\theta}_1. \tag{VI.13.34}$$

In the vicinity of $\bar{\theta}_1$,

$$\xi = \bar{\xi} + b(q - \bar{q})^2 + \cdots, \tag{VI.13.35}$$

with $\bar{\theta}_1$ and $\bar{q}$ connected by $\sin\bar{\theta}_1 = (\bar{q} + 1/2)\beta^{-1}$ and

$$b = (1/2)(d^2\xi/dq^2)|_{q=\bar{q}} = [1/(2\beta\cos\bar{\theta}_1)](d\Theta/d\bar{\theta}_1). \tag{VI.13.36}$$

We then have

$$S_1(\theta) \sim a\sum_{p,t}\varepsilon_p^h\left(\frac{\sin 2\bar{\theta}_1}{2\sin\theta|d\Theta/d\bar{\theta}_1|}\right)^{1/2}\exp\left[i\pi\left(\frac{s}{4} - \frac{p}{2} - \frac{t}{4} - m\right) + i\delta\right], \tag{VI.13.37}$$

where $s = \operatorname{sgn} d\Theta/d\bar{\theta}_1$ and $\delta \equiv 2\beta(pn\cos\bar{\theta}_2 - \cos\bar{\theta}_1)$. Finally, replacing ε_p^h with ε_p^e in Eq. (VI.13.37), we obtain $S_2(\theta)$.

The above treatment can be reformulated in a more rigorous way by applying the Watson–Regge method to the Mie series representing the scattering amplitude. The main difficulty encountered in pursuing this program is due to the vector character of the scattering process. By representing the electromagnetic process by a scalar one, the approach described in Section VI.5 for the circular cylinder can also be followed step by step for the sphere, as shown by Nussenzveig (Chapter IV [22]).

13.7 *Ray Optical Theory of the Glory Effect and the Rainbow*

For spheres significantly larger than the wavelength, the scattered field can be calculated by following either the ray optical approach discussed in Chapter II or the analysis of the above subsection. In particular, it can be easily shown by combining Eqs.(VI.13.33,34) that the scattering angle θ, measured with respect to the direction $\hat{z}$ of the incident wave, is related to the incidence angle θ_1 by the simple relation (see Fig. VI.7b)

$$\theta = (p - 1)\pi + 2\theta_1 - 2p\theta_2, \tag{VI.13.38}$$

$p - 1$ being the number of internal reflections and θ_2 the refraction angle.

The ray optical field $f(\theta)$ [see Eq. (II.12.42)] reflected by a sphere of radius a illuminated by a plane wave is given by $f(\theta) \propto (\rho'_1\rho'_2)^{1/2}$, where $\rho'_{1,2}$ are the curvature radii of the reflected wave front on the sphere surface. Then, taking into account Eq. (II.11.26), we obtain $f(\theta) \propto a/2$; that is, a sphere with radius large with respect to the wavelength scatters light by reflection isotropically.

In the more general case of rays entering the dielectric sphere, where they undergo $p - 1$ internal reflections, it follows from Eq. (VI.13.37) that the far-field intensity $i(\theta)$ is given by

$$i(\theta) \propto \frac{\sin 2\theta_1}{2\sin\theta(d\theta/d\theta_1)} = \frac{\sin 2\theta_1}{4\sin\theta[1 - (p/n)(\cos\theta_1/\cos\theta_2)]} \equiv D(\theta_1), \tag{VI.13.39}$$

where D is the so-called *divergence factor* [cf. Eq. (II.13.52)].

According to the above equation, the scattered intensity becomes infinite when either $\sin\theta = 0$ or $d\theta/d\theta_1 = 0$. The case $\sin\theta = 0$ with $\sin 2\theta_1 \neq 0$ gives rise to infinite intensities for direct backscattering (the *glory effect*) or direct forward scattering, while for $d\theta/d\theta_1 = 0$ we have the *rainbow effect*.

The glory effect occurs when the scattering angle is equal to a multiple of π for incidence angle $\theta_1 \neq 0$, $\pi/2$. The far field produced by glory rays backscattered by spherical particles is similar to that produced by a ring-shaped wave front. Therefore, there is a central spot of maximum brightness surrounded by rings of decreasing intensity.

The glory effect is well known in meteorology. It consists of concentric rings of color, with red the outermost and violet the innermost, encircling a bright

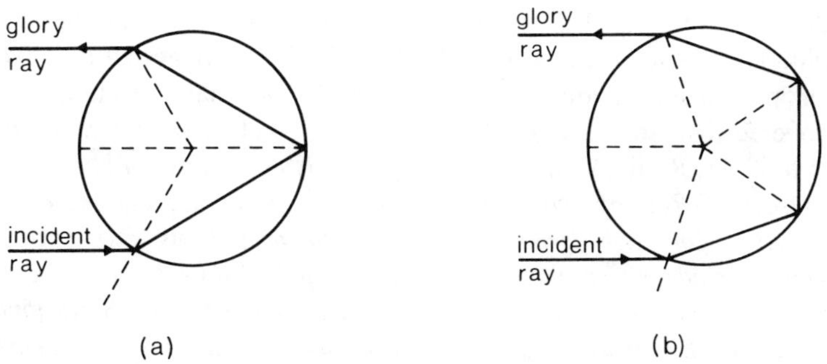

Fig. VI.21. Glory rays for (a) one and (b) two reflections.

central region in the direction opposite to that of the sun. It must not be confused with the diffraction coronas that surround the sun when covered by a thin veil of cloud. The glory effect appears when an observer stands on a high point or on a plane, looking at his own shadow, or that of the plane, projected on nearby thin clouds or mist. Under favorable conditions, the observer sees the shadow of his head, or that of the plane, surrounded by a bright halo.

From the point of view of the history of physics, it is noteworthy that in 1895 C. T. R. Wilson built the first cloud chamber in an attempt to study the glory effect experimentally.

The incidence angle of a glory ray (see Fig. VI.21) can be easily obtained by combining Eq. (VI.13.38) with the condition $\sin\theta = 0$. Then, taking into account that θ and θ_1 are both less than $\pi/2$, we obtain for a glory ray undergoing $p - 1$ reflections:

$$\theta_1 = p\theta_2 + (2 - p)(\pi/2). \qquad \text{(VI.13.40)}$$

Utilizing Snell's law for $p = 2$ (single reflection), the above relation gives $\cos\theta_2 = n/2$. Since θ_2 falls between 0 and $\pi/4$, n must be comprised between 2 and $2^{1/2}$. Analogous constraints can be derived for $p > 2$. Here we limit ourselves to observing that, according to geometric optics, no glory ray can exist for scattering from water drops ($n = 1.33$)! The failure to explain this effect in just the case that is most important from the physical point of view led van de Hulst to propose that the light of the glory is sent back by following paths consisting of rays that are reflected repeatedly within the droplet, together with small segments of surface waves.

The beautiful phenomenon of the rainbow [35], which sometimes accompanies the piercing of clouds by rays from the sun after a storm, has attracted the attention of naturalists since ancient times, when Aristotle proposed that the rainbow is an unusual kind of reflection of sunlight from clouds, which occurs at a fixed angle, giving rise to a circular cone of rainbow rays. The British philosopher and naturalist Roger Bacon [36] was the first to measure (in 1266) the angle of 138° formed by the rainbow rays and the incident sunlight. The German monk Theodorik of Freiberg suggested in 1304 that each drop of mist individually produces a rainbow. Analogous findings were rediscovered by Antonio de Dominis, Archbishop of Spalato, who described in his book "De Radiis Visus et Lucis" (1611) "*How the interior bow is made in round Drops of Rain by two Refractions of the Sun's light, and one Reflexion between them in each Drop of Water, and proves his Explications by Experiments made with a phial full of Water, and with Globes of Glass filled with Water, and placed in the Sun to make the Colours of the two Bows appear in them. The same Explication Des-Cartes hath pursued in his Meteors, and mended that of the exterior Bow*" (I. Newton, "Opticks," book one, part II, prop. IX, prob. IV). Newton completed the geometric optics theory by explaining the origin of the most conspicuous features of the rainbow—its colors. Guided by his findings on the dispersion of white light in prisms, he explained that what we observe is a collection of monochromatic rainbows, each one slightly displaced from the next. With his accurate measurements Newton calculated that the rainbow angle is 137°58′ for red light and 139°43′ for violet light.

In 1835 R. Potter pointed out that the crossing of various sets of light rays in a droplet gives rise to a caustic. Starting from these findings, George B. Airy succeeded in 1838 in determining the intensity distribution in a monochromatic rainbow by introducing the famous rainbow integral, which has become known as the Airy function. He based his reasoning on the Huygens principle by using a cubic wave front, as shown in Fig. VI.22.

According to Eq. (VI.13.39), the rainbow occurs for $d\theta/d\theta_1 = 0$, that is, in view of Eq. (VI.13.38), for incidence angle $\theta_{1,\mathrm{R}}$ given by

$$\theta_{1,\mathrm{R}} = \arcsin[(p^2 - n^2)/(p^2 - 1)]^{1/2}. \qquad \text{(VI.13.41)}$$

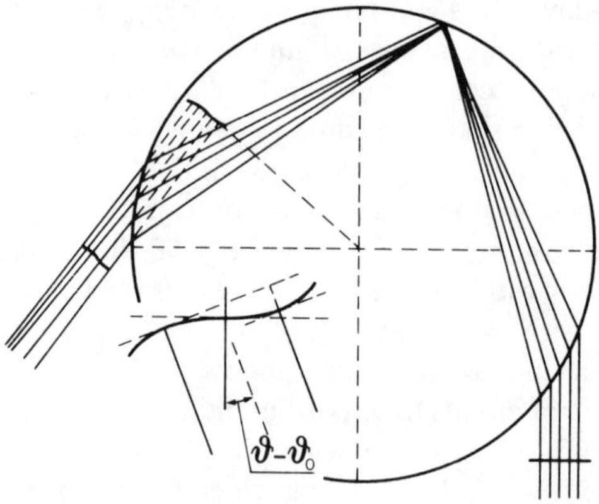

Fig. VI.22. Confluence of rays scattered by a droplet and giving rise to a congruence of backscattered rays characterized by a typical S-shaped wave front.

Since the argument of the function $\arcsin x$ must be less than unity, the rainbow ray must undergo a number of reflections larger than $n - 1$, n being the refractive index of the drops. In particular, for water drops ($n \cong 4/3$) the first rainbow ray occurs for $p = 2$ and is scattered with an angle $\theta_{\mathrm{R}}^{(1)}$ of

$$\theta_{\mathrm{R}}^{(1)} = \pi + 2\arcsin[(4 - n^2)/3]^{1/2} - 4\arcsin[(4 - n^2)/(3n^2)]^{1/2} \cong 138°. \tag{VI.13.42}$$

The second rainbow is due to rays with two reflections, and its geometric position is at $\theta_2^{(2)} \cong 128.7°$.

The amplitude near the rainbow angle can be obtained from the integral of Eq. (VI.13.32) by observing that, in the vicinity of θ_{R}, the quadratic expansion of Eq. (VI.13.35) is superseded by a cubic expression, owing to the form of the wave front shown in Fig. VI.22,

$$\xi = \xi_{\mathrm{R}} + (q - q_{\mathrm{R}})\,\partial\xi/\partial q + (1/6)(q - q_{\mathrm{R}})^3\,\partial^3\xi/\partial q^3. \tag{VI.13.43}$$

Then the gaussian integral leading to Eq. (VI.13.37) is replaced by the Airy function. Therefore, the monochromatic rainbow pattern exhibits the features of the field in proximity to a caustic, characterized by a series of fringes on the lit side. Simulations made by summing the Mie series have cast some doubts on the actual presence of sharp fringes. Recently [37], the Chester–Friedman–Ursell (CFU) method (Chapter V) was applied to an integral similar to that of Eq. (VI.13.32) obtained by calculating the scattering amplitude by the Watson–Regge method. Since the CFU method yields a representation based on a combination of the Airy function Ai and its derivative Ai′, the zeros of Ai are compensated by the presence of Ai′. The remarkable improvement obtained with these modifications is evident from Fig. VI.23.

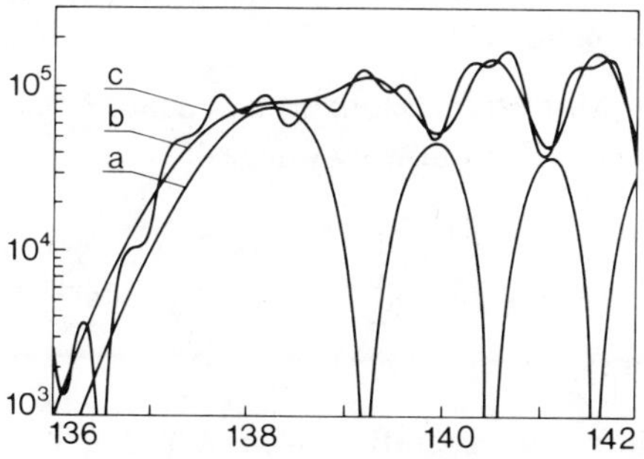

Fig. VI.23. Intensity of the scattered light corresponding to the rays of Fig. VI.22 for a size parameter $\beta = 1500$. Curve a represents the intensity calculated by Airy by evaluating the diffraction integral relative to the S-shaped wave front. Curve b represents the field calculated by taking into account the contribution of surface waves arising from the Watson–Regge representation of the exact scattered field in the scalar approximation. Curve c represents the solution obtained by adding up more than 1500 terms of the partial wave series representation of the scattered field. (From Nussenzveig [36].)

Problems

Section 2

1. Show that for a metallic half-plane the integral of Eq. (VI.2.2) reads for $\phi' > \pi$

$$u(\rho,\phi)=\frac{-i}{4\pi}\int_{\Gamma} e^{-ik\rho\cos\beta}\frac{\cos[(\beta+\phi)/2]}{\sin[(\beta+\phi)/2+\cos(\phi'/2)}d\beta$$

$$= e^{ik\rho\cos(\phi-\phi')}U(\phi-\phi'+\pi)+e^{-ik\rho\cos(\phi+\phi')}U(\phi+\phi'-3\pi)+\frac{i}{4\pi}$$

$$\times\int_{\mathrm{SDP}_2}\left\{\frac{-1}{\sin[(\beta+\phi)/2]+\cos(\phi'/2)}+\frac{1}{-\sin[(\beta+\phi)/2]+\cos(\phi'/2)}\right\}$$

$$+\cos\left(\frac{\beta+\phi}{2}\right)e^{-ik\rho\cos\beta}\,d\beta.$$

2. Starting from the above integral representation, derive Sommerfeld's expressions for the field relative to a metallic half-plane illuminated by a p- or an s-wave,

$$u_{\mathrm{p,s}}(\rho,\phi) = e^{ik\rho\cos(\phi-\phi')}F\{-(2k\rho)^{1/2}\cos[(\phi-\phi')/2]\}$$
$$\mp e^{ik\rho\cos(\phi+\phi')}F\{-(2k\rho)^{1/2}\cos[(\phi+\phi')/2]\},$$

where F is the Fresnel complex integral and the minus and plus signs apply, respectively, to the p- and s-polarizations.

3. Calculate the current on a metallic half-plane illuminated by an s-wave incident along the direction $\phi' = \pi/2$.

4. Show that the Maliuzhinets function M, defined by Eq. (VI.2.15), can be represented by means of the following integral:

$$M(\alpha) = \exp\left[\frac{i}{8\Phi}\int_0^{\alpha} d\mu \int_{-i\infty}^{+i\infty}\tan\left(\frac{\pi v}{4\Phi}\right)\frac{1}{\cos(v-\mu)}\,dv\right].$$

Hint: Consider the function $d\ln M/d\alpha$.

5. Using the above representation of $M(\alpha)$, prove the following relations:

$$M(\alpha+2\Phi)/M(\alpha-2\Phi) = \cot[(1/2)(\alpha+\pi/2)],$$
$$M(\alpha+\pi/2)M(\alpha-\pi/2) = M^2(\pi/2)\cos[(\pi\alpha)/(4\Phi)],$$
$$M(\alpha+\Phi)M(\alpha-\Phi) = M^2(\Phi)M'(\alpha),$$

where M' represents the Maliuzhinets function for $\Phi' = \Phi/2$.

6. Show that, for $\Phi = \pi$, M reduces to

$$M(\alpha) = \exp\left[\frac{1}{8\pi}\int_0^\alpha \frac{\pi\sin(u) - 2^{3/2}\pi\sin(u/2) - 2u}{\cos u}\,du\right].$$

7. Show that, for $\Phi = \pi$, M can be expressed by

$$M(\alpha) = \left(\frac{1}{\cos\alpha}\,\frac{1 - 2^{1/2}\alpha}{1 + 2^{1/2}\alpha}\right)^{1/8} \exp\left[-\frac{1}{4\pi}\sum_{n=0}\frac{|E_{2n}|\alpha^{2n+2}}{(2n+2)(2n)!}\right],$$

E_{2n} being Euler's numbers.

8. Calculate the function $g(\alpha)$ for $\Phi = \pi$ and $\theta_B \cong \pi/2$ by approximating $M(\alpha + \theta_B)$ with

$$\begin{aligned} M(\alpha + \theta_B) &\cong M(\alpha + \pi/2) + (\theta_B - \pi/2)\,dM/d\alpha \\ &= M(\alpha + \pi/2) - (\theta_B - \pi/2)M(\alpha + \pi/2) \\ &\quad \times [\cos(\alpha) - 2^{3/2}\sin(\alpha/2 + \pi/4) - 2\alpha/\pi - 1]/(8\cos\alpha) \end{aligned}$$

Hint: See, Problem 6.

9. Show that, for $\Phi = 3\pi/4$, M reads

$$M(\alpha) = \cos[(\alpha - \pi)/6]\cos[(\alpha + \pi)/6]/[\cos^2(\pi/6)\cos(\alpha/6)].$$

10. Prove Eq. (VI.2.11) by using Eq. (F2). *Hint*:

$$\int_{\mathrm{SDP}_2} [S(\beta + \phi) - S(\beta + \phi - 2\pi)]e^{-ik\rho\cos\beta}\,d\beta$$
$$\cong -(2\pi/k\rho)^{1/2}e^{i\pi/4}[S(2\pi + \phi) - S(\phi)]e^{-ik\rho}.$$

11. Calculate the diffraction coefficient D_S for a right-angled wedge by using Eq. (VI.2.18). In particular, discuss the dependence of D_S on θ_B.

12. Calculate the amplitude S_+ of the surface wave excited by a unit amplitude s-wave incident on a right-angled wedge ($\Phi = 3\pi/4$). *Hint*: Use the expression derived in Problem 9.

References

1. King, R. W. P., and Wu, T. T., "The Scattering and Diffraction of Waves." Harvard Univ. Press, Cambridge, Massachusetts, 1959.
2. Uslenghi, P. L. E., ed., "Electromagnetic Scattering." Academic Press, New York, 1978.
3. Maliuzhinets, G. D., *Sov. Phys.—Dokl.* (*Engl. Transl.*) **3**, 752 (1958).
4. Kouyoumjian, R. G., *in* "Numerical and Asymptotic Techniques in Electromagnetics" (R. Mittra, ed.), p. 166–215. Springer-Verlag, Berlin and New York, 1975.
5. Joo, C.-S., Ra, J.-W., and Shin, S.-Y., *Electron. Lett.* **16**, 934 (1980).

5a. Keller, J. B., *J. Appl. Phys.* **28**, 426 (1957).
6. Rhodes, D. R., "On the Theory of Scattering by Dielectric Bodies," Rep. 475-1. Antenna Lab., Ohio State Univ., Columbus, 1953.
7. Debye, P. J., *Phys. Z.* **9**, 775 (1908).
8. Poggio, A. J., and Miller, E. K., *in* "Computer Techniques for Electromagnetics" (R. Mittra, ed.), Chap. 4. Pergamon, New York, 1973.
9. Harrington, R. F., "Field Computation by Moment Method, "Macmillan, New York, 1968.
10. Baum, C. E., *Proc. IEEE* **64**, 1598 (1976).
11. Harrison, G. R., *Appl. Opt.* **4**, 1275 (1965).
12. Stroke, G. W., *in* "Handbuch der Physik" (S. Flugge, ed.), Vol. 29, p. 426–754. Springer-Verlag, Berlin and New York, 1967.
13. Denisyuk, *Dokl. Akad. Nauk. SSSR* **6**, 144 (1962).
14. Schmahl, G., and Rudolf, D. R., *Prog. Opt.* **14**, 195–244 (1976).
15. Labeyrie, A., and Flamand, J., *Opt. Commun.* **1**, 5 (1969).
16. Tamir, T., "Integrated Optics." Springer-Verlag, Berlin and New York, 1975.
17. Samson, J. A. R., "Techniques of Vacuum Ultraviolet Spectroscopy." Wiley, New York, 1967.
18. Petit, R., ed., "Electromagnetic Theory of Gratings." Springer-Verlag, Berlin and New York, 1980.
19. Zaki, K. A., and Neurenther, A. R., *IEEE Trans. Antennas Propag.* **AP-19**, 208 (1971).
20. Hessel, A., and Oliner, A. A., *Appl. Opt.* **4**, 1275 (1965).
21. Silver, S., ed., "Microwave Antenna Theory and Design," MIT Radiat. Lab. Ser. Vol. 12, Chap. 2. Cambridge, Massachusetts, 1949.
22. Newton, R. G., "Scattering Theory of Waves and Particles." McGraw-Hill, New York, 1966.
23. Kerker, M., "The Scattering of Light." Academic Press, New York, 1969.
24. Twersky, V., *Appl. Opt.* **3**, 1150 (1964).
25. Noble, B., *in* "Electromagnetic Waves" (R. Langer, ed.), p. 323–360. Univ. of Wisconsin Press, Madison, 1962.
26. Kleinman, R. E., *Proc. IEEE* **53**, 848 (1965); see also Ref. 2, Chap. 1.
27. Mie, G., *Ann. Phys. (Leipzig)* **25**, 377 (1908).
28. Debye, P., *Ann. Phys. (Leipzig)* **30**, 57 (1909).
29. Wickramasinghe, N. C., "Light Scattering Functions for Small Particles." Wiley, New York, 1973.
30. Landsberg, G. S., "Optica." Nauka, Moscow, 1976. (In Russ.)
31. Vauclair, S., *in* "Astrophysical Processes in Upper Main Sequence Stars" (B. Hauck and A. Maeder, eds.), p. 167. Geneva Observ. CH-1290, Sauverny, Switzerland, 1983.
32. Minogin, V. G., and Letokhov, V. S., *J. Opt. Soc. Am.* **69**, 413 (1979).
33. Irvine, W. M., *J. Opt. Soc. Am.* **55**, 16 (1965).
34. Ashkin, A., and Dziedzic, J. M., *Phys. Rev. Lett.* **38**, 1351 (1977).
35. Tricker, R. A. R., "Introduction to Meteorological Optics." Am. Elsevier, New York, 1970.
36. Nussenzveig, H. M., *in* "Light from the Sky" (J. Walker, ed.), p. 54–65. Freeman, San Francisco, California, 1980.
37. Khare, V., and Nussenzveig, H. M., *Phys. Rev. Lett.* **33**, 976 (1974).

Bibliography

Bohren, C. F., and Huffman, D. R., "Absorption and Scattering of Light by Small Particles." Wiley, New York, 1983.

Fabelinskii, I. L., "Molecular Scattering of Light." Plenum Press, New York (1968).

Ishimaru, A., "Wave Propagation and Scattering in Random Media." Academic Press, New York (1978).

Kazantsev, A. P., *Sov. Phys. Usp.* [*Engl. Transl.*] **21**, 58 (1978).
Lax, P. D., and Phillips, R. S., "Scattering Theory." Academic Press, New York, 1967.
Maystre, D., *Prog. Opt.* **21**, 3–57 (1984).
Reed, M., and Simon, B., "Methods of Modern Mathematical Physics," Vol. 3. Academic Press, New York, 1979.
Stenholm, S., Minogin, V. G., and Letokhov, V. S., *Opt. Commun.* **25**, 107 (1978).
Stroke, G. W., *Prog. Opt.* **2**, 1–72 (1963).
van de Hulst, H. C., "Multiple Light Scattering," Academic Press, New York, 1980.
Wilcox, C. H., "Scattering Theory for Diffraction Gratings," Springer-Verlag, Berlin and New York, 1984.
Welford, W. T., *Prog. Opt.* **4**, 241–280 (1965).

Chapter VII

Optical Resonators and Fabry–Perot Interferometers

1 Generalities on Electromagnetic Resonators

1.1 *Conventional Oscillators*

Physical systems displaced from their equilibrium state by an impulsive perturbation tend to resume their initial configuration by undergoing a series of damped oscillations. This behavior is of such a general nature that it is observed in the gigantic structure of a skyscraper struck by wind or in atoms hit by electromagnetic pulses, the only difference being the rapidity with which these oscillations tend to disappear. In some cases, in fact, no oscillation at all is observed owing to the predominance of the damping process, while in others the oscillations last so long that the system behaves as an *ideal oscillator*.

Readers with an elementary background in electronics know that by combining in parallel an inductance L with a capacitance C one obtains a system that, once excited, oscillates for an infinite time with an angular frequency $\omega_0 = 1/\sqrt{LC}$. In practice, due to the losses of inductance and capacitance, which can be represented by including the resistances R_L and R_C in their equivalent circuits (see insert of Fig. VII.I), these oscillations are damped. In fact, if we indicate with $i(t)$ the current flowing through the mesh formed by C and L we have

$$i(t) = (i_0/\sin\phi)e^{-Rt/2L}\sin\{[1/(LC) - [R/(2L)]^2]^{1/2}t + \phi\}, \qquad \text{(VII.1.1)}$$

where $R_L + R_C = R$ and i_0 and ϕ are, respectively, the initial current and phase of the oscillation. It is convenient to measure the damping of these oscillations by the *quality factor Q*

$$Q = \pi\tau/T = [L/(CR^2) - \tfrac{1}{4}]^{1/2} \sim (L/C)^{1/2}1/R, \qquad \text{(VII.1.2)}$$

$\tau = 2L/R$ being the time constant of the damping and $T = 2\pi/\omega_0$ the period of oscillation [see Eq. (VII.1.1)]. The factor Q is a measure of the departure

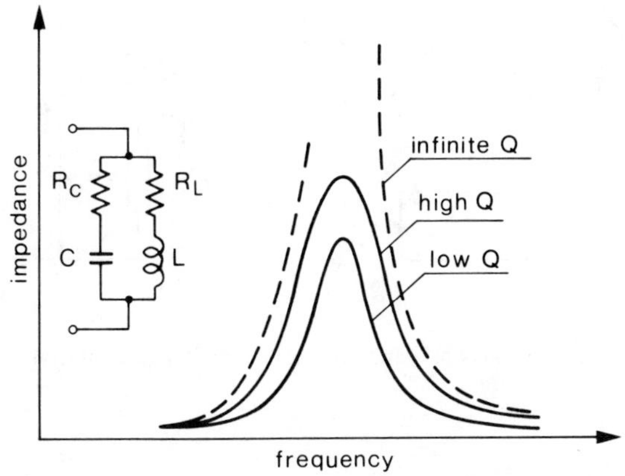

Fig. VII.1. Magnitude of the impedance of a resonant circuit versus frequency for different Q.

from the ideal case of a lossless system oscillating indefinitely; the larger Q, the more closely our oscillator approximates the ideal system. It can be easily shown by using the expression (VII.1.1) for the mesh current $i(t)$ that Q is also given by $Q = (1/2)\omega_0[Li^2 + (1/C)(\int i\,dt)^2]/Ri = 2\pi$ times energy stored in the LC circuit divided by energy lost per unit cycle $= 2\pi$ times energy stored in the LC circuit divided by energy lost per unit cycle.

Let us now look at the impedance $Z(\omega)$ of the LC circuit seen from the common terminals of L and C. For $R^2 \ll L/C$, in proximity to the resonance frequency $\omega_0 \cong 1/(LC)^{1/2}$, $Z(\omega)$ is given by

$$Z(\omega) = (L/C)^{1/2}Q/[1 + i2Q(\omega - \omega_0)/\omega_0] \equiv |Z(\omega)|e^{i\beta(\omega)}. \qquad \text{(VII.1.3)}$$

(We prefer in this introductory section to use the angular frequency ω; in the following section we will switch to the frequency ν.)

According to the above equation, for $Q \gg 1$ the modulus of Z is a very narrow peaked function of $\Delta\omega = \omega - \omega_0$ (see Fig. VII.1), while the phase $\beta(\omega)$ undergoes a 180° flip when ω passes ω_0.

Furthermore, we can easily infer from Eq. (VII.1.3) that the frequency band B over which the response is at least 71% of that at resonance (i.e., in the language of electrical engineers, within 3 dB of resonance) is $B = \omega_0/Q$.

When the resistances R_L and R_C are both negative, the current $i(t)$ is still given by Eq. (VII.1.1), with R a negative quantity. This means that the current undergoes oscillations whose amplitudes grow exponentially.

If we put in parallel two circuits having the same resonance frequency and generally different Q values and resonance impedances, the resultant impedance is given by

$$Z_{\text{tot}}(\omega) = Z_{\text{eq}}/[1 + i2Q_{\text{eq}}(\omega - \omega_0)/\omega], \qquad \text{(VII.1.4)}$$

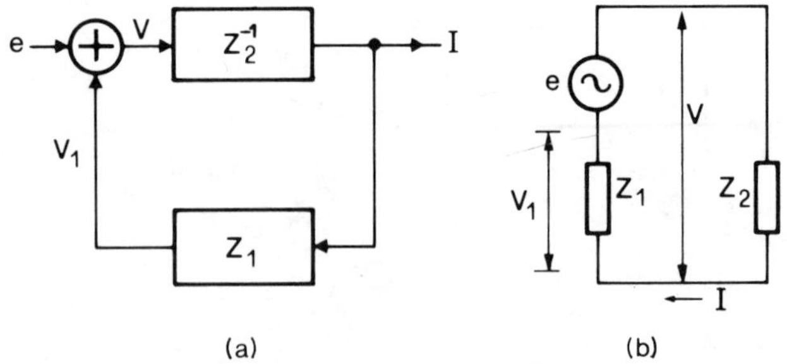

Fig. VII.2. (a) Feedback network and (b) mesh circuit representation of the current *I* produced by a voltage generator *e*.

where

$$Z_{eq} = Z_1 Z_2/(Z_1 + Z_2), \qquad Q_{eq} = (Q_1 Z_2 + Q_2 Z_1)/(Z_1 + Z_2) \qquad \text{(VII.1.5)}$$

with $Z_1 = Z_1(\omega_0)$ and $Z_2 = Z_2(\omega_0)$, so that Z_{eq} can become very large when $Z_1 + Z_2 = 0$. Therefore, by putting in parallel to a resonant circuit another circuit having a negative resonance impedance almost equal in modulus, we obtain a resonant circuit having an infinite resonance impedance. To achieve this result, the only necessary condition to be satisfied is the negative value of $Z_2(\omega_0)$.

An oscillator can be represented by a generator of noise of complex amplitude e in series with Z_1 and Z_2 (see Fig. VII.2a), which represent the impedances of the passive components (Z_1) accompanying the active element (Z_2). In the language of system theory, the relation between the current I circulating through the mesh and e can be represented by the closed loop of Fig. VII.2b. The response $Z_1(\omega)$ of the feedback filter may depend critically on the frequency, thus determining the oscillation frequency, while $Z_2(\omega)$ can be almost flat near the resonance frequency. In addition, while Z_1 generally represents a linear network, Z_2 exhibits a nonlinear dependence on the applied voltage, thus fixing the oscillation amplitude.

As an alternative to Fig. VII.2, we can represent an oscillator by the nonlinear feedback diagram of Fig. VII.3, consisting of a saturable amplifier with *amplification* $A(V)$, almost independent of ω in proximity to ω_0 and decreasing for increasing V, and a dephasing network with frequency response

$$F(\omega) = |F|e^{-i\beta(\omega)} = [1 + i2Q(\omega - \omega_0)/\omega_0]^{-1}, \qquad \text{(VII.1.6)}$$

where β represents the phase delay undergone by the input signal when it is transmitted through the closed loop. In other words, we have included in the saturable amplifier all the frequency-independent terms, so that the response F of the feedback network reduces to unity for $\omega = \omega_0$. To be more precise, we should also allow for a V dependence of Q in view of the nonlinear contribution of the active elements to the phase.

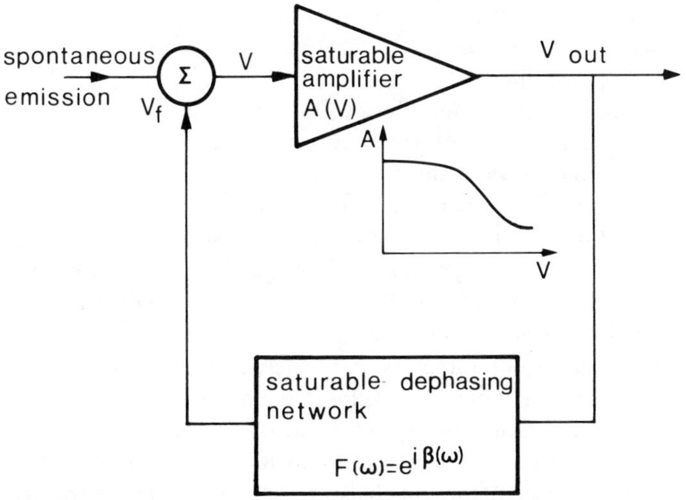

Fig. VII.3. Feedback network representing a nonlinear oscillator that includes a saturable amplifier with the characteristic amplification–voltage curve represented in the inset.

1.2 *Microwave and Optical Oscillators*

While for conventional low-frequency oscillators it is simple and unambiguous to identify the resonant circuit and the positive and negative resistances used in the previous discussion, the analysis of *microwave* [1] and *optical* [2–6] oscillators requires a certain amount of mathematical and physical ingenuity. To be specific, let us consider a modern microwave generator built by enclosing an inversely polarized IMPATT diode in a cavity (see Fig. VII.4). The diode can be represented as a polarization P acting as a source of the electric field, as described by Eq. (I.2.8). Accordingly, the field $\mathbf{E}(\mathbf{r}, t)$ can be interpreted as the response of the resonator to the dipoles distributed in the region of the diode junction, i.e.,

$$\mathbf{E}(\mathbf{r}, t) = \int_{-\infty}^{t} dt' \iiint_{\text{junction}} \mathbf{\Gamma}(\mathbf{r}, \mathbf{r}'; t - t') \cdot \mathbf{P}(\mathbf{r}', t') d\mathbf{r}' \qquad \text{(VII.1.7)}$$

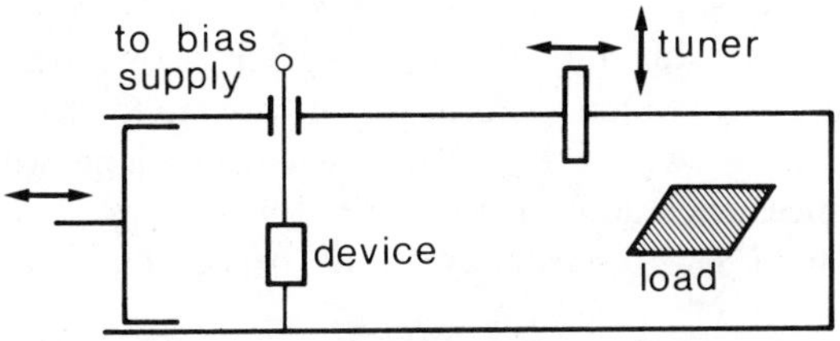

Fig. VII.4. Schematic diagram of a negative-resistance microwave oscillator.

where $\mathbf{\Gamma}$ is a dyadic representing the response to a dipole in $\mathbf{r}'$ turned on at t' for a very short interval, that is, the space–time Green's function of the cavity.

Since $\mathbf{P}$ is a (nonlinear) function of $\mathbf{E}$, Eq. (VII.1.7) [or equivalently, Eq. (1.2.9) supplied with the boundary conditions] can be used, in principle, to calculate the amplitude and the frequency of the oscillation. From a mathematical point of view, these calculations can be notably simplified by representing $\mathbf{E}$ and $\mathbf{P}$ as a combination of *modes of oscillation of the cavity*. According to this representation, which will be discussed in some detail, the cavity and consequently the whole oscillator can be modeled as an *infinite discrete sequence of simple oscillators having the simple properties of the LC circuits*. This, in turn, implies that *the system can oscillate at several frequencies*, so that, while sometimes it oscillates at a single frequency, under other conditions it can operate at several frequencies simultaneously. This is a completely new situation with respect to the traditional *LC* circuit and gives rise to the current distinction between *radio* and *microwave* oscillators.

A third class of oscillators consists of the *masers* and *lasers*, which are generally referred to as *quantum oscillators*. They are obtained by clever recourse to a *single quantum* of energy exchange between the laser field and the active medium, which, as originally suggested by Schawlow and Townes, allows the removal of the thermodynamic limit posed on the linewidth in conventional devices.

One of the main differences between optical and microwave oscillators is that the latter have an active region that is typically small compared to the emission wavelength λ and can be represented by a localized dipole. Conversely, in the laser oscillator the active region is large compared to λ, which implies that the problem of the interaction between the field and the dipole distribution becomes very involved.

Closed cavities can be used only in the microwave region. In fact, we know that the frequencies of a rectangular cavity are given by

$$\omega_{lmn} = \pi c[(l/a)^2 + (m/b)^2 + (n/d)^2]^{1/2}, \qquad \text{(VII.1.8)}$$

where a, b, and d are the dimensions of the prismatic enclosure and l, m, and n are nonnegative integers such that $l + m + n \geq 2$. If we consider that the dimensions would exceed 1 mm for technical reasons, it becomes impossible to use these devices to generate waves with a wavelength less than 1 mm.

To overcome this difficulty, Prokhorov, Schawlow, and Townes originally proposed using a Fabry–Perot (open) cavity, so that a laser oscillator has the form shown in Fig. VII.5, where a plane wave of average intensity I bounces back and forth between the mirrors M_1 and M_2 (having reflectivity R_1 and R_2, respectively) while being amplified by the active medium.

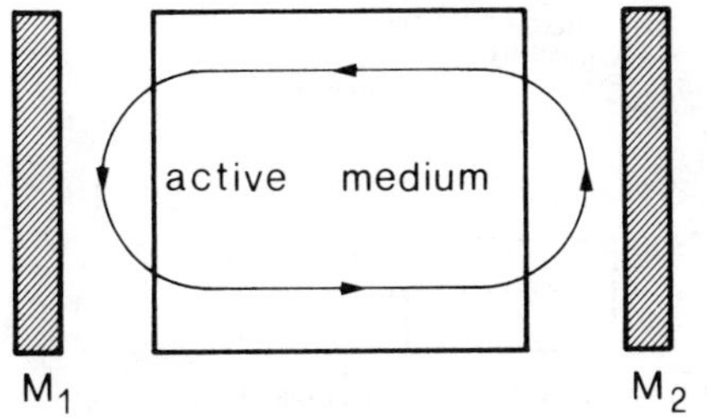

Fig. VII.5. Schematic of a laser oscillator.

Actually, the hypothesis that the oscillating beam is a plane wave is strictly connected with the assumption that the laser is supposed to operate in the stationary regime where losses and gains exactly compensate (*above threshold*). *Below threshold*, each elementary radiator emits incoherently from all the others, so that the total field consists of a superposition of wavelets leaving the active region along *all directions.*

Using the feedback network model, we can put

$$1 = R_1 R_2 e^{2G}, \tag{VII.1.9}$$

where G represents the amplification per pass. Furthermore, if we require that the plane wave undergoes a phase delay per round trip equal to $2m\pi$, with m an integer, when $\omega = \omega_0$ we have, in general, from Eq. (VII.1.6)

$$\beta_m(\omega) = 2m\pi + \arctan[2Q(\omega - \omega_0)/\omega_0], \tag{VII.1.10}$$

an equation that will be confirmed by the more accurate analysis of the following sections.

In some situations, it is important to take into account a detuning of the oscillation frequency of the cavity with respect to the central frequency ω_L of the gain curve of the active medium and accordingly allow for a frequency dependence of the gain G and the amplification A, thus having in general $A = A(I, \omega, \omega_L)$. Besides, it is not always possible to neglect the dephasing introduced by the active medium. In this case we must include a *dephasing network*, characterized by a function $F_L(I, \omega, \omega_L) = \exp[-i\beta_L(I, \omega, \omega_L)]$ in the feedback arm of the loop network of Fig. VII.3, where $\beta_L(I, \omega, \omega_L)$ is a nonlinear function of the intensity I.

The above discussion can be summarized by saying that we can relate the output $V(\omega)$ ($|V|^2 = I$ is the average intensity of the light inside the laser oscillator) to the input $e(\omega)$ of the feedback circuit, having the form of Fig. VII.6 and representing the laser oscillator of Fig. VII.5, by means of the equation

$$V(\omega) = e(\omega) A(I, \omega, \omega_L)/[1 - A(I, \omega, \omega_L) F_{\text{cav}}(\omega) F_L(I, \omega, \omega_L)]. \tag{VII.1.11}$$

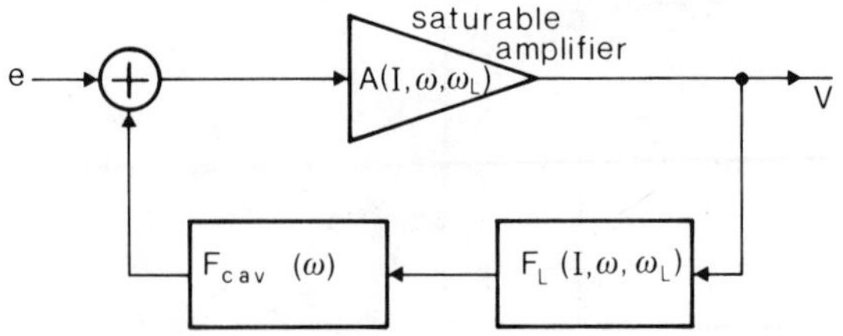

Fig. VII.6. Nonlinear feedback network equivalent to a laser oscillator.

The system will undergo free oscillation in the absence of the input signal $e(\omega)$ when it is possible to satisfy the *resonance condition*

$$A(I, \omega, \omega_L)|F_{\mathrm{cav}}(\omega)|\exp[-i\beta_{\mathrm{cav}}(\omega) - i\beta_L(I, \omega, \omega_L)] = 1, \qquad \text{(VII.1.12)}$$

which splits into the two nonlinear equations

$$\beta_{\mathrm{cav}}(\omega) + \beta_L(I, \omega, \omega_L) = 2m\pi, \qquad A(I, \omega, \omega_L)|F_{\mathrm{cav}}(\omega)| = 1, \qquad \text{(VII.1.13)}$$

which, solved simultaneously, provide the oscillation frequency ω and the intensity I of the oscillating beam.

In principle, a laser should oscillate on a line spectrum. However, in the model adopted up to now we have completely neglected the presence of noise. By using statistical arguments, Schawlow and Townes proved that quantum noise produces a broadening of the line given by (see Chapter I, Yariv [8])

$$\Delta\omega_{\mathrm{osc}} = [(\hbar\omega)/P](\Delta\omega_{\mathrm{cav}})^2\mu, \qquad \text{(VII.1.14)}$$

where $\Delta\omega_{\mathrm{cav}}$ is the half-width of the resonance at half-maximum, P is the power emitted by the oscillator, and $\mu = N_a/[N_a - N_b(g_a/g_b)]_{\mathrm{th}}$ accounts for the fact that the finite population N_b of the lower laser level requires a corresponding increase in the population N_a of the upper laser level at threshold in order for the gain to remain equal to the loss. In practice, it is almost impossible to reach the small bandwidth given above, since the oscillator is strongly affected by the fluctuations of the characteristics of the active medium and by vibrations of the cavity mirrors.

2 Generalities on Optical Resonators

Microwave cavity resonators were developed to be used mainly at centimetric wavelengths. An increase in λ calls for increasingly larger dimensions, until the point is reached where the devices turn into lumped-constant LC circuits. A reduction of λ means smaller and smaller cavities, to the detriment of their quality factor Q. Generally, in the millimetric band Q decreases as the inverse of the square root of the frequency. At very short wavelengths, say in the submillimetric range, cavities may no longer be designed for operation in a single low-order mode (e.g., TE_{011}), which would require prohibitively small dimensions. There is thus a need to operate with many modes of very high order, a single mode of operation no longer being achievable. In fact, as the order increases, the frequency separation between adjacent modes becomes comparable to the resonance linewidth, a circumstance that clearly detracts from the selectivity. Therefore, the use of such resonators is altogether unsuitable for forcing a system to oscillate at a predetermined frequency. In practice, the system oscillates simultaneously in several modes and produces an almost continuous emission spectrum centered on some average frequency.

For very high frequencies, according to the *Rayleigh–Jeans formula*, the number ΔN_m of modes comprised in a frequency interval $\Delta\nu$ is given, for a closed cavity of volume V and general shape, by [7]

$$\Delta N_m = [(8\pi)/c^3]\nu^2 \Delta\nu\, V, \tag{VII.2.1}$$

where c is the velocity of light. The consequence of this relation is that as the frequency rises the mode spectrum becomes far more dense. For $\nu = 5 \times 10^{14}$ Hz (the frequency of an He–Ne laser) and $\Delta\nu = 1.5$ GHz, Eq. (VII.2.1) gives about 1.7×10^8 modes per cubic centimeter. Under these conditions, quite obviously, a 10-cm^3 cavity consisting of a closed metallic enclosure—to which Eq. (VII.2.1) applies—will not be able to control the emission frequency of an active He–Ne plasma, which will come to oscillate on some billions of wavelengths, placed about 1 Hz from one another.

These considerations led Schawlow and Townes [8] and Prokhorov [9] to suggest the use of a different type of cavity that would favor certain so-called *fundamental modes* with respect to all others. They envisaged a suitable configuration of the cavity walls such that the Q of the fundamental modes would be very high while the remaining (much more numerous) modes would be so lossy as to preclude the onset of oscillations when the cavity is filled by an active medium.

The simplest example of a resonator satisfying the above requirements derives from the *Fabry–Perot interferometer* [10] made with a pair of plane-parallel mirrors (Fig. VII.7). This cavity can be thought of as obtained by eliminating the sidewalls of a cylindrical enclosure so that the only high-Q modes are those consisting of two quasi-plane waves traveling in opposite directions, perpendicular to the mirrors. These modes are found to be favored with respect to those made up by rays not parallel to the optic axis. The fact that some of the power carried by noncollimated rays is lost outside the cavity causes the fall of the relative Q.

Assuming that the fundamental modes of this structure are adequately treated as plane waves bouncing back and forth between two mirrors spaced a

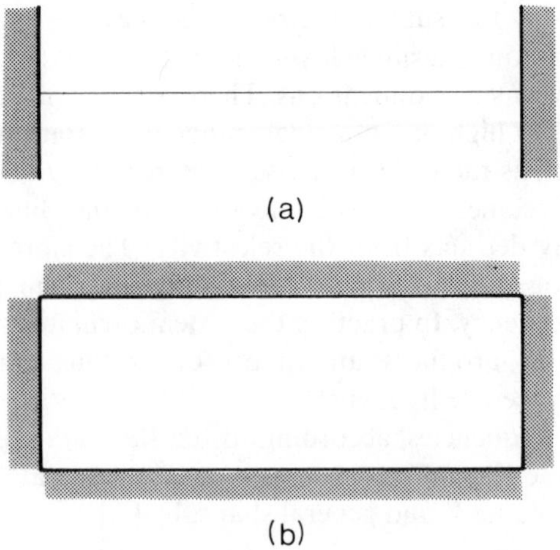

Fig. VII.7. Plane-parallel open resonator (a) obtained by eliminating the sidewalls of the cylindrical closed cavity (b).

distance d apart, the resonance frequencies ν_n will be given by the simple relation $\nu_n = \nu_0(1 + n)$, where $\nu_0 = c/2d$ represents the frequency separation between adjacent modes. Notice that when the cavity is used as a Fabry–Perot interferometer, the frequency of a wave may be measured to within an integer multiple of ν_0. In view of this, ν_0 is known as the *free spectral range* (FSR) (see Section III.15).

A characteristic of these resonators is the fact that each mode forms a *standing-wave pattern.* For some applications it is preferable to use configurations characterized by *traveling-wave modes.* As an example, the *ring resonator* introduced by Macek and Davis [11] (see Fig. VII.8) can be covered

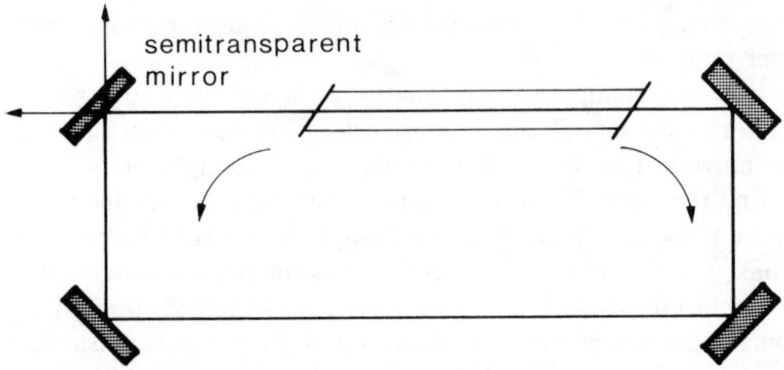

Fig. VII.8. Schematic representation of a square ring resonator. Other typical arrangements are triangular. The mirrors can be replaced with total-reflection prisms that have higher damage thresholds.

by wave beams circulating either counterclockwise or clockwise. When one of these waves is eliminated, the field reduces to a locally uniform traveling wave. Rotation of the system around an axis orthogonal to the ring laser plane produces a splitting of the frequencies of these two waves. By measuring their beating frequency the rotation rate can be inferred. This principle is used in *laser gyroscopes*.

When the losses produce a broadening of the resonances much wider than the intermode distance, the eigenfrequency spectrum becomes practically continuous. This characteristic can also be achieved by replacing a mirror with a scattering surface. Analogously, a cloud of scattering particles disseminated in an amplifying medium can provide sufficient feedback to operate a laser. In general, cavities exhibiting high mode degeneracy can be used to provide nonresonant feedback [12]. An important advantage of these resonators over the Fabry–Perot ones is that the mean frequency of the emission generated does not depend on the dimensions of the laser. This is the main prerequisite of a frequency standard, which, however, does not benefit from the space and time coherence of conventional lasers.

We cannot exclude from this overview some devices that use a distributed feedback mechanism to eliminate the mirrors. The idea of exploiting the Bragg scattering of two waves traveling in opposite directions through a periodic medium was first formulated by Kogelnik and Shank [13] (Fig. III.27), who were looking for a new cavity configuration to narrow the spectrum of injection lasers and make them compatible with the integrated optics technology.

3 Frequency Response of a Resonator

No field can be built up in an ideal cavity unless the exciting frequency coincides with the chosen resonance frequency. In practice, there will always be a narrow band of frequencies around the eigenfrequency over which appreciable excitation can occur.

If one excites the cavity with a monochromatic electric (or magnetic) dipole of constant strength, the field amplitude $|F(\nu)|$ at a generic point of the resonator will change with frequency in the way shown in Fig. VII.9. In contrast to the closed cavity, the resonance curve of the open resonator consists of many peaks, only a few of which exhibit a sharpness comparable with that of a closed cavity. This demonstrates the dramatic reduction of the effective number of modes resulting from sidewall removal.

The partial smearing out of the resonance peaks is due to leakage of radiation through the mirror itself and its edges and to dissipation of energy in the imperfectly reflecting walls. By virtue of the strong variation of radiation

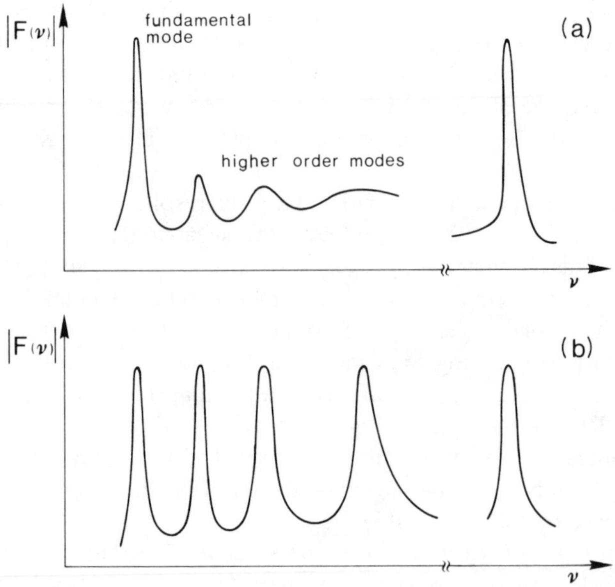

Fig. VII.9. Qualitative dependence of the frequency response amplitude $|F(\nu)|$ with respect to ν for (a) the plane-parallel resonator of Fig. VII.7a and (b) the closed cavity of Fig. VII.7b.

losses with the modal field distribution, the width of the resonance curve is an increasing function of the mode distance from the closest fundamental one.

Considering the cavity as the distributed version of a lumped-constant resonant circuit allows one to write the complex frequency response $F(\nu)$ in the form

$$F(\nu) = \sum_n \frac{F_n}{1 + i(\nu - \nu_n)2Q_n/\nu_n} + \int f(\nu, \nu')\, d\nu', \tag{VII.3.1}$$

where ν_n is the eigenfrequency of the nth mode, Q_n the quality factor measuring the sharpness of the nth resonance of the cavity, and F_n a weight factor depending on the resonator geometry. The integral on the right side of Eq. (VII.3.1) represents the contribution of the continuous part of the spectrum, which is always present in open systems. It is often useful to introduce a complex frequency $\tilde{\nu}_n = \nu_n + i\nu_n/(2Q_n)$, so that the denominators of the resonance factors in Eq. (VII.3.1) reduce to $i(\nu - \tilde{\nu}_n)2Q_n/\nu_n$.

The cavity spectrum can be considered discrete to the extent that the linewidth

$$\Delta\nu_n = \nu_n Q_n^{-1} \tag{VII.3.2}$$

of the nth mode is less than the distance of ν_n from the frequencies of the adjacent modes, apart from some occasional degeneracies.

According to Eq. (VII.3.1), by exciting the cavity with a time δ pulse, we generate a field proportional to $f(t)$, where (if we neglect the contribution of the continuous part of the spectrum)

$$f(t) = \sum_n F_n \exp\left(-\pi\nu_n \frac{t}{Q_n} + i2\pi\nu_n t\right). \tag{VII.3.3}$$

This shows that $Q_n/\nu_n\pi$ is the time constant of the nth mode oscillation damping. Assuming that only the nth mode is excited, Q_n can be written as

$$Q_n = 2\pi \frac{\text{energy stored}}{\text{energy lost per cycle}} = 2\pi\nu_n t_{\text{phot}}, \tag{VII.3.4}$$

where t_{phot} is the *lifetime of a photon in the cavity*. In a Fabry–Perot resonator the mode is made up of two planelike waves and, accordingly, the *energy lost per cycle* equals the *energy lost per pass* divided by the relative number of cycles performed. Consequently, if we indicate with α_n the loss per pass, Q_n can be expressed as

$$Q_n = k_n d/\alpha_n, \tag{VII.3.5}$$

where $k_n = 2\pi\nu_n/c$ is the wave number of the nth mode and d the cavity length.

Closely connected with the Q factor, the *finesse* F_n,

$$F_n = 1/\alpha_n = \nu_0/\Delta\nu_n, \tag{VII.3.6}$$

measures the resonance sharpness with respect to the free spectral range. All these quantities Q, F, t_{phot}, FSR, and α are interrelated. In particular, $Q = F\nu/\text{FSR}$ and $t_{\text{phot}} = F/\text{FSR} = 1/\alpha\ \text{FSR}$.

4 Ray Theory of a Closed Elliptic Resonator

In order to become acquainted with the basic features of the resonance field configurations, we consider in this section an exactly solvable model (in the ray optical limit) of an optical resonator consisting of a cylindrical cavity of elliptic cross section with semiaxes a and $d/2$ (Fig. VII.10). To keep the mathematics from becoming unwieldy, we examine a two-dimensional mode configuration in which $u(x, z)$ is independent of the y coordinate. In the ray optical limit, u can be expressed, as shown in Chapter II, by a combination of fields of the form $A\exp(-ikS)$, i.e.,

$$u(x, z) = A_1 e^{-ikS_1} + A_2 e^{-ikS_2} + \cdots + A_m e^{-ikS_m}. \tag{VII.4.1}$$

The splitting of u into m canonical functions, each representing a ray congruence, depends in general on the complexity of the boundaries. In the present case, because of the symmetry of the ellipse, u will be represented by

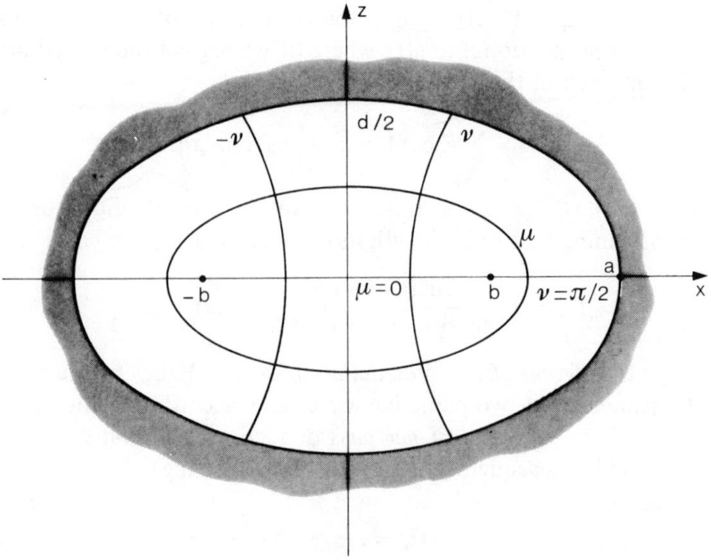

Fig. VII.10. Cross section of a cylindrical elliptic cavity. The coordinate curves μ = const and ν = const represent families of confocal hyperbolas and ellipses, respectively. The x axis is defined by $\nu = \pm\frac{1}{2}\pi$, $0 \leqslant \mu < \infty$ and $\mu = 0$, $-\pi < \nu < \pi$.

the sum of two functions (for example, for the ring cavity in Fig. VII.8 eight functions would be needed, two for each leg).

The S's can be obtained by solving the eikonal equation. In the present case, it is convenient to introduce a system of elliptic coordinates defined by Eq. (II.7.10), with $\mu \geq 0$ and $-\pi \leq \nu \leq \pi$, in which the curves μ = const represent ellipses with foci at $x = \pm b$ while the boundary of the cavity is defined by the curve $\mu = \bar{\mu}$. A simple way to solve the eikonal equation then is to look for solutions of the type (cf. Section II.12; do not confuse the coordinate ν with the symbol used to indicate a frequency)

$$S(\mu, \nu) = M(\mu) + E(\nu). \tag{VII.4.2a}$$

Proceeding in this way, we obtain (see Problem 1)

$$M(\mu) = \pm b \int_0^{\mu} (\sinh^2 \mu' + \cos^2 \nu_c)^{1/2}\, d\mu' + M(0), \tag{VII.4.2b}$$

$$E(\nu) = \pm b \int_0^{\nu} (\cos^2 \nu' - \cos^2 \nu_c)^{1/2}\, d\nu' + E(0). \tag{VII.4.2c}$$

Since the integrand of Eq. (VII.4.2c) becomes imaginary as $|\nu| > |\nu_c|$, the hyperbola of the equation $\nu = \pm \nu_c$ is the caustic of this ray congruence (Fig. VII.11c) and the mode configuration depends on the values taken on by $\cos \nu_c$. The field identified by each value of ν_c in the interval $(0, \pi/2)$ depends,

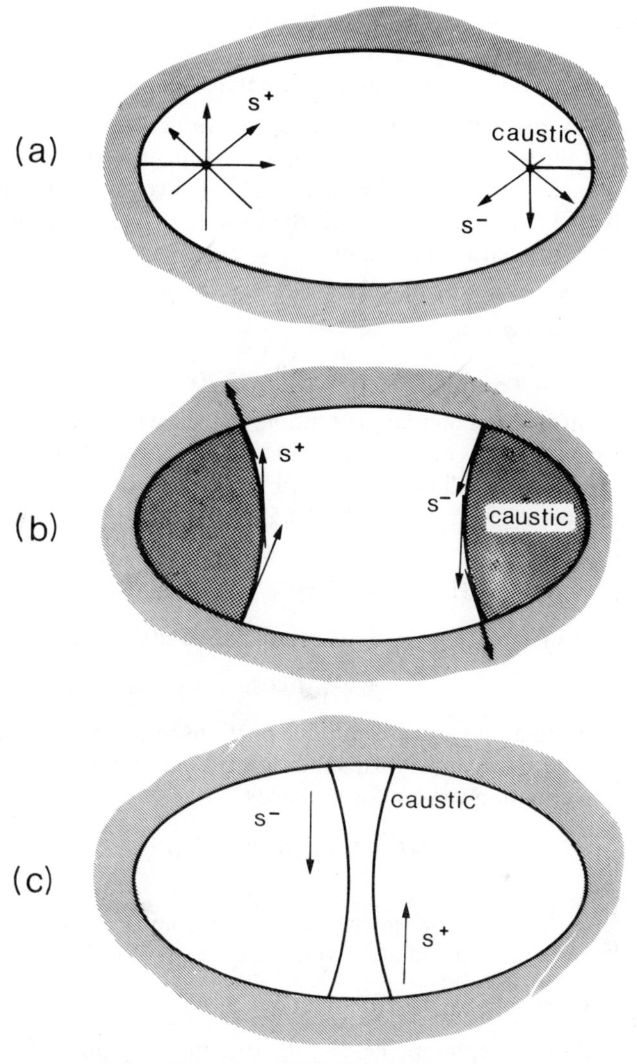

Fig. VII.11. Oscillation modes of an elliptic cavity obtained by superposition of two ray congruences moving upward and downward, respectively, and characterized by hyperbolic caustics having foci in $x = \pm b$. When the two caustics merge into the z axis, two imaginary foci appear in $z = \pm ib$ and the mode becomes gaussian. (a) Caustic merging into the z axis, (b) caustic with foci in $x = \pm b$, (c) caustic with foci in $z = \pm ib$.

as we will see below, on a discrete index n, so that the generic mode is labeled by the couple of continuous and discrete indices ν_c and n, respectively.

When $\nu_c = \pm\frac{1}{2}\pi$ the caustic collapses into straight half-lines departing from the foci of the ellipse and pointing toward $x = \pm\infty$ (Fig. VII.11a), while the wave fronts ($S =$ const) are circumferences centered at $x = \pm b$. The rays having a focus at $x = -b$ and pointing toward $z = +\infty$ are represented by an eikonal

$$S_-(\mu, \nu) = S_-(0, -\pi/2) - \operatorname{sgn}(z)b(\cosh\mu + \sin\nu), \qquad \text{(VII.4.3)}$$

sgn( ) being the sign function, which assumes the values $+1$ and -1 according to the sign of the argument and takes into account the fact that the wave fronts converge toward the focus for $z < 0$ and diverge from it for $z > 0$. Analogously, the rays passing through the focus $x = b$ and pointing toward $z = -\infty$ are associated with an eikonal

$$S_+(\mu, \nu) = S_+(0, \pi/2) + \operatorname{sgn}(z)b(\cosh\mu - \sin\nu), \qquad \text{(VII.4.4)}$$

so that we can finally write

$$u = A_+e^{-ikS_+} + A_-e^{-ikS_-} \equiv u_+ + u_-. \qquad \text{(VII.4.5)}$$

Since u must satisfy the boundary condition, the wave number k corresponding to the free oscillations of the cavity can take a discrete set of values, which will be indicated in the following by k_{res}. For the present example, imposing the vanishing of u, given by Eq. (VII.4.5), on the ellipse of the equation $\mu = \bar{\mu}$ yields

$$A_+(\bar{\mu}, \nu) = A_-(\bar{\mu}, \nu) \qquad \text{(VII.4.6a)}$$

$$\begin{aligned} k_{\mathrm{res}}[S_+(\bar{\mu}, \nu) - S_-(\bar{\mu}, \nu)] &= k_{\mathrm{res}}[S_+(0, \pi/2) - S_-(0, -\pi/2)] \\ &\quad \pm 2k_{\mathrm{res}}b\cosh\bar{\mu} = (2n_\pm + 1)\pi, \qquad \text{(VII.4.6b)} \end{aligned}$$

where the plus and minus signs refer to the portions of boundary with positive and negative z, respectively. Next, subtracting Eq. (VII.4.6b) written for n_- from the same equation written for n_+ yields

$$2k_{\mathrm{res}}(n, \pi/2)b\cos\bar{\mu} = (n_+ - n_-)\pi \rightarrow k_{\mathrm{res}} = n\pi/(2a), \qquad \text{(VII.4.7)}$$

with $n = n_+ - n_- = 1, \ldots$.

Consider, now, the case in which $\nu_c = 0$ (see Fig. VII.11c). Using Eqs. (VII.4.2) we can immediately verify that Eqs. (VII.4.3,4) modify into

$$S_\pm(\mu, \nu) = S_\pm(\mp i\pi/2, 0) + \operatorname{sgn}(z)b(\pm\sinh\mu + i\cos\nu). \qquad \text{(VII.4.8)}$$

The above eikonals represent fields with foci in the points of elliptic coordinates $\mu_f = \mp i/2$ and $\nu_f = 0$, which correspond to $x_f = 0$ and $z_f = \pm ib$. The corresponding fields $u_\pm$ read

$$\begin{aligned} u_\pm(\mu, \nu) &= \exp[-ikS_\pm(\mp i\pi/2, 0) \mp i\operatorname{sgn}(z)kb\sinh\mu + \operatorname{sgn}(z)kb\cos\nu] \\ &\cong \exp[-ikS_\pm(\mp i\pi/2, 0) \mp ikz + kb - \tfrac{1}{2}kb\,x^2/(b^2 + z^2)], \qquad \text{(VII.4.9)} \end{aligned}$$

where we have approximated $\cos\nu$ with $1 - \frac{1}{2}\nu^2$. While for a real focus the field amplitude $u_\pm$ is uniform on a phase front, for an imaginary focus it becomes gaussian; that is, it decays exponentially with the squared distance x^2 from the z-axis. The two fields of Eq. (VII.4.9) can be combined to generate an oscillation mode that satisfies a resonance condition similar to Eq. (VII.4.7), viz.

$$2k_{\mathrm{res}}(n,0)b\sinh\bar{\mu} = (n_+ - n_-)\pi \rightarrow k_{\mathrm{res}} = n\pi/d, \qquad n = 1,2,\ldots. \qquad \text{(VII.4.10)}$$

In general, each mode associated with a given value of v_c is characterized by a *hyperbolic caustic*. For $v_c = 0$ the two branches of the hyperbola tend to merge with the z axis and the field is confined along the minor axis of the cavity. The group of modes with $v_c > 0$ cover a large portion of the resonator volume and are termed *unstable*, while the modes with $v_c = 0$ are called *stable*. While in the latter case a large portion of the walls can be removed (Fig. VII.12a) without disturbing the field, the same cannot be done for the unstable modes (Fig. VII.12b), a fact that influences the *radiation losses*.

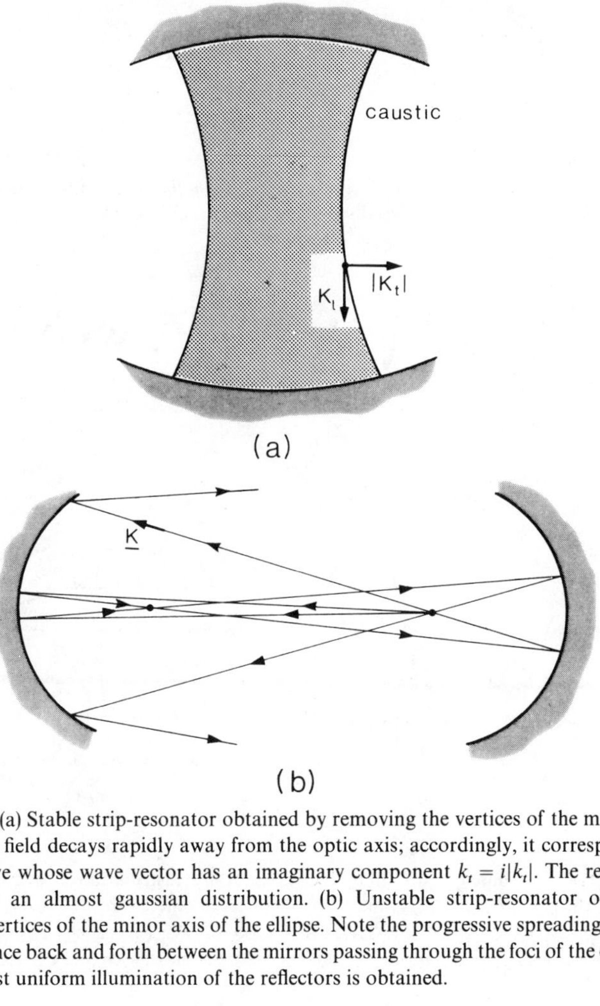

Fig. VII.12. (a) Stable strip-resonator obtained by removing the vertices of the major axis of the ellipse. The field decays rapidly away from the optic axis; accordingly, it corresponds to an evanescent wave whose wave vector has an imaginary component $k_t = i|k_t|$. The reflectors are illuminated by an almost gaussian distribution. (b) Unstable strip-resonator obtained by removing the vertices of the minor axis of the ellipse. Note the progressive spreading of the rays when they bounce back and forth between the mirrors passing through the foci of the ellipse. As a result, an almost uniform illumination of the reflectors is obtained.

5 Linear Resonators

The results we have obtained for the elliptic cavity are qualitatively true for most open resonators, which are generally composed of two mirrors facing each other (linear resonator; see Fig. VII.13). The field can be represented in

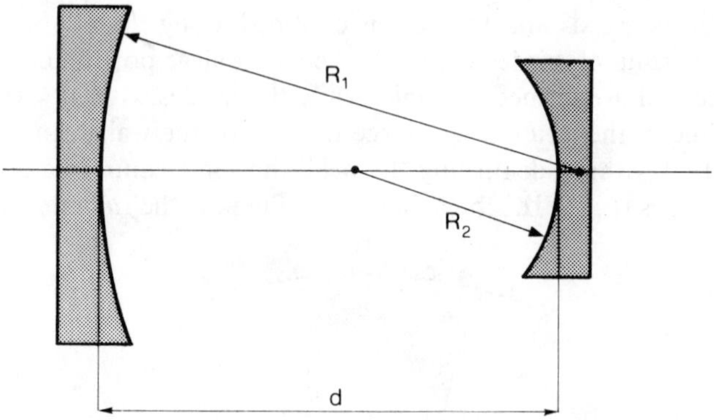

Fig. VII.13. Geometry of a simple *linear optical resonator*. The radiation is coupled outside the cavity by making one mirror partially transmitting, drilling a small hole on it, or using the rays scraping around the mirror edges (unstable cavities).

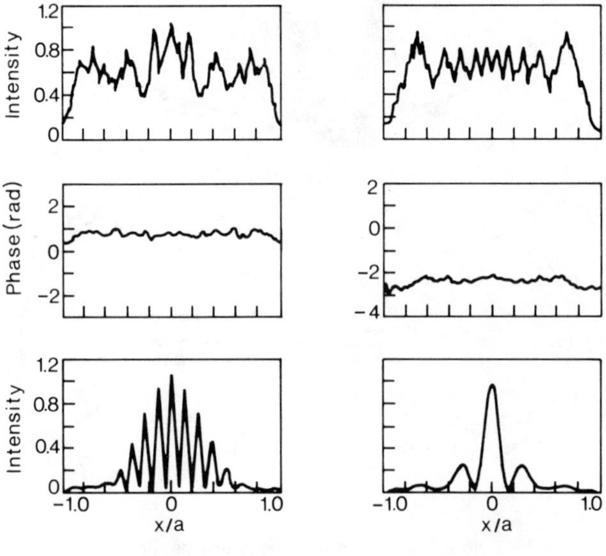

Fig. VII.14. Calculated near- and far-field intensity distributions for confocal unstable resonators. The two groups of diagrams were obtained for different outer Fresnel numbers N and magnifications M: (a) $N = 30$, $M = 1.42$, (b) $N = 60$, $M = 5$. For each group the top diagram represents the near field, the bottom one the fair field. (From Rensch and Chester [13a].)

the ray optical form $\exp(-ikS)$, and this assumption, together with the fact that actual resonators have one dimension, which defines the *optic axis*, much longer than the transverse ones, amounts to approximating the modes as *transverse electromagnetic* (*TEM*) *waves*. These circumstances will be instrumental in the following sections in permitting the use of the paraxial optics approximation on which the ray matrix formalism hinges. In addition, the modes can be classified into stable and unstable ones. While the closed elliptic cavity can support both kinds of modes, the open resonators generally admit only one kind of oscillation and they are accordingly termed *stable or unstable*. The modes of a stable resonator are concentrated near the optic axis and are, in general, notably insensitive to the mirror size. The modes of an unstable resonator fill the entire cavity volume and depend critically on the shape and size of the reflectors. A rippled output beam is a typical feature of these cavities used when a large mode volume is requested, as in high-power lasers (Fig. VII.14).

The component $\mathbf{k}_t$ of the wave vector perpendicular to the optic axis is imaginary in a stable cavity: in fact, a real $\mathbf{k}_t$ would imply a cross-section change under reflection (Fig. VII.12).

A stable beam appears limited by a caustic surface, which replaces the missing sidewalls in keeping the field confined near the axis. When the mirror diameter is sufficiently large, no edge effect occurs and the resonator is free of diffraction effects.

Among all the possible combinations of curvature radii and axial distances d, only a few configurations are currently used (Fig. VII.15). The *plane-parallel*

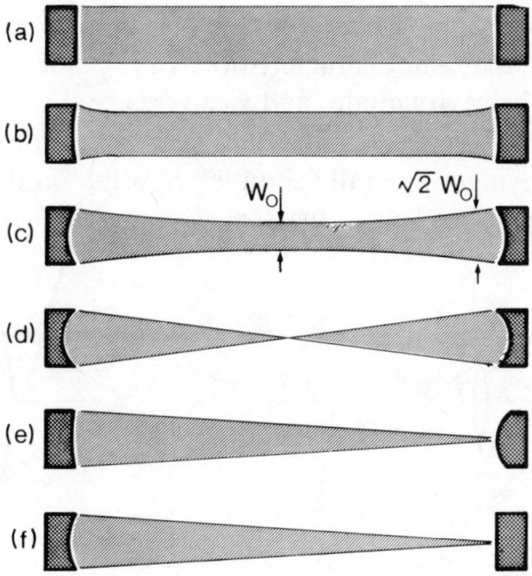

Fig. VII.15. Intracavity radiation pattern for (a) plane-parallel, (b) large-radius mirror, (c) confocal, (d) spherical, (e) concave–convex, and (f) hemispherical cavities. (From Bloom [13b].)

cavity, derived from the Fabry–Perot resonator, is the ancestor of all open resonators. The *symmetric confocal* cavity has two equal mirrors separated by a distance d equal to their curvature radius. By spacing the reflectors at a distance equal to twice the curvature radius, the *spherical configuration* is obtained. The *hemispherical* and *hemiconfocal configurations* have one plane mirror and half the length of the original spherical and confocal configurations. In particular, the plane-parallel and spherical cavities are *barely stable* in the sense that a slight deviation of the geometric parameters from the ideal ones makes these resonators unstable.

6 Characterization of Resonators by Means of Lens Sequences and g Parameters

In spite of the great variety of parameters characterizing the position and curvature radii of the mirrors, all resonators may be regarded as periodic focusing systems in which light rays, bouncing back and forth between the mirrors, have their trajectory periodically redirected by two curved surfaces. In the paraxial optics approximation a mirror and a thin lens of equal focal length are wholly equivalent, so that the cavity may be treated as an unlimited series of lenses whose focal lengths are alternately those of the two mirrors and whose spacing d equals the resonator length (Fig. VII.16) (the convention of considering positive the radius of curvature of a concave mirror and negative that of a convex one is usually adopted). Accordingly, the wave front behavior in the cavity may be studied by considering its evolution when traveling through a lumped-constant guide (consisting of an infinite sequence of lenses L_i, such that L_i has the same characteristics as L_{i+2}), the relative eigenmodes corresponding to the cavity modes and vice versa.

The characteristics of these modes can be inferred from the imaging of a point source occupying a generally complex position on the lens-guide optic axis. Proceeding from one lens to the next, L_i converts the ith image into the

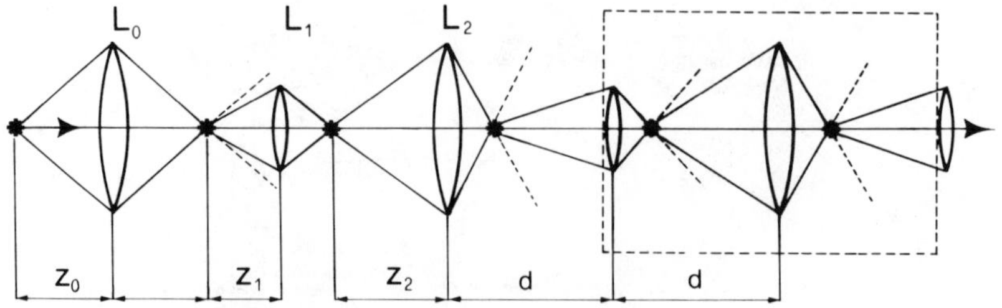

Fig. VII.16 Sequence of lenses equivalent to the resonator of Fig. VII.13. The distance z_i is taken positive when it corresponds to a focus lying on the left of the lens L_i.

$(i + 1)$st one at a distance z_{i+1} from L_{i+1}. In general, both z_i and z_{i+1} are different from z_0 (the distance of the point source from L_0), unless z_0 takes on certain particular values. In fact, a periodic image distribution is obtained when $z_2 = z_0$, so that, according to the lens equation, the condition for mutual imaging reads

$$1/z_0 + 1/(d - z_1) = 1/f_1, \qquad 1/z_1 + 1/(d - z_0) = 1/f_2. \tag{VII.6.1}$$

Solving for z_0, we get the two roots representing the distances of the two beam foci from mirror 1:

$$z_0^{(\pm)} = d\{1/2 + [g_1 - g_2 \pm 2(g_1^2 g_2^2 - g_1 g_2)^{1/2}]/[2(g_1 + g_2 - 2g_1 g_2)]\}, \tag{VII.6.2}$$

where we have introduced the *g parameters*, defined by

$$g_{1,2} = 1 - d/(2f_{1,2}). \tag{VII.6.3}$$

In the confocal configuration the foci of the two mirrors coincide ($d = f_1 + f_2$), i.e., in terms of g parameters

$$g_1 + g_2 = 2g_1 g_2. \tag{VII.6.4}$$

Therefore, if $g_1^2 g_2^2$ is greater than $g_1 g_2$, then one beam is collimated and the other shares the focus with the mirrors (Fig. VII.17), according to

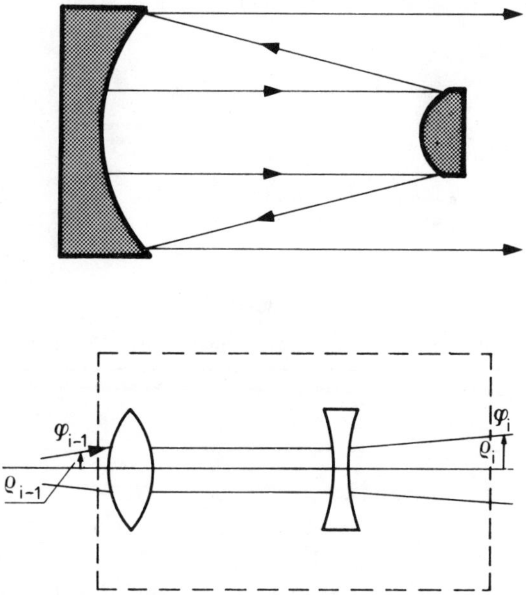

Fig. VII.17. Lens sequence equivalent to a confocal resonator. Note the beams collimated and confocal with the mirrors.

Eq. (VII.6.2), which in this case reads

$$z_0^{(-)} = d/[2(1 - g_1)] \qquad \text{and} \qquad z_0^{(+)} = \infty. \tag{VII.6.5}$$

The source position turns out to be either real or complex according to the sign of the radicand appearing in Eq. (VII.6.2); in particular, it is complex (which will be seen to correspond to stable resonators) if

$$1 \geq g_1 g_2 \geq 0, \tag{VII.6.6}$$

while it is real for unstable resonators.

It will be clear from the properties of the gaussian beams associated with a complex position of the source (see Section VII.7) that the present classification coincides with that given in Section VII.4. Hence, the cavity modes will be described by gaussian beams if and only if the product $g_1 g_2$ satisfies the inequality (VII.6.6).

The locations of the complex foci for the most commonly used resonators are listed in Table VII.1. Notice that plane-parallel and spherical resonators have real foci, in agreement with the denomination of *barely stable* used at the end of Section VII.5.

Within the approximation of paraxial optics, an arbitrary field distribution u on the plane $z = z_0$ generates a field that is periodically reproduced. Thus, we are led to look for fields having the form

$$\begin{aligned} u(\boldsymbol{\rho}, z) &= 2ik \iint_{-\infty}^{+\infty} u(\boldsymbol{\rho}_0, z_0) G(\boldsymbol{\rho} - \boldsymbol{\rho}_0, z - z_0)\, dx_0\, dy_0 \\ &= 2ik \sum_{lm} \frac{f_{lm}(z_0)}{l!\,m!} u_{lm}(\boldsymbol{\rho}, z - z_0), \end{aligned} \tag{VII.6.7a}$$

Table VII.1

Foci of the Spherical Waves Forming the Fundamental Modes

Configuration	$\dfrac{z_0^+}{d}$	$\dfrac{z_0^-}{d}$
Plane-parallel	∞	∞
Large-radius mirrors	$1/2 + i\dfrac{1}{[2(1-g)]^{1/2}}$	$1/2 - i\dfrac{1}{[2(1-g)]^{1/2}}$
Confocal	$1/2 + i/2$	$1/2 - i/2$
Spherical	$1/2$	$1/2$

where G is the Green's function of a homogeneous region,

$$f_{lm}(z_0) = \iint_{-\infty}^{\infty} u(\boldsymbol{\rho}_0, z_0) x_0^l y_0^m \, dx_0 \, dy_0, \qquad \text{(VII.6.7b)}$$

and we have expanded G in a power series in $x_0^l y_0^m$ by introducing the functions

$$u_{lm}(\boldsymbol{\rho}, z) \equiv (\partial^l/\partial x_0^l)(\partial^m/\partial y_0^m) G(\boldsymbol{\rho} - \boldsymbol{\rho}_0, z)|_{x_0 = y_0 = 0} \qquad \text{(VII.6.7c)}$$

which represent the field radiated by a multipole localized at z_0 of lth order in the x direction and mth order in the y direction. As a consequence, the most general field u can be decomposed into the superposition of modes u_{lm} having a focus at $z = z_0$ on the optic axis.

7 Fields Associated with Sources Located at Complex Points

We have seen above that, as long as the source occupies a real position, the field propagating along the lens cascade is completely defined by the series of focal points given by Eq. (VII.6.2) (see the example illustrated in Fig. VII.16). The case of a source lying at a point of complex abscissa does not admit an equally obvious physical interpretation. Actually, we would be tempted to discard such a possibility and conclude that for a complex z_0 the propagation along the lens sequence is interdicted, by analogy with metallic waveguides below cutoff. However, the existence in these structures of a stationary field decaying exponentially along the waveguide axis induces us to presume that a similar field exists in our case and decays in the directions transverse to the optic axis.

After these preliminaries, consider a source S lying at a point P with coordinates $(0, 0, -ib)$. The radiated field is proportional to the scalar Green's function G, which, assuming $\rho = (x^2 + y^2)^{1/2}$ to be negligible with respect to $|z + ib|$, is expressed by the Fresnel formula (see Section IV.10)

$$G = 1/[4\pi(z + ib)] \exp[-ik(z + ib + (1/2)\rho^2/(z + ib))]. \qquad \text{(VII.7.1)}$$

According to this definition, G represents a wave traveling from $z = -\infty$ to the source and from the source to $z = +\infty$; therefore, G *does not* satisfy uniformly Sommerfeld's radiation condition and cannot be considered as an outgoing Green's function.

If we set

$$1/q(z) \equiv 1/(z + ib) = z/(z^2 + b^2) - ib/(z^2 + b^2) \equiv 1/R(z) - i\lambda/[\pi w^2(z)], \qquad \text{(VII.7.2a)}$$

$$\psi(z) \equiv \arctan(z/b), \qquad \text{(VII.7.2b)}$$

we have

$$1/(z + ib) = -(i/b)[w_0/w(z)]e^{i\psi(z)}, \tag{VII.7.2c}$$

where $w_0 = w(0) = (\lambda b/\pi)^{1/2}$, so that, by choosing a positive b in order to guarantee a decrease of G away from the optic axis,

$$G \cong -\frac{i}{4\pi b}\frac{w_0}{w(z)}\exp\left[-i\left(kz - \psi + \frac{k\rho^2}{2R}\right) - \frac{\rho^2}{w^2(z)} + kb\right]$$

$$= \frac{1}{4\pi q(z)}\exp\left[-i\left(kz + \frac{\rho^2}{2q(z)}\right) + kb\right]. \tag{VII.7.3}$$

According to Eq. (VII. 7.3), $w(z)$ is the radius for which the field amplitude falls off to $1/e$ of its maximum value. In contrast to the case of a spherical wave, the field irradiated by a complex point exhibits a gaussian amplitude distribution on a spherical surface of constant phase and the radiation essentially flows through a section of radius $w(z)$ of the xy plane. If w is very small, the field may be conceived of as a pencil of rays, parallel to the z axis. In conclusion, the transition from real to complex positions transforms, so to speak, a spherical wave into a beam of rays, more or less well collimated.

The width $w(z)$ of the gaussian *beam spot*, which changes with z according to

$$w^2(z) = w_0^2(1 + z^2/b^2), \tag{VII.7.4}$$

has a minimum for $z = 0$, which justifies the use of the *beam waist* w_0 as a characteristic parameter. The beam is limited by a hyperboloid whose intersections with planes perpendicular to z are circumferences of radius $w(z)$ (Fig. VII.18). The quantity b may be interpreted as the distance from the waist within which the beam remains fairly well collimated. Beyond the range $z = b$ (also indicated by z_R), which is often called the *Rayleigh distance* of the beam, diffraction effects make the beam divergent.

Moving away from the waist, the beam contour hyperboloid $w(z)$ eventually takes on the shape of a cone whose half-angle equals the far-field

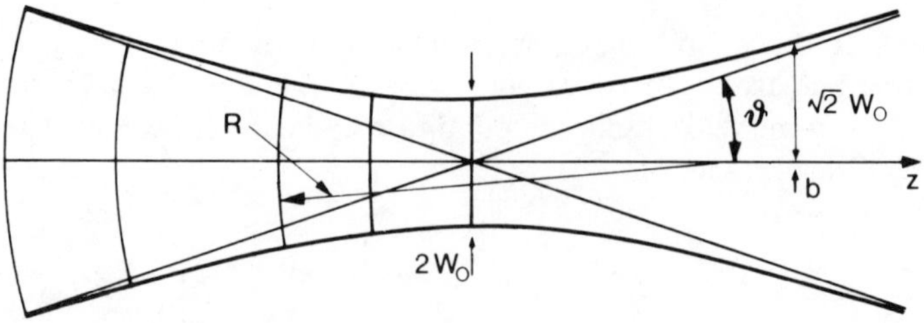

Fig. VII.18. Side view of a gaussian beam showing the wave fronts, waist, Rayleigh distance b, and diffraction angle θ.

half-divergence θ_{beam}, given by

$$\theta_{\text{beam}} = \arctan[\lambda/(\pi w_0)] \cong \lambda/(\pi w_0) = [\lambda/(\pi b)]^{1/2}. \qquad \text{(VII.7.5)}$$

According to Eq. (VII.7.5), the narrower the waist the wider the beam aperture, as implicit in the fact that the far-field is the Fourier transform of the waist field. This situation reproduces that of the main radiation lobe of an aperture antenna.

Referring to Eq. (VII.7.2a), it is immediately seen that the wave front radius R, given by $R(z) = z + b^2/z$, grows indefinitely large at the waist. Moving away from the waist, R at first decreases and attains its minimum value $2b$ at a distance b, then increases again and eventually reaches z.

The above analysis assumes that the scalar Green's function G provides a fully rigorous representation of the field vectors **E** and **H**. Without going into details, it is worth mentioning the conclusion of a vector analysis of wave beams developed by Goubau and Schwering [14]. Their electromagnetic treatment, proceeding from Maxwell's equations, shows that the wave beams are closely represented by expressions proportional to that provided by Eq. (VII.7.3) when they are slightly divergent (a few degrees). This result agrees with the fact that only well-collimated beams can be treated as TEM waves, to which the scalar theory applies correctly.

8 Hermite–Gauss and Laguerre–Gauss Beams

The analogy between gaussian beams and spherical waves can be further exploited to find out the expressions for the multipolar fields associated with wave beams. In fact, spherical waves belong to the wider family of multipolar fields, within which they represent the zeroth-order term. It may be conjectured that this is also the case with gaussian beams, as suggested by Eq. (VII.6.7).

Since the multipolar fields defined by Eq. (VII.6.7c) are not mutually orthogonal on the transverse plane, it is convenient to find an alternative family of multipolar fields, the so-called *beam modes* to be used as a basis for expanding a general field. To this end, let us write the field in the form $u(x, y, z) = A(x, y, z)e^{-ikz}$, where $k = k' + ik''$ is a generally complex propagation constant. Assuming that u propagates paraxially, we can easily prove that A satisfies the parabolic wave equation discussed in Section II.6. We look for a trial solution of the form

$$\begin{aligned} u(\boldsymbol{\rho}, z) &= [1/w(z)]f[2^{1/2}\rho/w(z)]\exp\{-i[\beta(z) + [k/(2q)]\rho^2 + kz]\} \\ &= A(\boldsymbol{\rho}, z)\exp(-ikz) \end{aligned} \qquad \text{(VII.8.1)}$$

with $\boldsymbol{\rho} = \hat{x}x + \hat{y}y$, where $w(z)$ and $q(z)$ are, respectively, the spot size and the complex curvature of a gaussian beam, defined by Eq. (VII.7.2) with $\lambda = 2\pi/k'$, and $\beta(z)$ is a phase factor (not to be confused with the β used to indicate the propagation constant) to be determined together with the unknown function f. Inserting Eq. (VII.8.1) into Eq. (II.6.12) yields

$$[\nabla_t^2 - (4/w^2)\boldsymbol{\rho} \cdot \boldsymbol{\nabla}_t]f = (4/w^2 + 2k\, d\beta/dz)f = \mu(w)f, \qquad \text{(VII.8.2)}$$

with $\boldsymbol{\nabla}_t = \hat{x}\,\partial/\partial x + \hat{y}\,\partial/\partial y$. Assuming f to be square integrable on the plane (x, y), Eq. (VII.8.2) uniquely defines f as an eigenfunction of the differential operator on the left (depending on the variables x and y only) with eigenvalue $\mu(w)$. By recalling the differential equation satisfied by the Hermite polynomials,

$$d^2H_m(t)/dt^2 - 2t\, dH_m(t)/dt = -2mH_m(t), \qquad \text{(VII.8.3)}$$

we can immediately prove that the normalized eigenfunctions of Eq. (VII.8.2) in cartesian coordinates are

$$f^{(H)}(x, y) = (\pi 2^{l+m-1} l!\, m!)^{-1/2} H_l(2^{1/2}x/w) H_m(2^{1/2}y/w), \qquad \text{(VII.8.4)}$$

with corresponding eigenvalues given by

$$\mu_{lm}(w) = -(4/w^2)(l + m). \qquad \text{(VII.8.5)}$$

A comparison of the expressions for $\mu(w)$ given by Eqs. (VII.8.2) and (VII.8.5) yields the differential equation

$$(1 + ik''/k')\, d\beta_{lm}/dz = -2(m + l + 1)/[k'w^2(z)] = -(m + l + 1)b/(z^2 + b^2), \qquad \text{(VII.8.6)}$$

whose solution provides for $\beta_{lm} = \beta'_{lm} - i\beta''_{lm}$ the complex expression

$$\beta_{lm}(z) = \frac{(l + m + 1)}{1 + ik''/k'} \arctan \frac{\lambda z}{\pi w_0^2} = \frac{(l + m + 1)}{1 + ik''/k'} \psi(z), \qquad \text{(VII.8.7)}$$

where $\psi(z)$ is defined by Eq. (VII.7.2b).

It should be noted that, according to Eq. (VII.8.1), the curvature of the phase fronts is the same for modes of all orders l and m, while the *phase velocity* v_{lm} is a decreasing function of the mode indices and an increasing function of the distance from the waist:

$$v_{lm}(z) = \frac{\omega}{k' + (d/dz)\beta'_{lm}}$$

$$= \frac{\omega/k'}{1 + [(l + m + 1)/(bk')]\{1/[1 + (z/b)^2]\}\{1/[1 + (k''/k')^2]\}}. \qquad \text{(VII.8.8)}$$

Table VII.2

Hermite and Generalized Laguerre Polynomials of Low Order

$H_0(t) = 1$	$L_0^l(t) = 1$
$H_1(t) = 2t$	$L_1^l(t) = l + 1 - t$
$H_2(t) = 4t^2 - 2$	$L_2^l(t) = \frac{(l+1)(l+2)}{2} - (l+2)t + \frac{t^2}{2}$
$H_3(t) = 8t^3 - 12t$	$L_3^l(t) = \frac{(l+1)(l+2)(l+3)}{6} - \frac{(l+2)(l+3)}{2}t$
$H_{n+1} = 2tH_n - 2nH_{n-1}$	$+\frac{l+3}{2}t^2 - \frac{t^3}{6}$
$H_n(t) = (-1)^n e^{t^2}\frac{d^n}{dt^n}(e^{-t^2})$	$L_n^l(t) = \frac{1}{n!}e^t t^{-l}\frac{d^n}{dt^n}(e^{-t}t^{n+l})$
$= 2^n t^n - 2^{n-1}\binom{n}{2}t^{n-2}$	$= \sum_{m=0}^{n}(-1)^m\binom{n+l}{n-m}\frac{t^m}{m!}$
$+ 2^{n-2}1\cdot 3\binom{n}{4}t^{n-4}$	$(n+1)L_{n+1}^l = (2n+l+1-t)L_n^l - (n+l)L_{n-1}^l$
$- 2^{n-3}1\cdot 3\cdot 5\binom{n}{6}t^{n-6}$	

We show in Table VII.2 some of the lowest-order modes $u_{lm}^{(H)} = A_{lm}^{(H)} \exp(-ikz)$, with

$$u_{lm}^{H}(\boldsymbol{\rho}, z) = \frac{(\pi 2^{l+m-1} l!m!)^{-1/2}}{w(z)} H_l\left(2^{1/2}\frac{x}{w}\right)H_m\left(2^{1/2}\frac{y}{w}\right) \times \exp\left[-i\left(-\frac{l+m+1}{1+ik''/k'}\psi(z) + \frac{k\rho^2}{2q} + kz\right)\right]. \quad \text{(VIII.8.9)}$$

For assigned k, the functions u_{lm}, regarded as eigenfunctions of the Laplacian ∇^2 of eigenvalue k^2, form a complete and orthonormal set of beam modes that replaces the multipolar fields of Eq. (VII.6.7c). The upper index (H) is used to distinguish these beams used in connection with cartesian coordinates, where Hermite polynomials are the natural choice, from those used in polar coordinates, where Laguerre polynomials come into play. In fact, Eq. (VII.8.2) also admits the following set of solutions, expressed as a combination of Laguerre polynomials and trigonometric functions:

$$f_{pl}^{(L)} = \sqrt{\frac{2p!}{\pi(l+p)!}}\left(2^{1/2}\frac{\rho}{w}\right)^l L_p^l\left(2\frac{\rho^2}{w^2}\right)\begin{Bmatrix}\sin l\phi \\ \cos l\phi\end{Bmatrix}, \quad \text{(VII.8.10)}$$

where L_p^l is the Laguerre polynomial of *radial index* p and *angular index* l, as

Table VII.3

Gaussian Beams for Rectangular $[u_{lm}^{(\mathrm{H})}(xyz)]$ and Cylindrical $[u_{pl}^{(\mathrm{L})}(\rho, z)]$ Geometries

$u_{00}^{(\mathrm{H})}(xyz)=u_{00}^{(\mathrm{L})}(\rho,z)=(\sqrt{2/\pi}/w)\exp[-i(kz+k\rho^2/2R-\psi)]\exp(-\rho^2/w^2)$	
$w(z)=w_0(1+z^2/b^2)^{1/2}$, $R(z)=z+b^2/z$, $\psi(z)=\arctan(z/b)/(1+ik''/k')$	
$u_{01}^{(\mathrm{H})}=2(y/w)e^{i\psi}u_{00}$	$u_{10}^{(\mathrm{L})}=(1-2\rho^2/w^2)e^{2i\psi}u_{00}$
$u_{10}^{(\mathrm{H})}=2(x/w)e^{i\psi}u_{00}$	$u_{11}^{(\mathrm{L})}=2(\rho/w)(1-\rho^2/w^2)e^{i3\psi}\begin{Bmatrix}\sin\phi\\ \cos\phi\end{Bmatrix}u_{00}$
$u_{11}^{(\mathrm{H})}=4(xy/w^2)e^{2i\psi}u_{00}$	$u_{20}^{(\mathrm{L})}=[1-4(\rho^2/w^2)+2(\rho^4/w^4)]e^{4i\psi}u_{00}$
$u_{20}^{(\mathrm{H})}=(1/\sqrt{2})[4(x^2/w^2)-1]e^{2i\psi}u_{00}$	$u_{21}^{(\mathrm{L})}=\sqrt{2/3}(\rho/w)[3-6(\rho^2/w^2)+2(\rho^4/w^4)]e^{5i\psi}\begin{Bmatrix}\sin\phi\\ \cos\phi\end{Bmatrix}u_{00}$
$u_{02}^{(\mathrm{H})}=(1/\sqrt{2})[4(y^2/w^2)-1]e^{2i\psi}u_{00}$	$u_{22}^{(\mathrm{L})}=(2/\sqrt{3})(\rho^2/w^2)(3-4\rho^2/w^2+\rho^4/w^4)e^{6i\psi}\begin{Bmatrix}\sin 2\phi\\ \cos 2\phi\end{Bmatrix}u_{00}$

can be easily proved by retracing the steps leading to Eq. (VII.8.4). As a consequence of the presence of the Laguerre polynomials in place of the Hermite ones, the phase factor β_{pl} becomes

$$\beta_{pl}^{(\mathrm{L})}(z) = [(2p + l + 1)/(1 + ik''/k')]\psi(z), \qquad \text{(VII.8.11)}$$

so that $u_{pl}^{(\mathrm{L})}$ reads

$$u_{pl}^{(\mathrm{L})} = \frac{1}{w(z)}\sqrt{\frac{2p!}{\pi(l+p)!}}\left(2^{1/2}\frac{\rho}{w}\right)^{l} L_p^l\left(2\frac{\rho^2}{w^2}\right)\begin{Bmatrix}\sin l\phi\\ \cos l\phi\end{Bmatrix}$$
$$\times \exp\left[-i\left(-\frac{2p+l+1}{1+ik''/k'}\psi(z)\right) + \frac{k\rho^2}{2q} + kz\right]. \qquad \text{(VII.8.9')}$$

Some of the lowest-order modal fields u_{pl} are shown in Table VII.3. As can be readily verified, all the $u_{pl}^{(\mathrm{L})}$ can be expressed as linear combinations of $u_{lm}^{(\mathrm{H})}$ [for example, $u_{10}^{(\mathrm{L})} = (u_{20}^{(\mathrm{H})} + u_{02}^{(\mathrm{H})})/4$]. Figure VII.19 shows the intensity distributions of a few TEM_{pl} modes and also that of TEM_{01*}, a hybrid mode with an annular distribution generated by the incoherent superposition of two TEM_{01} modes proportional to $\cos\phi$ and $\sin\phi$, respectively,

$$u_{0.1*}(x, y) \propto (a\sin\phi + b\cos\phi)\rho\exp(-\rho^2/w^2), \qquad \text{(VII.8.12)}$$

where a and b fluctuate independently of each other (due, for example, to laser instabilities). The average intensity I_{01*} of TEM_{01*} reads (see Figs. VII.19,20)

$$I_{01*} \propto \langle a^2\rangle\rho^2\exp(-2\rho^2), \qquad \text{(VII.8.13)}$$

when $\langle a^2\rangle = \langle b^2\rangle$.

The Laguerre–Gauss modes are axially symmetric for $l = 0$ and present

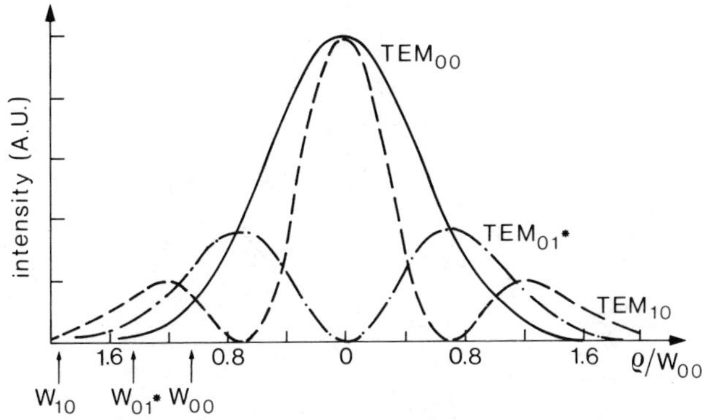

Fig. VII.19. Intensity distribution of some modes relative to cylindrical coordinates.

a number of dark rings equal to p. For $p = 0$, $u_{0l}^{(L)}$ has $2l$ radial nodal lines. As the radial and angular indices grow, so does the effective mode section. In the case of Hermite–Gauss modes, the dark rings are replaced by lines parallel to the symmetry axes, their number coinciding with the relative indices. Figure VII.21 shows the mode patterns and field configurations for linearly polarized Laguerre–Gauss modes (note the doughnut-shaped pattern of TEM_{01*}).

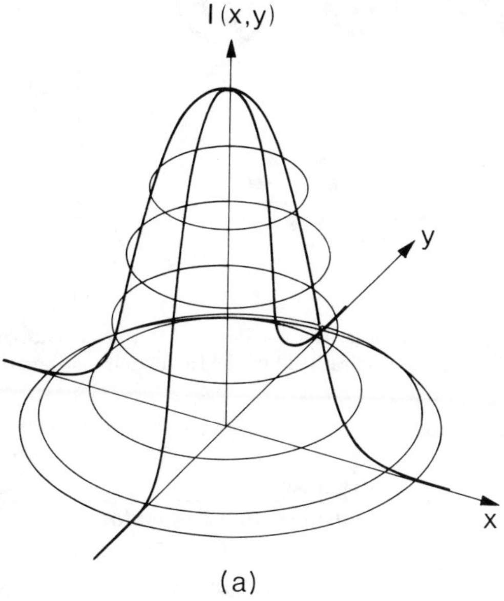

Fig. VII.20. Intensity distributions of (a) the fundamental and (b) the hybrid TEM_{01*} gaussian mode.

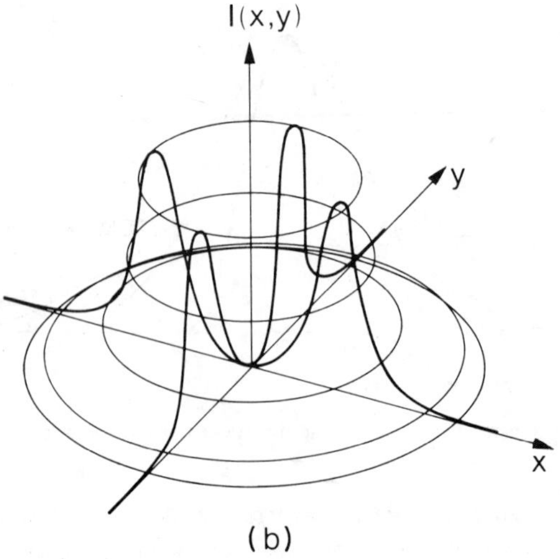

Fig. VII.20. (*continued*)

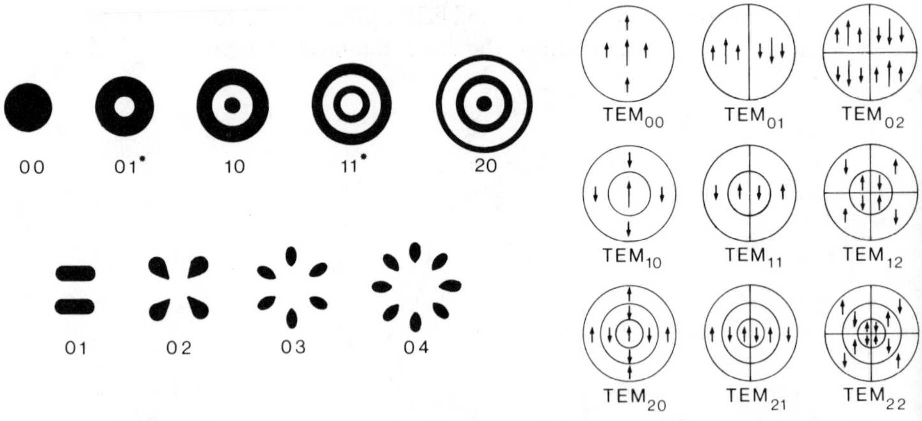

Fig. VII.21. Mode patterns and field configurations of some gaussian beam modes. The arrow lengths and directions are indicative of the field amplitudes and polarizations of Laguerre–Gauss modes.

9 Ray-Transfer Matrix Formalism for a Lens Waveguide Equivalent to a Resonator

As an alternative method for solving the system of equations for the successive images of an axial source [see Eq. (VII.6.1)], we can resort to the *ray-transfer matrix* formalism (see Section II.15), particularly when the cavity

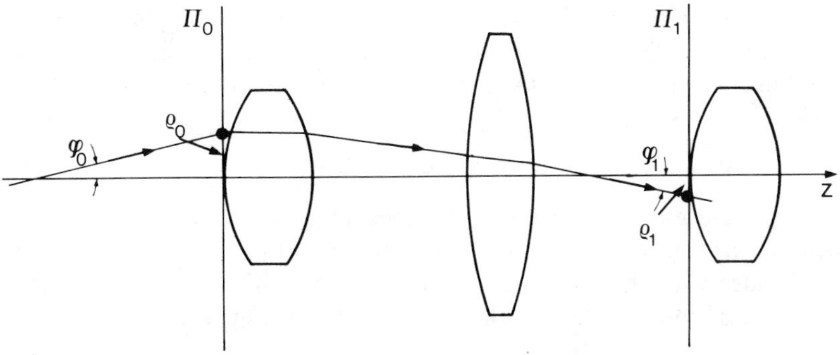

Fig. VII.22. Elements of a lens sequence. The planes tangent to the lens vertices represent the input and output of a generic block (cf. the rectangular region of Fig. VII.16).

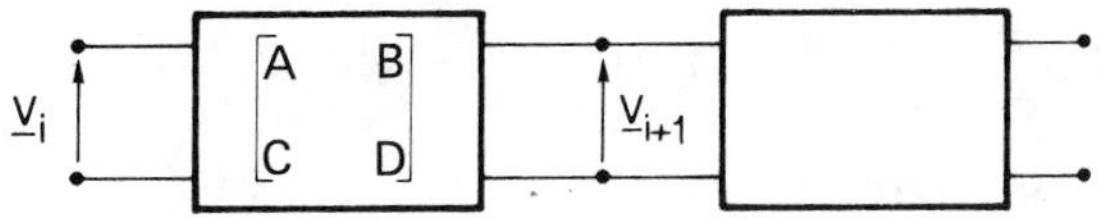

Fig. VII.23. Cascade of quadrupoles equivalent to a lens sequence. Here $\mathbf{v}_i = (\rho_i, \varphi_i)$ is a two-component vector representing the slope ϕ_i and the height ρ_i of a ray entering the ith block (cf. Section II.15).

consists of several optical elements. This method of analysis is essentially that developed by Pierce to analyze the passage of an electron beam through a sequence of electron lenses (see, e.g., Chapter II, Steffen [27]).

By grouping the lens sequence equivalent to the resonator into blocks formed by the least number of elements (Fig. VII.22), we set

$$\mathbf{v}_{i+1} = \gamma \mathbf{v}_i, \tag{VII.9.1}$$

where the vector $\mathbf{v}_i$ (see Fig. VII.23) depends on the height ρ_i and the slope ϕ_i of a meridional ray entering the ith block, while γ depends on the phase shift and attenuation of $\mathbf{v}_i$ in passing through the block.

9.1 *Magnification*

By introducing the ray-transfer matrix $\mathbf{S}$ of the blocks, Eq. (VII.9.1) can be rewritten as

$$\mathbf{S} \cdot \mathbf{v} = \gamma \mathbf{v}, \tag{VII.9.2}$$

which defines γ as an eigenvalue of $\mathbf{S}$. From the form of this matrix [see Eq. (II.15.21)], recalling that $\det|\mathbf{S}| = 1$, it follows that

$$\gamma_\pm = (A + D)/2 \pm i\{1 - [(A + D)/2]^2\}^{1/2} \equiv e^{\pm i\delta}, \tag{VII.9.3}$$

where $\delta = \arccos[\frac{1}{2}(A + D)]$. According to whether δ is complex or real, the amplitude of $\mathbf{v}$ changes or stays the same along the multipole cascade. Therefore, we can generalize the condition [Eq. (VII.6.6)] for a resonator to be *stable* to

$$|A + D| < 2. \tag{VII.9.4}$$

If γ is real and greater than one, it represents the *magnification* M of a beam when passing through one block. With reference to Figs. VII.17,23, we see that M coincides with the ratio ρ_{i+1}/ρ_i of the heights of a ray at the terminal sections of a block, so that for the lens sequence of Fig. VII.16,

$$\gamma = M = \rho_{i+1}/\rho_i = [z_1/(d - z_1)][z_2/(d - z_2)], \tag{VII.9.5}$$

which, using Eq. (VII.6.2), furnishes the *round-trip magnification* M of a cavity of parameters g_1 and g_2,

$$M = [1 + (1 - g_1^{-1}g_2^{-1})^{1/2}]/[1 - (1 - g_1^{-1}g_2^{-1})^{1/2}] \tag{VII.9.6}$$

9.2 *Complex Curvature*

Once γ is known, the vectors $\mathbf{v}$ can be expressed by means of the eigenvectors $\mathbf{v}_\pm$ of $\mathbf{S}$ in the form

$$\mathbf{v}_\pm = \begin{bmatrix} -B \\ A - \gamma_\pm \end{bmatrix}, \tag{VII.9.7}$$

so that the slopes $\phi_\pm$ of the rays represented by $\mathbf{v}_\pm$ are related to the heights through

$$\rho_\pm/\phi_\pm = -B/(A - \gamma_\pm). \tag{VII.9.8}$$

Since in the ray optical limit the field can be represented as $u(x, y) \propto e^{-ikS(x,y)}$, and the slope ϕ of a ray is related to S through $\phi = \partial S/\partial\rho$, Eq. (VII.9.8) gives

$$S_\pm(\rho) = \tfrac{1}{2}\rho^2(\gamma_\pm - A)/B = \tfrac{1}{2}\rho^2/q_\pm, \tag{VII.9.9}$$

where $q_\pm$, the generally complex curvature of the wave front, obeys the relation

$$1/q_\pm = -(1/2B)\{(A - D \pm i[4 - (A + D)^2]^{1/2}\}, \tag{VII.9.10}$$

which confirms Eq. (VII.9.4) as the condition for a resonator to support gaussian modes.

The curvature of a beam at the output of each block can be obtained by using the simple relation

$$q_{\text{out}} = (Aq_{\text{in}} + B)/(Cq_{\text{in}} + D), \tag{VII.9.11}$$

known as the *ABCD law*, which can be easily verified by observing that

$q = \rho/\phi$. When q_{in} assumes the value given by Eq. (VII.9.10), then $q_{\text{out}} = q_{\text{in}}$ as a consequence of the self-imaging of beams formed by rays satisfying Eq. (VII.9.10).

In conclusion, for a cavity described by the parameters g_1 and g_2, whose matrix $\mathbf{S} = \mathbf{S}_{12} \cdot \mathbf{S}_{21}$ can be shown to be given by

$$\mathbf{S} = \begin{bmatrix} 2g_2 - 1 & 2g_2 d \\ \frac{2}{d}(2g_1g_2 - g_1 - g_2) & 4g_1g_2 - 2g_2 - 1 \end{bmatrix}, \qquad \text{(VII.9.12)}$$

an analogous expression for $\mathbf{S}_{21} \cdot \mathbf{S}_{12}$ being obtained by exchanging the indexes. Equation (VII.9.10) gives for q_1 and q_2 relative to the two mirrors

$$1/q_1 = 1/R_1 - i(1/d)[(g_1/g_2)(1 - g_1g_2)]^{1/2}, \qquad \text{(VII.9.13a)}$$

$$1/q_2 = 1/R_2 - i(1/d)[(g_2/g_1)(1 - g_1g_2)]^{1/2}. \qquad \text{(VII.9.13b)}$$

When the cavity is stable the coefficients of the imaginary unit on the right side of Eqs. (VII.9.13) represent the spot size w_i, and read

$$w_1^2/(\lambda d) = N_1 = (1/\pi)[(g_1/g_2)(1 - g_1g_2)]^{-1/2}, \qquad \text{(VII.9.14a)}$$

$$w_2^2/(\lambda d) = N_2 = (1/\pi)[(g_2/g_1)(1 - g_1g_2)]^{-1/2}, \qquad \text{(VII.9.14b)}$$

N_1 and N_2 being the Fresnel numbers of the two spots as seen from the opposite mirrors. For a symmetric cavity the Fresnel number reduces to $N = \pi^{-1}(1 - g^2)^{-1/2}$ and correspondingly the spot size to $w = (\lambda R/\pi)^{1/2}[d/(2R - d)]^{1/4}$.

9.3 *Boyd–Kogelnik Stability Diagram*

The preceding discussion was designed to clarify the connection between the confinement condition (VII.6.6) and the transverse distribution of the modes inside the resonator (or the equivalent lens sequence). The *Boyd–Kogelnik stability diagram* [15], plotted on the basis of Eq. (VII.6.6) (Fig. VII.24), summarizes the mode properties of both optical resonators and periodic focusing systems. Any resonator is represented by a single point in the g_1–g_2 plane. Resonators located in the white regions are stable in the sense that the transverse displacement of a generic ray trajectory will oscillate periodically, but will remain bounded after many passes. Conversely, systems located in the shaded regions (unstable) admit divergent solutions of the equation $\mathbf{v}_{i+1} = \mathbf{S} \cdot \mathbf{v}_i = (\mathbf{S})^i \cdot \mathbf{v}_0$, which governs the transverse displacement of a generic ray. When the resonator lies well inside the stable region, the fundamental mode is narrowly confined along the optic axis, and the energy losses due to diffracted energy leakage past the edges of the mirrors are generally very small. Conversely, the losses of resonators lying well inside the

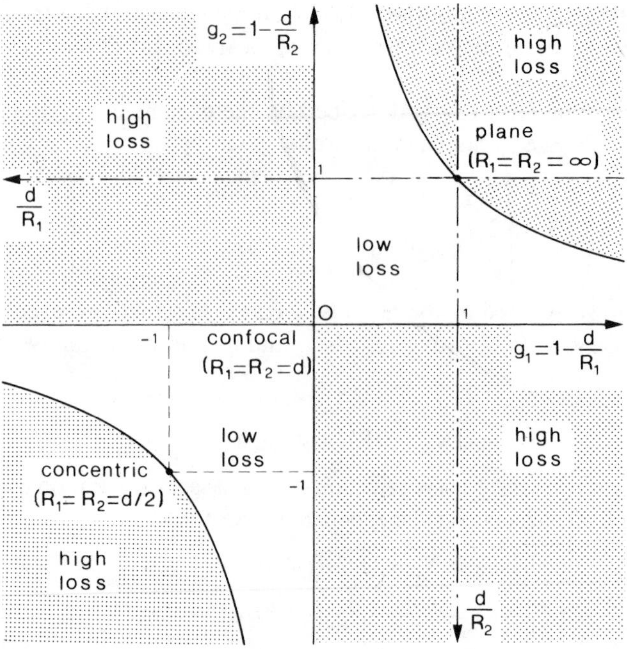

Fig. VII.24. Boyd-Kogelnik diagram showing stable (low-loss) and unstable (high-loss) values of the mirror radii of curvature and separation *d*.

unstable region are much higher, a defect that is largely compensated by a larger mode volume and an efficient diffraction coupling that eliminates the shortcomings of partial transmitting mirrors.

The wave analysis presented in Sections VII.14–19 shows that the transverse dimensions of the modes of a stable cavity increase when the stability boundaries are approached. In particular, a barely stable resonator (i.e., near the boundary) behaves as an open waveguide with walls coinciding with the mirrors and propagation directions perpendicular to the optic axis [2]. These waveguides operate very near cutoff, thus ensuring a nearly unitary reflection coefficient at the open ends.

10 Modal Representation of the Field Inside a Stable Resonator Free of Diffraction Losses

Let us consider a linear resonator delimited by two spherical mirrors of infinite extension and characterized by the amplitude reflection coefficients r_1 and r_2. In order to represent the field u excited by a time-harmonic source located inside the cavity, we find it convenient to represent u as a sum of eigenfunctions Φ_{lmn} (the reason for labeling them with three indices will become clear later on) of the Laplacian operator ∇^2 (see, e.g., Chapter IV, van Bladel [8]), i.e.,

$$\nabla^2\Phi_{lmn} = -k_{lmn}^2\Phi_{lmn}, \tag{VII.10.1}$$

which satisfy the orthogonality condition

$$(k_{l'm'n'}^2 - k_{lmn}^2)\iiint_{\text{cavity}} \Phi_{l'm'n'}\Phi_{lmn}\,dV = 0, \tag{VII.10.2}$$

(see Problem 7) and the impedance condition (cf. Section III.23)

$$\partial\Phi_{lmn}/\partial n_0 = (-ik_0/\zeta_0)Z_s\Phi_{lmn}, \tag{VII.10.3}$$

on the mirror surfaces, the surface impedance depending on the amplitude reflection coefficient, i.e., $Z_s = (r-1)/(r+1)$.

The field $\mathbf{E}(\mathbf{r}, t)$ produced by a polarization field $\mathbf{P}(\mathbf{r}, t)$, present inside the cavity, satisfies an inhomogenous Helmholtz equation together with the impedance conditions of Eq. (VII.10.3) on the mirrors. If we assume that $\mathbf{P}$ and $\mathbf{E}$ are linearly polarized and perpendicular to the optic axis, and $\nabla\cdot\mathbf{P} = 0$, Eq. (I.1.11) reduces to an inhomogeneous scalar Helmholtz equation, viz.

$$\nabla^2\mathbf{E}(\mathbf{r}) + (\omega^2/c^2)\mathbf{E} = -\mu_0\omega^2\mathbf{P}(\mathbf{r}). \tag{VII.10.4}$$

By expanding $\mathbf{E}$ in a series of cavity modes,

$$\mathbf{E}(\mathbf{r}) = \sum_n \mathbf{E}_n\Phi_n(\mathbf{r}), \tag{VII.10.5}$$

where we have summarized the three indices lmn by a single one, Eq. (VII.10.4) becomes

$$\sum_n \mathbf{E}_n(k^2 - k_n^2)\Phi_n = -\omega^2\mu_0\mathbf{P}, \tag{VII.10.6}$$

which allows us to specify the coefficients $\mathbf{E}_n$ in the form

$$\mathbf{E}_n = \frac{\omega^2\mu_0}{k_n^2 - k^2}\frac{\iiint \mathbf{P}\Phi_n\,dV}{\iiint \Phi_n^2\,dV} = \frac{\omega^2\mu_0}{k_n^2 - k^2}\mathbf{P}_n, \tag{VII.10.7}$$

where use has been made of Eq. (VII.10.2) with the additional hypothesis that different eigenfunctions having the same eigenvalue can be orthogonalized. The expansion (VII.10.5), then, becomes

$$\mathbf{E}(\mathbf{r}) = \omega^2\mu_0\sum_n \frac{\mathbf{P}_n}{k_n^2 - k^2}\Phi_n(\mathbf{r}). \tag{VII.10.8}$$

In particular, when $\mathbf{P}$ reduces to a delta function $[\mathbf{P} \propto \delta(\mathbf{r} - \mathbf{r}_0)]$, u represents the electromagnetic response of the cavity to a point source and, as such, coincides with the Green's function G_{cav} of the cavity, which, in consequence of Eq. (VII.10.8), reads

$$G_{\text{cav}}(\mathbf{r}, \mathbf{r}_s) = \omega^2 \mu_0 \sum_n \frac{1}{k_n^2 - k^2} \frac{\Phi_n(\mathbf{r})\Phi_n(\mathbf{r}_s)}{\iiint_{\text{cav}} \Phi_n^2 \, dV}. \tag{VII.10.9}$$

The reader should pay particular attention to the fact that the field is, in general, given by the contribution of the whole set of modes Φ_{lmn}, even when the cavity oscillates at a single frequency. In particular, when we say that a laser operates in a single mode, we mean simply that the field oscillates at a single frequency, but this does not automatically imply that the field is represented by a single eigenfunction Φ_{lmn}.

We are now left with the problem of determining the eigenfunctions Φ_{lmn}. If we suppose that the mirrors are sufficiently large to permit the total reflection of the gaussian beams of any order, we can put

$$\Phi_{lmn} \propto u_{lmn}^{(+)} + C_{lmn} u_{lmn}^{(-)}, \tag{VII.10.10}$$

where $u_{lmn}^{(+)}$ and $u_{lmn}^{(-)}$ represent either Hermite–Gauss or Laguerre–Gauss beams propagating, respectively, from left to right and from right to left, whose terminal wave fronts fit the mirror surfaces.

The coefficients C_{lmn} are determined by imposing the impedance conditions of Eq. (VII.10.3), which are equivalent to:

$$C_{lmn} u_{lmn}^{(-)}(x, y, z_2) = r_2 u_{lmn}^{(+)}(x, y, z_2), \tag{VII.10.11a}$$

$$u_{lmn}^{(+)}(x, y, z_1) = r_1 C_{lmn} u_{lmn}^{(-)}(x, y, z_1). \tag{VII.10.11b}$$

Since this system can be solved only if $u_{lmn}^{(-)}(x, y, z_2) u_{lmn}^{(+)}(x, y, z_1) = r_1 r_2 u_{lmn}^{(+)}(x, y, z_2) u_{lmn}^{(-)}(x, y, z_1)$, the boundary conditions transform into the following equation for the wave number k_{lmn} of Hermite–Gauss mode:

$$\exp\{-2i[k_{lmn} d - (l + m + 1)[\psi(z_2) - \psi(z_1)]]\} = 1/(r_1 r_2), \tag{VII.10.12}$$

which can be conveniently split into the equations

$$\begin{aligned} 2k''d - 2(k''/k')/[1 + (k''/k')^2](l + m + 1) & \\ \times [\arctan(z_2/b) - \arctan(z_1/b)] &= -\ln|r_1 r_2|, \end{aligned} \tag{VII.10.13a}$$

$$\begin{aligned} 2k'd - 2(l + m + 1)/[1 + (k''/k')^2] & \\ \times [\arctan(z_2/b) - \arctan(z_1/b)] &= 2\pi n + \arg(r_1 r_2), \end{aligned} \tag{VII.10.13b}$$

where for simplicity we have omitted the indices lmn from $k' = \operatorname{Re} k_{lmn}$ and $k'' = \operatorname{Im} k_{lmn}$.

It is convenient at this point to observe that the modes mostly excited by a source of frequency ω are those whose eigenvalues k_n are close to ω/c, as can be seen by inspecting the expression for the Green's function given in Eq. (VII.10.9). On the other hand, the reflectivity of the cavity mirrors is generally very high, which implies, in view of Eq. (VII.10.13a), that $k'' = O(1/d)$ while $k' = O(1/\lambda)$; since $d/\lambda \gg 1$ we have that $|k''/k'| \ll 1$ and Eqs. (VII.10.13) reduce to

$$2k''_{lmn}d = -\ln|r_1 r_2|, \tag{VII.10.14a}$$

$$k'_{lmn}d - (l + m + 1)[\arctan(z_2/b) - \arctan(z_1/b)] = n\pi + \tfrac{1}{2}\arg(r_1 r_2). \tag{VII.10.14b}$$

According to Eq. (VII.10.14a), $k''_{lmn} \cong (1 - |r_1 r_2|)/(2d)$ is independent of the particular mode considered, while the real part of k_{lmn} depends on the longitudinal index n and on the transverse ones l and m. For a Laguerre–Gauss mode, Eq. (VII.10.14b) must be replaced by

$$k'_{lpn}d - (2p + l + 1)[\arctan(z_2/b) - \arctan(z_1/b)] = n\pi + \tfrac{1}{2}\arg(r_1 r_2). \tag{VII.10.14c}$$

Finally, when the cavity is symmetric and $r_1 = r_2$, the combined use of Eqs. (VII.10.10,11) yields

$$\begin{aligned}\Phi_{lmn} &\propto \frac{w_0}{w} H_l H_m \exp\left(-i\frac{k'_{lmn}\rho^2}{2R} - \frac{\rho^2}{w^2}\right)\\ &\quad\times \sin\left[k_{lmn}z - (l + m + 1)\psi(z) + n\frac{\pi}{2}\right].\end{aligned} \tag{VII.10.15}$$

10.1 *Slowly Varying Amplitude (SVA) Approximation*

In spite of its formal simplicity, the modal representation of u by means of the series of Eq. (VII.10.5) does not lend itself to a simple description of the field. On the other hand, if we insert Eq. (VII.10.15) into Eq. (VII.10.5) and indicate with $\bar{n}$ the value of the index n for which $|k_{lmn} - \omega/c|$ is a minimum, we have

$$\begin{aligned}E &\propto \left[\sum_{lmn} e^{-i(n-\bar{n})\pi/2} E_{lmn} e^{-i(n-\bar{n})\pi z/d}\right] u^{(+)}_{lm\bar{n}} - \left[\sum_{lmn} e^{i(n-\bar{n})\pi/2} E_{lmn} e^{i(n-\bar{n})\pi z/d}\right] u^{(-)}_{lm\bar{n}}\\ &\equiv \sum_{lm} [E^{(+)}_{lm}(z) u^{(+)}_{lm\bar{n}} + E^{(-)}_{lm}(z) u^{(-)}_{lm\bar{n}}].\end{aligned} \tag{VII.10.16}$$

If the number of modes excited in the cavity is limited so that $|n - \bar{n}| \ll d/\lambda$, the $E^{(\pm)}_{lm}(z)$ are slowly varying functions of z over a distance λ, i.e., $|d^2 E^{(\pm)}_{lm}/dz^2| \ll k|dE^{(\pm)}_{lm}/dz|$.

Since the polarization **P** is the response of the medium to the field **E**, we can reasonably assume that **P** is represented by an expression similar to Eq. (VII.10.16), where the slowly varying functions $E_{lm}^{(\pm)}(z)$ are replaced by analogous functions $P_{lm}^{(\pm)}(z)$. Eventually, when we replace **E** and **P** in Eq. (VII.10.4) with these expressions and neglect the second derivatives with respect to z, we obtain the equation

$$\sum_{lm}\left[E_{lm}^{(+)}(k^2 - k_{\bar{n}}^2) - 2ik_{\bar{n}}\frac{dE_{lm}^{(+)}}{dz} + \omega^2\mu_0 P_{lm}^{(+)}\right]u_{lm\bar{n}}^{(+)}$$

$$+\sum_{lm}\left[E_{lm}^{(-)}(k^2 - k_{\bar{n}}^2) + 2ik_{\bar{n}}\frac{dE_{lm}^{(-)}}{dz} + \omega^2\mu_0 P_{lm}^{(-)}\right]u_{lm\bar{n}}^{(-)} = 0, \qquad \text{(VII.10.17)}$$

which, owing to the orthogonality condition between different $u_{lm\bar{n}}$, turns into a set of differential equations relating the amplitudes $E_{lm}^{(\pm)}(z)$ of the two nonuniform waves representing the TEM_{lm} mode of oscillation of the cavity to the analogous quantities related to the polarization. The solution of these equations will be discussed in Section VII.19.

11 Focus on Stable Resonators

According to the preceding analysis, stable cavities admit gaussian beams as their natural modes. Kogelnik and Rigrod [16] first experimentally verified the accuracy of this theoretical prediction by photographing with an image converter the single modes of a 1.15-μm He–Ne laser operating with a nearly spherical 230-cm-long cavity. Because of the difficulty of accurately measuring the intensity pattern, they turned to the measurement of the spacing between the zeros of the patterns and found close agreement with the values deduced in the above section. Mode purity was checked by verifying the absence of any modulation frequency in the intensity spectrum [17, 18].

After this short digression, let us go back to the evaluation of the mode parameters (spot size and waist location). They may be obtained in two different ways. One can determine the complex abscissa z_0 by using Eq. (VII.6.2) and then use Eq. (VII.7.2a) to calculate w and R on the mirrors, or one can determine the complex curvature on the mirrors by means of the ABCD matrix [see Eq. (VII.9.10)].

The waist w_0 of a symmetric cavity can be calculated by using the relation

$$w_0 = (\lambda/\pi)^{1/2}(d/2)^{1/4}(R - d/2)^{1/4} = [(\lambda d)/(2\pi)]^{1/2}[(1 + g)/(1 - g)]^{1/4}, \qquad \text{(VII.11.1)}$$

while the spot size on the mirrors is given by

$$w = w_0[2/(1 + g)]^{1/2}. \qquad \text{(VII.11.2)}$$

For an asymmetric cavity it can be shown that Eq. (VII.11.1) is replaced by

$$w_0 = (\lambda/\pi)^{1/2}\{[d(R_1 - d)(R_2 - d)(R_1 + R_2 - d)]/(R_1 + R_2 - 2d)^2\}^{1/4}, \tag{VII.11.3}$$

where R_1 and R_2 are positive if the mirrors are concave, negative otherwise. In addition, the distances d_1 and d_2 of the mirrors from the waist are given by

$$d_{1,2} = R_{1,2}/2 + (1/2)(R_{1,2}^2 - 4b^2)^{1/2}. \tag{VII.11.4}$$

The *resonance frequencies* may be obtained by requiring that the phase of the gaussian mode vary by an integer multiple of π when the wave front moves from one mirror to the opposite one [see Eqs. (VII.10.14b,c)]; thus we get

$$\nu_{lmn} = \nu_0\{n + [(l + m + 1)/\pi] \arccos(g_1 g_2)^{1/2}\}, \tag{VII.11.5a}$$

$$\nu_{lpn} = \nu_0\{n + [(2p + l + 1)/\pi] \arccos(g_1 g_2)^{1/2}\}, \tag{VII.11.5b}$$

for rectangular and cylindrical geometries, respectively.

For a confocal cavity the factor that multiplies the mode factors $l + m + 1$ and $2p + l + 1$ in the above equations reduces to 1/2. A strong frequency degeneracy is present in this case. Exact expressions for the resonance frequencies of cavities with mirrors of finite size will be discussed in the following (see Section VII.14); for the time being, Eqs. (VII.11.5) will be used except for the plane-parallel and spherical configurations, which, as already pointed out, are barely stable and have modes that deviate notably from the gaussian distribution.

The above discussion applies only to the resonance frequencies of an empty cavity. When the resonator is filled with an amplifying medium, anomalous dispersion causes the refractive index to vary with frequency close to the laser line center. As a result, the frequencies of the longitudinal modes are no longer exactly equally spaced. However, in gas lasers these frequency shifts are only of the order of 1 part in 1000.

11.1 *Diffraction Losses and Divergence*

The losses of a cavity can be calculated by taking into account the diffraction effects produced by the finite size of the mirrors. In Section VII.14 we will show that these effects are equivalent to the attenuation undergone by a beam traveling from one mirror to the opposite one through a medium having an imaginary component k'' of the propagation constant. In this way we are led to define the attenuation loss per transit α_{pl} for the plth mode as

$$\alpha_{pl} = 2k''_{pl}d. \tag{VII.11.6}$$

It will be seen that α_{pl} depends on the g parameters of the cavity and on the Fresnel number $N = a^2/\lambda d$, where a is the mirror radius. In general, α_{pl} must be calculated by solving the Fox–Li integral equations (see Section VII.14), as it is related to the eigenvalues γ_1 and γ_2 of these equations by

$$\alpha_{pl} = 1 - |\gamma_{1lp}\gamma_{2lp}|. \qquad \text{(VII.11.7)}$$

Vaynshteyn [2] has obtained some asymptotic expressions for the loss factor and the phase shift β per transit for the modes of *plane-parallel resonators* with circular mirrors, which, for large Fresnel numbers read

$$\alpha_{pl} = 8k_{lp}\delta(m + \delta)/[(m + \delta)^2 + \delta^2]^2, \qquad \beta_{lp} = [m/(4\delta)]\alpha_{lp}, \qquad \text{(VII.11.8)}$$

where $\delta = 0.824$ [see Section VII.16 and in particular Eq. (VII.16.23)], k_{lp} is the pth zero of the Bessel function J_l, and $m = (8\pi N)^{1/2}$. For confocal resonators Slepian [19] has found, for $N \gg 1$,

$$\alpha_{pl} = [2\pi(8\pi N)^{2p+l+1}e^{-4\pi N}]/[p!(p + l)!], \qquad \beta_{lp} = (2p + l + 1)\pi/2. \qquad \text{(VII.11.9)}$$

As shown by these expressions, the losses are a more rapidly increasing function of the modal indexes l and p in a confocal cavity than in a plane-parallel one. In Table VII.4 the ratio α_{pl}/α_{00} between the losses per transit of the higher modes and of the fundamental one is reported for the plane-parallel and confocal configurations. In particular, α_{10}/α_{00} is of the order of 20 in a confocal resonator with $N = 1$, that is, when the mirrors have the same diameter as the mode. It is thus clear that the fundamental mode of the plane-parallel resonator is much more susceptible to breaking into higher-order modes as a result of medium inhomogeneities (e.g., dust particles on the mirrors) and other imperfections than other types of resonators.

For stable cavities different from the plane-parallel, concentric, or confocal configurations, the modes tend to the gaussian profile. Then the diffraction loss per transit can be calculated to a first approximation by assuming that

Table VII.4

Ratio of Losses of Higher to Fundamental Modes

TEM_{pl}	α_{pl}/α_{00}, plane-parallel	α_{pl}/α_{00}, confocal
01	1.59	$4N$
02	2.13	$10N^2$
10	2.29	$16N^2$
03	2.65	$21N^3$
11	2.92	$85N^3$
04	3.15	$32N^4$
12	3.50	$160N^4$

only the fraction power hitting the mirrors is reflected back into the cavity [20]. Accordingly, we can put

$$\alpha_{pl} \cong 1 - \frac{\iint_{M_1} |u_{pl}|^2 \, dS_1}{\iint_{-\infty}^{+\infty} |u_{pl}|^2 \, dS_1} - \frac{\iint_{M_2} |u_{pl}|^2 \, dS_2}{\iint_{-\infty}^{+\infty} |u_{pl}|^2 \, dS_2}. \qquad \text{(VII.11.10)}$$

A parameter of great significance is the *full-width beam divergence* 2θ in the fundamental mode operation. For a symmetric cavity, it is given by [see Eq. (VII.7.5)]

$$2\theta = 2\lambda/(\pi w_0) = 2[2\lambda/(\pi d)]^{1/2}[(1-g)/(1+g)]^{1/4}. \qquad \text{(VII.11.11)}$$

11.2 *Effects of Cavity Misalignment*

Perfectly aligned cavities have been considered up to now. In case one of the mirrors, say M_1, is slightly tilted at an angle $\delta\phi_1$ (see Fig. VII.25), the beam is steered along a new axis passing through the new center of curvature O_1' of M_1. The new optic axis intersects the two mirrors at new vertices, which are displaced by the quantities δh_1 and δh_2 with respect to the original one. When the mirrors are rotated by $\delta\phi_1$ and $\delta\phi_2$, respectively, the displacements δh_1 and δh_2 are found to be

$$\begin{aligned} \delta h_1/d &= [g_2/(1-g_1g_2)]\,\delta\phi_1 + [1/(1-g_1g_2)]\,\delta\phi_2, \\ \delta h_2/d &= [1/(1-g_1g_2)]\,\delta\phi_1 + [g_2/(1-g_1g_2)]\,\delta\phi_2. \end{aligned} \qquad \text{(VII.11.12)}$$

Subtracting the first equation from the second one, the *steering angle* $\delta\phi$ of the new optic axis is obtained in the form

$$\delta\phi = [(1-g_2)/(1-g_1g_2)]\,\delta\phi_1 - [(1-g_1)/(1-g_1g_2)]\,\delta\phi_2. \qquad \text{(VII.11.13)}$$

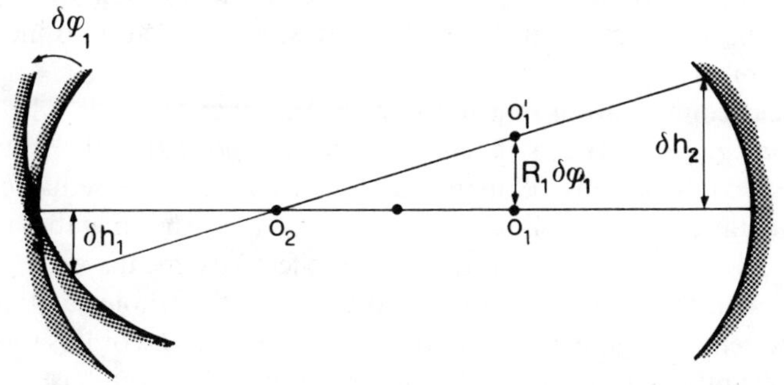

Fig. VII.25. Steering of the optic axis produced by mirror tilting.

12 Focus on Unstable Resonators

High-power laser systems can be classified into two types. In the so-called *MOPA* (*master oscillator power amplifier*), derived from electronic transmitters, a stable low-power laser drives some amplifying stages. In contrast, the generation and control of the oscillation are mixed with the amplification in the *power oscillator configuration*. The former system is more stable than self-excited oscillators. The latter configuration is more likely to be found in applications where small size and portability are more important than stability. Both systems, however, must meet the important requirement of forcing the laser to operate possibly in a single transverse mode (TEM_{00}) in order to produce an almost uniform wave front at the output; this will make it possible to focus the laser beam down to the diffraction limit, any departure from regularity of the wave front increasing the focal spot size and the far-field divergence. Multimode operation produces irregularities in the wave front, which fluctuates randomly so that it cannot be easily compensated by introducing suitably designed phase correctors at the output. In what follows no further consideration will be given to MOPA systems, but the discussion will concentrate on power oscillators.

A resonator may exhibit good optical homogeneity of the oscillation mode provided it is designed to compel all elementary radiators (excited atoms or molecules) to strongly interact among themselves, since this coupling ensures the necessary phase coherence among the oscillations of individual sources. Conversely, a beam opening up transversely as a result of diffraction hits many elementary sources. Since the cavity Fresnel number is the main factor controlling this effect, we may conclude that *a low value of N is a necessary condition for strong diffractive coupling among the particles of the cavity*. In fact, stable resonators are designed to have a low Fresnel number (~ 1) so that they will operate in a single mode. This condition can no longer be adopted for high-power lasers, which have a large active volume (for example, for a 1-liter active volume, to obtain a unit Fresnel number at the wavelength of 3 μm, the cavity should be 15 m long). Thus, the only approach left is to increase the Fresnel number.

The field configuration in a high-Fresnel-number cavity is schematically represented in Fig. VII.26a. The output beam (spotted) is the result of the incoherent contributions of many modes, which is a consequence of the inefficient diffraction coupling. On the other hand, the unstable resonator (Fig. VII.26b) compensates the diffraction effects by means of a geometric coupling, with the rays being compelled to sweep the whole cavity. In other words, the central region, whose section corresponds more or less to a Fresnel number of unity, acts as the mode core that produces a uniphase wave front that is steadily magnified after each round-trip. The beam is then coupled out of the cavity by scraping along the edges.

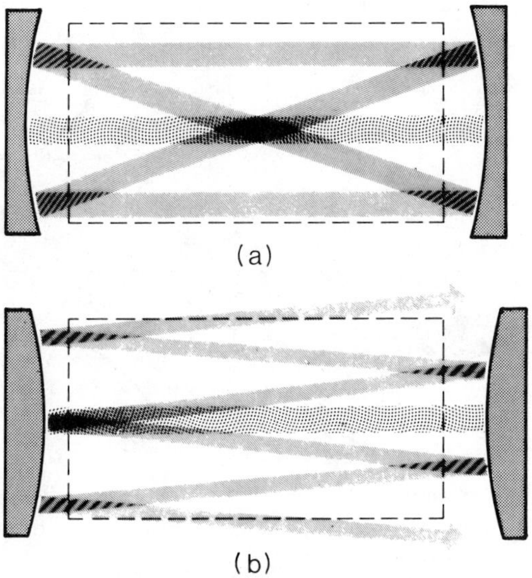

Fig. VII.26. Schematic representation of the fundamental and higher-order modes with $N \gg 1$ for (a) a stable and (b) an unstable configuration. (From Chodzko and Chester [5]. © John Wiley & Sons, Inc., 1976.)

The advantages of unstable resonators were first recognized in 1965 by Siegman [21], who also developed a simple geometric optics approach [22]. According to this approach, assuming that the fundamental modes of unstable resonators consist of two oppositely traveling spherical waves, let us denote by Γ_1 the fraction of total power of a wave of unit intensity, leaving mirror 1, that is reflected by mirror 2, and similarly define Γ_2. From Fig. VII.16 we can easily infer that

$$\Gamma_1 = \{z_1 a_1/[(d - z_1)a_2]\}^2, \qquad \Gamma_2 = \{z_2 a_2/[(d - z_2)a_1]\}^2, \qquad \text{(VII.12.1)}$$

where $z_{1,2}$ are the distances of the two spherical wave foci from the mirrors (cf. Section VII.6) and $a_{1,2}$ the mirror radii. The fractional intensity returning to either mirror after a round-trip is thus

$$\Gamma = \Gamma_1 \Gamma_2 = \left[\frac{z_1 z_2}{(d - z_1)(d - z_2)}\right]^2 = \left[\frac{1 - (1 - g_1^{-1} g_2^{-1})^{1/2}}{1 + (1 - g_1^{-1} g_2^{-1})^{1/2}}\right]^2 = \frac{1}{M^2}. \qquad \text{(VII.12.2)}$$

The main result of this geometric analysis is that *the losses per round-trip* $\alpha = 1 - \Gamma$ *are independent of the mirror diameters and depend on the resonator magnification* M only. We can draw on the $g_1 g_2$ plane a family of hyperbolas representing constant-loss configurations (Fig. VII.27).

Different kinds of unstable resonators are shown in Fig. VII.28; the first two are among the most frequently used because of the location of the foci

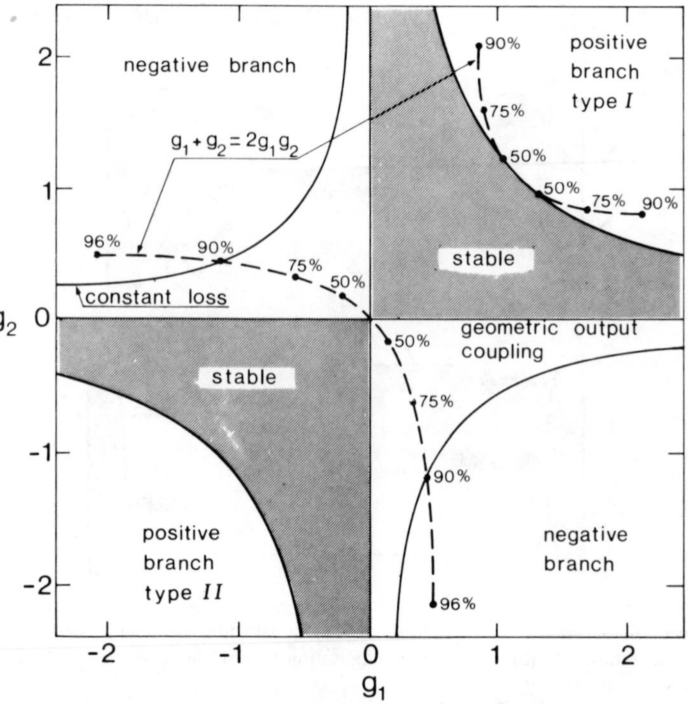

Fig. VII.27. The continuous hyperbola on the g plane represents configurations having constant magnification and loss (90%). The dashed hyperbola is the locus of the confocal configurations. (From Chodzko and Chester [5]. © John Wiley & Sons, Inc., 1976.)

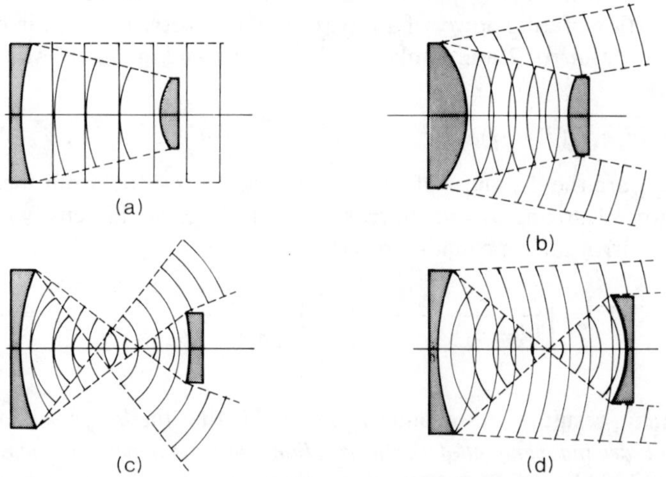

Fig. VII.28. Geometric wave front patterns in various unstable resonators: (a) and (b) positive branch (type I), $g_1g_2 > 1$, $g_1 > 0$; (c) positive branch (type II), $g_1g_2 > 1$, $g_1 < 0$; (d) negative branch, $g_1g_2 < 0$. (From Chodzko and Chester [5]. © John Wiley & Sons, Inc., 1976.)

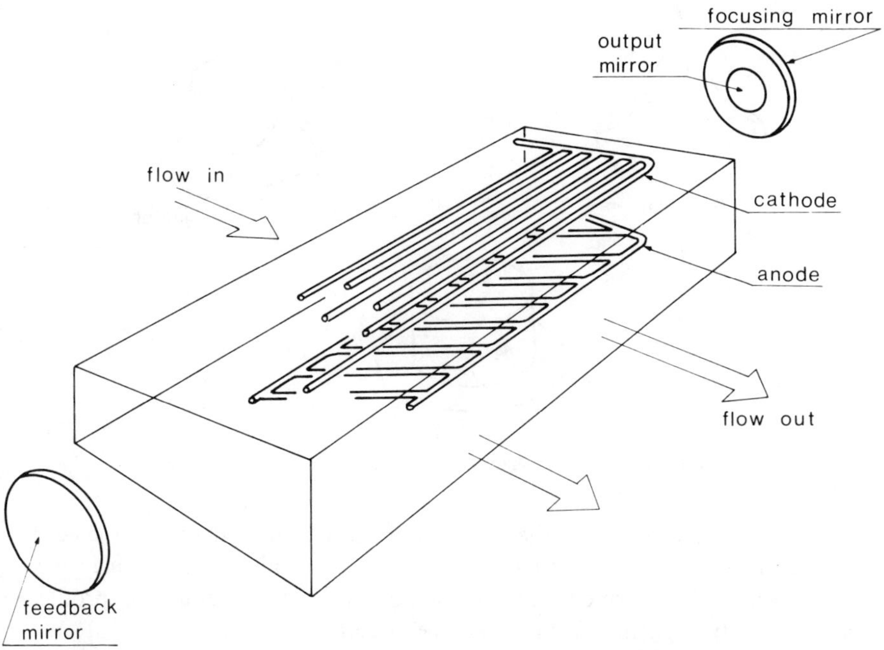

Fig. VII.29. Schematic of a telescopic resonator used in an electric discharge convection laser.

outside the cavity. In fact, their presence inside the active medium is not desirable because of the intense fields associated with them, which may produce nonlinear effects (even breakdown) that detract from the beam quality. However, the *negative* unstable configuration (Fig. VII.28d) has less stringent optical and mechanical tolerances. The *positive confocal* configuration (Fig. VII.28a), so termed because of the convex output mirror and its position in the upper right quadrant of the g plane, was originally proposed by Anan'ev [3] and is the most commonly used because it can produce a collimated output from one end only (see Fig. VII.29). The output mirror 2 (the smaller one) has radius a_2 and negative curvature radius R_2, while the reflector has radius a_1 and positive radius of curvature R_1. Since the foci of the two mirrors coincide, the g parameters must satisfy Eq. (VII.6.4). The loss per pass α is in the geometric limit equal to the coupling factor C, defined as

$$C = 1 - \Gamma = 1 - 1/M^2, \tag{VII.12.3}$$

while the magnification is simply related to R_1 and R_2 by

$$M = -R_1/R_2 \tag{VII.12.4}$$

and the g parameters obey the relations

$$g_1 = (M + 1)/(2M), \qquad g_2 = (M + 1)/2. \tag{VII.12.5}$$

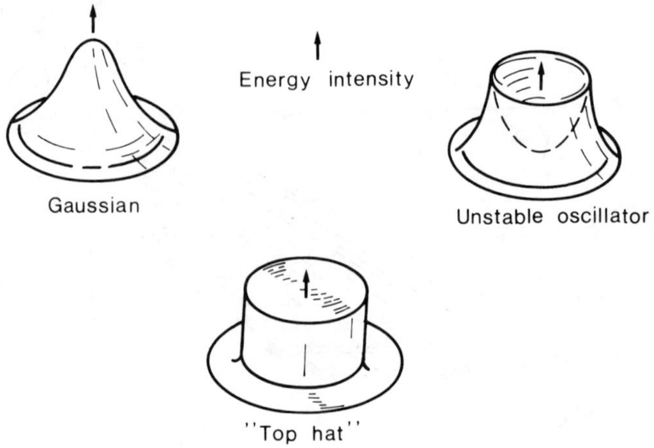

Fig. VII.30. Typical beam shapes used in high-power laser applications.

The beam associated with these resonators is doughnut-shaped in the near field because of the output mirror occluding the central region. *Fresnel rings* may appear in the intermediate field as a modulation to the doughnut profile, and fade away departing from the source, together with a small central spot that has extremely high power density and is potentially damaging to optical components. This spot, called both the *Poisson spot* and the *spot of Arago*, leads to self-focusing and potential material damage. In the far field, the beam profile resembles the Airy pattern.

The most frequently used beam distributions for laser applications are sketched in Fig. VII.30. In between the gaussian and the annular shapes is the so-called *top-hat distribution*, which can be thought of as a truncated gaussian. As to the misalignment, the unstable resonators are very sensitive to mirror tilting [23]. If the back mirror of a telescopic resonator is tilted through an angle $\delta\phi$, the angle ϕ through which the output beam is steered is equal to [cf. Eqs. (VII.11.13) and (VII.12.5)]

$$\phi = [2M/(M-1)]\,\delta\phi. \qquad \text{(VII.12.6)}$$

Because of the presence of the factor $M - 1$ in the denominator of this equation, a magnification larger than 1.5 is generally used.

13 Wave Theory of Empty Resonators

The finite size of the mirrors of a resonator introduces diffraction losses and mode conversion of the gaussian beams bouncing back and forth. As a result, the final spatial configuration of the modes can differ notably from the gaussian distribution. Ray optical theory is able to predict the configuration and the frequency of the modes of both stable and unstable resonators whenever the diffraction effects produced by the mirror edges and the apertures contained in the cavity are negligible. There are, however, parameters, such as the loss factor α, that cannot be calculated accurately in the frame of this theory.

A way to phase out this limitation is to start from the Maxwell equations, together with the boundary conditions represented by the finite mirrors of a linear resonator. The electromagnetic analysis of classical closed resonators was originally developed by J. J. Thomson and described in his "Recent Research in Electricity and Magnetism," published in 1893.

For ease of illustration let us assume that the field is described by a scalar function u, whose normal derivative vanishes at the cavity boundary. By taking advantage of Eq. (IV.2.10), we can express the field on the boundary as an integral, viz.

$$u(\mathbf{r}) = - \iint\limits_{\substack{\text{cavity}\\ \text{walls}}} \left[\frac{\partial}{\partial n_0} G(\mathbf{r} - \mathbf{r}') \right] u(\mathbf{r}')\, dS, \qquad \text{(VII.13.1)}$$

where $\mathbf{r}$ and $\mathbf{r}'$ belong to the cavity walls, while $\hat{n}_0$ is the outward normal. This relation can be regarded as an integral equation obeyed by each mode configuration, the oscillation frequency entering the equation through the Green's function G, which contains the wave number k in the exponent.

Physically, Eq. (VII.13.1) is a consequence of the fact that the field at every point of the boundary coincides with the field produced by the whole distribution of currents (proportional to u) smeared over the walls. The coupling between the field and the currents is instrumental in setting up the equilibrium configuration of the oscillation mode, while the frequency enters the process by fixing the phase delay by which each current element contributes to the field.

The integral equation (VII.13.1) admits solutions for real frequencies only when the walls of the cavity form a closed surface and the ohmic losses vanish identically.

An open resonator [2, 24] can be regarded as a closed cavity with a piece of surface coincident with the sphere of infinite radius. Since no radiation can be reflected from infinity, the modes are characterized by complex frequencies. In particular cases, the system may admit a set of low-loss modes forming a discrete spectrum in a given frequency range.

14 Fox–Li Integral Equations

Let us confine ourselves to a linear Fabry–Perot resonator. The fundamental modes may be assumed, whenever the Fresnel number is not very large, to be approximately of the TEM type, this hypothesis being necessary to justify recourse to the scalar diffraction theory. Furthermore, we can exploit the knowledge, gained from the ray optical analysis, that the modes are made of two opposite traveling waves, to infer that this description is still valid for finite wavelengths; we will denote by $u^{(+)}$ and $u^{(-)}$ the amplitudes of the forward and backward waves traveling, respectively, from mirror M_1 to M_2 and vice versa. By now applying the diffraction formulas derived in Section IV.2, we can relate $u^{(-)}$ on M_1 to $u^{(-)}$ on M_2. Following the same procedure for $u^{(+)}$ and assuming that the wave fronts coincide with the mirror surfaces, we end up with the equations [see Eq. (IV.2.14)]

$$u_2^{(+)}(x_2, y_2) = \frac{i}{2\lambda} \iint_{M_1} u_1^{(+)}(x_1, y_1) \frac{\exp[-ik(R + W_1 - W_2)]}{R} (1 + \cos\theta_1)\, dS_1, \tag{VII.14.1}$$

$$u_1^{(-)}(x_1, y_1) = \frac{i}{2\lambda} \iint_{M_2} u_2^{(-)}(x_2, y_2) \frac{\exp[-ik(R + W_2 - W_1)]}{R} (1 + \cos\theta_2)\, dS_2,$$

where the subscripts 1 and 2 refer to mirrors 1 and 2, respectively, R is the distance between the points P_1 and P_2 (see Fig. VII.31), θ_1 and θ_2 are the angles formed by $\mathbf{R}$ with the normals to M_1 and M_2 at P_1 and P_2, respectively, and W_1 and W_2 are the aberration functions of the mirrors. Equations (VII.14.1) also encompass the case of mirrors with output-coupling apertures, such as those used in high-power lasers [25].

When the mirror separation is much larger than the mirror dimension, the distance R can be approximated by

$$R \cong d + \tfrac{1}{2}(g_1/d)(x_1^2 + y_1^2) + \tfrac{1}{2}(g_2/d)(x_2^2 + y_2^2) - (1/d)(x_1 x_2 + y_1 y_2), \tag{VII.14.2}$$

where the g_i's are the cavity parameters.

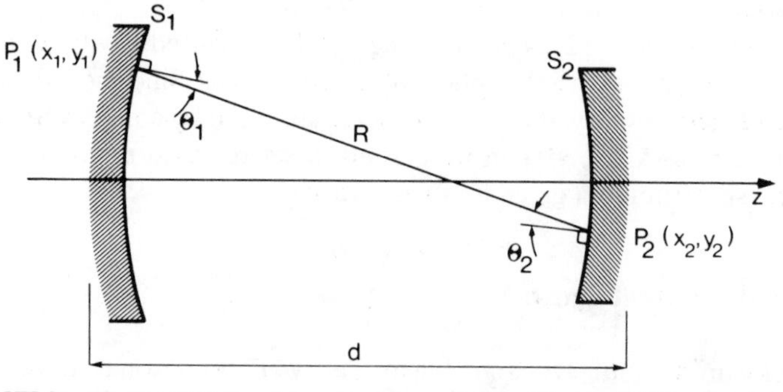

Fig. VII.31. Mode calculation for a resonator using the Kirchhoff diffraction integral.

The aberration functions W_1 and W_2 depend on the deviation of the mirror from ideality. For instance, when a mirror is slightly misaligned (see Section VII.11.2) by the angles $\delta\phi_x$ and $\delta\phi_y$, we have $W = x\,\delta\phi_x + y\,\delta\phi_y$. The Brewster angle windows are further sources of aberrations (astigmatism and coma), as is the laser material itself. The thermal load of Nd:YAG rods is at the origin of the *lensing effect*. In gas flow lasers, shock waves and turbulence inside the cavity are the major factors affecting the field, and their effects, too, can be simulated by introducing some ad hoc aberration functions.

Together with the integral equation we must consider the boundary conditions. If u is the magnetic field component tangent to mirrors having amplitude reflection coefficients r_1 and r_2, the boundary conditions reduce to

$$u_2^{(-)} = r_2 u_2^{(+)}, \qquad u_1^{(+)} = r_1 u_1^{(-)}. \tag{VII.14.3}$$

Replacing $u_2^{(-)}$ by $u_2^{(+)}$ and $u_1^{(+)}$ by $u_1^{(-)}$ in Eq. (VII.14.1) and approximating R by the right side of Eq. (VII.14.2), we obtain the system of Fredholm integral equations [26]

$$\gamma_2 u_1^{(-)} = \iint_{M_2} u_2^{(+)} K_{12} \frac{(1 + \cos\theta_2)}{2}\, dS_2, \tag{VII.14.4a}$$

$$\gamma_1 u_2^{(+)} = \iint_{M_1} u_1^{(-)} K_{21} \frac{(1 + \cos\theta_1)}{2}\, dS_1, \tag{VII.14.4b}$$

where

$$r_1\gamma_1 = r_2\gamma_2 = \exp(ikd),$$

$$K_{12}(x_1, y_1, x_2, y_2) = K(x_1, x_2)K(y_1, y_2)\exp[ik(W_2 - W_1)], \tag{VII.14.5}$$

$$K(u, v) = \left(\frac{i}{\lambda d}\right)^{1/2} \exp\left[-\frac{ik}{2d}(g_1 u^2 + g_2 v^2 - 2uv)\right],$$

Equations (VII.14.4) were first derived by Fox and Li [27] in 1960 for plane-parallel and confocal resonators and then extended to cavities with curved mirrors.

When the mirror sizes are small compared to the spacing d we can neglect the obliquity factors and set $\cos\theta_{1,2} = 1$. In that case, by replacing $u_2^{(+)}$ in Eq. (VII.14.4a) with the left side of Eq. (VII.14.4b), we obtain

$$\gamma_1\gamma_2 u_1^{(-)}(\mathbf{t}_1) = \iint_{M_1} Q_1(\mathbf{t}_1, \mathbf{t}'_1) u_1^{(-)}(\mathbf{t}'_1)\, dS, \tag{VII.14.6}$$

$\mathbf{t}_1$ and $\mathbf{t}_2$ being the transverse vectors ($\mathbf{t} = \hat{x}x + \hat{y}y$), while

$$Q_1(\mathbf{t}_1, \mathbf{t}'_1) = \iint_{M_2} K_{12}(\mathbf{t}_1 - \mathbf{t}_2)\mathbf{K}_{12}(\mathbf{t}_2 - \mathbf{t}'_1)\, dS_2, \qquad \text{(VII.14.7)}$$

an analogous equation being valid for $u_2^{(-)}$. In conclusion, the Fox–Li equations (VII.14.4) are equivalent to a couple of Fredholm integral equations having the common eigenvalue $\gamma_1\gamma_2$. From this it follows that Eqs. (VII.14.4) fix the product $\gamma_1\gamma_2$ only and not the single quantities γ_1 and γ_2.

14.1 *Losses per Transit*

Once the frequency around which we want to find the resonances has been chosen, the exact values of the resonance wave numbers will be obtained by finding the eigenvalues of Eq. (VII.14.6) and then applying Eq. (VII.14.5a). Proceeding in this way yields

$$k_{lmn} = k'_{lmn} + ik''_{lmn} = -i/(2d)\ln(\gamma_{1_{lmn}}\gamma_{2_{lmn}} r_1 r_2), \qquad \text{(VII.14.8)}$$

where the indices l, m, and n specify a mode. The real part $k'_{lmn} = 2\pi\nu_{lmn}/c$ of k_{lmn} determines the resonance frequency ν_{lmn}, and the imaginary part $k''_{lmn}\,(>0)$ fixes the required gain (cm^{-1}) for the cavity to oscillate. Because of the smallness of $k''d$, we may take, to a good approximation,

$$\begin{aligned} \alpha_{lm} &= 2k''_{lm}d \cong 1 - |\gamma_{1_{lm}}\gamma_{2_{lm}} r_1 r_2| \cong (1 - |\gamma_{1_{lm}}\gamma_{2_{lm}}|) + (1 - |r_1 r_2|) \\ &\equiv \alpha_{lm}^{(d)} + \alpha_{lm}^{(r)}, \end{aligned} \qquad \text{(VII.14.9)}$$

α_{lm} being the *power loss per transit* [see Eq. (VII.11.6)] (the longitudinal index n has been omitted since k''_{lmn} is practically independent of it). The above equation shows that α_{lm} depends on both the geometry of the cavity ($\alpha^{(d)}$) and the mirror reflectivity ($\alpha^{(r)}$), with the reflectivity contribution $\alpha^{(r)}$ simply adding to the diffraction term $\alpha^{(d)}$.

14.2 *Phase Shift*

To gain some insight into the dependence of ν_{lmn} [cf. Eq. (VII.11.5)] on the cavity geometry parameters, it is worth expressing k'_{lmn} in a way that recalls the resonance wave number of a prismatic cavity, given by Eq. (VII.1.8), as $n \to \infty$, i.e.,

$$k'_{lmn} = (\pi/d)n + \beta_{lm}/d, \qquad \text{(VII.14.10)}$$

where β_{lm} represents the mode phase shift per single pass measured with respect to an ideal plane wave having the same frequency. In this way, β_{lm} is a measure of the deviation of the ray trajectories from the optic axis. By using

Eqs. (VII.14.8, 10) we verify immediately that *$2\beta_{lm}$ is the phase of $\gamma_{1_{lm}}\gamma_{2_{lm}}r_1r_2$ between $-\pi$ and $+\pi$* and, accordingly, we find it convenient to introduce a complex quantity p_{lm} defined as

$$\gamma_{1_{lm}}\gamma_{2_{lm}} \equiv \exp(-4\pi i p_{lm}) \cong \exp(-2i\beta_{lm} - \alpha_{lm}). \qquad \text{(VII.14.11)}$$

Up to now we have taken for granted the existence of eigenfunctions of Eq. (VII.14.4). To prove this is not an easy matter. In fact, the kernel K_{12} of the integral equation is not Hermitian, which hinders recourse to the well-established theory of the Hermitian operators. This problem has challenged the ingenuity of many mathematicians, who have eventually succeeded in proving the existence of these eigenvalues [28].

14.3 *Symmetric Cavities*

The situation notably simplifies when the cavity possesses a plane of symmetry perpendicular to the optic axis. In this case the modes can be classified in two groups according whether $u_1^{(-)} = u_2^{(+)}$ $(\sigma = 1)$ or $u_1^{(-)} = -u_2^{(+)}$ $(\sigma = -1)$. Consequently, we have

$$\gamma u = \iint_M u K_{12}\, dS, \qquad \text{(VII.14.12)}$$

where $\gamma = (-1)^\sigma \gamma_1$ and $u = u_2^{(-)} = (-1)^\sigma u_1^{(+)}$. Equation (VII.14.12) exhibits a symmetric non-Hermitian kernel. Therefore, two solutions u_α and u_β, corresponding to the distinct eigenvalues γ_α and γ_β, will satisfy the orthogonality condition [cf. Eq. (VII.10.2) for an analogous condition for the cavity modes Φ_{lmn}]

$$\iint_M u_\alpha u_\beta\, dS = 0, \qquad \text{(VII.14.13)}$$

which differs from that holding in the case of Hermitian kernels in that it does not involve the conjugate of one of the two eigenfunctions. Consequently, when u_α and u_β are complex, they are not orthogonal in the power sense and the power associated with $u_\alpha + u_\beta$ does not resolve into the contributions of the individual modes u_α and u_β.

14.4 *Similarity Relations*

Before concluding this section, it will be useful to derive from Eq. (VII.14.4) some similarity relations between resonators of different dimensions but with mirrors of the same shape (e.g., both circular or rectangular). In fact, it is evident from Eq. (VII.14.4) that two resonators with the same values of the three parameters [29].

$$N = a_1 a_2/(\lambda d), \qquad G_1 = g_1 a_1/a_2 \qquad G_2 = g_2 a_2/a_1, \qquad \text{(VII.14.14)}$$

$a_{1,2}$ being a typical dimension of mirror 1 (or 2), have the same eigenvalues γ_i (and therefore the same diffraction losses and phase shifts) and, apart from a scale factor, the same field distribution for corresponding modes. In particular, for symmetric cavities ($a_1 = a_2$) N coincides with the Fresnel number, so that two resonators having the same g parameters and Fresnel number exhibit equal losses and phase shifts.

15 Overview of Mode Calculations

15.1 *Fox–Li Method*

The integral equation (VII.14.12) was first solved by Fox and Li [27, 29] by using an iterative calculation procedure [30]. They picked up an arbitrary initial field distribution on (say) mirror 1 and then used the integral of Eq. (VII.14.12) with $\gamma = 1$ to propagate it back and forth in the cavity. As a result of this process, the modes present in the initial distribution underwent an attenuation process that enhanced the fundamental one, having the lowest attenuation. The iteration was stopped when the shape of the field distribution did not vary from transit to transit and the amplitude decayed at an exponential rate. The steady-state distribution (apart from the exponential attenuation) was regarded as a normal mode. Supposing that this occurred at the ith iteration

$$u^{(i+1)} = \iint_M u^{(i)} K_{12} dS, \qquad \text{(VII.15.1)}$$

the γ coefficient having been omitted intentionally, then

$$u^{(i+1)}(x, y)/u^{(i)}(x, y) = \text{const} = \gamma. \qquad \text{(VII.15.2)}$$

The iterative method was successively used by Li[31] in 1965 to investigate nonconfocal symmetric cavities with circular mirrors. The results of these calculations are presented in Figs. VII.32 and 33, which give the loss per transit, the phase shift, and the field distribution for the TEM_{00} mode of symmetric cavities having various g parameters and Fresnel numbers. In this context, note that the horizontal portions of the phase-shift curves correspond to $\beta_{pl} = (2p + l + 1) \arccos g$ in accordance with Eq. (VII.11.5b).

A different approach was used by Checcacci *et al.* [32], who calculated the integral by means of the gaussian quadrature formulas and transformed the integral equation into a matrix equation.

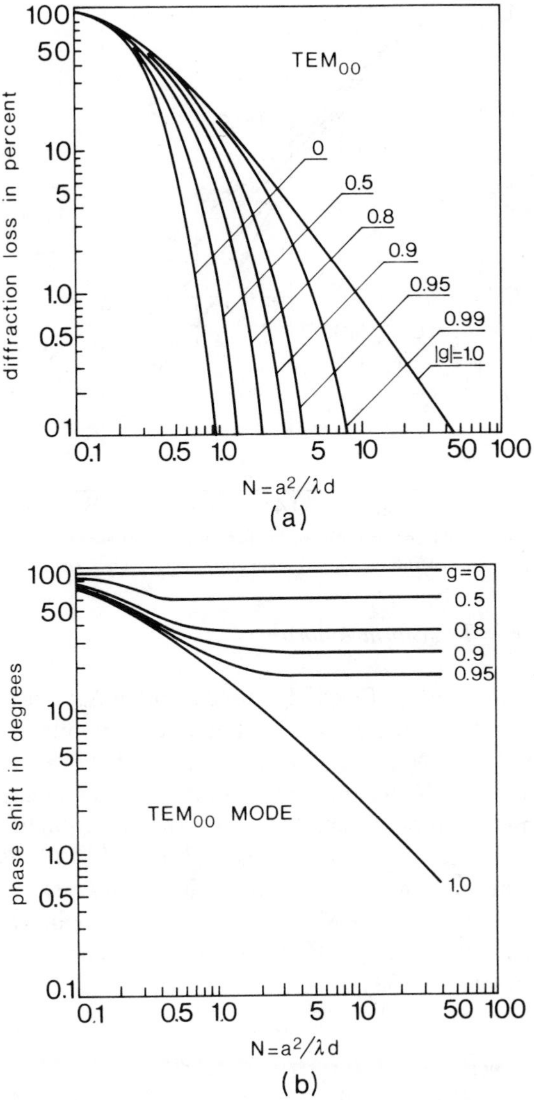

Fig. VII.32. (a) Diffraction loss α and (b) phase shift β per transit for the fundamental mode of a symmetrical cavity with circular mirrors of radius a versus the Fresnel number and the g parameters. Note that, for a TEM_{00} mode, α and β are the same for g and $-g$. (From Li [31]. © 1965 AT & T Bell Laboraties.)

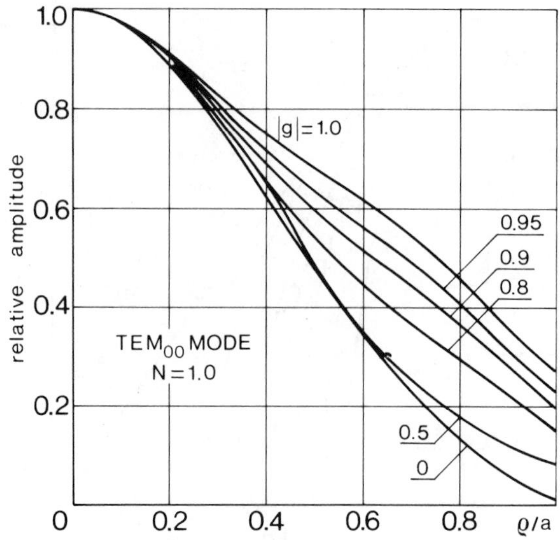

Fig. VII.33. Relative field distributions of the fundamental mode (TEM_{00}) of a symmetrical cavity with circular mirrors of radius a and Fresnel number $N = a^2/(\lambda d) = 1$.

15.2 *Fast Fourier Transform Algorithm*

To calculate the integrals, Fox and Li used a grid of N_p values of u. For one-dimensional diffraction integrals, the least number of points N_p is fixed by the Fresnel number N, i.e., $N_p > 4N$. For two-dimensional integrals $N_p \geq 16N^2$. Consequently, for high N the number of points becomes so large as to create storage problems in the computer and to make the execution times excessively long. Another significant objection to using Eq. (VII.14.12) as it stands for numerical calculations concerning situations in which $N \gg 1$ is that an array of N_p^2 values is required to store K_{12} if there are N_p grid points on M_1 and M_2. A way out of these difficulties can be found by simply rewriting Eq. (VII.14.12) as a convolution integral [33]

$$\gamma v(\mathbf{t}_1) = \iint_{-\infty}^{+\infty} P(\mathbf{t}_2)H(\mathbf{t}_1 - \mathbf{t}_2)f(\mathbf{t}_2)v(\mathbf{t}_2)\,dS_2, \tag{VII.15.3}$$

with $P(\mathbf{t}_2)$ the pupil function of the mirror, $v = uf^{1/2}$, and

$$f(\mathbf{t}) = \exp(-ikt^2/R), \qquad H(\mathbf{t}) = [i/(\lambda d)]^{1/2}\exp[-ikt^2/(2d)], \tag{VII.15.4}$$

R being the curvature radius of the mirrors.

Using Eq. (VII.15.3), two arrays of dimension N_p are required for H and f, respectively. The main advantage of using Eq. (VII.15.3) is that the right side is a convolution integral, which can be calculated very quickly by means of the fast Fourier transform (FFT) algorithms, viz.

$$\iint_{-\infty}^{+\infty} P(\mathbf{t}_2)H(\mathbf{t}_2 - \mathbf{t}_1)f(\mathbf{t}_2)v(\mathbf{t}_2)\,dS_2 = F^{-1}\{F\{H\} \cdot F\{Pfv\}\}. \qquad \text{(VII.15.5)}$$

Owing to the rapidity of the FFT algorithm, it does not particularly matter that one has to perform three Fourier transforms. For symmetric cavities with circular mirrors, the above integral becomes one-dimensional provided that a kernel containing a Bessel function J_l is introduced [cf. Eq. (VII.17.4)]. In this case one can use a special algorithm developed by Siegman [34] to calculate the Hankel transforms.

15.3 *Calculation of Higher-Order Modes*

The iterative approach is particularly suited for calculating the lowest symmetric and antisymmetric modes. If one wishes to calculate the higher modes, it is necessary to start with a field distribution that excludes the modes exhibiting an attenuation less than that of the mode one is looking for. This technique requires a considerable amount of work, which can be notably reduced by resorting to the *kernel expansion method* introduced by Streifer [35] and Bergstein and Marom [36]. It consists in expanding the kernel and the modes in a set of orthogonal functions, in such a way as to transform the integral equation into an infinite system of linear equations, whose eigenvalues reproduce the resonator phase delays and losses. In particular, Siegman and Miller [37] have proposed the *Prony method* to achieve higher efficiency in calculating the dominant modes for unstable resonators.

16 Stable Cavities with Rectangular Geometry

The following detailed considerations will be confined to symmetric cavities containing rectangular mirrors or apertures. The search for the modes is simplified notably by factorizing the field into the product of two functions, viz.

$$u(x, y) = f_x(x)f_y(y). \qquad \text{(VII.16.1)}$$

By using the property of K_{12} of factorizing into the product of two K functions, and assuming that the aberration functions vanish, the system of Eqs. (VII.14.4) reduces to the following homogeneous Fredholm equation of the second kind with symmetric continuous kernel:

$$\gamma f(v) = \int_{-a}^{a} f(u)K(u, v)\,du, \qquad \text{(VII.16.2)}$$

where f stands for either f_x or f_y and $2a$ is the width of the mirror. As it stands, this equation describes the resonances of an infinite-strip resonator (extending infinitely in one transverse direction).

If we now introduce [2] the new set of variables ξ and τ defined by

$$\xi = \arccos g, \qquad \tau = 2u/w = (2k \sin \xi/d)^{1/2} u, \tag{VII.16.3}$$

w being the spot size of the gaussian beam on the mirrors, Eq. (VII.16.2) becomes

$$\gamma f(\tau) = \int_{-\bar{\tau}}^{\bar{\tau}} G(\tau, \tau', \xi) f(\tau')\, d\tau' \tag{VII.16.4}$$

with $\bar{\tau} = 2(N_a/N_w)^{1/2} = 2a/w$ and

$$G(\tau, \tau', \xi) = \frac{e^{+i\pi/4}}{(4\pi \sin \xi)^{1/2}} \exp\left[-\frac{i}{4}\left(\frac{\tau^2 + \tau'^2}{\tan \xi} - \frac{2\tau\tau'}{\sin \xi}\right)\right], \tag{VII.16.5}$$

while $N_a = a^2/\lambda d$ is the cavity Fresnel number and $N_w = w^2/\lambda d = 1/(\pi \sin \xi)$ is the Fresnel number relative to the spot of the gaussian mode.

16.1 *Confocal Cavities*

In general, Eq. (VII.16.4) can only be solved numerically, although there are cases in which analytical solutions exist. An example is the case of a confocal cavity ($g = 0 \rightarrow \xi = \pi/2$), for which Eq. (VII.16.4) reduces to

$$\gamma f(t) = N_a^{1/2} e^{+i\pi/4} \int_{-1}^{1} e^{+ictt'} f(t')\, dt', \tag{VII.16.6}$$

where $t = \tau/\bar{\tau}$ and $c = 2\pi N_a$. The eigenfunctions of the above equation can be expressed by means of the angular and radial prolate spheroidal functions $S_{0m}(c, t)$ and $R_{0m}(c, t)$, as originally recognized by Landau, Slepian, and Pollack [38, 39] and by Boyd and Gordon [40]. These are solutions of the differential equation

$$(1 - t^2)\, d^2 f/dt^2 - 2t\, df/dt + (\chi - c^2 t^2) f = 0, \tag{VII.16.7}$$

which has continuous solutions in the closed t interval $[-1, 1]$ only for certain discrete real positive values $0 < \chi_0(c) < \chi_1(c) < \cdots < \chi_n(c)$ of the parameter χ. Corresponding to each eigenvalue $\chi_n(c)$, $n = 0, 1, 2, \ldots$, there is only one solution $S_{0n}(c, t)$ that is finite at $t = 0$ and such that $S_{0n}(c, 0) = P_n(0)$, where $P_n(t)$ is the nth Legendre polynomial. The functions $S_{0n}(c, t)$, called *angular prolate spheroidal functions*, constitute a complete orthogonal set of real functions in the interval $[-1, 1]$ and are continuous functions of c for $c \geq 0$. The function $S_{0n}(c, t)$ has exactly n zeros in $[-1, 1]$, reduces to $P_n(t)$ uniformly in $[-1, 1]$ as

$c \to \infty$, and is even or odd as n is even or odd. The eigenvalues $\chi_n(c)$ are continuous functions of c and $\chi_n(0) = n(n+1)$.

It can be shown that the eigenvalue γ_n of Eq. (VII.16.6) corresponding to the eigenfunction $S_{0n}(c, t)$ is given by

$$\gamma_n = 2N_a^{1/2} R_{0n}^{(1)}(2\pi N_a, 1) \exp[i(\pi/2)(n + 1/2)], \qquad \text{(VII.16.8)}$$

where $R_{0n}^{(1)}$, the so-called *radial prolate spheroidal* function, which differs from the S_{0n} only by a real scale factor $K_n(c)$, is given by

$$R_{0n}^{(1)}(c, t) = K_n(c) S_{0n}(c, t) \underset{t \to \infty}{\to} \cos[ct - \pi(n+1)/2]/ct. \qquad \text{(VII.16.9)}$$

According to Eq. (VII.16.8), the phase shift β_n [see Eq. (VII.14.10)] for a confocal strip resonator is a half-integer multiple of $\pi/2$, so that for the rectangular case β_{lm} is a multiple of $\pi/2$. Note that this result holds true for *finite-size* mirrors. This property of providing a phase-delay multiple of $\pi/2$ makes these resonators particularly suitable for use as interferometers. In fact, their resonance frequency is independent of both the Fresnel number and the order of the excited modes. This property can be explained by using a simple ray optical argument, noting that the optical path described by a general ray in going from one mirror to the other and back is always equal to $2d$, independent of the incidence angle. Thus, when the cavity is used as an interferometer, the resonance frequencies are independent of the ray directions, as pointed out originally by P. Connes (see Section VII.21.4).

16.2 *Vaynshteyn Theory of Concentric and Plane-Parallel Resonators*

A situation of particular interest occurs when the cavity is plane-parallel ($g = 1$) or concentric ($g = -1$). In this case $\xi = 0, \pi$ and the kernel Eq. (VII.16.5) is singular. However, by using as integration variable the quantity $t = \tau/(2 \sin \xi)^{1/2} = x(k/d)^{1/2}$, it can easily be shown that Eq. (VII.16.4) reduces to

$$\gamma f(t) = \frac{\exp(+i\pi/4)}{(2\pi)^{1/2}} \int_{-\bar{t}}^{\bar{t}} \exp\left[-\frac{1}{2}(t - t')^2\right] f(t')\, dt', \qquad \text{(VII.16.10)}$$

where $\bar{t} = (2\pi N_a)^{1/2} = (ka^2/d)^{1/2}$.

Vaynshteyn [2] has developed an elegant procedure for solving the above integral equation by using the eigenfunctions of the cavity with infinite mirrors. In this case the cavity can be seen as a waveguide extending from $t = -\infty$ to $t = +\infty$ and having height d. Analogously, the cavity with finite mirrors can be conceived as a truncated waveguide in which the modes travel toward the open ends, where they are partially reflected back as a result of the diffraction by the edges. Therefore, the eigenfunctions for assigned $\bar{t}$ can be expressed as a combination of the infinite waveguide modes subject to diffraction at the open ends.

If f represents the magnetic field component H_x relative to the cavity mirrors, the field inside $u_{q\varepsilon} = H_y$ inside the waveguide, related to the qth mode, reads

$$u_{q\varepsilon}(x, z) \propto \cos(q\pi z/d)\exp[i\varepsilon s_q x(k/d)^{1/2}], \qquad \text{(VII.16.11)}$$

where $\varepsilon = \pm 1$, the walls have coordinates $z = 0$, d, and

$$s_q = (k^2 - q^2\pi^2/d^2)^{1/2}(d/k)^{1/2}. \qquad \text{(VII.16.12)}$$

The wave function $u_{q\varepsilon}$ is the superposition of two plane waves that form an angle with the walls (see Fig. VII.34)

$$\phi_q = \pi/2 - \arctan[s_q/(kd)^{1/2}]. \qquad \text{(VII.16.13)}$$

In the present case we are interested in modes that are strongly reflected by the edges. According to the analysis of the diffraction by a half-plane, developed in Chapter VI, the field diffracted in the direction opposite to that of the incident beam becomes strong when the angle ϕ_q is almost equal to $\pi/2$. Consequently, we can restrict the analysis to those modes for which $s_q \ll (kd)^{1/2}$. Now, if k is represented in the form $k = \pi(n - 2p)/d$, with $-1/2 < p < 1/2$, and if we set $q = n - 2j$, with $j = 0, \pm 1, \pm 2, \ldots$, we have

$$s_j \cong [4\pi(j - p)]^{1/2}, \qquad \text{(VII.16.14)}$$

where we have replaced the index q with j and have made use of the approximate relation $k \cong q\pi/d$. In addition, we have considered only those modes whose index q differs from n by an even integer. The last assumption is due to the fact that the reflection at the open ends leaves the modes with even q uncoupled from those with odd q. This can be proved by observing that even

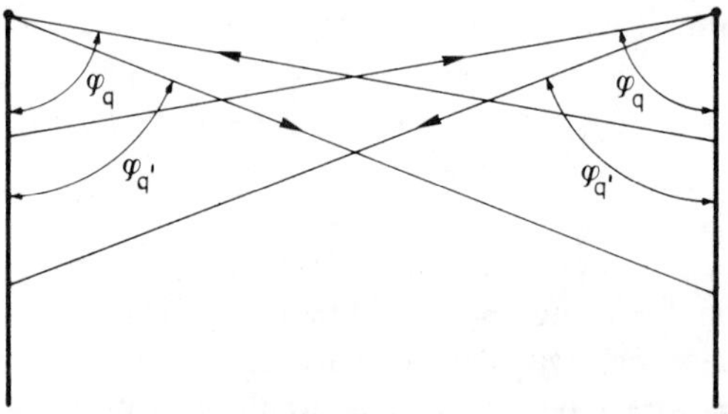

Fig. VII.34. Angles of incidence (ϕ_q) and diffraction ($\phi_{q'}$) relative to the conversion of the qth mode traveling upward into the q'th mode traveling downward induced by the diffraction at the waveguide edges.

and odd index modes correspond to equal and opposite field distributions, respectively, on the two mirrors. In addition, it is worth noting that s_j will be either real or imaginary according to whether $j - p > 0$ or <0. In the former case the mode propagates along the x axis, while in the latter case it is cut off.

For finite mirrors, we can follow Vaynshteyn [2] by expanding $f(t)$ as

$$f(t) = \sum_{j\varepsilon} F_{j\varepsilon} \exp(is_j\varepsilon t). \tag{VII.16.15}$$

Inserting the above series in Eq. (VII.16.10) yields

$$\sum_{j\varepsilon} F_{j\varepsilon} \exp(is_j\varepsilon t)\left\{\gamma + e^{is_j^2/2}\left[F\left(\frac{\bar{t} - t - s_j\varepsilon}{2^{1/2}}\right) - F\left(-\frac{\bar{t} + t + s_j\varepsilon}{2^{1/2}}\right)\right]\right\} = 0, \tag{VII.16.16}$$

where F is the complex Fresnel integral defined by Eq. (V.3.5). When we are far from the waveguide ends and $\bar{t} \gg 1$, the above equation reduces to an identity satisfied by choosing $\gamma = \exp(is_j^2/2) = \exp(-2\pi ip)$. This, in turn, implies that when they are far from either end, individual modes travel, as expected, without interfering with one another. In contrast, near the end $t = \bar{t}$ Eq. (VII.16.16) reduces to

$$\sum_{j\varepsilon} F_{j\varepsilon} \exp(is_j\varepsilon\bar{t}) F\left(-\frac{s_j\varepsilon}{2^{1/2}}\right) = 0, \tag{VII.16.17}$$

since $F(-2^{1/2}\bar{t}) \to 1$ as $\bar{t} \to \infty$.

Let us forget for a while that we are analyzing a resonator and consider a mode j' traveling toward the end $t = \bar{t}$ in a guide extending from $-\infty$ to $\bar{t}$ and giving rise by diffraction to a series of reflected modes. Clearly, Eq. (VII.16.10) must still hold, with the coordinate $-\bar{t}$ replaced by $-\infty$, for it prescribes that the fields on the two faces are identical.

Now, if we denote by $r_{jj'}$ the amplitude of the jth reflected mode ($\varepsilon = 1$) excited by the j'th incident one ($\varepsilon = -1$), and put $F_{j,-1} = 0$ for $j' \neq j$, $F_{j',-1} \exp(-is_{j'}\bar{t}) = 1$, and $F_{j,1} \exp(is_j\bar{t}) = r_{jj'}$, Eq. (VII.16.17) transforms into

$$\sum_j r_{jj'} F\left(-\frac{s_j}{2^{1/2}}\right) = -F\left(\frac{s_{j'}}{2^{1/2}}\right). \tag{VII.16.18}$$

The coefficients $r_{jj'}$ can be obtained by solving the system obtained by rewriting the above equation for all values of j' and putting $r_{jj'} = r_{j'j}$ by virtue of the reciprocity theorem. As a first approximation we have

$$r_{00} \cong -F(s_0/2^{1/2})/F(-s_0/2^{1/2}) = -\exp[i\delta(1 + i)s_0], \tag{VII.16.19}$$

where $\delta = 0.824$, while all other coefficients vanish. The last formula for r_{00} was derived by Vaynshteyn (see Chapter IV in [5]) as a limit for small p of the exact expression obtained by using the Wiener–Hopf–Fock factorization

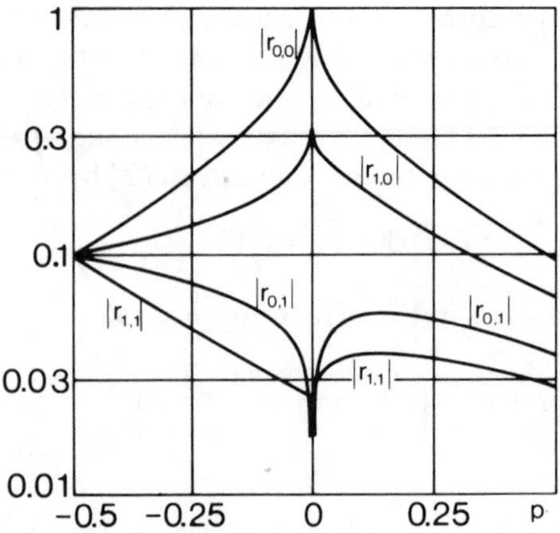

Fig. VII.35. Amplitudes of the reflection $|r_{jj}|$ and conversion $|r_{jj'}|$ coefficients for modes near cutoff versus the parameter $p = -s_j^2/(4\pi) + j$, where s_j is related to the angle $\phi_j \cong \pi/2 - [4\pi(j-p)/(kd)]^{1/2}$. (From Vaynshteyn [2].)

technique. A more accurate calculation of these coefficients, carried out by Vaynshteyn, yields (see Fig. VII.35)

$$r_{jj} \cong -\exp[i\delta(1+i)s_j], \qquad r_{jj'} \cong -\frac{2s_j}{s_j + s_{j'}}\exp[\tfrac{1}{2}i\delta(1+i)(s_j + s_{j'})]. \tag{VII.16.20}$$

(For a different approximate approach to the calculation of the reflection coefficients see Luchini and Solimeno [41].)

The cavity may now be represented as a set of transmission lines of length $2\bar{t}$, having propagation constants s_j and terminated by two n-ports characterized by a scattering matrix $[r_{jj'}]$. To a good approximation all $r_{jj'}$ coefficients may be assumed to vanish, except r_{00}, a fact that leads us to write f in the form

$$f(t) = \exp(is_0 t) \pm \exp(-is_0 t). \tag{VII.16.21}$$

Since for $t = \pm\bar{t}$, $\pm\exp(-is_0 t) = r_{00}\exp(is_0 t)$, we obtain by replacing r_{00} by its expression in Eq. (VII.16.19)

$$(2\bar{t} + \delta + i\delta)(4\pi p_m)^{1/2} = \pi(m+1), \qquad m = 0, 1, \ldots, \tag{VII.16.22}$$

where the index m labels the different values that can be taken by p_m. A different transverse mode corresponds to each p_m. By solving the above equation with respect to the real and imaginary parts of the complex parameter p, we obtain the phase shift β_m and the loss per transit α_m of the mth mode of the cavity, viz.

$$\beta_m = 2\pi p'_m = \frac{2\pi^2(m+1)^2\bar{t}(\bar{t}+\delta)}{[(2\bar{t}+\delta)^2 + \delta^2]^2}, \qquad \alpha_m = \pi p''_m = \frac{\pi^2(m+1)^2\delta(2\bar{t}+\delta)}{2[(2\bar{t}+\delta)^2+\delta^2]^2}. \tag{VII.16.23}$$

Now, combining the modes of two strip-resonators, we obtain the field configuration for a rectangular cavity. For $\bar{t} \gg 1$, to a good approximation we have

$$f_{lm}(x, y) \propto \begin{Bmatrix} \cos \\ \sin \end{Bmatrix}(k_x x)\begin{Bmatrix} \cos \\ \sin \end{Bmatrix}(k_y y), \qquad \text{(VII.16.24)}$$

where the cosine or sine function applies if $l(m)$ is even or odd, and

$$k_x = (l + 1)(\pi/2a)[1 + \delta(1 + i)/(8\pi N_a)^{1/2}]^{-1}, \qquad \text{(VII.16.25)}$$

while k_y is given by a similar expression obtained by replacing l with m and N_a with N_b.

Vaynshteyn also applied this approach to the case of circular mirrors, showing that the eigenfunctions are

$$f_{lm}(\rho, \phi) \propto J_l\left[\frac{k_{lm}\rho/a}{1 + \delta(1 + i)/(8\pi N)^{1/2}}\right]\cos l\phi, \qquad \text{(VII.16.26)}$$

where k_{lm} is the mth zero of $J_l(x)$ and $N = a^2/\lambda d$, a being the radius of the mirror. In this case the losses are given by Eq. (VII.11.8).

17 Rotationally Symmetric Cavities

Under the assumption of circular symmetry, it is expedient to seek a representation of the field relative to the circular mirrors of a symmetric cavity, of the form

$$u(\rho, \phi) = \sum_{l,p}(S_{lp}\sin l\phi + C_{lp}\cos l\phi)R_p^l(\rho), \qquad \text{(VII.17.1)}$$

where S_{lp} and C_{lp} are suitable coefficients and $R_p^l(\rho)$ a function to be determined by solving a Fox–Li integral equation, whose kernel can easily be found by averaging Eq. (VII.15.1) with respect to $\cos l\phi(\sin l\phi)$, that is,

$$\gamma_{lp}\int_0^{2\pi} u(\rho, \phi)\begin{Bmatrix} \cos l\phi \\ \sin l\phi \end{Bmatrix} d\phi = \int_0^a \rho'\,d\rho' \int_0^{2\pi} u(\rho', \phi')\,d\phi' \times \int_0^{2\pi} K_{12}(\rho, \rho', \phi, \phi')\begin{Bmatrix} \cos l\phi \\ \sin l\phi \end{Bmatrix} d\phi. \qquad \text{(VII.17.2)}$$

The kernel K_{12} coincides with that of Eq. (VII.14.5b) with $W_1 = W_2 = 0$, viz.

$$K_{12}(\rho, \rho', \phi, \phi') = [-i/(\lambda d)]\exp\{-ik/(2d)[g\rho^2 + g\rho'^2 - 2\rho\rho'\cos(\phi - \phi')]\}. \qquad \text{(VII.17.3)}$$

Finally, inserting the right sides of Eqs. (VII.17.1,3) into Eq. (VII.17.2) yields

$$\gamma_{lp}R_p^l(\rho) = 2\pi\frac{i^{l+1}}{\lambda d}\int_0^a J_l\left(\frac{k\rho\rho'}{d}\right)\exp\left[-\frac{ikg}{2d}(\rho^2+\rho'^2)\right]R_p^l(\rho')\rho'\,d\rho'$$
$$= i^{l+1}2\pi N\int_0^1 J_l\left(2\pi N\frac{\rho}{a}\eta\right)\exp\left[-i\pi Ng\left(\frac{\rho^2}{a^2}+\eta^2\right)\right]R_p^l(\eta)\eta\,d\eta, \qquad \text{(VII.17.4)}$$

where $N = a^2/\lambda d$ and $\eta = \rho'/a$. Note that when g is changed to $-g$, the eigenvalue γ_{lp} is changed to $(-1)^{l+1}\gamma_{lp}^*$. Consequently, two resonators having the same Fresnel number and opposite g parameters present the same losses.

18 Diffraction Theory of Unstable Resonators

In a telescopic resonator with circular mirrors M_1 and M_2, the radiation is taken out from M_1 (see Fig. VII.17) and we can assume for simplicity that M_2 has an infinite diameter. Consequently, the Fox–Li equation reduces to

$$\gamma R^l(\xi) = i^{l+1}\frac{2\pi N_1}{g_2}\int_0^1 J_l\left(\frac{\pi N_1\xi\eta}{g_2}\right)$$
$$\times\exp\left[-i\frac{\pi N_1}{2g_2}(2g_2g_1-1)(\xi^2+\eta^2)\right]R^l(\eta)\eta\,d\eta, \qquad \text{(VII.18.1)}$$

where $e^{\pm il\phi}R^l(\rho/a)$ is the field distribution on M_1, $N_1 = a_1^2/\lambda d$ is the Fresnel number of M_1, and ξ and η are radial coordinates normalized to a_1.

It is easy to show that Eq. (VII.18.1) describes as well a symmetric

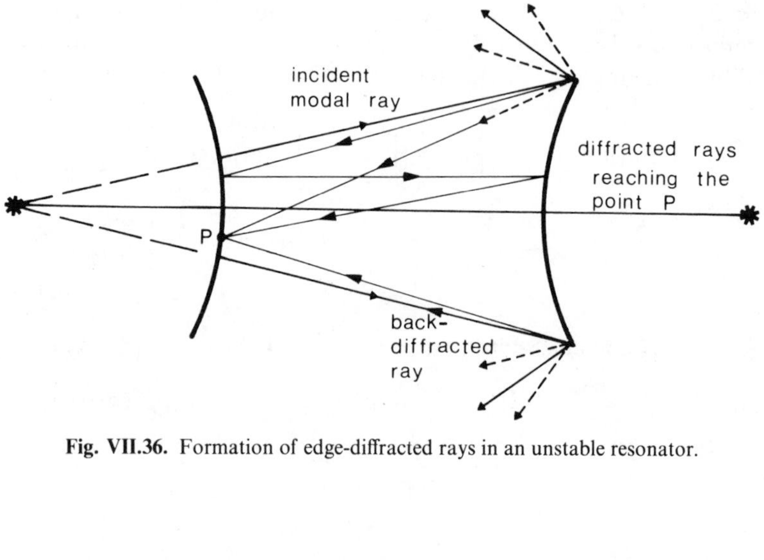

Fig. VII.36. Formation of edge-diffracted rays in an unstable resonator.

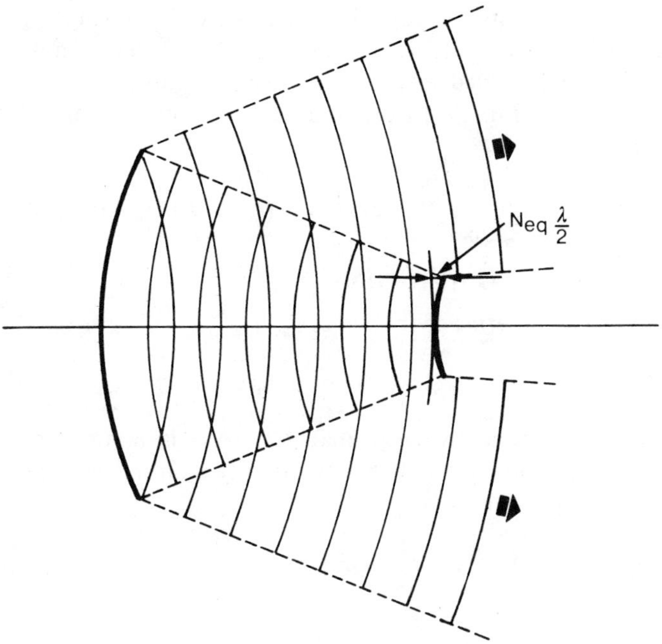

Fig. VII.37. Equivalent Fresnel number in an unstable resonator.

resonator (see Fig. VII.36) having a Fresnel number $N_{sym} = N_1/(2g_2)$ and a g-parameter $g_{sym} = (g_1 g_2)^{1/2}$. Introducing the *equivalent Fresnel number* N_{eq}, which represents the distance of the mirror edge from the wave front touching the center of the mirror divided by $\lambda/2$ (see Fig. VII.37),

$$N_{eq} = N_1[(g_1/g_2)(g_1 g_2 - 1)]^{1/2}(N_0/2) = (M - 1/M), \tag{VII.18.2}$$

M being the magnification of the resonator and $N = N_1/2g_2$, Eq. (VII.18.1) becomes

$$M\gamma f(M\xi) = i^{l+1}2t\int_0^1 J_l(t\xi\eta)\exp\left[-i\frac{t}{2}(\xi^2+\eta^2)\right]f(\eta)\eta\,d\eta \tag{VII.18.3}$$

where $t = 2\pi MN = 4\pi N_{eq}M^2/(M^2-1)$ and $f(\xi) = R^l(\xi)\exp(i\pi N_{eq}\xi^2)$. For Fresnel numbers sufficiently large, the above integral can be evaluated asymptotically by using Eqs. (V.8.1,5), thus obtaining for rotationally invariant modes ($l = 0$)

$$\begin{aligned} M\gamma f(M\xi) &= f(\xi) - \frac{e^{-i\pi/4}}{(2t)^{1/2}}f(1)\left[\frac{e^{-i(t/2)(1-\xi)^2}}{\sqrt{1-\xi}} + \frac{e^{-i(t/2)(1+\xi)^2}}{\sqrt{1+\xi}}\right] \\ &\equiv f(\xi) + f(1)F_1(\xi, t), \end{aligned} \tag{VII.18.4}$$

where $F_1(\xi, t)$ represents the edge-diffracted wave. Following a procedure introduced by Horwitz [42] for rectangular mirrors, let us now define a sequence of edge-diffracted waves F_n obtained by substituting F_{n-1} in the right-hand side of Eq. (VII.18.3) and subtracting the edge diffraction contribution,

$$F_n(M\xi, t) = i2t \int_0^1 J_0(t\xi\eta) \exp\left[-i\frac{t}{2}(\xi^2 + \eta^2)\right] F_{n-1}(\eta, t)\eta \, d\eta$$
$$- F_{n-1}(1, t) F_1(M\xi, t). \qquad \text{(VII.18.5)}$$

By expanding $f(\xi)$ in terms of the waves F_n,

$$f(\xi) = 1 + \sum_{n=1}^{m} a_n F_n(\xi, t), \qquad \text{(VII.18.6)}$$

m being chosen sufficiently large that $F_{m+1}(\xi, t)$ is almost constant for $0 \leq \xi \leq 1$, by inserting Eq. (VII.18.6) into Eq. (VII.18.3) and using the relations (VII.18.5), we finally obtain

$$M\gamma\left[1 + \sum_{n=1}^{m} a_n F_n(\xi, t)\right] = 1 + \sum_{n=1}^{m} a_n F_{n+1}(\xi, t) + \left[1 + \sum_{n=1}^{m} a_n F_n(1, t)\right] F_1(\xi, t). \qquad \text{(VII.18.7)}$$

Next, taking into account that $F_{m+1}(\xi, t)$ is almost constant for $0 < \xi < 1$ and equating the coefficients of $F_n(\xi, t)$ on either side of the above equation, it follows that $a_m = (\tilde{\gamma} - 1)/F_{m+1}(1, t)$, $a_n/a_{n-1} = \tilde{\gamma}^{-1}$, so that

$$a_n = [(\tilde{\gamma} - 1)/F_{m+1}(1, t)]\tilde{\gamma}^{m-n}, \qquad \text{(VII.18.8)}$$

where $\tilde{\gamma} \equiv M\gamma$. Finally, equating the coefficients of $F_1(\xi, t)$ in Eq. (VII.18.7) and using Eq. (VII.18.8) yields the polynomial equation in $\tilde{\gamma}$

$$\tilde{\gamma}^m(\tilde{\gamma} - 1) = F_{m+1}(1, t) + (\tilde{\gamma} - 1) \sum_{n=1}^{m} \tilde{\gamma}^{m-n} F_n(1, t), \qquad \text{(VII.18.9)}$$

whose solutions allow the calculation of the coefficients a_n by means of Eq. (VII.18.8).

For $t \to \infty$, (Butts and Avizonis [42a]) $F_n(\xi, t)$ can be expressed by the function

$$F(\xi, t) = [-e^{-it(1+\xi^2)}/(1 - \xi^2)][J_0(2t\xi) + i\xi J_1(2t\xi)], \qquad \text{(VII.18.10)}$$

where J_0 and J_1 are Bessel functions, by putting

$$F_n(\xi, t) = F(\xi/M^n, t/M_{n-1}), \qquad \text{(VII.18.11)}$$

where M^n is the nth power of M while

$$M_n = \sum_{k=0}^{m} M^{-2k}. \qquad \text{(VII.18.12)}$$

In particular,

$$F_m(0, t) \cong -\exp(-i4\pi N_{eq}),$$
$$F_m(1, t) = -\exp[-i4\pi N_{eq}(1 + M^{-2m})]J_0[8\pi N_{eq})/(M^m)], \qquad \text{(VII.18.13)}$$

where we have approximated M_m by $M_\infty = M^2/(M^2 - 1)$ and replaced t/M_∞ by $4\pi N_{eq}$. Since F_m must change slightly over the interval (0, 1), Eq. (VII.18.13) indicates that M^m must be larger than $8\pi N_{eq}$. Several numerical tests have shown that m must be fixed in accordance with the *Horwitz criterion*

$$M^m = 250N_{eq}. \qquad \text{(VII.18.14)}$$

The roots of Eq. (VII.18.9) also provide the loss factor α for the respective modes through the usual expression

$$\alpha = 1 - |\tilde{\gamma}/M|^2. \qquad \text{(VII.18.15)}$$

Then, for assigned M and N_{eq}, it is natural to order the eigenvalues $\tilde{\gamma}$ by decreasing modulus. In particular, the dominant eigenvalue is that associated with the fundamental mode. For $N_{eq} \to \infty$, the dominant eigenvalue tends to unity. In fact, being the fundamental mode of a resonator with unlimited mirrors, a spherical wave (see Section VII.12) $f(\xi)$ must be identically constant and the a_n must vanish. For finite mirror $\tilde{\gamma}$ deviates from unity and, if we plot α versus N_{eq} for several modes, we obtain a graph characterized by some sort of periodicity (see Fig. VII.38). We note that $|\tilde{\gamma}|$ reaches the maximum value for half-integer values of N_{eq}. This means that the losses of the fundamental mode reach their minimum values for half-integer values of N_{eq} and are maximum for integer values. This behavior is a consequence of the complex interference process among the unperturbed spherical wave and the waves diffracted by the edges [43].

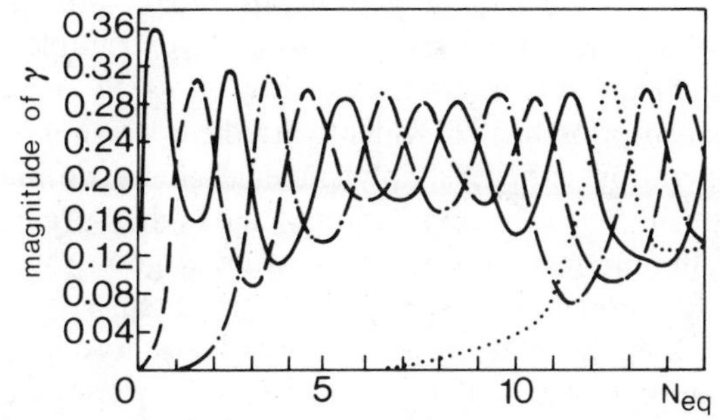

Fig. VII.38. Magnitude of the eigenvalues γ of the axially symmetric modes 1, 2, 3, and 4 of a symmetric unstable resonator with $M = 5$ and $g = 2.6$. Note that $|\gamma|$ oscillates around the geometric optics values. (From Siegman and Miller [37].)

19 Active Resonators

The considerations developed up to now have been confined to empty resonators so that no conclusion could be drawn about the amplitude and frequency of the laser oscillator. To gain information on these parameters we must solve the nonlinear equations (VII.1.13), which in turn requires knowledge of the analytic dependence of the laser amplification $A_L(I, \omega, \omega_L)$ and the phase delay $\beta_L(I, \omega, \omega_L)$ on the frequencies ω and ω_L and the intensity. Actually, we are quite imprecise when we refer to the intensity without specifying the section of the amplifying medium, since the field is in general nonuniform along the cavity axis. Therefore, the introduction of the parameter I, used before now for defining A_L and β_L, must be taken only as indicative of the fact that both these quantities depend on the general level of the oscillation mode.

A second assumption implicit in the use of a single oscillation frequency in the definition of the functions A_L and β_L is that the laser oscillates on a *single mode*, which amounts to a field oscillating at a *single frequency* almost coincident with that of a given longitudinal mode of the cavity. This assumption is more or less verified in some classes of lasers characterized by a *homogeneous linewidth*, while for gas lasers, which exhibit Doppler-broadened inhomogeneous lines, the oscillation on a single mode requires complex precautions.

The analysis of multimode operation is very complicated as it must account for the nonlinear coupling between different modes, which may produce conspicuous effect such as the emission of periodic sequences or trains of short pulses associated with a phase coherence between the several oscillation modes (see Chapter I, Yariv [8]). When *phase locking* is produced by the nonlinear behavior of the active medium, we speak of *self-locking* to stress the difference from the case in which the locking is obtained by inclusion in the optical cavity of a passive medium with a transmissivity that is an increasing function of the light intensity (*passive mode-locking*). Examples of such media, known as *saturable absorbers*, are provided by gases (SF_6), liquids (dyes), or solids with an absorption line coincident with the operation frequency of the laser. Their operation can be simply understood by noting that the absorptance drops off when the population of the higher level of the absorber equals that of the fundamental one. (Therefore, a good saturable adsorber must saturate for moderate values of the light intensity.) A more reliable technique for locking the modes is to include in the cavity a nonlinear medium whose losses can be modulated in time at the mode spacing frequency by the application of an electric signal. This technique, largely used to generate short pulses from the argon laser, is known as *active mode-locking*.

In the rest of this section we will dwell mainly on the establishment of continuous oscillations in single-mode lasers having a homogeneously broadened transition. Although this is not the most common situation, the required analysis is somewhat less involved than that necessary in the case of inhomogeneously broadened lines, to which we will refer occasionally [44].

19.1 *Nonlinear Dielectric Susceptivity of Laser Media*

In proximity to an isolated resonance, the susceptivity of the active medium can be represented in the simple form (cf. Eqs. (I.2.57)

$$\chi(\nu) = \chi_0 + i[f\,\Delta N/(1 + \mathscr{I})]\Gamma/[\Gamma + i(\nu - \nu_L)] = \chi' + i\chi''. \qquad \text{(VII.19.1)}$$

Here Γ is a damping factor having the dimension of a frequency (generally measured in reciprocal centimeters) which measures the linewidth of the gain curve, represented by $\chi''(\nu)$, at half its maximum height, f is the *oscillator strength*, and $\mathscr{I} = \langle E^2 \rangle / E_S^2$ is the time-averaged squared electric field normalized with respect to the quantity E_S, which plays the role of a *saturation field*. In writing χ in the above form we have introduced χ_0 to represent nonresonant contributions to the susceptivity, which can be taken as sensibly constant around the central frequency ν_L of the laser transition. Finally, $\Delta N = (e^2\,\Delta N_0)/(2\varepsilon_0 m\omega_L\Gamma)$ is proportional to the population inversion density ΔN_0.

In most cases χ deviates slightly from χ_0, so that the medium is characterized near a laser line by a complex refractive index $\tilde{n} = n + i\kappa$ given by

$$n + i\kappa \cong \chi_0^{1/2} + [i/(2\chi_0^{1/2}][f\,\Delta N/(1+\mathscr{I})]\Gamma/[\Gamma + i(\nu - \nu_L)] = n_0 + \Delta n + i\kappa. \qquad \text{(VII.19.2)}$$

In terms of the dimensionless *Lorentzian function* $\mathscr{L}$,

$$\mathscr{L}(\nu - \nu_L) = \Gamma^2/[\Gamma^2 + (\nu - \nu_L)^2] \equiv 1/(1 + \zeta^2), \qquad \text{(VII.19.3)}$$

$\zeta = (\nu - \nu_L)/\Gamma$ being the *detuning parameter*, we have

$$\Delta n = [1/(2n_0)][f\,\Delta N/(1 + \mathscr{I})]\mathscr{L}\zeta,$$
$$\kappa = [1/(2n_0)][f\,\Delta N/(1 + \mathscr{I})]\mathscr{L}. \qquad \text{(VII.19.4)}$$

For passive media $\kappa < 0$; for active media $\kappa > 0$.

According to Eq. (VII.19.4) the propagation constant $k = \beta + ig/2$ of the laser medium is

$$k = \beta + ig/2 = k_0\left(n_0 + \frac{1}{2n_0}\frac{f\,\Delta N}{1 + \mathscr{I}}\mathscr{L}\zeta\right) + ik_0\frac{1}{2n_0}\frac{f\,\Delta N\,\mathscr{L}}{1 + \mathscr{I}}, \qquad \text{(VII.19.5)}$$

where the gain constant g of the medium at frequency ν_L has been chosen so that the intensity of a plane wave traversing the medium grows with distance z as e^{gz}.

For a *Doppler-broadened line* the above expression for g must be replaced by the following formula, whose derivation goes beyond the scope of this book:

$$g(\nu) = \frac{1}{4\pi}\left(\frac{\ln 2}{\pi}\right)^{1/2} \frac{\lambda^2 \, \Delta N_0}{\Delta\nu_D \, \tau_{sp}} \exp\left\{-\left[\frac{2(\nu - \nu_L)}{\Delta\nu_D}(\ln 2)^{1/2}\right]^2\right\}. \qquad \text{(VII.19.6)}$$

Here τ_{sp} is the *natural* (radiative) *lifetime* for the transition and $\Delta\nu_D$ is the full width of the line at half-maximum, which depends on the atomic mass M in accordance with

$$\Delta\nu_D = 2\nu_L[2KT/(Mc^2)\ln 2]^{1/2}, \qquad \text{(VII.19.7)}$$

where K and T are Boltzmann's constant and the absolute temperature, respectively.

The change Δn in the refractive index at a frequency ν in the vicinity of ν_L can be calculated by means of the Kramers–Krönig dispersion relations (see problem I.9 and Bennett [18])

$$\Delta n(\nu) = \frac{c}{\pi\omega}\int_0^\infty \frac{\nu' g(\nu')}{\nu'^2 - \nu^2}\, d\nu' \cong \frac{cg(\nu_L)}{\omega\pi^{1/2}} e^{-\zeta^2} \int_0^\zeta e^{x^2}\, dx = -\frac{cg(\nu_L)}{\omega\pi^{1/2}} Z(\zeta), \qquad \text{(VII.19.8)}$$

where $\zeta = 2(\ln 2)^{1/2}(\nu - \nu_L)/\Delta\nu_D$ is the detuning factor for the Doppler line and $Z(\zeta)$ stands for *Dawson's integral* $Z(\zeta) = e^{-\zeta^2}\int_0^\zeta e^{x^2}\, dx$ (see Chapter II, Abramowitz and Stegun [4], p. 319). The integral $Z(\zeta)$, which resembles the function $\zeta/(1 + \zeta^2)$, presents a maximum $Z_{max} = 0.54$ for $\zeta = 0.924$ and an inflection point at $\zeta = 1.5$. Therefore, the change Δn in refractive index vanishes at $\nu = \nu_L$ and reaches its peak positive and negative values at $\zeta = \pm 0.924$.

19.2 *Single-Mode Operation*

For an accurate analysis of single-mode operation of lasers the reader is invited to read accurately Sargent *et al.* (Chapter I [6], Chap. 7). Here we limit ourselves to considering a resonator filled with a homogeneously broadened laser medium characterized by the susceptivity of Eq. (VII.19.1) and oscillating at a single frequency ν. In this case Eq. (VII.10.7) yields

$$P_n = \frac{if\varepsilon_0}{1 + i\zeta} \frac{\iiint_{\text{cavity}}[E\,\Delta N\Phi_n/(1 + \mathscr{I})]\, dV}{\iiint_{\text{cavity}} \Phi_n^2\, dV}, \qquad \text{(VII.19.9)}$$

the index n standing for lmn. Supposing now that $\mathscr{I}$ is so small that the saturation factor can be put equal to one, we have from Eqs. (VII.10.5),

(VII.10.7), and (VII.19.9)

$$(k_n^2 - k^2)E_n = [ifk_0^2\langle \Delta N \rangle_n/(1 + i\zeta)]E_n, \qquad \text{(VII.19.10)}$$

where

$$\langle \Delta N \rangle_n = \frac{\iiint \Delta N \Phi_n^2 \, dV}{\iiint \Phi_n^2 \, dV}.$$

This equation is satisfied either when $E_n = 0$ or when

$$k_n^2 - k^2 = ifk_0^2\langle \Delta N \rangle_n/(f + i\zeta). \qquad \text{(VII.19.11)}$$

This means that, apart from the undetermined multiplicative factor A_n, the cavity field will coincide with the mode Φ_n whose eigenvalue k_n satisfies the last relation, which, after we split k_n into its real part k_n' and imaginary part k_n'', can be rewritten in the form

$$(2\pi/c)(\nu - \nu_n) = -[fk_0\langle \Delta N \rangle_n/(2k)]\mathscr{L}\zeta, \qquad k_n'' = [fk_0\langle \Delta N \rangle_n/(2k)]\mathscr{L}. \qquad \text{(VII.19.12)}$$

Since ν_n is fixed by the resonance condition (VII.10.14b,c), Eq. (VII.19.12a) defines the detuning of the oscillation frequency with respect to the center ν_n of the cavity line. In particular, for $\Delta N = 0$ the oscillation frequency ν coincides with ν_n. For increasing population inversion $\langle \Delta N \rangle_n$, ν tends to deviate from the cavity resonance and takes values between ν_n and ν_L. This phenomenon is referred to as *frequency pulling*. Equation (VII.19.12b) establishes a relation between the losses of the resonator, represented by k'', and the population inversion $\langle \Delta N \rangle_n$. The value ΔN_{th} satisfying this condition is called the *threshold population inversion*. If we express $\langle \Delta N \rangle_n$ in the form $\langle \Delta N \rangle_n = \mathscr{N}_n \Delta N_{th}$, where $\mathscr{N}_n$ is called the *excitation parameter* and ΔN_{th} refers to the case in which $\nu_L = \nu_n$, the laser is below or above threshold according to whether $\mathscr{L}\mathscr{N} < 1$ or $\mathscr{L}\mathscr{N} > 1$.

If we neglect diffraction losses, we have to a good approximation [see Eq. (VII.10.14a)] $k_n'' = (1 - |r_1 r_2|)/(2d)$, so that Eq. (VII.19.12b) gives for the threshold inversion

$$\Delta N_{th} = [2k/(fk_0^2\mathscr{L})](1 - |r_1 r_2|)/(2d). \qquad \text{(VII.19.13a)}$$

For a Doppler-broadened gas laser, this equation must be replaced by [18]

$$\Delta N_{0,th} = [1/(\pi \ln 2)^{1/2}]k^2[(1 - |r_1 r_2|)/(2d)]\, \Delta\nu_D \tau_{sp} e^{\zeta^2}, \qquad \text{(VII.19.13b)}$$

and the expression for the laser detuning given by Eq. (VII.19.12a) becomes

$$\nu - \nu_n = -\frac{c}{2\pi}\frac{g(\nu_L)}{\pi^{1/2}} Z(\zeta), \qquad \text{(VII.19.14)}$$

where Z and ζ are defined in Eq. (VII.19.8) (see Bennett [18], Eq. (2.116)).

For $\mathscr{L}\mathscr{N}_n > 1$ the field becomes so large that we cannot neglect the saturation effects. In this case, by assuming that E can be represented by a simple cavity mode Φ_n, Eq. (VII.19.9) yields

$$P_n = i\frac{f\varepsilon_0}{1 + i\zeta}E_n\frac{\iiint \Delta N\Phi_n^2(1 + |\Phi_n|^2E_n^2\mathscr{L}/\bar{E}_S^2)^{-1}\,dV}{\iiint \Phi_n^2\,dV}, \tag{VII.19.15}$$

where we have taken explicitly into account the dependence of the saturation parameter $E_S^2 = \bar{E}_S^2/\mathscr{L}$ on the detuning parameter ζ. Since $|\Phi_n^2|$ is a rapidly varying function of z while the transverse distribution changes slowly with z, we can replace the integrands in the above equations with their average values over distances of a few wavelengths along the z axis, thus getting

$$P_n = i\frac{f\varepsilon_0}{1 + i\zeta}E_n\frac{\iiint \Delta N h(W)u_{lm}^2\,dV}{\iiint u_{lm}^2\,dV}, \tag{VII.19.16}$$

with

$$W = (E_n^2/\bar{E}_S^2)\mathscr{L}|u_{lm}^2| \tag{VII.19.17}$$

and

$$h(W) = 2\left\langle\frac{\sin^2 kz}{1 + W\sin^2 kz}\right\rangle = 1 - \frac{3}{4}W + \cdots. \tag{VII.19.18}$$

For W sufficiently small we can approximate $h(W)$ with $1 - \frac{3}{2}W$ and Eq. (VII.19.16) becomes

$$P_n = i\frac{f\varepsilon_0\langle\Delta N\rangle_n}{1 + i\zeta}\left(1 - \frac{3}{4}\frac{E_n^2}{\bar{E}_S^2}\mathscr{L}\frac{\iiint \Delta N|u_{lm}^2|u_{lm}^2\,dV}{\iiint \Delta N u_{lm}^2\,dV}\right)E_n. \tag{VII.19.19}$$

Accordingly, Eqs. (VII.19.12) are replaced by

$$(2\pi/c)(y - v_n) = -[f\langle\Delta N\rangle_n/(2k)]\mathscr{L}\zeta(1 - \tfrac{3}{2}E_n^2\alpha_n\mathscr{L}), \tag{VII.19.20a}$$

$$1 = \mathscr{L}\mathscr{N}_n(1 - \tfrac{3}{2}E_n^2\alpha_n\mathscr{L}), \tag{VII.19.20b}$$

with α_n replacing some factors of the second term in parentheses in Eq. (VII.19.19). Generally, the modes of the oscillators are represented by plane waves and ΔN is uniform over the entire volume of the cavity.

As a consequence of Eq. (VII.19.20b), the amplitude E_n is given by

$$E_n^2 = [2/(3\alpha_n)](\mathscr{L} - 1/\mathscr{N}_n)1/\mathscr{L}^2. \tag{VII.19.21}$$

A good approximation results from taking $v = v_n$ in the calculation of E_n^2.

When we change the value of v_n (for instance, by displacing one of the cavity mirrors) the amplitude of the cavity oscillation changes according to

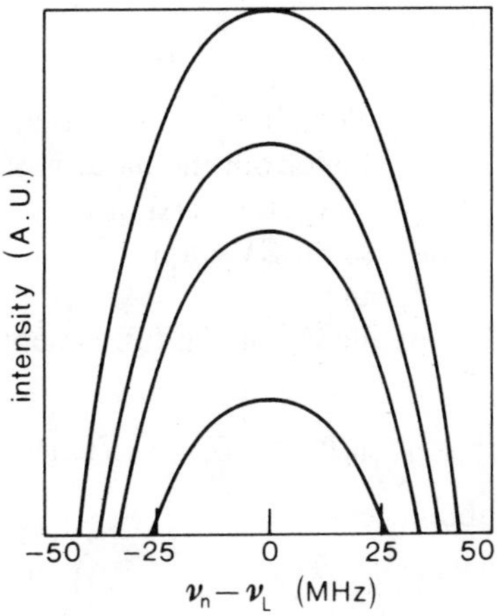

Fig. VII.39. Single-mode intensity versus detuning $\nu_n - \nu_L$ for a homogeneously broadened laser medium ($\Gamma = 100$ MHz) and increasing excitation parameter $\mathcal{N} = 1.05, 1.10, 1.15$, and 1.2.

Fig. VII.39. As expected, the intensity is a maximum for $\nu_n = \nu_L$ and vanishes for $\zeta^2 = \mathcal{N}_n - 1$. Accordingly, the maximum allowable detuning of the cavity depends on the laser linewidth Γ and on the amount by which the population inversion exceeds the threshold value ΔN_{th}, which is in turn a measure of the cavity losses. Broadly speaking, we can affirm that an increase of the laser linewidth and a reduction of the cavity bandwidth, together with an increase of the population inversion, produce a widening of the maximum allowable detuning of the cavity frequency with respect to the center of the laser line. In particular, when the detuning interval becomes larger than the intermode spacing, the laser output remains practically unchanged when we vary the cavity length. This condition is achieved when the relative excitation $\mathcal{N} > [c/(2d\Gamma) + 1]^2$.

When we slightly alter the length of the cavity of a single-mode laser with Doppler broadening (e.g., an He–Ne laser) we observe an output power versus length (or equivalently, cavity frequency) that increases to a definite maximum at the line center, as predicted by Lamb (see Chapter I, Sargent *et al.* [6]), and observed for the first time by McFarlane *et al.* [45] and Szöke and Javan [46]. This is due to the fact that each particle, because of its thermal speed, sees the two waves that form the standing-wave pattern of the cavity mode shifted up and down in frequency. The width of this dip is determined by natural linewidths rather than by a Doppler width. This effect can be observed whenever a standing wave can be set up in a gaseous absorbing or amplifying medium. It is currently used in *saturation absorption spectroscopy* (see, for instance, the monographs by Letokhov and Chebotayev [47] and Corney [48]) to measure the natural linewidths of lines broadened by the Doppler mechanism.

19.3 *Single-Frequency Multimode Oscillations*

When the gain per pass is larger than the threshold value, the saturation of the medium can induce a strong deviation of the field from the pattern of a single mode Φ_n. A significant variation of the population inversion over the cavity mode can produce similar effects. In this case the SVA approximation presented in Section VII.10.1 can be usefully adopted.

In particular, if we assume that the laser oscillates on the fundamental mode TEM_{00}, Eq. (VII.10.17) yields

$$E^{(\pm)}(k^2 - k_{\bar{n}}^2) \mp 2ik_{\bar{n}}\, dE^{(\pm)}/dz + \omega^2\mu_0 P^{(\pm)} = 0, \qquad \text{(VII.19.22)}$$

from which, by simple manipulations, we obtain

$$d|E^{(\pm)}|^2/dz = \pm\alpha(z, E^{(+)}, E^{(-)})|E^{(\pm)}|^2, \qquad \text{(VII.19.23)}$$

where α is the amplification relative to the intensity $|E^{(\pm)}|^2$,

$$\begin{aligned}\alpha(z, E^{(+)}, E^{(-)}) &= -\mathrm{Re}\{(i/k_{\bar{n}})[k^2 - k_{\bar{n}}^2 + \omega^2\mu_0 P^{(\pm)}/E^{(\pm)}]\}\\ &= -\mathrm{Re}[i(k^2 - k_{\bar{n}}^2)/k_{\bar{n}}] + g(z, E^{(+)}, E^{(-)}).\end{aligned} \qquad \text{(VII.19.24)}$$

Here g is the gain of the laser medium,

$$\begin{aligned}g(z, E^{(+)}, E^{(-)}) &= -\mathrm{Re}\left[\frac{i}{k\bar{n}}\omega^2\mu_0\frac{P^{(\pm)}}{E^{(\pm)}}\right]\\ &= \mathrm{Re}\left(\frac{k_0^2}{k_{\bar{n}}}\frac{f}{1 + i\zeta}\right)\\ &\quad\times\frac{\iint_{-\infty}^{+\infty}\langle|u_{00}|^2\,\Delta N/(1 + |E^{(+)}u_{00}^{(+)} + E^{(-)}u_{00}^{(-)}|^2/E_S^2)\rangle\,dx\,dy}{\iint_{-\infty}^{+\infty}|u_{00}|^2\,dx\,dy},\end{aligned} \qquad \text{(VII.19.25)}$$

where $\langle\ \rangle$ indicates the average value over an interval $\Delta z > \lambda$.

Analogously, we can show that the phases $\phi^{(\pm)}$ of $E^{(\pm)}$ $(E^{(\pm)} = |E^{(\pm)}|e^{i\phi^{(\pm)}})$ satisfy

$$\begin{aligned}d\phi^{(\pm)}/dz &= \pm\tfrac{1}{2}\mathrm{Re}\{(1/k_{\bar{n}})[k^2 - k_{\bar{n}}^2 + \omega^2\mu_0 P^{(\pm)}/E^{(\pm)}]\}\\ &= \mp\beta[z, E^{(+)}, E^{(-)}].\end{aligned} \qquad \text{(VII.19.26)}$$

From Eqs. (VII.10.11,16) we can easily infer that the generally complex amplitudes $E^{(\pm)}$ satisfy the following condition at the locations z_1 and z_2 of the cavity mirrors:

$$[E^{(+)}(z_1)E^{(-)}(z_2)]/[E^{(+)}(z_2)E^{(-)}(z_1)] = 1. \qquad \text{(VII.19.27)}$$

For amplitude negligible with respect to the saturation amplitude E_S, g and β

are independent of $E^{(\pm)}$ and the systems of Eqs. (VII.19.23) and (VII.19.26) can easily be integrated by quadrature. Besides, from the boundary condition of Eq. (VII.19.27) we obtain

$$\int_{z_1}^{z_2} \alpha(z)\,dz = \int_{z_1}^{z_2} \beta(z)\,dz = 0. \qquad \text{(VII.19.28)}$$

These equations fix the resonance value of $k_{\bar{n}}$, according to the relation

$$(k^2 - k_{\bar{n}}^2)d = -k_0^2 \frac{if}{1 + i\zeta} \int_{z_1}^{z_2} \left(\frac{\iint_{-\infty}^{+\infty} \Delta N |u_{00}|^2\,dx\,dy}{\iint_{-\infty}^{+\infty} |u_{00}|^2\,dx\,dy} \right) dz, \qquad \text{(VII.19.29)}$$

which replaces Eq. (VII.19.11). When the integral in parentheses in the above equation does not depend sensibly on z, both $\alpha(z)$ and $\beta(z)$ vanish and the fields $E^{(\pm)}$ are constant along z. Physically, this corresponds to the fact that a single mode $\Phi_{\bar{n}}$ is excited.

19.4 *Rigrod's Equations*

The equations presented above cannot be easily solved when saturation effects are included. However, in order to obtain a qualitative idea of these effects we will assume that the mode u_{00} is a plane wave and ΔN is constant over the cavity volume. In addition, we will replace $|E^{(+)}u_{00} + E^{(-)}u_{00}|^2$ with $|E^{(+)}|^2 + |E^{(-)}|^2$ in the saturation factor appearing in the integral in the numerator of Eq. (VII.19.25). Using these approximations, Rigrod [49] analyzed the solutions of Eq. (VII.19.23) by expressing the gain in the form

$$g = g_0/[1 + \mathscr{I}^{(+)} + \mathscr{I}^{(-)}], \qquad \text{(VII.19.30)}$$

where $\mathscr{I}^{(\pm)} = |E^{(\pm)}u_{00}|^2/E_S^2$ and g_0 represents the *small-signal gain* of the laser medium. The exponential variation of $|u_{00}| \propto e^{-k''z}$ allows us to recast Eq. (VII.19.23) in the form

$$d\mathscr{I}^{(\pm)}/dz = \pm\{g_0/[1 + \mathscr{I}^{(+)} + \mathscr{I}^{(-)}]\}\mathscr{I}^{\pm)}, \qquad \text{(VII.19.31)}$$

with the boundary conditions

$$\mathscr{I}^{(+)}(z_1) = |r_1|^2 \mathscr{I}^{(-)}(z_1), \qquad \mathscr{I}^{(-)}(z_2) = |r_2|^2 \mathscr{I}^{(+)}(z_2). \qquad \text{(VII.19.32)}$$

From Eqs. (VII.19.32) it follows that $\mathscr{I}^{(+)}(z)\mathscr{I}^{(-)}(z) = \text{const}$. Consequently, we have [cf. Eq. (VII.1.9)]

$$\{[\mathscr{I}^{(+)}(z_2)]/[\mathscr{I}^{(+)}(z_1)]\}^2 = \{[\mathscr{I}^{(-)}(z_1)]/[\mathscr{I}^{(-)}(z_2)]\}^2 = 1/|r_1 r_2|^2 = e^{2G}, \qquad \text{(VII.19.33)}$$

where $G = \ln[\mathscr{I}^{(+)}(z_2)/\mathscr{I}^{(+)}(z_1)]$ represents the *available gain per pass*. For the system to oscillate the small-signal gain g_0 must exceed G/d.

To calculate $\mathscr{I}^{(+)}(z_1)$ we must integrate Eq. (VII.19.31). By doing so we obtain

$$\mathscr{I}^{(+)}(z_1) = \frac{g_0 d + \ln|r_1 r_2|}{|r_1 r_2|^{-1} + |r_1|^{-2} - |r_2/r_1| - 1} \cong \frac{g_0 d - \frac{1}{2}(A_1 + A_2 + T_1)}{A_1 + A_2 + T_1}, \tag{VII.19.34}$$

where we have indicated with A and T the absorbance and the transmittance of the mirrors ($T_2 = 0$). The last expression is justified by the fact that for $|r| \sim 1$ we have $1/|r| \sim 1 + \frac{1}{2}A + \frac{1}{2}T$.

19.5 *Power Extraction*

In a resonator that uses a semitransparent mirror M_1 of transmittance T_1 to extract radiation from the cavity, the output intensity $\mathscr{I}_{\text{out}}$, normalized to the saturation intensity, can be derived from Eq. (VII.19.34):

$$\mathscr{I}_{\text{out}} = (T_1/|r_1|^2)\mathscr{I}^{(+)}(z_1) \cong g_0 d[T_1/(A_1 + A_2 + T_1)][1 - (A_1 + A_2 + T_1)/(2g_0 d)]. \tag{VII.19.35}$$

Consequently, $\mathscr{I}_{\text{out}}$ is maximum when the transmittance of the output mirror is equal to T_{opt}, given by

$$T_{\text{opt}} = (A_1 + A_2)\{[2g_0 d/(A_1 + A_2)]^{1/2} - 1\}. \tag{VII.19.36}$$

Thus, the maximum power that may be extracted from a plane cavity is given for $2g_0 d \gg A_1 + A_2$ by

$$P_{\text{out}} \cong V_{\text{mode}}\, I_S g_0\{1 - [(A_1 + A_2)/(2g_0 d)]^{1/2}\}, \tag{VII.19.37}$$

where $V_{\text{mode}} = dw^2$ is the volume of the oscillation mode and I_S is the saturation intensity (watts per square centimeter) of the active medium.

The diffraction losses, both intrinsic to the laser medium and produced by reflection, have been neglected so far. We may now allow for them by including them in the coefficient $A_1 + A_2$, which must be considered a measure of the total internal losses of the laser resonator. We may furthermore assume that Eq. (VII.19.37), established for plane mirror resonators, holds for a more general stable cavity, as long as the oscillation modes do not diverge too markedly.

Equation (VII.19.35) can also be used for unstable resonators having a moderate magnification (i.e., $M \sim 1$). In this case we can state for a telescopic resonator

$$P_{\text{out}} \cong V_{\text{mode}} I_S g_0 (M - 1)\frac{1 - (1/2g_0 d)\ln(M/|R_1 R_2|)}{(M/R_1 R_2)^{1/2} - (R_1 R_2/M)^{1/2}}. \tag{VII.19.38}$$

The factor $M - 1$ allows for the fact that the power is extracted through a cross section that is $(M - 1)$ times as large as the surface of the concave mirror. The reflection coefficients $R = 1 - A - A_d$ are defined in such a way as to include the diffraction effects A_d, which are particularly important in these cases. As these losses, and hence $A_1 A_2$, are appreciably dependent on the equivalent Fresnel number N_{eq}, P_{out} will be optimized by suitably selecting N_{eq} for a given value of M.

20 Frequency Control

The behavior of a cavity filled with an active medium depends strongly on the mechanisms controlling the broadening of the resonant gain curve [50]. When we have a homogeneous linewidth, the presence of an oscillation at a given frequency tends to prevent the onset of other oscillations at different frequencies. In contrast, when the line exhibits some amount of inhomogeneous broadening, then an almost resonant field does not interfere with all of the active molecules, and some other oscillations can be stimulated. This mechanism explains the multimode operation observed in many lasers, especially those characterized by high gain, which is sometimes desired in order to generate radiation having a bandwidth large enough to be modulated at high frequency.

20.1 *Control of the Cavity Length*

We saw in Section VII.19 that when the laser linewidth is comparable to the mode spacing ν_0 and the excitation factor $\mathcal{N}$ is almost unity, P_{out} depends appreciably on the detuning $\nu_{lmn} - \nu_L$. Optimizing the output power then requires adjusting the length d to obtain $\nu_{lmn} = \nu_L$. It is easily seen that a variation of d by $\lambda/2$ implies a resonance frequency variation by ν_0; therefore, the cavity length must be controlled to an accuracy much higher than λ. This is sometimes achieved by mounting the mirrors on two plates spaced by Invar bars or, more simply, by securing them on a granite bench. In addition, a piezoelectric transducer (PZT) whose thickness may be made to vary by a few micrometers by a suitable control voltage is introduced between one of the mirrors and the supporting plate.

20.2 *Frequency Control by Prisms, Gratings, Étalons, and Resonant Reflectors*

Another way to control the frequency is to introduce inside the cavity a dispersive element [50] in the form of a prism plus étalon (see Fig. VII.40) or to replace a mirror at one end of the resonator with a grating or a stack of low-

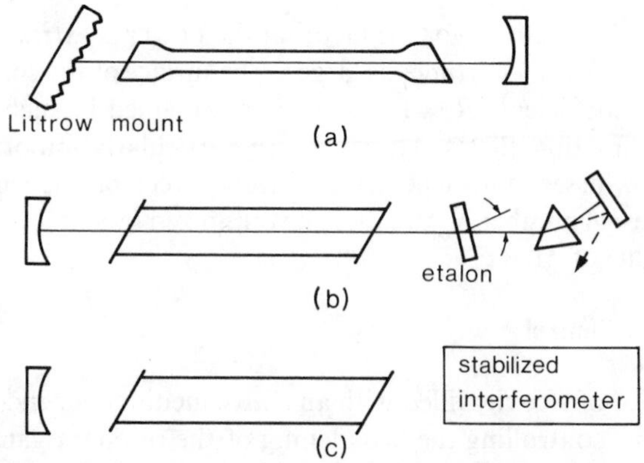

Fig. VII.40. Cavities containing dispersive elements for selecting the emission line of (a) CO_2 or (b) Ar lasers by means of a grating or a prism plus étalon, respectively. The resonator in (c) is equipped with an interferometer to reduce the linewidth of an He–Ne laser.

finesse étalons forming a so-called *resonant reflector*. A widely used technique employs a tilted Fabry–Perot étalon (see following section) inserted at a small angle in the laser resonator. The reflectivity of the composite mirror consisting of the tilted étalon and adjacent end mirror corresponds to the transmission curve of a simple Fabry–Perot resonator of high FSR. The reflectivity maxima of the composite mirror are adjusted by slightly tilting the étalon. The cavity will oscillate at a resonant frequency corresponding to an antiresonance of the étalon.

In molecular lasers the gain peaks at many frequencies spaced a few reciprocal centimeters apart. The transfer of energy by molecular collisions in gases tends to favor lasing of the system on the line having the highest gain (*line competition effect*). In other words, the set of laser lines behaves like the gain curve of a single homogeneous line. Sometimes we must force the system to lase on a particular line not exactly coincident with the most efficient one. This can be achieved by increasing the losses of the cavity corresponding to all the lines except the one of interest. Because the interline distance is of the order of a few reciprocal centimeters, we abandon the use of an étalon, which has an FSR of the same order of magnitude as the line spacing (a few reciprocal centimeters), in favor of a grating in the Littrow mount, as illustrated in Figs. VII.40a and VII.41. For a nonmonochromatic beam, only the portion of the spectrum having a wavelength satisfying Eq. (VI.10.2) with $\alpha = 0$ is retroreflected into the cavity, while the other spectral components are spread around the optic axis and are therefore lost to the cavity. When the grating is rotated, it changes the wavelength retroreflected into the resonator. The bandwidth of solid-state lasers [51] may be narrowed by replacing a mirror with a resonant reflector designed to have narrow regions of high reflectivity (say 1 Å).

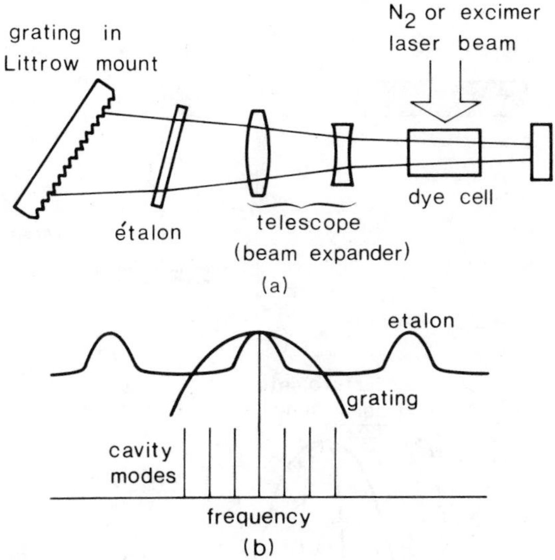

Fig. VII.41. (a) Grating preselection and (b) cavity mode selection of a dye laser with an intracavity étalon.

20.3 *Single-Frequency Operation by Means of a Fox–Smith Interferometer*

A number of interferometric mode selection techniques have been proposed. They are all based on the principle of replacing one mirror with a second cavity having a free spectral range comparable to the laser gain bandwidth, which can be thought of as a composite mirror of variable reflectivity.

In a cavity with longitudinal mode selection controlled by the Fox–Smith interferometer [52] illustrated in Fig. VII.42, a long laser cavity $d_1 + d_2$ is coupled to a relatively small folded cavity $d_2 + d_3$ by means of a beam splitter M_2. In this device, the reflectance presented to the laser by the folded cavity is high enough to permit oscillation over a small portion of its free spectral range

$$c/2(d_2 + d_3), \tag{VII.20.1}$$

which exceeds the oscillation bandwidth of the laser.

A wave traveling back and forth exhibits a fractional power loss given by

$$\alpha = 1 - \frac{R_1 T_2^2 R_3}{[1 - (R_3 R_4)^{1/2} R_2]^2 + 4(R_3 R_4)^{1/2} R_2 \sin^2[(2\pi/\lambda)(d_2 + d_3)]}, \tag{VII.20.2}$$

where the R_i are the mirror reflectivites and T_2 is the transmittance of the beamsplitter.

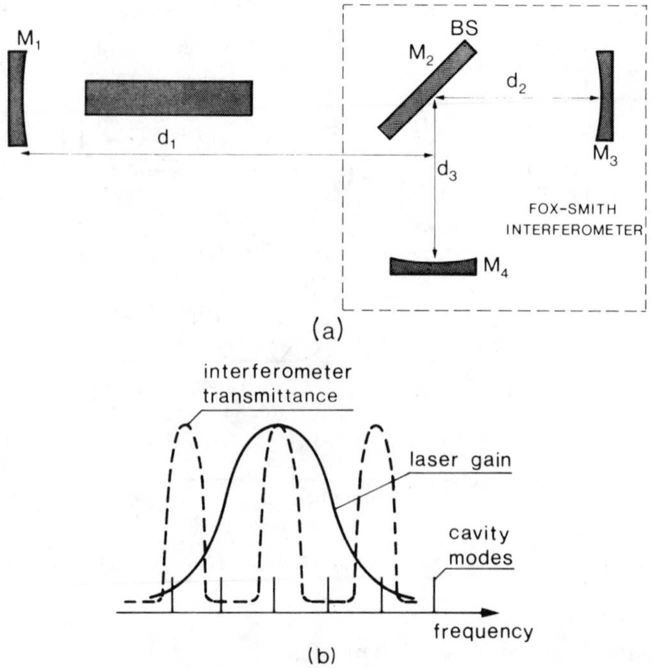

Fig. VII.42. Schematic of a stabilized single-mode laser using a Fox–Smith interferometer: (a) mirror geometry, (b) laser gain and interferometer transmittance curve. By changing d_3 a peak in the laser output occurs when there is coincidence of a cavity mode with a maximum of the interferometer transmittance.

When $R_i = 1$, $T_2 = 0.5$, $d_2 + d_3 = \lambda m/2$, and $d_1 + d_2 = \lambda n/2$ (with m and n integers), α vanishes and no power is lost. This is due to the fact that the standing-wave pattern of the oscillation mode has a node on the beamsplitter. In other words, by regulating the arms d_2 and d_3 in such a way as to satisfy the above conditions, the resonator will present zero losses at the assigned wavelength. The factor α increases rapidly when we pass to the next resonance and so on. The envelope of the losses of the different modes yields the transmittance curve sketched in Fig. VII.42b.

20.4 *Injection Locking*

An important approach to the frequency control of a laser is to impress on it an external signal generated by a well-stabilized low-power laser oscillating at the same frequency. This technique, known as *injection locking* [53, 54], exploits the nonlinear dependence of the gain on the oscillation amplitude, as originally discussed by Van der Pol in 1927 in his analysis of the frequency locking in triode oscillators (see ref.I.6, p. 52). (See Chapter I, Sargent *et al.* [6], p. 52).

The effect of a stabilized monochromatic signal $e(t) = e \sin \omega_{ext} t$ on the amplitude $V(t)$ of the field of a laser oscillator can be studied with the help of the feedback network of Fig. VII.6 by approximating the amplification characteristic $A = A(V^2)$ with the quadratic law $A = A_0 - A_3 V^2$ and neglecting the dependence of the dephasing factor $\beta(I, \omega, \omega_L)$ on the frequency ω, owing to the narrowness of the spectral range of $V(t)$ centered around ω_{ext}. Accordingly $V(t)$ satisfies the nonlinear *Van der Pol differential equation*

$$d^2 V/dt^2 - (d/dt)(A_0 V - A_3 V^3) + \omega_{cav}^2 V = V_{ext} \sin \omega_{ext} t, \qquad \text{(VII.20.3)}$$

where V_{ext} is proportional to the amplitude of the external signal. For $\omega_{cav} \sim \omega_{ext}$ the slow varying amplitude approximation can be used, so that

$$V(t) = \mathrm{Re}\{E(t) \exp[i\omega_{ext} t + i\Phi(t)]\}, \qquad \text{(VII.20.4)}$$

where $E(t)$ and $\Phi(t)$ are two slow functions of t such that we can neglect their second-order time derivatives. Then, capitalizing on this condition and inserting the right-hand side of Eq. (VII.20.4) into Eq. (VII.20.3) yields

$$dE/dt = \tfrac{1}{2}(A_0 - A_3 E^2)E - \tfrac{1}{2}\omega_{ext} V_{ext} \cos \Phi, \qquad \text{(VII.20.5a)}$$

$$d\Phi/dt = \omega_{cav} - \omega_{ext} + \tfrac{1}{2}\omega_{ext}(V_{ext}/E) \sin \Phi. \qquad \text{(VII.20.5b)}$$

For V_{ext} sufficiently small the term proportional to V_{ext} in Eq. (VII.20.5a) can be neglected by using the value $E = (A_0/A_3)^{1/2}$ in Eq. (VII.20.5b). Next, integrating the latter equation we obtain [55]

$$\Phi(t) = 2 \arctan\{1/K - [(K^2 - 1)^{1/2}/K] \tan[\tfrac{1}{2}L(t - t_0)(K^2 - 1)^{1/2}]\}, \qquad \text{(VII.20.6)}$$

where $L = \omega_{ext} V_{ext}/(2E)$ is the *locking coefficient*, $K = (\omega_{cav} - \omega_{ext})/L$ the detuning parameter normalized with respect to L, and t_0 an integration constant. We immediately observe that for $|K| > 1$, $d\Phi/dt$ never vanishes, so that the oscillator never reaches a steady state. The instantaneous oscillation frequency $\omega(t)$ is a periodic function of t, viz.

$$\omega(t) = \omega_{ext} + d\Phi/dt = \omega_{cav} + L \sin \Phi(t). \qquad \text{(VII.20.7)}$$

In contrast, for $|K| < 1$, the function tan( ) of Eq. (VII.20.6) turns into tanh and $\Phi(t)$ tends asymptotically to a value Φ_∞ simply related to K:

$$\Phi_\infty = \arcsin K. \qquad \text{(VII.20.8)}$$

In addition, the instantaneous frequency $\omega(t)$ tends asymptotically to the frequency of the external signal with a rate depending on the product $L(1 - K^2)^{1/2}$.

In conclusion, the laser oscillator can be synchronized with the frequency of an external source only if the locking coefficient L is larger than the detuning

$|\omega_{cav} - \omega_{ext}|$ of the external frequency ω_{ext} with respect to the laser mode frequency ω_{cav}. This condition fixes a lower bound for the minimum power injected into the oscillator to observe a steady-state frequency locking.

21 Fabry–Perot Interferometers

The plane-parallel optical resonator was devised by C. Fabry and A. Perot [10] in 1897 as an interferometric apparatus, and it has been extensively used since then for the spectral analysis of hyperfine structures or for accurate metrological measurements [56, 57]. The invention of the laser triggered a great many analytic and experimental studies on the properties and applications of these devices, which are now built with mirrors of any focal length and spacings ranging from a few millimeters to several meters. Accounts of the results of these analyses have been presented in numerous sections of this chapter. However, owing to the great importance of plane-parallel resonators as interferometers, we will discuss their main features in this section.

Although, in principle, we should apply modal analysis to the study of Fabry–Perot interferometers, a simpler approach is generally preferred, since diffraction effects can be neglected in many cases owing to their very large Fresnel numbers. In addition, the diffraction analysis of the field transmitted by an interferometer would imply the solution of inhomogeneous Fox–Li integral equations because of the presence of a source term representing the incident beam, an approach that would make the whole analysis very messy.

In their simplest form, Fabry–Perot interferometers consist of two parallel optical surfaces, between which multiple reflections occur. They may consist of a pair of semitransparent mirrors assembled a small distance d apart with the mirrored surfaces facing each other and accurately parallel (see Fig. VII.43a). The second surfaces of the mirrors are normally wedged (typically at 10–30′) with respect to the first surfaces in order to eliminate interference of the rays they reflect with those undergoing multiple reflections. These secondary reflections cannot be eliminated by simply adding an AR coating, since even the best "V" AR treatments have 0.1–0.2% reflectivity.

Devices with fixed spacing are generally referred to as *étalons*. In these instruments control of the transmittance curve can be achieved by changing the incidence angle. An étalon can also be made by depositing multidielectric layers acting as mirrors on the opposite faces of optical flats of quartz or other optical materials that are transparent in the region in which the interferometer must be used (see Fig. VII.43b). The quality of these *solid-state étalons* (SSE) depends critically on the parallelism of the two faces (typically, better than 0.2′) and on the variation of the refractive index n and the thickness d with

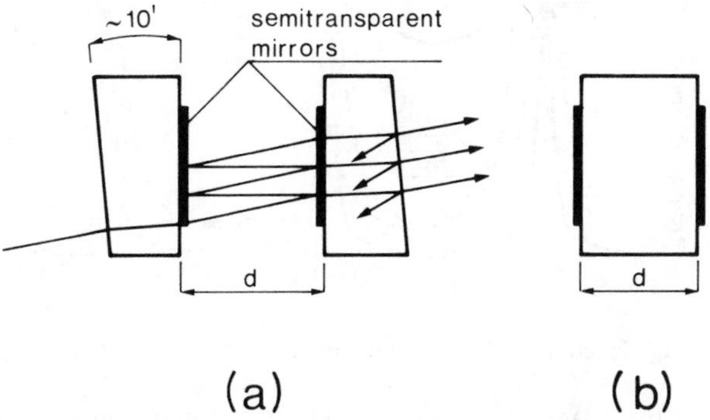

Fig. VII.43. Schematic of two Fabry–Perot interferometers. In (a) two semitransparent mirrors are assembled a distance d apart with the faces accurately parallel. In (b) a solid-state étalon (SSE) is shown.

temperature. These two parameters are usually controlled by placing the étalon in an oven in order to guarantee accurate temperature regulation (typically $\pm 0.01°C$).

21.1 *Transmittance of an Ideal Fabry–Perot Interferometer*

The Fabry–Perot interferometer derives its popularity from the fact that its transmittance $T(\lambda_0, \theta)$ varies periodically with the frequency, in the form of a series of peaks of nearly unit amplitude and very small width. In particular, for an ideal Fabry–Perot interferometer of infinite extension and illuminated by a plane wave the transmittance $T(\lambda_0, \theta)$ is given by the Airy function (see Section III.12.1)

$$T(\lambda_0, \theta) = \frac{T^2}{(1-R)^2} \frac{1}{1 + [4R/(1-R)^2 \sin^2(k_0 nd \cos\theta')}$$

$$\equiv \frac{T^2}{(1-R)^2} A(F_R, k_0 nd \cos\theta'), \qquad \text{(VII.21.1)}$$

where $R = (R_1 R_2)^{1/2}$, $T = (T_1 T_2)^{1/2}$, and A represents the Airy transmittance function (see Fig. VII.44) expressed in terms of the reflecting finesse F_R, which will be defined in Eq. (VII.21.6). Here $T_{1,2}$ and $R_{1,2}$ are the transmittance and the reflectance, respectively, of the two reflecting surfaces, while n and d are the refractive index and the thickness of the medium between the two surfaces. The incidence angle θ' of the two plane waves present in the spacer is related to the external incidence angle θ by Snell's law, $\sin\theta = n \sin\theta'$. In several cases, θ and θ' are so small that $\theta' = \theta/n$. The transmittances $T_{1,2}$ and the reflectances

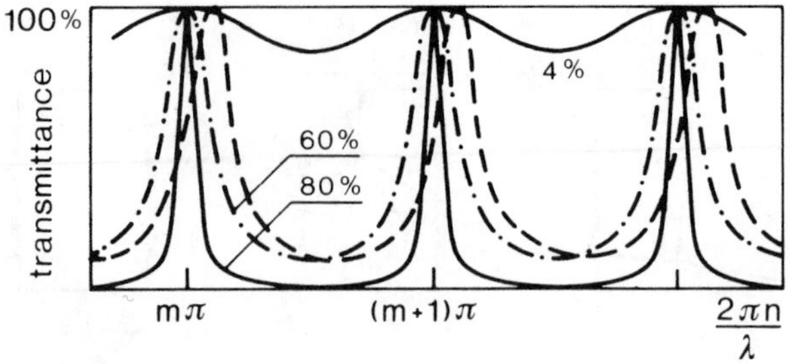

Fig. VII.44. Airy function versus $2\pi/\lambda$ for different reflectances of a Fabry–Perot interferometer. An increase of R (expressed in percent) produces a sharpening of the transmission peaks. The dashed (---) curves represent the transmittance of a tilted étalon.

$R_{1,2}$ are related through the equation

$$T_{1,2} + R_{1,2} + A_{1,2} = 1, \qquad \text{(VII.21.2)}$$

where $A_{1,2}$ represents the absorbance of face 1 (2) due to the absorption losses of the coating and the scattering produced by the unavoidable roughness of the surface ($A = A_{\text{abs}} + A_{\text{sc}}$). Here A_{sc} is given approximately by $2(k_0 n\sigma_w)^2$, where σ_w represents the standard deviation of the reflecting surface from an ideal plane.

According to the Airy formula, the transmission curve of a Fabry–Perot interferometer consists of a series of identical peaks equidistant on a frequency (or wave number λ_0^{-1}) scale. At normal incidence ($\theta' = 0$) the wavelengths transmitted with minimum attenuation are given by $\lambda_m = 2nd/m$, where m is an integer called the *order of interference*. At oblique incidence the wavelength of maximum transmission is changed to $\lambda_m = 2nd\cos\theta'/m$. For θ' sufficiently small, $\nu_m = c/\lambda_m$ is displaced from the corresponding frequency $\nu_{m\text{o}}$ for normal incidence by the quantity $\delta\nu_m$, which depends quadratically on θ', viz.

$$\delta\nu_m = \nu_{m\text{o}}\theta'^2/2. \qquad \text{(VII.21.3)}$$

Solid-state étalons suffer in thermal stability since both the spacing d and the refractive index n depend on the temperature. In fact, the transmitted frequencies ν_m are shifted by $\delta\nu_m$ when T changes by δT, viz.

$$\delta\nu_m = -\nu_m[(1/n)(dn/dT) + (1/d)(d(d)/dT)]\,\delta T \equiv -\eta\,\delta T. \qquad \text{(VII.21.4)}$$

The temperature coefficient η of the optical length of the étalon is of the order [58] of 2–10 × 10^{-6}. This means that to guarantee a shift $\delta\nu_m = \pm 20$ MHz for an étalon working at 500 nm, the temperature should be stabilized to within $\pm 0.01\,°$C. This condition is generally achieved by placing the étalon in a small enclosure whose temperature is actively stabilized.

The transmittance oscillates between T_{max} and T_{min}, whose ratio $C = T_{max}/T_{min}$, referred to as the *contrast* of the interferometer, is given by

$$C = \left\{\frac{[1 + (R_1 R_2)^{1/2}]}{[1 - (R_1 R_2)^{1/2}]}\right\}^2. \tag{VII.21.5}$$

Interferometers used for spectroscopic analysis are characterized by a high contrast, while the étalons used in laser cavities generally have a moderate C.

When the contrast is sufficiently high we find it convenient to characterize the étalon by means of the finesse F, whose inverse measures the full width at the half-maximum points of the transmission peak with respect to the free spectral range $\nu_0 = c\cos\theta'/(2nd)$. For the Airy transmisssion function, the finesse F becomes

$$F_R = \pi(R_1 R_2)^{1/4}/[1 - (R_1 R_2)^{1/2}], \tag{VII.21.6}$$

where the subscript R reminds us of the dependence of F_R on the reflectivities of the mirrors.

In the following discussion, we will derive a more complete expression for F by taking into account the effects due to the finite angular aperture of the light beam and the aberrations of the mirrors. Accordingly, F_R represents only the contribution of the reflectivity and is therefore referred to as the *reflecting finesse.*

The finesse is a measure of the interferometer's ability to resolve closely spaced lines in the spectrum of a plane incident wave. It can be thought of as the effective number of interfering beams involved in forming the étalon multiple-beam interference fringes and is proportional to the lifetime of a photon placed in the interferometer.

21.2 *Use of a Fabry–Perot Interferometer*

The Fabry–Perot interferometer is generally used [59] by illuminating its entrance face with a collimated beam and examining the transmitted beam in the focal plane of a lens placed on the exit side.

Originally, the Fabry–Perot interferometer was used as a spectrograph by recording on a photographic plate placed on the focal plane of the exit lens. If a monochromatic source has an extended area, the radiation falling on the étalon can be represented as a superposition of plane waves with wave vectors filling some solid angle. Consequently, those components forming a discrete sequence of incidence angles θ_m such that

$$k_0 nd\cos\theta_m = m\pi \tag{VII.21.7}$$

are transmitted.

When a uniform monochromatic field is viewed through a Fabry–Perot interferometer, the field is seen as a concentric narrow-ring pattern at infinity centered at normal incidence to the interferometer. The angular diameters of the jth and kth rings are related by $2(\cos\theta_j - \cos\theta_k) = (k-j)\lambda_0/(dn) \cong \theta_k^2 - \theta_j^2$. We may measure the angular width of a ring by its half-intensity points $\theta_{1/2} = \lambda_0/(F2dn\theta)$. Accordingly, such a ring subtends a solid angle Ω given by

$$\Omega = 2\pi\theta\theta_{1/2} = \pi\lambda_0/(Fdn) = 2\pi/\mathscr{R}, \qquad \text{(VII.21.8)}$$

where $\mathscr{R} = \lambda/\lambda_{1/2} = 2Fd/\lambda_0$ is the theoretical resolution of the interferometer. According to the expression for $\mathscr{R}$, the resolution of a Fabry–Perot interferometer may be varied over a wide range by varying the spacing d and, to a lesser extent, by varying the reflecting finesse.

The rings are imaged on the focal plane of the exit lens, where they produce a set of fine circular fringes of equal inclination, called *Haidinger rings* after the Austrian physicist Wilhelm Karl Haidinger (1795–1871). The radii ρ_m of these rings are given by

$$\rho_m = f\theta_m = 2^{1/2}f[1 - m\lambda_0/(2d)]^{1/2}, \qquad \text{(VII.21.9)}$$

f being the focal length of the exit lens.

When the source has a narrow (frequency) spectrum, each wavelength gives rise to a set of fringes. By observing the photographed pattern with a microdensitometer, the spectrum can be deduced with a certain limit of resolution. With this technique the hyperfine structures of atoms, which reveal the effect of the atomic nucleus on optical spectra, have been extensively studied. Modern Fabry–Perot interferometers are used in the study of Raman, Brillouin, or Rayleigh scattering from gases, liquids, or solids.

In 1948 Jacquinot and Dufour introduced a Fabry–Perot spectrometer in which the photographic plate was replaced by a photocell (which today would be replaced by a photomultiplier or a photodiode) behind a pinhole set in the focal plane of the exit lens. This method is called *central spot scanning*. By varying linearly in time the pressure of the gas in which the interferometer is placed, or by displacing the mirrors supported by piezoelectric spacers, we obtain a photodetector signal that is proportional to the spectral intensity of the source relative to the frequency at which the étalon is tuned at that instant. For example, by placing the spectrometer in a windowed gastight chamber containing SF_6 gas, for which the STP refractive index is roughly 1.00078, a scanning rate of 3.9 Å/atm can be achieved [60]. Whereas, for pressure scanning the total spectral range swept is independent of the spacing d, for mechanical scanning the range increases with decreasing d; an entire free spectral range is scanned when d changes by $\lambda/2$.

Spectroscopic instruments are characterized by the *throughput* U (see Section II.15 and IV.15), defined as the product of the area S and solid angle Ω

that the instrument accepts. A systematic treatment of the throughputs of various instruments was first given by Jacquinot [56]. For the Fabry–Perot interferometer we have from Eq. (VII.21.8),

$$U = S\Omega = \pi\lambda_0 S/(Fd) = 2\pi S/\mathscr{R}. \qquad \text{(VII.21.10)}$$

The throughput U, also called the *étendue* or *light-gathering power*, measures the ability of the interferometer to gather the maximum possible radiant power from a given source and to disperse and detect it appropriately.

If we indicate with L (luminosity) the ratio of the flux $\Phi(k_0)$ per unit wave number (erg centimeters per second) arriving at the detector to the brightness $B(k_0)$ per unit wave number (ergs per centimeter per steradian per second) of an extended source, we have

$$L = \Phi/B = \tau U = \tau 2\pi S/\mathscr{R}, \qquad \text{(VII.21.11)}$$

where τ is the peak transmittance of the interferometer.

It is noteworthy that the product of luminosity times resolution is a constant that depends on the étalon area S and the transmittance factor. Jacquinot showed that the product $L\mathscr{R}$ for Fabry–Perot interferometers is 70–350 times the analogous product for grating spectrometers. This high luminosity $\times$ resolution product can be easily explained by observing that at a given resolution the Fabry–Perot interferometer accepts light from a much larger solid angle than a grating spectrometer. For a more accurate account of the methods of recording and scanning the reader may consult the monograph by Cook [59].

21.3 *Multiple Fabry–Perot Spectrometers*

There are cases in which high resolution is required over a relatively large wavelength range. In particular, in the study of hyperfine structure the overall spread of the components may be many times the width of a single component. This requires an extension of the free spectral range to permit an unambiguous study of the spectrum, which is generally achieved by placing in series [60] with the main interferometer a second étalon having a shorter length d'. In general, regions of low transmittance of the short interferometer coincide with peaks of transmittance of the longer one. By adjusting d', peaks of the short one may be made to coincide with every qth peak of the longer one. In this way all but the qth maximum of transmittance of the longer interferometer are suppressed, but the resolution is the same as that of the longer étalon.

This principle may be extended to three or more Fabry–Perot interferometers in series. Of particular relevance for spectroscopic studies of planetary atmospheres, interstellar lines, or solar lines is the PEPSIOS spectrometer [61] combining three interferometers with an interference filter.

21.4 *Instrumental Function of an Interferometer*

When the source has a narrow spectrum of practically zero width, the spectral profile observed by means of the interferometer is represented by the so-called *instrumental function* $W(k_0)$. The form of W depends on the characteristics of the spectrometer and of the input beam. In the ideal case W coincides with the Airy function. In practice, the interferometer mirrors will have irregularities due to nonflatness that will broaden a fringe and reduce its peak height. In addition, the radiation collected by the photodetector contains plane waves comprised in a solid angle Ω having a half-apex aperture $\alpha = a/f$, a being the radius of the pinhole placed on the focal plane of the exit lens having focal length f.

The departure of W from the Airy function due to the above factors has been analyzed by Chabbal [62]. In particular, the nonflatness of the mirrors can be treated by considering the interferometer as a mosaic of microétalons making up the total plate area. The distribution of d can be characterized by a distribution function $D(x - \bar{d}) = dS/dx$, where dS is the area where the thickness has a value between x and $x + dx$ and $\bar{d}$ is the average thickness. For instance $D(x - \bar{d})$ is a rectangular function if the circular plates have a slight spherical curvature; it is a gaussian function in the case of randomly distributed microdefects due to the residual roughness of the surface. It is evident that D weighs the thickness d appearing in the Airy function. Consequently, we can express $W(k_0)$ as a convolution integral, viz.

$$W(k_0) = \tau \int_{x_{\min}}^{x_{\max}} D(x - \bar{d}) A(F, k_0 n x \cos\theta')\, dx = D * A. \qquad \text{(VII.21.16)}$$

The finite angular radius $\alpha = a/f$ of the pinhole placed in front of the photodetector produces a modification of the above expression for W, which reads

$$W(k_0) = \tau \int_0^{\alpha} d(\theta^2) \int_{x_{\min}}^{x_{\max}} D(x - \bar{d}) A[F, k_0 n x \cos(\theta/n)]\, dx, \qquad \text{(VII.21.17)}$$

where we have put $\theta' = \theta/n$. This last integral can be calculated either numerically or by expanding A in a Fourier series (see Problem 25 and Cook [59]).

Finally, we must take into account the finite diameter of the beam illuminating the interferometer.

A simple way to account for all the above effects consists of assuming that W is proportional to the Airy function corresponding to an *instrument finesse* F_I given by

$$1/F_I^2 = 1/F_R^2 + 1/F_F^2 + 1/F_P^2 + 1/F_D^2, \qquad \text{(VII.21.18)}$$

where $F_F = M/2$ is the *figure finesse*, M being the fractional wavelength deviation from planarity across the mirror aperture, $F_P = \lambda_0 f^2/(a^2 d)$ is the *pinhole finesse*, and $F_D = 2D^2/(\lambda_0 nd)$ the *diffraction finesse*, D being the diameter of the limiting aperture of the interferometer. Note that F_F also accounts for lack of parallelism of the mirrors. The best mirrors are made with a planarity of $\lambda/200$ at 500 nm. As a consequence, the instrumental finesse cannot be larger than 100.

As we have seen above, the instrumental profile of the Fabry–Perot interferometer depends on several factors. However, in every case, it is perfectly symmetric, in contrast to the asymmetric form of the instrumental profile of grating spectrometers. Because of this symmetry, the Fabry–Perot interferometer is particularly suited to making accurate measurements of the asymmetry of lines emitted by astronomical objects and providing important information about the hydrodynamic conditions of the surfaces of these objects.

The size of optical flats is limited by the area over which it is possible to make them flat to $\lambda/200$ (typically 2–3″). On the other hand, the solid angle of acceptance Ω is inversely proportional to the resolving power $\mathscr{R}$. This means that the étendue of these instruments decreases with $\mathscr{R}$. To overcome this limitation. Connes [63] introduced in 1958 an interferometer made of two spherical mirrors whose separation is equal to their radius of curvature. This device has the same instrumental profile, free spectral range, and resolving power as a plane interferometer of double spacing. However, it has the important property that the solid angle of acceptance is proportional to $\mathscr{R}$. As a consequence the étendue of this instrument can be much higher than that of a plane interferometer, an advantage important for separations greater than 0.1 m.

Problems

Section 4

1. Show that in the elliptic coordinate system μ, v, defined by $x = b\cosh\mu\sin v$, $z = b\sinh\mu\cos v$, the eikonal equation reads

$$|\nabla S|^2 = \{1/[b^2(\cosh^2\mu - \sin^2 v)]\}[(\partial S/\partial\mu)^2 + (\partial S/\partial v)^2].$$

Section 9

2. Consider a mirrorless cavity filled with a *lenslike medium* having a refractive index that changes radially according to the law $n(\rho) = n_0 - \frac{1}{2}n_2\rho^2$ and extending from $z = 0$ to $z = d$. The cavity axis coincides with the z axis. Find the parameters of the fundamental gaussian mode, assuming a refractive index exterior to the cavity equal to unity. *Hint*: Use the ABCD matrix method. To calculate the cavity matrix, first consider a thin section between z

and $z + dz$, so that $\mathbf{S}(z, z + dz)$ reduces to that of a thin lens having focal length $f = n_0/(n_2\,dz)$. Accordingly,

$$\mathbf{S}(z, z + dz) = \mathbf{1} - (n_2/n_0)\mathbf{A}\,dz$$

with

$$\mathbf{A} = \begin{vmatrix} 0 & 0 \\ 1 & 0 \end{vmatrix}.$$

Next, integrate the equation

$$\mathbf{S}(0, z + dz) - \mathbf{S}(0, z) = -(n_2/n_0)\mathbf{A} \cdot \mathbf{S}(0, z)\,dz$$

by obtaining

$$\mathbf{S}(0, z) = 1 - (n_2/n_0)z\mathbf{A},$$

in view of the equation $\mathbf{A}^2 = 0$.

3. Consider a far-infrared (FIR) laser using an HCN plasma as the active medium emitting at $\lambda = 337\ \mu$m. Assume an electron density profile in the positive column proportional to $J_0(2.4048\ \rho/a)$, a being the plasma tube radius and J_0 the Bessel function of zero order. Because of the presence of the electron gas, the refractive index n varies as $(1 - \omega_p^2/\omega^2)^{1/2}$, where $\omega_p = (e^2 n_e/m_e\varepsilon_0)^{1/2} = 5.6 \times 10^4\, n_e^{1/2}$ rad/s is the plasma frequency and n_e is the number of free electrons per cubic centimeter. As a consequence of the radial variation of n, the plasma forms a diverging lenslike medium that tends to defocus the cavity modes. Assuming $d = 3$ m, mirror radii $R = 10$ m, and $a = 5$ cm, calculate the maximum electron density for which the cavity is stable. *Hint*: Use the ray matrix of a lenslike medium with a parabolic profile derived in the preceding problem (approximate the Bessel function with a parabola) and apply the condition of stability $|A + D| < 2$.

4. Derive an expression for the spot size and the curvature radius of a gaussian beam focused by a thin lens into a medium of refractive index n. Find the position and the size of the waist.

5. Compute the field diffracted by a transparency

$$T(x, y) = \exp[-(x^2 + y^2)/a^2]$$

lit perpendicularly by a gaussian beam. Also calculate the relative ray matrix by applying the ABCD law.

6. Consider a symmetric cavity with gaussian profiles of the mirror reflectivities. Find the spot size and the curvature radii of the TEM_{00} mode in proximity to the mirrors and evaluate the field transmitted outside the cavity together with the integral reflection losses. *Hint*: Treat the cavity as a lens

sequence in which each lens is followed by a gaussian transparency. Note that the intensity of the gaussian mode traveling from left to right does not coincide with that of the mode traveling in the opposite direction. (For details see Zucker [64].)

Section 10

7. Prove the orthogonality relation of Eq. (VII.10.2) for the modes of a closed cavity with surface-impedance walls. *Hint*: Read van Bladel (Chapter IV [8], Chap. 10) and use Eq. (IV.2.1).

Section 11

8. Evaluate the diameter of an aperture in a plane-parallel or a confocal cavity, placed near a mirror in order to force a CO_2 laser to oscillate on the TEM_{00} mode. Assume a gain $g = 0.04\ \text{cm}^{-1}$ and a cavity length $d = 1$ m. *Hint*: Use the asymptotic expressions (VII.11.8,9) for the cavity loss factors.

9. A simple method for coupling radiation out of an infrared laser is to use an end mirror with a hole in it. Consider a resonator consisting of a plane and a concave mirror. Assume that the hole is smaller than the beam spot size, so that the mode patterns are not significantly perturbed. Calculate (a) the loss factor for the TEM_{00}, TEM_{10}, and TEM_{20} cylindrically symmetric modes and (b) the far field of the relative beams transmitted through the hole. Discuss in particular the case of $\lambda = 3\ \mu$m, $d = 1$ m, and the curvature radius R of the concave mirror equal to 20 m. (For details see D. E. McCumber [65].)

10. Consider a krypton laser emitting 0.5 W through a semitransparent mirror having a 2% transmittance. Compute the average number of photons in the cavity consisting of a plane mirror and a semitransparent output mirror having a curvature radius of 6 m. The mirrors are spaced 0.75 m apart and the laser operates at 415 nm. Compare the number of photons per unit volume with the gas number density and the presumed number of plasma electrons.

11. Consider a cavity 1.7 m long consisting of a plane reflector and a semitransparent concave mirror with curvature radius $R = 15$ m. Calculate the far-field divergence of the output beam by taking into account the effect on the beam of the output mirror, which acts as a plane–concave diverging lens. Assume a refractive index $n = 1.5$, a lens thickness $d = 1$ cm, and $\lambda = 500$ nm.

12. Design a pair of converging and diverging lenses suitable for focusing a gaussian beam of given parameters to an assigned point and changing the waist spot size into a given interval. Consider a beam having a spot size of 1.5 mm and far-field divergence of 0.7 mrad, and make the wavelength equal to 500 nm.

13. An optical resonator designed for use with the argon-ion laser transition at 5145 Å consists of two mirrors having reflectivities of 99.8 and 87.8%, respectively, which are spaced 60 cm apart. Calculate the width of the resonant modes of the passive cavity and the theoretical laser linewidth given by Eq. (VII.1.14) when the single-mode output power is 0.5 W. Then, using Eqs. (VII.11.5), estimate the change in the resonator length that would result in a shift of the frequency of oscillation equal to the theoretical linewidth of the laser output. (From Corney [48], p. 375.)

14. Calculate the fraction of the power in the TEM_{00} and TEM_{01} cylindrically symmetric modes that is lost when an aperture of diameter 3.5 w is inserted close to one of the mirrors of a symmetric mirror laser cavity. (From Corney [48], p. 375.)

Section 14

15. Show that two Laguerre–Gauss modes satisfy the orthogonality condition represented by Eq. (VII.14.13) on each mirror of a linear resonator.

Section 16

16. Show that for $\bar{\tau} = \infty$ Eq. (VII.16.4) is satisfied by a Hermite–Gauss mode. Calculate the relative eigenvalue γ.

17. Calculate the complex parameter p for a strip resonator by expressing the mode in the form

$$f(t) = \cos s_0 t + A \cos s_1 t_1 + B \cos s_{-1} t.$$

18. Use Eqs. (VII.16.19,20) for the reflection and mode conversion coefficients at the open ends. Discuss the values of the Fresnel number of the cavity for which the amplitudes A and B of Problem 17 become negligible.

Section 17

19. Show that for $a = \infty$, Eq. (VII.17.4) is satisfied by a Laguerre–Gauss mode.

Section 19

20. Show that

$$\frac{1}{2\pi}\int_0^{2\pi} \frac{dx}{1 + \varepsilon \cos x} = \sum_{n=0}^{\infty} \left(\frac{\varepsilon}{2}\right)^{2n} \frac{(2n)!}{(n!)^2} = \frac{1}{(1 - \varepsilon^2)^{1/2}}.$$

Then, using this result, prove Eq. (VII.19.8).

21. Show that for ΔN constant over the section of the fundamental mode of a laser resonator the gain $g(z, E^{(+)}, E^{(-)})$ defined by Eq. (VII.19.25) is given by

$$g(z, E^{(+)}, E^{(-)}) = k_0^2 f \,\mathrm{Re}\{[1/[k_{\bar{n}}(1 + i\zeta)]\}\, \Delta N(z)[\ln(1 + W)]/W,$$

where

$$W = \mathscr{I}^{(+)} + \mathscr{I}^{(-)} + 2[\mathscr{I}^{(+)}\mathscr{I}^{(-)}]^{1/2}\cos(2k_n' z + \phi),$$
$$\mathscr{I}^{(+)} = |E^{(\pm)}|^2 |u_{00}^{(\pm)}(0, 0, z)|^2/E_s^2.$$

22. With reference to Problem 21, show that

$$\frac{2}{\lambda}\int_z^{z+\lambda/2} \frac{\ln(1 + W)}{W}\,dz = 1 - \tfrac{1}{2}[\mathscr{I}^{(+)} + \mathscr{I}^{(-)}] + \tfrac{1}{3}[\mathscr{I}^{(+)} + \mathscr{I}^{(-)}]^2$$
$$+ \tfrac{2}{3}\mathscr{I}^{(+)}\mathscr{I}^{(-)} + 0[\mathscr{I}^{(+)^3}, \mathscr{I}^{(-)^3}],$$

where

$$\mathscr{I}^{(+)} = |E^{(\pm)}|^2 |u_{00}^{(\pm)}(0, 0, z)|^2/E_s^2.$$

23. Using the expressions derived in the above problems, show that Rigrod's equations [Eq. (VII.19.31)] are replaced for a gaussian mode of low intensity by

$$d\mathscr{I}^{(\pm)}/dz = \pm g_0(z)\{1 - \tfrac{1}{2}[\mathscr{I}^{(+)} + \mathscr{I}^{(-)}] + \tfrac{1}{3}[(\mathscr{I}^{(+)} + \mathscr{I}^{(-)}]^2 + \tfrac{2}{3}\mathscr{I}^{(+)}\mathscr{I}^{(-)}\}.$$

Next, express z as a function of $\mathscr{I}^{(+)}$ for $g_0(z) = \text{const}$ by exploiting the fact that $\mathscr{I}^{(+)}\mathscr{I}^{(-)} = \text{const}$.

Section 20

24. Consider a solid-state étalon forming an angle θ with a gaussian beam. Calculate the intensity of the transmitted beam. In particular, discuss the reduction of the maximum transmitted intensity occurring for large values of θ. This phenomenon is due to the partial superposition of the multiple reflected beams occurring when $\theta \neq 0$, and is at the origin of the *walk-off* losses.

25. Show that the Airy formula for the transmittance can be expanded in the Fourier series

$$\frac{1}{1 + [4R/(1 - R)^2]\sin^2\phi} = \frac{1 - R}{1 + R}\left[1 + 2\sum_{n=1}^{\infty} R^n \cos(2n\phi)\right].$$

Hint: Express $2\cos(2n\phi)$ in the form $e^{i2n\phi} + e^{-i2n\phi}$ and show that the right-hand side of the above equation is a geometric series which, when summed, reproduces the left side.

26. Show that when the retardation vector $\mathbf{D}_1$ of a double Fabry–Perot interferometer is nearly parallel to $\mathbf{D}_2$, the fringes of superposition in white light are hyperbolas. (See Cook [59], p. 132.)

27. Consider an étalon illuminated at normal incidence by a gaussian beam. Discuss the form of the image collected on the focal plane of the exit lens of the interferometer. In particular, examine the relation between the waist of the beam, the spacing of the étalon, and the number of rings observed.

28. Define the main parameters of a glass étalon ($n = 1.5$) intended to select a single frequency in the neighborhood of 4880 Å from the output of an argon ion laser. Consider that in this case the gain bandwidth is of the order of 6 GHz while the homogeneous linewidth is almost 500 MHz. (See Hercher [66].)

29. Consider a pulsed dye laser using a grating for tuning. Design a solid-state étalon to narrow the emitted line to 100 MHz. Assume that the linewidth without the étalon is about 2 GHz. Compute the tilting angle necessary to scan the interval of 2 GHz.

30. Show that the characteristic matrix $\mathbf{M}$ (cf. Chapter III) of an étalon made of an optical flat coated on the two faces with two identical multilayers is given by

$$\mathbf{M} = \begin{bmatrix} (AD + BC)\cos\phi - i\left(AC\hat{Z} + \dfrac{BD}{\hat{Z}}\right)\sin\phi & 2AB\cos\phi - i\left(A^2\hat{Z} + \dfrac{B^2}{\hat{Z}}\right)\sin\phi \\ 2CD\cos\phi - i\left(C^2\hat{Z} + \dfrac{D^2}{\hat{Z}}\right)\sin\phi & (BC + AD)\cos\phi - i\left(AC\hat{Z} + \dfrac{BD}{\hat{Z}}\right)\sin\phi \end{bmatrix},$$

where $\phi = k_0 nd\cos\theta'$, $\hat{Z} = (\zeta_0/n)\cos\theta'(1/\cos\theta')$ for TM (TE) waves, and A, B, C, and D are the components of the characteristic matrix of the multilayer coating.

31. Show that the amplitude transmission coefficient t relative to the exit face of the étalon discussed in Problem 30 is given by

$$t = \{[AD + BC + (AB/\hat{Z}_{\text{ext}} + CD\hat{Z}_{\text{ext}}]\cos\phi - i[AC\hat{Z} + (BD/\hat{Z}) + (A^2\hat{Z})/(2\hat{Z}_{\text{ext}}) + B^2/(2\hat{Z}\hat{Z}_{\text{ext}}) + (1/2)C^2\hat{Z}\hat{Z}_{\text{ext}} + (1/2)(D^2\hat{Z}_{\text{ext}}/\hat{Z})]\sin\phi\}^{-1},$$

where $\hat{Z}_{\text{ext}}$ is the impedance of the medium on the left and right sides of the two faces. *Hint*: Show that $t = (1 + r)/(A + B/Z_{\text{ext}})$, where r is the reflection coefficient of the entrance face and A, B are coefficients of the above matrix, and use Eq. (III.12.6).

32. Use the above expression for t to calculate the transmittance of a Fabry–Perot interferometer with silver- and aluminum-coated mirrors.

References

1. Liao, S. Y., "Microwave Devices and Circuits." Prentice-Hall, Englewood Cliffs, New Jersey, 1980.
2. Vaynshteyn, L. A., "Open Resonators and Open Waveguides." Golem Press, Boulder, Colorado, 1969.
3. Anan'ev, Y., "Résonateurs Optiques et Problème de Divergence du Rayonnement Laser." Editions M.I.R., Moscow, 1982.
4. Ronchi, L., *in* "Laser Handbook" (F. T. Arecchi and E. O. Schulz-Dubois, eds.), Vol. 1, p. 153. North-Holland Publ., Amsterdam, 1972.
5. Chodzko, R. A., and Chester, A. N., *in* "Handbook of Chemical Lasers" (R. W. G. Gross and J. F. Bott, eds.), p. 95–203. Wiley, New York, 1976.
6. Steier, W. H., *in* "Laser Handbook" (M. L. Stitch, ed.), Vol. 3, p. 5. North-Holland Publ., Amsterdam, 1979.
7. Courant, R., and Hilbert, D., "Methods of Mathematical Physics," Vol. 1, Theorem 18, p. 442. Wiley (Interscience), New York, 1953.
8. Schawlow, A. L., and Townes, C. H., *Phys. Rev.* **112**, 1940 (1958).
9. Prokhorov, A. M., *JETP (USSR)* **34**, 1658 (1958).
10. Fabry, C., and Perot, A., *Ann. Chim. Phys.* **16**, 115 (1899).
11. Macek, W. M., and Davis, D. T. M., *Appl. Phys. Lett.* **2**, 67 (1963).
12. Ambartsumyan, R. V., Basov, N. G., Kryukov, B. G., and Letokhov, V. S., *Prog. Quantum Electron.* **1**, Part 3 (1970).
13. Kogelnik, H., and Shank, C. V., *Appl. Phys. Lett.* **18**, 152 (1971).

13a. Rensch, D. B., and Chester, A. N., *Appl. Opt.* **12**, 997 (1973).

13b. Bloom, A. L., *Spectra-Phys. Laser Tech. Bull.* No. 2, Mountain View, California (1963).

14. Goubau, G., and Schwering, F., *IRE Trans. Antennas Propag.* **AP-9**, 248 (1961).
15. Boyd, G. D., and Kogelnik, H., *Bell Syst. Tech. J.* **41**, 1347 (1962).
16. Kogelnik, H., and Rigrod, W. W., *Proc. IRE* **50**, 220 (1962).
17. Herriott, D. R., *J. Opt. Soc. Am.* **52**, 31 (1962).
18. Bennett, W. R., Jr., "The Physics of Gas Lasers." Gordon and Breach, N.Y., 1977.
19. Slepian, D., *Bell Syst. Tech. J.* **43**, 3009 (1964).
20. Solimeno, S., and Torre, A., *Phys. Res.* **A237**, 298 (1985).
21. Siegman, A., *Proc. IEEE* **53**, 277 (1965).
22. Siegman, A., and Arrathon, R. W., *IEEE J. Quantum Electron.* **QE-3**, 156 (1967).
23. Krupke, W. F., and Sooy, W. R., *IEEE J. Quantum Electron.* **QE-5**, 575 (1969).
24. Nefiodov, E. I., "Otkritie Koaksialnie Resonansnie Strukturi." Nauka, Moscow, 1982. (In Russ.)
25. Patel, C. K. N., *Appl. Phys. Lett.* **7**, 15 (1965).
26. Tricomi, F., "Integral Equations." Wiley (Interscience), New York, 1957.
27. Fox, A. G., and Li, T., *Bell Syst. Tech. J.* **40**, 453 (1960).
28. Gordon, E. I., and White, A. D., *Proc. IEEE* **52**, 206 (1964).
29. Gordon, J. P., and Kogelnik, H., *Bell Syst. Tech. J.* **43**, 2873 (1964).
30. Wilkinson, J. H., "The Algebraic Eigenvalue Problem. Oxford Univ. Press (Clarendon), London and New York, 1965.
31. Li, T., *Bell Syst. Tech. J.* **44**, 917 (1965).
32. Checcacci, P. F., Consortini, A., and Scheggi, A., *Proc. IEEE* **54**, 1329 (1966).
33. Johnson, M. M., *Appl. Opt.* **13**, 2326 (1974).
34. Siegman, A. E., *Opt. Lett.* **1**, 13 (1977).
35. Streifer, W., *J. Opt. Soc. Am.* **55**, 868 (1965).

36. Bergstein, L., and Marom, E., *J. Opt. Soc. Am.* **56**, 16 (1966).
37. Siegman, A. E., and Miller, H. Y., *Appl. Opt.* **9**, 2729 (1970).
38. Slepian, D., and Pollack, H. O., *Bell Syst. Tech. J.* **40**, 43 (1960).
39. Landau, N. J., and Pollack, H. O., *Bell Syst. Tech. J.* **40**, 65 (1960); **42**, 1295 (1962).
40. Boyd, G. D., and Gordon, J. P., *Bell Syst. Tech. J.* **41**, 489 (1961).
41. Luchini, P., and Solimeno, S., *Opt. Lett.* **7**, 259 (1982).
42. Horwitz, P., *J. Opt. Soc. Am.* **63**, 1528 (1973).
42a. Butts, R. and Avizonis, P., *J. Opt. Soc. Am.* **68,** 1072 (1978).
43. Chen, L. W., and Felsen, L. B., *IEEE J. Quantum Electron.* **QE-9**, 1102 (1973).
44. Stenholm, S., *in* "Progress in Quantum Electronics" (J. H. Sanders and K. W. H. Stevens, eds.), Vol. 1, Part A, p. 189–271. Pergamon, Oxford, 1970.
45. McFarlane, R. A., Bennett, W. R., and Lamb, W. E., *Appl. Phys. Lett.* **2**, 189 (1963).
46. Szöke, A., and Javan, A., *Phys. Rev. Lett.* **10**, 521 (1963).
47. Letokhov, V. S., and Chebotayev, V. P., "Non-linear Laser Spectroscopy." Springer-Verlag, Berlin and New York, 1977.
48. Corney, A., "Atomic and Laser Spectroscopy," Chap. 13. Oxford Univ. Press (Clarendon), London and New York, 1977.
49. Rigrod, W. W., *J. Appl. Phys.* **34**, 2602 (1963); **36**, 2487 (1965).
50. Smith, P. W., *Proc. IEEE* **60**, 422 (1972).
51. Koechner, W. "Solid-state Laser Engineering." Springer-Verlag, Berlin and New York, 1976.
52. Smith, P. W., *IEEE J. Quantum Electron.* **QE-1**, 343 (1965).
53. Kurokawa, K., *Proc. IEEE* **61**, 1386 (1973).
54. Buczek, C. J., Freiberg, R. J., and Skolnick, M. L., *Proc. IEEE* **61**, 1411 (1973).
55. Adler, R., *Proc. IEEE* **61**, 1380 (1973).
56. Jacquinot, P., *Rep. Prog. Phys.* **23**, 267 (1960).
57. Girard, A., and Jacquinot, P., *in* "Advanced Optical Techniques" (A. C. S. van Heel, ed.), p. 71–121. North-Holland Publ., Amsterdam, 1967.
58. Danielmyer, H. G., *IEEE J. Quantum Electron.* **QE-6**, 101 (1970).
59. Cook, A. H., "Interference of Electromagnetic Waves." Oxford Univ. Press (Clarendon), London and New York, 1971.
60. Roesler, F. L., *in* "Astrophysics, Part A, Optical and Infrared Astronomy" (N. Carleton, ed.), Methods of Experimental Physics, Vol. 12, p. 531–569. Academic Press, New York, 1974.
61. Mack, J. E., McNutt, D. P., Roesler, F. C., and Chabral, R., *Appl. Opt.* **2**, 873 (1963).
62. Chabbal, R., *J. Phys. Radium* **19**, 295 (1958).
63. Connes, P., *J. Phys. Radium* **19**, 262 (1958).
64. Zucker, H., *Bell Syst. Tech. J.* **49**, 2349 (1970).
65. McCumber, D. E., *Bell Syst. Tech. J.* **44**, 333 (1965).
66. Hercher, M., *Appl. Opt.* **8**, 1103 (1969).

Bibliography

Harris, S. E., *Proc. IEEE* **57**, 2096 (1969).
Basov, N. G., Krokhin, O. N., and Popov, Yu. M., *Sov. Phys. Uspekhi* **3**, 702 (1961).
Bertolotti, M., "Masers and Lasers: an Historical Approach." Adam-Hilgher Ltd., Bristol, 1983.
Siegman, A. E., Belanger, P. A., and Hardy, A., *in* "Optical Phase Conjugation" (R. A. Fisher, ed.), p. 469–528. Academic Press, New York, 1983.
Svelto, O., "Principles of Lasers." Plenum Press, New York, 1976.

Chapter VIII

Propagation in Optical Fibers

1 Geometric Optics

Electromagnetic energy at optical frequencies can be transferred from one place to another by letting the corresponding field propagate in suitable dielectric waveguides. Basically, the guiding properties of these structures are associated with the process of *total reflection*, according to which a light beam traveling in a medium having an index of refraction n_1 can be totally reflected when impinging on a discontinuity surface separating the first medium from a second one having an index of refraction $n_2 < n_1$. More precisely, there exists an angle θ_c [critical angle; see Eq. (III.20.1)],

$$\theta_c = \arcsin(n_2/n_1), \tag{VIII.1.1}$$

such that total reflection occurs whenever

$$\theta \geqq \theta_c, \tag{VIII.1.2}$$

θ being the angle between the ray and the normal $\hat{n}$ to the separation surface between the two media. For $\theta < \theta_c$, the ray is partially reflected and partially transmitted (see Fig. VIII.1).

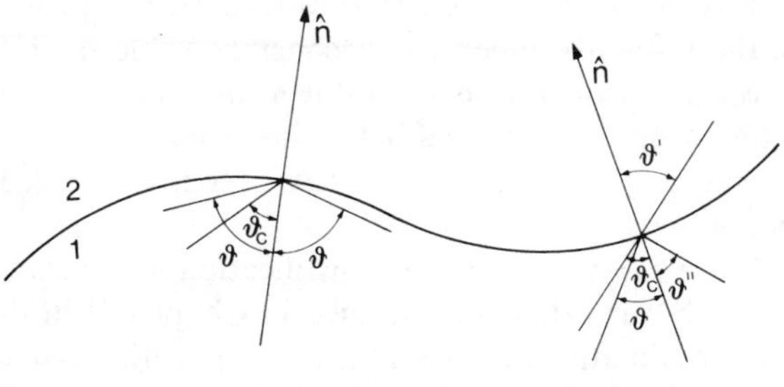

Fig. VIII.1. Reflection and refraction at the separation between two media.

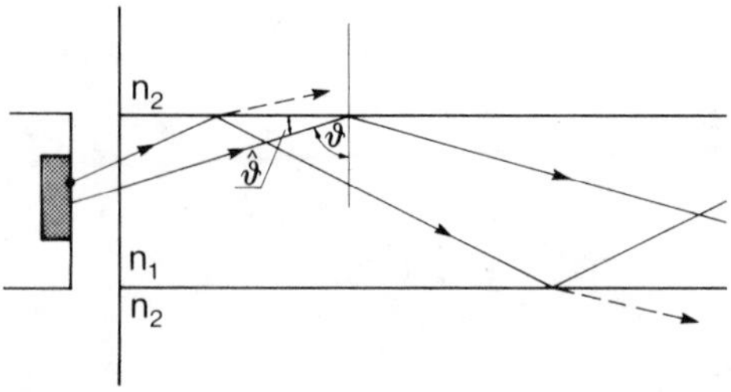

Fig. VIII.2. Schematic illustration of the guiding mechanism in a plane structure.

These processes are described by the laws of geometric optics (see Section II.11.1), according to which the incidence angle θ is equal to the reflection angle θ'' (that is, the angle between the reflected ray and $-\hat{n}$),

$$\theta = \theta'', \tag{VIII.1.3}$$

and the refraction angle θ' (that is, the angle between the transmitted ray and $\hat{n}$) is connected with θ by the relation [see Snell's law, Eq. (II.11.8)]

$$n_1 \sin\theta = n_2 \sin\theta'. \tag{VIII.1.4}$$

These preliminary considerations are sufficient to illustrate the guiding properties of a simple planar structure in which a dielectric layer of refractive index n_1 is surrounded by a dielectric material of refractive index $n_2 < n_1$ (see Fig. VIII.2). Light rays such that

$$\hat{\theta} = \pi/2 - \theta \leqq \pi/2 - \theta_c \tag{VIII.1.5}$$

are guided by means of multiple reflections, while those that do not satisfy Eq. (VIII.1.5) are attenuated due to successive refractions.

In practice, the waveguides most widely used for propagation over long distances have cylindrical symmetry. In this case, the ray trajectories become much more complicated, even if the elementary argument given above to explain their behavior generally applies. Furthermore, the decrease of the refractive index responsible for the ray confinement can be realized either in a discontinuous way (*step-index fibers*) or continuously (*graded-index fibers*). In the first case, the refractive index has a constant value n_1 in a cylindrical region of radius a (*core*) and a constant value n_2 in a concentric ringed region (*cladding*). In the second case, the refractive index continuously decreases in the core with distance ρ from the symmetry axis z to match the constant value n_2 in the cladding (see Fig. VIII.3).

The concept of a ray has its rigorous justification in the eikonal equation [Eq. (II.3.1)], which was extensively studied in Chapter II in the context of the ray optics approximation and has a field of application essentially limited to wavelengths much smaller than the waveguide dimensions.

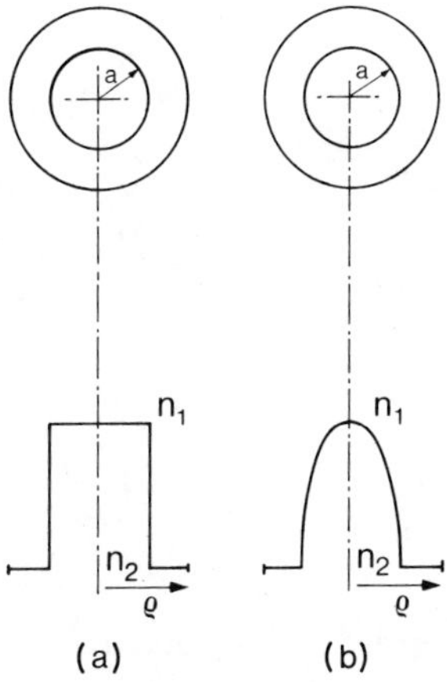

Fig. VIII.3. (a) Step-index and (b) graded-index profiles.

2 Step-Index Fibers

The elementary description of propagation given above applies, for a circular cylindrical step-index fiber, to the rays injected in a plane containing the z axis (*meridional rays*) [e.g., see Eq. (II.13.11)]. In fact, symmetry ensures that the trajectories remain in this plane, so that they can be described in terms of the two-dimensional scheme of Fig. (VIII.2). This allows us to introduce the parameter θ_M (*acceptance angle*), which is the largest angle a guided meridional ray can form with the z axis. Recalling Eqs. (VIII.1.1) and (VIII.1.4), we have (see Fig. VIII.4)

$$n_e \sin \theta_M = n_1 \sin \theta'_M = n_1 \cos \theta_c = (n_1^2 - n_2^2)^{1/2}, \qquad \text{(VIII.2.1)}$$

where n_e, the index of refraction of the medium in front of the fiber input, is in general $\simeq 1$. In practice, the difference between the values of the refractive

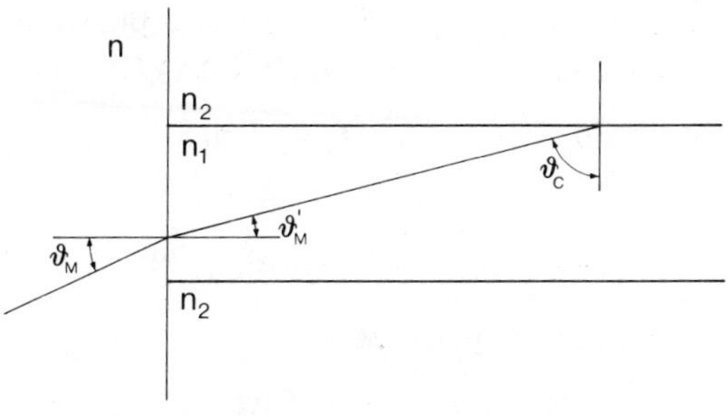

Fig. VIII.4. Acceptance angle in a step-index fiber.

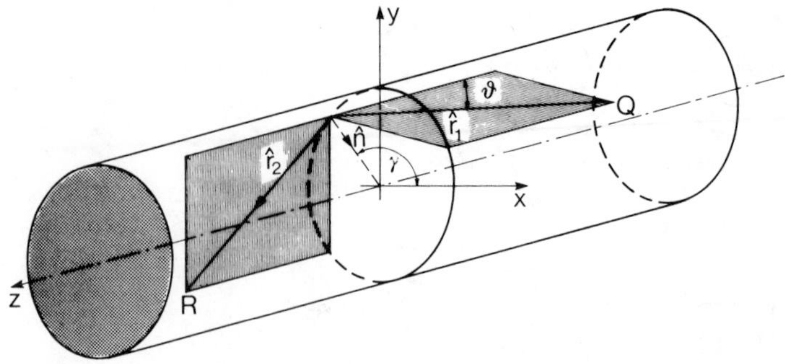

Fig. VIII.5. Skew rays in a step-index fiber.

index in the core and in the cladding is of the order of few percent,

$$\Delta \equiv (n_1^2 - n_2^2)/(2n_1^2) \ll 1, \qquad \text{(VIII.2.2)}$$

so that

$$\theta_M \cong \sin\theta_M = (n_1/n_e)(2\Delta)^{1/2}. \qquad \text{(VIII.2.3)}$$

It is also useful to define the *numerical aperture* NA (see also Section II.15.3) through the relation

$$\mathrm{NA} = n_e \sin\theta_M = (n_1^2 - n_2^2)^{1/2}. \qquad \text{(VIII.2.4)}$$

The description becomes more complicated for the nonmeridional rays (*skew rays*), since the plane containing the single ray segment and the two generatrices of the cylinder intercepted by it changes at every reflection (see Fig. VIII.5). However, it can be shown that the angle of incidence ψ between the generic ray segment $\hat{r}_j$ and the normal $\hat{n}$ to the core–cladding interface is a *constant of the motion* [see *skewness invariant* (Section II.13.1)], together with θ and γ (see Fig. VIII.5), and that

$$\cos\psi = -\sin\theta\cos\gamma. \qquad \text{(VIII.2.5)}$$

This relation allows us to extend the definition of acceptance angle to the case of skew rays. The guided rays are, according to Eq. (VIII.2.5), those for which

$$|\cos\psi| = \sin\theta|\cos\gamma| \le \cos\theta_c = (1/n_1)(n_1^2 - n_2^2)^{1/2}, \qquad \text{(VIII.2.6a)}$$

that is,

$$\sin\theta \le (n_1^2 - n_2^2)^{1/2}/(n_1|\cos\gamma|), \qquad \text{(VIII.2.6b)}$$

corresponding to a numerical aperture

$$\mathrm{NA} = (n_1^2 - n_2^2)^{1/2}/|\cos\gamma| \qquad \text{(VIII.2.7)}$$

larger than the numerical aperture for meridional rays. However, a rigorous modal analysis shows that skew rays may exhibit a loss mechanism at the core–cladding interface due to its curvature. In ray language, it is possible to distinguish between three classes or rays. The *guided rays*, either meridional or skew, satisfy the condition

$$\sin\theta \leqq (n_1^2 - n_2^2)^{1/2}/n_1 \tag{VIII.2.8}$$

and propagate without attenuation. The rays that do not satisfy Eq. (VIII.2.8) are subdivided into *leaky tunneling rays*, satisfying Eq. (VIII.2.6) (they are obviously all skew rays), and *leaky refracted rays*, which do not obey Eq. (VIII.2.6). In general, leaky tunneling rays attenuate much more slowly than refracted rays.

3 Graded-Index Fibers

The guiding effect can be achieved by means of structures constituted by a central region in which the index of refraction decreases continously with the distance from the z axis to match the constant value assumed in the cladding (see Fig. VIII.3). In order to justify this assertion, we recall the results obtained in the context of ray optics when studying propagation in the presence of refractive index profiles whose radial behavior in the core is of the kind just mentioned [e.g., see Eqs. (II.12.14) and (II.13.31)]. If we do this it is apparent [e.g., see [Eq. (II.13.33)] that the electromagnetic field tends to be confined near the z axis of symmetry while propagating in this direction, which explains the guiding properties of such structures.

We wish to give another illustration of this by considering a gaussian refractive index profile of the type

$$n^2(\rho) = n_0^2 \exp(-b\rho^2/n_0), \qquad 0 \leqq \rho \leqq a, \tag{VIII.3.1}$$

with $\rho^2 = x^2 + y^2$, x and y being cartesian transverse coordinates. This example is particularly interesting since the associated trajectories are very simple and it can be considered representative of a large class of index profiles. In fact, if, according to what happens in practice,

$$ba^2/n_0 \ll 1, \tag{VIII.3.2}$$

Eq. (VIII.3.1) represents a good approximation to the parabolic profile

$$n^2(x, y) = n_0^2 - n_0 b\rho^2, \tag{VIII.3.3}$$

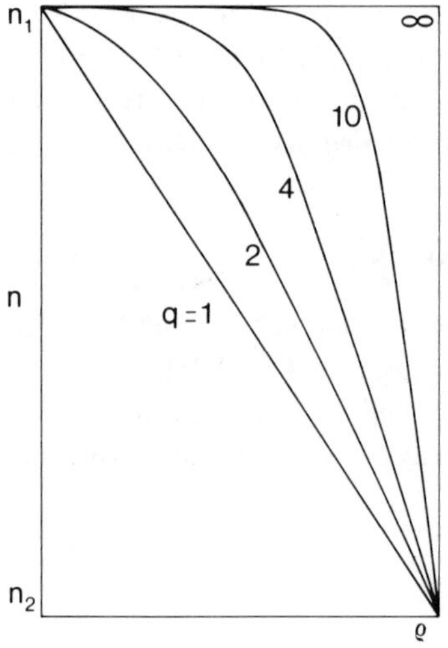

Fig. VIII.6. Refractive index profiles corresponding to Eq. (VIII.3.4).

which is a particular case ($q = 2$) of the general class of index profiles (see Fig. VIII.6)

$$n(\rho) = \begin{cases} n_0[1 - 2\Delta(\rho/a)^q]^{1/2} & \rho \leqq a, \\ n_0(1 - 2\Delta)^{1/2} & \rho > a. \end{cases} \tag{VIII.3.4}$$

In the paraxial approximation, dealing with rays having a small inclination with respect to the z axis, the ray trajectories are described by the equations (see Problem II.2)

$$d^2x/dz^2 = -(b/n_0)x, \qquad d^2y/dz^2 = -(b/n_0)y, \tag{VIII.3.5}$$

which admit the general solution

$$x(z) = A_x \sin[(b/n_0)^{1/2}z + \phi_x], \qquad y(z) = A_y \sin[(b/n_0)^{1/2}z + \phi_y]. \tag{VIII.3.6}$$

These equations can be compared, in the paraxial approximation, with Eq. (II.13.33), with which they share the property of having a spatial periodicity independent of the initial position and inclination of the ray (see Fig. VIII.7).

The guided rays are those for which

$$A_x^2 + A_y^2 < a^2, \tag{VIII.3.7}$$

while rays that do not satisfy Eq. (VIII.3.7) penetrate the cladding and cross it following a straight line (Fig. VIII.7). According to Eq. (VIII.3.7), the guided rays obey the relation

$$(dx/dz)^2 + (dy/dz)^2 \leqq (A_x^2 + A_y^2)b/n_0 \leqq ba^2/n_0, \tag{VIII.3.8}$$

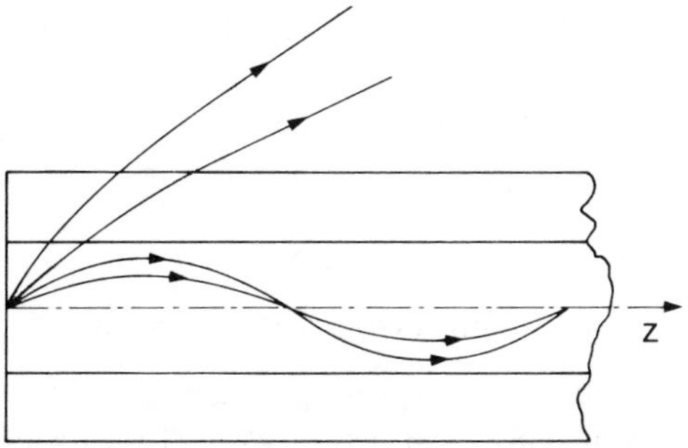

Fig. VIII.7. Guided and unguided rays in a graded-index fiber.

which justifies *a posteriori* the use of the paraxial approximation [see Eq. (VIII.3.2)]. Among the guided rays, the meridional rays are those for which $\phi_x = \phi_y$.

As in the case of the step-index fibers, it is possible to introduce for the graded-index fibers an acceptance angle that turns out to be a function of the distance ρ from the z axis. To this end, let us remember that, according to Eq. (II.2.5), the phase variation along the ray for an elementary path $d\mathbf{r}$ is given by

$$-k_0 \nabla S \cdot d\mathbf{r}, \tag{VIII.3.9}$$

which makes it possible to define, a local wave vector [see Eq. (II.4.2)]

$$\mathbf{k}(\mathbf{r}) = k_0 \nabla S, \tag{VIII.3.10}$$

which makes it possible to define a local wave vector [see Eq. (II.4.2)] vector is a constant of the motion, as can be seen immediately by projecting Eq. (II.4.2) along the z axis and taking the derivative with respect to the curvilinear abscissa s of both sides of the resulting equation; if $n(\mathbf{r})$ is assumed to be independent of z, we get

$$d(\nabla S)_z/ds = (d/ds)[n(x, y)\, dz/ds], \tag{VIII.3.11}$$

the right-hand side being zero according to Eq. (II.4.5), since $(\nabla n)_z = 0$. The angle $\theta(\rho)$ between the ray and the z axis at a distance ρ is then expressed through the relation

$$\cos\theta(\rho) = k_z/[\mathbf{k}(\rho) \cdot \mathbf{k}(\rho)]^{1/2}, \tag{VIII.3.12}$$

or, according to Eqs. (VIII.3.10) and (II.3.1),

$$\cos\theta(\rho) = k_z/[k_0 n(\rho)]. \tag{VIII.3.13}$$

For any given ρ, $\theta(\rho)$ attains its maximum value $\theta_M(\rho)$ in correspondence to the rays passing through ρ and tangent to the core–cladding interface which exhibit the minimum value $k_0 n(a)$ for k_z so that

$$\cos\theta_M(\rho) = n(a)/n(\rho). \tag{VIII.3.14}$$

Accordingly, the numerical aperture is given by

$$\mathrm{NA} = n(\rho)\sin\theta_M(\rho) = [n^2(\rho) - n^2(a)]^{1/2}, \tag{VIII.3.15}$$

which is formally the same expression as that for a step-index fiber [Eq. (VIII.2.4)], for which $n(\rho) \equiv n_1$ irrespective of the distance from the fiber axis of the point chosen on the fiber input.

The general arguments outlined in the preceding sections are extensively treated in specialized books [1–8], whose number is steadily increasing because of the growing relevance of this subject.

4 Mode Theory

The study of a general solution of Maxwell's equations in a cylindrical fiber naturally leads to the concept of a *propagation mode*, that is, of a particular field configuration characterized by a z dependence of the kind $\exp(-i\beta z)$. The *guided modes* are propagation modes whose electromagnetic energy is mainly confined in the core irrespective of the fiber length.

The distinction between guided and unguided modes, which is not present in the theory of propagation in metallic waveguides, is essential for dielectric waveguides. In order to describe the electromagnetic field inside the fiber, one has in general to consider, besides the discrete spectrum of guided modes (corresponding, in geometric optics, to rays confined by successive reflections inside the core), a continuous spectrum of radiation modes (corresponding to rays crossing the fibers through successive reflections; see Fig. VIII.8).

The propagation of electromagnetic waves in a homogeneous dielectric medium is described by Maxwell's equations (I.1.1)–(I.1.4) in the absence of

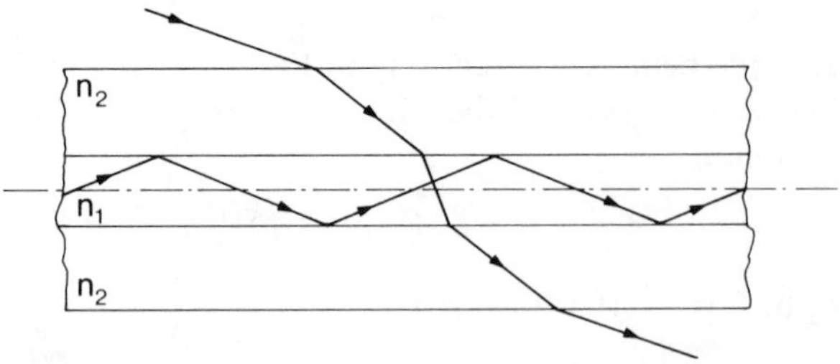

Fig. VIII.8. Rays corresponding to guided and radiation modes in a step-index fiber.

charges and currents, together with the *constitutive relations* expressed by Eqs. (I.1.6) and (I.2.3). Note that, in a nonstatic case ($\partial/\partial t \neq 0$), the last two of Maxwell's equations [Eqs. (I.1.3) and (I.1.4)] are a consequence of the first two, which can be immediately verified by applying the operator $\nabla \cdot$ on both sides of Eqs. (I.1.1) and (I.1.2) and taking advantage of the vector identity (A.15). Accordingly, in the following discussion it will suffice to consider only Eqs. (I.1.1) and (I.1.2), together with the continuity conditions for the tangential components of the electric and magnetic fields at any discontinuity surface of the dielectric constant $\varepsilon(\mathbf{r},\omega)$.

Let us look for monochromatic solutions of the form

$$\hat{\mathbf{E}}(\mathbf{r},t) = \mathbf{E}(\mathbf{r})e^{i\omega t}, \qquad \hat{\mathbf{H}}(\mathbf{r},t) = \mathbf{H}(\mathbf{r})e^{i\omega t}, \tag{VIII.4.1}$$

and let us consider the most general cylindrical structure [in which the z axis is parallel to the generatrices of the cylinder, on whose section a curvilinear coordinate system (q_1, q_2) has been introduced (see Fig. VIII.9)]. If the dielectric constant ε is a function of the transverse coordinates q_1 and q_2 alone, we are led by symmetry considerations to look for solutions of the form

$$\mathbf{E}(q_1,q_2,z) = \mathbf{E}_0(q_1,q_2)e^{-i\beta z}, \qquad \mathbf{H}(q_1,q_2,z) = \mathbf{H}_0(q_1,q_2)e^{-i\beta z}. \tag{VIII.4.2}$$

It is thus convenient to express the differential operator ∇ as the sum of its transverse and longitudinal components, that is, to write

$$\nabla = \nabla_{\mathrm{t}} + \hat{z}\,\partial/\partial z, \tag{VIII.4.3}$$

so that, taking into account Eqs. (I.1.1) and (I.1.2) and the constitutive relations, we obtain

$$\nabla_{\mathrm{t}} \times \mathbf{E}_0 - i\beta\hat{z} \times \mathbf{E}_0 = -i\omega\mu_0\mathbf{H}_0, \qquad \nabla_{\mathrm{t}} \times \mathbf{H}_0 - i\beta\hat{z} \times \mathbf{H}_0 = i\omega\varepsilon\mathbf{E}_0. \tag{VIII.4.4}$$

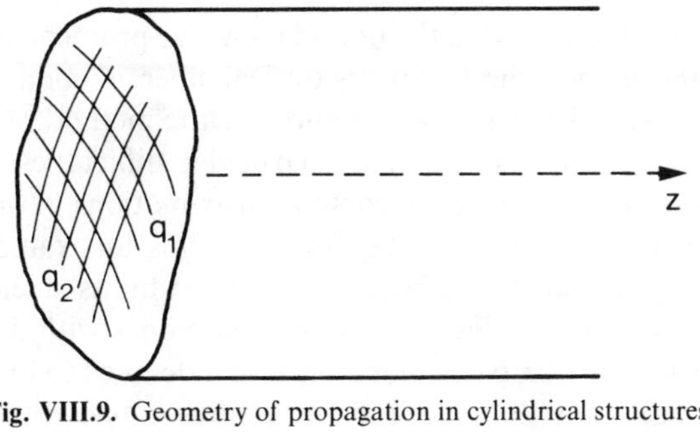

Fig. VIII.9. Geometry of propagation in cylindrical structures.

As will become apparent later, the constant β may only assume certain discrete values (*eigenvalues*) depending on the boundary conditions imposed by the guiding structure. If we write the electric and magnetic fields in terms of the transverse and longitudinal components

$$\mathbf{E}_0 = \mathbf{E}_t + \hat{z}E_z, \qquad \mathbf{H}_0 = \mathbf{H}_t + \hat{z}H_z, \tag{VIII.4.5}$$

it is possible, after some algebra, to express the transverse components of the fields in term of their longitudinal components according to the equations

$$\begin{aligned} \mathbf{H}_t &= -[i/(\omega^2\mu_0\varepsilon - \beta^2)](\beta \nabla_t H_z - \omega\varepsilon \nabla_t E_z \times \hat{z}), \\ \mathbf{E}_t &= -[i/(\omega^2\mu_0\varepsilon - \beta^2)](\beta \nabla_t E_z + \omega\mu_0 \nabla_t H_z \times \hat{z}), \end{aligned} \tag{VIII.4.6}$$

We are thus faced with the problem of finding the expressions for the longitudinal components H_z and E_z of the electric and magnetic fields. According to Eq. (II.8.1), the longitudinal component of the electric field obeys the equation

$$\nabla_t^2 E_z + (\omega^2\mu_0\varepsilon - \beta^2)E_z - i\beta\mathbf{E}_t \cdot \nabla_t \ln \varepsilon = 0, \tag{VIII.4.7}$$

while for the corresponding magnetic component

$$\nabla_t^2 H_z + (\omega^2\mu_0\varepsilon - \beta^2)H_z + i\omega\varepsilon \nabla_t(\ln \varepsilon) \times \mathbf{E}_t \cdot \hat{z} = 0. \tag{VIII.4.8}$$

These expressions reduce to

$$\nabla_t^2 E_z + (\omega^2\mu_0\varepsilon - \beta^2)E_z = 0, \tag{VIII.4.9}$$

$$\nabla_t^2 H_z + (\omega^2\mu_0\varepsilon - \beta^2)H_z = 0, \tag{VIII.4.10}$$

if we assume that ε does not vary appreciably over a distance of the order of the wavelength.

5 Mode Theory for Step-Index Fibers

The main difficulty in solving the boundary-value problem associated with the study of the guided modes in an optical fiber is connected with the integration of a partial differential equation by the method of separation of variables. Although this does not offer particular difficulties for step-index fibers, it is convenient to introduce some approximations in order to obtain simple expressions for the significant quantities. Thus, for example, we assume that the cladding extends to infinity, as suggested by its shielding role and justified by the exponential decay of the guided modes with distance ρ from the fiber axis. Furthermore, particular attention is devoted to the case in which the refractive indices of the core and the cladding differ by a few percent

($\Delta \ll 1$, *weakly guiding fibers*), the situation usually encountered in practice, since the smallness of Δ limits the distortion introduced by the fiber in a propagating pulse while still being compatible with its guiding property.

Equations (VIII.4.9) and (VIII.4.10) for the longitudinal components of the electric and magnetic fields, which are now rigorous since ε is constant in both the core and the cladding, read

$$\nabla_t^2 E_z + \chi^2 E_z = 0, \qquad \nabla_t^2 H_z + \chi^2 H_z = 0, \qquad \rho \leqq a, \tag{VIII.5.1}$$

$$\nabla_t^2 E_z - \gamma^2 E_z = 0, \qquad \nabla_t^2 H_z - \gamma^2 H_z = 0, \qquad \rho > a, \tag{VIII.5.2}$$

where

$$\chi^2 = \omega^2 \mu_0 \varepsilon_1 - \beta^2 = (\omega^2/c^2)\frac{\omega^2}{c^2} n_1^2 - \beta^2 \tag{VIII.5.3}$$

and

$$\gamma^2 = \beta^2 - \omega^2 \mu_0 \varepsilon_2 = \beta^2 - (\omega^2/c^2) n_2^2, \tag{VIII.5.4}$$

ε_1 and ε_2 (n_1 and n_2) being the values of the dielectric constant (of the refractive index) in the core and the cladding. It is customary to introduce, besides γ and χ, another parameter V (the so-called *normalized frequency*), defined as

$$\begin{aligned} V &= a(\chi^2 + \gamma^2)^{1/2} = a\omega[\mu_0(\varepsilon_1 - \varepsilon_2)]^{1/2} \\ &= a(\omega/c)(n_1^2 - n_2^2)^{1/2} = ak_0(n_1^2 - n_2^2)^{1/2}. \end{aligned} \tag{VIII.5.5}$$

For a guided mode, the propagation constant β is real and obeys the relation

$$(\omega/c)n_2 \leqq \beta \leqq (\omega/c)n_1 \tag{VIII.5.6}$$

(the positive values of β correspond to modes propagating in the positive direction of the z axis), as suggested by the facts that $(\omega/c)n_1 = k_0 n_1$ and $(\omega/c)n_2 = k_0 n_2$ are the values corresponding to a uniform plane wave propagating in homogeneous media with refractive indices n_1 and n_2 and that the guided case is an intermediate one. As a consequence, χ and γ are real and are assumed to be positive.

The symmetry of the problem suggests that we introduce polar coordinates ρ and ϕ, the operator ∇_t^2 in this case taking the form

$$\nabla_t^2 = \partial^2/\partial\rho^2 + (1/\rho)(\partial/\partial\rho) + (1/\rho^2)(\partial^2/\partial\phi^2), \tag{VIII.5.7}$$

and separate the variables by writing

$$E_z(\rho, \phi) = F_1(\rho)e^{i\nu\phi}, \qquad H_z(\rho, \phi) = F_2(\rho)e^{i\nu\phi}, \qquad \nu = 0, 1, 2, \ldots. \tag{VIII.5.8}$$

The set of Eqs. (VIII.5.1) and (VIII.5.2) then becomes

$$d^2F_{1,2}/d\rho^2 + (1/\rho)/(dF_{1,2}/d\rho) + (\chi^2 - v^2/\rho^2)F_{1,2} = 0, \qquad \rho \leqq a \tag{VIII.5.9}$$

and

$$d^2F_{1,2}/d\rho^2 + (1/\rho)/(dF_{1,2}/d\rho) - (\gamma^2 + v^2/\rho^2)F_{1,2} = 0, \qquad \rho > a, \tag{VIII.5.10}$$

which are, respectively, the *Bessel equation* and the *modified Bessel equation* and admit the solutions

$$F_{1,2}(\rho) = a_{1,2}J_v(\chi\rho) + b_{1,2}Y_v(\chi\rho), \qquad \rho \leqq a, \tag{VIII.5.11}$$

$$F_{1,2}(\rho) = c_{1,2}K_v(\gamma\rho) + d_{1,2}I_v(\gamma\rho), \qquad \rho > a, \tag{VIII.5.12}$$

where J_v and Y_v are *Bessel functions* (of the first and second kind) and K_v and I_v are *modified Hankel functions* (of the first and second kind (see Chapter II in [4] and Eqs. C.10 in the Appendix).

The functions Y_v and I_v diverge, respectively, for $\rho \to 0$ and (since $\gamma > 0$) for $\rho \to \infty$, so that they cannot be used to describe E_z and H_z (which are assumed to be finite everywhere), and accordingly we must set $b_{1,2} = d_{1,2} = 0$. Conversely, J_v is finite for $\rho = 0$ and K_v exhibits, for $\rho \to \infty$, the asymptotic behavior

$$K_v(\gamma\rho) \sim e^{-\gamma\rho}/(\gamma\rho)^{1/2}. \tag{VIII.5.13}$$

The continuity condition for the tangential components of the fields at the core–cladding interface then allows us to write

$$E_z(\rho, \phi) = \begin{cases} AJ_v(\chi\rho)e^{iv\phi}, & \rho \leqq a, \\ A[J_v(\chi a)/K_v(\gamma a)]K_v(\gamma\rho)e^{iv\phi}, & \rho > a, \end{cases} \tag{VIII.5.14}$$

and

$$H_z(\rho, \phi) = \begin{cases} BJ_v(\gamma\rho)e^{iv\phi}, & \rho \leqq a, \\ B[J_v(\chi a)/K(\gamma a)]K_v(\gamma\rho)e^{iv\phi}, & \rho > a, \end{cases} \tag{VIII.5.15}$$

where A and B are two arbitrary constants, the integer v representing the *azimuthal number.*

The transverse components of the fields can now be worked out by means of Eqs. (VIII.4.6), a task that is accomplished in the context of the cylindrical coordinate system depicted in Fig. VIII.10, and the continuity conditions

$$E_\phi(a^+) = E_\phi(a^-), \tag{VIII.5.16}$$

$$H_\phi(a^+) = H_\phi(a^-), \tag{VIII.5.17}$$

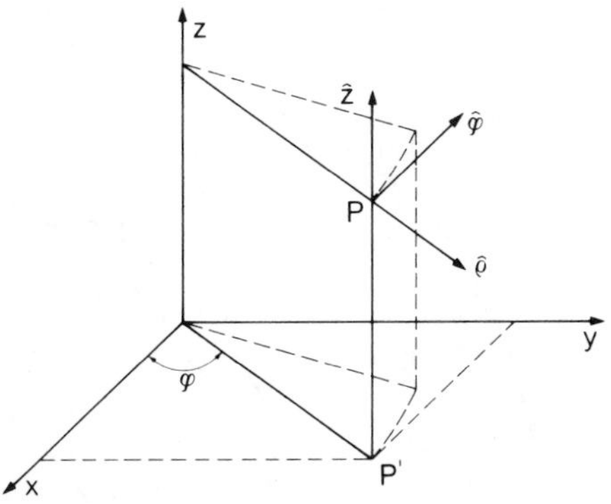

Fig. VIII.10. System of cylindrical coordinates.

can then be imposed. Actually, since it is possible to show that if $\varepsilon_1 \cong \varepsilon_2$ the relation

$$\mathbf{H}_t \cong (\varepsilon_1/\mu_0)^{1/2}\hat{z} \times \mathbf{E}_t \tag{VIII.5.18}$$

(similar to that valid for uniform plane-wave propagation in the core parallel to the z axis) connects the transverse components of the electric and magnetic fields, we can impose, instead of Eq. (VIII.5.17), the equivalent condition

$$E_\rho(a^+) = E_\rho(a^-). \tag{VIII.5.19}$$

Without going into the details of the calculation, we write the relation obtained from Eqs. (VIII.5.16) and (VIII.5.19), that is,

$$\left[\frac{J_{\nu-1}(\chi a)}{\chi a J_\nu(\chi a)} - \frac{K_{\nu-1}(\gamma a)}{\gamma a K_\nu(\gamma a)}\right]\left[\frac{J_{\nu+1}(\chi a)}{\chi a J_\nu(\chi a)} + \frac{K_{\nu+1}(\gamma a)}{\gamma a K_\nu(\gamma a)}\right] = 0. \tag{VIII.5.20}$$

Equation (VIII.5.20) is known as the *characteristic equation*, and its solution gives rise [with the help of Eqs. (VIII.5.3) and (VIII.5.4)] to a double infinity of discrete values for β which provide the admissible propagation constants $\beta_{\nu\delta}$ of the guided modes (δ being a positive integer distinguishing the various solutions for every fixed ν). Once the behavior of $\beta_{\nu\delta}$ as a function of the angular frequency ω is known, it is possible to introduce the concept of *cutoff frequency* $\omega_{\nu\delta}$, which determines the lower limit below which the mode ν, δ is no longer guided. In fact, according to Eq. (VIII.5.13), the mode is no longer confined in the core as $\gamma \to 0$, that is [see Eq. (VIII.5.4)], for a value $\omega_{\nu\delta}$ of ω such that

$$\beta(\omega_{\nu\delta}) = (\omega_{\nu\delta}/c)n_2, \tag{VIII.5.21}$$

since in this case the mode is spread over the whole space.

For $\omega < \omega_{\nu\delta}$, $\beta_{\nu\delta}$ becomes complex and, accordingly, the corresponding mode attenuates while propagating. It is possible to show that, unlike the case

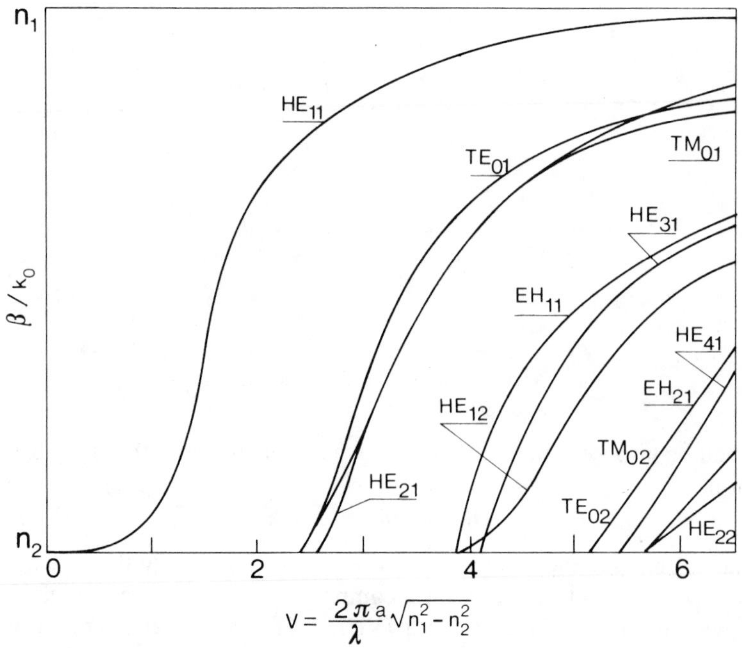

Fig. VIII.11. Normalized propagation constant as a function of normalized frequency V for various lowest-order modes in a step-index fiber.

of the metallic waveguides, there exists for dielectric waveguides a fundamental mode (indicated by the symbol HE_{11}) that does not admit cutoff [1] (see Fig. VIII.11).

6 Weakly Guiding Step-Index Fibers

A substantial simplification, for both the mode structure and the characteristic equation, is achieved whenever $\Delta \ll 1$, a situation always encountered in practice when dealing with fibers for telecommunications. The main advantage is the possibility of describing the electromagnetic field by means of a superposition of *linearly polarized modes* $(\mathrm{LP})_{\nu\delta}$ whose longitudinal components are negligible with respect to the transverse ones (of order $\Delta^{1/2}$) [1, 9].

After introducing two unit vectors $\hat{x}$ and $\hat{y}$ in the x and y directions (see Fig. VIII.10), we have for the electric transverse components of these modes

$$(\mathrm{LP}_x)_{\nu\delta} = \beta A_0[\sin(\nu\phi), -\cos(\nu\phi)]\hat{x} \begin{cases} J_\nu(\chi\rho)/J_\nu(\chi a), & \rho \leqq a, \\ K_\nu(\gamma\rho)/K_\nu(\gamma a), & \rho > a, \end{cases} \tag{VIII.6.1}$$

and

$$(\mathrm{LP}_y)_{\nu\delta} = \beta A_0[\cos(\nu\phi), \sin(\nu\phi)]\hat{y} \begin{cases} J_\nu(\chi\rho)/J_\nu(\chi a), & \rho \leqq a, \\ K_\nu(\gamma\rho)/K_\nu(\gamma a), & \rho > a, \end{cases} \tag{VIII.6.2}$$

where A_0 is an arbitrary constant. The [,] means that one can choose either the first or the second expression inside the brackets, β, χ, and γ being related to ν and δ through the characteristic equation

$$-J_\nu(\chi a)/\chi a J_{\nu+1}(\chi a) = K_\nu(\gamma a)/\gamma a K_{\nu+1}(\gamma a). \tag{VIII.6.3}$$

The transverse components of the magnetic field are obtained from Eqs. (VIII.6.1) and (VIII.6.2) by means of Eq. (VIII.5.18). Each of the LP modes exhibits, for $\nu \geq 1$, a fourth-order degeneracy, in the sense that two LP_x and two LP_y modes (corresponding to the "sin ϕ" and "cos ϕ" solutions) with the same $\beta_{\nu\delta}$ are present for any fixed ν and δ. The degeneracy is of order two for the modes $\nu = 0$.

It is possible to obtain a closed-form approximation for the solutions of Eq. (VIII.6.3) for frequencies much larger than the cutoff frequency. Far from cutoff ($\gamma a \to \infty$), Eq. (VIII.6.3) reduces, with the help of Eq. (VIII.5.13), to

$$\chi a J_{\nu+1}(\chi a) = \gamma a J_\nu(\chi a), \tag{VIII.6.4}$$

which, since $\gamma a \cong V$, is equivalent to

$$\chi a J_{\nu+1}(\chi a) = V J_\nu(\chi a). \tag{VIII.6.5}$$

By taking the derivative of both sides of Eq. (VIII.6.5) with respect to V and solving with respect to $d(\chi a)/dV$, we obtain

$$\frac{d(\chi a)}{dV} = \frac{J_\nu(\chi a)}{-\nu J_{\nu+1}(\chi a) + \chi a J_\nu(\chi a) - [\nu V/(\chi a)]J_\nu(\chi a) + V J_{\nu+1}(\chi a)}, \tag{VIII.6.6}$$

where use has been made of the identities [8]

$$(d/d\zeta)J_\nu(\zeta) = \tfrac{1}{2}[J_{\nu-1}(\zeta) - J_{\nu+2}(\zeta)], \tag{VIII.6.7}$$

$$(\nu/\zeta)J_\nu(\zeta) = \tfrac{1}{2}[J_{\nu+1}(\zeta) + J_{\nu-1}(\zeta)]. \tag{VIII.6.8}$$

By substituting Eq. (VIII.6.5) into Eq. (VIII.6.6), we finally get

$$\frac{d(\chi a)}{dV} = \frac{1}{[V/(\chi a)](V - 2\nu) + \chi a} \cong \chi a/[V(V - 2\nu)], \tag{VIII.6.9}$$

whose solution is

$$\chi a = (\chi a)_\infty (1 - 2\nu/V)^{1/(2\nu)} \tag{VIII.6.10}$$

for $\nu \neq 0$ and

$$\chi a = (\chi a)_\infty e^{-1/V} \tag{VIII.6.11}$$

for $\nu = 0$, $(\chi a)_\infty$ being the δth zero of the equation $J_\nu(\chi a) = 0$ [see Eq. (VIII.6.5)].

In a *multimode fiber*—that is, in a fiber supporting many guided modes—$\gamma a \gg 1$ for almost all of them (with the exception of the few near cutoff), so that Eqs. (VIII.6.10) and (VIII.6.11) provide to a good approximation, after using Eq. (VIII.5.3), the propagation constants for nearly all modes.

The cutoff situation, opposite to the one just described, is realized whenever $\gamma a \to 0$. By taking in Eq. (VIII.6.3) the asymptotic expansion of $K_\nu(\gamma a)$ for $\gamma a \to 0$, it is possible to obtain its solutions and from them to derive the cutoff frequencies $\omega_{\nu\delta}$, which can be written in the form

$$\omega_{\nu\delta} = (c/a)[1/(n_1^2 - n_2^2)^{1/2}]\xi_{\nu\delta}, \qquad \text{(VIII.6.12)}$$

where the $\xi_{\nu\delta}$ represent, for $\nu \geq 1$, the solutions of the equations $J_{\nu-1}(\xi) = 0$ (with the exclusion, for $\nu \geq 2$, of the solution $\xi = 0$), while the $\xi_{0\delta}$ are the solutions of the equation $J_1(\xi) = 0$. Since $J_1(0) = 0$, $\xi_{01} = 0$, and the mode $(\mathrm{LP})_{01}$, and it only, has no cutoff frequency. The lowest cutoff frequency is obtained by calculating the smallest of the remaining $\xi_{\nu\delta}$, which turns out to be the first zero of $J_0(\xi)$, that is, $\xi_{11} = 2.405\ldots$. By recalling Eqs. (VIII.5.5) and (VIII.6.12), we can express the condition of monomode operation for a step-index fiber by the relation

$$V \leqq 2.405. \qquad \text{(VIII.6.13)}$$

7 Parabolic-Index Fibers

The study of the guided modes of a fiber with a parabolic-index profile is simplified if the refractive index is supposed to vary slowly over the distance of a wavelength, and if it is assumed to maintain its parabolic behavior for any ρ (thus taking on unphysical values for ρ large enough) (see Fig. VIII.12). This

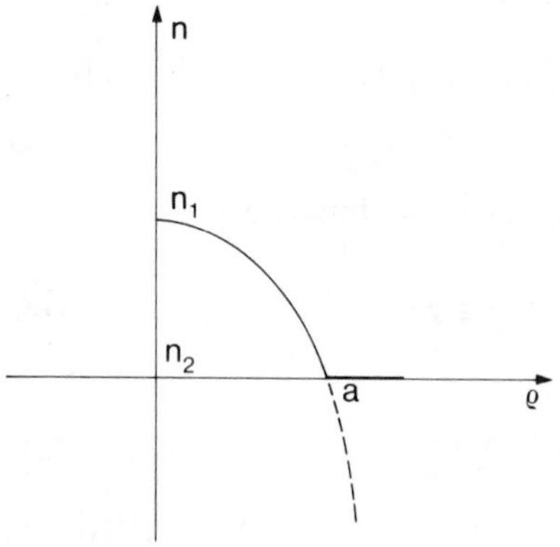

Fig. VIII.12. Actual and ideal parabolic refractive-index profiles.

last hypothesis is justified by the results of Section VIII.3 on the trajectories of the guided rays, from which it is reasonable to infer that at least the lower-order modes remain confined around the fiber axis, so that their behavior is insensitive to that of the refractive index for large ρ. In this way, it is possible to avoid the difficulties connected with the necessity of matching the tangential components of the field at the core–cladding interface and to resort directly to the scalar theory (polarized modes) and to the introduction of cartesian coordinates.

Recalling Eq. (I.1.9), each cartesian component $E_i(x, y)$ of the field obeys the equation (in the scalar approximation)

$$\partial^2 E_i/\partial x^2 + \partial^2 E_i/\partial y^2 + \{k_0^2[n_0^2 - n_0 b(x^2 + y^2)] - \beta^2\}E_i = 0, \qquad \text{(VIII.7.1)}$$

where use has been made of Eq. (VIII.3.3), that is,

$$n^2 = c^2\mu_0\varepsilon = n_0^2 - n_0 b(x^2 + y^2). \qquad \text{(VIII.7.2)}$$

By separating the variables, that is, by writing $E_i(x, y) = f(x)g(y)$, Eq. (VIII.7.1) becomes

$$(1/f)(d^2f/dx^2) + (n_0^2k_0^2 - \beta^2 - k_0^2 n_0 b x^2) + [(1/g)(d^2g/dy^2) - k_0^2 n_0 b y^2] = 0, \qquad \text{(VIII.7.3)}$$

which is equivalent to the two equations

$$d^2f/dx^2 + (n_0^2k_0^2 - R^2 - \beta^2 - k_0^2 n_0 b x^2)f = 0, \qquad \text{(VIII.7.4)}$$

$$d^2g/dy^2 + (R^2 - k_0^2 n_0 b y^2)g = 0, \qquad \text{(VIII.7.5)}$$

R^2 being the separation constant. After we introduce the dimensionless variables

$$\xi = k_0^{1/2}(n_0 b)^{1/4}x, \qquad \text{(VIII.7.6)}$$

$$\eta = k_0^{1/2}(n_0 b)^{1/4}y, \qquad \text{(VIII.7.7)}$$

the equations become

$$d^2f/d\xi^2 + (\sigma^2 - \xi^2)f = 0, \qquad \text{(VIII.7.8)}$$

$$d^2g/d\eta^2 + (\psi^2 - \eta^2)g = 0, \qquad \text{(VIII.7.9)}$$

with

$$\sigma^2 = (n_0^2k_0^2 - R^2 - \beta^2)/[k_0(n_0 b)^{1/2}], \qquad \text{(VIII.7.10)}$$

$$\psi^2 = R^2/[k_0(n_0 b)^{1/2}]. \qquad \text{(VIII.7.11)}$$

Equation (VIII.7.8) [or Eq. (VIII.7.9)] is a familiar one, since it is the Schrödinger equation for a one-dimensional harmonic oscillator and its solution can be found in any textbook on quantum mechanics. It can be shown that Eqs. (VIII.7.8) and (VIII.7.9) have solutions that are finite and continuous everywhere and that tend to zero for ξ and η tending toward $\pm\infty$, if and only if

$$\sigma^2 = 2p + 1, \qquad p = 0, 1, 2, \ldots, \tag{VIII.7.12}$$

$$\psi^2 = 2q + 1, \qquad q = 0, 1, 2, \ldots, \tag{VIII.7.13}$$

and that they read

$$f(\xi) = e^{-\xi^2/2} H_p(\xi), \tag{VIII.7.14}$$

$$g(\eta) = e^{-\eta^2/2} H_q(\eta), \tag{VIII.7.15}$$

H_m being the Hermite polynomials of mth order (see Section VII.8, Table VII.2, and Eqs. (C.4)).

According to these considerations, we can write the set of linearly polarized guided modes

$$E_{pq}(\xi, \eta) = [(2/\pi)^{1/2}/(2^{p+q} p!\, q!)^{1/2}] H_p(\xi) H_q(\eta) e^{-(\xi^2+\eta^2)}, \tag{VIII.7.16}$$

where the normalization factor has been chosen in such a way that

$$\int_{-\infty}^{+\infty} dx \int_{-\infty}^{+\infty} dy\, |E_{pq}|^2 = 1. \tag{VIII.7.17}$$

By introducing Eqs. (VIII.7.12) and (VIII.7.13) into Eqs. (VIII.7.10) and (VIII.7.11), we now obtain the propagation constant β_{pq} in the form

$$\beta_{pq} = [n_0^2 k_0^2 - 2k_0 (n_0 b)^{1/2} (p + q + 1)]^{1/2}. \tag{VIII.7.18}$$

For the lower-order modes, since $b^{1/2}/k_0 \ll 1$ due to the negligible variation of the refractive index over a wavelength, Eq. (VIII.7.18) reduces to

$$\beta_{pq} = n_0 k_0 - (b/n_0)^{1/2} (p + q + 1). \tag{VIII.7.19}$$

The results of this section can be compared with those obtained in the context of geometric optics in Section (II.12.1a) for a radial parabolic distribution of the refractive index profile [see Eq. (II.12.14)]. In particular, it is interesting to compare the modal distributions in Eq. (VIII.7.16) with the field distribution in Eq. (II.12.17) and to note the remarkable identity of the two expressions for the propagation constants β_{pq} [Eqs. (II.12.21) and (VIII.7.18)] worked out by two completely different approaches.

8 Nonguided Modes

The solutions of the characteristic equation for which β is real and obeys the relation [see Eq. (VIII.5.6)]

$$k_0 n_2 < \beta < k_0 n_1 \tag{VIII.8.1}$$

give rise (see Section VIII.5) to a discrete set of guided modes that propagate without attenuation along the fiber and vanish exponentially at large distances from the axis. In order to describe the electromagnetic field in a complete way, however, it is necessary to add a continuous set of modes to the previous one. From an analytical point of view, such configurations are obtained by letting $d_{1,2}$ [see Eq. (VIII.5.12)] be different from zero, which implies imaginary values for γ if the physical condition of a vanishing field for $\rho \to \infty$ must be satisfied. In fact, we have

$$I_\nu(\gamma\rho) \underset{\gamma\rho \to \infty}{\sim} e^{\gamma\rho}/(\gamma\rho)^{1/2}, \tag{VIII.8.2}$$

which, together with the asymptotic behavior of $K_\nu(\gamma\rho)$ [see Eq. (VIII.5.13)], shows that only imaginary values of γ are admissible. Accordingly, from Eq. (VIII.5.4) it is possible to deduce that β can assume real values in the interval

$$0 < \beta \leqq n_2 k_0, \tag{VIII.8.3}$$

and imaginary values in the interval

$$-i\infty < \beta < 0. \tag{VIII.8.4}$$

The choice of positive real values in Eq. (VIII.8.3) and negative *imaginary* values in Eq. (VIII.8.4) corresponds, respectively, to considering modes propagating in the positive z direction and modes decaying exponentially in the same direction. It is worth noting that, thanks to the further degree of freedom introduced by keeping I_ν in Eq. (VIII.5.12), the continuity condition at the core–cladding interface can be satisfied without giving rise to a characteristic equation. Thus, the values of β need no longer be discrete, but can take on any continuous value in the intervals defined by Eqs. (VIII.8.3) and (VIII.8.4).

In general, the electromagnetic field associated with a cylindrical fiber can be written as a suitable superposition of guided and continuous modes in the form

$$\begin{aligned}\mathbf{E}(\boldsymbol{\rho}, z, t) &= \sum_{\nu,\delta} A_{\nu,\delta}\mathbf{E}_{\nu,\delta}\exp(i\omega t - i\beta_{\nu,\delta}z)\\ &+ \sum_{\nu}\int_0^{k_0 n_2} A_\nu(\xi)\mathbf{E}_\nu(\xi)\exp[i\omega t - i\beta_\nu(\xi)z]\,d\xi\\ &+ \sum_{\nu}\int_{k_0 n_2}^{\infty} A_\nu(\xi)\mathbf{E}_\nu(\xi)\exp[i\omega t - i\beta_\nu(\xi)z]\,d\xi,\end{aligned} \tag{VIII.8.5}$$

where the first term represents the contribution of the guided modes, $\mathbf{E}_{\nu,\delta}$ being the relative modal configurations (the $\boldsymbol{\rho}$ and ω dependences are omitted for notational simplicity) and $A_{\nu,\delta}$ the corresponding amplitudes, and the second and the third terms represent the contributions of the continuous modes, for which, respectively,

$$0 \leqq \gamma < ik_0 \tag{VIII.8.6}$$

and

$$ik_0 n_2 < \gamma < i\infty, \tag{VIII.8.7}$$

that is, if $\xi = -i\gamma$,

$$0 \leqq \xi < k_0 n_2, \tag{VIII.8.8}$$

$$k_0 n_2 < \xi < \infty. \tag{VIII.8.9}$$

The continuous modes defined by Eq. (VIII.8.8), for which β is real, are called *radiation modes*, while those defined by Eq. (VIII.8.9) are called *evanescent modes.*

In practice, the representation of the electromagnetic field given by Eq. (VIII.8.5) is very complicated and, in the majority of cases concerning propagation, one prefers to adopt an approximate description in which the radiation modes are replaced by a suitable set of discrete modes (*leaky modes*) [7], which decay exponentially for z positive and, together with the guided modes, provide a good description of the field in the core and its vicinity (so that, for z sufficiently large, the field is represented by the guided modes alone). This does not contradict the fact that the individual radiation mode does not attenuate with z, since any realistic field carrying energy along the fiber is an integral over a finite interval of values of ξ, and it is this integral that goes to zero when z becomes very large.

The set of leaky modes is obtained in a natural way by choosing the solutions of the characteristic equation having a negative imaginary part β_2 of β and a real part β_1 satisfying Eq. (VIII.8.3), which, as we would expect, corresponds to propagation directions forming an angle larger than $\pi/2 - \theta_c$ with the z axis. In this way, a given mode is a guided one if the frequency of the field is larger than its cutoff frequency and a leaky one in the opposite case.

Notice that, according to Eqs. (VIII.5.4) and (VIII.8.3) and the fact the β_2 is negative, the real part of γ is negative, so that, recalling Eq. (VIII.5.13), we immediately see that the leaky modes diverge for $\rho \to \infty$. This implies that an electromagnetic field propagating along the fiber can be well represented by a superposition of guided and leaky modes (see also Section III.19), but not too far from the core–cladding interface.

9 Single-Mode Fibers

We saw in Section VIII.6 how a step-index fiber is capable of operating in a monomodal regime, that is, of propagating only the two degenerate orthogonally polarized states pertaining to the mode $(\mathrm{LP})_{01}$, provided that the normalized frequency V satisfies Eq. (VIII.6.13). When working in the 1.2–1.6-μm region, where silica fibers exhibit low losses and chromatic dispersion (Sections VIII.13 and VIII.14), single-mode fibers have a great potential for ultrahigh-bandwidth communications, which is a compelling reason for studying their propagation characteristics in some detail. However, this investigation cannot be limited to the step-index profile, for which an analytical if algebraically complicated solution already exists. In fact, in the significant wavelength range of operation (at least at present), monomodality is achieved for small core radii. This does not allow, during fabrication, a precise control of the fiber refractive index profile, which exhibits distortions or dips around the fiber axis. Luckily, it turns out that the modal field and the propagation constant of a single-mode fiber are quite insensitive to the precise details of the refractive index profile and that they can be characterized in a way independent of its fine structure [10].

In agreement with the analysis carried out in the previous sections, it is reasonable to assume that for weakly guiding fibers having an arbitrary refractive index profile $n(\rho)$ in the core, the field of each mode can be approximated by a transverse linearly polarized wave solution of the scalar wave equation. For example, by identifying the polarization direction with the x axis and by writing approximately (taking advantage of the weak-guidance hypothesis)

$$E_x = \psi e^{-i\beta z}, \qquad H_y \cong (\varepsilon/\mu_0)^{1/2} E_x, \tag{VIII.9.1}$$

the modal amplitude $\psi_\nu(\rho)$ corresponding to a mode of azimuthal number ν obeys the equation

$$d^2\psi_\nu/d\rho^2 + (1/\rho)(d\psi_\nu/d\rho) + [k_0^2 n^2(\rho) - \beta^2 - \nu^2/\rho^2]\psi_\nu = 0. \tag{VIII.9.2}$$

The direct approach to investigating the propagation characteristics of single-mode optical fibers is to specify a particular form of $n(\rho)$ and to look for some kind of exact numerical method [11], in general difficult to implement for arbitrary profiles, for solving Eq. (VIII.9.2). Actually, a much simpler approach has been developed that enables the important mode characteristics to be derived from much reduced profile data, adequately expressible through only two or three parameters corresponding to the moments of the index profile [12]. In this approach, it is customary to introduce the *profile shape function* $s(\rho)$, defined by the relation

$$n^2(\rho) = n_2^2 + (n_0^2 - n_2^2)s(\rho), \tag{VIII.9.3}$$

where n_0 and n_2 are, respectively, the maximum value $n(\rho)$ assumes in the core and the constant value it assumes in the cladding. The shape function $s(\rho)$ is obviously identically zero for ρ larger than the core radius ($\rho \geq a$; see Fig. VIII.13), and its moments Ω_l are defined through the relation

$$N_l \equiv \int_0^a [n^2(\rho) - n_2^2]\rho^{l+1}\, d\rho \equiv (n_0^2 - n_2^2)a^{l+2}\Omega_l, \tag{VIII.9.4}$$

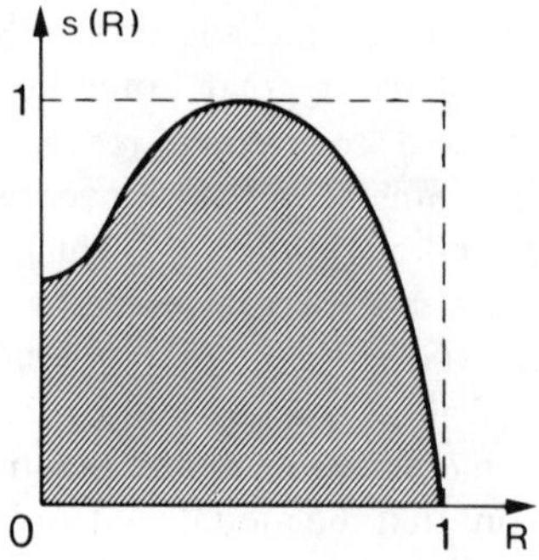

Fig. VIII.13. Typical behavior of the profile shape function $s(R)$ $(R = \rho/a)$.

so that

$$\Omega_l = \int_0^1 s(R) R^{l+1}\, dR, \tag{VIII.9.5}$$

with $R = \rho/a$. Note that only the even moments are sufficient for defining $s(R)$ completely, since it is defined only for $R \geq 0$. The moments N_0 and Ω_0 are connected in a simple way to the *effective normalized frequency* of the fiber, defined as

$$\bar{V}^2 = 2k_0^2 \int_0^a [n^2(\rho) - n_2^2]\rho\, d\rho, \tag{VIII.9.6}$$

by the relation

$$\bar{V}^2 = 2k_0^2 N_0 = 2V^2\Omega_0, \tag{VIII.9.7}$$

where

$$V = k_0 a(n_0^2 - n_2^2)^{1/2} \tag{VIII.9.8}$$

is the normalized frequency of a step-index fiber having the same radius a and n_0 as the core refractive index.

In the rest of this section we will closely follow the procedure adopted in Hussey and Pask [13]. After considering the fundamental mode, for which $v = 0$, we rewrite Eq. (VIII.9.2) in the form

$$[-D_0 + \tfrac{1}{2}\bar{V}^2\bar{s}(R)]\psi = \Gamma^2\psi, \tag{VIII.9.9}$$

where $R = \rho/a$, $\bar{s}(R) = s(R)/\Omega_0$,

$$D_0 = -d^2/dR^2 - (1/R)(d/dR), \tag{VIII.9.10}$$

and

$$\Gamma^2 = \beta^2 a^2 - k_0^2 n_2^2 a^2. \tag{VIII.9.11}$$

If we assume a reference profile $\hat{s}$ and a field $\hat{\psi}$ that satisfy the equation

$$(-D_0 + \tfrac{1}{2}\bar{V}^2\bar{\hat{s}})\hat{\psi} = \hat{\Gamma}^2\hat{\psi}, \tag{VIII.9.12}$$

ordinary first-order perturbation theory gives

$$\frac{\Gamma^2}{\bar{V}} = \frac{\hat{\Gamma}^2}{\bar{V}} + \frac{\bar{V}}{2}\frac{\int_0^\infty(\bar{s} - \bar{\hat{s}})\hat{\psi}^2 R\,dR}{\int_0^\infty \hat{\psi}^2 R\,dR}. \tag{VIII.9.13}$$

In particular, if $\hat{\psi}^2$ is expanded as a polynomial in R^2 [according to Eq. (VIII.9.9) $\hat{\psi}$ is an even function of R] with expansion coefficient b_i, then

$$\int_0^\infty (\bar{s} - \bar{\hat{s}})^2 R\,dR = \sum_{i=1}^{\infty} b_i(\bar{\Omega}_{2i} - \bar{\hat{\Omega}}_{2i}), \tag{VIII.9.14}$$

where $\bar{\Omega}_l = \Omega_l/\Omega_0$.

Usually, it is convenient to choose the reference profile as that pertaining to a step-index fiber, for which the field $\hat{\psi} = \psi_s$ and $\hat{\Gamma}/a = \gamma$ are well known (Section VIII.6), so that $V = \bar{V}$, its core radius a_s being assumed as an *a priori* unknown reference parameter. After evaluating the coefficients b_i, it is possible to obtain, in terms of the lower-order moments,

$$a_s/a = (2\bar{\Omega}_2)^{1/2}, \qquad \Gamma^2/a^2 \cong \gamma^2(\bar{V})/(2\bar{\Omega}_2), \tag{VIII.9.15}$$

for $x \le 0$, and

$$a_s/a = (3\bar{\Omega}_4)^{1/4}, \qquad \Gamma^2/a^2 \cong [\gamma^2(\bar{V})/(3\bar{\Omega}_4)^{1/2}]\{1 + x\chi^2 a_s^2/[4J_1^2(\chi a_s)]\}, \tag{VIII.9.16}$$

for $x > 0$, where the parameter

$$x = 1 - (2\bar{\Omega}_2/(3\bar{\Omega}_4)^{1/2}) \tag{VIII.9.17}$$

is connected to the profile type. In particular, $x = 0$ for a step-index profile, while it is positive for power-law profiles of the type $s_1(R) = 1 - R^q$ and negative for those of the type $s_2(R) = 1 - (1 - R)^q$ (which exhibit a dip on the fiber axis).

Equations (VIII.9.15) and (VIII.9.16) allow us to characterize single-mode optical fibers in terms of the first three moments Ω_0, Ω_2, and Ω_4 of the refractive index profile. Since they represent averaged quantities, this procedure automatically introduces a smoothing effect, which tends to cancel out the inessential fine details of $n(\rho)$.

In order to describe pulse propagation and chromatic dispersion (Section VIII.13), it is customary to introduce a parameter b defined by the relation [9]

$$\beta = k_0 n_2(1 + b\tilde{\Delta}), \tag{VIII.9.18}$$

where $\tilde{\Delta} = \Delta(n_0/n_2)^2$, which can be expressed in terms of Γ as

$$b(V) = \Gamma^2/V^2 = 2\Omega_0\Gamma^2/\bar{V}^2 \equiv 2\Omega_0\bar{b}(\bar{V}). \qquad \text{(VIII.9.19)}$$

Another important parameter, expecially in connection with source–fiber coupling and microbending losses (see Section VIII.15), is the *spot size* ρ_0. It can be defined by recalling that for step-index and power-law profiles, $\psi(\rho)$ can be well approximated by a guassian function, that is,

$$\psi(\rho) \cong \exp(-\tfrac{1}{2}\rho^2/\rho_0^2), \qquad \text{(VIII.9.20)}$$

or directly by the relation

$$\left(\frac{\rho_0}{a}\right)^2 = \frac{\int_0^\infty \psi^2 R^3\, dR}{\int_0^\infty \psi^2 R\, dR}. \qquad \text{(VIII.9.21)}$$

It can be explicitly evaluated according to the preceding approach, which yields

$$(\rho_0/a)_{\bar{V}} = (2\bar{\Omega}_2)^{1/2}(\rho_0/a)_{s,V=\bar{V}}. \qquad \text{(VIII.9.22)}$$

In particular, it can be shown [10] that, to a good approximation, one can write, for $V > 1$,

$$(\rho_0/a)_s^2 = 1/\ln V^2. \qquad \text{(VIII.9.23)}$$

It remains to evaluate the second-mode cutoff. This can be done by applying the preceding method to the second mode $\nu = 1$ and taking into account the cutoff condition $\Gamma = 0$. After some algebra, it is possible to express this condition through the relation

$$\bar{V} \leqq \bar{V}_{co} \cong 2.405/(1 - 1.1419x)^{1/2}. \qquad \text{(VIII.9.24)}$$

10 The Electromagnetic Field Inside the Fiber

In the previous sections, the fundamental aspects of propagation in optical fibers have been investigated by examining the behavior of the propagation modes at a fixed frequency in ideal dielectric waveguides. It is now necessary to consider the actual situation in which the field, consisting of a superposition of various modes and having a finite bandwidth, propagates in a real fiber exhibiting unavoidable deviations of the refractive index from ideality, which give rise to attenuation and coupling among the various modes. The simultaneous excitation of many modes and the dependence on the frequency of the mode propagation constant introduce a distortion of the signal that affects the fiber performance.

By indicating with $\mathbf{E}_m(\boldsymbol{\rho}, \omega)$ the spatial configuration of the generic mode identified by the index m (which, as already seen, corresponds in practice to a number of indices), and with $\beta_m(\omega)$ its propagation constant (real for guided modes and complex for leaky modes), we have for the analytic signal (see Section I.8) corresponding to a monochromatic electric field propagating in an ideal fiber

$$\hat{\mathbf{E}}(\boldsymbol{\rho}, z, t) = \sum_m c_m \mathbf{E}_m(\boldsymbol{\rho}; \omega) \exp[-i\beta_m(\omega)z + i\omega t], \qquad \text{(VIII.10.1)}$$

where the c_m are suitable expansion coefficients. If the field has a finite bandwidth $\delta\omega$, Eq. (VIII.10.1) is replaced by

$$\hat{\mathbf{E}}(\boldsymbol{\rho}, z, t) = \sum_m \int_0^\infty c_m(\omega)\mathbf{E}_m(\boldsymbol{\rho}; \omega) \exp[-i\beta_m(\omega)z + i\omega t]\, d\omega. \qquad \text{(VIII.10.2)}$$

As already noted, the field is not purely transverse, the ratio between the longitudinal and transverse components being of order $\Delta^{1/2}$. However, the significant quantity usually considered is the power $p^{(\sigma)}$ carried through a given transverse section σ of the fiber (see Fig. VIII.14), which is given by

$$P^{(\sigma)}(z, t) = \mathrm{Re} \iint_\sigma \tilde{\mathbf{S}} \cdot \hat{z}\, dx\, dy, \qquad \text{(VIII.10.3)}$$

where $\hat{z}$ is a unit vector in the positive direction of the z axis and $\tilde{\mathbf{S}}$ is the complex Poynting vector, defined as (see Section I.8)

$$\tilde{\mathbf{S}} = \tfrac{1}{2}\langle \hat{\mathbf{E}} \times \hat{\mathbf{H}}^* \rangle_t, \qquad \text{(VIII.10.4)}$$

the $\langle\ \rangle_t$ indicating a temporal average over few periods $(2\pi/\omega)$ of the field. Hence, only the transverse component of the field is necessary in order to evaluate this quantity. Taking into account this component, equations identical to Eqs. (VIII.10.1) and (VIII.10.2) hold true for the magnetic field, provided the substitutions $\hat{\mathbf{E}} \to \hat{\mathbf{H}}$ and $\mathbf{E}_m \to \mathbf{H}_m$ are made. The expansion coefficients c_m of the guided modes can be determined, in principle, by knowledge of the boundary condition $\mathbf{E}(\boldsymbol{\rho}, z = 0, t)$, taking advantage of the orthogonality relation

$$\int_{-\infty}^{+\infty} dx \int_{-\infty}^{+\infty} dy [\mathbf{E}_m(\boldsymbol{\rho}; \omega) \times \mathbf{H}_n^*(\boldsymbol{\rho}; \omega)] \cdot \hat{z} = 2P\delta_{mn}, \qquad \text{(VIII.10.5)}$$

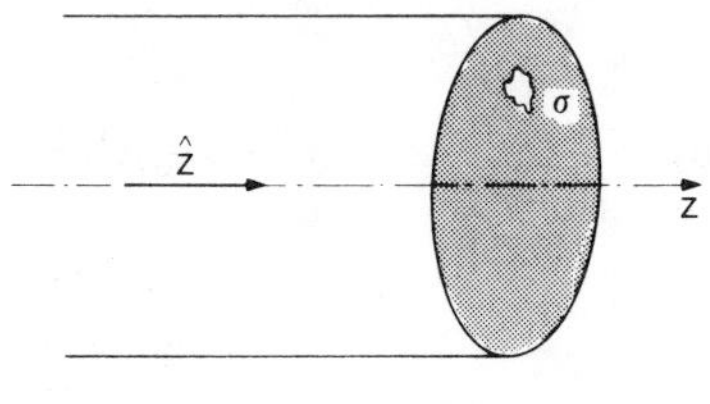

Fig. VIII.14. Transverse fiber section.

where P is a positive normalization constant and δ_{mn} is the Kronecker symbol. An equation similar to Eq. (VIII.10.5) is approximately valid for the *leaky tunneling modes* corresponding to leaky tunneling rays (see Section VIII.2), which can still contribute appreciably to the field after remarkably long distances.

If we assume that the bandwidth $\delta\omega$ of the field statisfies the relation (as happens in most practical cases)

$$\delta\omega/\omega_0 \ll 1, \qquad \text{(VIII.10.6)}$$

ω_0 being a typical frequency of the field, we can write $\mathbf{E}_m(\boldsymbol{\rho};\omega) \simeq \mathbf{E}_m(\boldsymbol{\rho};\omega_0)$ in Eq. (VIII.10.2), and thus obtain the simplified expression

$$\hat{\mathbf{E}}(\boldsymbol{\rho}, z, t) = \sum_m \mathbf{E}_m(\boldsymbol{\rho};\omega_0) \int_0^\infty c_m(\omega) \exp[-i\beta_m(\omega)z + i\omega t]\, d\omega. \qquad \text{(VIII.10.7)}$$

Notice that Eq. (VIII.10.2) or Eq. (VIII.10.7) can be immediately generalized to describe the case of a nonideal fiber with a refractive index profile slightly different from the ideal one by letting the coefficients $c_m(\omega)$ depend on z. This takes into account the mechanism of coupling among the various modes introduced by the irregularities in the refractive index, so that in general we have

$$\hat{\mathbf{E}}(\rho, z, t) = \sum_m \mathbf{E}_m(\boldsymbol{\rho};\omega) \int_0^\infty c_m(\omega, z) \exp[-i\beta_m(\omega)z + i\omega t]\, d\omega. \qquad \text{(VIII.10.8)}$$

The values of the $c_m(\omega, z = 0)$ are immediately obtained in terms of the boundary condition by using Eqs. (VIII.10.5) and (VIII.10.7), which yield

$$c_m(\omega, z = 0) = \frac{1}{4\pi P} \int_{-\infty}^{+\infty} dt\, e^{-i\omega t} \int_{-\infty}^{+\infty} dx \int_{-\infty}^{+\infty} dy\, \hat{z} \cdot [\hat{\mathbf{E}}(\boldsymbol{\rho}, z = 0, t) \times \mathbf{H}_m^*(\boldsymbol{\rho};\omega_0)] \qquad \text{(VIII.10.9)}$$

[in the ideal case $c_m(\omega, z) = c_m(\omega, 0) \equiv c_m(\omega)$]. By factorizing the rapidly varying (both in space and time) terms, Eq. (VIII.10.8) can be rewritten as

$$\hat{\mathbf{E}}(\boldsymbol{\rho}, z, t) = \sum_m \mathbf{E}_m(\rho) \exp[-i\beta_m(\omega_0)z + i\omega_0 t]\Phi_m(z, t), \qquad \text{(VIII.10.10)}$$

where $\mathbf{E}_m(\boldsymbol{\rho}) \equiv \mathbf{E}_m(\boldsymbol{\rho};\omega_0)$ and the Φ_m, which represent slowly varying amplitudes, are given by

$$\Phi_m(z, t) = \int_0^\infty c_m(\omega, z) \exp\{-i[\beta_m(\omega) - \beta_m(\omega_0)]z + i(\omega - \omega_0)t\}\, d\omega. \qquad \text{(VIII.10.11)}$$

By inserting Eqs. (VIII.10.4) and (VIII.10.10) and the analogous equations for the magnetic field into Eq. (VIII.10.3), we have

$$P^{(\sigma)}(z,t) = \frac{1}{2}\mathrm{Re}\sum_{m} F^{(\sigma)}_{mm}\langle|\Phi_m(z,t)|^2\rangle_t + \frac{1}{2}\mathrm{Re}\sum_{m\neq n}\sum F^{(\sigma)}_{mn}\exp\{i[\beta_n(\omega_0) - \beta_m(\omega_0)]z\}$$

$$\times\ \langle\Phi_m(z,t)\Phi_n^*(z,t)\rangle_t, \qquad \text{(VIII.10.12)}$$

where

$$F^{(\sigma)}_{mn} = \iint_{\sigma} dx\,dy\,[\mathbf{E}_m(\boldsymbol{\rho};\omega_0) \times \mathbf{H}_n^*(\boldsymbol{\rho};\omega_0)]\cdot\hat{\mathbf{z}}. \qquad \text{(VIII.10.13)}$$

If the area σ coincides with the whole fiber section, we obtain

$$P^{(\infty)}(z,t) = P\sum_{m}\langle|\Phi_m(z,t)|^2\rangle_t \equiv \sum_{m} P_m(z,t), \qquad \text{(VIII.10.14)}$$

the nondiagonal terms of interference among the various modes vanishing because of the orthogonality relation [see Eq. (VIII.10.5)]. The surviving diagonal terms $P_m(z,t)$ can be interpreted in a natural way as the total power carried by the mth mode through the whole fiber section at a given z.

11 Attenuation

The attenuation of the radiation field inside an optical fiber is due both to absorption by the fiber material (including scattering processes due to density fluctuations at microscopic and atomic levels) and to the very process of propagation inside the waveguide. The first attenuation mechanism concerns the material itself and can be investigated in a generic sample, while the second one depends specifically on the geometric form of the waveguide.The losses due to glass absorption can be subdivided into three parts: intrinsic material absorption, absorption by impurities unavoidably present in the material, and absorption due to atomic defects. These losses can be described phenomenologically in terms of a *loss factor* α, characteristic of the material under consideration, which expresses the relative decrement per unit length of the total energy carried by the electromagnetic field. One has of course to introduce two loss factors α_1 and α_2 related, respectively, to the core and the cladding materials. The relative decrement per unit length of the total energy $I_m(z)$ carried by the mth mode,

$$I_m(z) = \int_{-\infty}^{+\infty} P_m(z,t)\,dt, \qquad \text{(VIII.11.1)}$$

is then given by

$$\alpha_m = \alpha_1 f_m^{(1)} + \alpha_2 f_m^{(2)}, \tag{VIII.11.2}$$

where $f_m^{(1)}$ and $f_m^{(2)}$ represent, respectively, the fractions of the energy of the mth mode traveling in the core and in the cladding (evaluated for the ideal fiber).

Another source of attenuation is light scattering due to density fluctuations at the atomic level. If atoms and molecules were placed according to a perfectly homogeneous structure, the fields scattered by the single atoms would interfere destructively, and there would be no scattering. This is prevented by the occurrence of local time-dependent inhomogeneities due to thermal fluctuations. For a fiber, these inhomogeneities are static and are those present at the transition temperature T, which remain "frozen" in the glass when it solidifies. The whole process gives rise to an attenuation coefficient $\alpha^{(s)}$ due to this kind of scattering (*Rayleigh scattering*), which reads (see also Section VI.13.4)

$$\alpha^{(s)} = [(8\pi^3)/(3\lambda_0^4)](n^2 - 1)K_B T\beta_c, \tag{VIII.11.3}$$

where λ_0 is the radiation wavelength *in vacuo*, n the material refractive index, K_B Boltzmann's constant, and β_c the isothermal compressibility of the medium.

In Fig. VIII.15 the attenuation is shown as a function of λ_0. The loss mechanism associated with Rayleigh scattering represents a lower bound for attenuation, which decreases with $1/\lambda_0^4$.

Various kinds of losses associated with the geometric structure are present in the fiber; examples are those already mentioned concerning the leaky modes, which have a loss factor $\alpha = -2\beta_2$, and the finite fiber curvature.

The finite diameter of the cladding induces further losses associated with the fact that a fraction of the electromagnetic energy travels in the high-loss

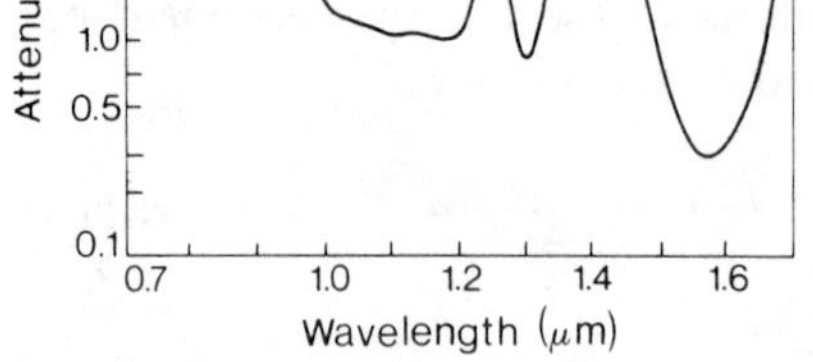

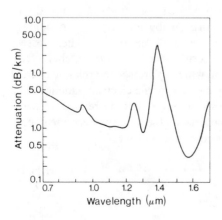

Fig. VIII.15. Spectral attenuation of a typical fiber.

jacket surrounding the fiber. Other losses are induced by nonlinear optical effects (*stimulated Raman* and *Brillouin*), which are usually negligible at the low powers employed for optical transmissions (see Section VIII.18).

Besides these loss mechanisms, which are essentially of a deterministic nature, there are others associated with the fact that each fiber unavoidably exhibits random deformations of its core and microbendings of its axis. These irregularities give rise to coupling among the various modes with mutual exchange of electromagnetic power. In this frame, the progressive power flow from guided to leaky modes and from leaky to refracted ones is a source of attenuation for the energy propagating in the fiber.

12 Modal Dispersion

By recalling Eqs. (VIII.10.10) and (VIII.10.11) and taking advantage of Eq. (VIII.10.6) we can write approximately

$$\begin{aligned}\hat{\mathbf{E}}(\boldsymbol{\rho}, z, t) = {} & \sum_m \mathbf{E}_m(\boldsymbol{\rho}) \exp[-i\beta_m(\omega_0)z + i\omega_0 t] \int_0^\infty c_m(\omega, z) \\ & \times \exp[-i\beta'_m(\omega_0)(\omega - \omega_0)z] \\ & \times \exp\left[-\frac{i}{2}\beta''_m(\omega_0)(\omega - \omega_0)^2 z + i(\omega - \omega_0)t\right] d\omega, \qquad \text{(VIII.12.1)}\end{aligned}$$

where $\beta'_m(\omega_0) = (d\beta_m/d\omega)_{\omega=\omega_0}$ and $\beta''_m(\omega_0) = (d^2\beta_m/d\omega^2)_{\omega=\omega_0}$.

Let us first investigate the case in which $\delta\omega$ is so small that the term in $(\omega - \omega_0)^2$ appearing in Eq. (VIII.12.1) can be neglected. By introducing the *group velocity* v_m of the mth mode,

$$v_m = 1/\beta'_m(\omega_0), \qquad \text{(VIII.12.2)}$$

we easily obtain for an ideal fiber

$$\hat{\mathbf{E}}(\boldsymbol{\rho}, z, t) = \sum_m \mathbf{E}_m(\boldsymbol{\rho}) \exp[-i\beta_m(\omega_0)z + i\omega_0 t]\Phi_m\left(0, t - \frac{z}{v_m}\right), \qquad \text{(VIII.12.3)}$$

which shows that the amplitudes of the various modes propagate with different velocities v_m. In particular, the power per mode $P_m(z, t)$ obeys the relation

$$P_m(z, t) = P_m(0, t - z/v_m), \qquad \text{(VIII.12.4)}$$

so that a signal formed by the superposition of various modes undergoes a distortion caused by their different velocities of propagation (*modal dispersion*). By defining the *modal delay* τ_m of the mth mode through the relation

$$\tau_m = L/v_m \qquad \text{(VIII.12.5)}$$

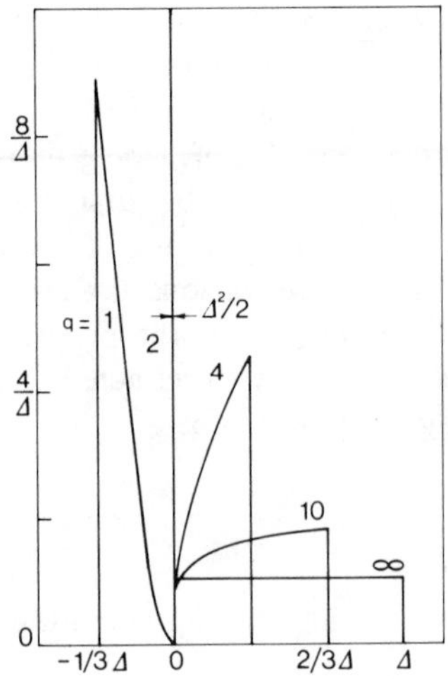

Fig. VIII.16. Impulse response function for different kinds of profiles versus normalized delay $(t - Ln_0/c)/(Ln_0/c)$.

(L being the fiber length), a quantitative determination of modal dispersion can be associated with the maximum possible difference between modal delays

$$T^{(\mathrm{mod})} = \max|\tau_m - \tau_n|. \tag{VIII.12.6}$$

For practical purposes, it is often convenient to consider the shape of a pulse after a given fiber length, provided its input width is negligible and all the modes are uniformly excited (*impulse response function*). This pulse shape is given in Fig. VIII.16 for different kind of profiles. It can be shown that the width of the exit pulse attains its minimum value $\Delta^2/8$ for [14]

$$q = 2 - 2.4\,\Delta, \tag{VIII.12.7}$$

where q is the exponent characterizing the power-law profile [see Eq. (VIII.3.4)].

13 Chromatic Dispersion

In the previous section, the distortion effect associated with the finite bandwidth $\delta\omega$ of the singal, contained in the terms quadratic in $(\omega - \omega_0)$ in Eq. (VIII.12.1), has been neglected. This effect, which turns out to be proportional to $\delta\omega$, is in practice negligible in multimode fibers if the exciting source is sufficiently monochromatic, while it represents the only source of distortion in an ideal single-mode fiber. It can be investigated in a simple way by considering the frequency dependence of the group velocity v_m (for a given mode m) and the associated delay between the "slowest" and the "fastest" frequencies ω' and ω''. More precisely, we have

$$T_m^{(\mathrm{cr})} = L\left[\frac{1}{v_m(\omega')} - \frac{1}{v_m(\omega'')}\right] = L\left(\frac{d\beta_m}{d\omega_{\omega=\omega'}} - \frac{d\beta_m}{d\omega_{\omega=\omega''}}\right), \qquad \text{(VIII.13.1)}$$

which, under the usual assumption contained in Eq. (VIII.10.6), can be written as

$$T_m^{(\mathrm{cr})} = L|d^2\beta_m/d\omega^2_{\omega=\omega_0}|\delta\omega, \qquad \text{(VIII.13.2)}$$

where, under standard conditions, $\delta\omega$ is the carrier bandwidth (usually larger than the bandwidth associated with signal modulation).

When performing the derivatives appearing in Eq. (VIII.13.2), various contributions arise due to the fact that the propagation constant β_m, besides depending on ω because of the dispersive nature of the fiber material (*material dispersion*), exhibits a frequency dependence that is due to the very presence of the waveguide structure and that would also be present for a fiber made up of an ideally nondispersive material (*waveguide dispersion*) and one associated with the variation of the refractive index profile with ω itself (*profile dispersion*). In general, these three effects are connected in a complicated way and it is not possible to separate their relative contributions.

Concerning the evaluation of $T_m^{(\mathrm{cr})}$, it is reasonable to assume that for the modes of a multimode fiber that are far from cutoff,

$$\beta_m(\omega) = n_1(\omega)\omega/c, \qquad \text{(VIII.13.3)}$$

where $n_1(\omega)$ is the refractive index of the core material, since they tend to be confined inside the core, which is regarded as infinite in extension. In terms of the wavelength λ_0 (*in vacuo*), it is not difficult to show that Eq. (VIII.13.2) yields in this case

$$T^{(\mathrm{cr})} = (L\lambda_0/c)|d^2n_1(\lambda_0)/d\lambda_0^2|\,\delta\lambda_0, \qquad \text{(VIII.13.4)}$$

where $\delta\lambda_0 = \lambda_0\delta\omega/\omega_0$, implying that chromatic dispersion reduces to material dispersion. According to Eq. (VIII.13.4), material dispersion becomes negligible in proximity to the wavelength $\bar{\lambda}_0$, such that

$$[d^2n_1(\lambda_0)/d\lambda_0^2]_{\lambda_0=\bar{\lambda}_0} = 0. \qquad \text{(VIII.13.5)}$$

In general, the most significant quantity is not $T^{(\mathrm{cr})}$ but $T^{(\mathrm{cr})}/L\,\delta\lambda_0$, that is, the delay per unit fiber length and unit $\delta\lambda_0$, which is the quantity plotted as a function of λ_0 in Fig. VIII.17.

For a single-mode fiber—or for a multimode fiber, when considering modes near cutoff or when working in a wavelength region such that $\lambda_0 \simeq \bar{\lambda}_0$ (where waveguide and profile dispersion can no longer be neglected)—Eq. (VIII.13.4) becomes questionable and it is necessary to resort to more sophisticated approaches on the lines indicated in Gloge [15] and fully

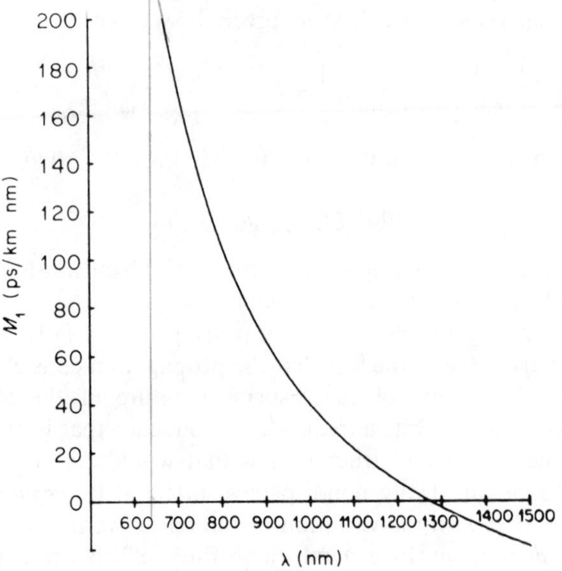

Fig. VIII.17. Material dispersion versus wavelength for silica. The quantity M_1 corresponds to $T^{(cr)}$ defined by Eq. (VIII.13.4).

pursued in Gambling *et al.* [16] for a single-mode fiber. More precisely, after recalling Eq. (VIII.9.18) and inserting it into Eq. (VIII.13.2), it is possible to obtain, after some algebra,

$$T^{(cr)} = L|T_{cmd} + T_{wd} + T_{cpd}|\,\delta\lambda_0, \tag{VIII.13.6}$$

where

$$T_{cmd} = (\lambda_0/c)\{A(V)\,d^2n_1/d\lambda_0^2 + [1 - A(V)]\,d^2n_2/d\lambda_0^2\}, \tag{VIII.13.7}$$

with

$$A(V) = \tfrac{1}{2}[d(bV)/dV + b], \tag{VIII.13.8}$$

$$T_{wd} = (n_2\,\Delta/c\lambda_0)BV\,d^2(bV)/dV^2, \tag{VIII.13.9}$$

with

$$B = [1 - (\lambda_0/n_2)(dn_2/d\lambda_0)]^2, \tag{VIII.13.10}$$

$$T_{cpd} = -(n_2/c)CD(V)\,d\tilde{\Delta}/d\lambda_0, \qquad \tilde{\Delta} = \Delta(n_0/n_2)^2, \tag{VIII.13.11}$$

with

$$C = 1 - (\lambda_0/n_2)\,dn_2/d\lambda_0 - (\lambda_0/4\tilde{\Delta})\,d\tilde{\Delta}/d\lambda_0, \tag{VIII.13.12}$$

and

$$D(V) = V\,d^2(bV)/dV^2 + d(bV)/dV - b \tag{VIII.13.13}$$

[for the definition of b, see Eq. (VIII.9.18)]. Here $T_{\rm cmd}$, $T_{\rm wd}$, and $T_{\rm cpd}$ are, respectively, the *composite material, waveguide*, and *profile dispersion.*

The regrouping of the various contributions in the form adopted above is dictated by the fact that each of them goes to zero whenever the corresponding dispersion parameter is assumed to be zero. Considering, for example, composite material dispersion, it is easy to see that for modes sufficiently far from cutoff, where $b \simeq 1$ and $V\,db/dV \simeq 0$, it is equivalent to the expression provided by Eq. (VIII.13.4), while near cutoff the dispersive properties of the cladding material become significant. For a step-index fiber, at $V = 2.402$, it is possible to derive the expression

$$T_{\rm cmd} = \frac{\lambda_0}{c}\,|0.83\,d^2n_1/d\lambda_0^2 + 0.17\,d^2n_2/d\lambda_0^2|. \qquad \text{(VIII.13.14)}$$

We note that, at a given wavelength, $T_{\rm cmd}$, $T_{\rm wd}$, and $T_{\rm cpd}$ do not always have the same sign, and that the value at which $T^{(\rm cr)}$ becomes negligible is in general different from $\bar{\lambda}_0$. The reader can get an idea of the relative role of the three dispersion times by looking at Fig. VIII.18.

14 Modal Noise

Let us return to the expression provided by Eq. (VIII.10.12) for the power $P^\sigma(z,t)$ crossing a given section of area $\sigma(\boldsymbol{\rho})$ centered on $\boldsymbol{\rho}$. Before proceeding, we observe that in practical situations the averaging operation indicated by an overbar corresponds to that performed by the detector employed over a time interval $T_{\rm D}$, which is usually long compared with the characteristic fluctuation time of the exciting source (*coherence time* $T_{\rm c}$). Under this hypothesis, we can replace the overbar by the symbol $\langle\ \rangle_{\rm av}$, which stands for *ensemble average* over the source fluctuations. Thus, the power P_m carried by the mth mode reads

$$P_m(z,t) = P\langle|\Phi_m(z,t)|^2\rangle_{\rm av}, \qquad \text{(VIII.14.1)}$$

while the nondiagonal terms in Eq. (VIII.10.12) contain the cross-correlation products

$$T_{nm}(z,t) = \langle\Phi_m(z,t)\Phi_n^*(z,t)\rangle_{\rm av}. \qquad \text{(VIII.14.2)}$$

In order to give an explicit expression for the T_{nm}'s, let us neglect mode coupling and chromatic dispersion so that it is possible to write (see Section 12) $\phi_m(z,t) = \phi_m(0, t - z/v_m)$ and, accordingly,

$$\begin{aligned} T_{nm}(z,t) &= \langle\Phi_m(0,t - z/v_m)\Phi_n^*(0,t - z/v_n)\rangle_{\rm av} \\ &= S(t - z/v_m)S^*(t - z/v_n)\langle F(t - z/v_n)F^*(t - z/v_n)\rangle_{\rm av} \\ &\equiv S(t - z/v_m)S^*(t - z/v_n)G(|\tau_{nm}|), \end{aligned} \qquad \text{(VIII.14.3)}$$

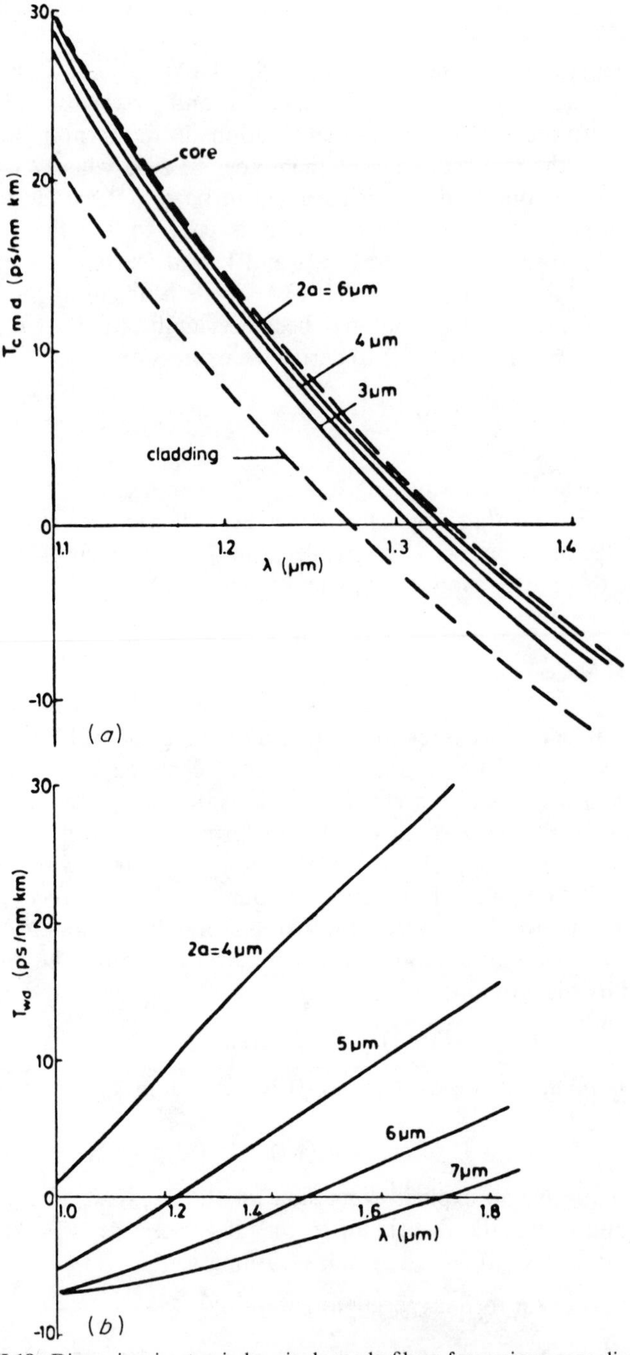

Fig. VIII.18. Dispersion in step-index single-mode fibers for various core diameters. The quantities (a) Tcmd, (b) Twd, and (c) Tcpd are defined by Eqs. (VIII.13.7), (VIII.13.9), and (VIII.13.11), respectively, while (d) shows the total dispersion T. (From Gambling, Matsumura, and Ragdale [16].)

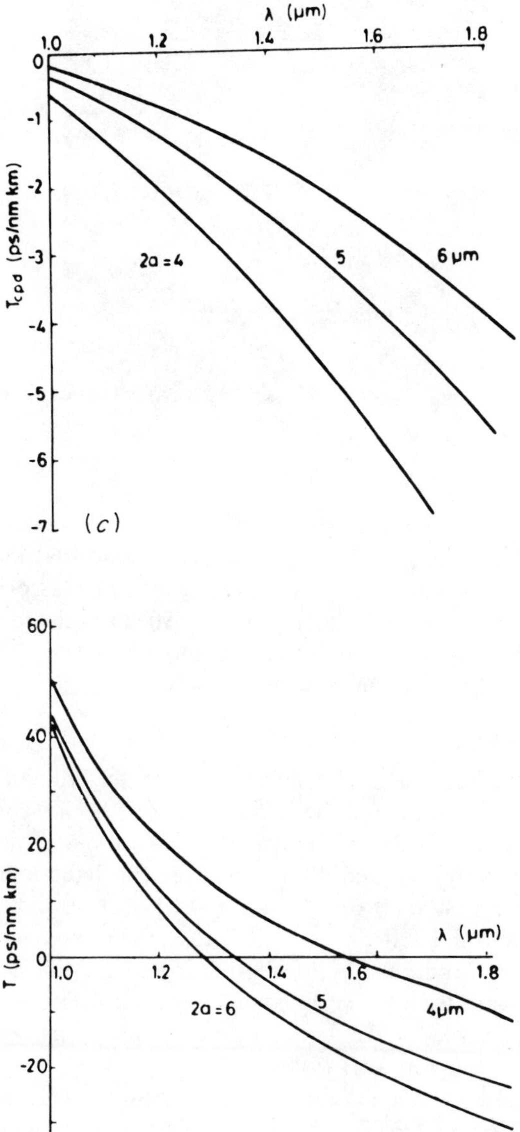

Fig. VIII.18. (*Continued*)

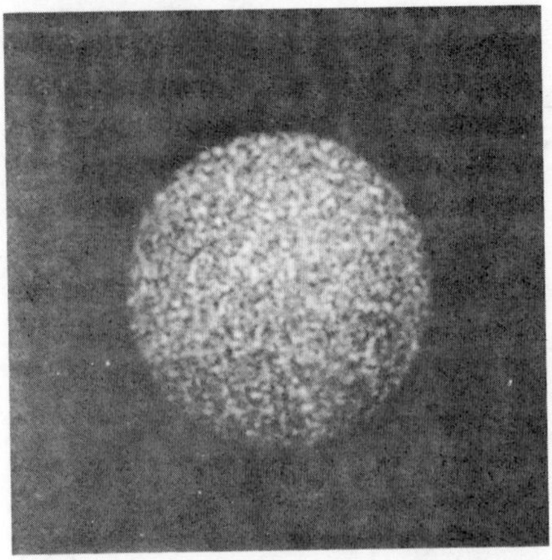

Fig. VIII.19. Typical speckle pattern on fiber exit face.

where we have assumed, for simplicity, that the excitation on the input face of the fiber is the same for all the guided modes and that it is possible to factorize it as the product of a fast fluctuating part $F(t)$ (associated with the source fluctuations) and a relatively slow one $S(t)$ (associated with environmental fluctuations and signal deterministic modulation).

The quantity $G(|\tau_{nm}|)$ becomes negligible whenever

$$|\tau_{nm}| = L|1/v_n - 1/v_m| \gg T_c, \qquad \text{(VIII.14.4)}$$

that is, when the mutual modal delay between the nth and mth modes exceeds the coherence time of the exciting source. If this condition is fulfilled for every possible pair of modes, then all the nondiagonal terms in Eq. (VIII.10.12) disappear and P^σ [as well as the light intensity $I(\rho, z, t)$ obtained by taking the limit of P^σ/σ for $\sigma \to 0$] is expressed as a sum over the diagonal terms alone. If this is not the case, then nondiagonal terms are present whose contributions, by locally adding to or subtracting from the background constituted by the diagonal terms according to their relative phases, are responsible for the appearance of a *speckle pattern* on the fiber transverse section (see Fig. VIII.19) [17].

In general, these speckle patterns undergo random dynamic changes due to the fact that $S(t)$ may exhibit, besides the deterministic modulation associated with the information-carrying signal, a stochastic time dependence connected with fiber vibrations or small changes in the emission wavelength of the exciting source. This continuous change in the speckle pattern may result in loss fluctuations, that is, in a stochastic amplitude modulation of the transmitted signal (*modal noise*) [18], if a speckle-dependent loss mech-

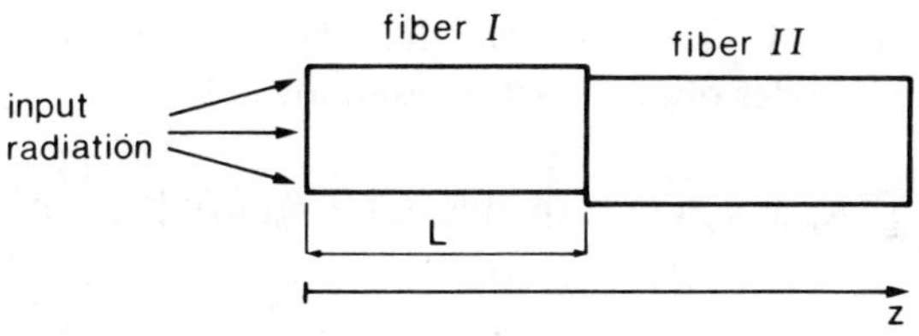

Fig. VIII.20. Example of speckle-dependent loss mechanism.

anism is present at some point of the fiber link, such as a misaligned fiber-to-fiber joint (see Fig. VIII.20).

From what we have stated above, it is apparent that modal noise is a problem associated with coherent sources. In fact, if a fiber is excited with a particularly incoherent source (like a light-emitting diode), Eq. (VIII.14.4) is verified, due to the smallness of T_c, for any pair of modes and for any reasonable fiber length L, so that no speckle pattern is present.

15 Coupled-Mode Theory

A coupling mechanism among the various modes [giving rise to the z dependence of the c_m's in Eq. (VIII.10.8)] is introduced by the deviations from the ideal cylindrical geometry of the fiber. This mechanism is responsible for an energy exchange among the propagation modes, so that the power P_m carried by each mode turns out to be a mixture of the initial powers of the various modes and to travel with a velocity that represents a kind of weighted average of the velocities. As we would expect, this effect is beneficial for modal dispersion, which tends to be reduced by coupling. There is, however, a negative effect concerning losses, due to progressive coupling among guided and radiation modes, which results in a net power flux out of the fiber.

In the large majority of cases, the coupling is of a random nature because fiber imperfections are unavoidably associated with the manufacturing process and the assembly of the waveguide structure. For this reason, and in view of the fact that the problem of propagation in the presence of a deterministic coupling, corresponding to a detailed knowledge of the perturbation, is analytically tractable in only a small number of cases, a statistical model is often necessary.

Before introducing a particular statistical model, we must determine the set of equations describing the evolution of the $c_m(\omega, z)$'s appearing in Eq. (VIII.10.8) in the presence of a deterministic coupling. If we consider weakly guiding fibers and neglect coupling between modes traveling in opposite directions, it can be shown that the amplitudes of the forward-traveling guided modes obey the set of equations [1]

$$\frac{dc_m(\omega, z)}{dz} = \sum_{n=1}^{N} K_{mn}(z) c_n(\omega, z) \exp\{i[\beta_m(\omega) - \beta_n(\omega)]z\} \qquad (m = 1, 2 \ldots, N), \tag{VIII.15.1}$$

N being the total number of guided modes, where

$$K_{mn}(z) = \frac{\omega\varepsilon_0}{4iP} \int_{-\infty}^{\infty} dx \int_{-\infty}^{\infty} dy [n^2(\boldsymbol{\rho}, z) - n_0^2(\boldsymbol{\rho})] \mathbf{E}_m(\boldsymbol{\rho}) \cdot \mathbf{E}_n^*(\boldsymbol{\rho}). \tag{VIII.15.2}$$

In Eq. (VIII.15.2), $n(\boldsymbol{\rho}, z)$ represents the refractive index of the real fiber, while $n_0(\boldsymbol{\rho})$, $\mathbf{E}_m$, and $\mathbf{E}_n$, respectively, represent the refractive index and the transverse components of the mth and nth modes of the ideal fiber. In writing Eqs. (VIII.15.1) the coupling to the continuum of the radiation modes has been neglected, so that in this approximation the total energy contained in the guided modes is conserved. Furthermore, it has been assumed that, for every z, the real fiber differs slightly from the ideal one (see Fig. VIII.21) so that the ideal modes furnish a suitable set for expansion of the electromagnetic field. In the more general case—for example, where deviations of the fiber axis from straightness are present (*microbendings*; see Fig. VIII.22)—this is no longer true and it is convenient to introduce a set of "local" modes relative to the ideal fiber that, for every z, most resembles the real one. It is possible to show that this amounts to considering the spatial configurations $\mathbf{E}_m(\boldsymbol{\rho}, z)$ and $\mathbf{H}_m(\boldsymbol{\rho}, z)$ and the relative propagation

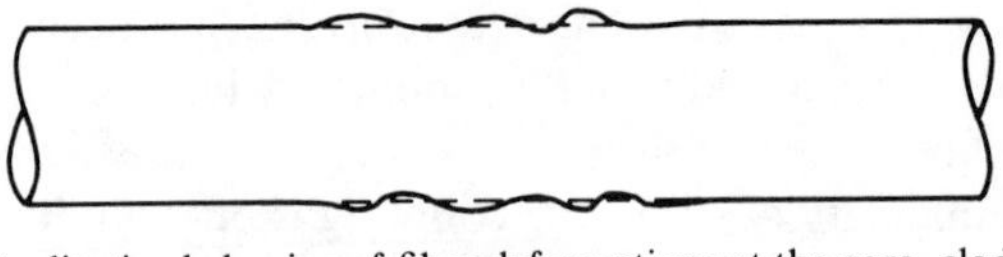

Fig. VIII.21. Qualitative behavior of fiber deformations at the core–cladding interface.

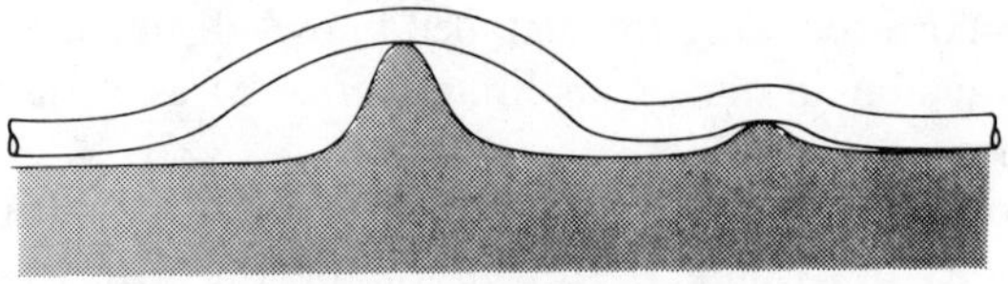

Fig. VIII.22. Microbending imperfections.

constant as z-dependent quantities and substituting for Eqs. (VIII.15.1) and (VIII.15.2)

$$\frac{dc_m(\omega,z)}{dz} = \sum_{n\neq m} R_{mn}(z) \times \exp\left\{i\int_0^z [\beta_m(z')-\beta_n(z')]\,dz'\right\} c_n(\omega,z) \qquad (m=1,2\ldots,N),$$

(VIII.15.3)

and

$$R_{mn}(z) = \frac{\omega\varepsilon_0}{4P(\beta_m-\beta_n)}\int_{-\infty}^{+\infty} dx \int_{-\infty}^{+\infty} dy \frac{\partial n^2(\boldsymbol{\rho},z)}{\partial z}\mathbf{E}_m(\boldsymbol{\rho},z)\cdot\mathbf{E}_n(\boldsymbol{\rho},z),$$

(VIII.15.4)

where $\boldsymbol{\rho}$ and z are relative to the coordinate system of the ideal fiber.

The perturbations associated with Fig. VIII.21 can be represented by writing the distance $D(\phi,z)$ between the core–cladding interface and the fiber axis in the form.

$$D(\phi,z) = a + \sum_l f_l\cos(l\phi + \psi_l), \qquad \text{(VIII.15.5)}$$

while microbendings are described in terms of the quantity $n^2(\boldsymbol{\rho},z)$ obtained from $n_0^2(\rho)$ by substituting for ρ

$$\rho' = \{[x - f(z)]^2 + [y - g(z)]^2\}^{1/2}, \qquad \text{(VIII.15.6)}$$

$f(z)$ and $g(z)$, respectively, specifying the distance of the axes x and y from the axis of the real fiber.

By taking advantage of the relation

$$\partial n^2(\boldsymbol{\rho},z)/\partial z = (\partial n_0^2(\rho')/\partial\rho')(\partial\rho'/\partial z) \qquad \text{(VIII.15.7)}$$

and of Eq. (VIII.15.6), we find, to the lowest order in the perturbation, that microbendings can only couple modes for which the *selection rule*

$$\nu' - \nu = \pm 1 \qquad \text{(VIII.15.8)}$$

is satisfied, ν being the azimuthal number. In the situation described by Eq. (VIII.15.5), we have the selection rules

$$\nu' - \nu = \pm l, \qquad \text{(VIII.15.9)}$$

according to which the lth term in Eq. (VIII.15.5) couples modes whose azimuthal numbers ν and ν' satisfy Eq. (VIII.15.9).

The systems of Eqs. (VIII.15.1) and (VIII.15.3) are analytically solvable only in some limiting cases in which the behavior of the imperfections is very simple and few modes are coupled. The problem of propagation in the presence of coupling becomes analytically tractable for a larger class of situations after the introduction of a statistical model, which is only able to describe "averaged" quantities. More precisely, we introduce an ensemble of macroscopically identical fibers differing among themselves for the random behavior of microscopic imperfections, and evaluate the significant physical quantities averaged over this ensemble, an operation indicated by the symbol $\langle\,\rangle$. If we perform the averaging operation on the system of Eqs. (VIII.15.1) or (VIII.15.3), we derive the following system of coupled equations for the average modal powers $\langle P_m \rangle = P\langle\langle\Phi_m(z,t)\rangle_{av}\rangle$ [see Eq. (VIII.14.1)] [1, 19]:

$$\frac{\partial\langle P_m(z,t)\rangle}{\partial z} + \frac{1}{v_m}\frac{\partial}{\partial t}\langle P_m(z,t)\rangle = -\alpha_m\langle P_m(z,t)\rangle + \sum_{n=1}^{N} h_{mn}[\langle P_n(z,t)\rangle - \langle P_m(z,t)\rangle] \qquad (m = 1, 2\ldots, N), \tag{VIII.15.10}$$

where α_m represents the total loss factor of the mth mode, introduced phenomenologically, and

$$h_{mn} = \int_{-\infty}^{+\infty} \langle K_{mn}(z)K_{mn}^*(0)\rangle \exp\{i[\beta_n(\omega_0) - \beta_m(\omega_0)]z\}\,dz, \tag{VIII.15.11}$$

an analogous expression being valid with the R_{mn}'s instead of the K_{mn}'s.

The *correlation length* L_c of the stochastic variable $K_{mn}(z)$ [or $R_{mn}(z)$], that is, the length of the interval over which the correlation function $\langle K_{mn}(z)K_{mn}^*(0)\rangle$ is nonvanishing, corresponds to the average spatial period of the irregularities in the single fiber. In deriving Eqs. (VIII.15.10), it has been assumed that the $c_m(z)$'s do not appreciably vary over the distance L_c (*weak-coupling hypothesis*).

The presence of modal coupling is responsible for a loss mechanism that can be interpreted in terms of coupling between guided and radiation modes. These losses are added to those associated with material absorption and light scattering, which are represented by the attenuation coefficients α_m in Eqs. (VIII.15.10). They can be investigated by solving the system of Eqs. (VIII.15.10) in a stationary case and looking for an asymptotic solution in the form

$$P_m(z) \propto \exp[-(\alpha + \gamma_1)z], \tag{VIII.15.12}$$

where γ_1 represents the loss due to coupling, under suitable hypotheses for the h_{mn}'s.

As already mentioned, coupling influences modal dispersion, in a way analytically described by Eqs. (VIII.15.10). Although the problem is not solvable in general, a simple description can be obtained by investigating the asymptotic behavior of a pulse of initially negligible width in terms of its temporal width $\sigma(z)$. If $\langle P(z,t)\rangle$ is the total power crossing a given fiber section, we define

$$\bar{t} = \frac{\int_0^\infty t\langle P(z,t)\rangle\,dt}{\int_0^\infty \langle P(z,t)\rangle\,dt} \tag{VIII.15.13}$$

and

$$\overline{t^2} = \frac{\int_0^\infty t^2\langle P(z,t)\rangle\,dt}{\int_0^\infty \langle P(z,t)\rangle\,dt}, \tag{VIII.15.14}$$

so that we can set

$$\sigma(z) = (\overline{t^2} - \bar{t}^2)^{1/2}. \tag{VIII.15.15}$$

For large z it is possible to derive

$$\sigma(z) \propto z^{1/2}, \tag{VIII.15.16}$$

while for small z

$$\sigma(z) \propto z. \tag{VIII.15.17}$$

The two limiting cases corresponding to Eqs. (VIII.15.17) and (VIII.15.16) describe, respectively, the situation over distances so small that coupling can be neglected, so that modal dispersion has the usual linear dependence on the fiber length, and that over long distances, where coupling becomes important. In the latter case, Eq. (VIII.15.16) ensures a dependence of dispersion on the distance traveled weaker than the one related to propagation without coupling (Fig. VIII.23). This square root of z dependence can be

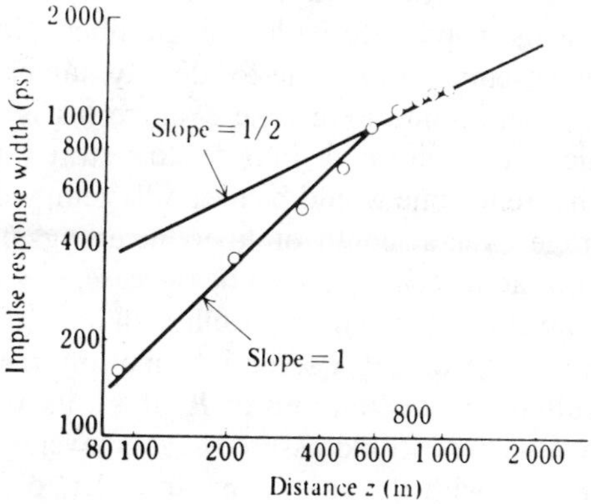

Fig. VIII.23. Pulse temporal width versus length traveled in the presence of mode coupling. (From Okoshi [6].)

heuristically explained by resorting to a simple model in which only two modes are present [20], coupled in such a way that the probability that a photon traveling in a given mode has to jump in the other one over a distance dz is $h\,dz$ (with h independent of z). The evolution of a pulse can then be described by a random-walk mechanism according to which the photon travels, on each interval of length $1/h$, with velocity v_1 or v_2, thus acquiring, with respect to the average transit time,

$$\bar{t} = (1/h)[2/(v_1 + v_2)], \tag{VIII.15.18}$$

a delay given by

$$|t - \bar{t}| = (1/h)|1/v_1 - 2/(v_1 + v_2)| \cong [1/(2h)]|v_2 - v_1|/v_1^2. \tag{VIII.15.19}$$

According to the general theory of diffusive processes, the mean square value of the total delay is given by the square of the delay over the single step of length $1/h$ times the number zh of steps in the distance z, that is,

$$(\overline{t^2} - \bar{t}^2)^{1/2} = \overline{(t - \bar{t})^2}^{1/2} \cong [|v_2 - v_1|/(2v_1^2)](z/h)^{1/2}, \tag{VIII.15.20}$$

in agreement with Eq. (VIII.15.16).

16 Statistical Theory of Propagation in an Ensemble of Fibers

The statistical approach to propagation inside an optical fiber in the presence of mode coupling, developed in the last section, concerns the evaluation of the average power per mode $\langle P_m \rangle$. The averaging procedure over the ensemble of macroscopically similar fibers is rather artificial, however, since one always deals in practice with a single fiber. Nor does the static nature of the fiber ensemble allow one to identify, under some hypothesis of ergodicity, $\langle P_m \rangle$ with some long time average (as is often the case in statistical mechanics). Also, although, in order to restore the validity of some kind of "ergodic theorem," one would be naturally tempted to identify $\langle P_m \rangle$ with a space average over a length of fiber including many perturbation periods, this does not actually turn out to be the case.

Under these conditions, the only possibility that is left, in order to extract from the knowledge of $\langle P_m \rangle$ some information relevant to practical situations, is to evaluate the fluctuations of P_m about its average value, since only their relative smallness would assure that the average value of P_m can be assumed to coincide with the actual one. In order to do this, it is necessary to write down a set of coupled differential equations for $\langle P_m^2 \rangle$ and

$\langle P_m P_n \rangle$, which, for a lossless fiber and monochromatic excitation, reads [19]

$$\frac{d}{dz}\langle P_m^2 \rangle = 2 \sum_{k \neq m} h_{mk}(\langle P_m^2 \rangle - 2\langle P_m P_k \rangle) \qquad (m = 1, 2 \ldots, N), \qquad \text{(VIII.16.1)}$$

$$\frac{d}{dz}\langle P_m P_n \rangle_{m \neq n} = 2h_{mn}\langle P_m P_n \rangle + \sum_k h_{nk}(\langle P_m P_n \rangle - \langle P_m P_k \rangle)$$
$$+ \sum_k h_{mk}(\langle P_m P_n \rangle - \langle P_n P_k \rangle) \qquad (m, n = 1, 2 \ldots, N). \qquad \text{(VIII.16.2)}$$

We now observe that Eqs. (VIII.15.10) furnish, in the stationary case ($\partial/\partial t = 0$) and for negligible losses ($\alpha_m = 0$), *asymptotic power equipartition*, that is, $\langle P_m(z \to \infty) \rangle$ independent of m. In the monochromatic case, this property allows us to obtain from the set of Eqs. (VIII.16.1) and (VIII.16.2) the asymptotic relation [19]

$$\langle P_m^2 \rangle = \frac{2N}{N+1}\langle P_m \rangle^2 \underset{N \gg 1}{\cong} 2\langle P_m \rangle^2 \qquad (z \to \infty), \qquad \text{(VIII.16.3)}$$

a result that, rather disappointingly, implies 100% uncertainty about the actual value of P_m, once its average value is known.

This situation changes in the case of a signal having finite bandwidth $\delta\omega$, where it can be shown [21] that the nomalized variance

$$\sigma_m^2(z) \equiv (\langle P_m^2 \rangle - \langle P_m \rangle^2)/\langle P_m \rangle^2 \qquad \text{(VIII.16.4)}$$

tends to vanish for $z \gg z_c^{(m,n)}$ $(n, m = 1, 2, \ldots, N)$, where $z_c^{(m,n)} = T_c/|1/v_n - 1/v_m|$ can be interpreted as a typical distance after which the mth and nth modes are practically decorrelated [see Eq. (VIII.14.4)]. Therefore, under this condition, no statistical uncertainty is present in the determination of P_m for a single fiber.

17 Polarization-Maintaining Optical Fibers

A circularly symmetric optical fiber working in a single-mode regime is actually a two-mode fiber, since it is able to propagate two orthogonally polarized eigenmodes [e.g., the $(\mathrm{LP}_x)_{01}$ and $(\mathrm{LP}_y)_{01}$ modes of a step-index fiber]. If the fiber possesses an ideal structure, these two polarization states are obviously degenerate; that is, their propagation constants β_x and β_y coincide (the principal axes $\hat{x}$ and $\hat{y}$ of the fiber being arbitrary). In practice, the values of β_x and β_y are very close, a circumstance that can induce a strong coupling mechanism between the two orthogonal polarization modes; this, in turn, causes a power exchange, with an associated polarization scrambling, taking place over very short distances traveled (from a few centimeters to some meters).

In order to understand this, one can rely on coupled-mode theory, which in particular describes the evolution of the two polarization states provided that in Eq. (VIII.15.2) the scalar product of the transverse parts of the eigenmodes is replaced by the scalar product of their longitudinal parts. In most practical cases, the coupling coefficient $h_{xy}(\chi)$ [$\chi = |\beta_x - \beta_y|$; see Eq. (VIII.15.11)], which represents the power spectrum of the random perturbation $K_{xy}(z)$ inducing the coupling, has a low-pass filter characteristic with a cutoff spatial frequency χ_c [22]. Thus, according to Eq. (VIII.15.11), the transfer of power between polarization states is small as long as

$$|\beta_x - \beta_y| \gg \chi_c. \tag{VIII.17.1}$$

while it becomes important in the opposite case and thus, *a fortiori*, when $\beta_x \simeq \beta_y$.

However, there are applications, like heterodyne-type optical communications and fiber sensors, that require the polarization state of the output signal to be kept constant. According to the above considerations, one method for achieving this is to make the difference between the propagation constants as large as possible. This can be done by breaking the circular symmetry of the fiber cross section, either by changing the geometric shape of the core from circular to elliptical or by a large transverse stress asymmetry in the core region [23].

It is customary to introduce a normalized quantity B called the *modal birefringence*, given by (let us assume, for definiteness, $\beta_x > \beta_y$)

$$B = 2(\beta_x - \beta_y)/(\beta_x + \beta_y) \cong (\beta_x - \beta_y)/n_0 k_0 \tag{VIII.17.2}$$

(typical values of B being in the ranges 10^{-5}–10^{-7} and 10^{-3}–10^{-4}, respectively, for commonly available and polarization-maintaining fibers), whose magnitude gives an estimate of the capacity of the fiber to hold the polarization once it is excited in a single-polarization state by light polarized along one of its principal axes. If the light is polarized at an angle θ with respect to the x axis, it will pass through various states of elliptical polarization as a function of the fiber length z, according to the values taken on by the phase retardation

$$\Phi(z) = (\beta_x - \beta_y)z \cong Bn_0 k_0 z, \tag{VIII.17.3}$$

the *beat length* L_b being defined by the relation

$$2\pi = Bn_0 k_0 L_b. \tag{VIII.17.4}$$

In terms of this quantity, the condition formulated by Eq. (VIII.17.1) can be expressed by saying that, in order to have a negligible energy transfer between the two polarization states, the beat length must be much shorter than the typical spatial period of geometric perturbations in the fiber.

It is apparent that removing the degeneracy in a single-mode fiber gives rise to a modal dispersion between the two polarization modes (*polarization dispersion*). According to Section VIII.12, the mutual modal delay between the two modes is given by

$$\tau_p = L(d/d\omega)(\beta_x - \beta_y), \tag{VIII.17.5}$$

or, after introducing the effective refractive index $n_x = c\beta_x/\omega$ ($n_y = c\beta_y/\omega$), by

$$\tau_p = L(n_x - n_y)/c + (L\omega/c)(dn_x/d\omega - dn_y/d\omega) \cong L(n_x - n_y)/c \cong LB/(c/n_0), \tag{VIII.17.6}$$

where we have taken advantage of the low dispersion ($dn/d\omega \ll n/\omega$) exhibited by ordinary silica glass in the near-infrared region relevant to fiber operation.

18 Nonlinear Effects in Optical Fibers

Many characteristics of extremely low-loss fibers, whose advent points toward the feasibility of long-distance wide band transmission systems, enhance the relevance of nonlinear phenomena in optical waveguide propagation. They are, besides the long interaction length provided by the fiber itself, the small core diameter pertinent to monomode operation and the use of narrow-linewidth single-frequency lasers. In particular, the product of the fiber length L and the intensity $P/(\pi a^2)$, associated with a core radius a and an input power P, can become large enough, compared with nonlinear propagation in an unbounded medium, to balance the intrinsically small nonlinearity of silica glass at relatively low power.

While nonlinear effects may be detrimental and place an ultimate limitation on the power level that can be used in fiber telecommunication systems, in some cases they can be employed to advantage, either for realizing special kinds of optical devices (e.g., *fiber Raman lasers* [24]), or for actually improving the fiber performances (e.g., solitons).

Let us recall (see Chapter II) that nonlinear optical processes are usually described in terms of the polarizability P, which can be formally expanded in a power series in the electric field as (see Section I.2.1)

$$P \sim \chi_1 E + \chi_2 E^2 + \chi_3 E^3 + \cdots. \tag{VIII.18.1}$$

The first nonlinear term $\chi_2 E^2$, associated, for example, with second-harmonic generation, is zero in glasses because of inversion symmetry, so that in practice all the nonlinear effects taking place inside a glass optical fiber are associated with the term $\chi_3 E^3$. They can be roughly divided into two classes, according whether the induced polarization vibrates with the same frequency as the incident field. The second class includes *stimulated Raman scattering* (SRS), *stimulated Brillouin scattering* (SBS), and *four-photon mixing*, while the first class contains the so-called *self-induced effects*, which can be formally described, as we will see in the following section, in terms of a nonlinear refractive index (*optical Kerr effect*).

In this section we will briefly summarize SRS and SBS because of the limits they impose on the peak power injected into the fiber while in the next section we will treat in greater detail self-induced effects for which a unified treatment is possible.

Both SRS and SBS can be classically viewed as a coupled three-wave interaction among an incident wave (*pump*), a *signal* wave (Stokes or Brillouin), and a vibrational or acoustical wave, respectively [25]. As a result, part of the energy initially contained in the pump is progressively transferred to the signal wave, in both the forward and backward directions for SRS, and only in the backward one for SBS. Also, in optical communication systems, if no signal field is deliberately injected into the fiber, a weak signal due to spontaneous emission is always present, and can be significantly amplified and subtract power from the information-carrying pump wave.

The two processes are usually characterized by the small-signal *gain coefficients* $G_R(\nu)$ and $G_B(\nu)$, which represent the exponential gain of the spectral intensity of the signal at frequency ν, provided the pump depletion is neglected, according to the relations

$$I_R(z,\nu) = I_R(0,\nu)\exp[G_R(\nu)z], \qquad \text{(VIII.18.2)}$$

$$I_B(z,\nu) = I_B(0,\nu)\exp[G_B(\nu)z]. \qquad \text{(VIII.18.3)}$$

Both coefficients are proportional to the pump intensity I_0. For the SRS we have

$$G_R(\nu) = g(\Delta\nu)I_0, \qquad \text{(VIII.18.4)}$$

where $\Delta\nu = \nu_0 - \nu_S$ is the difference between the pump frequency ν_0 and the Stokes line frequency ν_S, and $g(\Delta\nu)$, plotted in Fig. VIII.24, is given by

$$g(\Delta\nu) = \sigma_0(\Delta\nu)c^2/[h\nu_S^3 n^2(\nu_S)], \qquad \text{(VIII.18.5)}$$

where $\sigma_0(\Delta\nu)$ is the Stokes cross section per unit volume and per unit frequency, $n(\nu_S)$ indicating the refractive index at frequency ν_S.

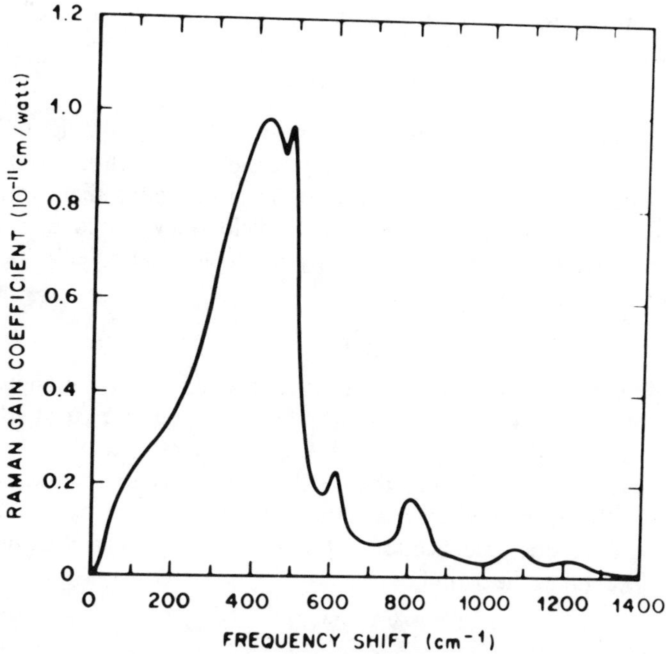

Fig. VIII.24. Raman gain versus frequency for fused silica at a pump wavelength of 1.0 μm. (From Stolen [24].)

Brillouin gain depends in a critical way on whether the pump linewidth $\Delta\nu_P$ is larger or smaller than the spontaneous Brillouin linewidth $\Delta\nu_B$ (typically around 15–40 MHz for fused silica at the wavelength relevant for fiber transmissions), the gain coefficient being reduced by the ratio $\Delta\nu_B/\Delta\nu_P$ when $\Delta\nu_P \gg \Delta\nu_B$ [26]. Whenever this last condition is fulfilled, SRS is the dominant nonlinear process (its threshold—that is, the maximum pump power that can be launched in the fiber before its effect becomes detectable—being in the region of 1 W or higher), while SBS turns out to be the leading one if a very narrowband laser source (≤ 1 MHz) is employed. In the latter case, the maximum input power [26] launchable in a multikilometer fiber is severely limited (to a few milliwatts) by SBS, which converts a significant portion of the forward-traveling optical power associated with the pump wave into a backward-traveling parasite signal wave.

19 Self-Induced Nonlinear Effects

The third-order polarizability $\mathbf{P}^{(3)}$ corresponding to the third term on the right of Eq. (VIII.18.1) can be written, by assuming the fiber material to be isotropic and its nonlinear response to be dominated by the fast-responding electronic processes, in the dispersionless form [27] (see also Chapter I, Owyoung [4]).

$$\mathbf{P}^{(3)} = \varepsilon_0 \chi^{(3)} \mathbf{E} \cdot \mathbf{E}\mathbf{E}, \tag{VIII.19.1}$$

where ε_0 is the electric permeability of the vacuum and $\chi^{(3)}$ the nonlinear susceptibility of the medium. In order to recognize the self-induced effects, it is now necessary to isolate in Eq. (VIII.19.1) the terms that vibrate approximately at the same frequency as the field (they would oscillate exactly at its frequency only for a perfectly monochromatic field), for which Eq. (VIII.19.1) can be expressed as

$$\mathbf{P}_\omega^{(3)} = \boldsymbol{\varepsilon}^{(3)} \cdot \mathbf{E}_\omega, \tag{VIII.19.2}$$

where $\mathbf{P}_\omega^{(3)}$ and $\mathbf{E}_\omega$ denote the time Fourier transforms at the angular frequency ω of $\mathbf{P}^{(3)}$ and $\mathbf{E}$ and $\boldsymbol{\varepsilon}^{(3)}$ is a suitable tensor [e.g., see Eq. (VIII.19.4)]. After introducing the analytic signal $\hat{\mathbf{E}}$ [28] [remember that $\mathbf{E} = (\hat{\mathbf{E}} + \hat{\mathbf{E}}^*)/2$; see Section I.8], it is possible to show that, inside a weakly guiding fiber where the transverse part of the field $\mathbf{E}_\mathrm{T}$ is large compared with the longitudinal one (see Section VIII.6), the transverse component of Eq. (VIII.19.2) takes the form [29]

$$\mathbf{P}_{\mathrm{T}\omega}^{(3)} = \boldsymbol{\varepsilon}_\mathrm{T}^{(3)} \cdot \mathbf{E}_{\mathrm{T}\omega}, \tag{VIII.19.3}$$

where

$$\boldsymbol{\varepsilon}_\mathrm{T}^{(3)} = \frac{\varepsilon_0 \chi^{(3)}}{2} \begin{bmatrix} \hat{E}_\mathrm{T} \cdot \hat{E}_\mathrm{T}^* + \frac{1}{2}|\hat{E}_x|^2 & \frac{1}{2}\hat{E}_y \hat{E}_x^* \\ \frac{1}{2}\hat{E}_y^* \hat{E}_x & \hat{E}_\mathrm{T} \cdot \hat{E}_\mathrm{T}^* + \frac{1}{2}|\hat{E}_y|^2 \end{bmatrix}. \tag{VIII.19.4}$$

For polarization-maintaining fibers (see Section VIII.17) we can, for example, set $E_y = 0$ (the nonlinear interaction is seen *a posteriori* not to couple orthogonal polarizations), and Eq. (VIII.19.3) reduces to

$$\mathbf{P}_{\mathrm{T}\omega}^{(3)} = \varepsilon_2 |\hat{E}_x|^2 \mathbf{E}_{\mathrm{T}\omega}, \tag{VIII.19.5}$$

where $\varepsilon_2 = \frac{3}{4}\varepsilon_0 \chi^{(3)}$. The same relation holds true for regular fibers, which scramble polarization over short distances and for which, on the average, $\hat{E}_y \hat{E}_x^* = 0$, provided that we set $\varepsilon_2 = \frac{5}{4}\varepsilon_0 \chi^{(3)}$. In both situations, we can formally introduce a nonlinear contribution to the refractive index through the usual relation between the Fourier components of the displacement vector $\mathbf{D}_{\mathrm{T}\omega}$ and of the electric field $\mathbf{E}_{\mathrm{T}\omega}$,

$$\mathbf{D}_\omega = \varepsilon_1(\omega) \mathbf{E}_\omega + \varepsilon^{(3)} \mathbf{E}_\omega, \tag{VIII.19.6}$$

which yields, according to Eq. (VIII.19.5),

$$\varepsilon = \varepsilon_1 + \varepsilon_2 |\hat{E}_x|^2, \tag{VIII.19.7}$$

so that we have, to a good approximation,

$$n(\omega) = n_1(\omega) + n_2 |\hat{E}_x|^2, \tag{VIII.19.8}$$

where $n = (\varepsilon/\varepsilon_0)^{1/2}$, $n_1 = (\varepsilon_1/\varepsilon_0)^{1/2}$ is the linear part of the refractive index, and $n_2 = \varepsilon_2/(2n_1\varepsilon_0)$. Equation (VIII.19.18) gives an example of a time-dependent (through $|\hat{E}_x|^2$) refractive index whose time scale is large compared with $1/\Delta\omega$, $\Delta\omega$ representing the typical scale of variation of $n_1(\omega)$ (see Section I.2 and Chapter I Ref. [3].) Note that Eq. (VIII.19.8) can also be rewritten in terms of the intensity I as $n = n_1 + \mathcal{N}_2 I$, where $\mathcal{N}_2 = 2(n_2/n_1)\zeta_0$. For silica, $n_1 \simeq 10^{-22}\,(\text{m}/V)^2$ and $\mathcal{N}_2 = 5 \times 10^{-16}\,\text{cm}^2/\text{Watt}$.

The so-called *optical Kerr effect* is associated with the presence of the nonlinear term, quadratic in the electric field, appearing on the right side of Eqs. (VIII.19.7) and (VIII.19.8). In order to investigate the influence of this contribution on the optical propagation inside the fiber, we can either deal directly with the solutions of Maxwell's equations in the presence of a medium characterized by the dielectric constant given in Eq. (VIII.19.7) [30] or, as we will do in the following, rely on the coupled-mode theory outlined in Section VIII.15 (see also Section III.17), by identifying the nonlinear part of the refractive index with a fiber perturbation.

Before proceeding, we must point out that, in order to extend the formalism to situations in which backward-traveling waves play a fundamental role, the field inside the fiber must be written as

$$\begin{aligned}\hat{E}_x(\boldsymbol{\rho}, z, t) &= \hat{E}_x^+(\boldsymbol{\rho}, z, t) + \hat{E}_x^-(\boldsymbol{\rho}, z, t)\\ &= \sum_m \tilde{E}_m(\boldsymbol{\rho})\{\exp[-i\beta_m(\omega_0)z + i\omega_0 t]\Phi_m^+(z,t)\\ &\quad + \exp[i\beta_m(\omega_0)z + i\omega_0 t]\Phi_m^-(z,t)\},\end{aligned} \qquad \text{(VIII.19.9)}$$

with

$$\Phi_m^\pm = \int_0^\infty \exp\{i(\omega - \omega_0)t \mp i[\beta_m(\omega) - \beta_m(\omega_0)]z\} c_m^\pm(\omega, z)\, d\omega, \qquad \text{(VIII.19.10)}$$

where $\tilde{E}_m(\boldsymbol{\rho}) = \mathbf{E}_m \cdot \hat{x}$, $\hat{x}$ being a unit vector parallel to the x axis. Equation (VIII.19.9) is the natural generalization of Eq. (VIII.10.10) to the case in which backward-traveling modes cannot be neglected. Coupled-mode theory must accordingly be modified to include coupling with backward-traveling waves, and the set of coupled equations describing the evolution of the $c_m^\pm$'s read [1]

$$\begin{aligned}\frac{dc_m^{(+)}(\omega,z)}{dz} &= \sum_n [K_{mn}^{+,+} \exp\{i[\beta_m(\omega) - \beta_n(\omega)]z\} c_n^+(\omega, z)\\ &\quad + K_{mn}^{+,-} \exp\{i[\beta_m(\omega) + \beta_n(\omega)]z\} c_n^-(\omega, z)],\\ \frac{dc_m^{(-)}(\omega,z)}{dz} &= \sum_n [K_{mn}^{-,+} \exp\{-i[\beta_m(\omega) + \beta_n(\omega)]z\} c_n^+(\omega, z)\\ &\quad + K_{mn}^{-,-} \exp\{-i[\beta_m(\omega) - \beta_n(\omega)]z\} c_n^-(\omega, z)],\end{aligned} \qquad \text{(VIII.19.11)}$$

the coupling coefficients turning out to be given, according to Eq. (VIII.19.7), by

$$K_{mn}^{p,q} = \frac{p\omega\varepsilon_0}{4iP}\int_{-\infty}^{+\infty} dx \int_{-\infty}^{+\infty} dy |\hat{E}_x(\boldsymbol{\rho}, z, t)|^2 \tilde{E}_m(\boldsymbol{\rho})\tilde{E}_n(\boldsymbol{\rho}).$$
$$= \frac{p\omega\varepsilon_2}{4iP}\int_{-\infty}^{+\infty} dx \int_{-\infty}^{+\infty} dy |\hat{E}_x(\boldsymbol{\rho}, z, t)|^2 \tilde{E}_m(\boldsymbol{\rho})\tilde{E}_n(\boldsymbol{\rho}). \tag{VIII.19.12}$$

Starting from Eqs. (VIII.19.11), it is possible to obtain a set of equations for the $\Phi_m^{\pm}$'s that read [31]

$$L_m^+\Phi_m^+ = -2i\left(\sum_{n\neq m} R_{mn}|\Phi_n^+|^2 + \frac{1}{2}R_{mm}|\Phi_m^+|^2\right)\Phi_m^+ - 2i\sum_n R_{mn}|\Phi_n^-|^2\Phi_m^+$$
$$- 2i\sum_{n\neq m} R_{mn}\Phi_n^+\Phi_n^-\Phi_m^{-*},$$
$$L_m^-\Phi_m^- = 2i\left(\sum_{n\neq m} R_{mn}|\Phi_n^-|^2 + \frac{1}{2}R_{mm}|\Phi_m^-|^2\right)\Phi_m^- + 2i\sum_n R_{mn}|\Phi_n^+|^2\Phi_m^-$$
$$+ 2i\sum_{n\neq m} R_{mn}\Phi_n^+\Phi_n^-\Phi_m^{+*} \qquad (n, m = 1, 2, \ldots, N) \tag{VIII.19.13}$$

where the coupling coefficients R_{mn} are expressed as superposition integrals of the mode spatial configurations

$$R_{mn} = k_0 n_2 \frac{\int_{-\infty}^{+\infty} dx \int_{\infty}^{+\infty} dy\, \tilde{E}_m^2(\boldsymbol{\rho})\tilde{E}_n^2(\boldsymbol{\rho})}{\int_{-\infty}^{+\infty} dx \int_{-\infty}^{+\infty} dy\, \tilde{E}_m^2(\boldsymbol{\rho})}, \tag{VIII.19.14}$$

while the differential operations $L_m^{\pm}$ are defined as

$$L_m^{\pm} = \frac{\partial}{\partial z} \pm \frac{1}{v_m}\frac{\partial}{\partial t} \mp \frac{i}{2A_m}\frac{\partial^2}{\partial t^2} \mp \frac{1}{3!B_m}\frac{\partial^3}{\partial t^3} \pm \frac{i}{4!C_m}\frac{\partial^4}{\partial t^4} + \cdots, \tag{VIII.19.15}$$

in terms of the group velocity v_m [Eq. (VIII.12.2)], of the *second-order group dispersion*

$$A_m = (d^2\beta_m/d\omega^2)^{-1}_{\omega=\omega_0}, \tag{VIII.19.16}$$

and of the *higher-order group dispersions*

$$B_m = (d^3\beta_m/d\omega^3)^{-1}|_{\omega=\omega_0}, \qquad C_m = (d^4\beta_m/d\omega^4)^{-1}|_{\omega=\omega_0}, \ldots. \tag{VIII.19.17}$$

In particular, if the $\tilde{E}_m(\boldsymbol{\rho})$'s are chosen in such a way that

$$\iint_{-\infty}^{+\infty} \tilde{E}_m^2(\boldsymbol{\rho})\, dx\, dy = 1, \tag{VIII.19.18}$$

the R_{mm}'s can be rewritten as

$$R_{mm} = k_0 n_2/\sigma_m \tag{VIII.19.19}$$

where σ_m represents the *effective area* of the mode [29].

The set of Eqs. (VIII.19.13) is the starting point for describing various kinds of nonlinear effects that can take place when propagating an optical signal inside a long fiber, such as *self-phase modulation, soliton propagation,* and *degenerate four-wave mixing*, which we are going to describe below.

19.1 *Self-Phase Modulation*

Let us assume a fiber of length L and an input field present only at $z = 0$, so that no backward-traveling mode is excited, as assured by the set of Eqs. (VIII.9.13), which implies $\Phi_m^-(z,t) = 0$ (for all m) if $\Phi_m^-(z = L, t) = 0$ (for all m). If we deal with a monomode fiber, Eqs. (VIII.19.13) reduce to the single nonlinear differential equation

$$(\partial/\partial z + (1/v_1)(\partial/\partial t) - [i/(2A_1)](\partial^2/\partial t^2))\Phi_1^+ = -iR_{11}|\Phi_1^+|^2\Phi_1^+, \tag{VIII.19.20}$$

where we have neglected dispersion effects of order higher than the second. The structure of the above relation clearly shows how the behavior of a pulse propagating inside the fiber depends on the relative weight of the diffusive term $[-i/(2A_1)]\,\partial^2/\partial t^2$ accounting for chromatic dispersion and the nonlinear term $-iR_{11}|\Phi_1|^2\Phi_1$. In fact [apart from the case in which they are both negligible and Eq. (VIII.19.20) admits a distortionless solution of the type $\Phi_1(z,t) = \Phi_1(z = 0, t - z/v_1)$], if the diffusive term dominates, a pulse injected at $z = 0$ undergoes a temporal spreading due to chromatic dispersion, while in the opposite situation its phase is subject to a modulation. This statement is easily understood by inspecting the solutions of Eq. (VIII.19.20), which read

$$\Phi_1^+(z,t) = \left(-\frac{iA_1}{2\pi z}\right)^{1/2}\int_{-\infty}^{+\infty}\exp\left\{i\left(\frac{A_1}{2z}\right)\left[t' - \left(t - \frac{z}{v_1}\right)\right]^2\right\}\Phi_1^+(z = 0, t)\,dt', \tag{VIII.19.21}$$

$$\Phi_1^+(z,t) = \exp\left[-iR_{11}\left|\Phi_1^+(z = 0, t - \frac{z}{v_1})\right|^2 z\right]\Phi_1^+\left(z = 0, t - \frac{z}{v_1}\right), \tag{VIII.19.22}$$

respectively, in the dispersion- and nonlinearity-dominated regimes.

In general, Eq. (VIII.19.20) does not admit analytic solutions if both the diffusive and nonlinear terms are included, and it is necessary to resort to approximate or numerical solutions, whose behavior depends on the interplay between these two terms. What we can say approximately, by inspecting Eq. (VIII.19.22), is that the instantaneous frequency of the field is given by

$$\omega(t) = \omega_0 - R_{11}(\partial/\partial t)|\Phi_1^+(z = 0, t - z/v_1)|^2 z, \qquad \text{(VIII.19.23)}$$

so that the leading edge of the pulse is downshifted in frequency, while the trailing edge is upshifted; since, for *normal dispersion* (see Section I.2) ($A_1 > 0$, $\lambda \leqq 1.3$ μm for fused silical, the group velocity v_1 is less for higher frequencies, the frequency sweep due to nonlinear *self-phase modulation* gives rise to pulse broadening, while for *anomalous dispersion* ($A_1 < 0$, $\lambda \geqq 1.3$ μm for fused silica) pulse narrowing occurs [32, 33].

19.2 *Envelope Solitons*

The considerations of the preceding subsection show that, in the anomalous-dispersion regime, the pulse narrowing induced by nonlinearity tends to neutralize the pulse broadening caused by chromatic dispersion; when exact balance occurs, the optical pulse can be shown to form an *envelope soliton* [34], which propagates without undergoing distortion. This is actually confirmed by the existence of an exact solution of Eq. (VIII.19.20) in the form

$$\Phi_1^+(z, t) = \Phi_0 \exp[iz/(2A_1\tau^2)] \operatorname{sech}[(t - z/v_1)/\tau], \qquad \text{(VIII.19.24)}$$

provided that the balance condition between the amplitude $|\Phi_0|$ and the temporal width τ at $z = 0$,

$$-1/(A_1\tau^2) = R_{11}|\Phi_0|^2, \qquad \text{(VIII.19.25)}$$

is satisfied. The envelope of the above *fundamental soliton* travels without changing its shape with a velocity equal to the group velocity v_1 of the mode under consideration.

Besides the fundamental soliton represented by Eq. (VIII.19.24), there exist *higher-order solitons* [35, 36] associated with a periodic z behavior of the pulse shape, which correspond to solutions of Eq. (VIII.19.20) satisfying the boundary condition

$$\Phi_1^+(z = 0, t) = N\Phi_0 \operatorname{sech}(t/\tau), \qquad \text{(VIII.19.26)}$$

where N is an integer number (≥ 2) and $|\Phi_0|$ and τ are still related by Eq. (VIII.19.25). By using numerical techniques it can be shown that, during propagation, there is first a narrowing of the pulse shape, followed by a more complex behavior consisting of a sequence of narrowings and splittings (see Fig. VIII.25), the period z_0 being given by [33, 37]

$$z_0 = (\pi/2)\tau^2|A_1|. \qquad \text{(VIII.19.27)}$$

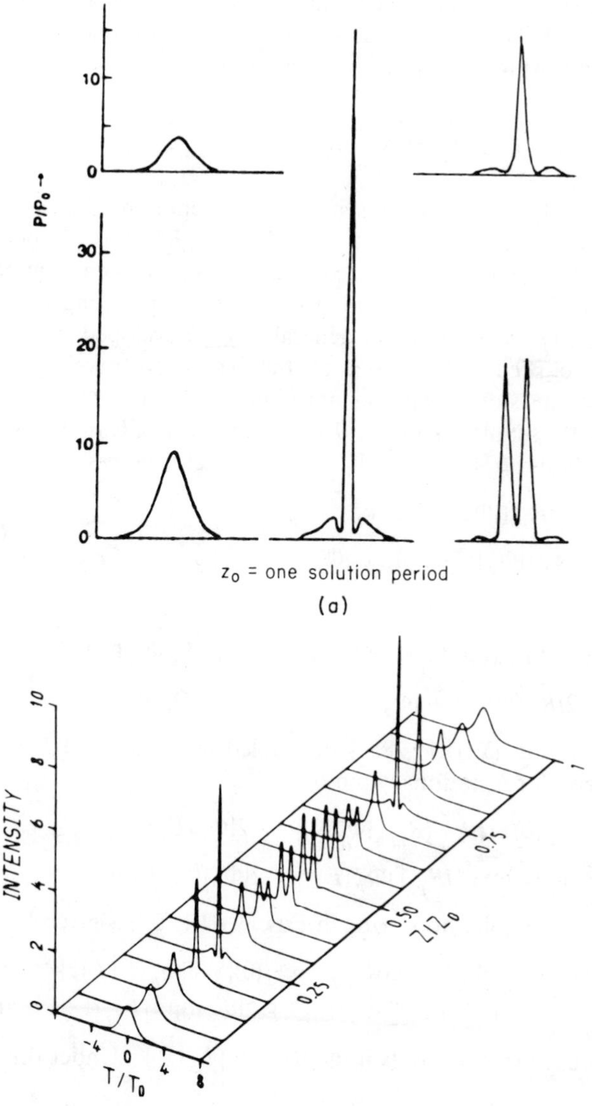

Fig. VIII.25. (a) Plots of $P/P_0 = |\Phi_1(z,t)|^2/|\Phi_0|^2$ for $N = 2$ (above) and $N = 3$ (below) solitons at fixed z/z_0 (respectively equal to 10^{-1} and 5×10^{-1} above and 10^{-1}, 25×10^{-2}, 5×10^{-1} below) as a function of $s = (t - z/v)/\tau$ (from Mollenauer and Stolen [1982], bibliography). (b) Perspective plot of P/P_0 as a function of $T/T_0 = s$ at various positions along the fiber [see Eqs. (VIII.19.24), (VIII.19.25) and (VIII.19.26)] (from Stolen [1984], bibliography).

It is hardly necessary to point out the implications that the possibility of propagating envelope solitons can have for the bit rate of transmission in a monomode fiber. The reader interested in the subject is referred to Hasegawa and Kodama [38] for an ample discussion.

19.3 *Degenerate Four-Wave Mixing*

Let us consider the situation schematically represented in Fig. VIII.26, in which the fiber input consists of three fields, E_l^+, E_p^+, and E_p^-, possessing the same central frequency ω_0, that excite, respectively, a given forward-traveling mode l at $z=0$ (*signal mode*) and two counterpropagating modes $p \neq l$ at $z=0$ and $z=L$ (*pump modes*). The general theory developed in Section VIII.19 allows one to predict the onset of the backward-traveling mode l, the evolution of the slowly varying amplitudes Φ_l^+, Φ_l^-, Φ_p^+, and Φ_p^- being governed by the set of Eqs. (VIII.19.13). After taking advantage of the strong-pump hypothesis $|\Phi_p^+|, |\Phi_p^-| \gg |\Phi_l^+|, |\Phi_l^-|$, the set becomes

$$\begin{aligned} L_p^+\Phi_p^+ &= -iR_{pp}(|\Phi_p^+|^2 + 2|\Phi_p^-|^2)\Phi_p^+, \\ L_p^-\Phi_p^- &= iR_{pp}(|\Phi_p^-|^2 + 2|\Phi_p^+|^2)\Phi_p^-, \end{aligned} \tag{VIII.19.28}$$

and

$$\begin{aligned} L_l^+\Phi_l^+ &= -2iR_{lp}(|\Phi_p^+|^2 + |\Phi_p^-|^2)\Phi_l^+ - 2iR_{lp}\Phi_p^+\Phi_p^-\Phi_l^{-*}, \\ L_l^-\Phi_l^- &= 2iR_{lp}(|\Phi_p^+|^2 + |\Phi_p^-|^2)\Phi_l^- + 2iR_{lp}\Phi_p^+\Phi_p^-\Phi_l^{+*}. \end{aligned} \tag{VIII.19.29}$$

The set of Eqs. (VIII.19.28) is decoupled from Eqs. (VIII.19.29), and its solution reads, for a stationary pump,

$$\begin{aligned} \Phi_p^+(z) &= \Phi_p^+(0)\exp\{-iR_{pp}[|\Phi_p^+(0)|^2 + 2|\Phi_p^-(L)|^2]z\}, \\ \Phi_p^-(z) &= \Phi_p^-(L)\exp\{iR_{pp}[|\Phi_p^-(L)|^2 + 2|\Phi_p^+(0)|^2](z-L)\}, \end{aligned} \tag{VIII.19.30}$$

so that the term $\Phi_p^+\Phi_p^-$ appearing in Eqs. (VIII.19.29) is given by

$$\begin{aligned} \Phi_p^+(z)\Phi_p^-(z) = \Phi_p^+(0)\Phi_p^-(L)&\exp\{iR_{pp}[|\Phi_p^+(0)|^2 - |\Phi_p^-(L)|^2]z\} \\ &\times \exp\{-iR_{pp}[|\Phi_p^-(L)|^2 + 2|\Phi_p^+(0)|^2]L\}, \end{aligned} \tag{VIII.19.31}$$

which is independent of z only if $|\Phi_p^+(0)|^2 = |\Phi_p^-(L)|^2$. Under this hypothesis,

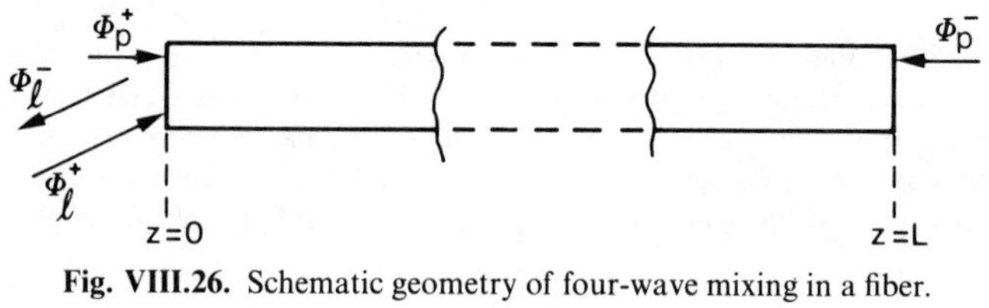

Fig. VIII.26. Schematic geometry of four-wave mixing in a fiber.

after putting

$$S_{lp} = 2R_{lp}[|\Phi_p^+(0)|^2 + |\Phi_p^-(L)|^2], \tag{VIII.19.32}$$

$$T_{lp} = 2R_{lp}\Phi_p^+(z)\Phi_p^-(z), \tag{VIII.19.33}$$

the set of Eqs. (VIII.19.29) can be written as

$$L_l^+\Phi_l^+ = iS_{lp}\Phi_l^+ - iT_{lp}\Phi_l^{-*}, \qquad L_l^-\Phi_l^- = iS_{lp}\Phi_l^- + iT_{lp}\Phi_l^{+*}, \tag{VIII.19.34}$$

which describes the process of *degenerate four-wave mixing* inside an optical fiber (the term "degenerate" referring to the common value ω_0 of the central frequency of signal and pump). In particular, in the stationary case ($L_l^+ = L_l^- = d/dz$), it reproduces the well-known equations usually found in the literature [30, 39], which admit the solution

$$\begin{aligned}\Phi_l^+(z) &= \exp(-iS_{lp}z)\frac{\cos[|T_{lp}|(z-L)]}{\cos(|T_{lp}|L)}\Phi_l^+(z=0),\\ \Phi_l^-(z) &= i\exp(iS_{lp}z)\frac{T_{lp}}{|T_{lp}|}\frac{\sin[|T_{lp}|(z-L)]}{\cos(|T_{lp}|L)}\Phi_l^{+*}(z=0).\end{aligned} \tag{VIII.19.35}$$

At $z = 0$, one has

$$\Phi_l^-(z=0) = Q_{lp}\Phi_l^{+*}(z=0), \tag{VIII.19.36}$$

with

$$Q_{lp} = -i(T_{lp}/|T_{lp}|)\tan(|T_{lp}|L), \tag{VIII.19.37}$$

which shows that the backward-traveling mode generated through the nonlinear interaction constitutes, at $z = 0$, a phase-conjugate replica of the forward-traveling mode, amplified or attenuated according whether $|\tan(|T_{lp}|L)| >$ or < 1.

The above result can be generalized to describe the situation in which an image-bearing beam excites a number of forward-traveling modes, whose amplitudes at $z = 0$ are connected to those of the corresponding backward-traveling modes by Eq. (VIII.19.36). In this case, one is naturally led to check whether a situation analogous to that encountered in the frame of *optical phase conjugation* [40] in bulk nonlinear crystals exists, where the backward-traveling field at $z = 0$ bears a phase-conjugate image of the input one, that is, in a monochromatic case,

$$E_x^-(\boldsymbol{\rho}, z=0) \propto E_x^{+*}(\boldsymbol{\rho}, z=0). \tag{VIII.19.38}$$

According to Eqs. (VIII.19.9), (VIII.19.33), and (VIII.19.36), we can write, in the weak-signal approximation $|T_{lp}|L \ll 1$,

$$E_x^-(\boldsymbol{\rho}, z=0) = -2i\Phi_p^+\Phi_p^- L\sum_l \tilde{E}_l(\boldsymbol{\rho})R_{lp}\Phi_l^{+*}(z=0), \tag{VIII.19.39}$$

to be compared with

$$E_x^+(\boldsymbol{\rho}, z = 0) = \sum_l \tilde{E}_l(\boldsymbol{\rho})\Phi_l^+(z = 0). \qquad \text{(VIII.19.40)}$$

From an inspection of Eqs. (VIII.19.39) and (VIII.19.40), it is evident that phase conjugation is not achieved unless the R_{lp}'s (that is, the overlap integrals between the lth and pth modes) are approximately independent of the mode index l for the group of modes under consideration.

If the input signal exhibits a time dependence, we must solve Eqs. (VIII.19.29) with L_l^+ and L_l^- given by Eq. (VIII.19.15). By limiting ourselves to taking into account first- and second-order group dispersions, that is, by writing

$$L_l^{\pm} = \partial/\partial z \pm (1/v_l)(\partial/\partial t) \mp [i/(2A_l)]/(\partial^2/\partial t^2), \qquad \text{(VIII.19.41)}$$

we find an analytic solution [41] that, in particular, generalizes Eq. (VIII.19.36) in the form

$$\Phi_l^-(z = 0, t) = \int_{-\infty}^{+\infty} e^{i\Omega t} R(\Omega) G(\Omega)\, d\Omega, \qquad \text{(VIII.19.42)}$$

where

$$G(\Omega) = \frac{1}{2\pi}\int_{-\infty}^{+\infty} e^{-i\Omega t}\phi_l^{+*}(z = 0, t)\, dt \qquad \text{(VIII.19.43)}$$

and

$$R(\Omega) = iT_{lp}\tan[f(\Omega)L]/\{i(\Omega/V_l)\tan[f(\Omega)L] - f(\Omega)\}, \qquad \text{(VIII.19.44)}$$

with $f(\Omega) = (|T|^2 + \Omega^2/v_l^2)^{1/2}$. The phase-conjugate response of the fiber is thus analogous to that of an optical bandpass filter [42]. Finally, we note that Eqs. (VIII.19.42), (VIII.19.43), and (VIII.19.44) do not contain A_l, so that the phase-conjugation process turns out to be independent of chromatic dispersion [41].

Problems

Section 2

1. Derive Eq. (VIII.2.5).

Section 3

2. Evaluate the ratio of the power guided by a graded-index fiber [see Eq. (VIII.3.4)] to that guided by an equivalent step-index fiber ($n_0 = n_1$ and same values of the core radius and the refractive index of the cladding), provided they are excited by two equal butt-jointed sources isotropically emitting inside the acceptance angle.

Section 4

3. Derive Eq. (VIII.4.8).

Section 5

4. Derive Eq. (VIII.5.20).

Section 6

5. By exploiting the properties of the zeros of the Bessel functions J_ν, show that the number of guided modes of a step-index fiber is given by $N = V^2/2$, in the limit $V \gg 1$.

6. Show that the number of guided modes of a graded-index fiber [see Eq. (VIII.3.4)] is $N = (V^2/2)q/(q+2)$. *Hint*: See [2], p. 257.

7. Derive Eq. (VIII.6.12). *Hint*: Use recurrence properties and asymptotic behavior of J_ν and K_ν.

Section 9

8. Evaluate the moments Ω_l for a generic graded-index fiber [see Eq. (VIII.3.4)].

9. Derive Eq. (VIII.9.13).

Section 10

10. Show the orthogonality between different modes of a parabolic fiber [see Eq. (VIII.7.16) *Hint*: Recall that, for a weakly guiding fiber, $\mathbf{H} \cong (\varepsilon/\mu_0)^{1/2}\hat{z} \times \mathbf{E}$.

11. Find the $c_m(\omega, z = 0)$'s of Eq. (VIII.10.9) for a parabolic fiber assuming a linearly polarized gaussian input $E(\rho, z = 0, t) = E_0 \exp[-\rho^2/(2R^2) + i\omega_0 t]$ ($R \ll a$). *Hint*: As for Problem 10.

Section 12

12. Relying on the result of Problem 5, show that, for $\nu \gg 1$, $\beta_{\nu\delta} = k_0 n_1 - \pi^2 M^2/(8k_0 n_1 a^2)$, with $M = \nu + 2\delta$, and evaluate $T^{(\mathrm{mod})}$.

13. Evaluate $|\tau_{p+1,q} - \tau_{pq}|$ for a parabolic fiber at the lowest significant order in Δ.

Section 13

14. Derive Eq. (VIII.13.2) for large L starting from Eq. (VIII.19.21) and assuming an input gaussian pulse of the form $\Phi_1^+(z = 0, t) = \Phi_{10} e^{-t^2/(2T_p^2)}$.

15. As in Problem 14, by assuming an input of the form $\Phi_1^+(z = 0, t) = \Phi_{10} e^{-t^2/(2T_p^2)} F(t)$, where F is a stochastic quantity with correlation function $\langle F(t')F^*(t'')\rangle = \exp(-|t' - t''|^2/T_c^2)$.

16. Show that the maximum transmissible bandwidth Δf achievable with a monomode fiber exhibiting only chromatic dispersion is proportional to $1/L$ and $1/L^{1/2}$, respectively, according to whether the source bandwidth or the modulation-frequency spread gives the dominant contribution to $\delta\omega$. *Hint*: $\Delta f \cong 1/T^{(\mathrm{cr})}$.

Section 15

17. Starting from Eqs. (VIII.15.10), find the steady-state ($\partial/\partial t = 0$) power distribution for a two-mode fiber. *Hint*: Look for particular solutions of the kind $\langle P_{1,2}\rangle = A_{1,2}e^{-\sigma z}$.

Section 16

18. Starting from Eqs. (VIII.15.10), show that asymptotic power equipartition among different guided modes takes place in a steady-state lossless case.

Section 17

19. Consider an elliptical monomode fiber and an input field linearly polarized in a direction forming an angle of $\pi/4$ with the principal axes. Evaluate the distance over which the initial power is completely transferred over the orthogonally polarized state. *Hint*: See Marcuse [1], p. 157.

Section 19

20. Derive Eq. (VIII.19.21).

21. Starting from Eq. (VIII.19.23), derive the approximate relation $(2\tau)^2 P_0 = 1.6\ \mathrm{W \cdot ps^2}$, which relates τ to the peak power P_0 necessary to excite a fundamental soliton; assume $n_2 = 1.2 \times 10^{-22}\ \mathrm{m^2/V^2}$, $A_1 = 10^{27}\ \mathrm{cm/s^2}$, $\pi a^2 = 20\ \mu\mathrm{m}^2$, $\lambda_0 = 1.3\ \mu\mathrm{m}$, and $n_0 = 1.5$. *Hint*: First show that, by normalizing the mode configuration to unity, $R_{11} \cong k_0 n_2/(\pi a^2)$.

22. Find the most general solution of Eq. (VIII.19.20) of the form $\Phi_1^+(z,t) = \exp[i(c_1 z + c_2 t)]f(t - z/v)$, with $v \neq v_1$ [38].

References

1. Marcuse, D., "Theory of Dielectric Optical Waveguides." Academic Press, New York, 1974.
2. Arnaud, J., "Beam and Fiber Optics." Academic Press, New York, 1976.
3. Unger, H. G., "Planar Optical Waveguides and Fibers." Oxford Univ. Press (Clarendon), London and New York, 1977.
4. Midwinter, J. E., "Optical Fibers for Transmission." Wiley, New York, 1979.
5. Adams, M. J., "An Introduction to Optical Waveguides." Wiley, New York, 1981.
6. Okoshi, T., "Optical Fibers." Academic Press, New York, 1982.
7. Snyder, A. W., and Love, J. D., "Optical Waveguide Theory." Chapman & Hall, London, 1983.

8. Jeunhomme, L. B., "Single-Mode Fiber Optics Principles and Applications." Dekker, New York, 1983.
9. Gloge, D., *Appl. Opt.* **10**, 2252, (1971).
10. Snyder, A. W., *Proc. IEEE* **69**, 6 (1981).
11. Gambling, W. A., and Matsumura, H., *Opt. Quantum Elect.* **10**, 31 (1978).
12. Stewart, W. J., *Electron. Lett.* **16**, 380 (1980).
13. Hussey, C. D., and Pask, C., *Proc. IEEE* **129**, 123 (1982).
14. Gloge, D., *Rep. Prog. Phys.* **42**, 1777 (1979).
15. Gloge, D., *Appl. Opt.* **10**, 2442 (1971).
16. Gambling, W. A., Matsumura, H., and Ragdale, C. M., *IEE Microwave Opt. Acoust.* **3**, 239 (1979).
17. Crosignani, B., Daino, B., and Di Porto, P., *J. Opt. Soc. Am.* **66**, 1312 (1976).
18. Epworth, R. E., *Laser Focus* **17**, 109 (1981).
19. Crosignani, B., Daino, B., and Di Porto, P., *IEEE Trans. Microwave Theory Tech.* **MTT-23**, 416 (1975).
20. Kawakami, S., and Ikeda, M., *IEEE J. Quantum Electron.* **QE-14**, 608 (1978).
21. Crosignani, B., Wabnitz, S., and Di Porto, P., *Opt. Lett.* **9**, 371 (1984).
22. Kaminow, I. P., *IEEE J. Quantum Electron.* **QE-17**, 15 (1981).
23. Okoshi, T., *IEEE J. Quantum Electron.* **QE-17**, 879 (1981).
24. Stolen, R. H., *in* "Fiber and Integrated Optics" (D. B. Ostrowsky, ed.), p. 157–182. Plenum, New York, 1979.
25. Smith, R. G., *Appl. Opt.* **11**, 2489 (1972).
26. Cotter, D., *J. Opt. Commun.* **4**, 10 (1983).
27. Bloembergen, N., "Nonlinear Optics." Benjamin, New York, 1965.
28. Gabor, D., *J. Inst. Electr. Eng.* **93**, 429 (1946).
29. Crosignani, B., Cutolo, A., and Di Porto, P., *J. Opt. Soc. Am.* **72**, 1136 (1982).
30. Yariv, A., AuYeung, J., Fekete, D., and Pepper, D., *Appl. Phys. Lett.* **32**, 635 (1978).
31. Crosignani, B., *in* "New Directions in Guided Wave and Coherent Optics" (D. B. Ostrowsky and E. Spitz, eds.), p. 23–41. Nijhoff, The Hague, 1984.
32. Stolen, R. H., *in* "Optical Fiber Telecommunications" (S. E. Miller aand A. G. Chynoweth, eds.), p. 125–150. Academic Press, New York, 1979.
33. Mollenauer, L. F., Stolen, R. H., and Gordon, J. P., *Phys. Rev. Lett.* **45**, 1095 (1980).
34. Hasegawa, A., and Tappert, F., *Appl. Phys. Lett.* **23**, 142 (1973).
35. Zacharov, V. E., and Shabat, A. B., *Zh. Eksp. Teor. Fiz.* **64**, 1627 (1973) [*Sov. Phys—JETP* (*Engl. Transl.*) **37**, 823 (1973)].
36. Satsuma, J., and Yajima, N., *Suppl. Prog. Theor. Phys.* No. 55, 284 (1974).
37. Stolen, R. H., Mollenauer, L. F., and Tomlinson, W. J., *Opt. Lett* **8**, 186 (1983).
38. Hasegawa, A., and Kodama, Y., *Proc. IEEE* **69**, 1145 (1981).
39. Hellwarth, R. W., *IEEE J. Quantum Electron.* **QE-15**, 101 (1979).
40. Fisher, R. A., ed., "Optical Phase Conjugation." Academic Press, New York, 1983.
41. Crosignani, B., and Di Porto, P., *Opt. Lett.* **7**, 489 (1982).
42. Pepper, D. M., and Abrams, R. L., *Opt. Lett* **3**, 212 (1978).

Bibliography

Keiser, G. E., "Optical Fiber Communications." McGraw-Hill, New York, 1983.

Mollenauer, L. F., and Stolen, R. H., *Laser Focus* **18**, 193 (1982).

Sharma, A. B., Halme, S. J., and Butusov, M. M., "Optical Fiber Systems and Their Components." Springer-Verlag, Berlin and New York, 1981.

Stolen, R. H., *in* "New Directions in Guided Wave and Coherent Optics" (D. B. Ostrowsky and E. Spitz, eds.), p. 1–22. Nijhoff, The Hague, 1984.

Appendix

A Vector Formulas

$$\mathbf{A}\cdot\mathbf{B}\times\mathbf{C}=\mathbf{A}\times\mathbf{B}\cdot\mathbf{C}=\mathbf{B}\cdot\mathbf{C}\times\mathbf{A}=\mathbf{B}\times\mathbf{C}\cdot\mathbf{A}=\mathbf{C}\cdot\mathbf{A}\times\mathbf{B}=\mathbf{C}\times\mathbf{A}\cdot\mathbf{B} \tag{A.1}$$

$$\mathbf{A}\times(\mathbf{B}\times\mathbf{C})=(\mathbf{A}\cdot\mathbf{C})\mathbf{B}-(\mathbf{A}\cdot\mathbf{B})\mathbf{C} \tag{A.2}$$

$$\mathbf{A}\times(\mathbf{B}\times\mathbf{C})+\mathbf{B}\times(\mathbf{C}\times\mathbf{A})+\mathbf{C}\times(\mathbf{A}\times\mathbf{B})=0 \tag{A.3}$$

$$(\mathbf{A}\times\mathbf{B})\cdot(\mathbf{C}\times\mathbf{D})=(\mathbf{A}\cdot\mathbf{C})(\mathbf{B}\cdot\mathbf{D})-(\mathbf{A}\cdot\mathbf{D})(\mathbf{B}\cdot\mathbf{C}) \tag{A.4}$$

$$(\mathbf{A}\times\mathbf{B})\times(\mathbf{C}\times\mathbf{D})=(\mathbf{A}\times\mathbf{B}\cdot\mathbf{D})\mathbf{C}-(\mathbf{A}\times\mathbf{B}\cdot\mathbf{C})\mathbf{D} \tag{A.5}$$

$$\nabla(fg)=\nabla(gf)=f\nabla g+g\nabla f \tag{A.6}$$

$$\nabla\cdot(f\mathbf{A})=f\nabla\cdot\mathbf{A}+\mathbf{A}\cdot\nabla f \tag{A.7}$$

$$\nabla\times(f\mathbf{A})=f\nabla\times\mathbf{A}+\nabla f\times\mathbf{A} \tag{A.8}$$

$$\nabla\cdot(\mathbf{A}\times\mathbf{B})=\mathbf{B}\cdot(\nabla\times\mathbf{A})-\mathbf{A}\cdot(\nabla\times\mathbf{B}) \tag{A.9}$$

$$\nabla\times(\mathbf{A}\times\mathbf{B})=\mathbf{A}(\nabla\cdot\mathbf{B})-\mathbf{B}(\nabla\cdot\mathbf{A})+(\mathbf{B}\cdot\nabla)\mathbf{A}-(\mathbf{A}\cdot\nabla)\mathbf{B} \tag{A.10}$$

$$\nabla(\mathbf{A}\cdot\mathbf{B})=\mathbf{A}\times(\nabla\times\mathbf{B})+\mathbf{B}\times(\nabla\times\mathbf{A})+(\mathbf{A}\cdot\nabla)\mathbf{B}+(\mathbf{B}\cdot\nabla)\mathbf{A} \tag{A.11}$$

$$\nabla^2 f=\nabla\cdot\nabla f \tag{A.12}$$

$$\nabla^2\mathbf{A}=\nabla(\nabla\cdot\mathbf{A})-\nabla\times\nabla\times\mathbf{A} \tag{A.13}$$

$$\nabla\times\nabla f=0 \tag{A.14}$$

$$\nabla\cdot\nabla\times\mathbf{A}=0 \tag{A.15}$$

$$\nabla^2(fg)=f\nabla^2 g+g\nabla^2 f+2\nabla f\cdot\nabla g \tag{A.16}$$

B Dyadics and their Properties

$$\mathbf{ab}=\begin{bmatrix} a_x b_x & a_x b_y & a_x b_z \\ a_y b_x & a_y b_y & a_y b_z \\ a_z b_x & a_z b_y & a_z b_z \end{bmatrix}=\begin{bmatrix} a_x \\ a_y \\ a_z \end{bmatrix}[b_x b_y b_z] \tag{B.1}$$

$$\widetilde{\mathbf{ab}} = \mathbf{ba} \tag{B.2}$$

$$\mathrm{Tr}(\mathbf{ab}) = \mathbf{a} \cdot \mathbf{b} \tag{B.3}$$

$$\mathbf{ab} \cdot \mathbf{c} = \mathbf{a}(\mathbf{b} \cdot \mathbf{c}) = (\mathbf{b} \cdot \mathbf{c}) = (\mathbf{b} \cdot \mathbf{c}) \begin{bmatrix} a_x \\ a_y \\ a_z \end{bmatrix} \tag{B.4}$$

$$\mathbf{c} \cdot \mathbf{ab} = (\mathbf{c} \cdot \mathbf{a})[b_x b_y b_z] \tag{B.5}$$

$$\mathbf{ab} \cdot \mathbf{cd} = (\mathbf{b} \cdot \mathbf{c})\mathbf{ad} = (\mathbf{b} \cdot \mathbf{c}) \begin{bmatrix} a_x d_x & a_x d_y & a_x d_z \\ a_y d_x & a_y d_y & a_y d_z \\ a_z d_x & a_z d_y & a_z d_z \end{bmatrix} \tag{B.6}$$

$$\mathbf{ab} : \mathbf{cd} = (\mathbf{b} \cdot \mathbf{c})(\mathbf{a} \cdot \mathbf{d}) = \mathrm{Tr}(\mathbf{ab} \cdot \mathbf{cd}) \tag{B.7}$$

$$\mathbf{A} \vdots \mathbf{B} = A_{ij} B_{ji} \tag{B8}$$

$$\nabla \mathbf{A} : \nabla \mathbf{B} = \left(\frac{\partial}{\partial x_q} A_{ij} \right) \frac{\partial}{\partial x_q} B_{jk} \tag{B9}$$

$$\nabla \mathbf{A} : \nabla \mathbf{B} = \left(\frac{\partial}{\partial x_q} A_{ij} \right) \frac{\partial}{\partial x_q} B_{ij} \tag{B10}$$

C Special Functions

Orthogonal Polynomials

Legendre polynomials

$$P_n(z) = \frac{2^{-n}}{n!} \frac{d^n}{dz^n} (z^2 - 1)^n, \qquad \int_{-1}^{+1} P_n(z) P_{n'}(z)\, dz = \frac{1}{n + \frac{1}{2}} \delta_{nn'} \tag{C.1}$$

Associate Legendre functions

$$P_n^m(z) = (1 - z^2)^{|m|/2} \frac{d^{|m|}}{dz^{|m|}} P_n(z) \begin{cases} \displaystyle\int_{-1}^{+1} P_n^m(z) P_{n'}^m(z)\, dz \\ \displaystyle\qquad = \frac{(n + \frac{1}{2})^{-1} (n + m)!}{(n - m)!} \delta_{nn'} \\ \\ \displaystyle\int_{-1}^{+1} P_n^m(z) P_n^{m'}(z) (1 - z^2)^{-1}\, dz \\ \displaystyle\qquad = \frac{(n + m)!}{(n - m)!} \delta_{mm'} \end{cases} \tag{C.2}$$

Laguerre polynomials

$$L_n^\alpha(z) = \frac{1}{n!}\frac{1}{z^\alpha}e^z\frac{d^n}{dz^n}(e^{-z}z^{n+\alpha})$$

$$\int_0^\infty e^{-z}z^\alpha L_n^\alpha(z)L_{n'}^\alpha(z)\,dz = \frac{\Gamma(\alpha+n+1)}{n!}\delta_{nn'} \tag{C.3}$$

Hermite polynomials

$$H_n(z) = (-1)^n e^{z^2}\frac{d^n}{dz^n}e^{-z^2}, \qquad \int_{-\infty}^{+\infty} e^{-z^2}H_n(z)H_{n'}(z)\,dz = 2^n\pi^{1/2}n!\,\delta_{nn'} \tag{C.4}$$

Zernike circle polynomials

$$R_n^{(m)}(z) = \sum_{s=0}^{(1/2)(n-|m|)}(-1)^s\frac{(n-s)!}{s![(n+m)/2-s]![(n-m)/2-s]!}z^{n-2s}$$

$$= \frac{1}{[(n-|m|)/2]!\,z^{|m|}}\left(\frac{d}{dz^2}\right)^{(n-|m|)/2}[z^{n+|m|}(z^2-1)^{(n-|m|)/2}]$$

$$R_n^{(m)}(1) = 1, \qquad R_n^{(n)}(z) = z^n$$

$$\int_0^1 zR_n^{(m)}(z)R_{n'}^{(m)}(z)\,dz = \frac{1}{2(n+1)}\delta_{nn'}$$

$$\int_0^1 zR_n^{(m)}(z)J_m(az)\,dz = (-1)^{(n-m)/2}\frac{J_{n+1}(a)}{a} \tag{C.5}$$

Scalar Spherical Harmonics

$$\begin{aligned}
Y_0^0 &= 1/(4\pi)^{1/2}\\
Y_1^0 &= (3/4\pi)^{1/2}\cos\theta\\
Y_1^{\pm1} &= \mp(3/8\pi)^{1/2}\sin\theta e^{\pm i\phi}\\
Y_2^0 &= (5/4\pi)^{1/2}(1/4)^{1/2}(3\cos^2\theta-1)\\
Y_2^{\pm1} &= \mp(5/4\pi)^{1/2}(3/2)^{1/2}\cos\theta\sin\theta e^{\pm i\phi}\\
Y_2^{\pm2} &= (5/4\pi)^{1/2}(3/8)^{1/2}\sin^2\theta e^{\pm 2i\phi}\\
Y_3^0 &= (7/4\pi)^{1/2}(1/4)^{1/2}(2\cos^3\theta-3\cos\theta\sin^2\theta)\\
Y_3^{\pm1} &= \mp(7/4\pi)^{1/2}(3/16)^{1/2}(4\cos^2\theta\sin\theta-\sin^3\theta)e^{\pm i\phi},
\end{aligned} \tag{C.6}$$

$$Y_3^{\pm 2} = (7/4\pi)^{1/2}(15/8)^{1/2} \cos\theta \sin^2\theta e^{\pm 2i\phi}$$
$$Y_3^{\pm 3} = \mp (7/4\pi)^{1/2}(5/16)^{1/2} \sin^3\theta e^{\pm 3i\phi}$$

Vector Spherical Harmonics

$$\begin{aligned}
\mathbf{Y}_1^1 &= -(1/2)^{1/2} Y_1^0 \hat{e}_{+1} + (1/2)^{1/2} Y_1^1 \hat{e}_0 \\
\mathbf{Y}_1^0 &= -(1/2)^{1/2} Y_1^{-1} \hat{e}_{+1} + (1/2)^{1/2} Y_1^1 \hat{e}_{-1} \\
\mathbf{Y}_1^{-1} &= -(1/2)^{1/2} Y_1^{-1} \hat{e}_0 + (1/2)^{1/2} Y_1^0 \hat{e}_{-1} \\
\mathbf{Y}_2^2 &= -(1/3)^{1/2} Y_2^1 \hat{e}_{+1} + (2/3)^{1/2} Y_2^2 \hat{e}_0 \\
\mathbf{Y}_2^1 &= -(1/2)^{1/2} Y_2^0 \hat{e}_{+1} + (1/6)^{1/2} Y_2^1 \hat{e}_0 + (1/3)^{1/2} Y_2^2 \hat{e}_{-1} \\
\mathbf{Y}_2^0 &= -(1/2)^{1/2} Y_2^{-1} \hat{e}_{+1} + (1/2)^{1/2} Y_2^1 \hat{e}_{-1} \\
\mathbf{Y}_2^{-1} &= (1/3)^{1/2} Y_2^{-2} \hat{e}_{+1} - (1/6)^{1/2} Y_2^{-1} \hat{e}_0 + (1/2)^{1/2} Y_2^0 \hat{e}_{-1} \\
\mathbf{Y}_2^{-2} &= -(2/3)^{1/2} Y_2^{-2} \hat{e}_0 + (1/3)^{1/2} Y_2^{-1} \hat{e}_{-1}
\end{aligned} \tag{C.7}$$

where

$$\begin{aligned}
\hat{e}_{+1} &= -(e^{i\phi}/2^{1/2})[(\sin\theta)\hat{\rho} + (\cos\theta)\hat{\theta} + i\hat{\phi}], \\
\hat{e}_0 &= (\cos\theta)\hat{\rho} - (\sin\theta)\hat{\theta} \\
\hat{e}_{-1} &= (e^{-i\phi}/2^{1/2})[(\sin\theta)\hat{\rho} + (\cos\theta)\hat{\theta} - i\hat{\phi}]
\end{aligned}$$

Maliuzhinet's Function

$$\begin{aligned}
M(\alpha) &= \prod_{n,m=1}^{\infty} \left\{1 - \left[\frac{\alpha/\pi}{N(2n-1) + m - 1/2}\right]^2\right\}^{(-1)^{m+1}} \\
&= \exp\left[\frac{i}{8\phi}\int_0^{\alpha} d\mu \int_{-i\infty}^{+i\infty} \tan\left(\frac{\pi v}{4\phi}\right)\frac{1}{\cos(v-\mu)}\,dv\right], \\
&\qquad \text{where} \quad \phi = \frac{2\pi}{N} \\
\frac{M(\alpha + 2\phi)}{M(\alpha - 2\phi)} &= \operatorname{cotan}\left[\frac{1}{2}\left(\alpha + \frac{\pi}{2}\right)\right] \\
M\left(\alpha + \frac{\pi}{2}\right)M\left(\alpha - \frac{\pi}{2}\right) &= M^2 \frac{\pi}{2}\cos\left(\frac{\pi\alpha}{4\phi}\right) \\
M(\alpha + \phi)M(\alpha - \phi) &= M^2(\phi)M'(\alpha)
\end{aligned} \tag{C.8}$$

Bessel Functions

$$J_\nu(z) = \left(\frac{z}{2}\right)^\nu \sum_{n=0}^{\infty} \frac{(-1)^n (z/2)^{2n}}{n!\,\Gamma(\nu+n+1)}$$

$$= \frac{1}{\pi}\int_0^\pi \cos(z\sin\theta - \nu\theta)\,d\theta - \frac{\sin\nu\pi}{\pi}\int_0^\infty e^{-z\sinh t - \nu t}\,dt$$

$$Y_\nu(z) = \frac{J_\nu(z)\cos(\nu\pi) - J_{-\nu}(z)}{\sin(\nu\pi)}$$

$$= \frac{1}{\pi}\int_0^\pi \sin(z\sin\theta - \nu\theta)\,d\theta - \frac{1}{\pi}\int_0^\infty [e^{\nu t} + e^{-\nu t}\cos(\nu\pi)]e^{-z\sinh t}\,dt$$

$$J_{-n}(z) = (-1)^n J_n(z)$$

$$H_\nu^{(1)}(z) = J_\nu(z) + iY_\nu(z) = -\frac{i}{\pi}\int_{-\infty}^{\infty+\pi i} e^{z\sinh t - \nu t}\,dt$$

$$H_\nu^{(2)}(z) = J_\nu(z) - iY_\nu(z) = \frac{i}{\pi}\int_{-\infty}^{\infty-\pi i} e^{z\sinh t - \nu t}\,dt \qquad \text{(C.9)}$$

Modified Bessel Functions

$$I_\nu(z) = e^{-i(\pi/2)\nu} J_\nu(e^{i\pi/2}z) = \left(\frac{z}{2}\right)^\nu \sum_{n=0}^{\infty} \frac{(z/2)^{2n}}{n!\,\Gamma(\nu+n+1)}$$

$$K_\nu(z) = \frac{\pi}{2}\csc(\pi\nu)[I_{-\nu}(z) - I_\nu(z)] = \frac{\pi^{1/2}(z/2)^\nu}{\Gamma(\nu+1/2)}\int_1^\infty e^{-zt}(t^2-1)^{\nu-1/2}\,dt \qquad \text{(C.10)}$$

Spherical Bessel Functions

$$j_n(z) = \left(\frac{\pi}{2z}\right)^{1/2} J_{n+1/2}(z) = z^n\left(-\frac{1}{z}\frac{d}{dz}\right)^n \frac{\sin z}{z}$$

$$y_n(z) = \left(\frac{\pi}{2z}\right)^{1/2} Y_{n+1/2}(z) = -z^n\left(-\frac{1}{z}\frac{d}{dz}\right)^n \frac{\cos z}{z} \qquad \text{(C.11)}$$

$$h_n^{(1)}(z) = j_n(z) + iy_n(z) = h_n^{(2)*}(z)$$

Riccati–Bessel Functions

$$\psi_n(z) = zj_n(z), \qquad \zeta_n^{(1)}(z) = zh_n^{(1)}(z), \qquad \zeta_n^{(2)}(z) = zh_n^{(2)}(z) \qquad \text{(C.12)}$$

Parabolic Cylinder Function

$$D_{-1/2}(z) = \left(\frac{z}{2\pi}\right)^{1/2} K_{1/4}\left(\frac{z^2}{4}\right) = \frac{z}{2}\frac{1}{\Gamma(3/4)}\int_1^\infty \frac{e^{-z^2t/4}}{(t^2-1)^{1/4}}\,dt \qquad \text{(C.13)}$$

Fresnel Integrals

$$C_1(z) = \left(\frac{2}{\pi}\right)^{1/2}\int_0^z \cos t^2\,dt$$

$$S_1(z) = \left(\frac{2}{\pi}\right)^{1/2}\int_0^z \sin t^2\,dt$$

$$F(z) = \left(\frac{1}{\pi}\right)^{1/2} e^{i\pi/4}\int_z^\infty e^{-it^2}\,dt = \frac{1}{2} - \frac{1}{2}[C_1(z) - iS_1(z)](1+i) \qquad \text{(C.14)}$$

$$\sim \begin{cases} U_{\mathrm{F}}(-z) + \dfrac{2}{2\pi^{1/2}}\dfrac{1}{z}e^{-i(z^2+\pi/4)} & (|z| \gg 1) \\ \dfrac{1}{2} - \dfrac{e^{i\pi/4}}{\pi^{1/2}}z\left(1 - \dfrac{iz^2}{1!3} - \dfrac{z^4}{2!5} + \dfrac{iz^6}{3!7} + \cdots\right) & (|z| \ll 1) \end{cases}$$

Note: see Eq. (C.19) for U_{F}.

Gudermannian Function

$$gd(z) = 2\arctan e^z - \pi/2 \qquad \text{(C.15)}$$

Airy Functions

$$\mathrm{Ai}(z) = \frac{1}{\pi}\int_0^\infty \cos\left(\frac{t^3}{3} + zt\right)dt = c_1 f(z) - c_2 g(z)$$

$$\mathrm{Bi}(z) = \frac{1}{\pi}\int_0^\infty \left[e^{-t^3/3+zt} + \sin\left(\frac{t^3}{3} + zt\right)\right]dt = 3^{1/2}\{c_1 f(z) + c_2 g(z)\}$$

$$f(z) = \sum_{k=0}^\infty 3^k \frac{\Gamma(\frac{1}{3}+k)}{\Gamma(\frac{1}{3})}\frac{z^{3k}}{(3k)!}, \qquad g(z) = \sum_{k=0}^\infty 3^k \frac{\Gamma(\frac{2}{3}+k)}{\Gamma(\frac{2}{3})}\frac{z^{3k+1}}{(3k+1)!}$$

$$c_1 = \frac{3^{-2/3}}{\Gamma(\frac{2}{3})} = 0.35503, \qquad c_2 = \frac{3^{-1/3}}{\Gamma(\frac{1}{3})} = 0.25882, \qquad \text{(C.16)}$$

$$\mathrm{Ai}(z) = \frac{z^{1/2}}{3}[I_{-1/3}(\zeta) - I_{1/3}(\zeta)] = \frac{z^{1/2}}{\pi 3^{1/2}}K_{1/3}(\zeta), \qquad z > 0, \qquad \zeta = \frac{2}{3}z^{3/2}$$

$$\mathrm{Ai}(-z) = \frac{z^{1/2}}{3}[J_{1/3}(\zeta) + J_{-1/3}(\zeta)],$$

$$\mathrm{Bi}(z) = \left(\frac{z}{3}\right)^{1/2}[I_{-1/3}(\zeta) + I_{1/3}(\zeta)] \qquad \mathrm{Bi}(-z) = \left(\frac{z}{3}\right)^{1/2}[J_{-1/3}(\zeta) - J_{1/3}(\zeta)].$$

Sign Function

$$\mathrm{sgn}(x) = \begin{cases} 1, & x > 0, \\ 0, & x = 0, \\ -1, & x < 0 \end{cases} \tag{C.17}$$

Gauss Gamma Function

$$\Gamma(z) = \int_0^\infty t^{z-1}e^{-t}\,dt, \qquad (\mathrm{Re}\, z > 0);$$

$$\Gamma(z+1) = z\Gamma(z), \qquad \Gamma(n+1) = n!;$$

$$\Gamma(z) \sim e^{-z}z^{z-1/2}(2\pi)^{1/2}\left(1 + \frac{1}{12z} + \cdots\right), \qquad z \to \infty,\ |\arg z| < \pi, \text{ (Stirling formula);} \tag{C.18}$$

$$\frac{1}{\Gamma(z)} = ze^{\gamma z}\prod_{n=1}^{\infty}\left\{\left(1 + \frac{z}{n}\right)e^{-z/n}\right\},$$

(Euler's infinite product) with $\gamma = .57721$, (Euler's constant);

Unit Step-Function $U_F(x)$ for x Complex

$$U_F(-x) = \begin{cases} 1, & \mathrm{Im}(xe^{-i\pi/4}) < 0, \\ 0, & \text{otherwise, (see Eq. V.3.12)} \end{cases} \tag{C.19}$$

Sinc Function

$$\mathrm{sinc}(x) = \frac{\sin x}{x} \tag{C.20}$$

Lommel Functions of Two Variables

$$U_\nu(w,z) = \sum_{n=0}^{\infty}(-1)^n\left(\frac{w}{z}\right)^{\nu+2n}J_{\nu+2n}(z)$$

$$V_\nu(w,z) = \sum_{n=0}^{\infty}(-1)^n\left(\frac{z}{w}\right)^{\nu+2n}J_{\nu+2n}(z) = \cos\left(\frac{w}{2} + \frac{z^2}{2w} + \pi\frac{\nu}{2}\right) + U_{2-\nu}(w,z) \tag{C.21}$$

D Series Expansions

Ordinary Series Expansion

$$f(\mathbf{r}, \lambdabar) = \sum_{n=0}^{\infty} \frac{1}{n!} \frac{d^n}{d\lambdabar^n} f(\mathbf{r}, \lambdabar)\bigg|_{\lambdabar=0} \lambdabar^n, \qquad \lambdabar = \frac{\lambda}{2\pi} = \frac{1}{k_0} \tag{D.1}$$

where $\lim_{N\to\infty}(f - \Sigma_{n=0}^{N}\cdots) = 0$, for $\lambdabar$ sufficiently small

Asymptotic Series Expansion

$$f(\mathbf{r}, \lambdabar) \sim \sum_{n=0}^{\infty} \frac{A_n(\mathbf{r}, \lambdabar)}{n!} \lambdabar^n, \tag{D.2}$$

where $\lim_{\lambdabar\to 0}(f - \Sigma_{n=0}^{N}\cdots)\lambdabar^N = 0$, for every N. For f satisfying the Helmholtz wave-equation, [Eq. (I.1.12)], the A_n's are obtained by integrating the system of Eqs. (II.2.8).

Expansion in Spherical Harmonics

$$\mathbf{f}(\mathbf{r}) = \sum C_{nm} \mathbf{Y}_n^m(\theta, \phi) \tag{D.3}$$

Jacobi Identity

$$e^{iz\cos\phi} = J_0(z) + 2\sum_{s=1}^{\infty} i^s J_s(z)\cos(s\phi) \tag{D.4}$$

Bauer's Formula

$$\begin{aligned} e^{iz\cos\phi} &= \left(\frac{\pi}{2z}\right)^{1/2} \sum_{s=0}^{\infty} i^s(2s+1) J_{s+1/2}(z) P_s(\cos\phi) \\ e^{-iaz^2/2} &= \left(\frac{2\pi}{a}\right)^{1/2} e^{-ia/4} \sum_{s=0}^{\infty} (-i)^s(2s+1) J_{s+1/2}\frac{a}{4} R_2^{(0)}(z) \end{aligned} \tag{D.5}$$

Addition Theorem for Legendre Polynomials

$$P_n(\cos\omega) = \frac{4\pi}{2n+1} \sum_{m=-n}^{+n} Y_n^{*m}(\theta, \phi) Y_n^m(\theta', \phi') \tag{D.6}$$

where ω is the angle formed by the two directions (θ, ϕ) and (θ', ϕ').

Product of Two Spherical Harmonics

$$Y_n^m(\theta,\phi)Y_{n}^{m'}(\theta,\phi) = \sum_{j=|n-n'|}^{n+n'} \sum_{m''=-j}^{j} C_{nn'j}^{mm'm''} Y_j^{m''}(\theta,\phi) \tag{D.7}$$

where the C's are variously called Wigner coefficients, $3j$ symbols, or Clebsch–Gordan coefficients. Properties and tables of these coefficients can be found, e.g., in M. Weissbluth, "Atoms and Molecules," Academic Press, New York (1978).

Graf's Addition Theorem For Bessel Functions

$$J_n(\chi w)e^{in\psi} = \sum_{m=-\infty}^{\infty} e^{im(\phi-\phi')}J_{n+m}(\chi\rho')J_m(\chi\rho), \tag{D.8}$$

where

$\rho<\rho'$, $w=\{\rho^2+\rho'^2-2\rho\rho'\cos(\phi-\phi')\}^{1/2}$, and $\sin\psi=(\rho'/w)\sin(\phi-\phi')$.

E Integral Transforms

Space Fourier Transforms

$$F[u(x);k] = \int_{-\infty}^{+\infty} u(x)e^{ikx}\,dx;$$

$$F[u(x,y);k_x,k_y] = \iint_{-\infty}^{+\infty} u(x,y)e^{ik_x x+ik_y y}\,dx\,dy \tag{E.1}$$

$$F[u(\mathbf{r});\mathbf{k}] = \iiint_{-\infty}^{+\infty} u(\mathbf{r})e^{i\mathbf{k}\cdot\mathbf{r}}\,d\mathbf{r} \equiv u_{\mathbf{k}}$$

Time Fourier Transforms

$$F[u(t);\omega] = \int_{-\infty}^{+\infty} u(t)e^{-i\omega t}\,dt \equiv u_\omega \tag{E.2}$$

Space-Time Fourier Transforms

$$F[u(\mathbf{r},t);\mathbf{k},\omega] = \iiiint_{-\infty}^{+\infty} u(\mathbf{r},t)e^{i\mathbf{k}\cdot\mathbf{r}-i\omega t}\,d\mathbf{r}\,dt = u_{\mathbf{k},\omega} \tag{E.3}$$

Fourier Antitransforms

$$F^{-1}[w(k);x] = \frac{1}{2\pi}\int_{-\infty}^{+\infty} w(k)e^{-ikx}\,dk;$$

$$F^{-1}[w(\mathbf{k});\mathbf{r}] = \frac{1}{(2\pi)^3}\iiint_{-\infty}^{+\infty} w(\mathbf{k})e^{-i\mathbf{k}\cdot\mathbf{r}}\,d\mathbf{k}; \tag{E.4}$$

$$F^{-1}[w(\mathbf{k},\omega);\mathbf{r},t] = \frac{1}{(2\pi)^4}\iiiint_{-\infty}^{+\infty} w(\mathbf{k},\omega)e^{-i\mathbf{k}\cdot\mathbf{r}+i\omega t}\,d\mathbf{k}\,d\omega$$

Fourier Integral Theorem

$$F^{-1}[F[u(\mathbf{r}',t');\mathbf{k},\omega];\mathbf{r},t] = u(r,t) \tag{E.5}$$

Convolution Theorem

$$F[u(\mathbf{r},t) * v(\mathbf{r},t);\mathbf{k},\omega] = F[u(\mathbf{r},t);\mathbf{k},\omega]F[v(\mathbf{r},t);\mathbf{k},\omega], \tag{E.6}$$

where * represents the *convolution operator*

$$u(\mathbf{r},t) * v(\mathbf{r},t) = \iiiint_{-\infty}^{+\infty} u(\mathbf{r}-\mathbf{r}',t-t')v(\mathbf{r}',t')\,dr'\,dt'$$

Hankel Transform of Order m

$$H_m[u(\rho);\chi] = \int_0^{\infty} u(\rho)J_m(\chi\rho)\rho\,d\rho \tag{E.7}$$

Hankel Integral Theorem

$$H_m[H_m[u(\rho');\chi];\rho] = u(\rho) \tag{E.8}$$

Relation Between Fourier and Hankel Transforms

Let $u(x, y)$ depend on $\rho = (x^2 + y^2)^{1/2}$ only, then

$$F[u(\rho);k_x,k_y] = 2\pi H_0[u(\rho);\rho] \tag{E.9}$$

$f(x)$	$g(y) = \int_0^\infty f(x)(xy)^{1/2} J_\nu(xy)\,dx = y^{1/2} H_\nu[x^{1/2} f; y]$
$x^{1/2}/(b^2 + x^2)$	$y^{1/2} K_0(by) \qquad (\nu = 0)$
$x^{1/2}(a^2 + x^2)^{-1/2}$	$\left(\frac{2a}{\pi}\right)^{1/2} K_{1/2}(ay) \qquad (\nu = 0)$
$x^{1/2} e^{iax^2}$	$\frac{\pi^{1/2}}{8}\left(\frac{y}{a}\right)^{3/2}\left[J_{\nu/2+1/2}\frac{y^2}{8a} + iJ_{(\nu-1)/2}\frac{y^2}{8a}\right] \times \exp\left[-i\left(\frac{y^2}{8a} - \frac{\nu\pi}{4}\right)\right]$
$x^{\nu+1/2} e^{-ax^2/2} L_n^\nu(ax^2)$	$(-1)^n \frac{1}{a^{\nu+1}} e^{-y^2/2a} y^{\nu+1/2} L_n^\nu\left(\frac{y^2}{a}\right)$
$x^{1/2} e^{iax^2}$	$\frac{i}{2a} y^{1/2} e^{-iy^2/(4a)} \qquad (\nu = 0)$
$x^{1/2} e^{-iax^2/2} \quad (x < 1)$; $0 \quad (x > 1)$	$e^{-ia/4}\left(\frac{2\pi}{a}\right)^{1/2} \sum_{s=0}^{\infty} i^s(2s+1) J_{s+1/2}\left(\frac{a}{4}\right)\frac{J_{2s+1}(y)}{y} = \frac{1}{a}\left[-ie^{iy^2/2a} + ie^{-ia/2} V_0(a, y) - e^{-ia/2} V_1(a, y)\right] \qquad (\nu = 0)$
$x^{1/2} e^{-iax^2/2} R_4^{(0)}(x) \quad (x < 1)$; $0 \quad (x > 1)$	$e^{-ia/4}\left(\frac{2\pi}{a}\right)^{1/2} \frac{1}{y} \sum_{s=0}^{\infty} i^s(2s+1) J_{s+1/2}\left(\frac{a}{4}\right) \times [a_s J_{2s+5}(y) + b_s J_{2s+1}(y) + c_s J_{2s-3}(y)] \qquad (\nu = 0)$, where $a_s = \frac{3}{2}\frac{(s+2)(s+1)}{(2s+3)(2s+1)}$, $b_s = \frac{s(s+1)}{(2s+3)(2s-1)}$, $c_s = \frac{3}{2}\frac{s(s-1)}{(2s+1)(2s-1)}$

[a] From F. Oberhettinger, "Tables of Bessel Transforms," Springer-Verlag, Berlin and New York (1972).

F Asymptotic Expansions of Diffraction Integrals

$$e^{i\pi/4}\left(\frac{k}{2\pi}\right)^{1/2}\int_0^\infty g(s)e^{-ikh(s)}\,ds \sim I_{\mathrm{sp}} + I_{\mathrm{ep}} \tag{F.1}$$

where

$$I_{\mathrm{sp}} = \frac{e^{i(1-\varepsilon)\pi/4 - ikh\bar{a}}}{|h''(\bar{a})|^{1/2}} g(\bar{a}) + O(k^{-1})$$

$$I_{\mathrm{ep}} = \frac{e^{-i\pi/4 - ikh(0)}}{(2\pi k)^{1/2}} \frac{g(0)}{h'(0)} + O(k^{-3/2}) \qquad (g(0) \neq 0)$$

$$- \frac{e^{i\pi/4 - ikh(0)}}{k(2\pi k)^{1/2}} \frac{g'(0)}{h'(0)^2} + O(k^{-5/2}) \qquad (g(0) = 0)$$

with $\varepsilon = \operatorname{sgn} h''(0)$, the symbols $'$ and $''$ indicating first and second derivative with respect to s, and $\bar{a}$ a stationary point.

$$e^{i\pi/4}\left(\frac{k}{2\pi}\right)^{1/2}\int_{\mathrm{SDP}} \frac{f(\beta)}{\beta - \beta_p} e^{-ik\rho\cos(\beta-\phi)}\,d\beta$$

$$\sim -\frac{i}{\rho^{1/2}} \frac{f(\phi)}{\phi - \beta_p} + O(k^{-1}) \qquad (\phi \neq \beta_p)$$

$$\sim (2\pi k)^{1/2} e^{i\pi/4} e^{ik\rho\cos(\phi-\beta_p)/2}$$

$$\times\, G\left\{(\beta_p - \phi)\left(\frac{k\rho}{2}\right)^{1/2}\right\} f(\phi) + O(k^{-1}) \qquad (\phi \cong \beta_p) \tag{F.2}$$

$$\left(\frac{k}{2\pi}\right)^{1/2} e^{i\pi/4}\int_\Gamma (s - s_b)^{1/2} g(s) e^{-ikh(s)}\,ds \sim I_{\mathrm{SDP}} + I_{\Gamma_{\mathrm{B}}} \tag{F.3}$$

where the saddle-point contribution is

$$I_{\mathrm{SDP}} = \frac{e^{i(1-\varepsilon)\pi/4 - ikh(s^*)}}{|h''(s^*)|^{1/2}} (s^* - s_b)^{1/2} g(s^*) + O(k^{-1})$$

with $\varepsilon = \operatorname{sgn} h''(s^*)$ and s^* = saddle point, and the branch-cut contribution is

$$I_{\Gamma_{\mathrm{B}}} = -\frac{i}{k} \frac{e^{-ikh(s_b)}}{2^{1/2}[h'(s_b)]^{3/2}} g(s_b) + O(k^{-2})$$

$$\left(\frac{k}{2\pi}\right)^{1/2} e^{i\pi/4}\int_\Gamma (s - s_b)^{-1/2} g(s) e^{-ikh(s)}\,ds \sim I_{\mathrm{SDP}} + I_{\Gamma_{\mathrm{B}}} \tag{F.4}$$

where

$$I_{\mathrm{SDP}} = \frac{e^{i(1-\varepsilon)\pi/4 - ikh(s^*)}}{|h''(s^*)|^{1/2}} (s^* - s_b)^{-1/2} g(s^*) + O(k^{-1})$$

and

$$\mathrm{I}_{\Gamma_{\mathrm{B}}} = 2\frac{e^{-ikh(s_b)}}{|2h'(s_b)|^{1/2}}g(s_b) + O(k^{-1})$$

$$\begin{aligned}
&\int_0^\infty g(s)J_0(ks)\,ds \sim \frac{g(0)}{k} - \frac{g''(0)}{2k^3} + \frac{g''''(0)}{8k^5} + O(k^{-7}) \\
&\int_0^\infty g(s)J_1(ks)\,ds \sim \frac{g(0)}{k} + \frac{g'(0)}{k^2} - \frac{g'''(0)}{2k^4} + O(k^{-6}) \qquad\qquad (\mathrm{F.5}) \\
&\int_0^\infty g(s)e^{-\alpha ks}\,ds \sim \frac{g(0)}{\alpha k} + \frac{g'(0)}{(\alpha k)^2} + \frac{g'''(0)}{(\alpha k)^3} + O(k^{-4}) \quad (\alpha \text{ generally complex}) \\
&\int_0^\infty g(s)e^{-k^2s^2}\,ds \sim \left(\frac{\pi}{4}\right)^{1/2}\frac{g(0)}{k} + \frac{g'(0)}{2k^2} + \left(\frac{\pi}{4}\right)^{1/2}\frac{g''(0)}{4k^3} + O(k^{-4})
\end{aligned}$$

INDEX

W

Z